T0211647

Lecture Notes in Computer Science 12137

More information about this series at http://www.springer.com/series/7407

Valeria V. Krzhizhanovskaya ·
Gábor Závodszky · Michael H. Lees ·
Jack J. Dongarra · Peter M. A. Sloot ·
Sérgio Brissos · João Teixeira (Eds.)

Computational Science – ICCS 2020

20th International Conference
Amsterdam, The Netherlands, June 3–5, 2020
Proceedings, Part I

Springer

Editors
Valeria V. Krzhizhanovskaya (iD)
University of Amsterdam
Amsterdam, The Netherlands

Michael H. Lees
University of Amsterdam
Amsterdam, The Netherlands

Peter M. A. Sloot (iD)
University of Amsterdam
Amsterdam, The Netherlands

ITMO University
Saint Petersburg, Russia

Nanyang Technological University
Singapore, Singapore

João Teixeira
Intellegibilis
Setúbal, Portugal

Gábor Závodszky (iD)
University of Amsterdam
Amsterdam, The Netherlands

Jack J. Dongarra (iD)
University of Tennessee
Knoxville, TN, USA

Sérgio Brissos
Intellegibilis
Setúbal, Portugal

ISSN 0302-9743   ISSN 1611-3349   (electronic)
Lecture Notes in Computer Science
ISBN 978-3-030-50370-3   ISBN 978-3-030-50371-0   (eBook)
https://doi.org/10.1007/978-3-030-50371-0

LNCS Sublibrary: SL1 – Theoretical Computer Science and General Issues

This Springer imprint is published by the registered company Springer Nature Switzerland AG
The registered company address is: Gewerbestrasse 11, 6330 Cham, Switzerland

Preface

Twenty Years of Computational Science

Welcome to the 20th Annual International Conference on Computational Science (ICCS – https://www.iccs-meeting.org/iccs2020/).

During the preparation for this 20th edition of ICCS we were considering all kinds of nice ways to celebrate two decennia of computational science. Afterall when we started this international conference series, we never expected it to be so successful and running for so long at so many different locations across the globe! So we worked on a mind-blowing line up of renowned keynotes, music by scientists, awards, a play written by and performed by computational scientists, press attendance, a lovely venue... you name it, we had it all in place. Then corona hit us.

After many long debates and considerations, we decided to cancel the physical event but still support our scientists and allow for publication of their accepted peer-reviewed work. We are proud to present the proceedings you are reading as a result of that.

ICCS 2020 is jointly organized by the University of Amsterdam, NTU Singapore, and the University of Tennessee.

The International Conference on Computational Science is an annual conference that brings together researchers and scientists from mathematics and computer science as basic computing disciplines, as well as researchers from various application areas who are pioneering computational methods in sciences such as physics, chemistry, life sciences, engineering, arts and humanitarian fields, to discuss problems and solutions in the area, to identify new issues, and to shape future directions for research.

Since its inception in 2001, ICCS has attracted increasingly higher quality and numbers of attendees and papers, and 2020 was no exception, with over 350 papers accepted for publication. The proceedings series have become a major intellectual resource for computational science researchers, defining and advancing the state of the art in this field.

The theme for ICCS 2020, "Twenty Years of Computational Science", highlights the role of Computational Science over the last 20 years, its numerous achievements, and its future challenges. This conference was a unique event focusing on recent developments in: scalable scientific algorithms, advanced software tools, computational grids, advanced numerical methods, and novel application areas. These innovative novel models, algorithms, and tools drive new science through efficient application in areas such as physical systems, computational and systems biology, environmental systems, finance, and others.

This year we had 719 submissions (230 submissions to the main track and 489 to the thematic tracks). In the main track, 101 full papers were accepted (44%). In the thematic tracks, 249 full papers were accepted (51%). A high acceptance rate in the thematic tracks is explained by the nature of these, where many experts in a particular field are personally invited by track organizers to participate in their sessions.

ICCS relies strongly on the vital contributions of our thematic track organizers to attract high-quality papers in many subject areas. We would like to thank all committee members from the main and thematic tracks for their contribution to ensure a high standard for the accepted papers. We would also like to thank Springer, Elsevier, the Informatics Institute of the University of Amsterdam, the Institute for Advanced Study of the University of Amsterdam, the SURFsara Supercomputing Centre, the Netherlands eScience Center, the VECMA Project, and Intellegibilis for their support. Finally, we very much appreciate all the Local Organizing Committee members for their hard work to prepare this conference.

We are proud to note that ICCS is an A-rank conference in the CORE classification.

We wish you good health in these troubled times and hope to see you next year for ICCS 2021.

June 2020

Valeria V. Krzhizhanovskaya
Gábor Závodszky
Michael Lees
Jack Dongarra
Peter M. A. Sloot
Sérgio Brissos
João Teixeira

Organization

Thematic Tracks and Organizers

Advances in High-Performance Computational Earth Sciences: Applications and Frameworks – IHPCES

Takashi Shimokawabe
Kohei Fujita
Dominik Bartuschat

Agent-Based Simulations, Adaptive Algorithms and Solvers – ABS-AAS

Maciej Paszynski
David Pardo
Victor Calo
Robert Schaefer
Quanling Deng

Applications of Computational Methods in Artificial Intelligence and Machine Learning – ACMAIML

Kourosh Modarresi
Raja Velu
Paul Hofmann

Biomedical and Bioinformatics Challenges for Computer Science – BBC

Mario Cannataro
Giuseppe Agapito
Mauro Castelli
Riccardo Dondi
Rodrigo Weber dos Santos
Italo Zoppis

Classifier Learning from Difficult Data – CLD^2

Michał Woźniak
Bartosz Krawczyk
Paweł Ksieniewicz

Complex Social Systems through the Lens of Computational Science – CSOC

Debraj Roy
Michael Lees
Tatiana Filatova

Computational Health – CompHealth

Sergey Kovalchuk
Stefan Thurner
Georgiy Bobashev

Computational Methods for Emerging Problems in (dis-)Information Analysis – DisA

Michal Choras
Konstantinos Demestichas

Computational Optimization, Modelling and Simulation – COMS

Xin-She Yang
Slawomir Koziel
Leifur Leifsson

Computational Science in IoT and Smart Systems – IoTSS

Vaidy Sunderam
Dariusz Mrozek

Computer Graphics, Image Processing and Artificial Intelligence – CGIPAI

Andres Iglesias
Lihua You
Alexander Malyshev
Hassan Ugail

Data-Driven Computational Sciences – DDCS

Craig C. Douglas
Ana Cortes
Hiroshi Fujiwara
Robert Lodder
Abani Patra
Han Yu

Machine Learning and Data Assimilation for Dynamical Systems – MLDADS

Rossella Arcucci
Yi-Ke Guo

Meshfree Methods in Computational Sciences – MESHFREE

Vaclav Skala
Samsul Ariffin Abdul Karim
Marco Evangelos Biancolini
Robert Schaback

Rongjiang Pan
Edward J. Kansa

Multiscale Modelling and Simulation – MMS

Derek Groen
Stefano Casarin
Alfons Hoekstra
Bartosz Bosak
Diana Suleimenova

Quantum Computing Workshop – QCW

Katarzyna Rycerz
Marian Bubak

Simulations of Flow and Transport: Modeling, Algorithms and Computation – SOFTMAC

Shuyu Sun
Jingfa Li
James Liu

Smart Systems: Bringing Together Computer Vision, Sensor Networks and Machine Learning – SmartSys

Pedro J. S. Cardoso
João M. F. Rodrigues
Roberto Lam
Janio Monteiro

Software Engineering for Computational Science – SE4Science

Jeffrey Carver
Neil Chue Hong
Carlos Martinez-Ortiz

Solving Problems with Uncertainties – SPU

Vassil Alexandrov
Aneta Karaivanova

Teaching Computational Science – WTCS

Angela Shiflet
Alfredo Tirado-Ramos
Evguenia Alexandrova

Uncertainty Quantification for Computational Models – UNEQUIvOCAL

Wouter Edeling
Anna Nikishova
Peter Coveney

Program Committee and Reviewers

Ahmad Abdelfattah
Samsul Ariffin
  Abdul Karim
Evgenia Adamopoulou
Jaime Afonso Martins
Giuseppe Agapito
Ram Akella
Elisabete Alberdi Celaya
Luis Alexandre
Vassil Alexandrov
Evguenia Alexandrova
Hesham H. Ali
Julen Alvarez-Aramberri
Domingos Alves
Julio Amador Diaz Lopez
Stanislaw
  Ambroszkiewicz
Tomasz Andrysiak
Michael Antolovich
Hartwig Anzt
Hideo Aochi
Hamid Arabnejad
Rossella Arcucci
Khurshid Asghar
Marina Balakhontceva
Bartosz Balis
Krzysztof Banas
João Barroso
Dominik Bartuschat
Nuno Basurto
Pouria Behnoudfar
Joern Behrens
Adrian Bekasiewicz
Gebrai Bekdas
Stefano Beretta
Benjamin Berkels
Martino Bernard

Daniel Berrar
Sanjukta Bhowmick
Marco Evangelos
  Biancolini
Georgiy Bobashev
Bartosz Bosak
Marian Bubak
Jérémy Buisson
Robert Burduk
Michael Burkhart
Allah Bux
Aleksander Byrski
Cristiano Cabrita
Xing Cai
Barbara Calabrese
Jose Camata
Mario Cannataro
Alberto Cano
Pedro Jorge Sequeira
  Cardoso
Jeffrey Carver
Stefano Casarin
Manuel Castañón-Puga
Mauro Castelli
Eduardo Cesar
Nicholas Chancellor
Patrikakis Charalampos
Ehtzaz Chaudhry
Chuanfa Chen
Siew Ann Cheong
Andrey Chernykh
Lock-Yue Chew
Su Fong Chien
Marta Chinnici
Sung-Bae Cho
Michal Choras
Loo Chu Kiong

Neil Chue Hong
Svetlana Chuprina
Paola Cinnella
Noélia Correia
Adriano Cortes
Ana Cortes
Enrique
  Costa-Montenegro
David Coster
Helene Coullon
Peter Coveney
Attila Csikasz-Nagy
Loïc Cudennec
Javier Cuenca
Yifeng Cui
António Cunha
Ben Czaja
Pawel Czarnul
Flávio Martins
Bhaskar Dasgupta
Konstantinos Demestichas
Quanling Deng
Nilanjan Dey
Khaldoon Dhou
Jamie Diner
Jacek Dlugopolski
Simona Domesová
Riccardo Dondi
Craig C. Douglas
Linda Douw
Rafal Drezewski
Hans du Buf
Vitor Duarte
Richard Dwight
Wouter Edeling
Waleed Ejaz
Dina El-Reedy

Amgad Elsayed
Nahid Emad
Chriatian Engelmann
Gökhan Ertaylan
Alex Fedoseyev
Luis Manuel Fernández
Antonino Fiannaca
Christos
 Filelis-Papadopoulos
Rupert Ford
Piotr Frackiewicz
Martin Frank
Ruy Freitas Reis
Karl Frinkle
Haibin Fu
Kohei Fujita
Hiroshi Fujiwara
Takeshi Fukaya
Wlodzimierz Funika
Takashi Furumura
Ernst Fusch
Mohamed Gaber
David Gal
Marco Gallieri
Teresa Galvao
Akemi Galvez
Salvador García
Bartlomiej Gardas
Delia Garijo
Frédéric Gava
Piotr Gawron
Bernhard Geiger
Alex Gerbessiotis
Ivo Goncalves
Antonio Gonzalez Pardo
Jorge
 González-Domínguez
Yuriy Gorbachev
Pawel Gorecki
Michael Gowanlock
Manuel Grana
George Gravvanis
Derek Groen
Lutz Gross
Sophia
 Grundner-Culemann

Pedro Guerreiro
Tobias Guggemos
Xiaohu Guo
Piotr Gurgul
Filip Guzy
Pietro Hiram Guzzi
Zulfiqar Habib
Panagiotis Hadjidoukas
Masatoshi Hanai
John Hanley
Erik Hanson
Habibollah Haron
Carina Haupt
Claire Heaney
Alexander Heinecke
Jurjen Rienk Helmus
Álvaro Herrero
Bogumila Hnatkowska
Maximilian Höb
Erlend Hodneland
Olivier Hoenen
Paul Hofmann
Che-Lun Hung
Andres Iglesias
Takeshi Iwashita
Alireza Jahani
Momin Jamil
Vytautas Jancauskas
João Janeiro
Peter Janku
Fredrik Jansson
Jirí Jaroš
Caroline Jay
Shalu Jhanwar
Zhigang Jia
Chao Jin
Zhong Jin
David Johnson
Guido Juckeland
Maria Juliano
Edward J. Kansa
Aneta Karaivanova
Takahiro Katagiri
Timo Kehrer
Wayne Kelly
Christoph Kessler

Jakub Klikowski
Harald Koestler
Ivana Kolingerova
Georgy Kopanitsa
Gregor Kosec
Sotiris Kotsiantis
Ilias Kotsireas
Sergey Kovalchuk
Michal Koziarski
Slawomir Koziel
Rafal Kozik
Bartosz Krawczyk
Elisabeth Krueger
Valeria Krzhizhanovskaya
Pawel Ksieniewicz
Marek Kubalcík
Sebastian Kuckuk
Eileen Kuehn
Michael Kuhn
Michal Kulczewski
Krzysztof Kurowski
Massimo La Rosa
Yu-Kun Lai
Jalal Lakhlili
Roberto Lam
Anna-Lena Lamprecht
Rubin Landau
Johannes Langguth
Elisabeth Larsson
Michael Lees
Leifur Leifsson
Kenneth Leiter
Roy Lettieri
Andrew Lewis
Jingfa Li
Khang-Jie Liew
Hong Liu
Hui Liu
Yen-Chen Liu
Zhao Liu
Pengcheng Liu
James Liu
Marcelo Lobosco
Robert Lodder
Marcin Los
Stephane Louise

Frederic Loulergue
Paul Lu
Stefan Luding
Onnie Luk
Scott MacLachlan
Luca Magri
Imran Mahmood
Zuzana Majdisova
Alexander Malyshev
Muazzam Maqsood
Livia Marcellino
Tomas Margalef
Tiziana Margaria
Svetozar Margenov
Urszula
    Markowska-Kaczmar
Osni Marques
Carmen Marquez
Carlos Martinez-Ortiz
Paula Martins
Flávio Martins
Luke Mason
Pawel Matuszyk
Valerie Maxville
Wagner Meira Jr.
Roderick Melnik
Valentin Melnikov
Ivan Merelli
Choras Michal
Leandro Minku
Jaroslaw Miszczak
Janio Monteiro
Kourosh Modarresi
Fernando Monteiro
James Montgomery
Andrew Moore
Dariusz Mrozek
Peter Mueller
Khan Muhammad
Judit Muñoz
Philip Nadler
Hiromichi Nagao
Jethro Nagawkar
Kengo Nakajima
Ionel Michael Navon
Philipp Neumann

Mai Nguyen
Hoang Nguyen
Nancy Nichols
Anna Nikishova
Hitoshi Nishizawa
Brayton Noll
Algirdas Noreika
Enrique Onieva
Kenji Ono
Eneko Osaba
Aziz Ouaarab
Serban Ovidiu
Raymond Padmos
Wojciech Palacz
Ivan Palomares
Rongjiang Pan
Joao Papa
Nikela Papadopoulou
Marcin Paprzycki
David Pardo
Anna Paszynska
Maciej Paszynski
Abani Patra
Dana Petcu
Serge Petiton
Bernhard Pfahringer
Frank Phillipson
Juan C. Pichel
Anna
    Pietrenko-Dabrowska
Laércio L. Pilla
Armando Pinho
Tomasz Piontek
Yuri Pirola
Igor Podolak
Cristina Portales
Simon Portegies Zwart
Roland Potthast
Ela Pustulka-Hunt
Vladimir Puzyrev
Alexander Pyayt
Rick Quax
Cesar Quilodran Casas
Barbara Quintela
Ajaykumar Rajasekharan
Celia Ramos

Lukasz Rauch
Vishal Raul
Robin Richardson
Heike Riel
Sophie Robert
Luis M. Rocha
Joao Rodrigues
Daniel Rodriguez
Albert Romkes
Debraj Roy
Katarzyna Rycerz
Alberto Sanchez
Gabriele Santin
Alex Savio
Robert Schaback
Robert Schaefer
Rafal Scherer
Ulf D. Schiller
Bertil Schmidt
Martin Schreiber
Alexander Schug
Gabriela Schütz
Marinella Sciortino
Diego Sevilla
Angela Shiflet
Takashi Shimokawabe
Marcin Sieniek
Nazareen Sikkandar
    Basha
Anna Sikora
Janaína De Andrade Silva
Diana Sima
Robert Sinkovits
Haozhen Situ
Leszek Siwik
Vaclav Skala
Peter Sloot
Renata Slota
Grazyna Slusarczyk
Sucha Smanchat
Marek Smieja
Maciej Smolka
Bartlomiej Sniezynski
Isabel Sofia Brito
Katarzyna Stapor
Bogdan Staszewski

Contents – Part I

ICCS Main Track

An Efficient New Static Scheduling Heuristic
for Accelerated Architectures 3
  Thomas McSweeney, Neil Walton, and Mawussi Zounon

Multilevel Parallel Computations for Solving Multistage Multicriteria
Optimization Problems. 17
  Victor Gergel and Evgeny Kozinov

Enabling Hardware Affinity in JVM-Based Applications:
A Case Study for Big Data 31
  Roberto R. Expósito, Jorge Veiga, and Juan Touriño

An Optimizing Multi-platform Source-to-source Compiler Framework
for the NEURON MODeling Language 45
  *Pramod Kumbhar, Omar Awile, Liam Keegan, Jorge Blanco Alonso,
  James King, Michael Hines, and Felix Schürmann*

Parallel Numerical Solution of a 2D Chemotaxis-Stokes System
on GPUs Technology 59
  Raffaele D'Ambrosio, Stefano Di Giovacchino, and Donato Pera

Automatic Management of Cloud Applications with Use of Proximal
Policy Optimization. 73
  Włodzimierz Funika, Paweł Koperek, and Jacek Kitowski

Utilizing GPU Performance Counters to Characterize GPU Kernels
via Machine Learning 88
  Bob Zigon and Fengguang Song

A Massively Parallel Algorithm for the Three-Dimensional
Navier-Stokes-Boussinesq Simulations of the Atmospheric Phenomena 102
  Maciej Paszyński, Leszek Siwik, Krzysztof Podsiadło, and Peter Minev

Reconstruction of Low Energy Neutrino Events with GPUs at IceCube 118
  Maicon Hieronymus, Bertil Schmidt, and Sebastian Böser

Cache-Aware Matrix Polynomials. 132
  Dominik Huber, Martin Schreiber, Dai Yang, and Martin Schulz

QEScalor: Quantitative Elastic Scaling Framework in Distributed
Streaming Processing.   147
   *Weimin Mu, Zongze Jin, Weilin Zhu, Fan Liu, Zhenzhen Li, Ziyuan Zhu,
   and Weiping Wang*

From Conditional Independence to Parallel Execution
in Hierarchical Models.   161
   Balazs Nemeth, Tom Haber, Jori Liesenborgs, and Wim Lamotte

Improving Performance of the Hypre Iterative Solver for Uintah
Combustion Codes on Manycore Architectures Using MPI Endpoints
and Kernel Consolidation.   175
   Damodar Sahasrabudhe and Martin Berzins

Analysis of Checkpoint I/O Behavior   191
   *Betzabeth León, Pilar Gomez-Sanchez, Daniel Franco,
   Dolores Rexachs, and Emilio Luque*

Enabling EASEY Deployment of Containerized Applications
for Future HPC Systems   206
   Maximilian Höb and Dieter Kranzlmüller

Reproducibility of Computational Experiments on Kubernetes-Managed
Container Clouds with HyperFlow   220
   Michał Orzechowski, Bartosz Baliś, Renata G. Słota, and Jacek Kitowski

GPU-Accelerated RDP Algorithm for Data Segmentation.   234
   Pau Cebrian and Juan Carlos Moure

Sparse Matrix-Based HPC Tomography.   248
   *Stefano Marchesini, Anuradha Trivedi, Pablo Enfedaque,
   Talita Perciano, and Dilworth Parkinson*

heFFTe: Highly Efficient FFT for Exascale   262
   Alan Ayala, Stanimire Tomov, Azzam Haidar, and Jack Dongarra

Scalable Workflow-Driven Hydrologic Analysis in HydroFrame   276
   *Shweta Purawat, Cathie Olschanowsky, Laura E. Condon,
   Reed Maxwell, and Ilkay Altintas*

Patient-Specific Cardiac Parametrization from Eikonal Simulations   290
   *Daniel Ganellari, Gundolf Haase, Gerhard Zumbusch, Johannes Lotz,
   Patrick Peltzer, Klaus Leppkes, and Uwe Naumann*

An Empirical Analysis of Predictors for Workload Estimation in Healthcare . . .   304
   *Roberto Gatta, Mauro Vallati, Ilenia Pirola, Jacopo Lenkowicz,
   Luca Tagliaferri, Carlo Cappelli, and Maurizio Castellano*

How You Say or What You Say? Neural Activity in Message
Credibility Evaluation   312
  Łukasz Kwaśniewicz, Grzegorz M. Wójcik, Andrzej Kawiak,
  Piotr Schneider, and Adam Wierzbicki

Look Who's Talking: Modeling Decision Making Based
on Source Credibility.   327
  Andrzej Kawiak, Grzegorz M. Wójcik, Lukasz Kwasniewicz,
  Piotr Schneider, and Adam Wierzbicki

An Adaptive Network Model for Burnout and Dreaming   342
  Mathijs Maijer, Esra Solak, and Jan Treur

Computational Analysis of the Adaptive Causal Relationships Between
Cannabis, Anxiety and Sleep   357
  Merijn van Leeuwen, Kirsten Wolthuis, and Jan Treur

Detecting Critical Transitions in the Human Innate Immune System
Post-cardiac Surgery   371
  Alva Presbitero, Rick Quax, Valeria V. Krzhizhanovskaya,
  and Peter M. A. Sloot

Using Individual-Based Models to Look Beyond the Horizon:
The Changing Effects of Household-Based Clustering of Susceptibility
to Measles in the Next 20 Years   385
  Elise Kuylen, Jori Liesenborgs, Jan Broeckhove, and Niel Hens

Modelling the Effects of Antibiotics on Gut Flora Using a Nonlinear
Compartment Model with Uncertain Parameters   399
  Thulasi Jegatheesan and Hermann J. Eberl

Stochastic Volatility and Early Warning Indicator   413
  Guseon Ji, Hyeongwoo Kong, Woo Chang Kim, and Kwangwon Ahn

Boost and Burst: Bubbles in the Bitcoin Market   422
  Nam-Kyoung Lee, Eojin Yi, and Kwangwon Ahn

Estimation of Tipping Points for Critical and Transitional Regimes
in the Evolution of Complex Interbank Network   432
  Valentina Y. Guleva

Modeling of Fire Spread Including Different Heat Transfer Mechanisms
Using Cellular Automata   445
  Jarosław Wąs, Artur Karp, Szymon Łukasik, and Dariusz Pałka

Narrow Passage Problem Solution for Motion Planning   459
  Jakub Szkandera, Ivana Kolingerová, and Martin Maňák

Fault Injection, Detection and Treatment in Simulated
Autonomous Vehicles   471
  Daniel Garrido, Leonardo Ferreira, João Jacob,
  and Daniel Castro Silva

Using Cellular Automata to Model High Density Pedestrian Dynamics   486
  Grzegorz Bazior, Dariusz Pałka, and Jarosław Wąs

Autonomous Vehicles as Local Traffic Optimizers   499
  Ashna Bhatia, Jordan Ivanchev, David Eckhoff, and Alois Knoll

Modeling Helping Behavior in Emergency Evacuations Using Volunteer's
Dilemma Game.   513
  Jaeyoung Kwak, Michael H. Lees, Wentong Cai, and Marcus E. H. Ong

Learning Mixed Traffic Signatures in Shared Networks   524
  Hamidreza Anvari and Paul Lu

A Novel Metric to Evaluate In Situ Workflows.   538
  Tu Mai Anh Do, Loïc Pottier, Stephen Thomas, Rafael Ferreira da Silva,
  Michel A. Cuendet, Harel Weinstein, Trilce Estrada, Michela Taufer,
  and Ewa Deelman

Social Recommendation in Heterogeneous Evolving Relation Network   554
  Bo Jiang, Zhigang Lu, Yuling Liu, Ning Li, and Zelin Cui

DDNE: Discriminative Distance Metric Learning for Network Embedding . . .   568
  Xiaoxue Li, Yangxi Li, Yanmin Shang, Lingling Tong, Fang Fang,
  Pengfei Yin, Jie Cheng, and Jing Li

Extracting Backbone Structure of a Road Network from Raw Data   582
  Hoai Nguyen Huynh and Roshini Selvakumar

Look Deep into the New Deep Network: A Measurement Study on the
ZeroNet.   595
  Siyuan Wang, Yue Gao, Jinqiao Shi, Xuebin Wang, Can Zhao,
  and Zelin Yin

Identifying Influential Spreaders On a Weighted Network Using
HookeRank Method   609
  Sanjay Kumar, Nipun Aggarwal, and B. S. Panda

Community Aware Models of Meme Spreading
in Micro-blog Social Networks   623
  Mikołaj Kromka, Wojciech Czech, and Witold Dzwinel

A Dynamic Vote-Rank Based Approach for Effective Sequential
Initialization of Information Spreading Processes Within
Complex Networks   638
  Patryk Pazura, Kamil Bortko, Jarosław Jankowski,
  and Radosław Michalski

On the Planarity of Validated Complexes of Model Organisms
in Protein-Protein Interaction Networks   652
  Kathryn Cooper, Nathan Cornelius, William Gasper,
  Sanjukta Bhowmick, and Hesham Ali

Towards Modeling of Information Processing Within Business-Processes
of Service-Providing Organizations   667
  Sergey V. Kovalchuk, Anastasia A. Funkner, Ksenia Y. Balabaeva,
  Ilya V. Derevitskii, Vladimir V. Fonin, and Nikita V. Bukhanov

A Probabilistic Infection Model for Efficient Trace-Prediction
of Disease Outbreaks in Contact Networks   676
  William Qian, Sanjukta Bhowmick, Marty O'Neill,
  Susie Ramisetty-Mikler, and Armin R. Mikler

Eigen-AD: Algorithmic Differentiation of the Eigen Library   690
  Patrick Peltzer, Johannes Lotz, and Uwe Naumann

Correction to: A Probabilistic Infection Model for Efficient Trace-
Prediction of Disease Outbreaks in Contact Networks C1
  William Qian, Sanjukta Bhowmick, Marty O'Neill,
  Susie Ramisetty-Mikler, and Armin R. Mikler

Author Index   705

ICCS Main Track

An Efficient New Static Scheduling Heuristic for Accelerated Architectures

Thomas McSweeney[1]([⊠]) [iD], Neil Walton[1] [iD], and Mawussi Zounon[1,2] [iD]

[1] University of Manchester, Manchester, UK
thomas.mcsweeney@postgrad.manchester.ac.uk
[2] The Numerical Algorithms Group (NAG), Manchester, UK

Abstract. Heterogeneous architectures that use *Graphics Processing Units* (GPUs) for general computations, in addition to multicore CPUs, are increasingly common in high-performance computing. However many of the existing methods for scheduling precedence-constrained tasks on such platforms were intended for more diversely heterogeneous clusters, such as the classic *Heterogeneous Earliest Finish Time* (HEFT) heuristic. We propose a new static scheduling heuristic called *Heterogeneous Optimistic Finish Time* (HOFT) which exploits the binary heterogeneity of accelerated platforms. Through extensive experimentation with custom software for simulating task scheduling problems on user-defined CPU-GPU platforms, we show that HOFT can obtain schedules at least 5% shorter than HEFT's for medium-to-large numerical linear algebra application task graphs and around 3% shorter on average for a large collection of randomly-generated graphs.

Keywords: High-Performance Computing · GPU computing · Scheduling · Precedence constraints · Directed Acyclic Graphs

1 Introduction

Modern *High-Performance Computing* (HPC) machines typically comprise hundreds or even thousands of networked nodes. These nodes are increasingly likely to be *heterogeneous*, hosting one or more powerful accelerators—usually GPUs—in addition to multicore CPUs. For example, Summit, which currently heads the Top500[1] list of the world's fastest supercomputers, comprises over 4000 nodes, each with two 22-core IBM Power9 CPUs and six NVIDIA Tesla V100 GPUs.

Task-based parallel programming is a paradigm that aims to harness this processor heterogeneity. Here a program is described as a collection of *tasks*—logically discrete atomic units of work—with *precedence constraints* that define the order in which they can be executed. This can be expressed in the form of a graph, where each vertex represents a task and edges the precedence constraints

---

[1] https://www.top500.org/.

Supported by the Engineering and Physical Sciences Research Council (EPSRC).

© Springer Nature Switzerland AG 2020
V. V. Krzhizhanovskaya et al. (Eds.): ICCS 2020, LNCS 12137, pp. 3–16, 2020.
https://doi.org/10.1007/978-3-030-50371-0_1

between them. We are interested only in the case when such task graphs are *Directed Acyclic Graphs* (DAGs)—directed and without any cycles.

The immediate question is, how do we find the optimal way to assign the tasks to a set of heterogeneous processing resources while still respecting the precedence constraints? In other words, what *schedule* should we follow? This DAG scheduling problem is known to be NP-complete, even for homogeneous processors [17], so typically we must rely on heuristic algorithms that give us reasonably good solutions in a reasonable time.

A fundamental distinction is made between *static* and *dynamic* scheduling. Static schedules are fixed before execution based on the information available at that time, whereas dynamic schedules are determined during runtime. There are generic advantages and disadvantages to both: static scheduling makes greater use of the data so is superior when it is sufficiently accurate, whereas dynamic scheduling uses more recent data. In practice task scheduling is usually handled by a *runtime system*, such as OmpSs [11], PaRSEC [6], or StarPU [4]. Most such systems use previous execution traces to predict task execution and data transfer times at runtime. On a single machine the latter is tricky because of shared buses and the possibility of asynchronous data transfers. Hence at present dynamic scheduling is typically preferred. However static schedules can be surprisingly robust, even when estimates are poor [1]. Furthermore, robustness can be improved using timing distribution information [18]. In addition, superior performance can be achieved in dynamic environments by modifying an existing static schedule, rather than computing a new one from scratch [1].

In this paper we therefore focus on the problem of finding good static schedules for multicore and GPU platforms. To facilitate this investigation, we developed an open-source software simulator which allows users to simulate the static scheduling of arbitrary task DAGs on arbitrary CPU-GPU platforms, without worrying about the time or energy usage constraints imposed by real systems.

The popular HEFT scheduling heuristic comprises two phases: a *task prioritization* phase in which the order tasks are to be scheduled is determined and a *processor selection* phase in which they are actually assigned to the processing resources. In this article we introduce HOFT, which follows the HEFT framework but modifies both phases in order to exploit accelerated architectures in particular, without significantly increasing the complexity of the algorithm. HOFT works by first computing a table of *optimistic* estimates of the earliest possible times all tasks can be completed on both processor types and using this to guide both phases. Simulations with real and randomly-generated DAGs on both single and multiple GPU target platforms suggest that HOFT is always at least competitive with HEFT and frequently superior.

Explicitly, the two main contributions of this paper are:

1. A new static scheduling heuristic that is optimized specifically for accelerated heterogeneous architectures;
2. Open-source simulation software that allows researchers to implement and evaluate their own scheduling algorithms for user-defined CPU-GPU platforms in a fast and reproducible manner.

The remainder of this paper is structured as follows. In Sect. 2 we summarize the relevant existing literature. Then in Sect. 3 we explicitly define the simulation model we use to study the static task scheduling problem. We describe HEFT in detail in Sect. 4, including also benchmarking results with our simulation model and a minor modification to the algorithm that we found performs well. In Sect. 5 we describe our new HOFT heuristic, before detailing the numerical experiments that we undertook to evaluate its performance in Sect. 6. Finally in Sect. 7 we state our conclusions from this investigation and outline future work that we believe may be useful.

2 Related Work

Broadly, static scheduling methods can be divided into three categories: *mathematical programming*, *guided-random search* and *heuristics*. The first is based on formulating the scheduling problem as a mathematical program; see, for example, Kumar's constraint programming formulation in [14]. However solving these is usually so expensive that they are restricted to small task graphs. Guided-random search is a term used for any method that generates a large population of potential schedules and then selects the best among them. Typically these are more general optimization schemes such as genetic algorithms which are refined for the task scheduling problem. As a rule, such methods tend to find very high-quality schedules but take a long time to do so [7].

Heuristics are the most popular approach in practice as they are often competitive with the alternatives and considerably faster. In turn *listing* heuristics are the most popular kind. They follow a two-phase structure: an ordered list of all tasks is first constructed (task prioritization) and they are then scheduled in this order according to some rule (processor selection). HEFT is the most prominent example: all tasks are prioritized according to their *upward rank* and then scheduled on the processor expected to complete their execution at the earliest time; a fuller description is given in Sect. 4. Canon et al. [8] compared twenty different task scheduling heuristics and found that HEFT was almost always among the best in terms of both schedule makespan and robustness.

Many modifications of HEFT have been proposed in the literature, such as HEFT *with lookahead* from Bittencourt, Sakellariou and Madeira [5], which has the same task prioritization phase but schedules all tasks on the resources estimated to minimize the completion time of their *children*. This has the effect of increasing the time complexity of the algorithm so, in an attempt to incorporate a degree of lookahead into the HEFT framework without increasing the cost, Arabnejad and Barbosa proposed *Predict Earliest Finish Time* (PEFT) [3]. The main innovation is that rather than just minimizing the completion time of a task during processor selection, we also try to minimize an *optimistic* estimate of the time it will take to execute the remaining unscheduled tasks in the DAG.

Like the majority of existing methods for static DAG scheduling in heterogeneous computing, HEFT was originally intended for clusters with diverse nodes. At least one extension specifically targeting accelerated architectures has

been proposed before, namely HEFT-NC (*No Cross*) from Shetti, Fahmy and
Bretschneider [15], but our new HOFT heuristic differs from this in both the
task prioritization and processor selection phases of the algorithm.

3   Simulation Model

In this paper we use a simulation model to study the static task scheduling
problem for multicore and GPU. This simulator follows the mathematical model
described in Sect. 3.1 and therefore facilitates the evaluation of scheduling algorithms for idealized CPU-GPU platforms. The advantage of this approach is
that it allows us to compare multiple algorithms and determine how intrinsically
well-suited they are for accelerated architectures. Although this model may not
capture the full range of real-world behavior, we gathered data from a single
heterogeneous node of a local computing cluster to guide its development and
retain the most salient features. This node comprises four octacore Intel (Skylake) Xeon Gold 6130 CPUs running at 2.10 GHz with 192 GB RAM and four
Nvidia V100-SXM2-16 GB (Volta) GPUs, each with 16 GB GPU global memory, 5120 CUDA Cores and NVLink interconnect. We used *Basic Linear Algebra
Subroutine* (BLAS) [10] and *Linear Algebra PACKage* (LAPACK) [2] kernels for
benchmarking as they are widely-used in scientific computing applications.

The simulator is implemented in `Python` and the complete source code is
available on Github[2]. All code used to generate results presented in this paper is
available in the folder `simulator/scripts` so interested researchers may repeat
our experiments for themselves. In addition, users may make modifications to
the simulator that they believe will more accurately reflect their own target
environment.

3.1   Mathematical Model

The simulator software implements the following mathematical model of the
problem. Suppose we have a task DAG G consisting of n tasks and e edges that
we wish to execute on a target platform H comprising P processing resources
of two types, P_C CPU resources and $P - P_C = P_G$ GPU resources. In keeping
with much of the related literature and based on current programming practices,
we consider CPU cores individually but regard entire GPUs as discrete [1]. For
example, a node comprising 4 GPUs and 4 octacore CPUs would be viewed as
4 GPU resources and $4 \times 8 = 32$ CPU resources.

We assume that all tasks $t_1, \ldots, t_n$ are atomic and cannot be divided across
multiple resources or aggregated to form larger tasks. Further, all resources can
only execute a single task at any one time and can in principle execute all tasks,
albeit with different processing times. Given the increasing versatility of modern
GPUs and the parallelism of modern CPUs, the latter is a much less restrictive
assumption than it once may have been.

---

[2] https://github.com/mcsweeney90/heterogeneous_optimistic_finish_time.

In our experiments, we found that the spread of kernel processing times was usually tight, with the standard deviation often being two orders of magnitude smaller than the mean. Thus we assume that all task execution times on all processing resources of a single type are identical. In particular, this means that each task has only two possible *computation costs*: a CPU execution time $w_C(t_i)$ and a GPU execution time $w_G(t_i)$. When necessary, we denote by w_{im} the processing time of task t_i on the specific resource p_m.

The *communication cost* between task t_i and its child t_j is the length of time between when execution of t_i is complete and execution of t_j can begin, including all relevant latency and data transfer times. Since this depends on where each task is executed, we view this as a function $c_{ij}(p_m, p_n)$. We assume that the communication cost is always zero when $m = n$ and that there are only four possible communication costs between tasks t_i and t_j when this isn't the case: $c_{ij}(C, C)$, from a CPU to a different CPU; $c_{ij}(C, G)$, from CPU to GPU; $c_{ij}(G, C)$, from GPU to CPU; and $c_{ij}(G, G)$ from GPU to a different GPU.

A *schedule* is a mapping from tasks to processing resources, as well as the precise time at which their execution should begin. Our goal is to find a schedule which minimizes the *makespan* of the task graph, the total execution time of the application it represents. A task with no successors is called an *exit* task. Once all tasks have been scheduled, the makespan is easily computed as the earliest time all exit tasks will be completed. Note that although we assume that all costs represent time, they could be anything else we wish to minimize, such as energy consumption, so long as this is done consistently. We do not however consider the multi-objective optimization problem of trading off two or more different cost types here.

3.2   Testing Environments

In the numerical experiments described later in this article, we consider two simulated target platforms: *Single GPU*, comprising 1 GPU and 1 octacore CPU, and *Multiple GPU*, comprising 4 GPUs and 4 octacore CPUs. The latter represents the node we used to guide the development of our simulator and the former is considered in order to study how the number of GPUs affects performance. We follow the convention that a CPU core is dedicated to managing each of the GPUs [4], so these two platforms are actually assumed to comprise 7 CPU and 1 GPU resources, and 28 CPU and 4 GPU resources, respectively. Based on our exploratory experiments, we make two further assumptions. First, since communication costs between CPU resources were negligible relative to all other combinations, we assume they are zero—i.e., $c_{ij}(C, C) = 0, \forall i, j$. Second, because CPU-GPU communication costs were very similar to the corresponding GPU-CPU and GPU-GPU costs, we take them to be identical—i.e., $c_{ij}(C, G) = c_{ij}(G, C) = c_{ij}(G, G), \forall i, j$. These assumptions will obviously not be representative of all possible architectures but the simulator software allows users to repeat our experiments for more accurate representations of their own target platforms if they wish.

We consider the scheduling of two different sets of DAGs. The first consists of ten DAGs comprising between 35 and 22, 100 tasks which correspond to the *Cholesky factorization* of $N \times N$ tiled matrices, where $N = 5, 10, 15, \ldots, 50$. In particular, the DAGs are based on a common implementation of Cholesky factorization for tiled matrices which uses GEMM (matrix multiplication), SYRK (symmetric rank-k update) and TRSM (triangular solve) BLAS kernels, as well as the POTRF (Cholesky factorization) LAPACK routine [10]. All task CPU/GPU processing times are means of 1000 real timings of that task kernel. Likewise, communication costs are sample means of real communication timings between the relevant task and resource types. All numerical experiments were performed for tile sizes 128 and 1024; which was used will always be specified where results are presented. Those sizes were chosen as they roughly mark the upper and lower limits of tile sizes typically used for CPU-GPU platforms.

The standard way to quantify the relative amount of communication and computation represented by a task graph is the *Computation-to-Communication Ratio* (CCR), the mean computation cost of the DAG divided by the mean communication cost. For the Cholesky DAGs, the CCR was about 1 for tile size 128 and about 18 for tile size 1024, with minor variation depending on the total number of tasks in the DAG.

We constructed a set of randomly-generated DAGs with a wide range of CCRs, based on the topologies of the 180 DAGs with 1002 tasks from the *Standard Task Graph* (STG) set [16]. Following the approach in [9], we selected GPU execution times for all tasks uniformly at random from $[1, 100]$ and computed the corresponding CPU times by multiplying by a random variable from a Gamma distribution. To consider a variety of potential applications, for each DAG we made two copies: *low acceleration*, for which the mean and standard deviation of the Gamma distribution was defined to be 5, and *high acceleration*, for which both were taken to be 50 instead. These values roughly correspond to what we observed in our benchmarking of BLAS and LAPACK kernels with tile sizes 128 and 1024, respectively. Finally, for both parameter regimes, we made three copies of each DAG and randomly generated communication costs such that the CCR fell into each of the intervals $[0, 10]$, $[10, 20]$ and $[20, 50]$. Thus in total the random DAG set contains $180 \times 2 \times 3 = 1080$ DAGs.

4   HEFT

Recall that as a listing scheduler HEFT comprises two phases, an initial *task prioritization* phase in which the order all tasks are to be scheduled is determined and a *processor selection* phase in which the processing resource each task is to be scheduled on is decided. Here we describe both in order to give a complete description of the HEFT algorithm.

The *critical path* of a DAG is the longest path through it, and is important because it gives a lower bound on the optimal schedule makespan for that DAG. Heuristics for homogeneous platforms often use the *upward rank*, the length of the critical path from that task to an exit task, including the task itself

[17], to determine priorities. Computing the critical path is not straightforward for heterogeneous platforms so HEFT extends the idea by using *mean* values instead. Intuitively, the task prioritization phase of HEFT can be viewed as an approximate dynamic program applied to a simplified version of the task DAG that uses mean values to set all weights.

More formally, we first define the *mean execution cost* of all tasks t_i through

$$\overline{w_i} := \sum_{m=1}^{P} \frac{w_{im}}{P} = \frac{w_C(t_i)P_C + w_G(t_i)P_G}{P}, \tag{1}$$

where the second expression is how $\overline{w_i}$ would be computed under the assumptions of our model. Likewise, the *mean communication cost* $\overline{c_{ij}}$ between t_i and t_j is the average of all such costs over all possible combinations of resources,

$$\overline{c_{ij}} = \frac{1}{P^2} \sum_{m,n} c_{ij}(p_m, p_n) = \frac{1}{P^2} \sum_{k,\ell \in \{C,G\}} A_{k\ell} c_{ij}(k, \ell), \tag{2}$$

where $A_{CC} = P_C(P_C - 1)$, $A_{CG} = P_C P_G = A_{GC}$, and $A_{GG} = P_G(P_G - 1)$. For all tasks t_i in the DAG, we define their upward ranks $rank_u(t_i)$ recursively, starting from the exit task(s), by

$$rank_u(t_i) = \overline{w_i} + \max_{t_j \in Ch(t_i)} (\overline{c_{ij}} + rank_u(t_j)), \tag{3}$$

where $Ch(t_i)$ is the set of t_i's immediate successors in the DAG. The task prioritization phase then concludes by listing all tasks in decreasing order of upward rank, with ties broken arbitrarily.

The processor selection phase of HEFT is now straightforward: we move down the list and assign each task to the resource expected to complete its execution at the earliest time. Let R_{m_i} be the earliest time at which the processing resource p_m is actually free to execute task t_i, $Pa(t_i)$ be the set of t_i's immediate predecessors in the DAG, and $AFT(t_k)$ be the time when execution of a task t_k is actually completed (which in the static case is known precisely once it has been scheduled). Then the *earliest start time* of task t_i on processing resource p_m is computed through

$$EST(t_i, p_m) = \max \left\{ R_{m_i}, \max_{t_k \in Pa(t_i)} (AFT(t_k) + c_{ki}(p_k, p_m)) \right\} \tag{4}$$

and the *earliest finish time* $EFT(t_i, p_m)$ of task t_i on p_m is given by

$$EFT(t_i, p_m) = w_{im} + EST(t_i, p_m). \tag{5}$$

HEFT follows an *insertion-based* policy that allows tasks to be inserted between two that are already scheduled, assuming precedence constraints are still respected, so R_{m_i} may not simply be the latest finish time of all tasks on p_m. A complete description of HEFT is given in Algorithm 1. HEFT has a time complexity of $O(P \cdot e)$. For dense DAGs, the number of edges is proportional to n^2, where n is the number of tasks, so the complexity is effectively $O(n^2 P)$ [17].

Algorithm 1: HEFT.
___
1   Set the computation cost of all tasks using (1)
2   Set the communication cost of all edges using (2)
3   Compute $rank_u$ for all tasks according to (3)
4   Sort the tasks into a priority list by non-increasing order of $rank_u$
5   **for** *task in list* **do**
6   **for** *each resource p_k* **do**
7 | Compute $EFT(t_i, p_k)$ using (4) and (5)
8   **end**
9      $p_m := \arg\min_k(EFT(t_i, p_k))$
10   Schedule t_i on the resource p_m
11 **end**
___

4.1 Benchmarking

Using our simulation model, we investigated the quality of the schedules computed by HEFT for the Cholesky and randomly-generated DAG sets on both the single and multiple GPU target platforms described in Sect. 3. The metric used for evaluation was the *speedup*, the ratio of the *minimal serial time* (MST)—the minimum execution time of the DAG on any single resource—to the makespan. Intuitively, speedup tells us how well the schedule exploits the parallelism of the target platform.

Figure 1a shows the speedup of HEFT for the Cholesky DAGs with tile size 128. The most interesting takeaway is the difference between the two platforms. With multiple GPUs the speedup increased uniformly with the number of tasks until a small decline for the very largest DAG, but for a single GPU the speedup stagnated much more quickly. This was due to the GPU being continuously busy and adding little additional value once the DAGs became sufficiently large and more GPUs therefore postponing this effect. Results were broadly similar for Cholesky DAGs with tile size 1024, with the exception that the speedup values were uniformly smaller, reaching a maximum of just over four for the multiple GPU platform. This was because the GPUs were so much faster for the larger tile size that HEFT made little use of the CPUs and so speedup was almost entirely determined by the number of GPUs available.

Figure 1b shows the speedups for all 540 high acceleration randomly-generated DAGs, ordered by their CCRs; results were broadly similar for the low acceleration DAGs. The speedups for the single GPU platform are much smaller with a narrower spread compared to the other platform, as for the Cholesky DAGs. More surprising is that HEFT sometimes returned a schedule with speedup less than one for DAGs with small CCR values, which we call a *failure* since this is obviously unwanted behavior. These failures were due to the greediness of the HEFT processor selection phase, which always schedules a task on the processing resource that minimizes its earliest finish time without considering the communication costs that may later be incurred by doing so.

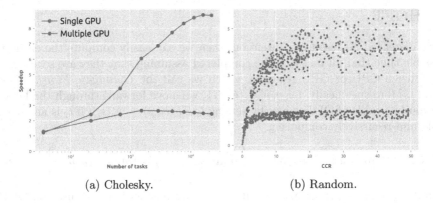

(a) Cholesky.   (b) Random.

Fig. 1. Speedup of HEFT for Cholesky (tile size 128) and random (high acceleration) DAG sets.

The effect is more pronounced when the CCR is low because the unforeseen communication costs are proportionally larger.

4.2 HEFT-WM

The implicit assumption underlying the use of mean values when computing task priorities in HEFT is that the probability of a task being scheduled on a processing resource is identical for all resources. But this is obviously not the case: if a task's GPU execution time is ten times smaller than its CPU execution time then it is considerably more likely to be scheduled on a GPU, even if not precisely ten times so. This suggests that we should weight the mean values according to each task's *acceleration ratio* r_i, the CPU time divided by the GPU time. In particular, for each task t_i we estimate its computation cost to be

$$\overline{w_i} = \frac{w_C(t_i)P_C + r_i w_G(t_i)P_G}{P_C + r_i P_G}, \tag{6}$$

and for each edge (say, between tasks t_i and t_j) the estimated communication cost is

$$\overline{c_{ij}} = \frac{A_{CC} \cdot c_{ij}(C,C) + A_{CG}\big(r_i c_{ij}(G,C) + r_j c_{ij}(C,G)\big) + r_i r_j A_{GG} \cdot c_{ij}(G,G)}{(r_i P_G + P_C) \cdot (r_j P_G + P_C)}. \tag{7}$$

We call the modified heuristic which follows Algorithm 1 but uses (6) instead of (1) and (7) instead of (2) HEFT-WM (for *Weighted Mean*). We were unable to find any explicit references to this modification in the literature but given its simplicity we do not think it is a novel idea and suspect it has surely been used before in practice.

5 HOFT

If we disregard all resource contention, then we can easily compute the earliest possible time that all tasks can be completed assuming that they are scheduled on either a CPU or GPU resource, which we call the *Optimistic Finish Time* (OFT). More specifically, for $p, p' \in \{C, G\}$, we move forward through the DAG and build a table of OFT values by setting $OFT(t_i, p) = w_p(t_i)$, if t_i is an entry task, and recursively computing

$$OFT(t_i, p) = w_p(t_i) + \max_{t_j \in Pa(t_i)} \left\{ \min_{p'} \{OFT(t_j, p') + \delta_{pp'} c_{ij}(p, p')\} \right\}, \quad (8)$$

for all other tasks, where $\delta_{pp'} = 1$ if $p = p'$ and 0 otherwise. We use the OFT table as the basis for the task prioritization and processor selection phases of a new HEFT-like heuristic optimized for CPU-GPU platforms that we call *Heterogeneous Optimistic Finish Time* (HOFT). Note that computing the OFT table does not increase the order of HEFT's time complexity.

Among several possible ways of using the OFT to compute a complete task prioritization, we found the most effective to be the following. First, define the weights of all tasks to be the ratio of the maximum and minimum OFT values,

$$\overline{w_i} = \frac{\max\{OFT(t_i, C), OFT(t_i, G)\}}{\min\{OFT(t_i, C), OFT(t_i, G)\}}. \quad (9)$$

Now assume that all edge weights are zero, $\overline{c_{ij}} \equiv 0, \forall i, j$, and compute the upward rank of all tasks with these values. Upward ranking is used to ensure that all precedence constraints are met. Intuitively, tasks with a strong preference for one resource type—as suggested by a high ratio—should be scheduled first.

We also propose a new processor selection phase which proceeds as follows. Working down the priority list, each task t_i is scheduled on the processing resource p_m with the smallest EFT as in HEFT except when p_m is not also the fastest resource type for that task. In such cases, let p_f be the resource of the fastest type with the minimal EFT and compute

$$s_m := EFT(t_i, p_f) - EFT(t_i, p_m), \quad (10)$$

the saving that we expect to make by scheduling t_i on p_m rather than p_f. Suppose that p_m is of type $T_m \in \{C, G\}$. By assuming that each child task t_j of t_i is scheduled on the type of resource T_j which minimizes its OFT and disregarding the potential need to wait for other parent tasks to complete, we estimate $E(Ch(t_i)|p_m)$, the earliest finish time of all child tasks, given that t_i is scheduled on p_m, through

$$E(Ch(t_i)|p_m) := \max_{t_j \in Ch(t_i)} \left(EFT(t_i, p_m) + c_{ij}(T_m, T_j) + w_{T_j}(t_j) \right). \quad (11)$$

Likewise for p_f we compute $E(Ch(t_i)|p_f)$ and if

$$s_m > E(Ch(t_i)|p_m) - E(Ch(t_i)|p_f) \quad (12)$$

we schedule task t_i on p_m; otherwise, we schedule it on p_f. Intuitively, the processor selection always chooses the resource with the smallest EFT unless by doing so we expect to increase the earliest possible time at which all child tasks can be completed.

Algorithm 2: HOFT.

1 Compute the OFT table for all tasks using (8)
2 Set the computation cost of all tasks using (9)
3 Set the communication cost of all edges to be zero
4 Compute $rank_u$ for all tasks according to (3)
5 Sort the tasks into a priority list by non-increasing order of $rank_u$
6 **for** *task in list* **do**
7 $\quad$ **for** *each resource p_k* **do**
8 $\quad\quad$ | Compute $EFT(t_i, p_k)$ using (4) and (5)
9 $\quad$ **end**
10 $\quad$ $p_m := \arg\min_k(EFT(t_i, p_k))$
11 $\quad$ **if** $w_{im} \neq \min(w_C(t_i), w_G(t_i))$ **then**
12 $\quad\quad$ | $p_f := \arg\min_k\big(EFT(t_i, p_k)|w_{ik} = \min(w_C(t_i), w_G(t_i))\big)$
13 $\quad\quad$ Compute s_m using (10)
14 $\quad\quad$ Compute $E(Ch(t_i)|p_m)$ and $E(Ch(t_i)|p_f)$ using (11)
15 $\quad\quad$ **if** (12) *holds* **then**
16 $\quad\quad\quad$ | Schedule t_i on p_m
17 $\quad\quad$ **else**
18 $\quad\quad\quad$ | Schedule t_i on p_f
19 $\quad\quad$ **end**
20 $\quad$ **end**
21 **end**

---

6   Simulation Results

Figure 2 shows the reduction in schedule makespan, as a percentage of the HEFT schedule, achieved by HOFT and HEFT-WM for the set of Cholesky DAGs, on both the single and multiple GPU target platforms. The overall trend for the multiple GPU platform is that HOFT improves relative to both HEFT variants as the number of tasks in the DAG grows larger; it is always better than standard HEFT for tile size 1024. For the single GPU platform, HOFT was almost always the best, except for the smallest DAGs and the largest DAG with tile size 128 (for which all three heuristics were almost identical). Interestingly, we found that the processor selection phase of HOFT never actually differed from HEFT's for these DAGs and so the task prioritization phase alone was key.

HOFT achieved smaller makespans than HEFT on average for the set of randomly-generated DAGs, especially for those with high acceleration, but was slightly inferior to HEFT-WM, as can be seen from Table 1. However also

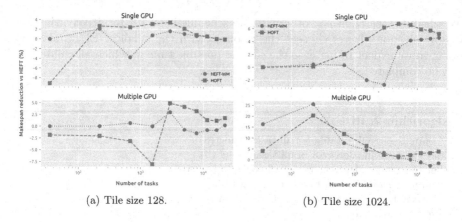

(a) Tile size 128.   (b) Tile size 1024.

Fig. 2. HEFT-WM and HOFT compared to HEFT for Cholesky DAGs.

included in the table is HOFT-WM, the heuristic obtained by using the HEFT-WM task prioritization with the HOFT processor selection. HOFT-WM was identical to HEFT-WM for the Cholesky DAGs but improved on both heuristics for the randomly-generated DAG set, suggesting that the HOFT processor selection phase is generally more effective than HEFT's, no matter which task ranking is used. The alternative processor selection also reduced the failure rate for DAGs with very low CCR by about half on the single GPU platform, although it made little difference on the multiple GPU platform.

Table 1. Makespan reduction vs. HEFT. Shown are both the average percentage reduction (*APR*) and the percentage of (540) DAGs for which each heuristic improved on the HEFT schedule (*Better*).

Heuristic	Single GPU				Multiple GPU			
	Low acc.		High acc.		Low acc.		High acc.	
	APR	*Better*	*APR*	*Better*	*APR*	*Better*	*APR*	*Better*
HEFT-WM	0.8	74.8	2.3	69.6	1.6	84.8	2.4	79.8
HOFT	−0.2	50.3	3.8	83.1	1.4	69.2	2.3	76.5
HOFT-WM	0.8	70.9	4.6	76.9	1.4	78.1	3.7	81.1

7   Conclusions

Overall our simulations suggest that HOFT is often superior to—and always competitive with—both standard HEFT and HEFT-WM for multicore and GPU platforms, especially when task acceleration ratios are high. The processor selection phase in particular appears to be more effective, at least on average, for any task prioritization. It should also be noted that HEFT-WM was almost always superior to the standard algorithm in our simulations, suggesting that it should perhaps be the default in accelerated environments.

Although the number of failures HOFT recorded for DAGs with low CCRs was slightly smaller than for HEFT, the failure probability was still unacceptably high. Using the lookahead processor selection from [5] instead reduced the probability of failure further but it was still nonzero and the additional computational cost was considerable. We investigated cheaper *sampling-based* selection phases that consider only a small sample of the child tasks, selected either at random or based on priorities. These did reduce the failure probability in some cases but the improvement was usually minor. Alternatives to lookahead that we intend to consider in future are the duplication [13] or aggregation of tasks.

Static schedules for multicore and GPU are most useful in practice as a foundation for superior dynamic schedules [1], or when their robustness is reinforced using statistical analysis [18]. This paper was concerned with computing the initial static schedules; the natural next step is to consider their extension to real—i.e., dynamic—environments. This is often called *stochastic* scheduling, since the costs are usually modeled as random variables from some—possibly unknown—distribution. The goal is typically to find methods that bound variation in the schedule makespan, which is notoriously difficult [12]. Experimentation with existing stochastic scheduling heuristics, such as *Monte Carlo Scheduling* [18], suggested to us that the main issue with adapting such methods for multicore and GPU is their cost, which can be much greater than computing the static schedule itself. Hence in future work we intend to investigate cheaper ways to make effective use of static schedules for real platforms.

References

1. Agullo, E., Beaumont, O., Eyraud-Dubois, L., Kumar, S.: Are static schedules so bad? A case study on Cholesky factorization. In: 2016 IEEE International Parallel and Distributed Processing Symposium (IPDPS), pp. 1021–1030, May 2016. https://doi.org/10.1109/IPDPS.2016.90
2. Anderson, E., et al.: LAPACK Users' Guide, 3rd edn. Society for Industrial and Applied Mathematics, Philadelphia (1999)
3. Arabnejad, H., Barbosa, J.G.: List scheduling algorithm for heterogeneous systems by an optimistic cost table. IEEE Trans. Parallel Distrib. Syst. **25**(3), 682–694 (2014). https://doi.org/10.1109/TPDS.2013.57
4. Augonnet, C., Thibault, S., Namyst, R., Wacrenier, P.A.: StarPU: a unified platform for task scheduling on heterogeneous multicore architectures. Concurr. Comput. Pract. Exp. **23**(2), 187–198 (2011). https://doi.org/10.1002/cpe.1631
5. Bittencourt, L.F., Sakellariou, R., Madeira, E.R.M.: DAG scheduling using a lookahead variant of the Heterogeneous Earliest Finish Time algorithm. In: 2010 18th Euromicro Conference on Parallel, Distributed and Network-based Processing, pp. 27–34, February 2010. https://doi.org/10.1109/PDP.2010.56
6. Bosilca, G., Bouteiller, A., Danalis, A., Faverge, M., Hérault, T., Dongarra, J.J.: PaRSEC: exploiting heterogeneity to enhance scalability. Comput. Sci. Eng. **15**(6), 36–45 (2013). https://doi.org/10.1109/MCSE.2013.98

7. Braun, T.D., et al.: A comparison of eleven static heuristics for mapping a class of independent tasks onto heterogeneous distributed computing systems. J. Parallel Distrib. Comput. **61**(6), 810–837 (2001). https://doi.org/10.1006/jpdc.2000.1714

8. Canon, L.C., Jeannot, E., Sakellariou, R., Zheng, W.: Comparative evaluation of the robustness of DAG scheduling heuristics. In: Gorlatch, S., Fragopoulou, P., Priol, T. (eds.) Grid Computing, pp. 73–84. Springer, Cham (2008). https://doi.org/10.1007/978-0-387-09457-1_7

9. Canon, L.-C., Marchal, L., Simon, B., Vivien, F.: Online scheduling of task graphs on hybrid platforms. In: Aldinucci, M., Padovani, L., Torquati, M. (eds.) Euro-Par 2018. LNCS, vol. 11014, pp. 192–204. Springer, Cham (2018). https://doi.org/10.1007/978-3-319-96983-1_14

10. Dongarra, J.J., Du Croz, J., Hammarling, S., Duff, I.S.: A set of level 3 basic linear algebra subprograms. ACM Trans. Math. Softw. **16**(1), 1–17 (1990). https://doi.org/10.1145/77626.79170

11. Duran, A., et al.: OmpSs: a proposal for programming heterogeneous multi-core architectures. Parallel Process. Lett. **21**(02), 173–193 (2011). https://doi.org/10.1142/S0129626411000151

12. Hagstrom, J.N.: Computational complexity of PERT problems. Networks **18**(2), 139–147. https://doi.org/10.1002/net.3230180206

13. Ahmad, I., Kwok, Y.-K.: On exploiting task duplication in parallel program scheduling. IEEE Trans. Parallel Distrib. Syst. **9**(9), 872–892 (1998). https://doi.org/10.1109/71.722221

14. Kumar, S.: Scheduling of dense linear algebra kernels on heterogeneous resources. Ph.D. thesis, University of Bordeaux, April 2017. https://tel.archives-ouvertes.fr/tel-01538516/file/KUMAR_SURAL_2017.pdf

15. Shetti, K.R., Fahmy, S.A., Bretschneider, T.: Optimization of the HEFT algorithm for a CPU-GPU environment. In: PDCAT, pp. 212–218 (2013). https://doi.org/10.1109/PDCAT.2013.40

16. Tobita, T., Kasahara, H.: A standard task graph set for fair evaluation of multiprocessor scheduling algorithms. J. Scheduling **5**(5), 379–394. https://doi.org/10.1002/jos.116

17. Topcuoglu, H., Hariri, S., Wu, M.Y.: Performance-effective and low-complexity task scheduling for heterogeneous computing. IEEE Trans. Parallel Distrib. Syst. **13**(3), 260–274 (2002). https://doi.org/10.1109/71.993206

18. Zheng, W., Sakellariou, R.: Stochastic DAG scheduling using a Monte Carlo approach. J. Parallel Distrib. Comput. **73**(12), 1673–1689 (2013). https://doi.org/10.1016/j.jpdc.2013.07.019

Multilevel Parallel Computations for Solving Multistage Multicriteria Optimization Problems

Victor Gergel$^{(\boxtimes)}$ and Evgeny Kozinov

Lobachevsky State University of Nizhni Novgorod, Nizhni Novgorod, Russia
`gergel@unn.ru, evgeny.kozinov@itmm.unn.ru`

Abstract. In the present paper, a novel approach for solving the computationally costly multicriteria optimization problems is considered. Within the framework of the developed approach, the obtaining of the efficient decisions is ensured by means of several different methods for the scalarization of the efficiency criteria. The proposed approach provides an opportunity to alter the scalarization methods and the parameters of these ones in the course of computations that results in the necessity of multiple solving the time-consuming global optimization problems. Overcoming the computational complexity is provided by reusing the computed search information and efficient parallel computing on high-performance computing systems. The performed numerical experiments confirmed the developed approach to allow reducing the amount and time of computations for solving the time-consuming multicriteria optimization problems.

Keywords: Multicriteria optimization · Criteria scalarization · Global optimization · Search information · Parallel computations · Numerical experiment

1 Introduction

The paper discusses a novel approach for solving the time-consuming multicriteria optimization (MCO) problems. Such problems arise in many applications that is confirmed by a wide spectrum of research on this subject – see, for example, monographs [1–5] and reviews of scientific and practical results [6–8].

The solving of the MCO problems is usually reduced to finding the efficient (non-dominated) decisions[1]. In the limiting case, it may appear to be necessary to obtain the whole set of the efficient decisions (the Pareto set) that may

---

[1] The solutions, which cannot be improved with respect to any criteria without worsening of the efficiency values with respect to other criteria are understood as the efficient (non-dominated) decisions.

This research was supported by the Russian Science Foundation, project No 16-11-10150 "Novel efficient methods and software tools for time-consuming decision making problems using supercomputers of superior performance.".

require a large amount of computations. Another approach used widely consists in finding a relatively small set of efficient decisions only. As a rule, to find particular efficient decisions, the vector criterion is transformed to a single (scalar) criterion, which can be optimized by some algorithms of nonlinear programming. Among such approaches, one can outline various types of criteria convolutions, the lexicographic optimization methods, the reference point algorithms, etc. [1–3].

In the present paper it is supposed that the efficiency criteria can be multiextremal and computing the values of criteria and constraints can be time-consuming. But the computational complexity is even higher as it is supposed that the applied scalarization method can be varied in the course of computations that leads to a necessity of multiple solving of the global optimization problems [9–11]. In the framework of developed approach efficient global optimization methods are proposed for solving such computationally expensive MCO problems and different multilevel parallel computation schemes are investigated for executing these methods on high-performance supercomputer systems.

Further structure of the paper is organized as follows. In Sect. 2, the statement of the multicriteria optimization problem is given and a general scheme for the criteria scalarization is proposed. In Sect. 3, efficient global optimization methods utilizing the computed search information are considered. In Sect. 4, the multilevel schemes of parallel computations for the multistage solving of the computational-costly MCO problems are given. Section 5 presents the results of numerical experiments confirming the developed approach to be promising. In Conclusion, the obtained results are summarized and main directions of further investigations are outlined.

2 Multicriteria Optimization Problem Statement

Multicriteria optimization (MCO) problem can be formulated as follows [1–5]

$$f(y) \to \min, y \in D,\qquad(1)$$

where $y = (y_1, y_2, \ldots, y_N)$ is a *vector of varied parameters*,
$f(y) = (f_1(y), f_2(y), \ldots, f_s(y))$ is a vector efficiency criterion, and $D \subset R^N$ is a search domain

$$D = \{y \in R^N : a_i \le y_i \le b_i, 1 \le i \le N\},\qquad(2)$$

for the given vectors a and b. Without loss of generality, in further consideration the criteria $f_i(y)$, $1 \le i \le s$ are suggested to be non-negative and the decrease of these ones corresponds to the increase of the decision efficiency.

In the most general case, the criteria $f_i(y)$, $1 \le i \le s$ can be multiextremal, and the procedure of computing the values of these ones can appear to be time consuming. Also, the criteria $f_i(y)$, $1 \le i \le s$ are supposed to satisfy the Lipschitz condition

$$|f_i(y_1) - f_i(y_2)| \le L_i \|y_1 - y_2\|, 1 \le i \le s,\qquad(3)$$

where L_i, $1 \leq j \leq s$ are the Lipschitz constants for the criteria $f_j(y)$, $1 \leq j \leq s$ and $\| * \|$ denotes the Euclidean norm in R^N.

The efficiency criteria of the MCO problems are usually contradictory, and the decisions $y^* \in D$ with the best values with respect to all criteria simultaneously may be absent. In such situations, for the MCO problems it is appropriate to find the efficient (non-dominated) decisions, for which the improvement of the values with respect to any criterion results in worsening the efficiency values with respect to other criteria. Obtaining of the whole set of efficient decisions (the Pareto set) may require to perform a large amount of computations. As a result, another approach is applied for solving the MCO problems often – finding only a relatively small set of efficient decisions defined according to the requirements of the decision maker.

An approach to obtaining particular efficient decisions used widely consists in the transformation of a vector criterion into some general scalar criterion of efficiency[2] [1–5]

$$\min \varphi(y) = F(\alpha, y), y \in D, \tag{4}$$

where F is the objective function generated as a result of scalarization of the criteria f_i, $1 \leq i \leq s$, α is a vector of parameters of the criteria convolution applied, and D is the search domain from (2). Because of (3), the function $F(\alpha, y)$ also satisfies the Lipschitz condition with some constant L i.e.

$$|F(\alpha, y') - F(\alpha, y'')| \leq L\|y_1 - y_2\|. \tag{5}$$

To construct a general scalar efficiency function $F(\alpha, y)$ from (4), one can use, in particular, the following methods of the criteria scalarization.

1. One of the scalarization methods used often consists in the use of the minmax convolution of criteria [1–3]:

$$F_1(\lambda, y) = \max(\lambda_i f_i(y), 1 \leq i \leq s),$$
$$\lambda = (\lambda_1, \lambda_2, \ldots, \lambda_s) \in \Lambda \subset R^s : \textstyle\sum_{i=1}^s \lambda_i = 1, \lambda_i \geq 0, 1 \leq i \leq s. \tag{6}$$

2. Another approach used widely is applied if there are some *a priori* estimates of the criteria values for the required decision (for example, based on some ideal decision or any existing prototype). In such cases, the solving of a MCO problem may consist in finding an efficient decision corresponding to given criteria values most completely. The scalar criterion $F_2(\lambda, y)$ can be presented as root-mean-square deviation of a point $y \in D$ from the ideal decision y^* [3]:

$$F_2(\theta, y) = \frac{1}{s} \sum_{i=1}^s \theta_i(f_i(y) - f_i(y^*))^2, y \in D, \tag{7}$$

where the parameters $0 \leq \theta_i \leq 1$, $1 \leq i < s$ are the indicators of importance of the approximation precision with respect to each varied parameter y_i, $1 \leq i \leq N$ separately.

---

[2] It is worth noting that such an approach provides an opportunity to use a wide set of already existing global optimization methods for solving the MCO problems.

3. In the case when the criteria can be arranged in the importance, the method of successive concessions (MSC) [2–4] is applied often, according to which the solving of a MCO problem is reduced to a multistage solving of the global optimization problems with nonlinear constraints:

$$f_1^* = \min_{y \in D} f_1(y),$$
$$f_i^* = \min_{y \in D} f_i(y), f_j(y) \le f_j^* + \delta_j, 1 \le j < i, 1 < i \le s, \quad (8)$$
$$P_{lex}^{\delta}(f, \delta, D) = Arg\min_{y \in D} f_s(y), f_j(y) \le f_j^* + \delta_j, 1 \le j < s,$$

where the notation Arg means the set of all points $y \in D$, at which the minimum value of the optimized criterion is achieved and δ_i, $1 \le i \le s$ are the feasible concessions from the minimum values of the efficiency criteria. At that, the multi-step computations in the scheme (8) may be reduced to the minimization of the general scalar criterion $F_3(\lambda, y)$ [11]:

$$F_3(\delta, y) = f_s(y), f_i(y) \le f_i^{min} + \delta_i(f_i^{max} - f_i^{min}), 1 \le i < s, y \in D, \quad (9)$$

where f_i^{min}, f_i^{max}, $1 \le i < s$ are the minimum and maximum values of the criteria[3] in the domain D from (2) and $0 \le \delta_i \le 1$, $1 \le i < s$ are the concessions normalized to the interval $[0, 1]$.

It is worth noting that due to the possibility of the changing of the requirements to the optimality in the course of computations, the form of the scalar criterion $F(\alpha, y)$ from (4) may vary. Thus, it may turn to be necessary to alter the scalarization method used (6)–(9) and/or to change the convolution parameters λ, θ, and δ [9–11]. Such variations form a set of scalar global optimization problems (4)

$$\mathbb{F}_T = \{F_k(\alpha_i, y) : 1 \le i \le T, k = 1, 2, 3\}. \quad (10)$$

This set of problems may be formed progressively in the course of computations; the problems from the set may be solved strictly sequentially or simultaneously in the time-share mode. Besides, the problems from the set $\mathbb{F}_T$ may be solved in parallel using high-performance computer systems. An opportunity of forming the set $\mathbb{F}_T$ allows to formulate a *new approach to the multistage solving of the multicriteria optimization problems* (MMCO).

3   Methods of Multistage Solving of Multicriteria Optimization Problems

In the general case, the problems of the set $\mathbb{F}_T$ from (10) are the global optimization problems, the solving of which implies constructing some grids covering the search domain D – see, for example, [12–17]. The necessity to construct the coverage of the search domain D leads to the "curse of dimensionality" – the

---

[3] Since the magnitudes f_i^{min}, f_i^{max}, $1 \le i < s$, may be unknown *a priori*, the values of these ones may be replaced by some numerical estimates, which may be obtained using the available search information.

computational complexity of solving the global optimization problems increases exponentially with increasing dimensionality. This computational complexity can be decreased by reducing the dimensionality of the optimization problems being solved with the use of Peano *space-filling curves or evolvents* $y(x)$ mapping the interval $[0,1]$ onto a N-dimensional hypercube D unambiguously and continuously – see, for example, [12,15]. As a result of such a reduction, initial multidimensional problem of multicriteria optimization (4) is reduced to a one-dimensional problem:

$$f(y(x)) = (f_1(y(x)), f_2(y(x)), \ldots, f_s(y(x))) \to min, x \in [0,1]. \qquad (11)$$

It is worth noting that the one-dimensional functions obtained as a result of the reduction satisfy the uniform Hölder condition (see [12,15]) i. e.

$$|f_i(y(x')) - f_i(y(x''))| \le H_i |x' - x''|^{1/N}, x', x'' \in [0,1], 1 \le i < s \qquad (12)$$

where the constants H_i are defined by the relation $H_i = 2L_i\sqrt{N+3}$, $1 \le i \le m$, L_i are the Lipschitz constants from (4) and N is the dimensionality of the optimization problem (1).

As a result of the dimensionality reduction, the search information obtained in the course of computations can be represented in the form

$$A_k = \{(x_i, z_i, f_i = f(y(x_i)) : 1 \le i \le k\}, \qquad (13)$$

where x_i, $1 \le i \le k$ are the points of performed global search iterations, z_i, f_i, $1 \le i \le k$ are the values of scalar criterion $F(\alpha, y(x))$ from (4) and of the criteria $f_i(y)$ from (11), $1 \le i \le s$ computed in the points x_i, $1 \le i \le k$. Note that the data in the set A_k are arranged in the order[4] of increasing of the point coordinates x_i, $1 \le i \le k$ i.e.

$$x_1 < x_2 < \cdots < x_k \qquad (14)$$

for more efficient execution of the global search algorithm.

The availability of the set A_k from (13) allows recalculating the results of all computations of the criteria values performed earlier to the values of the current optimization problem $F(\alpha, y(x))$ from (4) being solved without repeating the time-consuming computations of the criteria values i.e.

$$(x_i, f_i) \to z_i = F(\alpha, y(x_i)), 1 \le i \le k. \qquad (15)$$

In this way, the search information A_k from (13) recalculated according to (15) can be reused to continue the solving of the MCO problem. Such an opportunity can provide an essential decrease of the amount of computations performed to solve every next problem of the set $\mathbb{F}_T$ from (10) down to performing some limited set of the global search iterations.

In the proposed approach, the Multidimensional Algorithm of Global Search (MAGS) developed within the framework of the information-statistical theory

---

[4] The arrangement of the data is reflected by the use of the lower index in (14).

of global search [12,15,18,19] is applied to solve the multiextremal optimization problems of the $\mathbb{F}_T$ from (10). The general computational scheme of MAGS can be presented as follows.

At the initial iteration of MAGS, a minimized function value $\varphi(y(x^0))$ from (4) is computed in some arbitrary point x^0 from the interval $(0, 1)$ (the computing of the function value will further be called a *trial*). Then, let us assume k, $k > 1$ global search iterations to be completed. The choice of the trial point of the next $(k + 1)^{th}$ iteration is determined by the following rules.

Rule 1. For each interval (x_{i-1}, x_i), $1 < i \leq k$ compute a magnitude $R(i)$ called further a *characteristic* of the interval.

Rule 2. Determine the interval (x_{t-1}, x_t), which the maximum characteristic corresponds to[5]

$$R(t) = \max \{R(i) : 1 < i \leq k\}. \tag{16}$$

Rule 3. Perform the new trial in the interval with the maximum characteristic

$$x^{k+1} \in (x_{t-1}, x_t). \tag{17}$$

The stopping condition, according to which the trials are terminated, is defined by the condition

$$(x_t - x_{t-1})^{1/N} \leq \varepsilon, \tag{18}$$

where t is from (15), N is the dimensionality of the problem being solved from (1), and $\varepsilon > 0$ is the predefined accuracy of the problem solution. If the stopping condition is not fulfilled, the number of iterations k is incremented by unity, and new global search iteration is performed.

The convergence conditions for the algorithms developed within the framework of the information-statistical theory of global search were considered in [12]. Thus, at appropriate numerical estimates of the Hölder constants H_i, $1 \leq i \leq m$ from (12), MAGS converges to all available global minima points of the minimized function $\varphi(y(x))$.

It is worth noting also that the application of the MAGS algorithm after solving the current problem $F(\alpha, y)$ from (4) to solving the next problems from the set $\mathbb{F}_T$ from (10) allows reusing the search information A_k from (13) obtained in the course of all preceding computations.

4   Multilevel Parallel Computations for Solving Multistage Multicriteria Optimization Problems

The applied general approach to parallel solving the global optimization problems is the following – parallel computations is provided by simultaneous computing the minimized function values $F(\alpha, y)$ from (4) in several different points of the search domain D – see, for example, [12,20]. Such an approach provides

---

[5] The characteristics $R(i)$, $1 < i \leq k$ may be interpreted as some measures of importance of the intervals with respect to the probability to find the global minimum point in respective intervals.

the parallelization of the most time-consuming part of the global search process and is a general one – it can be applied to almost all global search methods for a variety of global optimization problems [18, 19, 21, 22].

This approach can be applied at different computation levels – either at the level of solving of one of the problems of the set $\mathbb{F}_T$ from (10) or at the level of parallel solving of several problems of this set. These methods of parallel computations will be considered below in relation to multiprocessor computer systems with shared memory.

4.1 Parallel Computations in Solving Single Multicriteria Optimization Problem

Since the characteristics $R(i)$, $1 < i \le k$ of the search intervals (x_{i-1}, x_i), $1 < i \le k$ play the role of the measures of importance of the intervals with respect to the probability to find the global minimum points, the MAGS algorithm can be extended for the parallel execution at the following generalization of the rules (16)–(17) [12, 18, 20, 23]:

Rule 2′. Arrange the characteristics of the intervals in the decreasing order

$$R(t_1) \ge R(t_2) \ge \cdots \ge R(t_{k-2}) \ge R(t_{k-1}) \tag{19}$$

and select p intervals with the indices t_j, $1 \le j \le p$ having the maximum values of the characteristics (p is the number of processors (cores) employed in the parallel computations).

Rule 3′. Perform new trials (computing of the minimized function values $F(\alpha, y(x))$ in the points x^{k+j}, $1 \le j \le p$ placed in the intervals with the maximum characteristics from (19) in parallel.

The stopping condition for the algorithm (18) should be checked for all intervals, in which the scheduled trials are performed

$$(x_{t_j} - x_{t_j-1})^{1/N} \le \varepsilon, 1 \le t_j \le p. \tag{20}$$

As before, if the stopping condition is not satisfied, the number of iterations k is incremented by p, and new global search iteration is performed.

This extended version of the MAGS algorithm will further called *Parallel Multidimensional Algorithm of Global Search for solving Single* MCO problems (PMAGS-S).

4.2 Parallel Computations in Solving Several Multicriteria Optimization Problems

Another possible method of parallel computations consists in simultaneous solving several problems $F(\alpha, y)$ of the set $\mathbb{F}_T$ from (10). In this approach, the number of problems being solved simultaneously is determined by the number of processors (computational cores) available. The solving of each particular problem $F(\alpha, y)$ is performed using the MAGS algorithm. Then, taking into account that all the problems of the set $\mathbb{F}_T$ are generated from the same MCO problem (the values of the scalar criterion $F(\alpha, y)$ are computed on the basis of the criteria values $f_i(y)$, $1 \leq i \leq s$ from (1)), it is possible to provide the interchange of the computed search information. For this purpose, the computational scheme of the MAGS algorithm should be appended by the following rule:

Rule 4. After completing a trial (computing the values of the function $F(\alpha, y)$ and criteria $f_i(y)$, $1 \leq i \leq s$) by a processor, the point of current trial $y^{k+1} \in D$ and the values of criteria $f(y^{k+1})$) are transferred to all processors. Then the availability of the data transferred from other processors is checked, and new received data is included into the search information A_k from (13).

Such a mutual use of the search information A_k from (13) obtained when solving particular problems of the set $\mathbb{F}_T$ from (10) allows to reduce significantly the number of global search iterations performed for each problem $F(\alpha, y)$ – see Sect. 5 for the results of the numerical experiments.

This version of the MAGS algorithm will be further called *Parallel Multidimensional Algorithm of Global Search for solving Multiple* MCO problems (PMAGS-M).

4.3 Parallel Computations in Joint Solving of Several Multicriteria Optimization Problems

The computational scheme of the PMAGS-S algorithm can be applied for the parallel solving of several problems of the set $\mathbb{F}_T$ as well. In this case, the choice of the intervals with the maximum characteristics $R(i)$, $1 < i \leq k$ from (16) must be performed taking into account all simultaneously solved problems $F(\alpha, y)$:

$$R_{l_1}(t_1) \geq R_{l_2}(t_2) \geq \cdots \geq R_{l_{K-2}}(t_{K-2}) \geq R_{l_{K-1}}(t_{K-1}), 1 \leq l_i \leq p, 1 \leq i \leq K-1, \tag{21}$$

where l_i, $1 \leq i \leq K - 1$ is the index of the problem, which the characteristic R_{l_i} belongs to and K is the total number of trials for all problems being solved simultaneously.

In this approach, the problems, for which the trials are performed, are determined dynamically in accordance with (21) – at each current global search iteration for a problem $F(\alpha, y)$, the trials may be absent or all p trials may be performed.

This version of the MAGS algorithms will further be called *Parallel Multidimensional Algorithm of Global Search for Joint solving of Multiple* MCO problems (PMAGS-JM).

5   Results of Numerical Experiments

The numerical experiments were carried out using the "Lobachevsky" supercomputer at University of Nizhny Novgorod (operating system – CentOS 6.4, managing system – SLURM). Each supercomputer node had 2 Intel Sandy Bridge E5-2660 processors 2.2 GHz, 64 GB RAM. Each processor had 8 cores (i.e. total

16 CPU cores per node were available). To obtain the executable program code, Intel C++ 17.0 compiler was used. The numerical experiments were performed using the Globalizer system [24].

The comparison of the efficiency of the sequential version of the developed approach with other approaches to solving the MCO problems was performed in [10,11]. This paper presents the results of numerical experiments for the evaluation of the efficiency of the parallel generalization of the developed approach. Each experiment consisted of the solving of 100 two-dimensional bi-criterial test MCO problems, in which the criteria were defined as the multiextremal functions [12]

$$f(y_1, y_2) = -(AB + CD)^{1/2}$$
$$AB = \left(\sum_{i=1}^{7} \sum_{j=1}^{7} [A_{ij}a_{ij}(y_1, y_2) + B_{ij}b_{ij}(y_1, y_2)] \right)^2 \tag{22}$$
$$CD = \left(\sum_{i=1}^{7} \sum_{j=1}^{7} [C_{ij}a_{ij}(y_1, y_2) - D_{ij}b_{ij}(y_1, y_2)] \right)^2$$

where

$$a_{ij}(y_1, y_2) = \sin(\pi i y_1)\sin(\pi j y_2), \ b_{ij}(y_1, y_2) = \cos(\pi i y_1)\cos(\pi j y_2)$$

were defined in the ranges and the parameters $-1 \leq A_{ij}, B_{ij}, C_{ij}, D_{ij} \leq 1$ were independent random numbers distributed uniformly. The functions of this kind are multiextremal essentially and are used often in the evaluation of the efficiency of the global optimization algorithms [10–12,18,19,21].

When performing the numerical experiments, the construction of a numerical approximation of the Pareto domain was understood as a solution of a MCO problem. To construct an approximation of a Pareto domain for each MCO problem with the criteria from (22), 64 scalar global optimization subproblems $F(\alpha, y)$ from (4) were solved with different values of the criteria convolution coefficients (i.e. total 6400 global optimization problems were solved in each experiment). The obtained results of experiments were averaged over the number of solved MCO problems. It should be noted that since the developed approach is oriented onto the MCO problems, in which computing the criteria values requires a large amount of computations, in all tables presented below the computational costs of solving the MCO problems is measured in the numbers of global search iterations performed.

In carrying out the numerical experiments, the following values of parameters of the applied algorithms were used: the required accuracy of the problem solutions $\varepsilon = 0.01$ from (18) and (20), the reliability parameter[6] $r = 2.3$. The experiments were carried out using a single supercomputer node (two processors, 16 computational cores with shared memory). In Table 1, the indicators of the computational costs (the numbers of the performed global search iterations) for all considered schemes of parallel computations (see Sect. 4) are presented (Fig. 1).

---

[6] The reliability parameter is used in the construction of the numerical estimate of the constants L_j, $1 \leq j \leq s$ from (3), L from (5) and H from (12).

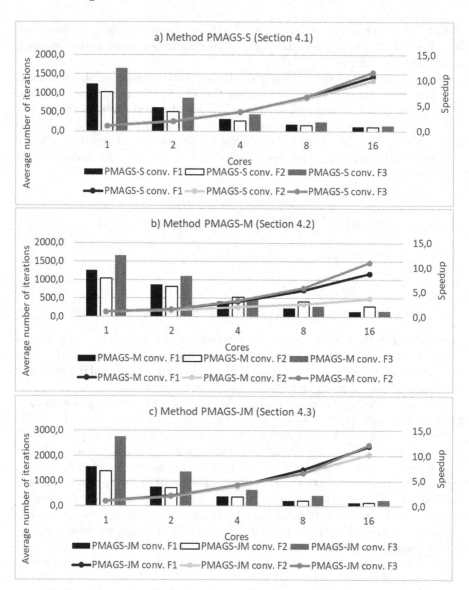

Fig. 1. Comparison of the averaged numbers of iterations executed for solving the MCO problems using various efficiency criteria convolutions

The results of experiments demonstrate the lowest computational costs (the number of performed global search iterations) to be achieved when using 16 cores for the PMAGS-S algorithm and the convolution F_2 from (7). Also, one can see from Table 1 that almost all computational schemes have a high efficiency from the viewpoint of parallelization. When using 16 cores, all algorithms except PMAGS-M with the convolutions F_1 from (6) and F_2 from (7) have demonstrated the speedup greater than 9.9.

Table 1. Comparison of performance of various schemes of parallel computations (the second column indicates the criteria convolution schemes F_1 from (6), F_2 from (7), F_3 from (9))

Method\Cores	Conv.	Average number of iterations					Speedup			
		1	2	4	8	16	2	4	8	16
Method	F_1	1238.3	625.7	324.5	186.7	115.3	**2.0**	3.8	6.6	10.7
PMAGS-S	F_2	**1018.6**	**512**	**264.2**	**158.5**	**102.9**	2.0	3.9	6.4	9.9
(Section 4.1)	F_3	1657.6	868.3	443.6	245.4	143.4	1.9	3.7	6.8	11.6
Method	F_1	1257.2	868.4	408.6	231.9	143.4	1.4	3.1	5.4	8.8
PMAGS-M	F_2	1035.9	811.2	532.1	407.6	284.3	1.3	1.9	2.5	3.6
(Section 4.2)	F_3	1653.3	1102.3	509.8	282.4	150.2	1.5	3.2	5.9	11.0
Method	F_1	1552.1	762.4	390.5	214	130.7	**2.0**	4.0	**7.3**	11.9
PMAGS-JM	F_2	1387.9	721.9	362.8	209.8	135.9	1.9	3.8	6.6	10.2
(Section 4.3)	F_3	2760.3	1371.7	652	421.4	227.6	**2.0**	**4.2**	6.6	**12.1**

Table 2. Comparison of the efficiency of the developed methods in solving the applied problem (the second row indicates the criteria convolution schemes F_1 from (6), F_2 from (7), F_3 from (9))

	Parallel scheme 4.1			Parallel scheme 4.2			Parallel scheme 4.3		
Convolution	F_1	F_2	F_3	F_1	F_2	F_3	F_1	F_2	F_3
Iterations	125	145	118	168	138	149	81	**82**	79

    In order to demonstrate the efficiency of the proposed approach, a problem of vibration isolation for a system with several degrees of freedom consisting of an isolated base and an elastic body has been solved. In the considered problem statement, the protected object was represented as multi-mass mechanical system consisting of several material points connected by vibration damping elements. As the criteria, the maximum deformation and maximum displacement of the object relative to the base were minimized (for details, see [25]). The dimensionality of the space of the optimized parameters was selected to be 3.

    The problem was solved by all considered methods using the Globalizer system. When solving the problem, the parameter $r = 3$ and the number of cores 16 were used. The number of convolutions was selected to be 16, and the accuracy of the method was set to $\varepsilon = 0.05$. The comparison of the efficiency of the methods by solving the applied problem is presented in Table 2.

    The results of the numerical experiments demonstrate that all methods have found the sufficient approximation of the Pareto domain. The lowest number of iterations performed the PMAGS-JM method. In Fig. 2, the computed approximation of the Pareto domain obtained by the PMAGS-JM method with the convolution F_2 from (7) is presented.

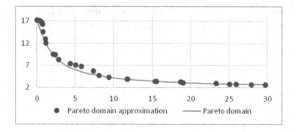

Fig. 2. Approximation of the Pareto domain for the problem of vibration isolation using PMAGS-JM method with the convolution F_2 from (7)

6   Conclusion

In the present paper, a novel approach for solving the time-consuming multicriteria optimization problems has been considered. Within the framework of the developed approach, obtaining the efficient decisions is provided by using several different efficiency criteria scalarization methods. The proposed approach allows altering the scalarization methods used and the parameters of these ones in the course of computations. In turn, such a variation of the problem statement leads to the need for solving multiple time-consuming global optimization problems. Overcoming the computational complexity is provided by means of the reuse of the whole search information obtained in the course of computations and efficient parallel computations on high-performance computational systems. The proposed methods of parallel computations can be used both for solving single MCO problems and for joint solving several ones.

The performed numerical experiments (total 6400 global optimization problems have been solved) and the example of solving the applied problem of vibration isolation confirm the developed approach to allow reducing the amount and time of computations for solving time-consuming MCO problems. In order to obtain more reliable evaluation of efficiency of the parallel computations, it is intended to continue carrying out the numerical experiments on solving the MCO problems with more efficiency criteria and for larger dimensionality.

Future research will also include investigations how to select the best parallel method automatically by using different computational platforms. Finally problems how to generalize the proposed approach for applying on multi-node cluster with GPU processors will be considered.

References

1. Ehrgott, M.: Multicriteria Optimization, 2nd edn. Springer, Heidelberg (2010)
2. Collette, Y., Siarry, P.: Multiobjective Optimization: Principles and Case Studies (Decision Engineering). Springer, Hiedelberg (2011). https://doi.org/10.1007/978-3-662-08883-8
3. Marler, R.T., Arora, J.S.: Multi-Objective Optimization: Concepts and Methods for Engineering. VDM Verlag (2009)

4. Pardalos, P.M., Žilinskas, A., Žilinskas, J.: Non-Convex Multi-Objective Optimization. Springer Optimization and Its Applications, vol. 123. Springer, New York (2017). https://doi.org/10.1007/978-3-319-61007-8
5. Parnell, G.S., Driscoll, P.J., Henderson, D.L. (eds.): Decision Making in Systems Engineering and Management, 2nd edn. Wiley, New Jersey (2011)
6. Figueira, J., Greco, S., Ehrgott, M. (eds.): Multiple Criteria Decision Analysis: State of the Art Surveys. Springer, New York (2005). https://doi.org/10.1007/b100605
7. Hillermeier, C., Jahn, J.: Multiobjective optimization: survey of methods and industrial applications. Surv. Math. Ind. **11**, 1–42 (2005)
8. Cho, J.-H., Wang, Y., Chen, I.-R., Chan, K.S., Swami, A.: A survey on modeling and optimizing multi-objective systems. IEEE Commun. Surv. Tutor. **19**(3), 1867–1901 (2017)
9. Gergel, V.P., Kozinov, E.A.: Accelerating multicriterial optimization by the intensive exploitation of accumulated search data. In: AIP Conference Proceedings, vol. 1776, p. 090003 (2016). https://doi.org/10.1063/1.4965367
10. Gergel, V., Kozinov, E.: Efficient multicriterial optimization based on intensive reuse of search information. J. Global Optim. **71**(1), 73–90 (2018). https://doi.org/10.1007/s10898-018-0624-3
11. Gergel, V., Kozinov, E.: Multistage global search using various scalarization schemes in multicriteria optimization problems. In: Le Thi, H.A., Le, H.M., Pham Dinh, T. (eds.) WCGO 2019. AISC, vol. 991, pp. 638–648. Springer, Cham (2020). https://doi.org/10.1007/978-3-030-21803-4_64
12. Strongin, R., Sergeyev, Y.: Global Optimization with Non-Convex Constraints: Sequential and Parallel Algorithms, 2nd edn. Kluwer Academic Publishers, Dordrecht (2013). (3rd edn. 2014)
13. Zhigljavsky, A., Žilinskas, A.: Stochastic Global Optimization. Springer, Boston (2008). https://doi.org/10.1007/978-0-387-74740-8
14. Locatelli, M., Schoen, F.: Global optimization: theory, algorithms, and applications. In: SIAM (2013)
15. Sergeyev, Y.D., Strongin, R.G., Lera, D.: Introduction to Global Optimization Exploiting Space-Filling Curves. Springer Briefs in Optimization. Springer, New York (2013). https://doi.org/10.1007/978-1-4614-8042-6
16. Paulavičius, R., Žilinskas, J.: Simplicial Global Optimization. Springer Briefs in Optimization. Springer, New York (2014). https://doi.org/10.1007/978-1-4614-9093-7
17. Floudas, C.A., Pardalos, M.P.: Recent Advances in Global Optimization. Princeton University Press, Princeton (2016)
18. Gergel, V.P., Strongin, R.G.: Parallel computing for globally optimal decision making. In: Malyshkin, V.E. (ed.) PaCT 2003. LNCS, vol. 2763, pp. 76–88. Springer, Heidelberg (2003). https://doi.org/10.1007/978-3-540-45145-7_7
19. Sergeyev, Y.D.: An information global optimization algorithm with local tuning. SIAM J. Optim. **5**(4), 858–870 (1995)
20. Strongin, R.G., Sergeyev, Y.D.: Global multidimensional optimization on parallel computer. Parallel Comput. **18**(11), 1259–1273 (1992)
21. Sergeyev, Y.D., Kvasov, D.E.: A deterministic global optimization using smooth diagonal auxiliary functions. Commun. Nonlinear Sci. Numer. Simul. **21**(1–3), 99–111 (2015)
22. Barkalov, K., Gergel, V., Lebedev, I.: Solving global optimization problems on GPU cluster. In: AIP Conference Proceedings, vol. 1738, p. 400006 (2016) https://doi.org/10.1063/1.4952194

23. Sergeyev, Y.D., Grishagin, V.A.: Parallel asynchronous global search and the nested optimization scheme. J. Comput. Anal. Appl. **3**(2), 123–145 (2001). https://doi.org/10.1023/A:1010185125012
24. Sysoyev, A., Barkalov, K., Sovrasov, V., Lebedev, I., Gergel, V.: Globalizer – a parallel software system for solving global optimization problems. In: Malyshkin, V. (ed.) PaCT 2017. LNCS, vol. 10421, pp. 492–499. Springer, Cham (2017). https://doi.org/10.1007/978-3-319-62932-2_47
25. Balandin, D.V., Kogan, M.M.: Multi-objective generalized H2 control for optimal protection from vibration. In: 2018 UKACC 12th International Conference on Control, vol. 8516721, pp. 205–210 (2018)

Enabling Hardware Affinity
in JVM-Based Applications: A Case
Study for Big Data

Roberto R. Expósito$^{(\boxtimes)}$, Jorge Veiga, and Juan Touriño

Universidade da Coruña, CITIC, Computer Architecture Group, A Coruña, Spain
{roberto.rey.exposito,jorge.veiga,juan}@udc.es

Abstract. Java has been the backbone of Big Data processing for more
than a decade due to its interesting features such as object orientation, cross-platform portability and good programming productivity. In
fact, most popular Big Data frameworks such as Hadoop and Spark
are implemented in Java or using other languages designed to run on
the Java Virtual Machine (JVM) such as Scala. However, modern computing hardware is increasingly complex, featuring multiple processing
cores aggregated into one or more CPUs that are usually organized as
a Non-Uniform Memory Access (NUMA) architecture. The platform-
independent features of the JVM come at the cost of hardware abstraction, which makes it more difficult for Big Data developers to take advantage of hardware-aware optimizations based on managing CPU or NUMA
affinities. In this paper we introduce jhwloc, a Java library for easily managing such affinities in JVM-based applications and gathering
information about the underlying hardware topology. To demonstrate
the functionality and benefits of our proposal, we have extended Flame-
MR, our Java-based MapReduce framework, to provide support for setting CPU affinities through jhwloc. The experimental evaluation using
representative Big Data workloads has shown that performance can be
improved by up to 17% when efficiently exploiting the hardware. jhwloc
is publicly available to download at https://github.com/rreye/jhwloc.

Keywords: Big Data · Java Virtual Machine (JVM) · Hardware
affinity · MapReduce · Performance evaluation

1 Introduction

The emergence of Big Data technologies offers great opportunities for researchers
and scientists to exploit unprecedented volumes of data sources in innovative
ways, resulting in novel insight discovery and better decisions. Distributed processing frameworks are the great facilitators of the paradigm shift to Big Data,
as they enable the storage and processing of large datasets and the application
of analytics techniques to extract valuable information from such massive data.

The MapReduce paradigm [5] introduced by Google in 2004 and then pop-
ularized by the open-source Apache Hadoop project [19] has been considered

© Springer Nature Switzerland AG 2020
V. V. Krzhizhanovskaya et al. (Eds.): ICCS 2020, LNCS 12137, pp. 31–44, 2020.
https://doi.org/10.1007/978-3-030-50371-0_3

as a game changer to the way massive datasets are processed. During the past decade, there has been a huge effort in the development of Big Data frameworks. Some projects focus on adapting Hadoop to take advantage of specific hardware (RDMA-Hadoop [16]), or to provide improved performance for iterative algorithms by exploiting in-memory computations (Twister [7], Flame-MR [22]). Other frameworks have been designed from scratch to overcome other Hadoop limitations such as the lack of support for streaming computations, real-time processing and interactive analytics. Many of these frameworks are developed under the umbrella of the Apache Software Foundation: Storm [10], Spark [24], Flink [2], Samza [14]. Despite the large amount of existing frameworks, Hadoop and its ecosystem are still considered the cornerstone of Big Data as they provide the underlying core on top of which data can be processed efficiently. This core consists of the Hadoop Distributed File System (HDFS) [17], which allows to store and distribute data across a cluster of commodity machines, and Yet Another Resource Negotiator (YARN) [20], for the scalable management of the cluster resources. New frameworks generally only replace the Hadoop MapReduce data engine to provide faster processing speed.

Unlike the C++-based original MapReduce implementation by Google, the entire Hadoop stack is implemented in Java to increase portability and ease of setup. As Big Data users are generally non-expert programmers, the use of Java provides multiple appealing features for them: object orientation, automatic memory management, built-in multithreading, easy-to-learn properties, good programming productivity and a wide community of developers. Moreover, later Java releases adopt concepts from other paradigms like functional programming. A core feature of Java is cross-platform portability: programs written on one platform can be executed on any combination of software and hardware with adequate runtime support. This is achieved by compiling Java code to platform-independent bytecode first, instead of directly to platform-specific native code. The bytecode instructions are then executed by the Java Virtual Machine (JVM) that is specific to the host operating system and hardware combination. Furthermore, modern JVMs integrate efficient Just-in-Time (JIT) compilers that can provide near-native performance from Java bytecode.

Apart from Hadoop, most state-of-the-art Big Data frameworks are also implemented in Java (e.g., Flink, Storm 2, Flame-MR). Other frameworks rely on Scala (Spark, Samza), whose source code is compiled to Java bytecode so that the resulting executable runs on a JVM. However, the platform-independent feature provided by the JVM is only possible by abstracting most of the hardware layer away from developers. This makes it difficult or even impossible for them to access interesting low-level functionalities in JVM-based applications such as setting hardware affinities or gathering topology information for performing hardware-aware optimizations. The increasing complexity of multicore CPUs and the democratization of Non-Uniform Memory Access (NUMA) architectures [11] raise the need for exposing a portable view of the hardware topology to Java developers, while also providing an appropriate API to manage CPU and NUMA affinities. JVM languages in general, and Big Data frameworks in particular, may take advantage of such API to exploit the hardware more efficiently without having to resort to non-portable, command-line tools that offer limited functionalities. The contributions of this paper are:

- We present jhwloc, a Java library that provides a binding API to manage hardware affinities straightforwardly in JVM-based applications, as well as a means to gather information about the underlying hardware topology.
- We implement the support for managing CPU affinity using jhwloc in Flame-MR, a Java-based MapReduce framework, to demonstrate the functionality and benefit of our proposal.
- We analyze the impact of affinity on the performance of in-memory data processing with Flame-MR by evaluating six representative Big Data workloads on a 9-node cluster.

The remainder of the paper is organized as follows. Section 2 provides the background of the paper and summarizes the related work. Section 3 introduces the jhwloc library and its main features. Section 4 describes a case study of integrating jhwloc in Flame-MR, and presents the experimental evaluation. Finally, our concluding remarks are summarized in Sect. 5.

2 Background and Related Work

Exploiting modern hardware platforms requires in-depth knowledge of the underlying architecture together with appropriate expertise from the application behaviour. Current architectures provide a complex multi-level cache hierarchy with dedicated caches (one per core), a global cache (one for all the cores) or even partially shared caches. Moving a computing task from one core to another can cause performance degradation because of cache affinities. Simultaneous Multithreading (SMT) technologies [6] such as Intel Hyper-Threading (HT) [13] involve sharing computing resources of a single core between multiple logical Processing Units (PUs). This fact also means to share cache levels so that performance may be even reduced in some particular cases. Furthermore, NUMA architectures [11] are currently widely extended, in which memory is transparently distributed among CPUs connected through a cross-chip interconnect. Hence, an access from one CPU to the memory of another CPU (i.e., a remote memory access) incurs additional latency overhead due to transferring the data through the network. As an example, Fig. 1 shows the hierarchical organization of a typical NUMA machine with two octa-core CPUs that implement two-way SMT, so 8 cores and 16 PUs are available per CPU. All this complexity of the hardware topology of modern computing platforms is considered a critical aspect of performance and must be taken into account when trying to optimize parallel [18] and distributed applications [23].

Nowadays, the hardware locality (hwloc) project [9] is the most popular tool for exposing a static view of the topology of modern hardware, including CPU, memory and I/O devices. This project solves many interoperability issues due to the amount and variety of the sources of locality information for querying the

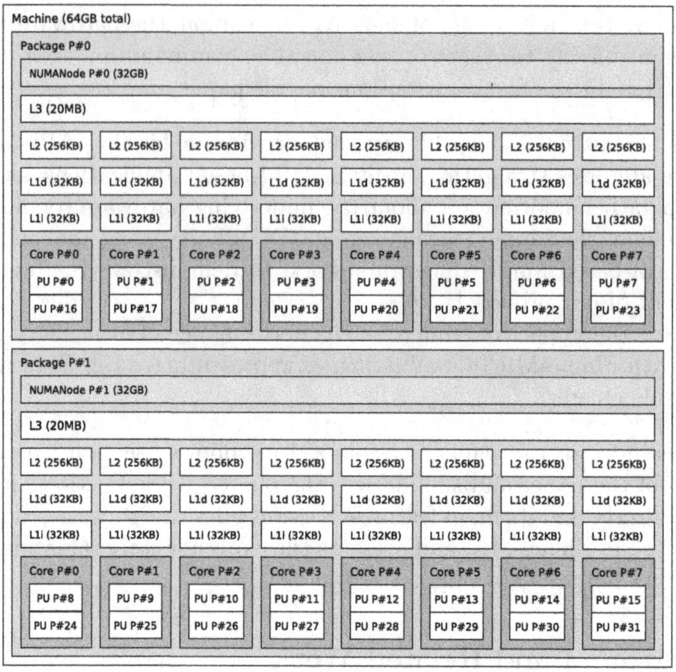

Fig. 1. Overview of a NUMA system with two octa-core CPUs

topology of a system. To do so, hwloc combines locality information obtained from the operating system (e.g., /sys in Linux), from the hardware itself (cpuid instruction in x86), or using high-level libraries (numactl) and tools (lscpu). Moreover, hwloc offers APIs to interoperate with device-specific libraries (e.g., libibverbs) and allows binding tasks (e.g., process, thread) according to hardware affinities (CPU and memory) in a portable and abstracted way by exposing a unified interface, as different operating systems have diverse binding support. As a consequence, hwloc has become the de facto standard software for modeling NUMA systems in High Performance Computing (HPC) environments. In fact, it is used by most message-passing implementations, many batch queueing systems, compilers and parallel libraries for HPC. Unfortunately, hwloc only provides C-based APIs for gathering topology information and binding tasks, whereas the JVM does not offer enough support for developing hardware-aware applications as it only allows to obtain the number of cores available in the system.

Overseer [15] is a Java-based framework to access low-level data such as performance counters, JVM internal events and temperature monitoring. Among its other features, Overseer also provides basic information about hardware topology through hwloc by resorting to the Java Native Interface (JNI). Moreover, it allows managing CPU affinity by relying on the Linux Kernel API (sched.h). However, the support provided by Overseer for both features is very limited. On the one hand, topology information is restricted to a few Java methods that only allow to obtain the number of available hardware resources of a certain type (e.g., cores per CPU), without providing any further details about cache hierarchy (e.g., size, associativity), NUMA nodes (e.g., local memory) or additional functionality to manipulate topologies (e.g., filtering, traversing). On the other hand, support for CPU binding is limited to use Linux-specific affinity masks that determine the set of CPUs on which a task is eligible to run, without providing convenient methods to operate on such CPU set (e.g., logical operations) or utility methods for performing more advanced binding operations in an easy way (e.g., bind to one thread of the last core of the machine). Furthermore, no kind of memory binding is provided by Overseer. Our `jhwloc` library allows to overcome all these limitations by currently providing support for more than 100 methods as the Java counterparts of the hwloc ones.

There exist few works that have evaluated the impact of managing hardware affinities on the performance of Big Data frameworks. In [1], authors analyze how NUMA affinity and SMT technologies affect Spark. They manage affinities using numactl and use hwloc to obtain the identifier of hardware threads. Their results reveal that performance degradation due to remote memory accesses is 10% on average. Authors in [4] characterize several TPC-H queries implemented on top of Spark, showing that NUMA affinity is slightly advantageous in preventing remote memory accesses. To manage NUMA affinity, they also rely on numactl. However, both studies are limited to evaluating a single machine, they lack the assessment of the impact of CPU binding on performance and they do not provide any useful API for gathering topology information and managing hardware affinities for JVM-based languages. Hence, our work extends the current state-of-the-art in all these directions.

3 Java Hardware Locality Library

The Java Hardware Locality (`jhwloc`) project has been designed as a wrapper library that consists of: (1) a set of standard Java classes that model hwloc functionalities using an object-oriented approach, and (2) the necessary native glue code that interfaces with the C-based hwloc API. In order to do so, `jhwloc` uses JNI to invoke the native methods implemented in C. Hence, `jhwloc` acts as a transparent bridge between a client application written in any JVM-based language and the hwloc library, but exposing a more friendly and object-oriented Java API to developers. One important advantage provided by `jhwloc` is that it frees Java developers from interacting directly with JNI calls, which is considered a cumbersome and time-consuming task. So, our library can be easily employed in Java applications to perform hardware-oriented performance tuning.

3.1 Java API

Currently, `jhwloc` provides Java counterparts for more than 100 hwloc functions, covering a significant part of its main functionality. A complete Javadoc documentation that describes all the public methods together with their parameters is publicly available at the jhwloc website. Basically, the jhwloc API enables Java developers to: (1) obtain the hierarchical hardware topology of key computing elements within a machine such as: NUMA nodes, shared/dedicated caches, CPU packages, cores and PUs (I/O devices are not yet supported); (2) gather various attributes from caches (e.g., size, associativity) and memory information; (3) manipulate hardware topologies through advanced operations (e.g., filtering, traversing); (4) build "fake" or synthetic topologies that allow querying them without having the underlying hardware available; (5) export topologies to XML files to reload them later; (6) manage hardware affinities in an easy way using bitmaps, which are sets of integers (positive or null) used to describe the location of topology objects on the CPU (CPU sets) and NUMA nodes (node sets); and (7) handle such bitmaps by providing advanced methods to operate over them through the hwloc bitmap API.

It is important to remark that, unlike C, the JVM provides an automatic memory management mechanism through the built-in Garbage Collector (GC), which is in charge of performing memory allocation/deallocation without interaction from the programmer. Hence, the memory binding performed by the hwloc functions that manage memory allocation explicitly (e.g., *alloc_membind*) or migrate already-allocated data (*set_area_membind*) cannot be supported in jhwloc. NUMA affinity management is thus restricted to those functions for performing implicit memory binding: *set_membind* and *get_membind*. These functions allow to define the current binding policy that will be applied to the subsequent calls to malloc-like operations performed by the GC.

3.2   Usage Example

As an illustrative usage example, Listing 1 presents a Java code snippet that shows the simplicity of use of the jhwloc API. Basic hardware topology information such as the number of cores and PUs and the available memory is obtained (lines 6–11) after creating and initializing an instance of the *HwlocTopology* class (lines 1–4), which represents an abstraction of the underlying hardware. Most of the jhwloc functionality is provided through this class with more than 50 Java methods available that allow to manipulate, traverse and browse the topology, as well as to perform CPU and memory binding operations. Next, the example shows how to manage CPU affinities by binding the current thread to the CPU set formed by the first and last PU of the machine. As can be seen, the Java objects that represent those PUs, which are instances of the *HwlocObject* class, can be easily obtained by using their indexes (lines 13–15). The CPU set objects from both PUs are then operated using a logical **and** (lines 16–17), and the returned CPU set is used to perform the actual CPU binding of the current thread (lines 18–19). The logical **and** operation is an example of the more than 30 Java methods that are provided to conform with the hwloc bitmap API, supported in jhwloc through the *HwlocBitmap* abstract class. This class is extended by the *HwlocCPUSet* and *HwlocNodeSet* subclasses that provide concrete implementations for representing CPU and NUMA node sets, respectively.

```
1  // Create, initialize and load a topology object
2  HwlocTopology topo = new HwlocTopology ();
3  topo.init ();
4  topo.load ();
5
6  // Get the number of cores, PUs, and the available memory
7  int nc = topo.get_nbobjs_by_type (HWLOC.OBJ_CORE);
8  int np = topo.get_nbobjs_by_type (HWLOC.OBJ_PU);
9  long mem = topo.get_root_obj ().getTotalMemory ();
10 System.out.println (''#Cores/PUs: ''+nc+''/''+np);
11 System.out.println (''Total memory: ''+mem);
12
13 // Get first and last PU objects
14 HwlocObject fpu = topo.get_obj_by_type (HWLOC.OBJ_PU, 0);
15 HwlocObject lpu = topo.get_obj_by_type (HWLOC.OBJ_PU, np-1);
16 // Logical 'and' over the CPU sets of both PUs
17 HwlocCPUSet cpuset = fpu.getCPUSet ().and (lpu.getCPUSet ());
18 // Bind current thread to the returned CPU set
19 topo.set_cpubind (cpuset, EnumSet.of (HWLOC.CPUBIND_THREAD));
```

Listing 1. Getting hardware topology information and managing CPU affinities

More advanced usage examples are provided together with the jhwloc source code. These examples include NUMA binding, manipulating bitmaps, obtaining cache information, traversing topologies, exporting topologies to XML and building synthetic ones, among other jhwloc functionalities.

4 Impact of CPU Affinity on Performance: Flame-MR Case Study

This section analyzes the impact of setting CPU affinity on the performance of Flame-MR, our Big Data processing framework. First, Flame-MR is briefly introduced in Sect. 4.1. Next, Sect. 4.2 describes how jhwloc has been integrated into Flame-MR to manage CPU affinities, explaining the different affinity levels that are supported. Section 4.3 details the experimental testbed, and finally Sect. 4.4 discusses the results obtained.

4.1 Flame-MR Overview

Flame-MR [22] is a Java-based MapReduce implementation that transparently accelerates Hadoop applications without modifying their source code. To do so, Flame-MR replaces the underlying data processing engine of Hadoop by an optimized, in-memory architecture that leverages system resources more efficiently.

The overall architecture of Flame-MR is based on the deployment of several *Worker* processes (i.e., JVMs) over the nodes of a cluster (see Fig. 2). Relying on an event-driven architecture, each *Worker* is in charge of executing the computational tasks (i.e., map/reduce operations) to process the input data from HDFS. The tasks are executed by a thread pool that can perform as many concurrent operations as the number of cores configured for each *Worker*. These operations are efficiently scheduled in order to pipeline data processing and data movement steps. The data pool allocates memory buffers in an optimized way, reducing the amount of buffer creations to minimize garbage collection overheads. Once the buffers are filled with data, they are stored into in-memory data structures to be processed by subsequent operations. Furthermore, these data structures can be cached in memory between MapReduce jobs to avoid writing intermediate results to HDFS, thus providing efficient iterative computations.

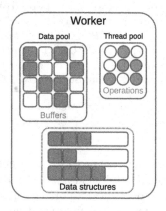

Fig. 2. Flame-MR *Worker* architecture

4.2 Managing CPU Affinities in Flame-MR

Flame-MR has been extended to use the functionalities provided by jhwloc to enable the binding of computational tasks to the hardware processing elements available in the system (i.e., CPU/cores/PUs). To do so, the software components that manage such tasks, *Worker* and thread pool classes, have been modified to make them aware of the hardware affinity level that is set by the user through the configuration file of Flame-MR.

When Flame-MR starts a computational task, it first checks the configuration file to determine if the jhwloc library must be called and, if so, the specific affinity level that must be enforced. Then, the *Worker* initializes a *HwlocTopology* object and uses the set_cpubind method provided by jhwloc to bind computational tasks to the appropriate hardware. The configuration of the *Workers* is affected by the affinity level being used, as the number of threads launched by them to execute map/reduce operations should be adapted to the hardware characteristics of the underlying system. The intuitive recommendation would be to create as many *Workers* as CPUs, and as many threads per *Worker* as cores/PUs available in the nodes. The CPU affinity levels currently supported by Flame-MR through jhwloc are:

- NONE: Flame-MR does not manage hardware affinities in any way (i.e., jhwloc is not used).
- CPU: *Workers* are bound to specific CPUs. Each JVM process that executes a *Worker* is bound to one of the available CPUs in the system using the jhwloc flag *HWLOC.CPUBIND_PROCESS*. Hence, each thread launched by a *Worker* is also bound to the same CPU, which means that the OS scheduler can migrate threads among its cores. The mapping between *Workers* and CPUs is done cyclically by allocating each *Worker* to a different CPU until all the CPUs are used, starting again if there are remaining *Workers*.
- CORE: map/reduce operations are bound to specific cores. Each thread launched by a *Worker* to perform such operations is bound to one of the available cores in the system using the flag *HWLOC.CPUBIND_THREAD*. So, the OS scheduler can migrate threads among the PUs of a core (if any). The mapping between threads and cores is done by allocating a group of cores from the same CPU to each *Worker*. Note that the number of threads used by all *Workers* executed in a node should not exceed the number of cores to avoid resource oversubscription.
- PU: map/reduce operations are bound to specific PUs. Each thread launched by a *Worker* to perform such operations is bound to one of the available PUs in the system using the flag *HWLOC.CPUBIND_THREAD*. The mapping between threads and PUs is done by allocating a group of cores to each *Worker*, and then distributing its threads over the PUs of those cores in a cyclic way. Note also that the number of threads used by all *Workers* on a node should not exceed the number of PUs to avoid oversubscription.

It is important to note that all the threads launched by a *Worker* process are created during the Flame-MR start-up phase. So, jhwloc is only accessed once to set the affinity level, avoiding any JNI overhead during data processing.

4.3 Experimental Configuration

Six MapReduce workloads from four domains that represent different Big Data use cases have been evaluated: (1) data sorting (Sort), (2) machine learning (K-Means), (3) graph processing (PageRank, Connected Components), and (4) genome sequence analysis (MarDRe, CloudRS). Sort is an I/O-bound micro-benchmark that sorts an input text dataset generated randomly. K-Means is an iterative clustering algorithm that classifies an input set of N samples into K clusters. PageRank and Connected Components are popular iterative algorithms for graph processing. PageRank obtains a ranking of the elements of a graph taking into account the number and quality of the links to each one, and Connected Components explores a graph to determine its subnets. MarDRe [8] and CloudRS [3] are bioinformatics tools for preprocessing genomics datasets.

MarDRe removes duplicate and near-duplicate DNA reads, whereas CloudRS performs read error correction.

In the experiments conducted in this paper, Sort processes a 100 GB dataset and K-Means performs a maximum of five iterations over a 35 GB dataset ($N = 360$ million samples) using 200 clusters ($K = 200$). Both PageRank and Connected Components execute five iterations over a 40 GB dataset (60 million pages). MarDRe removes duplicate reads using the SRR377645 dataset, named after its accession number in the European Nucleotide Archive (ENA) [12], which contains 241 million reads of 100 base pairs each (67 GB in total). CloudRS corrects read errors using the SRR921890 dataset, which contains 16 million reads of 100 base pairs each (5.2 GB in total). Both genomics datasets are publicly available to download at ENA website.

The experiments have been carried out on a 9-node cluster with one master and eight slave nodes running Flame-MR version 1.2. Each node consists of a NUMA system with two Intel Xeon E5-2660 octa-core CPUs. This CPU model features two-way Intel HT, so 8 cores and 16 logical PUs are available per CPU (i.e., 16 and 32 per node, respectively). Each node has a total of 64 GiB of memory evenly distributed between the two CPUs. The NUMA architecture just described is the one previously shown in Fig. 1, which also details the cache hierarchy. Additionally, each node has one local disk of 800 GiB intended for both HDFS and intermediate data storage during the execution of the workloads. Nodes are interconnected through Gigabit Ethernet (1 Gbps) and InfiniBand FDR (56 Gbps). The cluster runs Linux CentOS 6.10 with kernel release 2.6.32–754.3.5, whereas the JVM version is Oracle JDK 10.0.1.

To deploy and configure Flame-MR on the cluster, the Big Data Evaluator (BDEv) tool [21] has been used for ease of setup. Two *Workers* (i.e., two JVM processes) have been executed per slave node, since our preliminary experiments proved it to be the best configuration for Flame-MR on this system. Regarding the number of threads per *Worker*, two different configurations have been evaluated: (1) using as many threads per *Worker* as cores per CPU (8 threads), and (2) using as many threads per *Worker* as PUs per CPU (16 threads), thus also evaluating the impact of Intel HT on performance. Finally, the metric shown in the following graphs corresponds to the median runtime for a set of 10 executions for each experiment, clearing the OS buffer cache of the nodes between each execution. Variability is represented in the graphs by using error bars to indicate the minimum and maximum runtimes.

4.4 Performance Results

Figure 3 presents the measured runtimes of Flame-MR for all the workloads when using different affinity levels and *Worker* configurations as previously described. When running 8 threads per *Worker* (i.e., not using Intel HT), all the workloads benefit from enforcing some hardware affinity, although the best level to use varies for each workload. On the one hand, the performance improvements for Sort (Fig. 3a) and K-Means (Fig. 3b) with respect to the baseline scenario (i.e., without using affinity) are lower than for the remaining workloads. The main reason is that both workloads are the most I/O-intensive codes under evaluation, being clearly bottlenecked by disk performance (only one disk per slave node is available). This fact limits the potential benefits of using an enforced hardware placement. Nevertheless, the improvements obtained are up to 3% and 4% using the PU and CPU affinity levels for Sort and K-Means, respectively. On the other hand, the performance improvements for the remaining workloads (see Figs. 3c–3f) are significantly higher: up to 13%, 17%, 16% and 11% for PageRank, Connected Components, MarDRe and CloudRS, respectively. Although the best affinity level depends on each particular workload (e.g., CPU for PageRank, CORE for CloudRS), it can be concluded that the performance differences between different levels are generally small. Note that the workloads benefit not only from accelerating a single *Worker* (i.e., JVM) using CPU binding, but also from reducing the impact of the synchronizations among all *Workers*, performed between the global map and reduce phases for each MapReduce job. Moreover, this reduction in the runtimes is obtained as a zero-effort transparent configuration for Flame-MR users, without recompiling/modifying the source code of the workloads.

Regarding the impact of Intel HT on performance (i.e., running 16 threads per *Worker*), note that the results for the CORE affinity level cannot be shown when using two *Workers* per node since slave nodes have 16 physical cores, as mentioned in Sect. 4.3. In the HT scenario, K-Means, MarDRe and CloudRS take clear advantage of this technology. In the case of K-Means (see Fig. 3b), execution times are reduced with respect to the non-HT counterparts by 16% and 12% for the baseline and CPU affinity scenarios, respectively. The improvements for these two scenarios increase up to 26% and 32% for CloudRS (Fig. 3f), and 48% and 41% for MarDRe (Fig. 3e). However, the impact of HT can be considered negligible for Sort, as shown in Fig. 3a, whereas the performance of PageRank and Connected Components is even reduced in most scenarios. Note that using HT technology implies that the two logical PUs within a physical core must share not only all levels of the cache hierarchy, but also some of the computational units. This fact can degrade the performance of CPU-bound workloads due to increased cache miss rates and resource contention, as it seems to be the case with PageRank. Finally, the impact of enforcing hardware affinity when using HT can only be clearly appreciated for CloudRS, providing a reduction in the execution time of 14% when using the CPU affinity level.

We can conclude that there is no one-size-fits-all solution, since the best affinity level depends on each workload and its particular resource characterization. Furthermore, the impact of managing CPU affinities is clearly much more beneficial when HT is not used. However, the results shown in this section reinforce the utility and performance benefits of managing hardware affinities in Big Data JVM-based frameworks such as Flame-MR.

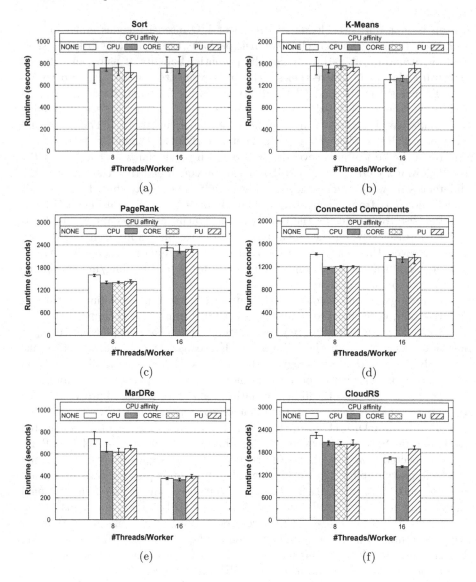

Fig. 3. Runtimes of Flame-MR for different workloads and CPU affinity levels

5 Conclusions

The complexity of current computing infrastructures raises the need for carefully placing applications on them so that affinities can be efficiently exploited by the hardware. However, the standard class library provided by the JVM lacks support for developing hardware-aware applications, preventing them from taking advantage of managing CPU or NUMA affinities. As most popular distributed processing frameworks are implemented using languages executed by the JVM, having such affinity support can be of great interest for the Big Data community.

In this paper we have introduced `jhwloc`, a Java library that exposes an object-oriented API that allows developers to gather information about the underlying hardware and bind tasks according to it. Acting as a wrapper between the JVM and the C-based hwloc library, the de facto standard in HPC environments, `jhwloc` enables JVM-based applications to easily manage affinities to perform hardware-aware optimizations. Furthermore, we have extended our Java MapReduce framework Flame-MR to include support for setting hardware affinities through `jhwloc`, as a case study to demonstrate the potential benefits. The experimental results, running six representative Big Data workloads on a 9-node cluster, have shown that performance can be transparently improved by up to 17%. Other popular JVM-based Big Data frameworks such as Hadoop, Spark, Flink, Storm and Samza could also take advantage of the features provided by `jhwloc` in a similar way to Flame-MR.

The source code of the `jhwloc` library is released under the open-source GNU GPLv3 license and is publicly available together with the Javadoc documentation at https://github.com/rreye/jhwloc. As future work, we aim to explore the impact of setting NUMA affinities on JVM performance when using different garbage collection algorithms. We also plan to extend `jhwloc` to provide other functionalities such as gathering information about the network topology.

Acknowledgments. This work was supported by the Ministry of Economy, Industry and Competitiveness of Spain and FEDER funds of the European Union [ref. TIN2016-75845-P (AEI/FEDER/EU)]; and by Xunta de Galicia and FEDER funds [Centro de Investigación de Galicia accreditation 2019–2022, ref. ED431G2019/01, Consolidation Program of Competitive Reference Groups, ref. ED431C2017/04].

References

1. Awan, A.J., Vlassov, V., Brorsson, M., Ayguade, E.: Node architecture implications for in-memory data analytics on scale-in clusters. In: Proceedings of 3rd IEEE/ACM International Conference on Big Data Computing, Applications and Technologies (BDCAT 2016), Shanghai, China, pp. 237–246 (2016)
2. Carbone, P., Katsifodimos, A., Ewen, S., Markl, V., Haridi, S., Tzoumas, K.: Apache Flink: stream and batch processing in a single engine. Bull. IEEE Tech. Comm. Data Eng. **38**(4), 28–38 (2015)
3. Chen, C.C., Chang, Y.J., Chung, W.C., Lee, D.T., Ho, J.M.: CloudRS: an error correction algorithm of high-throughput sequencing data based on scalable framework. In: Proceedings of IEEE International Conference on Big Data (IEEE Big-Data 2013), Santa Clara, CA, USA, pp. 717–722 (2013)
4. Chiba, T., Onodera, T.: Workload characterization and optimization of TPC-H queries on Apache Spark. In: Proceedings of IEEE International Symposium on Performance Analysis of Systems and Software (ISPASS 2016), Uppsala, Sweden, pp. 112–121 (2016)
5. Dean, J., Ghemawat, S.: MapReduce: simplified data processing on large clusters. Commun. ACM **51**(1), 107–113 (2008)
6. Eggers, S.J., Emer, J.S., Levy, H.M., Lo, J.L., Stamm, R.L., Tullsen, D.M.: Simultaneous multithreading: a platform for next-generation processors. IEEE Micro **17**(5), 12–19 (1997)

7. Ekanayake, J., et al.: Twister: a runtime for iterative MapReduce. In: Proceedings of 19th ACM International Symposium on High Performance Distributed Computing (HPDC 2010), Chicago, IL, USA, pp. 810–818 (2010)
8. Expósito, R.R., Veiga, J., González-Domínguez, J., Touriño, J.: MarDRe: efficient MapReduce-based removal of duplicate DNA reads in the cloud. Bioinformatics **33**(17), 2762–2764 (2017)
9. Goglin, B.: Managing the topology of heterogeneous cluster nodes with hardware locality (HWLOC). In: Proceedings of International Conference on High Performance Computing & Simulation (HPCS 2014), Bologna, Italy, pp. 74–81 (2014)
10. Iqbal, M.H., Soomro, T.R.: Big Data analysis: Apache Storm perspective. Int. J. Comput. Trends Technol. **19**(1), 9–14 (2015)
11. Lameter, C.: NUMA (Non-Uniform Memory Access): an overview. ACM Queue **11**(7), 40:40–40:51 (2013)
12. Leinonen, R., et al.: The European nucleotide archive. Nucleic Acids Res. **39**, D28–D31 (2011)
13. Marr, D.T., et al.: Hyper-threading technology architecture and microarchitecture. Intel Technol. J. **6**(1), 1–12 (2002)
14. Noghabi, S.A., et al.: Samza: stateful scalable stream processing at LinkedIn. Proc. VLDB Endow. **10**(12), 1634–1645 (2017)
15. Peternier, A., Bonetta, D., Binder, W., Pautasso, C.: Tool demonstration: overseer - low-level hardware monitoring and management for Java. In: Proceedings of 9th International Conference on Principles and Practice of Programming in Java (PPPJ 2011), Kongens Lyngby, Denmark, pp. 143–146 (2011)
16. Wasi-ur Rahman, M., et al.: High-performance RDMA-based design of Hadoop MapReduce over InfiniBand. In: Proceedings of IEEE 27th International Symposium on Parallel & Distributed Processing, Workshops & PHD Forum (IPDPSW 2013), Boston, MA, USA, pp. 1908–1917 (2013)
17. Shvachko, K., Kuang, H., Radia, S., Chansler, R.: The Hadoop distributed file system. In: Proceedings of IEEE 26th Symposium on Mass Storage Systems and Technologies (MSST 2010), Incline Village, NV, USA, pp. 1–10 (2010)
18. Terboven, C., an Mey, D., Schmidl, D., Jin, H., Reichstein, T.: Data and thread affinity in OpenMP programs. In: Proceedings of Workshop on Memory Access on Future Processors: A Solved Problem? (MAW 2008), Ischia, Italy, pp. 377–384 (2008)
19. The Apache Hadoop project: http://hadoop.apache.org. Accessed 31 Mar 2020
20. Vavilapalli, V.K., et al.: Apache Hadoop YARN: yet another resource negotiator. In: Proceedings of 4th Annual Symposium on Cloud Computing (SOCC 2013), Santa Clara, CA, USA, pp. 5:1–5:16 (2013)
21. Veiga, J., Enes, J., Expósito, R.R., Touriño, J.: BDEv 3.0: energy efficiency and microarchitectural characterization of Big Data processing frameworks. Future Gener. Comput. Syst. **86**, 565–581 (2018)
22. Veiga, J., Expósito, R.R., Taboada, G.L., Touriño, J.: Flame-MR: an event-driven architecture for MapReduce applications. Future Gener. Comput. Syst. **65**, 46–56 (2016)
23. Wang, L., Ren, R., Zhan, J., Jia, Z.: Characterization and architectural implications of Big Data workloads. In: Proceedings of IEEE International Symposium on Performance Analysis of Systems and Software (ISPASS 2016), Uppsala, Sweden, pp. 145–146 (2016)
24. Zaharia, M., et al.: Apache spark: a unified engine for Big Data processing. Commun. ACM **59**(11), 56–65 (2016)

An Optimizing Multi-platform
Source-to-source Compiler Framework
for the NEURON MODeling Language

Pramod Kumbhar[1], Omar Awile[1], Liam Keegan[1], Jorge Blanco Alonso[1],
James King[1], Michael Hines[2], and Felix Schürmann[1(✉)]

[1] Blue Brain Project, Ecole Polytechnique Fédérale de Lausanne (EPFL), Campus
Biotech, 1202 Geneva, Switzerland
`felix.schuermann@epfl.ch`
[2] Department of Neuroscience, Yale University, New Haven, CT 06510, USA

Abstract. Domain-specific languages (DSLs) play an increasingly
important role in the generation of high performing software. They allow
the user to exploit domain knowledge for the generation of more efficient
code on target architectures. Here, we describe a new code generation
framework (NMODL) for an existing DSL in the NEURON framework,
a widely used software for massively parallel simulation of biophysically detailed brain tissue models. Existing NMODL DSL transpilers
lack either essential features to generate optimized code or capability
to parse the diversity of existing models in the user community. Our
NMODL framework has been tested against a large number of previously published user models and offers high-level domain-specific optimizations and symbolic algebraic simplifications before target code generation. NMODL implements multiple SIMD and SPMD targets optimized for modern hardware. When comparing NMODL-generated kernels with NEURON we observe a speedup of up to 20×, resulting
in overall speedups of two different production simulations by ∼7×.
When compared to SIMD optimized kernels that heavily relied on auto-
vectorization by the compiler still a speedup of up to ∼2× is observed.

Keywords: NEURON · HPC · DSL · Code generation · Neuroscience

1 Introduction

The use of large scale simulation in modern neuroscience is becoming increasingly important (e.g. [2,24]) and has been enabled by substantial performance
progress in neurosimulation engines over the last decade and a half (e.g.
[14,17,19,20,27,28]). While excellent scaling has been achieved on a variety of
platforms with the conversion to vectorized implementations, domain specific
knowledge expressed in the models is not yet optimally used. In other fields,
the use of DSLs and subsequent code-to-code translation have been effective
in generating efficient codes and allowing easy adaptation to new architectures

© Springer Nature Switzerland AG 2020
V. V. Krzhizhanovskaya et al. (Eds.): ICCS 2020, LNCS 12137, pp. 45–58, 2020.
https://doi.org/10.1007/978-3-030-50371-0_4

[8,9,30,31]. This is becoming more important as the architectural diversity of hardware is increasing.

Motivated by these observations, we have revisited the widely adopted NEURON simulator [11], which enables simulations of biophysically detailed neuron models on computing platforms ranging from desktop to petascale supercomputers, and which has over 2,000 reported scientific studies using it. One of the key features of the NEURON simulator is extendability via a domain specific language (DSL) layer called the NEURON Model Description Language (NMODL) [12]. NMODL allows the neuroscientist to extend NEURON by incorporating a wide range of membrane and intracellular submodels. The domain scientist can easily express these channel properties in terms of algebraic and ordinary differential equations, kinetic schemes in NMODL without worrying about lower level implementation details.

The rate limiting aspect for performance of NEURON simulations is the execution of channels and synapses written in the NMODL DSL. The code generated from NMODL often accounts for more than 80% of overall execution time. There are more than six thousand NMODL files that are shared by the NEURON user community on the ModelDB platform [29]. As the type and number of mechanisms differ from model to model, hand-tuning of the generated code is not feasible. The goal of our NMODL Framework is to provide a tool that can parse all existing models, and generate optimized code from NMODL DSL code, which is responsible for more than 80% of the total simulation time. Here we present our effort in building a new NMODL source-to-source compiler that generates C++, OpenMP, OpenACC, CUDA and ISPC targeting modern SIMD hardware platforms. We also describe several techniques we employ in the NMODL Framework, which we believe to be useful beyond the immediate scope of the NEURON simulation framework.

2   Related Work

The reference implementation for the NMODL DSL specification is found in *nocmodl* [13], a component in the NEURON simulator. Over the years *nocmodl* underwent several iterations of development and gained support for a number of newer language constructs. One of the major limitations of *nocmodl* is its lack of flexibility. Instead of constructing an intermediate representation, such as an *Abstract Syntax Tree* (AST), it performs many code generation steps on the fly, while parsing the input. This leaves little room for performing global analysis, optimizations, or targeting a different simulator altogether. The CoreNEURON [19] library uses a modified version of *nocmodl* called *mod2c* [3], which duplicates most of the legacy code and has some of the same limitations as *nocmodl*. *Pynmodl* [23] is a Python based parsing and post-processing tool for NMODL. The primary focus of *pynmodl* is to parse and translate NMODL DSL to other computational neuroscience DSLs but does not support code generation for a particular simulator. The *modcc* source-to-source compiler is being developed as part of the Arbor simulator [1]. It is able to generate from NMODL DSL code, optimized C++/CUDA to be used with the Arbor simulator. It only implements a subset of the NMODL DSL specification and hence is only able to process a modest number of existing models available in ModelDB [29]. For a more comprehensive review of current code-generator techniques in computational neuroscience we refer the reader to Blundell et al. [4]. Other fields have adopted DSLs and code generation techniques [9,25,31] but they are not yet fully exploited in the context of NMODL. We conclude that current NMODL tools either lack support for the full NMODL DSL specification, lack the necessary flexibility to be used as a generic code generation framework, or are unable to adequately take advantage of modern hardware architectures, and thus are missing out on available performance from modern computing architectures.

3   NMODL DSL

In most simple terms, the NEURON simulator deals with two aspects of neuronal tissue simulations: 1) the exchange of spiking events between neuronal cells and 2) the numerical integration of a spatially discretized cable equation that is equipped with additional terms describing transmembrane currents resulting from ion channels and synapses. The NMODL DSL allows the modelers to efficiently express these transmembrane mechanisms. As an example, many models of neurons use a non-linear combination of the following basic ordinary differential equation to describe the kinetics of ion channels first developed by Hodking and Huxley [15]:

$$\frac{dV}{dt} = \left[I - \bar{g}_{Na} m^3 h (V - V_{Na}) - \bar{g}_K n^4 (V - V_K) - g_L (V - V_L) \right] / C \quad (1)$$

$$\frac{dn}{dt} = \alpha_n(V)(1 - n) - \beta_n(V) n \quad (2)$$

$$\frac{dm}{dt} = \alpha_m(V)(1 - m) - \beta_m(V) m \quad (3)$$

$$\frac{dh}{dt} = \alpha_h(V)(1 - h) - \beta_h(V) h \quad (4)$$

where V is the membrane potential, I is the membrane current, g_i is conductance per unit area for ith ion channel, n, m, and h are dimensionless quantities between 0 and 1 associated with channel activation and inactivation, V_i is the reversal potential of the ith ion channel, C is the membrane capacitance, and α_i and β_i are rate constants for the ith ion channel, which depend on voltage but not time.

Figure 1 shows a simplified NMODL DSL fragment of a specific ion channel, a voltage-gated calcium ion channel, published in Traub et al. [32]. Our example highlights the most important language constructs and serves as an example for the DSL-level optimizations presented in the following sections. The NMODL language specification can be found in [12].

At DSL level a lot of information is expressed implicitly that can be used to generate efficient code and expose more parallelism, e.g.:

```
1  TITLE T-calcium channel (adapted)         25  DERIVATIVE states {
2                                             26    rates(v)
3  NEURON {                                   27    m' = (minf - m)/mtau
4    SUFFIX cat                               28    h' = (hinf - h)/htau
5    USEION ca READ cai, cao WRITE ica        29  }
6    RANGE gcatbar, ica, gcat, hinf, minf, mtau, htau   30
7  }                                          31  PROCEDURE rates(v(mV)) {
8                                             32    LOCAL a, b, qt
9  PARAMETER {                                33    qt= q10^((celsius-25)/10)
10   celsius = 25 (degC)                      34    a = 0.2*(-1.0*v+19.26)
11   cao = 2 (mM)                             35    b = 0.009*exp(-v/22.03)
12 }                                          36    minf = a/(a+b)
13                                            37    mtau = betmt(v)/(qt*a0m*(1+alpmt(v)))
14 STATE { m h }                              38  }
15                                            39
16 ASSIGNED {                                 40  FUNCTION ghk(v(mV), ci(mM), co(mM)) (mV) {
17   ica gcat hinf htau minf mtau             41    LOCAL nu,f
18 }                                          42    f = KTF(celsius)/2
19                                            43    nu = v/f
20 BREAKPOINT {                               44    VERBATIM
21   SOLVE states METHOD cnexp                45    // C code implementation
22   gcat = gcatbar*m*m*h                     46    ENDVERBATIM
23   ica = gcat*ghk(v,cai,cao)                47    ghk = -f*(1.0-(ci/co)*exp(nu))*efun(nu)
24 }                                          48  }
```

Annotation labels: channel dependency; memory footprint; constant variables with limited precision; modifiable variables; often memory bound; often compute bound; if inlined into DERIVATIVE minf, mtau could become local variables; elemental function; unsafe for optimisations, need C lexer/parser

Fig. 1. NMODL example of a simplified model of a voltage-gated calcium channel showing different NMODL constructs and summary of optimization information available at DSL level. Keywords are printed in uppercase and marked with boldface.

- USEION statement describes the dependency between channels and can be used to build the runtime dependency graph to exploit micro-parallelism [22]
- PARAMETER block describes the variables that are constant, often can be stored with limited precision.
- ASSIGNED statement describes modifiable variables and can be allocated in fast memory.
- DERIVATIVE, KINETIC and SOLVE describes ODEs which can be analyzed and solved analytically to improve the performance as well as accuracy.
- BREAKPOINT describes current and voltage relation. If this is ohmic then one can use analytical expression instead of numerical derivatives to improve the accuracy as well as performance.
- PROCEDURE can be inlined at DSL level to eliminate RANGE variables and thereby significantly reduce memory access cost as well as memory footprint.

To use this information and perform such optimizations, often a global analysis of the NMODL DSL is required. For example, to perform inlining of a PROCEDURE one needs to find all function calls and recursively inline the function bodies. As *nocmodl* lacks the intermediate AST representation, this type of analysis is difficult to perform and such optimizations are not implemented. The NMODL Framework is designed to exploit such information from DSL specification and perform optimizations.

4   Design and Implementation

The implementation of the NMODL Framework can be broken down into four main components: lexer/parser implementation, DSL level optimisation passes, ODE solvers, and code generation passes. Figure 2 summarizes the overall architecture of NMODL Framework. As in any compiler framework, lexing and parsing are the

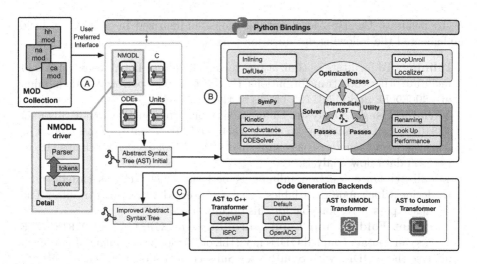

Fig. 2. Architecture of the NMODL Code Generation Framework showing: A) Input NMODL files are processed by different lexers & parsers generating the AST; B) Different analysis and optimisation passes further transform the AST; and C) The optimised AST is then converted to low level C++ code or other custom backends

first two steps performed on an input NMODL. The lexer implementation is based on the popular flex package and bison is used as the parser generator. The ODEs, units and inline C code need extra processing and hence separate lexers and parsers are implemented. DSL level optimizations and code generation are main aspects of this framework and discussed in detail in subsequent sections.

4.1   Optimization Passes

Modern compilers implement various passes for code analysis, code transformation, and optimized code generation. Optimizations such as constant folding, inlining, and loop unrolling are commonly found in all of today's major compilers. For example, the LLVM compiler framework [21] features more than one hundred compiler passes. In the context of the NMODL Framework, we focus on a few optimization passes with very specific objectives. By taking advantage of domain-specific and high-level information that is available in the DSL but later lost in the lower level $C++$ or $CUDA$ code, we are able to provide additional significant improvements in code performance compared to native compiler optimizations. For example, all NMODL `RANGE`, `ASSIGNED`, and `PARAMETER` variables are translated to *double* type variables in $C++$. Once this transformation is done, $C/C++$ compilers can no longer infer these high-level semantics from these variables. Another example is `RANGE` to `LOCAL` transformations with the help of `PROCEDURE` inlining discussed in Sect. 3. All `RANGE` variables in the NMODL DSL are converted to array variables and are dynamically allocated in $C++$. Once this transformation is done, the $C/C++$ compiler can only do limited optimizations.

To facilitate the DSL level optimizations summarized in Sect. 3, we have implemented the following optimization passes.

Inlining: To facilitate optimizations such as RANGE to LOCAL conversion and facilitate other code transformations, the *Inlining* pass performs code inlining of PROCEDURE and FUNCTION blocks at their call sites.

Variable Usage Analysis: Different variable types such as RANGE, GLOBAL, ASSIGNED can be analysed to check where and how often they are used. The *Variable Usage Analysis* pass implements *Definition-Use (DU)* chains [18] to perform data flow analysis.

Localiser: Once function inlining is performed, *DU* chains can be used to decide which RANGE variables can be converted to LOCAL variables. The *Localiser* pass is responsible for this optimization.

Constant Folding and Loop Unrolling: The KINETIC and DERIVATIVE blocks can contain coupled ODEs in WHILE or FOR loop statements. In order to analyse these ODEs with SymPy (see Subsect. 4.4), first we need to perform constant folding to know the iteration space of the loop and then perform *loop unrolling* to make all ODE statements explicit.

4.2 Code Generation

Once DSL and symbolic optimizations (see Subsect. 4.3) are performed on the AST, the NMODL Framework is ready to proceed to the code generation step (cf. Fig. 2). The *C++* code generator plays a special role, since it constitutes the base code generator extended by all other implementations. This allows easy implementation of a new target by overriding only necessary constructs of the base code generator.

To better leverage specific hardware platform features such as SIMD, multi-threading or SPMD we have further implemented code generation targets for OpenMP, OpenACC, ISPC (*Intel SPMD Program Compiler*) and experimentally CUDA. We chose ISPC for its performance portability and support for all major vector extensions on x86 (SSE4.2, AVX2, AVX-512), ARM NEON and NVIDIA GPUs (using NVPTX) giving us the ability to generate optimized SIMD code for all major computing platforms.

We have, furthermore, extended the *C++* target with an OpenMP and an OpenACC backend. These two code generators emit code that is largely identical to the *C++* code generator but add appropriate pragma annotations to support OpenMP shared-memory parallelism and OpenACC GPU acceleration. Finally, our code-generation framework supports CUDA as a main backend to target NVIDIA GPUs.

4.3 ODE Solvers

NMODL allows the user to specify the equations that define the system to be simulated in a variety of ways.

- The KINETIC block describes the system using a mass action kinetic scheme of reaction equations.
- The DERIVATIVE block specifies a system of coupled ODEs (note that any kinetic scheme can also be written as an equivalent system of ODEs.)
- Users can also specify systems of algebraic equations to be solved. The LINEAR and NONLINEAR blocks respectively specify systems of linear and nonlinear algebraic equations (applying a numerical integration scheme to a system of ODEs typically also results in a system of algebraic equations to solve.)

To reduce duplication of functionality for dealing with these related systems of equations, we implemented a hierarchy of transformations as shown in Fig. 3. First, any KINETIC blocks of mass action kinetic reaction statements are translated to DERIVATIVE blocks of the equivalent ODE system. Linear and independent ODEs are solved analytically. Otherwise a numerical integration scheme such as implicit Euler is used which results in a system of algebraic equations equivalent to a LINEAR or NONLINEAR block. If the system is linear and small, it is solved analytically at compile time using symbolic Gaussian elimination. Optionally, Common Subexpression Elimination (CSE) [7] can then be applied.

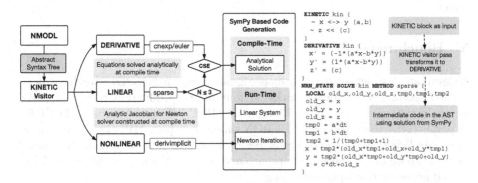

Fig. 3. On the left, unified ODE solver workflow showing ODEs from different NMODL constructs either produces compile–time analytical solutions, or run–time numerical solutions. On the right, example of KINETIC block and its transformation to SymPy based solution.

If the system is linear and large, it is solved (at run time) using a lower–upper (LU) matrix decomposition. Non-linear systems of equations are solved at run time by Newton iteration, which makes use of the analytic Jacobian calculated at compile time. The numerical ODE solver uses the Eigen [10] numerical linear algebra $C++$ template library, which produces highly optimized and vectorized routines for solving systems of linear equations.

4.4 SymPy

The analytic ODE solver uses SymPy [26], a Python library for symbolic calculations, which can simplify, differentiate and integrate symbolic mathematical expressions. Our analytical solver replaces the purely numerical approach used in other NMODL source-to-source compilers and simulators. It allows us to perform some computations analytically at compile time that were previously carried out at run time at each time step using approximate numerical differentiation.

Linear and independent ODEs have been typically replaced by an analytic solution that treats the coefficients as constant over a time step. NMODL increases the runtime performance by performing algebraic simplification and optionally replacing computationally expensive exponential calculations with the (1, 1) *Pade approximant* [5], consistent with the overall second order correct simulation accuracy (as suggested in [6], and implemented in [1]).

For coupled ODEs, the implicit Euler numerical integration scheme is applied which results in a set of simultaneous algebraic equations. For a linear systems of equations, the `sparse` solver method is used. For non-linear systems, the `derivimplicit` solver method is used. The `sparse` solver chooses from two solution methods, depending on the size of the system to be solved. For small systems (three or less equations), the system is solved by symbolic Gaussian elimination at compile time. The `derivimplicit` solver constructs a system of non–linear equations, which we solve using Newton's method at run time. We therefore compute the system's Jacobian, which is then used in the iterative solver.

5 Benchmarks

To evaluate the achieved performance gains through NMODL, we have performed comprehensive benchmarks on four major production hardware platforms, Intel Skylake 6140, KNL 7230, AMD EPYC 7451 and NVidia Tesla V100. In parallel NEURON and CoreNEURON simulations pure MPI execution expose more parallelism and achieve better performance. To provide a realistic benchmark setup we follow the same approach and perform our measurements on fully loaded nodes (process per physical core).

Benchmarks performed on the Intel platforms to compare auto-vectorization performance with ISPC were compiled with Intel Parallel Studio 2018.1, while all others were compiled with GCC. All benchmarks have been compiled with `-O2 -xHost` and `-O3` flags respectively. For GPU benchmarking we compared performance of Intel Skylake node with a NVidia Tesla V100 GPU.

We selected two brain tissue models: a somatosensory cortex and a hippocampus region model. The first, a somatosensory cortex microcircuit of a young rat published by the Blue Brain Project has 55 layer-specific morphological types and 207 morpho-electrical types [24]. The second, a model of a rat hippocampus CA1 [16] is built as part of the European Human Brain Project and has 13 morphological types and 17 morpho-electrical types. These models are selected

because they are computationally expensive and have a large number of mecha-nisms which allow us to assess performance benefits for different types of kernels used in production simulations. Based on these two models, we presented results for two benchmarks:

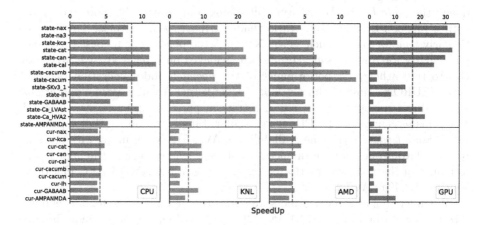

Fig. 4. Speedups of the 14 representative channels from the neocortex and hippocam-pus models, built using the NMODL Framework with ISPC target and run using CoreNEURON over *nocmodl* and NEURON. The dotted lines denote average speedups of the respective group of kernels

Channel Benchmark: This benchmark consists of 21 different morpho-electrical types selected to include all the unique mechanisms from somatosen-sory cortex and hippocampus CA1 models. A total of 4,608 cells are cre-ated without network connectivity as this benchmark is designed to measure performance of code generation for individual mechanisms. To benchmark GPU performance we created 3,000 artificial cells using mechanisms from this benchmark.

Simulation Benchmark: This benchmark measures the overall performance improvement for whole production simulations. We used 1,000 cells from the somatosensory cortex and hippocampus models simulating one second of bio-logical time using a timestep of 0.025 ms.

6   Results

The code generated from the NMODL DSL accounts for more than 90% of the simulation time in the above-described benchmarks. Each channel is writ-ten in the NMODL language and typically contains two kernels: *State Update* (denoted as *state-channel-name*) and *Current Update* (denoted as *cur-channel-name*). For calculating the speedup, we compare the runtimes of these two ker-nels translated using NMODL Framework with ISPC backend and simulated using CoreNEURON with the same mechanisms compiled with the NEURON simulator using *nocmodl*. The results of these benchmarks are shown in Fig. 4. We restrict ourselves to most expensive kernels from 14 representative channels. The four columns correspond to the four tested hardware platforms. The dotted lines show the average speedups achieved for the channels. Generally we observe a higher speedup on *State Update* kernels than on *Current Update* kernels. This is due to the *State Update* kernels typically being computationally more expensive, with a higher FLOP per byte ratio than *Current Update* kernels. This is particularly true for GPUs, such as the NVidia V100 platform. The best speedups, particularly for *State Update* kernels, are achieved on Intel KNL. We attribute this to the rather poor performance of the *nocmodl* code generation backend with NEURON. Most of the *Current Update* kernels require atomic operations with indirect memory access, which results in poor kernel performance on all platforms in general, and GPUs in particular. Finally, we notice that especially in the *State Update* kernel the availability of AVX-512 vector units, with optimal memory layout offers a performance advantage as can be seen in the higher performance of the two Intel platforms compared with the AMD EPYC platform, which only offers AVX2.

When looking at top performers we notice that several of the high-level optimizations described in Sects. 4.1 and 4.3 are at least equally if not more important than the generation of vectorized code. We observe that particularly our optimizations on the ODE statements using SymPy based solvers (e.g. *state-cacum*) can lead to speedups of more than 12× on AMD EPYC.

Table 1. Absolute runtime in seconds and speedup of 1 s simulated biological time of the hippocampus and somatosensory cortex simulations on Intel Skylake and Intel KNL platform using NEURON with *nocmodl* (NRN-NOCMODL), CoreNEURON with MOD2C (CN-MOD2C) and CoreNEURON with NMODL Framework (CN-NMODL). The total time is further broken down into *State* and *Current Update*, which represent the two main groups of computational kernels generated by the transpiler. Speedup is shown with respect to NEURON.

| | | Intel Skylake | | | Intel KNL | | |
| | Component | NRN | CN | | NRN | CN | |
		NOCMODL	MOD2C	NMODL	NOCMODL	MOD2C	NMODL
Hippocampus	State Update	1089.01 s	310.92 s	145.89 s	3260.16 s	525.89 s	251.81 s
	Current Update	866.81 s	239.52 s	171.99 s	1129.13 s	143.2 s	223.93 s
	Other	157.51 s	84.27 s	67.29 s	869.86 s	348.95 s	266.02 s
	Total	2113.34 s	634.71 s	385.17 s	5259.14 s	1018.04 s	741.76 s
	Speedup	—	3.33×	5.49×	—	5.17×	7.09×
Cortex	State Update	173.29 s	32.81 s	20.39 s	556.63 s	45.09 s	41.73 s
	Current Update	106.86 s	32.38 s	27.34 s	154.29 s	37.18 s	64.56 s
	Other	43.51 s	29.69 s	24.11 s	222.9 s	108.43 s	102.66 s
	Total	323.66 s	94.88 s	71.84 s	933.81 s	190.7 s	208.95 s
	Speedup	—	3.41×	4.51×	—	4.9×	4.47×

Table 1 shows the absolute time and speedup achieved for full simulations of the hippocampus CA1 and somatosensory cortex models. We compare the performance with three different configurations. The first configuration (*NRN-NOCMODL*) uses the *NEURON* simulator with *nocmodl* as the code generation backend. The second configuration (*CN-MOD2C*) uses the *CoreNEURON* library with *MOD2C* as code generation backend. The third configuration (*CN-NMODL*) uses the *CoreNEURON* library with the here-presented NMODL Framework as a code generation backend. Total times are broken down into the above-discussed *State Update* and *Current Update* kernels where the majority of the time is spent in NMODL generated code. The rest of the time is shown as *Other*.

The NMODL Framework shows up to 5.5× speedup on Skylake and up to 7× speedup on the KNL platform. The hippocampus model shows a larger speedup compared to the somatosensory cortex model because it uses *cacum*, *cacumb* and *kca* mechanisms with the *derivimplicit* integration scheme. The Eigen based solver implementation in NMODL offers therefore additional performance improvements. When compared with *CN-MOD2C*, *CN-NMODL* shows a up to 2× performance improvement with NMODL generated *State Update* kernels and up to 1.6× for whole simulation. Note that *CN-MOD2C* is heavily dependent on the auto-vectorization capabilities of the compiler. For example, if the GCC compiler is used instead of Intel, *CN-NMODL* becomes up to 3× and 6× faster compared to *CN-MOD2C* on Intel Skylake and Intel KNL platforms respectively. On the Intel KNL platform the *Current Update* kernels are ∼2× slower in *CN-NMODL* compared to *CN-MOD2C* (highlighted in red). These kernels require indirect addressing due to use of USEION constructs. We have found possible performance issues in ISPC when *gather-scatter* instructions and *atomic reductions* are used in the kernels with the target *avx512knl-i32x16*. These issues will be addressed in a future release of the NMODL Framework.

7   Discussion

Many scientific applications do not encode only a single mathematical problem, but scientific users provide the governing equations that need to be integrated by the solvers on a case by case basis. This can impact the success of auto-vectorization and thus strategies are required to allow the user to express the problem at hand on a high level, e.g. through DSLs, that help in producing optimized code.

In this paper we presented a novel NMODL code generation framework for the DSL of the widely used NEURON simulator. The DSL is translated into an AST that lends itself to specific optimization passes before it is handed off to different backends for generation of optimized code for the target platform. We have implemented optimization passes that relate to straight-forward transformation of the DSL code, but also more advanced optimization passes that intercept ODE statements for which an analytical solution can be used instead of having to resort to numerical integration. This functionality is built on top of the SymPy and Eigen libraries.

For code generation we have developed backends for C++ and OpenMP targeting CPUs and ISPC to target a wide variety of CPU architectures providing optimal SIMD performance and reducing the dependency on auto-vectorization capabilities of the compiler. Furthermore, we have developed both a CUDA backend specifically with NVidia GPUs in mind as well as a more generic OpenACC backend.

We have benchmarked kernels from production simulations of two different large-scale brain tissue models on Intel SKL, Intel KNL and AMD EPYC platforms. On those individual kernels, we saw performance improvements from 5× to 20×. In order to test how those kernel improvements translate into speedup of the entire simulations (which use the kernels in different ratios or not at all), we benchmarked production simulations on Intel KNL and Intel SKL platforms. Compared to the regular NEURON simulation environment, we observed a speedup of 6–10×. Compared to an optimized version of the NEURON simulator, CoreNEURON, which heavily relies on auto-vectorization of the compiler, the work presented here nonetheless resulted in a speedup of up to 2×.

Beyond the performance gains, a central goal of our effort is the ability to parse all previously published models. By using the grammar specification from the original NEURON NMODL language, we were able to demonstrate compatibility with 6,370 channels from the public model repository ModelDB. We furthermore took care to maintain the language semantics of the DSL in the AST, providing great flexibility to use NMODL as a generic NMODL parsing framework through its Python API and build new tools on top of it.

More generally, DSL have been used by many other fields to close the gap between the domain scientists and efficient code. Our study brings this capability of automatic generation of efficient and multi-platform code to the computational neuroscience community. Importantly, it is doing so by making extensive use of an abstract intermediate representations to perform optimizations using domain knowledge before it is lost during the translation into a general purpose programming language. This makes it possible to perform symbolic simplifications and equation solving on the level of the intermediate representation. We believe that this approach is applicable to other fields and other DSLs and thus transcends the problem at hand.

Acknowledgements. This study was supported by funding to the Blue Brain Project, a research center of the École Polytechnique Fédérale de Lausanne (EPFL), from the Swiss government's ETH Board of the Swiss Federal Institutes of Technology. The research was also funded by NIH grant number R01NS11613 to the Department of Neuroscience, Yale University, and the European Union's Horizon 2020 Framework Programme for Research and Innovation under Specific Grant Agreement number 785907 (Human Brain Project SGA2). We would like to thank Antonio Bellotta, Francesco Cremonesi, Ioannis Magkanaris, Matthias Wolf, Samuel Melchior and Tristan Carel for fruitful discussions and their contributions to the NMODL development. The AMD system for benchmarking was provided by Erlangen Regional Computing Center (RRZE).

References

1. Akar, N.A., et al.: Arbor — a morphologically-detailed neural network simulation library for contemporary high-performance computing architectures. In: 2019 27th Euromicro International Conference on Parallel, Distributed and Network-Based Processing (PDP), pp. 274–282, February 2019. https://doi.org/10.1109/EMPDP. 2019.8671560
2. Arkhipov, A., et al.: Visual physiology of the layer 4 cortical circuit in silico. PLOS Comput. Biol. **14**(11), 1–47 (2018). https://doi.org/10.1371/journal.pcbi.1006535
3. Blue Brain Project: MOD2C - CoreNEURON's converter for mod files to C code (2015). http://github.com/BlueBrain/mod2c
4. Blundell, I., et al.: Code generation in computational neuroscience: a review of tools and techniques. Front. Neuroinformatics **12** (2018). https://doi.org/10.3389/fninf.2018.00068
5. Cambridge, W.H. (ed.): Padé Approximants (Section 5.12): Numerical Recipes: The Art of Scientific Computing, 3rd edn. Cambridge University Press, Cambridge (2007). oCLC: ocn123285342
6. Casalegno, F., et al.: Error analysis and quantification in NEURON simulations. In: Proceedings of the VII European Congress on Computational Methods in Applied Sciences and Engineering (ECCOMAS Congress 2016), Crete Island, Greece (2016). https://doi.org/10.7712/100016.1892.7366
7. Cocke, J.: Global common subexpression elimination. ACM SIGPLAN Not. **5**(7), 20–24 (1970). https://doi.org/10.1145/390013.808480
8. Deutsch, A., et al.: Morpheus: a user-friendly modeling environment for multiscale and multicellular systems biology. Bioinformatics **30**(9). https://doi.org/10.1093/bioinformatics/btt772
9. DeVito, Z., et al.: Liszt: a domain specific language for building portable mesh-based PDE solvers. In: Proceedings of 2011 International Conference for High Performance Computing, Networking, Storage and Analysis, SC 2011. ACM (2011). https://doi.org/10.1145/2063384.2063396
10. Guennebaud, G., et al.: Eigen v3 (2010). http://eigen.tuxfamily.org
11. Hines, M.L., Carnevale, N.T.: The NEURON simulation environment. Neural Comput. **9**, 1179–1209 (1997)
12. Hines, M.L., Carnevale, N.T.: Expanding NEURON's repertoire of mechanisms with NMODL. Neural Comput. **12**, 995–1007 (2000). https://doi.org/10.1162/089976600300015475
13. Hines, M.: nocmodl - NEURON's converter for mod files to C code (1989). https://github.com/neuronsimulator/nrn/tree/master/src/nmodl
14. Hines, M.: Comparison of neuronal spike exchange methods on a Blue Gene/P supercomputer. Front. Comput. Neurosci. **5** (2011). https://doi.org/10.3389/fncom.2011.00049
15. Hodgkin, A.L., Huxley, A.F.: A quantitative description of membrane current and its application to conduction and excitation in nerve. J. Physiol. **117**(4), 500–544 (1952). https://doi.org/10.1113/jphysiol.1952.sp004764
16. Human Brain Project: Community Models of Hippocampus. https://www.humanbrainproject.eu/en/brain-simulation/hippocampus/
17. Jordan, J., et al.: Extremely scalable spiking neuronal network simulation code: from laptops to exascale computers. Front. Neuroinformatics **12**. https://doi.org/10.3389/fninf.2018.00034

18. Kennedy, K.: Use-definition chains with applications. Comput. Lang. **3**(3), 163–179 (1978). https://doi.org/10.1016/0096-0551(78)90009-7

19. Kumbhar, P., et al.: CoreNEURON: An Optimized Compute Engine for the NEURON Simulator. arXiv:1901.10975 [q-bio], January 2019

20. Kunkel, S., et al.: Spiking network simulation code for petascale computers. Front. Neuroinformatics **8**. https://doi.org/10.3389/fninf.2014.00078

21. Lattner, C., Adve, V.: LLVM: a compilation framework for lifelong program analysis & transformation. In: Proceedings of the International Symposium on Code Generation and Optimization: Feedback-Directed and Runtime Optimization, CGO 2004, p. 75. IEEE Computer Society (2004)

22. Magalhaes, B., Sterling, T., Schuermann, F., Hines, M.: Exploiting flow graph of system of odes to accelerate the simulation of biologically-detailed neural networks, pp. 176–187, May 2019. https://doi.org/10.1109/IPDPS.2019.00028

23. Marin, B.: Pynmodl: Python infrastructure for parsing and generating code from NMODL (2018). https://github.com/borismarin/pynmodl

24. Markram, H., et al.: Reconstruction and simulation of neocortical microcircuitry. Cell **163**(2), 456–492 (2015). https://doi.org/10.1016/j.cell.2015.09.029

25. Membarth, R., Hannig, F., Teich, J., Köstler, H.: Towards domain-specific computing for stencil codes in HPC. In: 2012 SC Companion: High Performance Computing, Networking Storage and Analysis, pp. 1133–1138 (2012)

26. Meurer, A., et al.: SymPy: symbolic computing in Python. PeerJ Comput. Sci. **3**, e103 (2017). https://doi.org/10.7717/peerj-cs.103

27. Migliore, M., et al.: Parallel network simulations with NEURON. J. Comput. Neurosci. **21**(2). https://doi.org/10.1007/s10827-006-7949-5

28. Morrison, A., et al.: Advancing the boundaries of high-connectivity network simulation with distributed computing. Neural Comput. **17**(8). https://doi.org/10.1162/0899766054026648

29. Nadkarni, P.M., et al.: ModelDB: an environment for running and storing computational models and their results applied to neuroscience. J. Am. Med. Inform. Assoc. **3**(6), 389–398 (1996). https://doi.org/10.1136/jamia.1996.97084512

30. Rathgeber, F., et al.: PyOP2: a high-level framework for performance-portable simulations on unstructured meshes. In: Proceedings of the 2012 SC Companion: High Performance Computing, Networking Storage and Analysis, SCC 2012. IEEE Computer Society (2012). https://doi.org/10.1109/SC.Companion.2012.134

31. Schmitt, C., et al.: ExaSlang: a domain-specific language for highly scalable multigrid solvers. In: 2014 Fourth International Workshop on Domain-Specific Languages and High-Level Frameworks for High Performance Computing (2014). https://doi.org/10.1109/WOLFHPC.2014.11

32. Traub, R., Wong, R., Miles, R., Michelson, H.: A model of CA3 hippocampal pyramidal neuron incorporating voltage-clamp data on intrinsic conductances. J. Neurophysiol. **66**, 635–650 (1991). https://doi.org/10.1152/jn.1991.66.2.635

Parallel Numerical Solution of a 2D Chemotaxis-Stokes System on GPUs Technology

Raffaele D'Ambrosio$^{(\boxtimes)}$, Stefano Di Giovacchino, and Donato Pera

Department of Information Engineering and Computer Science and Mathematics,
University of L'Aquila, L'Aquila, Italy
{raffaele.dambrosio,donato.pera}@univaq.it,
stefano.digiovacchino@graduate.univaq.it

Abstract. The aim of this paper is the numerical solution of a 2D chemotaxis model by a parallel numerical scheme, implemented on a GPU technology. The numerical discretization relies on the utilization of a finite difference scheme for the spatial part and the explicit Euler method for the time integration. Accuracy and stability properties are provided. The effectiveness of the approach, as well as the coherence of the results with respect to the modeled phenomenon, is provided through numerical evidence, also giving a performance analysis of the serial and the parallel implementations.

Keywords: Chemotaxis · GPU computing · Parallel numerical method

1 Introduction

Chemotaxis [3,10,11,14,18,23,24] is a very common phenomenon consisting in the movement of an organism in response to a chemical stimulus. For example, in order to find food, bacteria swim toward highest concentration of food molecules [10]. Another example is given by the motion of sperm towards the egg during fertilization in which chemotaxis phenomena are very crucial. Sometimes, as we can read in [23], the mechanism that allows chemotaxis phenomena in animals can be subverted; this is the case, for example, of cancer metastasis.

The model we deal with was first derived in [24] in order to describe the swimming of bacteria and oxygen transport near contact lines. Subsequently, this model was modified and completed by Cao in [3], where he described the motion of oxygen consumed by bacteria in a drop of water. The model is given by

This work is supported by GNCS-INDAM project and PRIN2017-MIUR project. The authors Raffaele D'Ambrosio and Stefano Di Giovacchino are members of the INdAM Research group GNCS. The authors are thankful to the Department of Information Engineering and Computer Science and Mathematics of the University of L'Aquila for the usage of the HPC Caliban Cluster (https://caliband.disim.univaq.it).

© Springer Nature Switzerland AG 2020
V. V. Krzhizhanovskaya et al. (Eds.): ICCS 2020, LNCS 12137, pp. 59–72, 2020.
https://doi.org/10.1007/978-3-030-50371-0_5

the chemotaxis-Stokes system with a rotational flux term, in a three-dimensional domain. The equations for an incompressible Navier-Stokes fluid are coupled with two parabolic equations, in which the first one presents a chemotactic term. In [3], it is proved in the two dimensional case and three dimensional case the existence and uniqueness of classical solution under a smallness assumption in the initial concentration. We will recall these results and we will give a wider description of this model in Sect. 2.

Numerical analysis plays a crucial role in computing solutions for PDEs system especially when it is quite difficult to find the analytical one or it is proved under some restrictive assumptions on the data of the problem. For this reason, our goal is to develop a numerical scheme to compute the solution of this system and to simulate it. In particular, we expect that there exists a time t after which the bacteria start the chemotaxis and move toward the oxygen.

Numerical solutions often require high spatial resolution to capture the detailed biophysical phenomena. As a consequence, long computational times are often required when using a serial implementation of a numerical scheme. Parallel computation can strongly improve the time efficiency of some numerical methods such as finite differences algorithms, which are relatively simple to implement and apply to a generic PDEs system. The Graphics Processing Units (GPUs) are perfect to use when we want to execute a numerical code based of a very large number of grid points, since the larger is the number of the grid points, the higher is the accuracy of the our numerical solution.

The codes used to study the performance of GPUs presented in this article were programmed using CUDA. The CUDA platform (Compute Unified Device Architecture), introduced by NVIDIA in 2007, was designed to support GPU execution of programs and focuses on data parallelism [12]. With CUDA, graphics cards can be programmed with a medium-level language, that can be seen as an extension to C/C++, without requiring a great deal of hardware expertise. We refer to [15,19] for a comprehensive introduction to GPU-based parallel computing, including details about the CUDA programming model and the architecture of current generation NVIDIA GPUs. As regards the application of GPU computing to partial differential equations, see [1,5,13] and references therein.

It is important to point out that, although the model is set in the three dimensional case we will perform our numerical analysis in a two dimensional setting. This assumption is not too restrictive since this is the most treated case in the literature concerning chemotaxis models (see [11] and reference therein). Indeed in many models, because of the microscopic third dimension, without loss of generality, cells are considered bidimensional.

This paper is organized as follows. In the next section, Sect. 2, a short description of the biological phenomenon and the equations of the model are presented. We present the numerical scheme in Sect. 3.

In the Sect. 4, the analysis of consistency and stability, for our numerical scheme, is given. Moreover, a set of numerical experiments are presented in Sect. 5. Section 6 contains the comparative performance evaluation between

GPUs and CPUs implementations of the numerical scheme. We summarize our work and we give some possible future developments in the final section, Sect. 7.

2   Mathematical Model

In this paper, we study the motion of oxygen consumed by bacteria in a drop of water. In particular, the model describes the motion of bacteria towards the zone of highest concentration of oxygen. However, the bacteria don't move directly toward these areas by using some rotations that can be completely random. The following initial boundary problem model, has been introduced in [3] and it is given by the following set of equations,

$$
\begin{cases}
n_t = \Delta n - \nabla \cdot (nS(x,n,c) \cdot \nabla c) - u \cdot \nabla n, & (x,t) \in \Omega \times (0,T), \\
c_t = \Delta c - nc - u \cdot \nabla c, & (x,t) \in \Omega \times (0,T,), \\
u_t = \Delta u + \nabla P + n \nabla \phi, & (x,t) \in \Omega \times (0,T), \\
\nabla \cdot u = 0, & (x,t) \in \Omega \times (0,T), \\
\nabla c \cdot \nu = (\nabla n - S(x,n,c)\nabla c) \cdot \nu = 0, \ u = 0, & (x,t) \in \partial\Omega \times (0,T), \\
n(x,0) = n_0(x), \ c(x,0) = c_0(x), \ u(x,0) = u_0(x), & x \in \Omega,
\end{cases}
\tag{1}
$$

where Ω is a bounded smooth domain in 2D or 3D, ν is the outward normal vector to the boundary $\partial\Omega$, n is the density of bacteria, c the oxygen's concentration and u and P are the velocity and the pressure of the fluid respectively.

The equation $(1)_1$ describes the density of bacteria. As we can see, this equation is a parabolic equation that admits a diffusion term Δn and a chemotactic term $\nabla \cdot (nS(x,n,c) \cdot \nabla c)$, that says that bacteria always move towards the higher oxygen's areas.

The parabolic equation $(1)_2$ describes the motion of oxygen concentration where the diffusion term is represented by Δc. The equations $(1)_3$ and $(1)_4$ are the well known Navier-Stokes equations for an incompressible fluid, subjected to an external force, without the convective term. This choice is due to the fact that, as we know from [2], the uniqueness of the solution is not yet guaranteed for a three dimensional Navier-Stokes problem. The tensor S is a rotational tensor, that takes into account the rotations of bacteria, and the function ϕ is a potential function that can be associated to an external force, therefore the term $n\nabla\phi$ can be seen as buoyant or electric force of bacterial mass. As in [3], we assume the following regularity conditions for the tensor S

$$
s_{ij} \in C^2 \left(\overline{\Omega} \times [0,\infty) \times [0,\infty) \right),
\tag{2}
$$

$$
|S(x,n,c)| := \max_{i,j\in\{1,2\}} \{|s_{ij}(x,n,c)|\} \le S_0 \text{ for all } (x,n,c) \in \overline{\Omega} \times [0,\infty) \times [0,\infty).
\tag{3}
$$

In order to describe the functional setting for the initial data we need to define the following operators and spaces.

Definition 1 (Stokes operator). *The Stokes operator on $L_\sigma^p(\Omega)$ is defined as $A_p = -\mathcal{P}\Delta$ with domain $D(A_p) = W^{2,p}(\Omega) \cap W_0^{1,p} \cap L_\sigma^p(\Omega)$, where $\mathcal{P}$ is the so-called Helmholtz projection. Since A_{p_1} and A_{p_2} coincide on the intersection of their domain for $p_1, p_2 \in (1, \infty)$, we will drop the index p.*

We will denote the first eigenvalue of A by λ_1', and by λ_1 the first nonzero eigenvalue of $-\Delta$ on Ω under Neumann boundary conditions.

The conditions on the initial data are as follows:

$$\begin{cases} n_0 \in L^\infty(\Omega), \\ c_0 \in W^{1,q}(\Omega), \quad q > N, \\ u_0 \in D(A^\alpha), \quad \alpha \in (\frac{N}{4}, 1), \end{cases} \tag{4}$$

$$n_0 \geq 0, \; c_0 \geq 0 \text{ on } \Omega. \tag{5}$$

As we will see, we also assume that $\|c_0\|_{L^\infty(\Omega)}$ is small. This assumption will be crucial for the existence of classical solution of (1), indeed the smallness of the initial concentration of bacteria can force the stability of system. With these assumptions, in [3], it has been proved that there exists a unique global classical solution for problem (1), for completeness, we report here these results.

Theorem 1. *Let $\Omega \subset \mathbb{R}^3$ be a bounded domain with smooth boundary. Assume that S fulfills (2) and (3). There is δ_0 with the following property: if the initial data fulfill (4) and (5), and*

$$\|c_0\|_{L^\infty(\Omega)} < \delta_0 \tag{6}$$

then (1) admits a global classical solution (n, c, u, P) which is bounded, and satisfies

$$\begin{cases} n \in C^{2,1}(\overline{\Omega} \times (0, \infty)) \cap C_{loc}^0((\overline{\Omega} \times [0, \infty)), \\ c \in C^{2,1}(\overline{\Omega} \times (0, \infty)) \cap C_{loc}^0((\overline{\Omega} \times [0, \infty)) \cap L^\infty((0, \infty); W^{1,q}(\Omega)), \\ u \in C^{2,1}(\overline{\Omega} \times (0, \infty)) \cap L^\infty((0, \infty); D(A^\alpha)) \cap C_{loc}^0([0, \infty); L^2(\Omega)), \\ P \in L^1((0, \infty); W^{1,2}(\Omega)). \end{cases} \tag{7}$$

3 Numerical Scheme

We now aim to discretize the differential problem (1), by a suitable finite difference numerical scheme, according to the classical method-of-lines [4,6–9,20,21]. As we said in the introduction, a simplifying assumption, we consider our problem in a two dimensional setting, therefore our spatial variable are given by (x, y). Moreover, the functions s_{ij} defined in (2) are given by $s_{i,j} = s_{i,j}(x, y, t)$, $i, j = 1, 2$. We assume that the domain Ω has the form $\Omega = [0, 1] \times [0, 1]$ and is discretized as follows. Given an integer N, we denote by $h = 1/(N+1)$ the spatial stepsize and accordingly define the grid

$$\Omega_h = \{(x_i, y_j) \in \Omega : x_i = ih, \; i = 0, 1, \ldots, N+1; \; y_j = jh, j = 0, 1, \ldots, N+1\}.$$

For a given function $u = u(x,y,t)$, in correspondence of a generic point $(x_i, y_j) \in \Omega_h$, we recall the following finite differences for the approximation of the first derivates

$$
\begin{aligned}
\frac{d}{dx}u(x_i,y_j,t) &= \frac{u(x_{i+1},y_j,t) - u(x_{i-1},y_j,t)}{2h} + O(h^2), \\
\frac{d}{dy}u(x_i,y_j,t) &= \frac{u(x_i,y_{j+1},t) - u(x_i,y_{j-1},t)}{2h} + O(h^2).
\end{aligned}
\tag{8}
$$

As regards the finite difference approximation of the second derivative, we adopt the following usual central finite difference discretization

$$
\begin{aligned}
\frac{d^2}{dx^2}u(x_i,y_j,t) &= \frac{u(x_{i+1},y_j,t) - 2u(x_i,y_j,t) + u(x_{i-1},y_j,t)}{h^2} + O(h^2), \\
\frac{d^2}{dy^2}u(x_i,y_j,t) &= \frac{u(x_i,y_{j+1},t) - 2u(x_i,y_j,t) + u(x_i,y_{j-1},t)}{h^2} + O(h^2).
\end{aligned}
\tag{9}
$$

Therefore, we have

$$
\begin{aligned}
\Delta u(x_i,y_j,t) = \frac{1}{h^2}\Big(&u(x_{i+1},y_j,t) + u(x_{i-1},y_j,t) + u(x_i,y_{j+1},t) \\
&+ u(x_i,y_{j-1},t) - 4u(x_i,y_j,t)\Big) + O(h^2).
\end{aligned}
$$

For any fixed time t, we denote by u_{ij} an approximate value of $u(x_i,y_j,t)$, with $i,j = 0,1,\ldots,N+1$. Then, for $i,j = 1,2,\ldots,N$, we obtain

$$
\begin{aligned}
\frac{du}{dx}(x_i,y_j,t) &\approx \frac{u_{i+1,j} - u_{i-1,j}}{2h}, & \frac{du}{dy}(x_i,y_j,t) &\approx \frac{u_{i,j+1} - u_{i,j-1}}{2h}, \\
\frac{d^2u}{dx^2}(x_i,y_j,t) &\approx \frac{u_{i+1,j} - 2u_{i,j} + u_{i-1,j}}{h^2}, & \frac{d^2u}{dy^2}(x_i,y_j,t) &\approx \frac{u_{i,j+1} - 2u_{i,j} + u_{i,j-1}}{h^2}
\end{aligned}
$$

and the five-point stencil

$$
\Delta u(x_i,y_i,t) \approx \frac{u_{i+1,j} + u_{i-1,j} + u_{i,j+1} + u_{i,j-1} - 4u_{i,j}}{h^2}
$$

for the Laplacian operator.

For the time discretization, we divide the time interval $[0,T]$, in M equidistant parts of length

$$
h_t = \frac{T}{M}.
$$

Then, we define the grid

$$
\mathcal{I}_{h_t} = \{t_k \in [0,T], \ t_k = kh_t, \ k = 0,1,\ldots,M\}
\tag{10}
$$

and denote by $u_{ij}^{(k)}$ an approximation of $u(x_i,y_j,t_k)$, with $(x_i,y_j) \in \Omega_h$, $t_k \in \mathcal{I}_{h_t}$.

In view of a parallel implementation, we adopt as time discretization that arises from the forward Euler scheme because it is directly parallelizable. The fully discretized problem reads as follows: as regards equation $(1)_2$, we have

$$c_{i,j}^{(k+1)} = c_{i,j}^{(k)} + h_t \left(\frac{c_{i+1,j}^{(k)} - 4c_{i,j}^{(k)} + c_{i-1,j}^{(k)} + c_{i,j-1}^{(k)} + c_{i,j+1}^{(k)}}{4h^2} \right.$$
$$\left. - n_{i,j}^{(k)} c_{i,j}^{(k)} - u_{1_{i,j}} \frac{c_{i+1,j}^{(k)} - c_{i-1,j}^{(k)}}{2h} - u_{2_{i,j}} \frac{c_{i,j+1}^{(k)} - c_{i,j-1}^{(k)}}{2h} \right).$$

For equation $(1)_3$, we have, for the component u_1,

$$u_{1_{i,j}}^{(k+1)} = u_{1_{i,j}}^{(k)} + h_t \left(\frac{u_{1_{i+1,j}}^{(k)} - 4u_{1_{i,j}}^{(k)} + u_{1_{i-1,j}}^{(k)} + u_{1_{i,j-1}}^{(k)} + u_{1_{i,j+1}}^{(k)}}{4h^2} \right.$$
$$\left. + \frac{P_{i+1,j}^{(k)} - P_{i-1,j}^{(k)}}{2h} + n_{i,j}^{(k)} \frac{\phi_{i+1,j} - \phi_{i-1,j}}{2h} \right)$$

and a similar formula also holds true for the component u_2. For equation $(1)_4$, we have

$$\frac{u_{1_{i+1,j}}^{(k)} - u_{1_{i-1,j}}^{(k)} + u_{2_{i,j+1}}^{(k)} - u_{2_{i,j-1}}^{(k)}}{2h} = 0.$$

As regards equation $(1)_1$, we have

$$n_{i,j}^{(k+1)} = n_{i,j}^{(k)} + \frac{h_t}{4h^2} \left[\alpha_{i,j}^{(k)} - 2h \left(u_{1_{i,j}}^{(k)} (n_{i+1,j}^{(k)} - n_{i-1,j}^{(k)}) + u_{2_{i,j}}^{(k)} (n_{i,j+1}^{(k)} - n_{i,j-1}^{(k)}) \right) \right],$$

for $i, j = 1, ..., N$, where $\alpha_{ij}^{(k)}$ contains all the terms of discretization independent on $u_{1_{i,j}}^{(k)}$ and $u_{2_{i,j}}^{(k)}$.

We finally provide a discretized equation for the pressure P. Indeed, by the incompressibility assumption $(1)_4$, the pressure P satisfies, at any time t, the equation

$$\Delta P = -\nabla \cdot (n\nabla\phi) = -\nabla n \cdot \nabla\phi - n\Delta\phi, \tag{11}$$

whose discretization leads to

$$\frac{P_{i+1,j}^{(k)} + P_{i-1,j}^{(k)} + P_{i,j+1}^{(k)} + P_{i,j-1}^{(k)} - 4P_{i,j}^{(k)}}{h^2} =$$
$$-\frac{(n_{i+1,j}^{(k)} - n_{i-1,j}^{(k)})(\phi_{i+1,j} - \phi_{i-1,j}) - (n_{i,j+1}^{(k)} - n_{i,j-1}^{(k)})(\phi_{i,j+1} - \phi_{i,j-1})}{4h^2}$$
$$- n_{i,j}^{(k)} \frac{\phi_{i+1,j} + \phi_{i-1,j} + \phi_{i,j+1} + \phi_{i,j-1} - 4\phi_{i,j}}{h^2}, \tag{12}$$

for $i, j = 1, 2, \ldots, N$.

We observe that, as regards the boundary conditions, we will always use the Dirichlet ones in the remainder of the treatise, that will give the values of the unknown functions when $(i, j) = (0, 0)$ and $(i, j) = (N+1, N+1)$.

4   Consistency and Stability Analysis

In this section, we want to analyze the consistency and stability of the numerical scheme introduced in the previous section. For the sake of clarity, here we distinguish the contribution to the global error arising from the spatial discretization and that coming from the time discretization. We observe that our analysis is given for problems having sufficient regularity in order to make the application of Taylor series arguments possible.

4.1   Analysis of the Spatial Discretization

Let us consider the system (1) and analyze the contribution to the global error associated to the space discretization of each equation. Let us first focus our attention on Eq. $(1)_2$. Replacing the exact solution in the right-hand side of the spatially discretized equation referred to the generic point (x, y) of the grid gives

$$
\begin{aligned}
&\Delta c(x, y) - n(x, y)c(x, y) - u(x, y) \cdot \nabla c(x, y) \\
&\approx \frac{c(x + h, y) + c(x - h, y) + c(x, y + h) + c(x, y - h) - 4c(x, y)}{h^2} \\
&\quad - n(x, y)c(x, y) - u_1(x, y)\frac{c(x + h, y) - c(x - h, y)}{2h} \\
&\quad - u_2(x, y)\frac{c(x, y + h) - c(x, y - h)}{2h},
\end{aligned} \tag{13}
$$

where we have denoted the spatial stepsize by h and neglected the time dependence for the sake of brevity. Expanding $c(x + h, y)$, $c(x - h, y)$, $c(x, y + h)$ and $c(x, y - h)$ in Taylor series around (x, y) and collecting the resulting expressions in (13), we obtain

$$
\begin{aligned}
&c_{xx}(x, y) + c_{yy}(x, y) - n(x, y)c(x, y) - u_1(x, y)c_x(x, y) - u_2(x, y)c_y(x, y) \\
&\approx c_{xx}(x, y) + c_{yy}(x, y) + \frac{h^2}{12}\left(c_{xxxx}(x, y) + c_{yyyy}(x, y)\right) - n(x, y)c(x, y) \\
&\quad - u_1(x, y)\left(c_x(x, y) + \frac{h^2}{6}c_{xxx}(x, y)\right) - u_2(x, y)\left(c_y(x, y) + \frac{h^2}{6}c_{yyy}(x, y)\right).
\end{aligned}
$$

This implies that the deviation between the exact operator and its spatial discretization is $\{\mathcal{O}\}(h^2)$. For all the other equations in (1), the analysis proceeds exactly in the same way. Therefore, the residuum obtained by replacing the exact solution in the spatially discretized problem is $\mathcal{O}(h^2)$. Then, the spatial discretization is consistent of order 2.

4.2   Analysis of the Time Discretization

Let us rewrite the numerical scheme introduced in Sect. 3 in the following form

$$
\begin{cases}
\dot{W}(t) = f(W(t)), \\
W(0) = W_0,
\end{cases} \tag{14}
$$

where $W^{(k)} \in \mathbb{R}^{4N^2}$ is the vector

$$W^{(k)} = \begin{bmatrix} c^{(k)} \\ n^{(k)} \\ u_1^{(k)} \\ u_2^{(k)} \end{bmatrix},$$

with $c^{(k)}$ the vectorization of $\left(c_{ij}^{(k)}\right)_{i,j=1}^{N}$ similarly for the other entries of $W^{(k)}$.
By applying the explicit Euler method, we have, at time $t_k \in \mathcal{I}_{h_t}$ defined in (10),

$$W^{(k)} = W^{(k-1)} + h_t f(W^{k-1}). \tag{15}$$

Clearly, the numerical scheme (15) is consistent with problem (1). We now aim to provide the conditions ensuring its stability. To this purpose, it is useful to rewrite the vector field f of (14) in the form

$$f(W^{(k)}) = GW^{(k)} + F(W^{(k)}),$$

where the matrix $G \in \mathbb{R}^{4N^2 \times 4N^2}$ contains the linear part of numerical scheme and the nonlinear function $F(W^{(k)}) \in \mathbb{R}^{4N^2}$ is the vector containing the nonlinear part of f. In this regard, following the lines drawn in [4,6,7,22], the following result holds.

Theorem 2. *The numerical method (15) is stable if*

$$\|I + h_t G\|_\infty + h_t F_{max} \leq 1,$$

being I the identity matrix and F_{max} an upper bound for the norm of the gradient of F.

Proof. Let us consider a perturbation of the solution at the step k, denoted by $\widetilde{W}^{(k)}$, that is,

$$\widetilde{W}^{(k)} = W^{(k)} + E^{(k)}.$$

By applying the method (15) to $\widetilde{W}^{(k)}$, we obtain

$$\widetilde{W}^{(k+1)} = \widetilde{W}^{(k)} + h_t[G\widetilde{W}^{(k)} + F(\widetilde{W}^{(k)})].$$

Therefore, we have

$$\begin{aligned} E^{(k+1)} &= W^{(k+1)} - \widetilde{W}^{(k+1)} \\ &= W^{(k)} + h_t[GW^{(k)} + F(W^{(k)})] - \widetilde{W}^{(k)} - h_t[G\widetilde{W}^{(k)} + F(\widetilde{W}^{(k)})] \\ &= E^{(k)} + h_t G E^{(k)} + h_t[F(W^{(k)}) - F(\widetilde{W}^k)] \\ &= (I + h_t G)E^{(k)} + h_t[F(W^{(k)}) - F(W^k + E^{(k)})]. \end{aligned} \tag{16}$$

By Taylor expansion arguments for $F(W^{(k)} + E^{(k)})$ around $W^{(k)}$, we have

$$E^{(k+1)} \leq (I + h_t G)E^{(k)} + h_t[\nabla F(W^{(k)})E^{(k)}].$$

Therefore, passing to the norm, we obtain

$$\left\|E^{(k+1)}\right\|_{\infty} \leq \left\|I + h_t G\right\|_{\infty} \left\|E^{(k)}\right\|_{\infty} + h_t \left\|\nabla F(W^{(k)})\right\|_{\infty} \left\|E^{(k)}\right\|_{\infty}$$
$$\leq \left\|I + h_t G\right\|_{\infty} \left\|E^{(k)}\right\|_{\infty} + h_t F_{max} \left\|E^{(k)}\right\|_{\infty}.$$

Thus, we obtained the following stability inequality

$$\left\|E^{(k+1)}\right\|_{\infty} \leq (\left\|I + h_t G\right\|_{\infty} + h_t F_{max}) \left\|E^{(k)}\right\|_{\infty}, \tag{17}$$

leading to the thesis.   □

5   Simulations and Numerical Results

In this section, we present our main numerical results. In particular, we are interested in observing the chemotactic effect of bacteria towards the oxygen. For simplicity, we have considered the rotational tensor S to be the identity. Moreover, we have supposed the vector field u to be null at the initial time and we have considered the following initial data for the function n, c and P:

$$P(x,y) = n(x,y) = \frac{100}{\sqrt{2\pi\sigma_n^2}} e^{-\frac{1}{2\sigma_n^2}[(x-\mu_n)^2 + (y-\mu_n)^2]},$$

$$c(x,y) = \frac{65}{\sqrt{2\pi\sigma_c^2}} e^{-\frac{1}{2\sigma_c^2}[(x-\mu_c)^2 + (y-\mu_c)^2]},$$

where $\sigma_n^2 = 0.01$, $\sigma_c^2 = 0.025$, $\mu_n = 0.5$ and $\mu_c = 0.8$. Moreover, we have assumed the potential to have the following form

$$\phi(x,y) = -\frac{70}{\sqrt{2\pi\sigma_\phi^2}} e^{-\frac{1}{2\sigma_\phi^2}[(x-\mu_{\phi_x})^2 + \frac{(y-\mu_{\phi_y})^2}{2}]},$$

where $\sigma_\phi^2 = 1$, $\mu_{\phi_x} = 1$ and $\mu_{\phi_y} = 0.5$ In order to see the phenomenon and to preserve the stability, we have to choose as time step $h_t = 2e-10$. The numerical pattern for the density at various time is depicted in Fig. 1. In our simulations, at the initial time, the bacteria are concentrated on the centre of the domain. Therefore, the action of potential is not well visible. At time $t = 0.2$ ms, we can see that the external force exerts a braking action on the bacteria in their motion toward the oxygen.

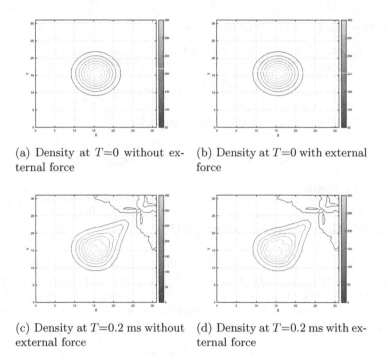

(a) Density at T=0 without external force   (b) Density at T=0 with external force

(c) Density at T=0.2 ms without external force   (d) Density at T=0.2 ms with external force

Fig. 1. Time evolution of the density towards the oxygen with grid size 32×32, where $T = T_{fin}$.

6   GPU Programming and Performance Evaluation

In this section we describe the basic logical steps required to implement the GPU codes and also the performance evaluation metrics used to evaluate the computing performance. As discussed in Sect. 3, the numerical scheme relies on a finite differences method for the spatial discretization and a time integration based on the explicit Euler method. From the programming point of view this mathematical approach leads to design a code where a *for loop* defines the clock time steps and where spatial values at each time iteration are updated in parallel by the GPU using the aforementioned numerical scheme.

This basic idea is to use the CPU (host) as owner of time clock activities and the GPU (device) as the owner of the massive computing activities related to the spatial part of the equations. This will lead to a *master-slave* model in which the CPU is the master because it controls the parallel executions on GPU and, therefore, the GPU works only on the spatial part of our scheme. The implemented code employs only the *global memory* in the CUDA kernel codes, while further optimizations related to the implentation of a code able to use the *shared memory* and/or CUDA dynamic parallelism [12,16,17] able to reduce the data transfer activity between host and device will be subjects of a further work.

According to all the principles above described, the code follows the following logical steps:

1. the CPU (host) loads the initial data from the its memory to the GPU (device) memory, global memory;
2. the GPU provides the massive computing activities, that is, the GPU has to execute the code related to spatial discretization because it is the parallelizable part of the code since, at each time step, the values referring to the current step only depends on those already computed in the previous one;
3. the GPU sends back to the CPU the partial/final results;
4. the CPU checks the time step and according to the maximum time value defined by the user restarts/stops the parallel computing process.

We have executed the code on two distinct architectures. The first has the following specifications: HP DL 585 G7 PROLIANT, with processors 4x AMD 6128 (8 core), with clock's frequency 2.0 GHz and RAM 64 GB, in which a GTX GeForce 1080, 8 GB RAM, is integrated. The GPU is the only difference between this architecture and the second one. Here, there are 3x GeForce GTX 670, 4 GB RAM. The operative system used is Linux CentOS 6.5. Finally, we have compiled the serial code with gcc 4.4.7 and the CUDA-C code with CUDA 9.1 in the first machine and CUDA 8.0 in the second one. In order to evaluate the performances of the two machines, it is very reasonable to compute the number of floating point operations executed for unity of time on GPUs, as a function of the dimension of the grid. Therefore, for any fixed size of the grid, if n is number of floating point operations and T is the CUDA code execution time on the GPUs, we have computed the number $f_{op} = n/T$ of floating point operations for unity of time (seconds).

We remark that the following results have been obtained with a single precision and that we have estimated the number of operations using the NVIDIA *nvprof* tool.

In Table 1, we can see the time comparison between the serial execution, the GPU parallel on two different devices and the parallel openMP version executed on a CPU based *shared memory* architecture with different number of threads, We report the corresponding graphs in Figs. 2. In particular, we can observe a good scaling of the code moving from the CPU technology to the GPU with the GTX GeForce 670 and GTX GeForce 1080 that provide reduced execution times.

Table 1. Computation times, in seconds, for the serial execution, parallel execution on GTX GeForce 1080, parallel execution on GTX GeForce 670 and parallel execution by using shared memory with openMP with different number of threads.

Dim	Serial kernel	GTX Force 1080	GeForce GTX 670
32	1.156	13.920×10^{-6}	18.677×10^{-3}
64	4.134	25.120×10^{-6}	21.203×10^{-3}
128	18.629	124.06×10^{-6}	50.671×10^{-3}
256	75.635	741.16×10^{-6}	183.45×10^{-3}
Dim	OpenMP(8)	OpenMP(16)	OpenMP(32)
32	0.524	0.686	6.918
64	1.708	1.637	1.487
128	11.565	9.455	9.547
256	46.065	38.635	41.320

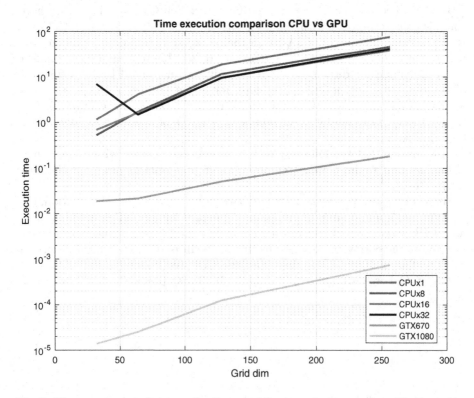

Fig. 2. Time comparison between the three architectures in the semilogarithmic scale. We can see the improvement that we have obtained by passing from CPU technology to GTX GeForce 670, until to GTX GeForce 1080.

7 Conclusions and Future Works

We have developed a parallel numerical scheme, implemented on GPU, to compute the solution of a chemotaxis system. We have made use of the central finite differences to approximate the spatial derivatives and of the explicit Euler method to discretize the time evolution of the system. We have analyzed accuracy and stability issues, implemented the code on CPU and GPU architectures and compared their performances in terms of time execution getting a good scalability for the GPU implentation. For the GPU kernel design, we have used the global memory and implemented a master-slave model, in which the CPU controls the time evolution while the GPU works exclusively on the spatial derivatives of our scheme. Future issues of this research are oriented to providing a 3D model with a deep optimized CUDA kernel code implemented by using dynamic parallelism and shared memory.

References

1. Aissa, M., Verstraete, T., Vuik, C.: Toward a GPU-aware comparison of explicit and implicit CFD simulations on structured meshes. Comput. Math. Appl. **74**(1), 201–217 (2017)
2. Boyer, F., Fabrie, P.: Mathematical Tools for the Study of the Incompressible Navier-Stokes Equations and Related Models. Springer, New York (2013). https://doi.org/10.1007/978-1-4614-5975-0
3. Cao, X.: Global classical solutions in chemotaxis(-Navier)-Stokes system with rotational flux term. J. Differ. Equations **261**(12), 6883–6914 (2016)
4. Cardone, A., D'Ambrosio, R., Paternoster, B.: Exponentially fitted IMEX methods for advection-diffusion problems. J. Comput. Appl. Math. **316**, 100–108 (2017)
5. Conte, D., D'Ambrosio, R., Paternoster, B.: GPU acceleration of waveform relaxation methods for large differential systems. Numer. Algorithms **71**(2), 293–310 (2016)
6. D'Ambrosio, R., Moccaldi, M., Paternoster, B.: Adapted numerical methods for advection-reaction-diffusion problems generating periodic wavefronts. Comput. Math. Appl. **74**(5), 1029–1042 (2017)
7. D'Ambrosio, R., Moccaldi, M., Paternoster, B.: Parameter estimation in IMEX-trigonometrically fitted methods for the numerical solution of reaction-diffusion problems. Comput. Phys. Commun. **226**, 55–66 (2018)
8. D'Ambrosio, R., Paternoster, B.: Numerical solution of reaction-diffusion systems of lambda-omega type by trigonometrically fitted methods. J. Comput. Appl. Math. **294**(C), 436–445 (2016)
9. D'Ambrosio, R., Paternoster, B.: Numerical solution of a diffusion problem by exponentially fitted finite difference methods. SpringerPlus **3**(1), 1–7 (2014). https://doi.org/10.1186/2193-1801-3-425
10. de Oliveira, S., Rosowski, E.E., Huttenlocher, A.: Neutrophil migration in infection and wound repair: going forward in reverse. Nat. Rev. Immunol. **16**(6), 378–391 (2016)
11. Di Francesco, M., Donatelli, D.: Singular convergence of nonlinear hyperbolic chemotaxis systems to Keller-Segel type models. Discrete Contin. Dyn. Syst. Ser. B **13**(1), 79–100 (2010)

12. Kirk, D.B., Hwu, W.M.W.: Programming Massively Parallel Processors: A Hands-on Approach, (third ed.). Morgan Kaufmann Publishers Inc., San Francisco (2016)
13. Magee, D.J., Niemeyer, K.E.: Accelerating solutions of one-dimensional unsteady PDEs with GPU-based swept time-space decomposition. J. Comput. Phys. **357**, 338–352 (2018)
14. Málaga, C., Minzoni, A.A., Plaza, R.G., Simeoni, C.: A chemotactic model for interaction of antagonistic microflora colonies: front asymptotics and numerical simulations. Stud. Appl. Math. **130**(3), 264–294 (2013)
15. Nvidia CUDA C Programming Guide, Version 9.1, NVIDIA Corporation
16. Nvidia TechBrief Dynamic Parallelism in CUDA
17. Pera, D.: Parallel numerical simulations of anisotropic and heterogeneous diffusion equations with GPGPU, PhD Thesis (2013)
18. Pera, D., Málaga, C., Simeoni, C., Plaza, R.G.: On the efficient numerical simulation of heterogeneous anisotropic diffusion models for tumor invasion using GPUs. Rend. Mat. Appl. **7**(40), 233–255 (2019)
19. Sanders, J., Kandrot, E.: CUDA by Example: An Introduction to General-Purpose GPU Programming. Addison-Wesley Professional, Boston (2010)
20. Schiesser, W.E.: The Numerical Method of Lines: Integration of Partial Differential Equations. Academic Press, San Diego (1991)
21. Schiesser, W.E., Griffiths, G.W.: A Compendium of Partial Differential Equation Models: Method of Lines Analysis with Matlab. Cambridge University Press, Cambridge (2009)
22. Smith, G.D.: Numerical solution of partial differential equations: Finite Difference Methods. Clarendon Press, Oxford (1985)
23. Stuelten, C.H., Parent, C.A., Montell, D.J.: Cell motility in cancer invasion and metastasis: insights from simple model organisms. Nat. Rev. Cancer **18**(5), 296–312 (2018)
24. Tuval, I., Cisneros, L., Dombrowski, C., Wolgemuth, C.W., Kessler, J.O., Goldstein, R.E.: Bacterial swimming and oxygen transport near contact lines. Proc. Natl. Acad. Sci. U.S.A. **102**(7), 2277–2282 (2005)

Automatic Management of Cloud Applications with Use of Proximal Policy Optimization

Włodzimierz Funika[1]([⊠]) [ID], Paweł Koperek[1] [ID], and Jacek Kitowski[1,2] [ID]

[1] Department of Computer Science, Faculty of Computer Science, Electronics and Telecommunication, AGH, al. Mickiewicza 30, 30-059 Kraków, Poland
{funika,kito}@agh.edu.pl, pkoperek@gmail.com
[2] ACC CYFRONET AGH, AGH, ul. Nawojki 11, 30-950 Kraków, Poland

Abstract. Reinforcement learning is a very active field of research with many practical applications. Success in many cases is driven by combining it with Deep Learning. In this paper we present the results of our attempt to use modern advancements in this area for automated management of resources used to host distributed software. We describe the use of an autonomous agent that employs a policy trained with use of Proximal Policy Optimization algorithm. The agent is managing a cloud infrastructure used to process a sample workload. We present the design and architecture of a complete autonomous management system and explain how the management policy was trained. Finally, we compare the performance to the traditional automatic management approach exploited in AWS stack and discuss feasibility to use the presented approach in other scenarios.

Keywords: Reinforcement learning · Cloud resources · Automatic management · Proximal policy optimization

1 Introduction

Leveraging the cloud computing infrastructures is one of the currently dominating trends in the design of modern software systems. The main advantages of this approach include high availability, increased security and flexibility in resource allocation. In many cases the ability to adjust the amount of used resources to the actual needs is the driver of the adoption. The promise to reduce the costs of resources is very compelling. Unfortunately it requires implementing special measures. The application needs in support for adding or removing more cores, RAM, hard drives (*vertical scaling*) or more virtual or physical machines (*horizontal scaling*) without breaking the core functionality. Furthermore one needs to develop a policy under which the resources will be added or removed. In certain scenarios, where the environment renders strong and stable seasonal behavior patterns, the configurations and resources can be approximated by human experience in advance. Nevertheless, in many other cases elasticity can

© Springer Nature Switzerland AG 2020
V. V. Krzhizhanovskaya et al. (Eds.): ICCS 2020, LNCS 12137, pp. 73–87, 2020.
https://doi.org/10.1007/978-3-030-50371-0_6

only be enabled by automatic scaling, which we can define as *a dynamic process [...] that adapts software configurations [...] and hardware resources provisioning [...] on-demand, according to the time-varying environmental conditions* [1].

Reinforcement learning techniques have been known for a long time [2,3]. Until recently they were mostly applicable to relatively simple problems, where observing the whole environment was easy and the number of possible actions was small. The new advancements in the area allow to tackle domains which are much more complicated, like computer games [4], control of robots [5] or the game of Go [6]. The state-of-the-art results can be obtained thanks to application of Deep Learning, e.g. in form of Deep Q Learning [7], Asynchronous Actor-Critic Agents (A3C) [8] or more recently Proximal Policy Optimization [9]. Those methods enable learning complex behaviors by directly observing an environment and interacting with it through pre-defined actions. In some cases this approach allowed to achieve results surpassing human decisions based performance.

Such successes encourage experimenting with applying Deep Reinforcement Learning (*DRL*) to other domains. One particular area, where this technique might deliver a lot of benefits, is the auto-scaling of applications deployed to compute clouds. The infrastructure used to host the application becomes the environment in which the automatic agent operates. The state of the application becomes the state which the agent alters and the operations which can be executed through a cloud vendor API become the agent's actions. Available measurements and metrics are well defined: usually the technology choices help to define which elements of the system should be observed and how. There is a variety of available monitoring software which can be used. In this type of task there are usually well defined goals (e.g. reducing request latency, average CPU load, memory consumption or monetary cost of hosting). Such an optimization goal can be translated into a reward function, which can be used by the agent as a feedback mechanism. The agent can discover management policies without any prior knowledge, by conducting experiments in the underlying infrastructure through a process of trials and errors.

The main disadvantage of the described approach is the cost of creating of a DRL policy. The policy needs to go through multiple iterations where it interacts with the automatically scaled system and observes its reactions. Unfortunately, since the decisions made at the beginning might be quite random, there is a substantial risk of destabilizing the observed application and making it unusable by the end users. Such a situation is unacceptable in a production system, what suggests that creating a special, duplicate training environment might be helpful. Unfortunately, this introduces additional resource requirements and therefore increases the overall cost. This situation might be mitigated by modelling the auto-scaled application and simulating it in an artificial environment. This enables replaying the observed workloads multiple times and much faster than they happened in reality. Since DRL usually can benefit from increasing the number of training iterations, the simulation-based approach can help to create more efficient policies.

In this paper we present a novel auto-scaling software system which leverages the simulation of a compute cloud. We present the Semantic-Based Automatic Monitoring and Management (SAMM) system [10] extended by a decision making component, which employs a DRL method for training - the Proximal Policy Optimization. SAMM observes a cloud-based application, passes its observations (metric measurements and information about the scheduled jobs) to the policy. In case the policy decides to execute an action, that information is used by SAMM to adjust the environment by e.g. creating or removing resources through cloud API function invocations. Before being deployed to the live system, the policy is trained in a simulator, what enables us to avoid potentially very costly bad decisions made during the training phase. To the best of our knowledge, this is the first attempt to use the PPO algorithm to control resources of a distributed application deployed in a real-world cloud computing environment. We present a complete and functioning implementation and share the results of experiments.

The paper is organized as follows: in Sect. 2 we present related work, Sect. 3 describes the design and architecture of the environment and the training of the decision policy. Section 4 provides details of the experiments and the environment in which they were being run. Section 5 summarizes our research and outlines next steps.

2   Related Work

Managing an infrastructure in a way, which minimizes the monetary cost while maintaining the business requirements, sometimes defined as Quality-of-Service (*QoS*) objectives, is a very complex subject. There have been numerous attempts to tackle this problem, which differ in many ways. One possible taxonomy has been proposed by the authors of [1], where the following features were chosen to classify auto-scaling systems:

- *self-awareness* - the ability to acquire and maintain knowledge about the system's state. It can be implemented in the form of: *stimulus awareness, time awareness, goal awareness, interaction awareness, meta awareness.*
- *self-adaptivity* - the ability to change own behavior depending on the requirements. We include the variants of *self-healing, self-configuring, self-optimizing, self-protecting.*
- *architectural patterns* - the structure of the auto-scaling process which includes the definition of interactions between components and specification of modules. The dominant approaches are: *Feedback loop* [11], *Observe-Decide-Act* (ODA) [12], *Monitor-Analysis-Plan-Execute* (MAPE) [13].
- *QoS modeling* - the control primitives which are used to change the environment and the model which connects QoS metrics to the control primitives. There are different types of models: static (the relationship between metrics and how resources are reallocated is defined before the system begins to work) [14], dynamic (resource allocation is determined based on statistical analysis of workload traces) [15], semi-dynamic [16], based on machine learning [17] or simulation [18].

- *granularity of control* - specification of what is the basic entity to control: virtual machine [19], container [20], application [21].
- *decision making* - definition of the process which produces a control decision, including:
  - the mechanism of reasoning and searching for the decisions
  - objectives and their representations
  - which control primitives need to be changed.

There are numerous different approaches, most notably:

  - rule-based control [22,23] - classic approach in which the scaling actions are executed when some pre-defined conditions occur, e.g. when the average CPU load reaches 0.9 a new VM is added to the pool of resources;
  - control theory based - the process of making a decision to change the pool of resources uses mechanisms described by the control theory, e.g. [24];
  - search based optimization - the possible decisions are treated as a part of a large but finite search space. The process of choosing among them becomes a search problem. There were many different attempts which have used this approach: [17,25,26]. The first attempts to exploit Deep Reinforcement Learning [27] also fit in this category.

The system presented in this paper can be classified as a *self-optimizing* and *interaction-aware*, with a *feedback-loop* based architecture. It operates on the container granularity level. The decision making process is based on the Deep Reinforcement Learning (PPO algorithm) approach. To our best knowledge our contribution is a first example of the complete system which implements such an approach in a widely available public cloud environment.

2.1 Reinforcement Learning

Reinforcement Learning (RL) [2,28] is a machine learning paradigm which focuses on discovering a policy for autonomous agents, which take actions within a specific environment. The goal of the policy is to maximize a specific reward whose value can be presented to the agent with a delay. The RL approach differs from *supervised learning*: training the agent is based on the fact that knowledge is gathered in a trial and error process, where the agent interacts with the environment and observes results of its actions. There is no *supervising entity*, which would be capable of providing feedback on whether certain actions are better than others. RL is also different from unsupervised learning: it focuses on maximizing the reward signal instead of finding the structure hidden in collections of unlabeled data.

There are multiple variants of Reinforcement Learning algorithms:

- *Online* and *offline*: the former approach assumes the agent's policy is updated based on the most recent data, after every step (e.g. every monitoring and management iteration), in the latter one - after a full *episode* (i.e. after interactions cease, the environment needs to restart and a complete reward is given).

– *Model-based* and *model-free*: the former approach assumes an explicit model of the environment (state transitions and reward estimations) are created, in the latter it is assumed a decision can be based only on a sample of information about transitions.

The *model-free* approach has become very popular recently thanks to combining it with Deep Learning and creating so called *Deep Reinforcement Learning* (DRL). Using this technique allowed to create autonomous agents which are capable of achieving human-level performance across many domains.

Policy gradient methods are an approach to DRL, which is believed to render good experimental results. The training process improves a vector of the policy parameters Θ based on the gradient of some estimated scalar performance objective $J(\Theta)$ in respect to the policy parameters. These methods seek to maximize performance (measured as a reward obtained from interactions with the environment) and as such they change the parameters according to an iterative process in which the changes to parameters approximate a *gradient ascent* in J:

$$\Theta_{k+1} = \Theta_k + \alpha \nabla_\Theta J(\Theta_k) \tag{1}$$

where Θ_k denotes policy's parameters in the k-th iteration of the training process.

There are many variants of policy gradient optimization methods, however in this paper we focus on the *Proximal Policy Optimization* (PPO) [9].

The algorithm aims to compute a parameter update at each step, that on the one hand minimizes the cost function, while at the same time ensures the difference to the previous policy to be relatively small. This is achieved by modifying the objective in such a way that it ensures the updates to parameters are not too big. The objective is therefore defined by the following function:

$$J(\Theta) = L^{CLIP}(\Theta) = \mathbb{E}_t \left[min(r_t(\Theta)A_t, clip(r_t(\Theta), 1 - \epsilon, 1 + \epsilon)A_t) \right] \tag{2}$$

where $\mathbb{E}_t$ denotes calculating average over a batch of samples at timestamp t, A_t is an estimator of the advantage function which helps to evaluate which action is the most beneficial in a given state. r_t marks probability ratio $r_t(\Theta) = \frac{\pi_\Theta(a_t|s_t)}{\pi_{\Theta_{old}}(a_t|s_t)}$ in which $\pi_\Theta(a_t|s_t)$ denotes the probability of taking an action a in state s by a stochastic policy and Θ_{old} are the policy parameters before the update. The $clip(r_t(\Theta), 1 - \epsilon, 1 + \epsilon)$ function keeps the value of $r_t(\Theta)$ within some specified limits (*clips* it at the end of the range) and ϵ is a hyperparameter with a typical value between 0.1 and 0.3.

In our previous research [29] we experimentally verified that PPO gave the best empirical results in automated resources management among the policy gradient methods. Therefore have chosen this algorithm for experiments presented in this paper.

3   Proposed System Design

In this section we present the design of an experimental system, which allows to employ DRL to create automatic management agents. These agents are meant to manage resources of a distributed application deployed in a cloud infrastructure. The system attempts to adapt itself to ever-changing environment conditions by continuously improving the DL model used as an agent's decision policy. To avoid the costs of poor decisions while training a policy, a simulation is used as a training environment.

We begin with presenting the components of the system's architecture and their responsibilities. In the following section we discuss details of how we trained the policy used to make resource management decisions.

3.1   Architecture

In our research we focused on utilizing some well known and tested frameworks to reduce the amount of time needed to implement our ideas. As a foundation we have chosen SAMM [10], a prototype monitoring system enabling experimenting with automatic management of computer resources. It allows to easily extend its monitoring capabilities by new types of resources to observe and integrate with different technologies and new algorithms and observe their impact on the observed system.

In our use case, SAMM is used to integrate other elements of the system together to form a feedback loop:

- Periodically polls measurements which describe the current state of the system (e.g. the average CPU utilization in the computation cluster, amount of used memory etc.),
- Aggregates measurements into metrics used by the decision policy,
- Communicates with the *Policy Evaluation Service*. Provides the current state of the system in a form of metric values and retrives decisions,
- Executes decisions through the cloud vendor API (e.g. Amazon Web Services API) taking into account environment constraints (e.g. warm-up and cooldown periods).

The raw measurements are being collected with use of the Graphite monitoring tool [30]. Every machine executing the computations is expected to automatically start sending frequent reports through that tool as soon as it starts operating. The measurements can be reported at different intervals, e.g. CPU and memory statistics are sent every 10 s while the number of running virtual machines (VM) is reported once per minute. To introduce a coherent view of the environment, Graphite aggregates the values within a common interval, which in our case is set to one minute.

The role of *Policy Evaluation Service* is straightforward: it evaluates the state of the observed system with the use of the policy trained using the Proximal Policy Optimization. There are three possible results of the evaluation: *do nothing* (according to the policy a proper amount of resources is being used, there is no need to change the environment), *add another VM* (there are not enough resources in the current state of the system), *remove a VM* (there are too many machines used, one of the currently running ones should be stopped).

The decisions of the agent are not always translated directly to changes of the environment. Taking an action is always subject to the environment constraints: starting or stopping a virtual machine takes some time. In order to observe the effects of the previous interaction, we need to wait for some time, i.e. wait for a *warm-up* (starting a new VM) or *cool-down* (stopping a VM) of the system. We might also need to simply wait until the previous request gets fulfilled or handle a request failure.

The presented system does not make many assumptions about the workload for which the infrastructure is being controlled. It is required though that: a) the work can be split into multiple, independent tasks, b) there is a concept of a queue which can be monitored to check how many tasks are waiting for processing, c) the VMs used for processing can be stopped, the tasks which would be processed by them would be retried, d) executing the same job multiple times does not carry a risk of system malfunction.

It is assumed that the process which generates workload is being run on a machine which is not subject to automatic scaling. In order to fulfill monitoring requirements it might be necessary to instrument that machine and the software package which is responsible for generating tasks.

The complete diagram of the architecture, including the relationships of the discussed components, is presented in Fig. 1.

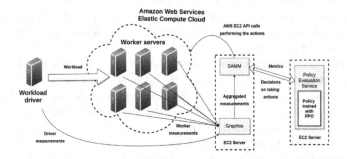

Fig. 1. Components of the discussed system. Arrows denote interactions between them.

3.2   Training the Policy

Autonomous management systems face a serious risk of introducing unnecessary costs while they are being trained. If they would start experimenting with executing the actions with regard to a real live application, they could significantly degrade its performance. This would in turn lead to business losses. A common solution to such a problem is to simulate the cloud resources [31]. Thanks to this approach, we can train the model in a safe, isolated environment, where even catastrophic events have no real consequences. There is also a number of other advantages:

- The computational cost of the simulation is orders of magnitude lower than running the actual system. We can potentially parallelize this process and evaluate multiple agents in parallel.
- The time in the simulation can be easily controlled. In a relatively short amount of time we can expose the agent to events from within a long period of time.
- Results are repeatable: if we need to replicate the training of a policy, we can rerun the simulation with the same set of parameters and we can expect the simulator to behave exactly in the same way.
- We can safely try to fine-tune the policy by changing the training algorithm parameters and re-running the simulation.

Unfortunately, there is one major downside: the simulator differs from the real environment. It is very hard to include every factor which might potentially affect how the real system behaves. Part of our research is to evaluate whether this problem limits the ability of the agent to make optimal decisions, or the effect can be reduced due to generalization done by the DNN.

We have decided to train the policy using a simulated environment which has been implemented following the results of our prior research [29,32]. The main simulation process utilizes the CloudSim Plus simulation framework [33] which has been used in a wide range of studies [34]. It is wrapped with the interface provided by the Open AI Gym framework [35]. This allows for decoupling the simulation from other elements of the system, which in turn allows to easily reuse the same environment in experiments with different algorithms. This also helps to parallelize the execution of the simulation in situations where multiple simulations need to be run simultaneously.

The agent observes the environment through the following metrics: number of running virtual machines, average CPU utilization, 90th percentile of CPU utilization, average RAM utilization, 90th percentile of RAM utilization, ratio of all the jobs waiting in the queue to all the jobs submitted, ratio of the jobs waiting in the queue submitted in the last monitoring interval to all the jobs submitted in the last monitoring interval.

The architecture of our training environment is presented in Fig. 2.

During the training, the simulated environment consisted of a single data-center which could host up to a 1000 virtual machines. Each virtual machine could provide a 4 core processor with 16 GB of RAM, similar to *xlarge* Amazon EC2 instances. The initial number of virtual machines was pre-configured to 100. Each simulation was run until all tasks were processed.

The workload simulation was based on logs of the actual jobs executed on IBM SP2 cluster working in the Swedish Royal Institute of Technology. It consisted of 28490 batch jobs executed between October 1996 thru August 1997. Jobs had varying time of execution and used different amounts of resources. The configuration of the simulated CPU cores was adjusted to the configuration of

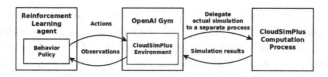

Fig. 2. Components of the training system; arrows denote interactions between them.

the simulation execution environment. To reduce the training time, the time in our simulation experiments was speeded up 1000 times.

The reward function was set up as the negative cost of running the infrastructure including some SLA penalties. This enabled us to formulate the training task as maximization of the objective function (minimizing the cost of the running infrastructure). The cost of running virtual machines was set to $0.2 per hour of their work. The SLA penalty was set to $0.00001 for every second of delay in task execution (e.g. waiting in the queue for execution).

As discussed in [29], the policy resulting from such a training can be applied to managing resources of workloads which are similar to the one used in simulations.

4 Experiments

We chose to use the *pytorch-dnn-evolution* tool [36] as an application for which the infrastructure is automatically managed. The tool implements a co-evolutionary algorithm discovering an optimal structure of a Deep Neural Network to solve a given problem [37]. This approach can be used for problems that can be solved using supervised learning, i.e. there is a training set available. In many such problems, however, due to the size of that dataset, evolutionary methods for discovering the network structure are very costly to apply. Evolution requires evaluating each candidate network by training it over this large dataset. In order to workaround this problem, we assume that by training on a small subset of the initial training dataset, neural networks can be still compared to each other.

Such an approach leads to an emergence of high number of relatively small tasks, which can be processed independently on a cluster of machines. These tasks are idempotent: they can be recalculated multiple times without a risk of corrupting the main evolutionary process. This means that each virtual machine used to conduct the training can be safely stopped at any point in time. Furthermore, the *pytorch-dnn-evolution* tool explicitly creates a queue which can be observed for monitoring the progress of the evolutionary process. Each iteration may have a different number of tasks. This allows to potentially reduce the amount of resources used and the cost of the evolution, if only we are able to design a dynamic resource allocation policy which will be able to add and remove VMs when necessary.

As a workload to which we have applied automatic scaling policies, we have chosen a simple evolutionary experiment which improves the architecture of a network which recognizes hand written numbers (the MNIST dataset [38]). The experiment consisted of executing 20 iterations of evolution over a population of 32 neural networks and 16 fitness predictors (subsets of 2000 samples from the original training set). Every network evaluation included 20 training iterations of a given fitness predictor.

As our cloud environment we have decided to use Amazon Web Services Elastic Compute Cloud (AWS EC2) [39]. The setup consisted of up to 20 instances of *m5a.large* virtual machines from US East 2 (Ohio) region. All VMs have been executed in the same availability zone to reduce the risk of introducing random delays caused by the network communication latency. The first VM has been used to host the workload driver, together with SAMM and Graphite. The Policy Evaluation Service has been deployed to a separate machine because of its higher resource requirements. The management agent could run between 1 and 18 VMs as workers which actually performed the calculations.

To provide a comparison for the results of work of our policy, we also tried to automatically manage the set of used virtual machines with an Auto Scaling Group. This AWS EC2's feature allows to automatically start and stop VMs based on the average CPU usage of the already started machines. If that metric becomes higher than a defined threshold, a new machine is started. Similarly, if the metric drops below this threshold, one of the machines is stopped. For our workload, the value of 80% average CPU usage allowed to achieve the best results.

5   Experiment Results

Below we present the results of the automatic management of the resources with use of the policy trained using the PPO algorithm, on the one hand and the threshold-based AWS policy, on the other hand. Figure 3 presents the number of virtual machines in the context of jobs waiting in the queue. The *step* shape of the charts comes from the fact that after performing an action through the API, the system needs to wait for 180 s. This is necessary to allow for startup of newly added machines or shutdown of the stopped ones.

Under the management of the PPO-trained policy, the overall experiment time was 598 min and the total cost of managed resources was $13.24. In the case of the threshold-based approach it was respectively 642 min and $14.21. The use of the first policy led to faster execution (by 42 min - 6.5%) and lower resources cost (by $0.97–6.8%) execution. For a fair cost comparison in the case of the first policy we need to include also the cost of the additional VM used to generate decisions ($0.86). This reduces the cost difference to $0.11 (0.7%). If the experiment would run for a longer period of time or more expensive resources would be used, such an additional cost would become negligible. The resources required to run the policy are constant, so they would become a very small percentage of the overall cloud resources cost.

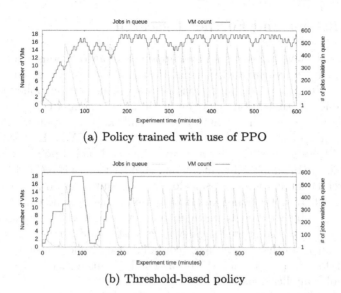

(a) Policy trained with use of PPO

(b) Threshold-based policy

Fig. 3. Number of started VMs in context of jobs waiting in the queue.

The PPO policy was quite conservative in introducing changes and maintained a similar amount of resources most of the time. From time to time it tried to reduce the number of virtual machines after a slightly smaller number of tasks were submitted in an iteration. Those drops were quickly compensated.

The threshold-based policy was more aggressive: after the first iteration it reduced the number of used VMs to 1. Unfortunately, this leads to a slow-down of the whole processing. The final task in this iteration was being picked up subsequently by machines, which would get stopped after a couple of minutes. This delayed its execution until there was only a single machine, which would not get stopped. That single situation delayed the overall evolution process. It is worth noting that after the initial attempts to reduce the amount of resources used, that policy set the number of used resources to a maximum of 18 and maintained this number till the completion of the experiment.

We acknowledge that it is not fully a fair comparison. It might be possible to set the threshold in a way which will allow to avoid the slow-down described above. Using a more complex policy using the same paradigm (a multiple-threshold policy) might help to achieve even better results. Those experiments prove however that the use of the PPO-trained policy renders results on-par with other approaches. At the same time, in the new approach, one can take multiple factors into account in the decision making process (e.g. amount of free RAM), there is no need to set a fixed threshold. Furthermore, as demonstrated in this experiment, it is possible to create a generic policy which can be used also to manage resources for other workloads in which the resources are used in a similar way.

84 W. Funika et al.

6 Conclusions and Further Research

In this paper we have presented a novel approach to autonomous resource management which uses recent advancements in the Deep Reinforcement Learning area. We explained how to train a cloud resource management policy using the Proximal Policy Optimization algorithm with use of a simulated cloud environment. Furthermore we demonstrated how to implement a system which enables deploying such a policy to a real cloud infrastructure - the AWS Elastic Compute Cloud. Finally, we showed that for a sample workload such a policy can manage the infrastructure in a more efficient manner comparing to a threshold-based policy. The careful examination of reasons for such a result revealed issues which led us to a different overall conclusion. Given more fine-tuning the threshold-based policy might be able to allow to achieve even better results. On the other hand, the DRL based approach offered slightly lower costs of infrastructure, while also having a number of other advantages (considering multiple decision factors, no requirement for setting the thresholds manually, re-using the policy across a range of similar applications).

Our approach to the training of the policy delivered good results. Even though we trained it by simulating a different, speeded up workload, our approach enabled to successfully manage the infrastructure for a real, sample machine-learning application. Simulations enabled us to avoid high costs of training. We were able to reduce the amount of time which was needed for a single simulation. At the same time we were able to avoid the cost of poor management decisions made by an untrained policy.

The presented system has a few limitations as well. Due to the grace period and ability to start or stop only a single VM, our policy could not react to environment changes fast enough. Furthermore, our policy was only able to efficiently handle situations, which it was exposed to in a prior training. When new jobs were being issued, the management decisions seemed reasonable, however after the workload had stopped, the number of used resources was not reduced immediately. In contrast, the threshold-based policy stopped all the machines within a couple of minutes once the evolution had stopped.

We plan to continue extending the approach discussed in this paper and mitigate the mentioned issues. We would like to extend the range of available actions in order to allow the policy to add or remove more virtual machines at once. Furthermore, we aim to modify the system to introduce a continuous policy improvement loop.

Acknowledgement. The research presented in this paper was supported by the funds assigned to AGH University of Science and Technology by the Polish Ministry of Science and Higher Education. The experiments have been carried out on the PL-Grid infrastructure resources of ACC Cyfronet AGH and on the Amazon Web Services Elastic Compute Cloud.

References

1. Chen, T., Bahsoon, R., Yao, X.: A survey and taxonomy of self-aware and self-adaptive cloud autoscaling systems. ACM Comput. Surv. **51**(3), 61:1–61:40 (2018)
2. Sutton, R.S.: Temporal credit assignment in reinforcement learning. PhD thesis (1984)
3. Kaelbling, L.P., et al.: Reinforcement learning: a survey. CoRR, cs.AI/9605103 (1996)
4. Mnih, V., et al.: Human-level control through deep reinforcement learning. Nature **518**(7540), 529–533 (2015)
5. Gu, S., et al.: Deep reinforcement learning for robotic manipulation with asynchronous off-policy updates. In: Proceedings 2017 IEEE International Conference on Robotics and Automation (ICRA), Piscataway, NJ, USA, May 2017. IEEE (2017)
6. Silver, D., et al.: Mastering the game of go without human knowledge. Nature **550**, 354–359 (2017)
7. Mnih, V., et al.: Playing atari with deep reinforcement learning. CoRR, abs/1312.5602 (2013)
8. Mnih, V., et al.: Asynchronous methods for deep reinforcement learning. In: Proceedings of The 33rd International Conference on Machine Learning, vol. 48, pp. 1928–1937. PMLR, 20–22 June 2016
9. Schulman, J., et al.: Proximal policy optimization algorithms. CoRR, abs/1707.06347 (2017)
10. Funika, W., et al.: Towards autonomic semantic-based management of distributed applications. Comput. Sci. **11**, 51 (2013)
11. Brun, Y., et al.: Engineering self-adaptive systems through feedback loops. In: Cheng, B.H.C., de Lemos, R., Giese, H., Inverardi, P., Magee, J. (eds.) Software Engineering for Self-Adaptive Systems. LNCS, vol. 5525, pp. 48–70. Springer, Heidelberg (2009). https://doi.org/10.1007/978-3-642-02161-9_3
12. Hoffman, H.: Seec: a framework for self-aware management of goals and constraints in computing systems (power-aware computing, accuracy-aware computing, adaptive computing, autonomic computing). PhD thesis, Cambridge, MA, USA (2013). AAI0829261
13. An architectural blueprint for autonomic computing. Technical report, IBM, June 2005
14. Huber, N., et al.: Model-based self-adaptive resource allocation in virtualized environments. In: Proceedings of the 6th International Symposium on Software Engineering for Adaptive and Self-Managing Systems, SEAMS 2011, pp. 90–99. ACM, New York (2011)
15. Kim, S., et al.: An allocation and provisioning model of science cloud for high throughput computing applications. In: Proceedings of the 2013 ACM Cloud and Autonomic Computing Conference, CAC 2013, pp. 27:1–27:8. ACM, New York (2013)
16. Kateb, D., et al.: Generic cloud platform multi-objective optimization leveraging models@run.time, March 2014
17. Minarolli, D., Freisleben, B.: Distributed resource allocation to virtual machines via artificial neural networks. In: Proceedings of the 2014 22Nd Euromicro International Conference on Parallel, Distributed, and Network-Based Processing, PDP 2014, pp. 490–499. IEEE Computer Society, Washington, DC (2014)

18. Wickremasinghe, B., et al.: Cloudanalyst: a cloudsim-based visual modeller for analysing cloud computing environments and applications. In: 2010 24th IEEE International Conference on Advanced Information Networking and Applications, pp. 446–452, April 2010

19. Qu, C., et al.: A reliable and cost-efficient auto-scaling system for web applications using heterogeneous spot instances. CoRR, abs/1509.05197 (2015)

20. Rodriguez, M.A., et al.: Containers orchestration with cost-efficient autoscaling in cloud computing environments. CoRR, abs/1812.00300 (2018)

21. Fernandez, H., et al.: Autoscaling web applications in heterogeneous cloud infrastructures. In: Proceedings of the 2014 IEEE International Conference on Cloud Engineering, IC2E 2014, pp. 195–204, Washington, DC, USA (2014)

22. Koperek, P., Funika, W.: Dynamic business metrics-driven resource provisioning in cloud environments. In: Wyrzykowski, R., Dongarra, J., Karczewski, K., Waśniewski, J. (eds.) PPAM 2011. LNCS, vol. 7204, pp. 171–180. Springer, Heidelberg (2012). https://doi.org/10.1007/978-3-642-31500-8_18

23. Ferretti S., et al.: Qos–aware clouds. In: 2010 IEEE 3rd International Conference on Cloud Computing, pp. 321–328, July 2010

24. Ashraf, A., et al.: Cramp: cost-efficient resource allocation for multiple web applications with proactive scaling. In: 4th IEEE International Conference on Cloud Computing Technology and Science Proceedings, pp. 581–586, December 2012

25. Xu, C.-Z., et al.: Url: a unified reinforcement learning approach for autonomic cloud management. J. Parallel Distrib. Comput. **72**, 95–105 (2012)

26. Xiong, P., et al.: Smartsla: cost-sensitive management of virtualized resources for CPU-bound database services. IEEE Trans. Parallel Distrib. Syst. **26**, 1 (2014)

27. Wang, Z., et al.: Automated cloud provisioning on AWS using deep reinforcement learning. CoRR, abs/1709.04305 (2017)

28. Kitowski, J., et al.: Computer simulation of heuristic reinforcement learning system for nuclear plant load changes control. Comput. Phys. Commun. **18**, 339–352 (1979)

29. Funika, W., Koperek, P.: Evaluating the use of policy gradient optimization approach for automatic cloud resource provisioning. In: Wyrzykowski, R., Deelman, E., Dongarra, J., Karczewski, K. (eds.) PPAM 2019. LNCS, vol. 12043, pp. 467–478. Springer, Cham (2020). https://doi.org/10.1007/978-3-030-43229-4_40

30. Graphite Project. https://graphiteapp.org/. Accessed 28 Nov 2019

31. Rząsa, W.: Predicting performance in a paas environment: a case study for a web application. Comput. Sci. **18**(1), 21 (2017)

32. Funika, W., et al.: Repeatable experiments in the cloud resources management domain with use of reinforcement learning. In: Cracow Grid Workshop 2018, pp. 31–32. ACC Cyfronet AGH, Kraków (2018)

33. Filho, M.C.S., et al.: Cloudsim plus: a cloud computing simulation framework pursuing software engineering principles for improved modularity, extensibility and correctness. In: 2017 IFIP/IEEE Symposium on Integrated Network and Service Management (IM), pp. 400–406, May 2017

34. Hussain, A., et al.: Investigation of cloud scheduling algorithms for resource utilization using cloudsim. Comput. Inform. **38**, 525–554 (2019)

35. Brockman, G., et al.: OpenAI Gym (2016). arxiv:1606.01540

36. PyTorch DNN Evolution. https://gitlab.com/pkoperek/pytorch-dnn-evolution. Accessed 01 Dec 2019

37. Funika, W., Koperek, P.: Co-evolution of fitness predictors and deep neural networks. In: Wyrzykowski, R., Dongarra, J., Deelman, E., Karczewski, K. (eds.) PPAM 2017. LNCS, vol. 10777, pp. 555–564. Springer, Cham (2018). https://doi.org/10.1007/978-3-319-78024-5_48
38. LeCun, Y., Cortes, C.: MNIST handwritten digit database (2010)
39. Amazon Web Services Elastic Compute Cloud. https://aws.amazon.com/ec2/. Accessed 30 Dec 2019

Utilizing GPU Performance Counters
to Characterize GPU Kernels
via Machine Learning

Bob Zigon[1]([✉]) and Fengguang Song[2]([✉])

[1] Beckman Coulter Inc., Indianapolis, IN, USA
bob.zigon@gmail.com
[2] Department of Computer Science, Indiana University-Purdue University
Indianapolis, Indianapolis, IN, USA
fgsong@cs.iupui.edu

Abstract. GPU computing kernels are relatively simple to write if achieving the best performance is not of the highest priority. However, it can quickly become a much more daunting task when users try to tune and optimize their kernels to obtain the highest performance. This is due to GPUs' massive degree of parallelism, complex memory hierarchy, fine grain synchronization, and long memory access latency. Hence, users must carry out the complex tasks of profiling, analyzing, and tuning to reduce performance bottlenecks. Today's GPUs can generate hundreds of performance events that comprehensively quantify the behavior of a kernel. Instead of relying on experts' manual analysis, this paper targets using machine learning methods to generalize GPU performance counter data to determine the characteristics of a GPU kernel as they will reveal possible reasons for low performance. We choose a set of problem-independent counters as our inputs to design and compare three machine learning methods to automatically classify the execution behavior of a kernel. The experimental results on stencil computing kernels and sparse matrix multiplications show the machine learning models' good accuracy, and demonstrate a feasible approach that is capable of classifying a kernel's characterizations and suggesting changes to a skilled user, who can subsequently improve kernel performance with less guessing.

Keywords: GPU computing · Hardware performance counters · Automatic performance analysis · Machine learning for HPC

1 Introduction

When writing high performance kernels for a modern GPU, several guidelines must be followed. For instance, utilization of both host and GPU memory bandwidth should be maximized. The idle time of parallel computing resources within the GPU should be minimized. Finally, instruction and memory access latency should be either hidden or minimized.

© Springer Nature Switzerland AG 2020
V. V. Krzhizhanovskaya et al. (Eds.): ICCS 2020, LNCS 12137, pp. 88–101, 2020.
https://doi.org/10.1007/978-3-030-50371-0_7

After successfully writing a kernel, most often a GPU programmer is faced with optimizing the kernel's performance. GPU code, however, is notoriously difficult to optimize given the complexity of the underlying hardware. This begs an interesting research question: *"Can machine learning be used to aid in improving GPU kernel performance?"* We have seen that optimizing GPU kernels is sufficiently difficult that only experts engage in the activity. For the average researchers it is both tiresome and tedious. In this paper, we design and develop a process that utilizes different machine learning (ML) techniques to generate insight into GPU performance tuning.

Our approach carefully creates ten classes of GPU kernels which have various performance patterns. Great numbers of executions with the ten classes using different parameters are then used as a dataset to train three distinct machine learning (ML) methods. The three methods include a deep neural network, a random forest, and a naive Bayes classifier. As for the particular machine learning *inference* stage, we use three other new kernels that the ML models have never seen so that the inference results can be compared and a good or poor ML model can be identified.

To the best of our knowledge, the main contributions of this paper are presented as follows.

1. A comprehensive machine learning software framework dedicated to evaluating the characteristics of GPU kernels using GPU hardware counters.
2. A software toolkit implemented for generating large amounts of training data for performance modeling.
3. Different machine learning models tailored and optimized to classify GPU kernels to help users tune ernel performance.

The rest of this paper is organized into the following sections. Section 2 describes related work while Sect. 3 discusses background and GPU performance bottlenecks. Section 4 describes our kernel classification and optimization process. Section 5 describes the three machine learning models and Sect. 6 presents experimental results. Finally, Sect. 7 discusses our conclusions.

2   Related Work

The process of program optimization requires first identifying the type of optimization to be performed. We choose to optimize execution time, that is, we strive to produce kernels that execute the fastest. However, optimization can also be performed with respect to power consumption. For instance, Antz [2] focused on optimizing matrix multiplication with respect to energy for GPUs.

A tool named BEAST (Bench-testing Environment for Automated Software Tuning) [9] was designed to autotune dense matrix-matrix multiplication. Using BEAST requires manually annotating a user's GPU program with the new language (The BEAST project has since been renamed BONSAI [12]). BEAST uses the following recipe for code optimization. First, a computational kernel is parameterized and implemented with a set of tunable parameters (e.g., tile sizes, compiler options, hardware switches), which generally defines a search space. Next, a number of pruning constraints are applied to trim the search space to a manageable size [6,9]. Finally, those kernel variants that have passed the pruning process are compiled, run, benchmarked, and then the best performers are identified.

Abe [1] studied the problem of modeling performance and power by using multiple linear regression. They quantified the impact of voltage and frequency scaling on multiple GPU architectures. Their approach used power and performance as the dependent variables while using the statistical data obtained from performance counters as the source of the independent variables.

Lai [8] took a different approach. An analytical tool called TEG (Timing Estimation tool for GPU) was developed to estimate GPU performance. TEG takes the disassembled CUDA kernel opcodes as input, along with an instruction trace from a GPU simulator named *Barra*, and generates a predicted execution time. TEG is a static analysis tool and does not need to execute the user code.

In reflection, our work is different from the previous work. Instead of predicting execution time or power consumption. We aim to classify GPU kernels in order to aid users in improving performance. We achieve the goal by running a target GPU kernel, extracting all performance counters, classifying the kernel to a similar kernel class by ML models, and finally suggesting an optimization strategy.

3   Background on Performance Metrics

Given our use of performance counters and performance bounds, we start with some definitions. Modern GPUs provide between 100 and 200 hardware counters that can be collected during kernel execution. These counters are also referred to as *events*. On the other hand, another term called *metric* is used to represent a kernel's characteristic, which is calculated based upon one or multiple *events*. In this paper, we use the *metrics* as the training data input to our ML methods.

In general, the first step in analyzing a GPU kernel is to determine if its performance is bounded by memory bandwidth, computation, or instruction/memory latency.

A *memory bound* kernel reaches the physical limits of a GPU device in terms of accesses to the global memory.

A *compute bound* kernel is one in which computation dominates the kernel time, under the assumption that there is no issue feeding the kernel with memory and there is good overlap of arithmetic and latency.

Finally, a *latency bound* kernel is one whose predominant stall reason is due to memory access latency. The global memory bus is still not saturated. An example is when a kernel has to wait to retrieve an operand due to an inadequate number of kernel threads [7].

4   The Process of GPU Kernel Classification and Optimization Recommendation

Our paper focuses on three commonly used kernel optimization methods (as our first step). The methods are: 1) efficient use of memory bandwidth, 2) efficient use of compute resources, and 3) efficient hiding and reduction of instruction and memory latency. Our future work plans to add more kernel optimization methods.

The whole process consists of six steps.

1. Design ten different kernel classes that exhibit different properties of GPU bottlenecks.
2. Generate the ten kernel classes' code samples.
3. Collect GPU performance counter data from running the generated code samples.
4. Create three ML classifiers.
5. Train the three ML classifiers using the performance counters.
6. Apply the best ML classifier to a new GPU kernel that needs to be optimized.

Steps 1–3 are mainly used for our training data generation. Each of the first three steps is described in the subsections that follow. Steps 4–6 are for our ML design and implementation and will be described in Sect. 5.

4.1 Designing Different Kernel Classes for Classification

Many problems in scientific computing can be described in terms of matrix computations. Popular operations include matrix-matrix multiplication, matrix-vector multiplication, and vector normalization, as well as stencil processing for Laplace partial differential equations.

The ten mathematical functions listed in Table 1 represent a list of commonly used GPU kernels, present in both matrix computations and scientific computing, that can be used to demonstrate memory bound, compute bound and latency bound behaviors.

The first six kernels (i.e., K1–K6) are revised from the BLAS Level 1 SAXPY functions but with increasing computation intensities. BLAS is a collection of functions, grouped into three sets, called Level 1, Level 2, and Level 3. Level 1 functions perform scalar, vector and vector-vector operations and have a computational complexity of $\mathcal{O}(n)$. BLAS Level 2 functions perform matrix-vector operations of the form $y \leftarrow \alpha Ax + \beta y$, and have a computational complexity of $\mathcal{O}(n^2)$. Finally, BLAS Level 3 performs matrix-matrix operations of the form $C \leftarrow \alpha AB + \beta C$, and have a computational complexity of $\mathcal{O}(n^3)$.

The seventh kernel K7 in Table 1, naive matrix multiplication, is an example of a function that exhibits a memory access pattern more complex than the previous six. This kernel does not use shared memory (SMEM) on the GPU.

The eighth kernel K8 in Table 1 is an example of a BLAS Level 3 function that exhibits a memory access pattern more complex than the first six. This kernel tiles the shared memory over the matrix for higher performance than kernel K7.

Finally, kernels K9 and K10 in Table 1 are examples of a 5-point stencil. They represent an $\mathcal{O}(\Delta x^2)$ finite difference approximation to the Laplacian of a function. Kernel K9 does not use shared memory while kernel K10 does.

The last column in Table 1, *Strategy Id*, shows the recommended strategies that can be used to improve kernel performance. Table 2 lists the strategy IDs and their corresponding descriptions.

Table 1. Functions of the ten classes of kernels that are used to train ML models along with the number of *training, validation* (val) and *test* samples. *SM* represents the percentage of the streaming multiprocessors utilized, *Mem* represents the percentage of memory bandwidth consumed, and *Strategy Id* describes the technique recommended to apply to a kernel to improve performance. The *SM* and *Mem* percentages were reported by NVidia's NSight Compute tool.

Kernel num	Kernel function	Train	Val	Test	SM	Mem	Strategy Id
K1	$z_i = x_i + y_i$	609	63	672	43%	84%	S4, S10, S20
K2	$z_i = K_1 x_i + K_2 y_i$	600	72	672	49%	83%	S4, S10, S20
K3	$z_i = \sin(K_1)x_i + \cos(K_2)y_i$	604	68	672	50%	24%	S4, S200
K4	$z_i = \sin(K_1)x_i + \cos(K_2)y_i$, ILP2, 32 bit read	606	66	672	49%	22%	S4, S200
K5	$z_i = \sin(K_1)x_i + \cos(K_2)y_i$, ILP2, 64 bit read	612	60	672	52%	52%	S1, S200
K6	$z_i = \sin(K_1)x_i + \cos(K_2)y_i$, ILP4, 128 bit read	611	61	672	45%	60%	S1, S200
K7	$C = AB$, not using shared memory (smem)	543	64	529	95%	22%	S100
K8	$C = AB + C$, uses smem, a tiled GEMM	523	63	528	92%	6%	S100
K9	$U_c^{t+1} = 0.25 \cdot \{U_n^t + U_s^t + U_e^t + U_w^t\}$ no smem	450	50	621	80%	43%	S20
K10	$U_c^{t+1} = 0.25 \cdot \{U_n^t + U_s^t + U_e^t + U_w^t\}$ smem	667	81	548	60%	63%	S1, S10, S30

Table 2. Suggested modifications that can be applied to a kernel to improve its performance.

Strategy ID	Strategy description
S1	Low GPU occupancy – vary thread block size, vary shared memory usage
S4	Unroll loops and improve ILP
S10	Improve read and write memory coalescing
S20	Improve cache locality using registers or shared memory
S30	Reduce bank conflicts and cache into registers
S100	Save results to shared memory or global memory
S200	Use low occupancy and high ILP

Design Rationale Behind Our Kernel Classes: The first kernel class K1 in Table 1 adds two single precision floating point vectors and stores the result. Note that K1 can be used to generate many K1 sample instances that use different parameters such as thread block size, input vector size and number of thread blocks. The simplicity of the function combined with the low arithmetic intensity causes the kernel to be memory bound.

The second kernel class K2 extends K1 by multiplying the vectors by two floating point constants. Since K1 is memory bound, the multiplications in K2 come at virtually no cost because they can be hidden by the memory accesses.

K2 demonstrates how to get more work done per clock cycle [13]. If the PTX (GPU virtual machine language) for kernel K2 is generated as in Listing 1.1, the opportunity for overlapping instruction execution is lost. The GPU stalls when an operand is not available and not on a memory read. This means that after line 1 issues, line 2 will stall because T1 is not available yet. The same pattern exists for lines 3 and 4.

Listing 1.1. Non Overlapping PTX

```
1.      T1 = X[ i ]
2.      T2 = T1*K1
3.      T3 = Y[ i ]
4.      T4 = T3*K2
5.      T5 = T2+T4
6.      Z[ i ] = T5
```

Listing 1.2. Overlapping PTX

```
10.     T1 = X[ i ]
20.     T2 = Y[ i ]
30.     T3 = T1*K1
40.     T4 = T2*K2
50.     T5 = T3+T4
60.     Z[ i ] = T5
```

To increase instruction throughput, the PTX for K2 needs to look like Listing 1.2. Line 10 will issue the memory read. Since the GPU does not block on memory reads, line 20 will then issue. Line 30 will stall because T1 is not likely to be available after line 20 issues. However, when T1 is available, it is likely that T2 is also available, so lines 30 and 40 will issue in parallel and line 50 will stall. As numeric expressions become more complex there can be many opportunities to overlap memory reads with arithmetic instruction execution.

In the third kernel class K3, the multiplicative constants K_1 and K_2 are replaced with the evaluation of two trigonometric functions. The result is that K3 stresses the floating point hardware with an arithmetic intensity level higher than kernel K2.

The fourth kernel class K4 extends K3 by exposing loop unrolling to instruction level parallelism (ILP). The loop in this SAXPY-like K4 kernel is first unrolled twice. The two vector accesses are then interleaved while performing 32-bit operand fetches. In fact, kernels K3–K6 each run faster than the previous kernel by exploiting different degrees of ILP and utilizing more memory bandwidth [13].

The fifth kernel class K5 is similar to K4 by way of loop unrolling. However, kernel K5 issues half as many memory instructions as K4 because the hardware is forced to generate 64-bit reads. This twice reduction in instruction count is important to instruction-bound or latency-bound kernels.

Kernel class K6 is similar to K5. However, K6 issues half as many memory instructions as K5 because the hardware is forced to generate 128-bit reads. Like kernel K5, this 2x reduction in instruction count is important to instruction-bound or latency-bound kernels because it increases memory bandwidth utilization.

Kernel K7, the naive matrix multiplication kernel, is needed because the memory access pattern aids in differentiating the kernel from the SAXPY like kernels. The addition of this kernel is consistent with the vector and matrix processing needs of scientific applications.

Kernel K8, the blocked BLAS Level 3 generalized matrix multiplication kernel, is added because it gave us an opportunity to use tile shared memory over the kernel arguments. The resulting memory access pattern is similar to kernel K7, but with much higher throughput, because the DRAM reads are cached in shared memory. Like kernel K7, this kernel is consistent with the matrix processing needs of scientific applications.

The ninth kernel class K9, a 5-point stencil kernel, is a memory bound kernel with no reuse of its operands. For that reason, we added the tenth kernel class K10, which is a 5-point stencil that caches its operands in the GPU shared memory.

Our initial work in this paper studied ten classes of kernels related to general matrices, but new classes can be added to support other problem domains. For example, if users were to explore social relationships between individuals, we may add graph kernels to classify graph domain related kernels.

4.2   Generating the Training Dataset

All three of our machine learning models (to be described in Sect. 5) require $(\bar{X}_i, Y_i)$ pairs in order to be trained on. The vector $\bar{X}_i$ consists of the GPU performance counters from the ith run of some kernel sample. The value Y_i is the label associated with $\bar{X}_i$. Each label Y_i is one of the ten kernel numbers: K1 to K10.

Our ML models will be trained by using the performance counters collected from hundreds of runs of the ten kernel classes. To automate the process, we created a toolkit called *Datagen* that performed two functions. First, when Datagen was launched with the -q parameter, a function was called to write all of the different ways a given kernel could be launched, to a file. Table 3 shows an example of the output from that function. Lines 1 through 3 show *kernel_one* being launched with different block sizes (**-bs**) and grid sizes (**-gs**). The **-numele** parameter describes the size of the vector that will be manipulated by the kernel.

The second function of Datagen is to automatically execute a kernel with the parameters that were passed in as shown in Table 3. The GPU profiler would launch Datagen, which in turn launched the target kernel, so that the performance counters could be collected.

Table 3. Example output from the Datagen query functions.

```
kernel_one, params -bs 16,16 -gs 10,1 -numele 10000
kernel_one, params -bs 16,18 -gs 10,1 -numele 10000
kernel_one, params -bs 16,20 -gs 20,2 -numele 10000
```

4.3   Collecting Performance Counter Data

Figure 1 shows a flow chart that describes our process to collect the performance counter data. First, all of the kernels were queried so that their launch parameters could be written to a file called TestConfig.txt. Then, a python script opened TestConfig.txt, parsed each line, and launched *nvprof* with the Datagen executable, the kernel name, and the kernel configuration.

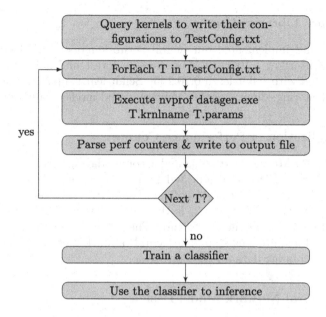

Fig. 1. Flow chart for performance counter data collection process.

5   Designing Machine Learning Models

In this section, we introduce the libraries we have used, and the three different ML methods we designed to perform GPU kernel classifications, which consist of the neural network model, the random forest model, and the naive Bayes model.

5.1   Performance Counters Used

In our experiments, we use the TITAN V GPU, which exposes 158 performance counters. Our Datagen can collect all of the counters, but the nature of our ten kernel classes means that only 95 of them would be non-zero. For example, our kernels operate on single precision floating point numbers and do not call atomic operations. As a result, the double precision floating point counters and atomic transaction counters are all zero.

There are three types of performance counters [7]. The first type is *absolute count*. These counters, for example, will record the total number of memory read and write requests. The second type is efficiency or *utilization counters*. These counters typically have a value between 0 and 1. They summarize the behavior of the hardware. For example, the `sm_efficiency` counter measures the overall activity of the SM's. Finally, there are *throughput counters*. These throughput counters typically measure the throughput of the various memory subsystems within the GPU. For example, the `gld_requested_throughput` counter measures the throughput of requested global loads in gigabytes per second.

5.2 Software Libraries Used

We designed and implemented our neural network model with Python 3.6.8, Keras, and Tensorflow 1.10.0 [10]. Our random forest and naive Bayes model are implemented with scikit-learn 0.20.2 [5]. Scikit-learn supports classification, regression, clustering, dimensionality reduction, and model evaluation.

5.3 Normalization of Performance Counters

In preparation for ML model training and accommodating various scales of metric values, we normalized our collected performance counters to the interval $[-1, 1]$ using the transform $F(x_{i,j}) = 2(x_{i,j} - \mu_j)/\sigma_j - 1$, where $x_{i,j}$ is the ith measurement of the jth feature, μ_j is the mean of the jth feature, and σ_j is the standard deviation of the jth feature. The random forest and naive Bayes methods do not require this step. The neural network method, however, must have all of the training data on the same scale.

5.4 The Neural Network Model Design

When doing experiments, we found that the efficiency and throughput performance counters produced better modeling results while the absolute counters negatively impacted our models. In other words, summarization information that efficiency and throughput counters offered caused ML models to generalize better. The absolute counters, on the other hand, mislead the ML classifiers. We therefore modified our methods to train on the union of the non-zero counters that only belong to the efficiency or throughput type. The result is a neural network with 48 inputs instead of 158.

The designed neural network is shown in Fig. 2. The input layer, consisting of 48 performance counters, was followed by a fully connected layer of 48 neurons and a ReLU activation function, another fully connected layer of 24 neurons and a ReLU, and finally a softmax activation function that generates the probability for each of the 10 classes [3, 11].

We elected to train against all of the performance counter data using 7-fold cross validation for each of 7 different optimizers: RMSprop, SGD (stochastic gradient descent), Adagrad, Adadelta, Adam, Adamax and Nadam. Figure 3 shows that Nadam produced the best results.

5.5 The Random Forest Model Design

Our second machine learning model is a random forest classifier [4], which is based upon a collection, or *ensemble*, of binary decision trees where the probability of each class is the average of the probabilities across all trees.

Decision trees are easy to interpret and understand. They can be visualized, and unlike neural networks, the training data does not need to be normalized before being processed by the algorithm. The drawback to a decision tree is that

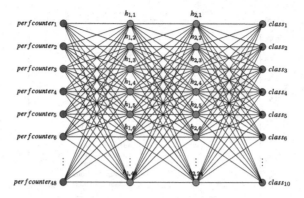

Fig. 2. Our neural network consists of 48 inputs, 2 hidden layers fully connected, and an output layer with 10 classes.

they can become overly complex which results in overfitting and poor generalization. Random forests manage the overfitting by utilizing hundreds or thousands of relatively simple decision trees. Our random forest consisted of 500 decision trees, each with a maximum depth of 6.

5.6 The naive Bayes Model Design

The naive Bayes classifier uses Bayes theorem to predict the class for a new test instance $\mathbf{x}$. The predicted class for $\mathbf{x}$ is given as

$$\hat{y} = \arg \max_{c_i}\{P(c_i|\mathbf{x})\}, \tag{1}$$

where a training dataset $\mathbf{D}$ consists of n points $\mathbf{x_i}$ in a d-dimensional space, and y_i denote the class for each point, with $y_i \in \{c_1, c_2, \ldots, c_k\}$. The posterior probability $P(c_i|\mathbf{x})$ for each class c_i is given by

$$P(c_i|\mathbf{x}) = \frac{P(\mathbf{x}|c_i) \cdot P(c_i)}{P(\mathbf{x})} \tag{2}$$

where $P(\mathbf{x}|c_i)$ is the *likelihood*, $P(c_i)$ is the *prior probability* of class c_i, and $P(\mathbf{x})$ is the probability of observing $\mathbf{x}$ from any of the k classes, given as

$$P(\mathbf{x}) = \sum_{j=1}^{k} P(\mathbf{x}|c_j) \cdot P(c_j). \tag{3}$$

6   Experimental Results with the Three Different ML Methods

In this section we show the results from the experiments conducted using the three models. Table 1 shows the distribution of training, validation and testing

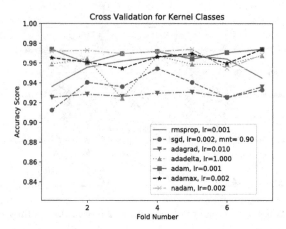

Fig. 3. 7-fold cross validation with 7 optimizers shows that the Nadam optimizer produced the best results.

samples that were used. Each of the GPU kernels was compiled using Visual Studio 2015 and CUDA 10.0 on a quadcore, Lenovo D30 ThinkStation running Windows 7/64. NSight Visual Studio 6.0.0.18227 was the visual profiler used. The GPU kernels were subsequently executed on an NVIDIA TITAN V GPU with device driver version 417.01.

For each ML model we generate a *confusion matrix*. The confusion matrix shows the ways in which a classification model fails when it makes predictions. The confusion matrices are shown in Figs. 4, 5a and 5b. Based on the confusion matrices, we observe that all three models are nearly identical with the exception of kernels K3 and K4. In this case, the random forest has the largest normalized score.

In Table 4 the F1-score for the random forest is mostly larger when compared to the other classifiers. The F1-score [11] measures a model's accuracy as function of precision and recall.

6.1 Evaluation of the Inference Results

Inferencing is the process of using a ML model to generate a category for an input the ML model has never seen. In our experiments, we use three test kernels that our models have never seen.

The first test kernel is a 9-point stencil shown in Eq. 4. This equation represents a $\mathcal{O}(\Delta x^4)$ finite difference approximation to the Laplacian.

$$U^{t+1}_{center} = \frac{1}{120}\{4U^t_n + 4U^t_s + 4U^t_e + 4U^t_w + U^t_{ne} + U^t_{nw} + U^t_{se} + U^t_{sw}\} \quad (4)$$

The second test kernel implements a sparse matrix vector multiplication (SpMV). The matrix is in compressed sparse row format and has a sparsity of 5% (i.e. 95% of the values are non-zero).

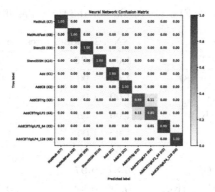

Fig. 4. Neural network confusion matrix.

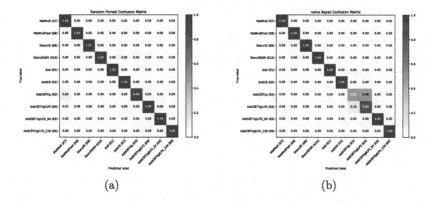

Fig. 5. Random forest and naive Bayes confusion matrices.

The third test kernel is a version of the 9-point stencil kernel that uses shared memory to cache variables from DRAM. If the models are doing a good job of generalizing, then we are postulating that the 9-point kernels will be mapped to the 5-point kernel class used in training.

Table 5 shows the inference results for our three ML models given the three test kernels. Each ML model was presented with ten different instances of the three kernels. All three models classified the 9-point stencil as the 5-point stencil, and the 9-point stencil with shared memory as the 5-point stencil with shared memory.

For the third test kernel, SpMV, the behavior of the models starts to diverge. The naive Bayes model classified SpMV as the 5-point stencil with shared memory. The random forest model classified SpMV as a matrix multiplication and a 5-point stencil with shared memory. Finally, the neural network model classified SpMV as being predominantly similar to kernel K3, $z_i = \sin(K_1)x_i + \cos(K_2)y_i$. The divergent results on SpMV may be caused by the specialty of the SpMV kernel, which is not only similar to matrix multiplication (K7), but also similar

Table 4. F1 score results for the three classifiers.

Kernel name	NN	Random forest	naive Bayes
MatMult (K7)	1.00	**0.96**	1.00
MatMultFast (K8)	1.00	1.00	1.00
Stencil5 (K9)	1.00	1.00	1.00
Stencil5SM (K10)	1.00	1.00	1.00
Add (K1)	1.00	1.00	1.00
AddCB (K2)	1.00	1.00	1.00
AddCBTrig (K3)	**0.87**	**0.97**	**0.64**
AddCBTrigILP2 (K4)	**0.86**	**0.99**	**0.76**
AddCBTrigILP2_64 (K5)	1.00	1.00	1.00
AddCBTrigILP4_128 (K6)	1.00	1.00	1.00

Table 5. Inference results for our three ML models on three new test kernels. The italic values show where the ML algorithms agree in their classification. The bold values show where they disagree.

ML model	Test kernel	MM (K7)	K8	Sten5 (K9)	Sten5SM (K10)	K1	K2	AddCBTrig (K3)	AddCBTrigILP2 (K4)	K5	K6
naive Bayes	Stencil9			*10*							
Random forest	Stencil9			*10*							
Neural net	Stencil9			*10*							
naive bayes	Stencil9SM				*10*						
Random forest	Stencil9SM				*10*						
Neural net	Stencil9SM				*10*						
naive bayes	SpMV05				10						
Random forest	SpMV05	**5**			**5**						
Neural net	SpMV05							**8**	**2**		

to K3 because SpMV consists of a collection of vector-vector dot products. Based on the result of SpMV, we can say that the neural network and random forest models perform better than the naive Bayes model.

7   Conclusion

In this paper, we have described a machine learning framework that was used to classify unseen GPU kernels into one of ten classes using machine learning and GPU hardware performance counters. With that classification information, users are able to pursue optimization strategies for their target kernel based on the strategies for the learned kernels in Tables 1 and 2. What is critical in this paper is that each of the ten kernel classes we selected for training purposes was each slightly more complex than the previous kernel. This devised additional complexity gradually utilized more and more of memory bandwidth and parallel resources, while minimizing the instruction and memory access latencies.

We applied our framework to three new GPU kernels that are widely used in many domains, i.e. the 9-point stencil (with and without shared memory) used to discretize the Laplacian operator in differential equations, and the sparse matrix vector multiplication procedure. We also found that the random forest model and the neural network model performed similarly with respect to the confusion matrix, the F1-score and the actual classification results.

Finally, this framework can be extended by adding more kernel classes in our training dataset to support classifying larger, more complex GPU kernels or kernel sections. Our future work, on the other hand, will seek to improve and extend the three ML models to support new types of GPU applications.

References

1. Abe, Y., Sasaki, H., Kato, S., Inoue, K., Edahiro, M., Peres, M.: Power and performance characterization and modeling of GPU-accelerated systems. In: IPDPS, pp. 113–122 (2014)
2. Anzt, H., Haugen, B., Kurzak, J., Luszczek, P., Dongarra, J.: Experiences in autotuning matrix multiplication for energy minimization on GPUs. Concurr. Comput. Pract. Exp. **27**(17), 5096–5113 (2015)
3. Bishop, C.M., et al.: Neural Networks for Pattern Recognition. Oxford University Press, Oxford (1995)
4. Bowles, M.: Machine Learning in Python: Essential Techniques for Predictive Analysis. Wiley, Hoboken (2015)
5. Géron, A.: Hands-On Machine Learning with Scikit-Learn and TensorFlow: Concepts, Tools, and Techniques to Build Intelligent Systems. O'Reilly Media Inc., Sebastopol (2017)
6. Haugen, B., Kurzak, J.: Search space pruning constraints visualization. In: 2014 Second IEEE Working Conference on Software Visualization (VISSOFT), pp. 30–39. IEEE (2014)
7. Kim, H., Vuduc, R., Baghsorkhi, S., Choi, J., Hwu, W.m.: Performance analysis and tuning for general purpose graphics processing units (GPGPU). Synth. Lect. Comput. Arch. **7**(2), 1–96 (2012)
8. Lai, J., Seznec, A.: Break down GPU execution time with an analytical method. In: Proceedings of the 2012 Workshop on Rapid Simulation and Performance Evaluation: Methods and Tools, pp. 33–39. ACM (2012)
9. Luszczek, P., Gates, M., Kurzak, J., Danalis, A., Dongarra, J.: Search space generation and pruning system for autotuners. In: 2016 IEEE International Parallel and Distributed Processing Symposium Workshops, pp. 1545–1554. IEEE (2016)
10. McClure, N.: TensorFlow Machine Learning Cookbook. Packt Publishing Ltd., Birmingham (2017)
11. Murphy, K.P.: Machine Learning: A Probabilistic Perspective. MIT Press, Cambridge (2012)
12. Tsai, Y.M., et al.: Bonsai. In: ISC 2018, Frankfurt Am Main, Germany, 24–28 June 2018. https://ssl.linklings.net/conferences/isc_hpc/assets/2018/posters/post101.pdf
13. Volkov, V.: Better performance at lower occupancy. In: NVidia GPU Conference 2010. NVidia, September 2010

A Massively Parallel Algorithm
for the Three-Dimensional
Navier-Stokes-Boussinesq Simulations
of the Atmospheric Phenomena

Maciej Paszyński[1]([✉]), Leszek Siwik[1], Krzysztof Podsiadło[1], and Peter Minev[2]

[1] Department of Computer Science, AGH University of Science and Technology,
Krakow, Poland
`paszynsk@agh.edu.pl`
[2] Applied Mathematics Institute, Mathematical and Statistical Sciences,
University of Alberta, Edmonton, Canada

Abstract. We present a massively parallel solver using the direction splitting technique and stabilized time-integration schemes for the solution of the three-dimensional non-stationary Navier-Stokes-Boussinesq equations. The model can be used for modeling atmospheric phenomena. The time integration scheme utilized enables for efficient direction splitting algorithm with finite difference solver. We show how to incorporate the terrain geometry into the simulation and how to perform the domain decomposition. The computational cost is linear $\mathcal{O}(N)$ over each sub-domain, and near to $\mathcal{O}(N/c)$ in parallel over 1024 processors, where N is the number of unknowns and c is the number of cores. This is even if we run the parallel simulator over complex terrain geometry. We analyze the parallel scalability experimentally up to 1024 processors over a PROMETHEUS Linux cluster with multi-core processors. The weak scalability of the code shows that increasing the number of subdomains and processors from 4 to 1024, where each processor processes the subdomain of $49 \times 49 \times 99$ internal points ($50 \times 50 \times 100$ box), results in the increase of the total computational time from 120 s to 178 s for a single time step. Thus, we can perform a single time step with over 1,128,000,000 unknowns within 3 min. The number of unknowns results from the fact that we have three components of the velocity vector field, one component of the pressure, and one component of the temperature scalar field over 256,000,000 mesh points. The computation of the one time step takes 3 min on a Linux cluster. The direction splitting solver is not an iterative solver; it solves the system accurately since it is equivalent to Gaussian elimination. Our code is interfaced with the mesh generator reading the NASA database and providing the Earth terrain map. The goal of the project is to provide a reliable tool for parallel, fully three-dimensional computations of the atmospheric phenomena.

Keywords: Massive parallel computations · Alternating direction solver · Navier-Stokes Boussinesq · Finite difference method

© Springer Nature Switzerland AG 2020
V. V. Krzhizhanovskaya et al. (Eds.): ICCS 2020, LNCS 12137, pp. 102–117, 2020.
https://doi.org/10.1007/978-3-030-50371-0_8

1   Introduction

Air pollution is receiving a lot of interest nowadays. It is visible, especially in the Kraków area in Poland (compare Fig. 1), as this is one of the most polluted cities in Europe [1]. People living there are more and more aware of the problem, which causes the raising of various movements that are trying to improve air quality. Air pollution grows because of multiple factors, including traffic, climate, heating in the winter, the city's architecture, etc. The ability to model atmospheric phenomena such as thermal inversion over the complicated terrain is crucial for reliable simulations of air pollution. Thermal inversion occurs when a layer of warm air stays over a layer of cool air, and the warm air holds down the cool air and it prevents pollutants from rising and scattering.

Fig. 1. Pollution with fog and thermal inversion over the same area near Kraków between October 2019 and January 2020 (photos by Maciej Paszyński)

We present a massively parallel solver using the direction splitting technique and stabilized time-integration schemes for the solution of the three-dimensional non-stationary Navier-Stokes-Boussinesq equations.

The Navier-Stokes-Boussinesq system is widely applied for modeling the atmospheric phenomena [2], oceanic flows [3] as well as the geodynamics simulations [4]. The model can be used for modeling atmospheric phenomena, in particular, these resulting in a thermal inversion. It can be used as well for modeling several other important atmospheric phenomena [5,6]. It may even be possible to run the climate simulation of the entire Earth atmosphere using the approach presented here. The time integration scheme utilized results in a Kronecker product structure of the matrices, and it enables for efficient direction splitting algorithm with finite difference solver [7], since the matrix is a Kronecker product of three three-diagonal matrices, resulting from discretizations along x, y, and z axes. The direction splitting solver is not an iterative solver; it is equivalent to the Gaussian elimination algorithm.

We show how to extend the alternating directions solver into non-regular geometries, including the terrain data, still preserving the linear computational cost of the solver. We follow the idea originally used in [8] for sequential computations of particle flow. In this paper, we focus on parallel computations, and we describe how to compute the Schur complements in parallel with linear cost, and how to aggregate them further and still have a tri-diagonal matrix that can be factorized with a linear computational cost using the Thomas algorithm. We also show how to modify the algorithm to work over the complicated non-regular terrain structure and still preserve the linear computational cost.

Thus, if well parallelized, the parallel factorization cost is near to $\mathcal{O}(N/c)$ in every time step, where N is the number of unknowns and c is the number of cores. We analyze the parallel scalability of the code up to 1024 multi-core processors over a PROMETHEUS Linux cluster [9] from the CYFRONET supercomputing center. Each subdomain is processed with $50 \times 50 \times 100$ finite difference mesh. Our code is interfaced with the mesh generator [10] reading the NASA database [11] and providing the Earth terrain map. The goal of the project is to provide a reliable tool for parallel fully three-dimensional computations of the atmospheric phenomena resulting in the thermal inversion and the pollution propagation.

In this paper, we focus on the description and scalability of the parallel solver algorithm, leaving the model formulation and large massive parallel simulations of different atmospheric phenomena for future work. This is a challenging task itself, requiring to acquire reliable data for the initial state, forcing, and boundary conditions.

2   Navier-Stokes Boussinesq Equations

The equations in the strong form are

$$\frac{\partial \mathbf{u}}{\partial t} + (\mathbf{u} \cdot \nabla)\mathbf{u} + \nabla p + \mathrm{Pr}\Delta \mathbf{u} = g\mathrm{Pr}\mathrm{Ra}T + f \text{ in } \Omega \times (0, T_f] \tag{1}$$

$$\nabla \cdot \mathbf{u} = 0 \text{ in } \Omega \times (0, T_f] \tag{2}$$

$$\mathbf{u} = 0 \text{ in } \partial\Omega \times (0, T_f] \tag{3}$$

$$\frac{\partial T}{\partial t} + (u \cdot \nabla)T + \Delta T = 0 \text{ in } \Omega \times (0, T_f] \tag{4}$$

$$T = 0 \text{ in } \partial\Omega \times (0, T_f] \tag{5}$$

where u is the velocity vector field, p is the pressure, $Pr = 0.7$ is the Prandt number, $g = (0, 0, -1)$ is the gravity force, $Ra = 1000.0$ is the Rayleigh number, T is the temperature scalar field.

We discretize using finite difference method in space and the time integration scheme resulting in a Kronecker product structure of the matrices.

We use the second-order in time unconditionally stable time integration scheme for the temperature equation and for the Navier-Stokes equation, with the predictor-corrector scheme for pressure. For example we can use the Douglass-Gunn scheme [13], performing an uniform partition of the time interval $\bar{I} = [0, T]$ as

$$0 = t_0 < t_1 < \ldots < t_{N-1} < t_N = T,$$

and denoting $\tau := t_{n+1} - t_n, \ \forall n = 0, \ldots, N - 1$. In the Douglas-Gunn scheme, we integrate the solution from time step t_n to t_{n+1} in three substeps as follows:

$$\begin{cases} (1 + \dfrac{\tau}{2}\mathcal{L}_1)u^{n+1/3} = \tau f^{n+1/2} + (1 - \dfrac{\tau}{2}\mathcal{L}_1 - \tau\mathcal{L}_2 - \tau\mathcal{L}_3)u^n, \\[2mm] (1 + \dfrac{\tau}{2}\mathcal{L}_2)u^{n+2/3} = u^{n+1/3} + \dfrac{\tau}{2}\mathcal{L}_2 u^n, \\[2mm] (1 + \dfrac{\tau}{2}\mathcal{L}_3)u^{n+1} = u^{n+2/3} + \dfrac{\tau}{2}\mathcal{L}_3 u^n. \end{cases} \tag{6}$$

For the Navier-Stokes equations, $\mathcal{L}_1 = \partial_{xx}$, $\mathcal{L}_2 = \partial_{yy}$, and $\mathcal{L}_3 = \partial_{zz}$, and the forcing term represents $g\mathrm{PrRa}T^{n+1/2}$ plus the convective flow and the pressure terms $(u^{n+1/2} \cdot \nabla u^{n+1/2}) + \nabla \tilde{p}^{n+1/2}$ treated explicitly as well. The pressure is computed with the predictor/corrector scheme. Namely, the predictor step

$$\tilde{p}^{n+\frac{1}{2}} = p^{n-\frac{1}{2}} + \phi^{n-\frac{1}{2}}, \tag{7}$$

with $p^{-\frac{1}{2}} = p_0$ and $\phi^{-\frac{1}{2}} = 0$ computes the pressure to be used in the velocity computations, the penalty steps

$$\begin{cases} \psi - \partial_{xx}\psi = -\dfrac{1}{\tau}\nabla \cdot u^{n+1}, \\[2mm] \xi - \partial_{yy}\xi = \psi, \\[2mm] \phi^{n+1/2} - \partial_{zz}\phi^{n+1/2} = \xi, \end{cases} \tag{8}$$

and the corrector step updates the pressure field based on the velocity results and the penalty step

$$p^{n+\frac{1}{2}} = p^{n-\frac{1}{2}} + \phi^{n+\frac{1}{2}} - \chi\nabla \cdot \left(\dfrac{1}{2}(u^{n+1} + u^n)\right). \tag{9}$$

These steps are carefully designed to stabilize the equations as well as to ensure the Kronecker product structure of matrix, resulting in the linear computational cost solver. The mathematical proofs of the stability of the formulations, motivating such the predictor/corrector (penalty) steps, can be found in [7,12] and the references there.

For the temperature equation, $\mathcal{L}_1 = \partial_{xx}$, $\mathcal{L}_2 = \partial_{yy}$, and $\mathcal{L}_3 = \partial_{zz}$, and the forcing term represents the advection term treated explicitly $(u^{n+1/2} \cdot \nabla T^{n+1/2})$.

For mathematical details on the problem formulation and its mathematical properties, we refer to [12].

Each equation in our scheme contains only derivatives in one direction, so they are of the following form

$$
\begin{aligned}
(1 + \alpha \partial_{xx}) \, u^{n+1/3} &= RHS_x \\
(1 + \alpha \partial_{yy}) \, u^{n+2/3} &= RHS_y \\
(1 + \alpha \partial_{zz}) \, u^{n+1} &= RHS_z
\end{aligned}
\tag{10}
$$

or the update of the pressure scalar field. Thus, when employing the finite difference method, we either endup with the Kronecker product matrices with sub-matrices being three-diagonal, or the point-wise updates of the pressure field

$$
\begin{aligned}
u_{ijk}^{n+1/3} + \alpha \frac{u_{(i-1)jk}^{n+1/3} - 2u_{ijk}^{n+1/3} + u_{(i+1)jk}^{n+1/3}}{dt} &= RHS_x \\
u_{ijk}^{n+2/3} + \alpha \frac{u_{i(j-1)k}^{n+2/3} - 2u_{ijk}^{n+2/3} + u_{i(j+1)k}^{n+1/3}}{dt} &= RHS_y \\
u_{ijk}^{n+1} + \alpha \frac{u_{ij(k-1)}^{n+1} - 2u_{ijk}^{n+1} + u_{ij(k+1)}^{n+1}}{dt} &= RHS_z,
\end{aligned}
\tag{11}
$$

where $\alpha = \tau/2$ or $\alpha = 1$, depending on the equation, which is equivalent to

$$
\begin{aligned}
\alpha u_{(i-1)jk}^{n+1/3} + (dt - 2\alpha)u_{ijk}^{n+1/3} + \alpha u_{(i+1)jk}^{n+1/3} &= dt * RHS_x \\
\alpha u_{i(j-1)k}^{n+2/3} + (dt - 2\alpha)u_{ijk}^{n+2/3} + \alpha u_{i(j+1)k}^{n+1/3} &= dt * RHS_y \\
\alpha u_{ij(k-1)}^{n+1} + (dt - 2\alpha)u_{ijk}^{n+1} + \alpha u_{ij(k+1)}^{n+1} &= dt * RHS_z,
\end{aligned}
\tag{12}
$$

These systems have a Kronecker product structure $\mathcal{M} = \mathcal{A}^x \otimes \mathcal{B}^y \otimes \mathcal{C}^z$ where the sub-matrices are aligned along the three axis of the system of coordinates, one of these sub-matrices is three-diagonal, and the other two sub-matrices are scalled identity matrices. From the parallel matrix computations point of view, discussed in our paper, it is important that in every time step, we have to factorize in parallel the system of linear equations having the Kronecker product structure.

3   Factorization of the System of Equations Possessing the Kronecker Product Structure

The direction splitting algorithm for the Kronecker product matrices implements three steps, which result is equivalent to the Gaussian elimination algorithm [14], since

$$
(\mathcal{M})^{-1} = (\mathcal{A}^x \otimes \mathcal{B}^y \otimes \mathcal{C}^z)^{-1} = (\mathcal{A}^x)^{-1} \otimes (\mathcal{B}^y)^{-1} \otimes (\mathcal{C}^z)^{-1}
\tag{13}
$$

Each of the three systems is three-diagonal,

$$
\begin{bmatrix}
A_{11}^x & A_{12}^x & \cdots & 0 \\
A_{21}^x & A_{22}^x & \cdots & 0 \\
\vdots & \vdots & \ddots & \vdots \\
0 & 0 & \cdots & A_{kk}^x
\end{bmatrix}
\begin{bmatrix}
z_{111} & z_{121} & \cdots & z_{1lm} \\
z_{211} & z_{221} & \cdots & z_{2lm} \\
\vdots & \vdots & \ddots & \vdots \\
z_{k11} & z_{k21} & \cdots & z_{klm}
\end{bmatrix}
=
\begin{bmatrix}
y_{111} & y_{121} & \cdots & y_{1lm} \\
y_{211} & y_{221} & \cdots & y_{2lm} \\
\vdots & \vdots & \ddots & \vdots \\
y_{k11} & y_{k21} & \cdots & y_{klm}
\end{bmatrix}
\tag{14}
$$

and we can solve it in a linear $\mathcal{O}(N)$ computational cost. First, we solve along x direction, second, we solve along y direction, and third, we solve along z direction.

4 Introduction of the Terrain

To obtain a reliable three-dimensional simulator of the atmospheric phenomena, we interconnect several components. We interface our code with mesh generator that provides an excellent approximation to the topography of the area [10], based on the NASA database [11]. The resulting mesh generated for the Krakow area is presented in Fig. 2.

In our system of linear equations, we have several tri-diagonal systems with multiple right-hand-sides, factorized along x, y and z directions. Each unknown in the system represents one point of the computational mesh. In the first system, the rows are ordered according to the coordinates of points, sorted along x axis. In the second system, the rows are ordered according to the y coordinates of points, and in the third system, according to z coordinates. When simulating the atmospheric phenomena like the thermal inversion over the prescribed terrain with alternating directions solver and finite difference method, we check if a given point is located in the computational domain. The unknowns representing points that are located inside the terrain (outside the atmospheric domain) are removed from the system of equations. This is done by identifying the indexes of the points along x, y, and z axes, in the three systems of coordinates. Then, we modify the systems of equations, so the corresponding three rows in the three systems of equations are reset to 0, the diagonal is set to 1, and the corresponding rows and columns of the three right-hand-sides are set 0.

For example, if we want to remove point (r, s, t) from the system, we perform the following modification in the first system.

The rows in the first system they follow the numbering of points along x axis. The number of columns corresponds to the number of lines along x axis perpendicular to OYZ plane. Each column of the right-hand side correspond to yz coordinates of a point over OYZ plane. We select the column corresponding to the "st" point. We factorize the system with this column separately, by replacing the row in the matrix by the identity on the diagonal and zero on the right-hand side. The other columns in the first system are factorized in a standard way.

$$
\begin{bmatrix}
A_{11}^x & A_{12}^x & \cdots & 0 \\
A_{21}^x & A_{22}^x & \cdots & 0 \\
\vdots & \vdots & \ddots & \vdots \\
0 & 0 & A_{rr}^x = 1.0 & 0 \\
\vdots & \vdots & \ddots & \vdots \\
0 & 0 & \cdots & A_{kk}^x
\end{bmatrix}
\begin{bmatrix}
z_{1st} \\
z_{2st} \\
\vdots \\
z_{rst} \\
\vdots \\
z_{kst}
\end{bmatrix}
=
\begin{bmatrix}
y_{1st} \\
y_{2st} \\
\vdots \\
y_{rst} = 0.0 \\
\vdots \\
y_{kst}
\end{bmatrix}
\tag{15}
$$

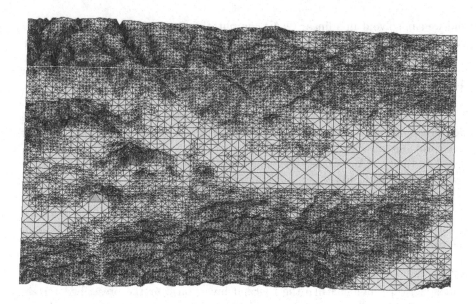

Fig. 2. The computational mesh generated based on the NASA database, representing the topography of the Krakow area.

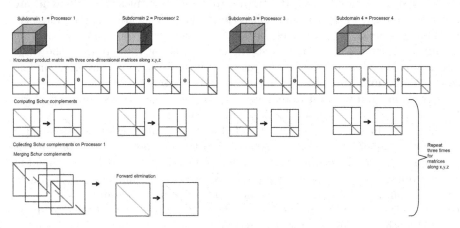

Fig. 3. Illustration on the parallel solver algorithm.

Analogous situation applies for the second system, this time with right-hand side columns representing lines perpendicular to OXZ plane. We factorize the "rt" column in the second system separately, by setting the row in the matrix as the identity on the diagonal, and using 0.0 on the right-hand side. The other columns in the second system are factorized in the standard way.

$$
\begin{bmatrix}
B_{11}^y & B_{12}^y & \cdots & & & & 0 \\
B_{21}^y & B_{22}^y & \cdots & & & & 0 \\
\vdots & \vdots & \ddots & & & & \vdots \\
\vdots & \vdots & & B_{ss}^y = 0 & & & \vdots \\
\vdots & \vdots & & & \ddots & & \vdots \\
0 & 0 & \cdots & & & & B_{ll}^y
\end{bmatrix}
\begin{bmatrix}
y_{r1t} \\
y_{r2t} \\
\vdots \\
y_{rst} \\
\vdots \\
y_{rlt}
\end{bmatrix}
=
\begin{bmatrix}
z_{r1t} \\
z_{r2t} \\
\vdots \\
z_{rst} = 0.0 \\
\vdots \\
z_{rlt}
\end{bmatrix}
\tag{16}
$$

Similarly, in the third system we factorize the "rs" column separately. The other columns in the third system are factorized in a standard way.

$$
\begin{bmatrix}
C_{1,1}^z & C_{1,2}^z & \cdots & & & & 0 \\
C_{2,1}^z & C_{2,2}^z & \cdots & & & & 0 \\
\vdots & \vdots & \ddots & & & & \vdots \\
\vdots & \vdots & & C_{tt}^z = 1.0 & & & \vdots \\
\vdots & \vdots & & & \ddots & & \vdots \\
0 & 0 & \cdots & & & & C_{m,m}^z
\end{bmatrix}
\begin{bmatrix}
x_{rs1} \\
x_{rs2} \\
\vdots \\
x_{rst} \\
\vdots \\
x_{rsm}
\end{bmatrix}
=
\begin{bmatrix}
b_{rs1} \\
b_{rs2} \\
\vdots \\
b_{rst} = 0.0 \\
\vdots \\
b_{rsm}
\end{bmatrix}
\tag{17}
$$

Using this trick for all the points in the terrain, we can factorize the Kronecker product system in a linear computational cost over the complex terrain geometry.

5 Parallel Factorization with Domain Decomposition Preserving the Linear Computational Cost

The computational domain is decomposed into several cube-shape sub-domains. We generate systems of linear equations over each sub-domain separately, and we enumerate the variables in a way that interface unknowns are located at the end of the matrices. We compute the Schur complement of the interior variables with respect to the interface variables. We do it in parallel over each of the subdomains. The important observation is that the Schur complement matrices will also be three-diagonal matrices. This is because the subdomain matrix is three-diagonal, and the Schur complement computation can be implemented as forward eliminations, performed in the three sub-systems, each of them stopped after processing the interior nodes in the particular systems. Later, we aggregate the Schur complements into one global matrix. We do it by global gather operation. This matrix is also tri-diagonal and can be factorized in a linear cost. Later, we scatter the solution, and we use the partial solutions from the global matrix to backward substitute each of the systems in parallel. These operations are illustrated in Fig. 3.

We perform this operation three times, for three submatrices of the Kronecker product matrix, defined along three axes of the coordinate system. We provide algebraic details below.

Thus, assuming we have $r-1$ rows to factorize in the first system ($r-1$ rows in the interior, and $k - r + 1$ rows on the interface), we run the forward elimination over the first matrix, along the x direction, and we stop it before processing the r-th row (denoted by red color). This partial forward elimination stopped at the r-th row ensures that below that row we have the Schur complement of the first $r - 1$ rows related with the interior points in the domain with respect to the next $k - r + 1$ rows related with the interface points (the Schur complement is denoted by blue color). This Schur complement matrix is indeed tri-diagonal:

$$
\begin{bmatrix}
A_{11}^x & A_{12}^x & 0 & & \cdots & & \cdots & & \cdots & & \cdots & & \cdots \\
A_{21}^x & A_{22}^x & A_{23}^x & & 0 & & \cdots & & \cdots & & \cdots & & \cdots \\
\vdots & \vdots & & \ddots & & \ddots & & \ddots & & \vdots & & \vdots & \\
\cdots & 0 & A_{(r-1)(r-2)}^x & A_{(r-1)(r-1)}^x & A_{(r-1)r}^x & & 0 & & \cdots & & \cdots & \\
\cdots & \cdots & & 0 & A_{r(r-1)}^x & A_{rr}^x & A_{r(r+1)}^x & & 0 & & \cdots & \\
\vdots & \vdots & & \ddots & & \vdots & & \ddots & & \ddots & & \vdots & \\
\cdots & \cdots & & \cdots & & \cdots & 0 & A_{(k-1)(k-2)}^x & A_{(k-1)(k-1)}^x & A_{(k-1)k}^x & \\
\cdots & \cdots & & \cdots & & \cdots & & 0 & A_{k(k-1)}^x & A_{kk}^x &
\end{bmatrix}
$$

Partial forward eliminations $\longrightarrow$

$$
\begin{bmatrix}
\tilde{A}_{11}^x & \tilde{A}_{12}^x & 0 & & \cdots & & \cdots & & \cdots & & \cdots \\
0 & \tilde{A}_{22}^x & \tilde{A}_{23}^x & & 0 & & \cdots & & \cdots & & \cdots \\
\vdots & \vdots & & \ddots & & \ddots & & \vdots & & \vdots & \\
\cdots & 0 & 0 & \tilde{A}_{(r-1)(r-1)}^x & \tilde{A}_{(r-1)r}^x & & 0 & & \cdots & & \cdots \\
\cdots & \cdots & 0 & 0 & \tilde{A}_{rr}^x & \tilde{A}_{r(r+1)}^x & & 0 & & \cdots \\
\cdots & \cdots & 0 & 0 & \tilde{A}_{r(r+1)}^x & \tilde{A}_{(r+1)(r+1)}^x & \tilde{A}_{(r+1)(r+2)}^x & & \cdots \\
\vdots & \vdots & & \ddots & & \vdots & & \ddots & & \ddots & & \vdots \\
\cdots & \cdots & \cdots & & \cdots & & \cdots & & 0 & \tilde{A}_{k(k-1)}^x & \tilde{A}_{kk}^x
\end{bmatrix}
\tag{18}
$$

$$
\begin{bmatrix}
z_{111} & z_{121} & \cdots & z_{1lm} \\
z_{211} & z_{221} & \cdots & z_{2lm} \\
\vdots & \vdots & \ddots & \vdots \\
z_{(r-1)11} & z_{(r-1)21} & \cdots & z_{(r-1)lm} \\
z_{r11} & z_{r21} & \cdots & z_{rlm} \\
\vdots & \vdots & \ddots & \vdots \\
z_{k11} & z_{k21} & \cdots & z_{klm}
\end{bmatrix}
=
\begin{bmatrix}
\tilde{y}_{111} & \tilde{y}_{121} & \cdots & \tilde{y}_{1lm} \\
\tilde{y}_{211} & \tilde{y}_{221} & \cdots & \tilde{y}_{2lm} \\
\vdots & \vdots & \ddots & \vdots \\
\tilde{y}_{(r-1)11} & \tilde{y}_{(r-1)21} & \cdots & \tilde{y}_{(r-1)lm} \\
\tilde{y}_{r11} & \tilde{y}_{r21} & \cdots & \tilde{y}_{rlm} \\
\vdots & \vdots & \ddots & \vdots \\
\tilde{y}_{k11} & \tilde{y}_{k21} & \cdots & \tilde{y}_{klm}
\end{bmatrix}
\tag{19}
$$

We perform this operation on every sub-domain, and then we gather on processor one the tri-diagonal Schur complements, we aggregate them into one matrix along x direction. The matrix is still a tri-diagonal matrix, and we solve the matrix using linear $\mathcal{O}(N)$ computational cost Gaussian elimination procedure with the Thomas algorithm.

Next, we scatter and substitute the partial solutions to sub-system over sub-domains. We do it by replacing the last $r - k + 1$ rows by the identity matrix and placing the solutions into the right-hand side blocks. Namely, on the right-hand side we replace rows from $r + 1$ (denoted by blue color) by the solution obtained in the global phase, to obtain:

$$\begin{bmatrix} \tilde{A}_{11}^x & \tilde{A}_{12}^x & 0 & \cdots & & \cdots & \cdots \\ 0 & \tilde{A}_{22}^x & \tilde{A}^x 23 & 0 & & \cdots & \cdots \\ \vdots & \vdots & \ddots & & \ddots & \vdots & \vdots \\ \cdots & 0 & 0 & \tilde{A}_{(r-1)(r-1)}^x & \tilde{A}_{(r-1)r}^x & 0 & \cdots \\ 0 & \cdots & \cdots & 0 & 1 & 0 & \cdots & 0 \\ \vdots & \vdots & \ddots & \vdots & & \ddots & \ddots & \vdots \\ 0 & \cdots & \cdots & & \cdots & 0 & 1 & 0 \\ 0 & \cdots & \cdots & & \cdots & & 0 & 1 \end{bmatrix} \quad (20)$$

$$\begin{bmatrix} z_{111} & z_{121} & \cdots & z_{1lm} \\ z_{211} & z_{221} & \cdots & z_{2lm} \\ \vdots & \vdots & \ddots & \vdots \\ z_{(r-1)11} & z_{(r-1)21} & \cdots & z_{(r-1)lm} \\ z_{r11} & z_{r21} & \cdots & z_{rlm} \\ z_{(k-1)11} & z_{(k-1)21} & \cdots & z_{(k-1)lm} \\ z_{k11} & z_{k21} & \cdots & z_{klm} \end{bmatrix} = \begin{bmatrix} \tilde{y}_{111} & \tilde{y}_{121} & \cdots & \tilde{y}_{1lm} \\ \tilde{y}_{211} & \tilde{y}_{221} & \cdots & \tilde{y}_{2lm} \\ \vdots & \vdots & \ddots & \vdots \\ \tilde{y}_{(r-1)11} & \tilde{y}_{(r-1)21} & \cdots & \tilde{y}_{(r-1)lm} \\ \hat{z}_{r11} & \hat{z}_{r21} & \cdots & \hat{z}_{rlm} \\ \vdots & \vdots & \ddots & \vdots \\ \hat{z}_{(k-1)11} & \hat{z}_{(k-1)21} & \cdots & \hat{z}_{(k-1)lm} \\ \hat{z}_{k11} & \hat{z}_{k21} & \cdots & \hat{z}_{klm} \end{bmatrix} \quad (21)$$

and running backward substitutions over each subdomain in parallel.

Next, we plug the solutions to the right-hand side of the second system along y axis, and we continue with the partial factorization. Now, we have $s - 1$ rows in the interior and $l - s + 1$ rows on the interface.

We compute the Schur complements in the same way as for the fist sub-system, thus we skip the algebraic details here. We perform this operation on every sub-domain, then we collect on processor one and aggregate the Schur complements into the global matrix along y directions. The global matrix is three-diagonal, and we solve it with Thomas algorithm. Next, we scatter and substitute the partial solutions to sub-system on each subdomain, and we solve by backward substitutions.

Finally, we plug the solution to the right-hand side of the third system along z axis, and we continue with the partial factorization. Now, we have $t - 1$ rows in the interior and $m - t + 1$ rows on the interface. The partial eliminations follow the same lines as for the two other directions, thus, we skip the algebraic details.

We repeat the computations for this third direction, computing the Schur complements on every sub-domain, collecting them into one global system, which is still three-diagonal, and we can solve it using the linear computational cost Thomas algorithm.

Next, we substitute the partial solution to sub-systems. We replace the last $t - m + 1$ rows by the identity matrix, and place the solutions into the right-hand side, and run the backward substitutions over each subdomain in parallel.

6 Parallel Scalability

The solver is implemented in fortran95 with OpenMP (see Algorithm 1) and MPI libraries used for parallelization. It does not use any other libraries, and it is a highly optimized code. We report in Fig. 4 and Table 1 the weak scalability for three different subdomain sizes, $50 \times 50 \times 100$, $25 \times 100 \times 100$, and $50 \times 50 \times 50$. The weak scalability for the subdomains of $49 \times 49 \times 99$ internal points, shows that increasing the number of processors from 4 to 1024, simultaneously increasing the number of subdomains from 4 to 1024, and the problem size from $50 \times 50 \times 400$ to $800 \times 800 \times 400$, results in the increase of the total computational time from 120 s to 178 s for a single time step. Thus, we can perform a single time step with over 1,128,000,000 unknowns (three components

---

Algorithm 1. OpenMP loop parallelization

```
!!$OMP PARALLEL DO DEFAULT(PRIVATE) PRIVATE(start)
SHARED(inverse,nrhs,nint) SHARED(s_mult,d)
DO i = 1,nrhs
   start = (i-1)*(nint+2)+2; s_mult(i,1) = d(start)*inverse
END DO
!!$OMP END PARALLEL DO
```

Algorithm 2. Loop unrolling to optimize cache usage

```
loop_end =nrhs/10; i=1; inverse =1.0/(mat%b(n)-cp(n-1)*mat%a(n));
dmult =mat%a(n)
DO j = 1,loop_end
   IF(i<nrhs-10)THEN
      finish = i*(nint+2)-1
      dp(n,i) = (d(finish) - dp(n-1,i) * dmult)*inverse
      dp(n,i+1) = (d(finish+(nint+2)) - dp(n-1,i+1) * dmult)*inverse
      ...
      dp(n,i+9) = (d(finish+(nint+2)*9) - dp(n-1,i+9) * dmult)*inverse
      i=i+10
   END IF
END DO
```

of the velocity vector field, and one component of the pressure and the temperature scalar fields over 256,000,000 mesh points) within 3 min on a cluster. For the numerical verification of the code, we refer to [15].

We report in Figure 5 and Table 2 the strong scalability for six different simulations, each one with box size $50 \times 50 \times 50$, with 8, 16, 32, 64, 128 and 256 subdomains. Since the number of nodes is multiplied by the number of unknowns (three components of the velocity vector field, one component of the pressure scalar field and one component of the temperature scalar field), we obtained between $8 \times 50 \times 50 \times 5 = 5$ millions, to $256 \times 50 \times 50 \times 5 = 160$ millions of unknowns. We can read the superlinear speedup for these plots, which is related to the optimization of cache usage on smaller subdomains, with optimizing the memory transfers to the computational kernel and loop unrolling technique, as illustrated in Algorithm 2.

In Fig. 6, we show some snapshots from the preliminary simulations. In here, we focused on the description and scalability of the parallel solver algorithm,

Table 1. Weak scalability up to 1024 processors (subdomains). Each grid box contains one subdomain with $49 \times 49 \times 99$, $24 \times 99 \times 99$, or $49 \times 49 \times 49$ internal points, respectively, one subdomain per processor.

Subdomains = Processors	Grid	$50 \times 50 \times 50$ Time [s]	$25 \times 100 \times 100$ Time [s]	$50 \times 50 \times 100$ Time [s]
1	(1, 1, 1)	19	58	–
2	(1, 1, 2)	23	63	–
4	(1, 1, 4)	23	66	120
8	(2, 1, 4)	63	85	157
16	(2, 2, 4)	36	97	152
32	(4, 2, 4)	42	100	150
64	(4, 4, 4)	49	115	157
128	(8, 4, 4)	63	129	160
256	(8, 8, 4)	72	144	166
512	(16, 8, 4)	–	–	170
1024	(16, 16, 4)	–	–	178

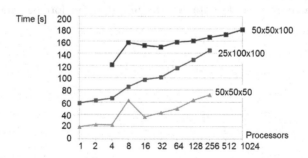

Fig. 4. Weak scalability for subdomains with $49 \times 49 \times 99$, $24 \times 99 \times 99$, and $49 \times 49 \times 49$ internal points, one subdomain per processor, up to 1024 processors (subdomains). We increase the problem size with the number of processors. For the ideal parallel code, the execution time remains constant.

Table 2. Strong scallability up to 256 processors.

Processors ndofs * 1,000,000	4	8	16	32	64	128	256
5	120	63	–	–	–	–	–
10	–	157	36	–	–	–	–
20	–	–	152	42	–	–	–
40	–	–	–	150	49	–	–
80	–	–	–	–	157	63	–
160	–	–	–	–	–	160	72

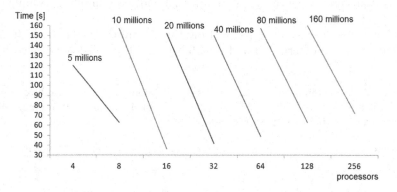

Fig. 5. Strong scallability for meshes with different sizes, for different numbers of processors. For larger meshes, it is only possible to run them on maximum number of
processors.

leaving the model formulation and large massive parallel simulations of different
atmospheric phenomena for the future work. This will be a challenging task itself,
requiring to acquire reliable data for the initial state, forcing, and boundary
conditions.

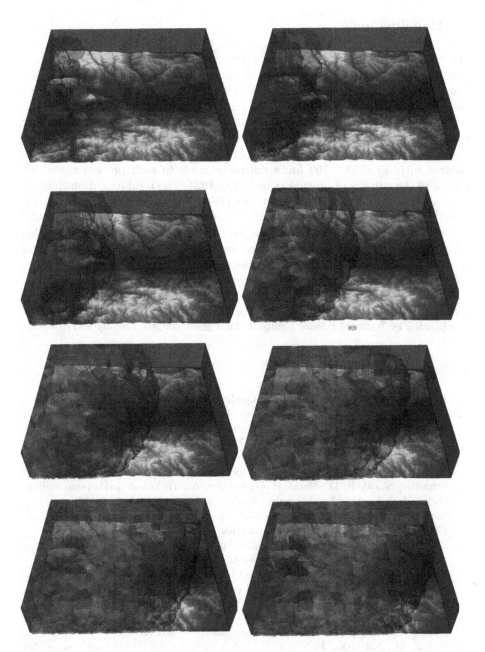

Fig. 6. Snapshots from the simulation

7   Conclusions

We described a parallel algorithm for the factorization of Kronecker product matrices. These matrices result from the finite-difference discretizations of the Navier-Stokes Boussinesq equations. The algorithm allows for simulating over the non-regular terrain topography. We showed that the Schur complements are tri-diagonal, and they can be computed, aggregated, and factorized in a linear computational cost. We analyzed the weak scalability over the PROMETHEUS Linux cluster from the CYFRONET supercomputing center. We assigned a sub-domain with $50 \times 50 \times 100$ finite difference mesh to each processor, and we increased the number of processors from 4 to 1024. The total execution time for a single time step increased from 120 s to 178 s for a single time step. Thus, we could perform computations for a single time step with around 1,128,000,000 unknowns within 3 min on a Linux cluster. This corresponds to 5 scalar fields over 256,000,000 mesh points. In future work, we plan to formulate the model parameters, initial state, forcing, and boundary conditions to perform massive parallel simulations of different atmospheric phenomena.

Acknowledgments. This work and the visit of prof. Petar Minev in AGH University is supported by National Science Centre, Poland grant no. 2017/26/M/ ST1/ 00281.

References

1. European Environment Agency: Air Quality in Europe - 2017 report. 13/2017
2. Marras, S., et al.: A review of element-based Galerkin methods for numerical weather prediction: finite elements, spectral elements, and discontinuous Galerkin. Arch. Comput. Methods Eng. **23**, 673–722 (2016)
3. Song, Y., Hou, T.: Parametric vertical coordinate formulation for multiscale, Boussinesq, and nonBoussinesq ocean modeling. Ocean. Model. **11**, 298–332 (2006)
4. Schaeffer, N., Jault, D., Nataf, H.-C., Furnier, A.: Turbulent geodynamo simulations: a leap towards Earth's core. Geophys. J. Int. **211**(1), 1–29 (2017)
5. Zhang, Z., Moore, J.C.: Mathematical and Physical Fundamentals of Climate Change, Chap. 11 - Atmospheric Dynamics, pp. 347–405 (2015)
6. Zeytounian, R.: Asymptotic Modeling of Atmospheric Flows. Springer, Heidelberg (1990). https://doi.org/10.1007/978-3-642-73800-5
7. Guermond, J.-L., Minev, P.D.: High-order time stepping for the Navier-Stokes equations with minimal computational complexity. J. Comput. Appl. Math. **310**, 92–103 (2017)
8. Keating, J., Minev, P.: A fast algorithm for direct simulation of particulate flows using conforming grids. J. Comput. Phys. **255**, 486–501 (2013)
9. Bubak, M., Kitowski, J., Wiatr, K. (eds.): eScience on Distributed Computing Infrastructure. Achievements of PLGrid Plus Domain-Specific Services and Tools, vol. 8500. Springer, Cham (2014). https://doi.org/10.1007/978-3-319-10894-0
10. https://github.com/Podsiadlo/terrain
11. Farr, T.G., et al.: The Shuttle Radar Topography Mission, Reviews of Geophysics, vol. 45, no. 2 (2005)

12. Guermond, J.L., Minev, P.D.: A new class of massively parallel direction splitting for the incompressible Navier–Stokes equations. Comput. Methods Appl. Mech. Eng. **200**, 2083–2093 (2011)
13. Douglas, J., Gunn, J.E.: A general formulation of alternating direction methods. Numerische Mathematik **6**(1), 428–453 (1964)
14. Golub, G.H., Van Loan, C.: Matrix Computations, 3rd edn. John Hopkins University Press, Baltimore (1996)
15. A. Takhirov, R. Frolov, P. Minev, Direction splitting scheme for Navier-Stokes-Boussinesq system in spherical shell geometries. arXiv:1905.02300 (2019)

Reconstruction of Low Energy Neutrino Events with GPUs at IceCube

Maicon Hieronymus[1][(✉)], Bertil Schmidt[1], and Sebastian Böser[2]

[1] Institute of Computer Science, Johannes Gutenberg University, Mainz, Germany
mhieronymus@uni-mainz.de
[2] Institute of Physics, Johannes Gutenberg University, Mainz, Germany

Abstract. IceCube is a cubic kilometer neutrino observatory located at the South Pole that produces massive amounts of data by measuring individual Cherenkov photons from neutrino interaction events in the energy range from few GeV to several PeV. The actual reconstruction of neutrino events in the GeV range is computationally challenging due to the scarcity of data produced by single events. This can lead to run times of several weeks for the state-of-the-art reconstruction method – `Pegleg` – on CPUs for typical workloads of many ten-thousand events. We propose a GPU version of `Pegleg` that probes the likelihood space with several hypotheses in parallel while adapting the amount of parallel sampled hypotheses dynamically in order to reduce computation time significantly. Our results show an average speedup of 14 (with a maximum of over 200) for 5262 reconstructed neutrino events of different flavors on a Titan V GPU compared to the multithreaded CPU version, which enables quicker and broader analysis of IceCube events.

Keywords: Neutrino oscillation · Neutrino physics · MultiNest · Reconstruction · GPU

1 Introduction

IceCube [10] is a cubic kilometer neutrino observatory located at the South Pole. Neutrinos are elementary particles that exist in three different flavors ν_e, ν_μ and ν_τ. Unlike any other particles, neutrinos can change their flavor during propagation [8]. This phenomenon, the so-called neutrino oscillations, implies that neutrino flavours have different masses. The question which of the three neutrino flavors is the heaviest is being investigated under the term "neutrino mass ordering". While most neutrino oscillation experiments are insensitive to this question, IceCube can address it through precision measurements of the ubiquitous flux of atmospheric neutrinos and the subtle effects the very dense matter in the earth core has on their flavor oscillations. These matter effects only appear for relatively low energy neutrinos below 15 GeV [1,13]. With such low energies only very few photon hits per event are detected in IceCube, heavily decreasing the signal-to-noise ratio. Neutrino events are reconstructed by comparing the observed hit-pattern to the expected hit pattern given an 8-dimensional event

V. V. Krzhizhanovskaya et al. (Eds.): ICCS 2020, LNCS 12137, pp. 118–131, 2020.
https://doi.org/10.1007/978-3-030-50371-0_9

hypothesis using a maximum likelihood method. The expected light intensity at any position and time for a given hypothesis is obtained from spline approximations to tabulated simulation data (so called *photosplines* [17]) that exploit various approximate symmetries of the light emission and propagation process. Even without noise, the reconstruction of such an event from the signal in IceCube is suffering from causality requirements on the photon arrival times, causing a likelihood function that is neither globally convex nor continuous and has many local minima. In particular gradient descent minimizers and algorithms only using local information struggle with this likelihood function, which increases the complexity of the event reconstruction problem. Efficient and fast evaluation of a suitable event hypothesis has thus become a limiting factor for the analysis of IceCube data in order to study neutrino properties. The algorithm that is currently used, `Pegleg` [16], partitions the exploration of the event likelihood space into two: (i) The vertex of the neutrino interaction and the direction of secondary particles emerging from this are optimized using MultiNest [6,7] – a multi modal nested sampling algorithm that handles degenerated likelihoods. (ii) Track length of the emerging muon and its energy depositions caused by the event are optimized separately. The reconstruction takes up to 10 mins per event on a typical workstation. This in turn leads to run times of several weeks for a typical analysis with tens of thousands of events [11,12].

In this paper, we present a GPU version of `Pegleg`, which features a massively parallel MultiNest probing algorithm (explained in Sect. 3) and massively parallel spline evaluation (explained in Sect. 4). Previous work on accelerating neutrino oscillation data analyses on GPUs has mainly focused on direct neutrino propagation to calculate oscillation probabilities [3,15] or direct photon propagation for a given hypothesis. This work is the first to reconstruct neutrinos given measurements from an interaction event by evaluating splines on a GPU.

2 Background

2.1 Millipede Likelihood

Consider the light yield at different positions and times of an interaction event of a neutrino with ice. We search for the parameters of the underlying event, i.e. the source vertex of the event, the energy and the direction of the neutrino (which mostly coincides with the direction of the secondary particles). If a muon neutrino ν_μ interacts in a charged-current interaction, it will also create a muon that can travel several ten meters through the ice, allowing for a) differentiation of ν_μ from other flavors and b) significantly improved directional resolution. The event hypothesis thus has eight free parameters $(\boldsymbol{y}, t, E, L)$ with $\boldsymbol{y} = (y_1, y_2, y_3, y_4 = \theta, y_5 = \phi)^{\mathrm{T}}$ the coordinates and direction of the neutrino event, t the time it occurred, E the energy of the neutrino and L the track length of an outgoing muon if present. The muon track is divided into many segments of fixed length, hence the name Millipede. The light detected by the i-th DOM (Digital Optical Modules) at the position x_i is a superposition of the light emitted at each segment (see Fig. 1) [18]. Equation 1 describes the amount

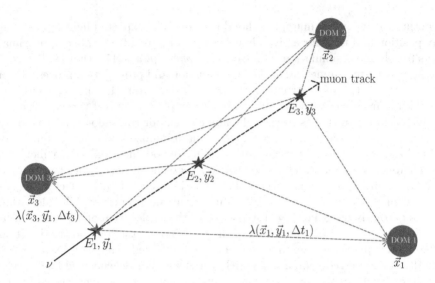

Fig. 1. A sketch of the millipede hypothesis with an incoming neutrino from the bottom left. The red star marks the origin of the neutrino event, with the blue stars marking track segments of an outgoing track. Each star is handled as a distinct photon source. The colored dotted lines depict the photons that arrive at each DOM given a photon source. (Color figure online)

of expected photons μ_l for a time bin k of a DOM i, where l is the index over all time bins of all DOMs.

$$\mu_l = \rho_l + \sum_{j=1}^{J} \lambda(\boldsymbol{x}_i, \boldsymbol{y}_j, \Delta t_{i,k}) \cdot E_j. \tag{1}$$

In Eq. 1, λ denotes how the expected light intensity at a position $\boldsymbol{x}_i$ scales with the energy loss E_j at a position $\boldsymbol{y}_j$. $\Delta t_{i,k}$ is the time delay between the photon production and detection at sensor i during time bin k. ρ_l is the noise which is specific for every DOM, such that $\rho_l = \rho_{l'}$ for all $l \neq l'$ that belong to the same DOM. λ is typically evaluated by using spline tables that describe a B-spline surface over a rectangular d-dimensional knot grid [17]. Each DOM i can have a different amount n_i of valid time bins k that are used during reconstruction such that we have a total of M DOMs, J energy losses and $M' \geq M$ total time bins. We can summarize Eq. 1 for all DOMs by Eq. 2.

$$\begin{pmatrix} \mu_1 - \rho_1 \\ \mu_2 - \rho_2 \\ \vdots \\ \mu_{M'} - \rho_{M'} \end{pmatrix} = \begin{pmatrix} \lambda_{1,1} & \lambda_{1,2} & \dots & \lambda_{1,J} \\ \lambda_{2,1} & \lambda_{2,2} & \dots & \lambda_{2,J} \\ \vdots & \vdots & \ddots & \vdots \\ \lambda_{M',1} & \lambda_{M',2} & \dots & \lambda_{M',J} \end{pmatrix} \cdot \begin{pmatrix} E_1 \\ E_2 \\ \vdots \\ E_J \end{pmatrix}. \tag{2}$$

Given a loss pattern we calculate the likelihood of photon counts, such that the predicted photon counts μ can be compared to the actual signal. The amount of detected photons N_l at one time bin of one DOM follows a Poisson distribution [9] with mean μ_l, which expands for N_l photons to [4]:

$$\mathcal{L}_l = \frac{\mu_l}{N_l!} e^{-\mu_l}. \tag{3}$$

The overall likelihood is the product of all contributions:

$$\mathcal{L} = \prod_{l=1}^{M'} \mathcal{L}_l = \prod_{l=1}^{M'} \frac{\mu_l}{N_l!} e^{-\mu_l}. \tag{4}$$

In order to reduce numerical instabilities, it is common to use the logarithm and work with the log-likelihood (llh) or the negative log-likelihood, which in turn needs to be minimized. Using the continuous charge variable instead of discrete photon counts [16] we can approximate the factorial with the gamma function, resulting in Eq. 5.

$$\ln \mathcal{L} = \sum_{l=1}^{M'} (N_l \cdot \ln \mu_l - \ln(\Gamma(N_l + 1)) - \mu_l). \tag{5}$$

The evaluation of the non-algebraic function λ for each μ_l makes the calculation of an analytical maximum of Eq. 5 infeasible. Thus, a numerical maximum likelihood approach is used instead.

2.2 Pegleg

Pegleg [16] employs a modified Millipede likelihood where a possible muon track is divided into segments of fixed energy losses – a good approximation for low-energy events with minimum ionizing muons. The event likelihood is minimized by using three different layers: Layer 1 optimizes the six parameters $(\boldsymbol{y}, t) = (y_1, y_2, y_3, y_4 = \theta, y_5 = \phi, t)$ using the MultiNest algorithm. Layer 2 receives fixed parameters $(\boldsymbol{y}, t)$ from Layer 1 and optimizes the track length $L = N \cdot l$ by fitting the number of track segments N with a given spacing l. Finally Layer 3 internally optimizes the energy loss $E_c = E_1$ of Eq. 2 of the first segment, i.e. the initial cascade.

Note that the response matrix Λ (see Eq. 2) contains the intensity changes for every DOM time bin and photon source, where every column corresponds to a different photon source as shown in Fig. 2. The first column is the initial cascade source and every other column is another track segment. We start with a single column and calculate the energy loss E_c and likelihood in Layer 3. All energy losses of the track segments are approximated by a minimum ionizing muon. We add muon segments successively and evaluate the likelihood of the new response matrix for every added segment until a predefined amount of subsequent columns does not improve the likelihood [16]. Evaluating a column of Λ is computationally expensive due to the spline evaluations via BSPLVB [2]. Thus, there is a negligible overhead when calculating J segments iteratively compared to calculating J segments at once.

$$
\left(
\begin{array}{cccc}
\boxed{\lambda_{1,1}} & \boxed{\lambda_{1,2}} & \boxed{\lambda_{1,3}} & \boxed{\lambda_{1,4}} \\
\lambda_{2,1} & \lambda_{2,2} & \lambda_{2,3} & \lambda_{2,4} \\
\vdots & \vdots & \vdots & \vdots \\
\lambda_{M',1} & \lambda_{M',2} & \lambda_{M',3} & \lambda_{M',4}
\end{array}
\right)
\cdot
\begin{pmatrix}
E_c \\
E_2 \\
E_3 \\
E_4
\end{pmatrix}
$$

$$\text{Cascade} \quad \text{Track 1} \quad \text{Track 2} \quad \text{Track 3}$$

Fig. 2. The first column of the response matrix Λ corresponds to the initial cascade of photons. All other columns correspond to a track segment of an outgoing muon. Increasing the track length translates directly to adding columns and energy losses.

For Layer 3 the parameters (y, t, L) are fixed and the last remaining free parameter is the energy loss E_c of the cascade segment. Using Eq. 2 we can describe the amount of expected photons for every muon track by Eq. 6.

$$\boldsymbol{\mu}_{\text{eff}} = \boldsymbol{\rho} + \Lambda \cdot (0, E_2, E_3, \ldots, E_J)^{\mathrm{T}}. \tag{6}$$

With every energy loss $E_2, E_3, \ldots, E_J$ fixed to minimum ionizing muon, we can solve Eq. 6 directly for the effective noise term $\boldsymbol{\mu}_{\text{eff}} = (\mu_{\text{eff},1}, \mu_{\text{eff},2}, \ldots, \mu_{\text{eff},M'})^{\mathrm{T}}$. To calculate the energy loss E_c, we take the first column of Λ and apply the Newton method [19], i.e. we calculate the gradient of the likelihood from Eq. 4 with respect to E_c and follow it's second gradient until we find the maximum likelihood with the iterative scheme shown in Eq. 7

$$\frac{\mathrm{d}\tilde{\mathcal{L}}^{(k+1)}}{\mathrm{d}E_c} = \frac{\mathrm{d}\tilde{\mathcal{L}}^{(k)}}{\mathrm{d}E_c} - \frac{\mathrm{d}\tilde{\mathcal{L}}^{(k)}}{\mathrm{d}E_c}\frac{\mathrm{d}^2\tilde{\mathcal{L}}^{(k)}}{\mathrm{d}^2E_c} \quad \text{with } k = 1, 2, \ldots, \tag{7}$$

In Eq. 7 $\frac{\mathrm{d}\tilde{\mathcal{L}}^{(1)}}{\mathrm{d}E_c}$ is an initial estimate, using either $20\,\text{GeV}$ which leads to $\mathcal{O}(10)$ steps until convergence or the result of a previous iteration which reduces the number of steps to $\mathcal{O}(3)$ with a tolerance of $10^{-6}\,\text{GeV}$. The gradient of $\mathcal{L}$ can be calculated by Eq. 8.

$$\frac{\mathrm{d}\mathcal{L}}{\mathrm{d}E_c} = \sum_{l=1}^{M'} \frac{N_l}{\lambda_{l,1}E_c + \mu_{\text{eff},l}} - \lambda_{l,1}. \tag{8}$$

Considering that the second term $\lambda_{l,1}$ in Eq. 8 is constant and the other part is strictly monotonically decreasing with E_c, there is at most one positive value of E_c that maximizes the likelihood.

3 Parallelizing MultiNest on a GPU

In this section we explain our GPU parallelization of Multinest. The spline approximations of the photon expectation tables do not vary throughout the minimization process. Thus, we need to copy them only once to the GPU device. The corresponding data transfer becomes negligible for the $\mathcal{O}(10000)$ calls for the likelihood in MultiNest. DOMs that are too far away or didn't see any charge are excluded in each evaluation of the splines, i.e. their entries in the response matrix Λ (Eq. 2) are 0, leaving us with $\mathcal{O}(100)$ evaluations for a column of the response matrix for low energy events.

To achieve high GPU occupancy, we evaluate n_p points in parallel, where each point represents an event hypothesis. Since we sample multiple points in parallel, we do not expect a sampling efficiency of 1, but allow undersampling of likelihood regions. This is reasonable as long as the sampling space includes the global minimum or as long as the ellipsoids that enclose the current sampling space are allowed to grow large enough.

3.1   Generating Initial Live Points

MultiNest starts with a number of live points n_{live}. We use the GPU to generate n_p many points in parallel, where $n_{\text{live}} \leq n_p$. We initialize the live points by generating n_p many points and taking the best n_{live} ones as start. The points are generated inside a hypercube in parallel on the CPU using OpenMP, whereby the likelihood is calculated on the GPU. In case the seed for the parameter space boundaries may not be good enough to generate points with different likelihoods at first try, we generate a new set if the difference of the highest and lowest found likelihood is lower than 10^{-4} which occurs when all points are far away from the minimum.

3.2   Sampling the Complete Space

In this stage MultiNest samples points within the hard constraint $\mathcal{L}_{points_i} > \mathcal{L}_{\text{lowest}}$, i.e. a new point has to have a higher likelihood than the lowest likelihood of the live points. We switch to ellipsoidal sampling after a certain sampling efficiency cannot be reached within five iterations in a row. For parallel sampling we keep that condition but we do not necessarily calculate new likelihoods in every iteration. Instead, we sample n_p many points in iteration i and iterate over that list of points until one satisfies the hard constraint. The inverse of the number of samples to find such a point is the sampling efficiency in iteration i. Algorithm 1 shows the pseudocode where we sample n_p points in parallel and process those as they were sampled individually. As long as we do not find a new point inside the hard constraint $\mathcal{L}_{points_i} > \mathcal{L}_{\text{lowest}}$ we do the following: We check if any sampled points are left from previous iterations. If this is not the case, we sample new points from the hypercube in parallel using OpenMP and calculate the likelihoods using the GPU. Subsequently, we iterate over the remaining points until one of them satisfies the hard constraint. If it is the last remaining point, we note that we need more points and exit the loop. If no point satisfies the hard constraint, we repeat the loop. Note, the approach presented in [6] applies a similar sampling scheme on a compute cluster using MPI with the difference that each process evaluates a single point. In contrast, our approach takes advantage of the compute power of a GPU by dynamically changing the number of parallel sampled points as outlined in the next subsection.

Algorithm 1. Sampling the complete space

```
1  do
2      if not remaining_points then
3          for j=1,n_p do in parallel                    ▷ Using OpenMP
4              points_j ← get_random(thread_id)
5          L_points ← get_llh(points)                    ▷ GPU
6      for i=remain_idx, n_p do
7          if L_{points_i} >lowlike then
8              accept points_i
9              if i == n_p then
10                 remaining_points ← False
11                 remain_idx ← 1
12             else
13                 remaining_points ← True
14                 remain_idx getsi + 1
15             exit
16         if i == n_p then
17             remaining_points ← False
18             remain_idx ← 1
19     if point accepted then exit
20 enddo
```

3.3 Sampling in Ellipsoids

Once the sampling efficiency consecutively falls below a threshold (e.g. five times as default), we switch to ellipsoidal sampling. The clustering and the overall algorithm remain the same, except for the sampling itself. We sample $n_{p,i}$ points at iteration i in parallel and check if any of them satisfies the hard constraint and then continue with possibly remaining points for the next iteration. Now we might have one or several distinct clusters, where each cluster is enclosed by one or more (overlapping) ellipsoids [5]. Hence we use parallel sampling within an isolated cluster. For each point to be sampled we randomly choose one ellipsoid and generate a point within this ellipsoid in parallel with OpenMP. After $n_{p,i}$ points have been generated, the GPU evaluates their likelihood. We choose a point as in Sect. 3.2 but if a point lies in k ellipsoids, we accept it with probability $1/k$ as does the CPU version (see Algorithm 2). During ellipsoidal sampling, the space to be explored varies and therefore we scale the number of points $n_{p,i}$ to sample in parallel by means of function $\delta(V; \alpha) \in [0, 1]$ that depends on the volume fraction V of the isolated cluster we want to sample from and a user-defined value α that controls the behaviour of that function. The overhead of sampling many points in parallel is negligible, whereas it dominates the calculation time if few points are sampled in parallel. Therefore we decrease the number of points sampled in parallel faster when the ellipsoids cover only a small fraction of the parameter space and decrease slower for bigger ellipsoids. To find an appropriate number of points for a volume fraction $V \in [0, 1]$ that is covered by the ellipsoids,

---

Algorithm 2. Sampling in ellipsoids

```
1   do
2       if not remaining_points then
3           for j=1,n_p do in parallel                          ▷ Using OpenMP
4               ell_j ← get_random_ellipsoid(thread_id)
5               points_j ← get_random(thread_id, ell_j)
6       L_points ← get_llh(points)                              ▷ GPU
7       for i=remain_idx, n_p do
8           if L_points_i >lowlike then
9               k ← 0
10              for j=1,n_ellipsoids do
11                  if point_in_ell(points_i, j then) k ← k + 1
12              accept points_i with prob 1/k
13              if point accepted then
14                  if i == n_p then
15                      remaining_points ← False
16                      remain_idx ← 1
17                  else
18                      remaining_points ← True
19                      remain_idx ← i + 1
20                  exit
21          if i == n_p then
22              remaining_points ← False
23              remain_idx ← 1
24      if point accepted then exit
25  enddo
```

---

we use a simple linear interpolation scheme with 5 knots, which depend on the user-defined value α:

$$f_1(V; \alpha) = \frac{4}{3} \cdot \frac{\alpha}{1-\alpha} \cdot V$$

$$f_2(V; \alpha) = \frac{\alpha}{3} + \frac{2}{3} \cdot \frac{\alpha}{1-\alpha} \cdot V$$

$$f_3(V; \alpha) = \alpha\left(1 - \frac{4}{3}\right) - \frac{4}{3\alpha} + \frac{8}{3} + V\left(\frac{4}{3 \cdot \alpha} - \frac{4}{3}\right) \qquad (9)$$

$$f_4(V; \alpha) = \frac{5}{3} - \frac{2}{3\alpha} + V\left(\frac{2}{3 \cdot \alpha} - \frac{2}{3}\right)$$

The function δ is shown in Fig. 3 for different values of α.

4   Evaluating the Response Matrix on a GPU

We want to evaluate several hypotheses in parallel on the GPU device, where the evaluation itself is not different to the CPU version, which uses SuiteSparse to

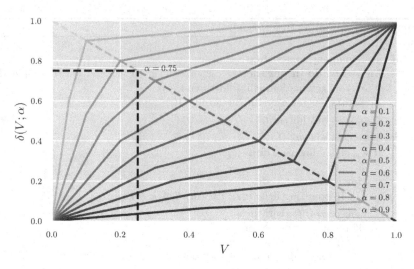

Fig. 3. A visualization of the scaling function $\delta(V;\alpha)$ to scale the number of parallel sampled points given a user-defined variable α and the volume fraction V that the isolated cluster covers from which we want to sample from. By setting α the user can define the intersection of the function with the diagonal.

handle the response matrix. However, different hypothesis might have its optimum at various amounts of track segments and hence converge at different time points. To balance the workload we thus overlap computation and communication using CUDA streams.

For n_p hypotheses and n_{streams} streams, we divide the work into chunks of size $n_{\text{chunk}} = \lfloor n_p/n_{\text{streams}} \rfloor$ where the last stream contains $n_p \bmod n_{\text{streams}}$ many hypotheses. On the CPU, we create n_p many threads to process each stream in parallel. Each hypothesis and therefore each kernel call is assigned to one CUDA thread block. The CPU checks every k column and corresponding likelihood evaluations if a stream has finished its work and sets it to inactive. Hence each stream uses n_{chunk} many CUDA thread blocks from which some might end directly if their hypothesis has already converged.

The workflow for one stream is as follows: (i) Evaluate k columns of an hypothesis. (ii) Calculate the cascade energy loss for each of the additional columns and the likelihood. (iii) Synchronize the stream and copy an array of convergence checks for all hypotheses to CPU. (iv) If all converged, set the stream to inactive and start an asynchronous copy of the found likelihoods from device to host. If at least one hypothesis has not converged, launch the kernels for evaluating columns and calculating the energy loss and likelihoods for the next k columns.

Evaluating the response matrix of n_p hypotheses is done with a CUDA kernel consisting of 512 threads per CUDA thread block, one CUDA thread block per hypothesis, n_{streams} streams and 40944 Bytes of shared memory and 128 registers per thread. Given the sparse nature of a low energy neutrino event, only few

DOMs, typically 10 to 60, and few time bins of a DOM, about 1 to 6 bins, deliver a non-zero entry of the response matrix which is why we settle with 512 threads in favour of more register memory per thread. According to [14], at least 128 threads have to be invoked to fully utilize processing units on Volta. Adding a response column involves a check for the number of valid bins in this DOM and a check if the DOM is too far away for the given hypothesis. Once a thread processes a DOM with valid bins, it calculates mean photon counts for the DOM given the hypothesis and evaluates an amplitude splinetable. In case there are more time bins, we also need the probability quantiles for every bin using time splinetables. Each thread evaluates several splines for an entry with BSPLVB, where we do not have a memory layout for our splines that ensures coalesced memory access, marking the evaluation process as the most time consuming part of the overall reconstruction. Layer 3 has least impact on run time. Here we start a CUDA kernel consisting of 256 threads per CUDA thread block, one CUDA thread block per hypothesis, n_{streams} streams and 96 registers per thread and 40872 Bytes of shared memory to calculate the gradient from Eq. 12 in parallel. The size, here the amount of doubles, of shared memory is determined via

$$\text{size} = \max\left(\frac{n_{\mathrm{threads}}}{2 \cdot w} + n_{\mathrm{valid\_bins}}, m_s + n_{\mathrm{valid\_bins}}\right), \tag{10}$$

where $w = 32$ is the typical size of a warp and the number of valid bins depends on the seen photons by all DOMs. m_s is the maximum amount of segments that can be used. The number of valid bins is the number of all bins in worst case, which can be a multiple of 5160 in case of the IceCube detector. Therefore the number of threads is restricted to a small number. We calculate the gradient of $\mathcal{L}$ with respect to the cascade energy E_c from Eq. 8. Since $\sum_{l=1}^{M'} \lambda_{l,1} = \lambda_{\mathrm{sum}}$ remains constant, we only need to update the first part of the term. To calculate the amount of seen photons for the cascade, we use

$$N_1 = \sum_{l=1}^{M'} \mu_{\mathrm{data},l} \cdot \lambda_{l,1}, \tag{11}$$

where $\mu_{\mathrm{data},l}$ is the data vector of seen charges. The second derivative is

$$\frac{d^2\mathcal{L}}{dE_c^2} = \left(\sum_{l=1}^{M'} \frac{N_l \cdot \lambda_{l,1}}{(\lambda_{l,1}E_c + \mu_{\mathrm{eff},l})^2}\right) - \lambda_{\mathrm{sum}}. \tag{12}$$

We leave $\mu_{\mathrm{eff},l}$ in shared memory and every row is being processed by another thread, such that all memory reads are done in a coalesced manner. The partial results of each thread is distributed with sum_and_broadcast in every iteration that utilizes warp shuffles and shared memory. Calculating the likelihood after the Newton method converged is done by evaluating Eq. 5.

5   Performance Evaluation

We compare the run time and the accuracy of our GPU version with the CPU version of Pegleg. We have measured run times depending on the energy of the neutrino events for both versions and discuss the achieved accuracy, where we use a fixed seed for generating the initial live points. The used parameters for the GPU version are shown in Table 1. The CPU version (compiled using GCC V6.3.0) runs on an Intel Xeon E5-2680@2.70 GHz with 16 threads and with 8 GB DDR3 RAM. The GPU version (compiled using CUDA V9.2.148) runs on an Intel Core i5-3450@3.10 GHz with 8 GB DDR3 RAM and a NVidia Titan V with 12 GB HBM2 VRAM. We use the GRECO (GeV Reconstructed Events with Containment for Oscillations) sample [16] which is a simulated set from different Monte Carlo generators. It provides a low energy sample with high statistics at E_ν ~5 GeV. These low energy events deposit photons in $\mathcal{O}(10)$ DOMs, where most DOMs feature a single time bin due to the few number of photons (~1) that are detected per DOM. This leads to a low signal to noise ratio and hence a poor resolution for the zenith angle, the neutrino energy and track against cascade separation.

Table 1. Settings of the GPU MultiNest algorithm that delivers the best run time while reconstructing all tested events with a low error.

Parameter	Used value
Number of streams	4
Minimum parallel evaluations per calls	96
Parallel evaluations per calls	480
α	0.3
n_{nlive}	30
Efficiency	2.0
Maximum modes	10
Tolerance	1.1
Importance sampling	False

5.1 Speedup

Figure 4 shows a heatmap of run times over energy ranges from 1.00 GeV to 980.71 GeV for 5262 events of different flavors reconstructed by the GPU and CPU. The average speedup is 14.09 with a minimum at 1.29 and a maximum at 248.40 and an interquantile range of 10.47. The median run times for the GPU and CPU versions slightly increase with the energy of the event. Calculating Pearson's correlation coefficient, a value of 0.76 (0.64) for the GPU (CPU) run time and the number of activated DOMs and 0.46 (0.37) for the run times and the number of time bins for each DOM. The GPU version scales best with the track length (correlation value of 0.07 compared to 0.33 of the CPU version) thanks to the usage of CUDA streams. The speedup correlates with the number of DOMs with a value of 0.39, with the number of time bins with a value of 0.06 and with the track length with a value of 0.73. The wide variety of speedups can be explained as follows:

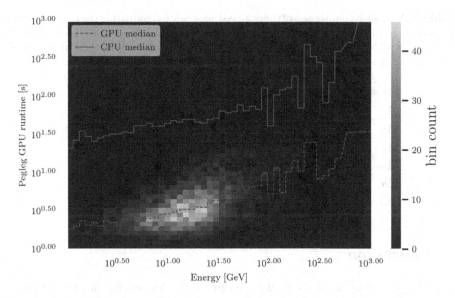

Fig. 4. 5262 events reconstructed with the GPU and CPU. The energy of an event correlates directly to the signal-to-noise ratio and the number of activated DOMs.

- Either one of the versions stops prematurely in a local minimum.
- The initial live points are favourable for one of the versions, such that it converges after fewer iterations.
- The event can be reconstructed within few iterations for both versions, such that the overhead of the GPU version of `Pegleg` dominates the runtime.

5.2 Accuracy

High speedups of the GPU version could mean a reconstruction prematurely terminated and did not reach the same accuracy as the CPU version. In order to investigate this in more detail, we compare between the calculated reconstruction and the provided ground truth on an event-by-event basis in Table 2 for 5262 events. With the exception of the time parameter, there is no systematic difference visible, i.e. all differences are centered around or close to zero. Overall, the CPU version yields slightly better reconstructions and has smaller interquartile ranges than our GPU version with the chosen setup that yields speedups of up to 250. This indicates for events that take especially long on CPU, that the GPU version stops early. For those events the CPU version usually yields a hypothesis with at least one parameter outside the interquantile range, rendering the result barely useful for further processing. Additional parameter tuning of the GPU version could enhance the accuracy especially for events in the tails which could lead to speedups closer to the mean.

Table 2. A detailed comparison of the achieved accuracy by comparing the reconstructed parameters to the truth. The CPU version is most of the time closer to the truth and has shorter tails with the GPU version not far behind. The interquantile range is similar, such that one can expect similar behavior of the GPU implementation.

Parameter	CPU median	GPU median	CPU IQR	GPU IQR
Δt	7.87	-14.57	50.40	54.35
Δy_1	-0.31	0.29	14.81	16.33
Δy_2	0.34	0.64	16.28	19.58
Δy_3	-0.13	0.92	7.88	8.47
$\Delta\theta$	-0.03	-0.11	0.64	0.76
$\Delta\phi$	0.01	-0.04	1.65	2.17

6 Conclusion

The computational analysis of data produced by low energy neutrino events is a major computational challenge for neutrino physicists at IceCube. In this paper we have shown how GPUs can be efficiently used to accelerate this task by an order-of-magnitude while achieving comparable accuracy. Our GPU version scales best with larger track lengths and achieves further speedup with more activated DOMs, whereas the number of time bins has no effect on the speedup. An interesting direction for further acceleration could be achieved by replacing the B-spline surface with neural networks that describe the light yield λ for a given light source and detector position. This can potentially achieve an even higher efficiency on GPUs compared to the B-spline approach.

Acknowledgements. Parts of this research were conducted using the supercomputer MOGON and auxiliary services offered by Johannes Gutenberg University Mainz (hpc.uni-mainz.de) which is a member of the AHRP and the Gauss Alliance e.V. This paper is supported by the NVidia GPU Grant Program, which donated a Titan V for this research.

References

1. Aartsen, M.G., et al.: Measurement of atmospheric neutrino oscillations with Ice-Cube. Phys. Rev. Lett. **111**(8) (2013). https://doi.org/10.1103/physrevlett.111.081801
2. de Boor, C.: On "best" interpolation. J. Approx. Theory **16**, 28–42 (1976). https://doi.org/10.1016/0021-9045(76)90093-9
3. Calland, R.G., Kaboth, A.C., Payne, D.: Accelerated event-by-event neutrino oscillation reweighting with matter effects on a GPU. J. Instrum. **9**(04), P04016–P04016 (2014). https://doi.org/10.1088/1748-0221/9/04/p04016
4. Cook, R.J., Lawless, J.: The statistical analysis of recurrent events. In: Statistics for Biology and Health. Springer, New York (2007). https://doi.org/10.1007/978-0-387-69810-6

5. Feroz, F., Hobson, M.P.: Multimodal nested sampling: an efficient and robust alternative to markov chain monte carlo methods for astronomical data analyses. Mon. Not. R. Astron. Soc. **384**(2), 449–463 (2008). https://doi.org/10.1111/j.1365-2966.2007.12353.x

6. Feroz, F., Hobson, M.P., Bridges, M.: MultiNest: an efficient and robust bayesian inference tool for cosmology and particle physics. Mon. Not. R. Astron. Soc. **398**(4), 1601–1614 (2009). https://doi.org/10.1111/j.1365-2966.2009.14548.x

7. Feroz, F., Hobson, M.P., Cameron, E., Pettitt, A.N.: Importance nested sampling and the MultiNest algorithm. arXiv:1306.2144 [astro-ph, physics:physics, stat], 10 June 2013. https://doi.org/10.21105/astro.1306.2144

8. Giunti, C., Kim, C.W.: Fundamentals of Neutrino Physics and Astrophysics. Oxford University Press, Oxford (2007). https://doi.org/10.1093/acprof:oso/9780198508717.001.0001

9. IceCube Collaboration: Energy reconstruction methods in the IceCube neutrino telescope. J. Instrum. **9**(03), P03009–P03009 (2014). https://doi.org/10.1088/1748-0221/9/03/p03009

10. IceCube Collaboration: The IceCube neutrino observatory: instrumentation and online systems. J. Instrum. **12**(3), P03012–P03012 (2017). https://doi.org/10.1088/1748-0221/12/03/P03012

11. IceCube Collaboration: Computational techniques for the analysis of small signals in high-statistics neutrino oscillation experiments (2018)

12. IceCube Collaboration: Measurement of atmospheric tau neutrino appearance with IceCube DeepCore. Phys. Rev. D **99**, 032007 (2019). https://doi.org/10.1103/PhysRevD.99.032007

13. IceCube Collaboration: Development of an analysis to probe the neutrino mass ordering with atmospheric neutrinos using three years of IceCube DeepCore data. Eur. Phys. J. C **80**(1), 9 (2020). https://doi.org/10.1140/epjc/s10052-019-7555-0

14. Jia, Z., Maggioni, M., Staiger, B., Scarpazza, D.P.: Dissecting the NVIDIA volta GPU architecture via microbenchmarking (2018). http://arxiv.org/abs/1804.06826

15. Kallenborn, F., Hundt, C., Böser, S., Schmidt, B.: Massively parallel computation of atmospheric neutrino oscillations on CUDA-enabled accelerators. Comput. Phys. Commun. **234**, 235–244 (2019). https://doi.org/10.1016/j.cpc.2018.07.022

16. Leuermann, M.: Testing the neutrino mass ordering with IceCube DeepCore. Ph.D. thesis, RWTH Aachen University (2018). https://doi.org/10.18154/RWTH-2018-231554

17. van Santen, J., Whitehorn, N.: Photospline: smooth, semi-analytic interpolation for photonics tables. Technical report, University of Wisconsin-Madison, 19 May 2011

18. Verpoest, S.: Search for particles with fractional charges in IceCube based on anomalous energy loss. Ph.D. thesis, Ghent University (2018). https://lib.ugent.be/fulltxt/RUG01/002/479/620/RUG01-002479620_2018_0001_AC.pdf

19. Ypma, T.: Historical development of the Newton-Raphson method. SIAM Rev. **37**(4), 531–551 (1995). https://doi.org/10.1137/1037125

Cache-Aware Matrix Polynomials

Dominik Huber[1], Martin Schreiber[1($\boxtimes$)] [iD], Dai Yang[2], and Martin Schulz[1]

[1] Department of Informatics, Technical University of Munich, Munich, Germany
{domi.huber,martin.schreiber}@tum.de, schulzm@in.tum.de
[2] NVIDIA, Munich, Germany
daiy@nvidia.com

Abstract. Efficient solvers for partial differential equations are among the most important areas of algorithmic research in high-performance computing. In this paper we present a new optimization for solving linear autonomous partial differential equations. Our approach is based on polynomial approximations for exponential time integration, which involves the computation of matrix polynomial terms $(M^p v)$ in every time step. This operation is very memory intensive and requires targeted optimizations. In our approach, we exploit the cache-hierarchy of modern computer architectures using a temporal cache blocking approach over the matrix polynomial terms.

We develop two single-core implementations realizing cache blocking over several sparse matrix-vector multiplications of the polynomial approximation and compare it to a reference method that performs the computation in the traditional iterative way. We evaluate our approach on three different hardware platforms and for a wide range of different matrices and demonstrate that our approach achieves time savings of up to 50% for a large number of matrices. This is especially the case on platforms with large caches, significantly increasing the performance to solve linear autonomous differential equations.

Keywords: Cache-blocking in time dimension · Matrix exponentiation · Higher-order time integration

1 Introduction

Solving time-depending partial differential equations (PDEs) on large-scale supercomputers is extremely resource demanding, yet applications demand the ability to operate on increasingly larger and more complex systems. Consequently, the development of efficient parallel PDE solvers from the mathematical side, as well as their efficient implementation on high-performance computing (HPC) systems is an active area of research. In this work, we investigate optimizations along the time dimension combining new approaches from mathematics and HPC research.

Our main application focus lies on linear autonomous PDEs that occur frequently, e.g., in full waveform inversion problems [9] or as part of splitting methods that incorporate non-linear parts in a separate way [8]. In general, such

© Springer Nature Switzerland AG 2020
V. V. Krzhizhanovskaya et al. (Eds.): ICCS 2020, LNCS 12137, pp. 132–146, 2020.
https://doi.org/10.1007/978-3-030-50371-0_10

PDEs are given by $\frac{\partial U(t)}{\partial t} = \mathcal{L}U(t)$ with $\mathcal{L}$ being the linear operator and $U(t)$ the solution at time t.

In order to solve such systems numerically for a given initial condition $U(0)$, we must apply a discretization. In particular, our presented HPC algorithms target commonly used discretization methods leading to a linear operator directly and explicitly given by a sparse matrix L. This is, e.g., the case when using discretizations based on finite differences or radial basis functions.

Furthermore, the discrete state of the solution at time t is given by $U(t)$, leading to $\frac{\partial U(t)}{\partial t} = LU(t)$. Such a discretization typically results in sparse matrices that are then used in matrix-vector-like computations LU, as it is common in off-the-shelf time integration methods. To provide an example, explicit Runge-Kutta (RK) methods rely on computations of the form $k_i = L\left(t_n + \Delta t c_i, U^n + \Delta t \sum_j a_{i,j} k_j\right)$ with U^n being the approximated solution at time t_n, k_j related to the j-th RK stage, $a_{i,j}$ an entry in the Butcher table (e.g., see [1]) and Δt the time step size as part of the time discretization. However, such a formulation targets non-autonomous systems with the assumption of $\mathcal{L}(t)$ varying over time, e.g., by external time-varying forces, hence involving the dependency on time via $t_n + \Delta t c_i$.

In contrast, the linear PDEs we target in this paper do not involve any time-depending terms and this is indeed the case for many other PDEs (Seismic waves, Tsunami simulations, etc.). This opens up a new branch of matrix-polynomial-based time integration methods of the form $U(t + \Delta t) = \sum_n \alpha_n (\Delta t L)^n U(t)$, which we explore in this paper as the target for our algorithmic HPC optimizations. Similarly to the RK-based methods, these methods rely on matrix-vector products.

For their efficient implementation, though, we need to take modern HPC architectures into account, in particular their cache and memory hierarchy. We, therefore, design and implement a novel temporal cache-blocking scheme over the linear operators L as part of such a matrix polynomial computation. This increases both spatial and temporal locality and leads to a high utilization of the cache resources, leading to a speed-up of up to 50% on some architectures.

Our main contributions are the development of these caching strategies in Sect. 3, an analytical performance model which is presented in Sect. 4 as well as the performance assessment in Sect. 5.

2   Related Work

The growing gap between computational performance and memory access latencies and bandwidth, commonly referred to as the memory wall [12], is one of the fundamental bottlenecks in modern computer architectures. Caches are commonly used to mitigate this problem, but require careful algorithm design to achieve the needed temporal and spatial locality that makes their use efficient. This is particularly true for PDE solvers, which we target in this paper. Algorithm design and optimization is, therefore, an active and wide field of research.

In the following we point out the most relevant related work and contrast it to our approach.

We first discuss very common optimization approaches for spatial dimensions. For matrix-vector and matrix-matrix multiplications, cache-blocking [6] is a well established technique and considered an essential optimization on today's architectures with deep memory hierarchies. For regular grid structures, this technique can be combined with tiling approaches, like spatial tiling [7], to further increase its efficiency. However, so far such optimizations only targeted the execution of a single operation, ignoring potential optimizations across multiple operators.

When considering the time dimension, temporal tiling and wavefront computations, as a generalization of it, has been shown to provide significantly improved performance on modern architectures [2, 11, 13]. In our work we build on this approach of temporal tiling, used for individual SpMvs, and apply it to a series of successive SpMvs, as they occur during the calculation of the matrix potentials $M^p v$ needed for our class of targeted PDEs.

Contrary to stencil computations, our algorithms do not perform blocking over several time steps, but rather several sparse matrix-vector multiplications (SpMvs) computing the polynomial terms (vectors) in every time step. Furthermore, our approach can also be applied out-of-the-box to non-uniform grids. For temporal tiling, this would pose new requirements on data dependencies, as it is based on the explicit use of the regular grid structure.

Within the scope of the project to develop "communication-avoiding Krylov subspace methods" several publications focus on comparable approaches (see Hoemmen [4] and references therein). One particular difference of our work is the application of this technique in polynomial time integration. We also provide two different implementations of this technique, which enable the cache blocking naturally with a very small preprocessing overhead.

3   Cache-Aware Matrix Polynomials

In this section we present two cache-aware algorithms for the calculation of the matrix polynomial terms $M^p v$: the Forward Blocking Method (FBM) and the Backward Blocking Method (BBM). In particular, matrix-polynomial-based time integration demands not only the vector $M^p v$ to be calculated, but rather all vectors $M^k v$, $k \in \{1, \cdots, p\}$, which makes it infeasible to explicitly precompute the matrices M^k before multiplying them with v. Therefore, these vectors are computed by typically successive matrix-vector multiplications with the same matrix M. With y_n denoting the result vector of the calculation of $M^n v$, the vector y_{n+1} is derived as $y_{n+1} = M^{n+1} v = M y_n$. This leads to the intuitive way to calculate $M^p v$ by successive matrix-vector multiplications, i.e., the vectors y_1 to y_p are calculated one after the other. We refer to this as the naive approach.

However, for sufficiently large problem sizes this results in no data reuse between the matrix-vector products, as the data is already evicted from the cache before it can be used again in the next multiplication. To avoid this situation our two methods use a blocking technique that enables the reuse of data over multiple matrix-vector calculations, which borrows some ideas from wavefront strategies in stencil computations. We interpret the vectors y_1 to y_p as one-dimensional domains at time steps 1 to p, similar to one-dimensional stencil computations. While in such stencil computations the dependencies between the time steps are given by the defined stencil. for our calculations these dependencies are defined by the positions of nonzero entries in every row of the matrix. For matrices arising from finite differences or radial basis function discretizations, these positions are usually regionally similar in neighboring rows. Based on this observation, we apply a blocking scheme to the matrix to describe dependencies between whole blocks of the vectors. Our algorithms then construct two-dimensional space-time tiles over the vectors that fit into cache, while simultaneously respecting the dependencies between all blocks of the vectors.

To achieve this, our two methods use two different concepts: FBM starts at the first vector y_1 and calculates successive blocks of a vector y_n until the dependencies for a block on the next vector y_{n+1} are fulfilled. BBM, on the other hand, starts at the last vector y_p and is stepping backwards through the dependency graph in a recursive way to calculate exactly the blocks needed to resolve the dependencies for the current block of y_p. To realize these two concepts, both methods demand distinct information about the dependencies between the vector blocks. Therefore, we use different data structures for the FBM and BBM, as discussed next.

3.1 Extended CSR Matrix Formats

As a basis for our cache-aware matrix polynomial scheme, we extended the CSR matrix storage format to provide additional information about the non-zero block structure of the matrix. The CSR format uses three arrays: the non-zero entries of the matrix in the array `val`, the corresponding column indices in the array `colInd` and the row information as pointers into the two other arrays in the array `rowPtr`. We extended this format by (conceptually) partitioning the matrix into blocks of size $B \times B$, while keeping the underlying data layout of the CSR format. The information about the non-zero block structure is then stored in additional arrays. However, we use different formats for the two methods: while we store the positions of all non-zero blocks for the BBM, only the position of one non-zero block per *blockRow* has to be stored for FBM.

Therefore, for FBM the CSR format is extended by only one additional array of size $\lceil \frac{n}{B} \rceil$ for an $M^{n \times m}$ matrix. In this array the maximum *block-column* index of every *block-row* is stored (see `maxBlockColInd` array in Fig. 1). Hence, only a relatively small overhead of additional data has to be stored and loaded.

The format used by BBM, on the other hand, provides the full information about the non-zero block structure of the matrix. This information is stored in two arrays similarly to the `colInd` and `rowPtr` arrays of the CSR format, but by dealing with all block rows and columns instead of single ones (see Fig. 2). Thus, the `blockRowPtr` array consists of offsets into the `blockColumnIndex` array, indicating the start of a *block row*. The `blockColumnIndex` array contains the

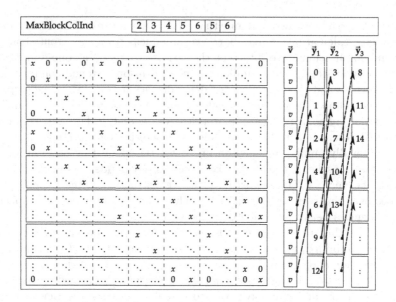

Fig. 1. Forward Blocking Method: This example shows the concept of the FBM for the calculation of $M^3 v$ with sparse matrix M (left), the dense source vector v/y_0 (most left vector) and a block size of $B = 2$. The following vectors y_1, y_2 and y_3 are the destination vectors of the successive SpMv operations needed to calculate $M^3 v$. The *numbers* inside the blocks of the destination vectors denote the order in which they are calculated by the FBM. The arrows indicate the dependencies between the vector blocks as encoded by the `MaxBlockColInd` array.

Algorithm 1. Forward Blocking Method: Calculates $y_p = M^p y_0$, where M is partitioned into *numBlocks* slices represented in the format described in Sec. 3.1

Require: y_0 is the source vector → lastBlockOf(y_0) = numBlocks **and** $y_i, i \in [1, p]$ are
    empty vectors → lastBlockOf(y_i)= −1, neededBlockFor(y_i) = maxBlockColInd[0]
1: **function** FBM(numBlocks)
2:     p0 ← 1
3:     **while** neededBlockFor(y_p) ! = −1 **do**
4:         **for** $i = p0$ to p **do**
5:             **while** lastBlockOf(y_{i-1}) ≥ neededBlockFor(y_i) **and not** $(i + 1 <= p$
  and lastBlockOf(y_i) >= neededBlockFor(y_{i+1})) **do**
6:                 SpMv(y_i, lastBlockOf(y_i)+1)
7:                 lastBlockOf(y_i)++
8:                 **if** lastBlockOf(y_i) < numBlocks −1 **then**
9:                     neededBlockFor(y_i) = maxBlockColInd[lastBlockOf(y_i)+1]
10:                **else**
11:                   p0 + +
12:                   neededBlockFor(y_i)=-1
13:                   **break**

block-column indices of non-zero blocks in a *block-row*. If B_r denotes the number of non-zero $B \times B$ blocks of an $M^{n \times m}$ matrix, the size of the two arrays is given by $\lceil \frac{n}{B} \rceil$ and B_r.

3.2 The Forward Blocking Method

We implemented FBM according to the pseudo code in Algorithm 1. For a better understanding of the underlying concept of this method, we present an example in Fig. 1. Based on this example we describe the algorithm while referring to the corresponding lines of code.

For each vector $\boldsymbol{y}_1$, $\boldsymbol{y}_2$ and $\boldsymbol{y}_3$ we track the information about its last calculated block (starting at -1) and the maximum index of the block of the predecessor vector that is needed to calculate the next block. For simplicity, in Algorithm 1 we compute these values by the function calls $\texttt{lastBlockOf}(\boldsymbol{y}_n)$ and $\texttt{neededBlockFor}(\boldsymbol{y}_n)$, respectively.

Starting at $\boldsymbol{y}_1$, FBM loops through the vectors (Line 2 & 3), thereby calculating blocks of vector $\boldsymbol{y}_n$ by an arbitrary SpMv kernel (Line 5) until one of the following two conditions is reached (Line 4):

- The forward pointer in the last calculated block points to an unfilled block of $\boldsymbol{y}_{n+1}$: this indicates that the currently calculated data can be used to calculate a block of vector $\boldsymbol{y}_{n+1}$ and, therefore, the loop jumps to the next vector to propagate the new data forward.
- The forward pointer to the next block of $\boldsymbol{y}_n$ to be calculated originates from an unfilled block of y_{n-1}: this indicates that there are more blocks of the previous vector(s) needed, so the loop starts again at vector y_1.

When a block of vector $\boldsymbol{y}_n$ with index B_i is calculated, the value of $\texttt{lastBlockOf}(\boldsymbol{y}_n)$ has to be incremented (Line 7) and the new value of $\texttt{neededBlockFor}(\boldsymbol{y}_n)$ can be read from $\texttt{maxBlockColInd}[B_i+1]$ (Line 9). Completely filled vectors are excluded from the loop (Line 11). This loop is repeated until the last block of $\boldsymbol{y}_3$ is filled (Line 1).

The numbers in the vectors in Fig. 1 illustrate the order in which the blocks of the vectors would be calculated by FBM in this example. This order exhibits improved temporal locality on both the matrix and the vectors, compared to successive matrix-vector products, as it traverses the $dim \times p$-plane of the vectors in wavefronts with a certain (constant) wavefront angle, resembling those in stencil computation. It can be observed that the minimum tile size is dependent on the distances between the lowest and highest column index in every row. Hence, for very large distances FBM produces large space-time tiles to respect these dependencies, which impedes cache usage and, among other issues, excludes periodic boundary problems from the application domain of this method.

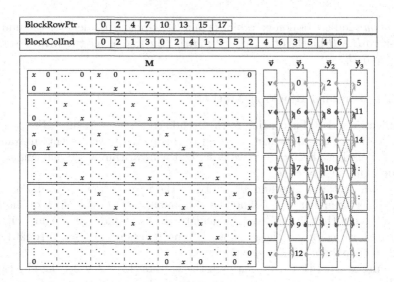

Fig. 2. Backward Blocking Method: This example shows the concept of the BBM for the calculation of M^3v with sparse matrix M (left), the dense source vector $\boldsymbol{v}/\boldsymbol{y_0}$ (most left vector) and a block size of $B = 2$. The following vectors $\boldsymbol{y}_1$, $\boldsymbol{y}_2$ and $\boldsymbol{y}_3$ are the destination vectors of the successive SpMv operations needed to calculate $M^3\boldsymbol{v}$. The numbers inside the blocks of these destination vectors denote the order in which they are calculated by the BBM. The arrows indicate the dependencies between the vector blocks as encoded by the arrays `blockRowPtr` and `blockColInd`. The BBM computes a block of $\boldsymbol{y}_3$ by recursively computing all the (not already computed) blocks of the previous vectors it depends on, e.g., to calculate the first block of $\boldsymbol{y}_3$ ($\boldsymbol{y}_3[0]$) the order of calculated blocks is $\boldsymbol{y}_1[0]$, $\boldsymbol{y}_1[2]$, $\boldsymbol{y}_2[0]$, $\boldsymbol{y}_1[4]$, $\boldsymbol{y}_2[2]$, $\boldsymbol{y}_3[0]$.

Algorithm 2. Calculates block B_i of vector y_n recursively

1: **function** LOOKUPREC(y_n, B_i)
2:    **if** $n > 1$ **then**
3:        **for** index=blockRowPtr[B_i] **to** blockRowPtr[B_{i+1}]−1 **do**
4:            B_{rec} =blockColInd[index]
5:            **if** $\boldsymbol{y}_{n-1}[B_{rec}]$.isEmpty() **then**
6:                lookupRec($vecy_{n-1}$, B_{rec})
7:    SPMV(y_n, B_i)

3.3   Concept of Backward Blocking

Figure 2 shows the concept of BBM. It loops over the blocks of the final result vector y_p ($\boldsymbol{y}_3$ in our example) and calculates the necessary blocks of the previous vectors recursively by calling the functions shown in Algorithm 2. As input parameters, this function takes a vector $\boldsymbol{y}_n$ and the index B_i of the block of this vector that will be calculated. To calculate this block $\boldsymbol{y}_n[B_i]$, all the blocks of $\boldsymbol{y}_{n-1}$ from which pointers lead to $\boldsymbol{y}_n[B_i]$ are needed. The indices of these blocks can simply be read from the entries `blockColInd[blockRowPtr[$B_i$]]` to `blockColInd[blockRowPtr[`B_{i+1}`]-1]` (Line 3). If such a block of $\boldsymbol{y}_{n-1}$ is not calculated, yet, a recursive function call is performed for this block index and vector $\boldsymbol{y}_{n-1}$ (Line 5 & 6). When all necessary blocks are filled, $\boldsymbol{y}_n[B_i]$ can finally be calculated using a SpMv kernel (Line 10). The algorithm is effectively stepping backwards through the dependence graph in a depth first traversal to reach the needed filled blocks and is calculating exactly the required data on the way backtracking forward through the dependence graph.

As above, the correct order of the calculations of the vector blocks improves the temporal locality of the data accesses. For the shown example of a regular grid, BBM calculates blocks of vectors (after a short initialization phase) in the same order as FBM. However, the additional information of the dependencies between the blocks leads to a decisive advantage. As discussed above FBM degenerates to nearly successive SpMv calculations for large distances between non-zeros in one or multiple rows. BBM can "compensate" a small number of such rows, if for a majority of rows these distances are small enough for the space-time tiles to fit into cache. For such cases, BBM breaks the wavefront analogy and only calculates exactly the needed data blocks. This is contrary to FBM, which calculates all blocks of a vector up to the maximum needed block.

4   Analytical Best-Case Performance Model

In order to understand the quality of our proposed solution, we derive an analytical model showing the upper bound for the performance improvements possible with our blocking methods. When calculating $M^p\boldsymbol{v}$ for large problem sizes without out cache blocking, the values of the matrix and vectors have to be loaded from memory for every matrix-vector multiplication. Thus, the time for the naive calculation is given by $T_{naive}(p) = p \times T_{mem}$, where T_{mem} denotes the time needed for an SpMv with no values cached.

Our approaches use cache blocking, hence—in the ideal case—the matrix and vector values are only loaded once from memory and then reside in cache for the rest of the computation. Following this observation, we model the computation time as $T_{blocked}(p) = T_{mem} + (p - 1) \times T_{cache}$, where T_{cache} is the time needed for in-cache SpMvs. The time savings through blocking can then be calculated as $T_{save}(p) = 1 - \frac{T_{blocked}(p)}{T_{naive}(p)}$. Consequently, for increasing exponents of the matrix (p), the expected time savings through blocking converge to $\lim_{p\to\infty} 1 - \frac{T_{mem}+(p-1)\times T_{cache}}{p\times T_{mem}} \approx 1 - \frac{T_{cache}}{T_{mem}}$.

The size of the data that has to fit into the cache to achieve full cache blocking (S_C) is heavily dependent on the specific matrix structure and the exponent of the matrix. I.e., the relation can be described as $S_C \propto R_{nz} \times B_w \times p$, where R_{nz} is the number of nonzero values per row, B_w the distance between the lowest and highest column index per row and p the exponent of the matrix. For regular grid based matrices, the computation order in which the two methods compute the blocks lead to a more accurate approximation of $S_C \approx \frac{B_w}{2}(p(R_{nz}(S_{val} + S_{ind}) + 3S_{val} + S_{ind}) + S_{val})$, where S_{val} is the size of the data type of the matrix/vector values and S_{ind} the size of the data type used for the index and pointer array of the CSR format.

5 Evaluation

5.1 Targeted Hardware Architectures

To analyze the effectiveness of our approach, we evaluate our approaches on three different hardware platforms: XeonBronze, XeonSilver and AMD. Both XeonBronze and XeonSilver are based on the Intel Skylake Architecture; the XeonBronze platform features an Intel Xeon Bronze 3106 8-core processor with a total of 80 GiB of DRAM, and the XeonSilver platform is equipped with two Intel Xeon Silver 4116 12-core processors and a total of 96 GiB of DRAM, arranged equally across all available memory slots to allow for optimal bandwidth. Our AMD platform is built with a single AMD Ryzen Threadripper 2990WX 32-core processor. It is based on the 2nd-generation AMD Zen architecture and features a total of 64 GiB of main memory.

The Intel Skylake [3] features a classical 3-layer cache design (L1I/D, L2 and L3), with each of the layers (L1I/D, L2 and L3) being non-inclusive. L1D and L1I caches are 32 KiB large and 8-way associative, the L2 cache has a size of 1 MiB and is 4-way associative, and all three caches are exclusive to a particular core. The L3, on the other hand, is a shared cache and has a size of 1.375 MiB per core on our reference systems, resulting in a total of 11 MiB (Xeon Bronze) and 16.5 MiB (Xeon Silver) L3 cache shared between the cores of a processor.

On AMD's 2nd-generation Zen architecture (Zen+) the L1I caches are 64 KiB and the L1D are 32 KiB per core and each core also has its own 256 KiB L2 cache. Unlike on Skylake, the L1 caches are full inclusive with respect to the L2 caches. A special design of Zen+ is the so-called CCX consisting of a cluster of 4 cores, which each shares an 8 MiB L3 cache. Two CCXs are located on one die and our reference platform (2990WX) features a total of 4 dies in its package. The dies are interconnected with a high-speed interconnect named *Infinity Fabric*. On the 2990WX, two memory controllers are attached to two of the dies, resulting in 4 NUMA domains, in which two of the domains do not have direct memory access and need to route accesses through another core.

5.2 Matrix Test Suite

The structure of the generated matrices depends on the particular grid, the finite difference order and the boundary condition. To identify the interplay between these parameters and our developed algorithms, we construct two matrix test suites that cover a wide range of combinations of these parameters. We give an overview of these matrices we used for our tests in Tables 1 and 2.

Table 1. Overview of matrices in test suite 1

Test suite 1 (TS 1)		
PDE	$\frac{\partial u}{\partial t} = \alpha \left(\frac{\partial^2 u}{\partial x_1^2} + \cdots + \frac{\partial^2 u}{\partial x_n^2} \right) = \alpha \nabla^2 u$	
FD orders	2, 4, 6, 8	
Boundary condition	Homogenous Dirichlet	
	TS 1a	**TS 1b**
Matrix IDs	0–27	28–55
Dimensionality	2D	3D
Grid dimensions	224, 316, 447, 632, 775, 894, 1000	37, 46, 58, 74, 84, 93, 100

Table 2. Overview of matrices in test suite 2

Test suite 2 (TS 2)			
Matrix IDs	56–136		
PDE	$U(x, t) = c \nabla U(x, t), c = 1$		
FD orders	2, 4, 6, 8		
Boundary condition	Periodic, $U(x, 0) = sin(2\pi(x - x_0))$		
Dimensionality	1D	2D	3D
Grid dimensions	$2^n, n \in [5, 10]$	$2^n \times 2^{n-1}, n \in [6, 10]$	$2^n \times 2^{n-1} \times 2^{n-2}, n \in [7, 10]$

5.3 Benchmark Description and Configuration

We investigate the behavior of our two methods for matrices of TS 1a and 1b. As FBM is not suited for problems with periodic boundary conditions (see Sect. 3.2), we test only BBM for matrices of TS 2.

We run tests using SSE4.2, AVX, AVX2 and AVX512 on the Intel systems and an AVX2 implementation on the AMD system, using block sizes of $B = 2^i, i \in \{6, \cdots, 12\}$. We further use an affinity of the single-threaded program to the core closest to the memory controller on each architecture. Our findings show that the differences in the vector instruction sets and the underlying micro architecture realizing them have a great impact on the performance of SpMvs with the matrices of our test suites: using AVX512 consistently leads to lower performance. Further, the performance of SSE, AVX and AVX2 instructions seem to be highly dependent on the specific matrix, making it difficult to get to a general conclusion on which vector extensions to use. Hence, if not further specified, we use the results of the best performing vector extension for our implementation and the reference method, respectively. We compare the results of our approaches to an implementation of the naive approach, which performs the matrix-vector multiplications sequentially (see Sect. 3). For this, we use the best performing block size evaluated per matrix and exponent. For all occurrences of SpMv calculations, we use the routine `mkl_sparse_d_mv()` of the Intel Math Kernel Library (MKL) [5].

We also use the same library on the AMD processor, although it often is reported to not reach high performance on non-Intel CPU types. Several factors led to this decision: by setting the environment variable MKL_DEBUG_CPU_TYPE=5, the library can be forced to choose the AVX2 code path instead of the default SSE path to which it normally falls back to on non-Intel CPUs. Comparing the performance of the AVX2 code path to other libraries on our AMD architecture, e.g., OSKI [10], we found that for our cases the optimal library is dependent on the specific problem type and size. Moreover, this paper focuses on exploring the general potential of cache-aware algorithms for this type of calculation, rather than achieving overall maximum performance in using SpMvs directly, motivated further by the results in the following section. We therefore stick with MKL on all architectures.

5.4 Results

In this section we present our results of FBM and BBM introduced in Sect. 3. We measure the time needed for the calculation of $M^p v$, $p \in \{2, 3, 4, 5\}$ and compare it to the reference implementation without cache blocking. Our results show improved performance of our blocking methods on all three architectures for a large number of matrices in the test suites.

For the matrices of TS 1a and 1b, both FBM and BBM produce quite similar performance behavior; consequently, these results are shown interchangeably in Fig. 3. For most of the matrices in TS 1a, our approaches outperform the reference method. We achieve time savings of up to 25%/15% on the Intel Xeon Bronze/Silver, respectively, and up to 50% on the AMD Ryzen Threadripper. The matrices in TS 1b lead to slightly less improvements on the Intel processors, but still produced time savings of up to ±15% and ±5%, respectively. On the AMD, we measure greater performance improvements of 40% to 50% for many of these matrices.

Regarding the periodic boundary problems of TS 2, BBM still achieves the same kind of performance improvements as for some matrices resulting from 2D FD grids (Figs. 4).

5.5 Evaluation Compared to the Analytical Model

On all three hardware platforms, we measure the in-L2/L3-cache and in-memory performance for SpMvs with matrices similar to those in the test suites and then use these values in our analytical model as described in Sect. 4. The upper bounds for the time savings of our blocking approach derived from the model are 20%/12% on Intel Xeon Silver, 30%/15% on Intel Xeon Bronze and 55%/50% on the AMD Ryzen Threadripper, which closely resembles our real measured performance.

The performance of our approaches depends on size and structure of the matrix as discussed in Sect. 4. Using cache blocking, these algorithms naturally can only provide significant performance improvements if the matrix itself does not fit into cache. Moreover, the matrix properties B_w and R_{nz} have to be small

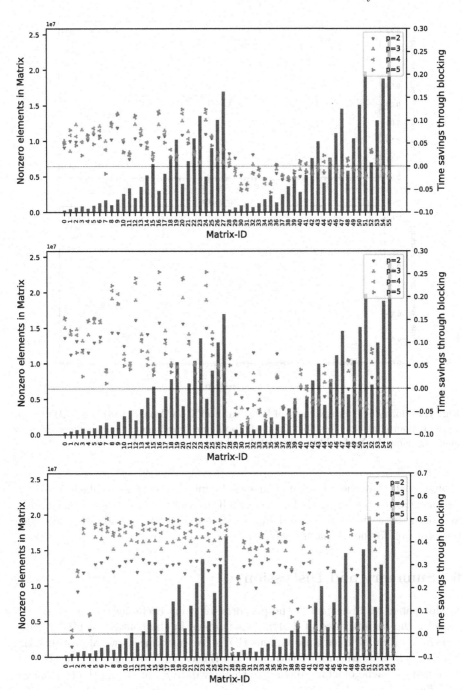

Fig. 3. Performance improvement of BBM/FBM on Xeon Silver (top), Xeon Bronze (mid) and AMD Ryzen Threadripper (bottom) for matrices in TS 1a and 1b (Dirichlet boundary condition): time reduction through blocking using best performing ISA for both, BBM/FBM and the reference method. The number of floating-point operations is $2 \times p \times$ *number of non-zeros.*

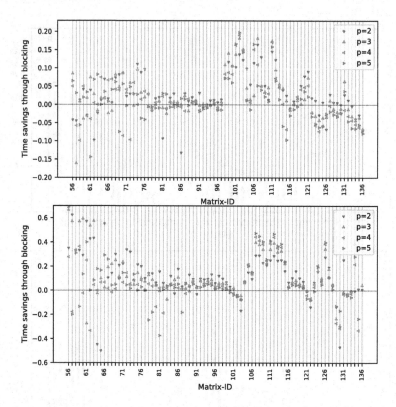

Fig. 4. Performance improvement of BBM on Intel Xeon Bronze (top) and AMD Ryzen Threadripper (bottom) for TS 2 (periodic boundary condition): time savings through blocking using AVX2 for both, the BBM and reference method.

enough such that the space-time tiles do fit into cache. This explains the poor performance of the methods for very small matrices (e.g., matrices 0, 1 and 2) and matrices with large B_w and R_{nz} (e.g., matrices 53, 54 and 55), while they perform well for large sparse matrices.

6   Summary and Discussion

In this paper we investigated the potential of using cache aware algorithms to increase the performance of matrix polynomials in the context of higher-order time integration of linear autonomous PDEs. We introduced two algorithms: the Forward Blocking Method (FBM) and the Backward Blocking Method (BBM), both using a cache blocking technique to allow data reuse during the calculation of $M^p v$.

Our evaluation on three different architectures showed both methods profit from larger and faster caches. Further, our approaches showed improved performance for a large number of matrices of our test suites. These are matrices not fitting into cache, while the generated space-time tiles do. We showed that the ratio of in-cache and in-memory SpMv is a good indicator for upper bounds of the performance improvements of our method to be expected on a specific architecture. This is also the deciding factor, why better results can be observed especially on AMD by blocking for the L3 cache.

Our experiments showed that BBM is the more flexible approach. While FBM is (by design) not suited for periodic boundary problems, our results showed BBM also achieved improved performance for such matrices.

Overall, our results showed promising time savings of both methods compared to the standard approach of successive sparse matrix-vector multiplications. Therefore, our approach is a significant step towards further reducing the wallclock time of higher-order time integrators for linear autonomous partial differential equations.

Future work will extend these algorithms to exploit multi-core architectures. Here, various new challenges will arise, such as possible data races, which ultimately show up for such unstructured problems. However, also opportunities such as the exploitation of additional caches can lead to further performance boosts. Additionally, future work will leverage the performance boosts of the presented algorithms in the context of time integrating PDEs.

References

1. Butcher, J.C.: Implicit runge-kutta processes. Math. Comput. **18**, 86 (1964)
2. Datta, K., Kamil, S., Williams, S., Oliker, L., Shalf, J., Yelick, K.: Optimization and performance modeling of stencil computations on modern microprocessors. SIAM Rev. **51**(1), 129–159 (2009). http://www.jstor.org/stable/20454196
3. Doweck, J., et al.: Inside 6th-generation intel core: new microarchitecture code-named skylake. IEEE Micro **37**(2), 52–62 (2017)
4. Hoemmen, M.F.: Communication-avoiding Krylov subspace methods. Ph.D. thesis, EECS Department, University of California, Berkeley (2010). http://www2.eecs.berkeley.edu/Pubs/TechRpts/2010/EECS-2010-37.html
5. Intel: Intel ® Math Kernel Library Developer Reference, revision: 023 edn. (2019)
6. Nishtala, R., Vuduc, R., Demmel, J., Yelick, K.: When cache blocking sparse matrix vector multiply works and why. Appl. Algebr. Eng. Commun. Comput. **18**, 297–311 (2007). https://doi.org/10.1007/s00200-007-0038-9
7. Rivera, G., Tseng, C.W.: Tiling optimizations for 3D scientific computations. In: Proceedings of the 2000 ACM/IEEE Conference on Supercomputing, SC 2000, p. 32-es. IEEE Computer Society, USA (2000)
8. Ruprecht, D., Speck, R.: Spectral deferred corrections with fast-wave slow-wave splitting. SIAM J. Sci. Comput. **38**(4), A2535–A2557 (2016)
9. Virieux, J., Asnaashari, A., Brossier, R., Métivier, L., Ribodetti, A., Zhou, W.: 6. An introduction to full waveform inversion (2014)
10. Vuduc, R., Demmel, J.W., Yelick, K.A.: OSKI: a library of automatically tuned sparse matrix kernels. J. Phys. Conf. Ser. **16**, 521–530 (2005)
11. Wellein, G., Hager, G., Zeiser, T., Wittmann, M., Fehske, H.: Efficient temporal blocking for stencil computations by multicore-aware wavefront parallelization, vol. 1, pp. 579–586 (2009)

12. Wulf, W., McKee, S.A.: Hitting the memory wall: implications of the obvious. Technical report, USA (1994)
13. Yount, C., Duran, A., Tobin, J.: Multi-level spatial and temporal tiling for efficient HPC stencil computation on many-core processors with large shared caches. Future Gener. Comput. Syst. **92**, 903–919 (2019)

QEScalor: Quantitative Elastic Scaling Framework in Distributed Streaming Processing

Weimin Mu[1,2], Zongze Jin[1(✉)], Weilin Zhu[1], Fan Liu[1,2], Zhenzhen Li[1,2], Ziyuan Zhu[1], and Weiping Wang[1]

[1] Institute of Information Engineering, Chinese Academy of Sciences, Beijing, China
{muweimin,jinzongze,zhuweilin,liufan,lizhenzhen,zhuziyuan,
wangweiping}@iie.ac.cn
[2] School of Cyber Security, University of Chinese Academy of Sciences,
Beijing, China

Abstract. Recently, researchers usually use the elastic scaling techniques as a powerful means of the distributed stream processing systems to deal with the high-speed data stream which arrives continuously and fluctuates constantly. The existing methods allocate the same amount of resources to the instances of the same operator, but they ignore the correlation between the operator performance and resource provision. It may lead to the waste of the resources caused by the over-provision or the huge overhead of the scheduling caused by the under-provision. To solve the above problems, we present a quantitative elastic scaling framework, named QEScalor, to allocate resources for the operator instances quantitatively based on the actual performance requirements. The experimental results show that compared with the existing works, the QEScalor can not only achieve resource-efficient elastic scaling with lower cost, but also it can enhance the total performance of the DSPAs.

Keywords: Data stream processing · Elastic scaling · Random forest

1 Introduction

Recently, the distributed stream processing systems (DSPSs) [1–4] offer a powerful means to extract the valuable information from the data streams in time. We usually use the directed acyclic graph (DAG) [5] to model the data stream processing application (DSPA) in DSPSs. In the DAG, each vertex represents a kind of operations, named as the operator, and each edge represents a data stream between two operators. At the run time, the DSPS initiates a certain number of operator instances for each operator and deploy them on the runtime environment.

Considering the constant fluctuating data stream, we adopt the elastic scaling in the DSPSs, which adjusts the number of the operators dynamically, to satisfy the QoS requirements. There have been many researches on the elastic scaling.

Zacheilas et al. [6] adjusts the number of operators based on the state transition graph. Hidalgo et al. [7] evaluates the processing power of the operator through the benchmarking, and adjusts the number of the operator instances based on the threshold and the workload prediction. Wei et al. [8] only adjusts the CPU frequency of the virtual machines as the workload fluctuates to reduce the energy cost. Marangozova-Martin et al. [9] presents the method to allocate three levels of resources to the operator instances including virtual machines, processes and threads. The above methods allocate the same amount of resources to the instances of the same operator, but they ignore the correlation between the operator performance and resource provision. Actually, the unreasonable resources provision for the operator instance will cause some severe problems. For example, over-provision will result in a waste of resources. Besides, the under-provision means that the DSPSs will create a lot of instances to achieve high processing performance, which results in the huge overhead of the scheduling and the state transition.

In this paper, we present a quantitative elastic scaling framework, named QEScalor, to allocate the resources for the operator instances quantitatively based on the actual performance requirements. This framework firstly builds the operator performance and resource provision model (OPRPM), then it generates the low-cost elastic scaling plan based on the OPRPM. The contributions of this paper are as follows:

- We use the QEScalor, which first considers the correlation between the operator performance and resource provision, to enhance the performance of the resource provision.
- We propose an online algorithm, named DSA. It learns the correlation samples of the operator performance and resource provision based on the gradient strategy and re-sampling mechanism. Besides, we use the random forest regression model (RFR) [10] to build the OPRPM to get the suitable resource provision options for the operator performance requirement.
- We present a quantitative and cost-based elastic scaling algorithm, QCESA. It refers to the prediction of the workload [11] and the operator performance [12] to generate the low-cost scaling plan based on the OPRPM to achieve the resource-efficient elastic scaling to improve the performance.
- We implement the QEScalor as a key part of our DataDock [11]. The experiment results show that our QEScalor can enhance the performance with the lower scaling cost on the real-world datasets.

We organize the rest of our paper as follows. Section 2 describes the design of QEScalor. Section 3 shows the experimental results of our framework. Finally, Sect. 4 concludes our paper.

2 System Design

2.1 Overview

In this subsection, we describe our framework QEScalor in detail in Fig. 1, which contains three modules: the online operator performance sampler (OOPSer), the operator performance and resource provision modeler (OPRPMer) and the quantitative elasticity controller (QECer). As is shown in Fig. 1, the QEScalor is an important module in the DataDock. In our previous work [11], we present our distributed stream processing system, DataDock, mainly aiming at processing the heterogeneous data in real-time.

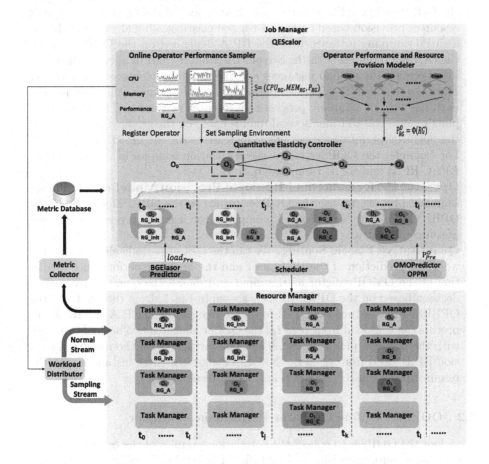

Fig. 1. QEScalor architecture

We use the OOPSer to learn the correlation samples of the operator performance and resource provision online. Then we use these samples as the input of the OPRPMer to build the operator performance and resource provision model (OPRPM). At last, we use the QECer to adjust the scaling plan according to the workload prediction of the BGElasor [11], the operator performance prediction of the OMOPredictor [12] and the OPRPM.

As is shown in Fig. 1, we take the DSPA including the new operator O_1 as an example to describe the work process of the QEScalor. The process contains three main stages.

- **Correlation Samples Online Learning.** When we use the QECer to execute the DSPA including the new operator O_1 at time t_0, it registers the O_1 on the OOPSer. The QECer starts enough instances of the O_1 with the default resources provision based on the cost-based elastic-scaling algorithm [11] and a single sampling instance with the RP_A depended on the OOPSer. Then the OOPSer interacts with the Stream Distributor (SD) to allocate the workload between the normal instances and the sampling instance. The OOPSer continues to collect the operator performance and resource utilization metrics until the current sampling process convergences at time t_i. The above sampling process will continue several rounds based on the gradient strategy until the operator performance no longer increases with the growth of resources. For example, the OOPSer learns the correlation samples of resource group RP_A, RP_B and RP_C from time t_0 to t_k.
- **Operator Performance and Resource Provision Modeling.** During the sampling process, we use the OOPSer to invoke the OPRPMer to build the OPRPM using the RFR, when it completes the learning process of one kind of source provision.
- **Quantitative Elastic Scaling.** We run the QECer periodically. It takes the workload prediction of the BGELasor and the operator performance prediction of the OMOPredictor as the input and makes the quantitative scaling decision based on the OPRPM. As is shown in Fig. 1, from time t_0 to t_j, the OPRPMer has learned the OPRPM of resource provision RP_A and RP_B. To process the workload from t_j to t_k, the QECer allocates two normal instances with the resource provision RP_A and one instance with the resource provision RP_B. And from t_k to t_l, the QECer allocates three instances with the resource provision RP_A, RP_B and RP_C respectively.

2.2 OOPSer: Online Operator Performance Sampler

We adopt the OOPSer to collect the correlation samples of the operator performance and resource provision to build the OPRPM. We focus on the two types of resources: the CPU and the memory. We do not consider the network bandwidth in our work, because the network bandwidth is more sufficient and cheaper than the CPU and the memory in the data center. Therefore, we only consider the correlation samples learning of two types of operators [12]: the compute-intensive Operator (COperator) and the compute-intensive operator mixed with the memory I/O (CMOperator).

We propose the Dynamic-Sampling-Algorithm (DSA) to collect the correlation samples online. We show it in Algorithm 1. The sampling process consists of three steps: Firstly, we create the sampling operator instance with specific resource provision. Secondly, we do not stress test on the sampling operator instance until the performance of the operator instance converges. Thirdly, we continuously collect the correlation samples during the test. Since we spend some time in stress testing, it will take a long time to complete the sampling. In order to speed upsampling, we dynamically adjust the sampling step according to the gradient of the performance change of the sampling operator instance. Meanwhile, we use the re-sampling method to add sampling points when the operator performance fluctuates to improve accuracy.

We are processing the online workload while sampling with the normal running of the DSPAs. We dynamically allocate the workload between the normal instances and the single sampling instance. It can reduce the time and resource overhead obviously compared with running the sampling alone.

When we complete the sampling, we get the correlation sample set $CS_o = (cs_o^0, cs_o^1, ..., cs_o^i)$, where $cs_o^i = (cpu_o^i, mem_o^i, p_o^i)$ is the correlation sample of the resource provision (cpu_o^i, mem_o^i) and the corresponding operator performance p_o^i. We use the CS_o to build the OPRPM of the operator o. However, the performance of operators fluctuates during the life cycle of the DSPA. With the online correlation samples learning mechanism, we can continuously collect the samples and update the OPRPM when the performance of the operator is inconsistent with the OPRPM.

Algorithm 1. DSA

---

sampling(start, step, delta)

1: stopCount = 0
2: step = step.min
3: cur = start
4: sample(cur)
5: **while** stopCount < maxStopCount **do**
6:    cur += step
7:    sample(cur)
8:    rate = (curload - lastload) * minStep / (cur - last) / lastload
9:    **if** abs(rate) < delta.stop **then**
10:       stopCount.increase()
11:    **else**
12:       stopCount = 0
13:    **end if**
14:    **if** rate < 0 or rate > upDelta **then**
15:       re-sample (sample.last , cur)
16:    **else**
17:       step = min(step.min * delta.up / rate), step.max)
18:    **end if**
19: **end while**

2.3   OPRPMer: Operator Performance and Resource Provision Modeler

We use the OPRPMer to build the model of the operator performance and the resource provision, with which we can predict the operator performance based on the given resource provision.

The correlation between operator performance and resource provision is commonly complex and nonlinear. The linear regression model can not capture the latent features of the correlation well, resulting in bad prediction. Besides, in our scenario, the correlation samples set CS_O is commonly small. Using the single nonlinear regression model, like the SVR [13], leads to overfitting easily. The ensemble learning model can improve the robustness of prediction by integrating many weak classifiers, which is more suitable for small sample learning. We adopt the random forest regression (RFR) model in the OPRPMer to capture the nonlinear correlation between the operator performance and the resource provision. According to the experiments, compared to the boosting models, such as the Adaboost [14], GBDT [15] and XGBoost [16], the RFR model performs better. Because the bootstrap strategy adopted by RFR model can avoid overfitting effectively when the sample set is small.

We take the correlation sample set CS_O learned by the OOPSer as the input to build the model in the OPRPMer. When invoked by the QECer, the OPRPMer takes $r = (cpu, mem)$ as input to get the operator performance prediction p_o corresponding to the r.

2.4   QECer: Quantitative Elasticity Controller

In this section, we build the QECer, which can ensure the end-to-end latency with the minimum elastic-scaling cost. It contains two parts: the Cost Model and Quantitative & Cost-based Elastic Scaling Algorithm (QCESA).

Cost Model. We build a cost model to evaluate the total cost of all elastic-scaling actions for an operator from the current epoch S to the future epoch F. The total cost $W_o(\mathbf{Ins})$, the startup times $C^u_{o \cdot t}(Ins)$ and the shutdown times $C^d_{o \cdot t}(Ins)$ are defined as:

$$\min \quad W_o(\mathbf{Ins}) = \min(\sum_{t_S}^{t_F} \sum_{r \in Res} |Ins^t_{o \cdot r}| p^r_o + \sum_{t_S}^{t_{F-1}} (p^u_o C^u_{o \cdot t}(Ins) + p^d_o C^d_{o \cdot t}(Ins)))$$

$$\text{s.t.} \quad \sum_{r \in Res} |Ins^t_{o \cdot r}| Perf_{o \cdot r} \geq Workload_t, \quad \forall t \in [t_S, t_F]$$

$$\tag{1}$$

$$C^u_{o \cdot t}(\mathbf{Ins}) = \sum_{r \in Res} \max(0, Ins^{t+1}_{o \cdot r} - Ins^t_{o \cdot r}) \tag{2}$$

$$C^d_{o \cdot t}(\mathbf{Ins}) = \sum_{r \in Res} \max(0, Ins^t_{o \cdot r} - Ins^{t+1}_{o \cdot r}) \tag{3}$$

where p_o^r is the cost of system resources used by the single instance with resource r for the operator o. $|Ins_{o \cdot r}^t|$ is the instance number of operator o with resource r at time t. In addition, $Perf_{o \cdot r}$ denotes the performance of each operator and $\sum_{r \in Res} |Ins_{o \cdot r}^t| Perf_{o \cdot r}$ denotes the total performance of operator o at time t. $Workload_t$ is the workload at epoch t. In order to satisfy the end-to-end latency, we ensure that the performance of each operator is not less than the workload. In other word, $\sum_{r \in Res} |Ins_{o \cdot r}^t| Perf_{o \cdot r} \geq Workload_t$ at any time. And p_o^u is the startup-cost of a single o instance. p_o^d is the shutdown-cost of a single o instance.

QCESA. To solve this expression $min(W_o)$, we propose the Quantitative and Cost-based Elastic Scaling Algorithm (QCESA). We show it in Algorithm 2. The QCESA considers not only the cost of instance startup and shutdown, but also the correlation of operator performance and resource provision. We use the QCESA to balance these parts of the cost to guarantee a low cost.

At first, we use the QCESA to compute the max workload $workload_{max}$ during $t \in [t_C, t_F]$. Then use it to calculate all candidates at all time $t \in [t_C, t_F]$. Each candidate is a combination of instances with different resource provision and instance number, of which the total performance $Perf_{cand \cdot total}^t \in [Workload_t, Workload_{max}]$. At last, we use dynamic programming to calculate the minimal cost.

3   Experiments

3.1   Environment

Settings. Our experiments run on Kubernetes (K8S) cluster, which we use as the Resource Manager on the DataDock, including eight servers. The version of K8S is 1.14.1. There are two types of servers in the K8S cluster: two GPU servers and six CPU servers. Each GPU server comprises 36 cores Intel Xeon CPU E5-2697 v4 2.30 GHz, 256 GB memory, two NVIDIA GeForce GTX 1060ti cards, and 500 GB disks. Each CPU server comprises 36 cores Intel Xeon CPU E5-2697 v4 2.30 GHz, 256 GB memory, and 500 GB disks. We use the GPU servers to run the JobManager, conducting the training and evaluation. We adopt the CPU servers to run the Task Manager, in which the operator instance runs. Besides, we conduct the evaluation of the OPRPMer with sklearn 0.22.1 running on python 3.7.

Datasets. In Table 1, we show our datasets and the intermediate results of our model at different stages. Firstly, we present the We use the real online workload processed by the DataDock in a day as the original dataset (OriWL-1day). In OOPser, we use two sampling algorithms which are FSSA and DSA to sample OriWL-1day. F1, F2, F3, F4 and F5 denote different steps of FSSA. A, B, C and D represent that the original dataset is processed by these operators. CO1 denotes CPU operator and CMO1 denotes CPU-Memory operator. In OPRPMer, we use DSA-A/B/C/D as the train set and use F1-A/B/C/D as the test set. Then we obtain the output of the random forest regression (RFR),

---

Algorithm 2. QCESA

---

schedule(load, res)

1: cand = loopCandidatesAtEachTime(load.max, load)
2: **if** load.size == 1 **then**
3:     plan.cost = res.start.cost
4:     plan.setplan(res.start.index, res.start)
5:     return plan
6: **end if**
7: **if** load.size == 2 **then**
8:     **if** load[0] < load[1] **then**
9:         plan.cost = res.end.cost + calcWarmup(res.start , res.end)
10:        plan.setplan(res.start.index, res.end)
11:        return plan
12:    **else**
13:        plan.cost = res.start.cost
14:        plan.setplan(res.start.index, res.start)
15:        return res.end
16:    **end if**
17: **end if**
18: plan.cost=max
19: **for** cur : load **do**
20:    **for** cand :cur.candidates **do**
21:        lplan = schedule(load.before(cur), res.with(cand))
22:        rplan = schedule(load.after(cur), res.with(cand))
23:        **if** plan.cost > lplan.cost+rplan.cost **then**
24:            plan.setplan(lplan,cur,rplan)
25:            plan.cost = lplan.cost+rplan.cost
26:        **end if**
27:    **end for**
28: **end for**
29: return plan

loopCandidatesAtEachTime(maxload, load)

1: **for** curload : load **do**
2:    **for** cand : res **do**
3:        **if** curload < cand.load < maxload **then**
4:            candidates.add(curload, cand)
5:        **end if**
6:    **end for**
7: **end for**
8: return candidates

DSA-RFR-A/B/C/D. In QECer, we should evaluate the system performance. From OPRMER, we use DSA-RFR-A/B/C/D as the input. From OMOPredictor, we adopt OP-PM-30 as the input, which represents the operator performance on Datadock online for 30 days. For the input of the BGElasor, we use OriWL-60days-FlowStat which represents the flow statistics for 60 days of data load on the DataDock online.

Table 1. The datasets description

Stage	IN or OUT		Datasets						
OOPSer	IN		OriWL-1day						
	OUT	CO1[A]	F1-A	F2-A	F3-A	F3-A	F5-A	DSA-A
		CO2[B]	F1-B	F2-B	F3-B	F3-B	F5-B	DSA-B
		CMO1[C]	F1-C	F2-C	F3-C	F3-C	F5-C	DSA-C
		CMO2[D]	F1-D	F2-D	F3-D	F3-D	F5-D	DSA-D
OPRPMer	IN	Train	DSA-A/B/C/D						
		Test	F1-A/B/C/D						
	OUT		DSA-RFR-A/B/C/D						
QECer	IN	From OPRMer	DSA-RFR-A/B/C/D						
		From OMOPredictor	OP-PM-30						
		From BGElasor	OriWL-60days-FlowStat						
	OUT		Resource Allocation Plan						

3.2 Performance Evaluation

In our algorithm, we guarantee the latency to reduce the total cost. Thus, we evaluate the Quantitative Elasticity Controller from two aspects: the total cost and the end-to-end latency guarantee. We use the Cost-Balance-Algorithm (CBA) [11] as the baseline algorithm. The CBA considers the running cost and the operation cost. Compared to the CBA, the QCESA takes the operator resource provision into account.

The performance of QCESA depends on the sampling of OOPSer and the predicted results of OPRPMer. For the OOPSer, we compare our method, DSA, with the Fixed-Step-Sampling-Algorithm (FSSA) to demonstrate that DSA is more accurate in the sampling stage to enhance scheduling accuracy and reduce the cost. For evaluating the OPRPMer, we compare the random forest regression model (RFR) with the following methods: Adaboost, GBDT and XGBoost, to demonstrate that RFR is more suitable for the current application scenarios. It can get more accurate prediction results and affect the overall performance of scheduling.

Total Cost. In this part, we take the CMO1 as an example to compare by using the total cost of elastic scaling. We use the workload prediction to generate the scaling plan for the CMO1 and calculate the total cost.

Moreover, to evaluate the effectiveness of the QCESA, we use four different resource provisions to test the CBA respectively. The four resource provision granularities are as follows: 1) $r_1 = (cpu = 0.6 * core, mem = 33.2\,\text{MB})$, 2) $r_2 = (cpu = 1.2 * core, mem = 33.3\,\text{MB})$, 3) $r_3 = (1.8 * core, mem = 33.5\,\text{MB})$, 4) $r_4 = (cpu = 2.4 * core, mem = 33.8\,\text{MB})$.

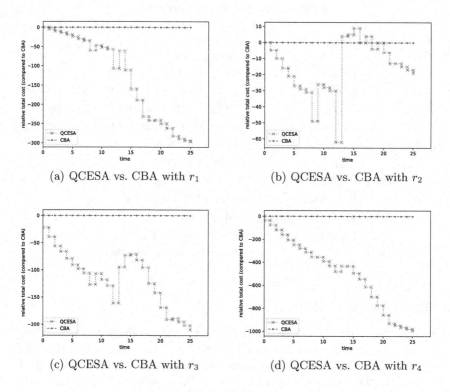

(a) QCESA vs. CBA with r_1 (b) QCESA vs. CBA with r_2

(c) QCESA vs. CBA with r_3 (d) QCESA vs. CBA with r_4

Fig. 2. Total cost

In Fig. 2, we can find that most of the time, the total cost of the QCESA is less than that of the CBA. Besides, as the system runs, the performance of the QCESA is becoming much higher than the CBA.

End-to-End Latency Guarantee. In this part, we focus on the end-to-end latency guarantee. We still take the CMO1 as an example to run on the DataDock and monitor the end-to-end latency.

In Fig. 3, we can see that both the QCESA and the CBA can guarantee that the performance of the operator is no less than the workload. And the end-to-end latency always stays stable and satisfies the requirement of the QoS. The reason is that both algorithms start the instances before the workload rises. Thus they can process the workload timely.

Impact of Sampling. To measure the impact of sampling, we compare the DSA with the Fixed-Step-Sampling-Algorithm (FSSA) in OOPSer. We run the FSSA and the DSA separately to collect the correlation samples of the four operators. For each operator, we run the DSA with the sampling step set to 1 and use the sampling result as the baseline. Besides, we also run the FSSA with the sampling step set to 2, 3, 4 and 5 as the contrast evaluation.

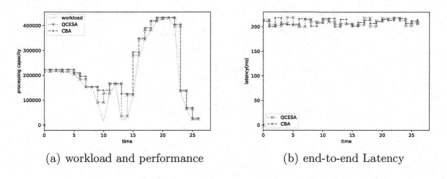

(a) workload and performance   (b) end-to-end Latency

Fig. 3. QoS guarantee

After we get all correlation samples of four operators with the FSSA and the DSA, we use the OPRPMer to build the OPRPMs. Then, we predict the operator performance based on the minimum sampling step using the OPRPMs. We use the Root Mean Square Errors (RMSE), the Mean Absolute Errors (MAE) and the Sampling Number (SN) to evaluate the effectiveness of the DSA. $RMSE = \sqrt{\frac{1}{n}\sum_{i=1}^{n}(x_i - \hat{x}_i)^2}$, $MAE = \frac{1}{n}\sum_{i=1}^{n}|x_i - \hat{x}_i|$, where x_i is the operator performance of baseline, and $\hat{x}_i$ is the predicted operator performance.

Table 2. The performance of each sampling method

Algorithm	Operator type											
	COperator						CMOperator					
	CO1			CO2			CMO1			CMO2		
	RMSE	MAE	SN	RMSE	MAE	SN	RMSE	MAE	SN	RMSE	MAE	SN
$FSSA_2$	0.0047	0.0596	16	0.0031	0.0393	17	0.0058	0.0519	12	0.0199	0.1231	19
$FSSA_3$	0.0049	0.0455	11	0.0015	0.0305	12	0.0074	0.0613	9	0.0067	0.0563	13
$FSSA_4$	0.0066	0.0599	9	0.0033	0.0521	9	0.0247	0.1434	7	0.0233	0.0924	10
$FSSA_5$	0.0237	0.1131	7	0.0026	0.0423	8	0.0192	0.1157	6	0.0161	0.0938	9
DSA	**0.0046**	**0.0568**	**14**	**0.0022**	**0.0402**	**11**	**0.0056**	**0.0525**	**15**	**0.0112**	**0.0911**	**9**

As is shown in Table 2, we can observe that the performance of the FSSA is not stable. When the step of the FSSA is 2, the CO1 and the CMO1 get the best performance. But when the step of the FSSA is 3, the CO2 and the CMO2 get the best performance. And the DSA performs well for all the four operators. Its performance is close to or even reaches the best performance. Moreover, the DSA has fewer sampling numbers when reaching the same performance. It benefits from the dynamical sampling strategy.

We show the sampling result of the CO1 and the CMO2 in Fig. 4. We can see where the performance fluctuates obviously, the sampling step is close to the minimum sampling step. Instead, when the performance changes smoothly, the DSA only uses a few sampling points to capture the main characteristics of performance changes.

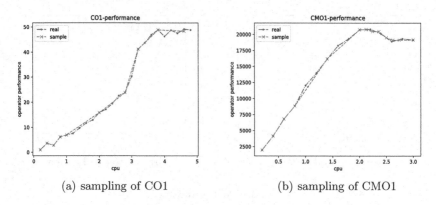

(a) sampling of CO1   (b) sampling of CMO1

Fig. 4. Sampling result

Comparison of Prediction Methods. To enhance the total performance, we should select the better prediction method for the OPRPMer. So we compare the random forest regression model with the following methods: Adaboost, GBDT and XGBoost, to demonstrate the effectiveness of the RFR model in this scenario.

We use the RMSE and the MAE to evaluate the performances of each model. There are several hyper-parameters in these approaches, we use the grid search and 10-folds cross-validation to select the key hyper-parameters. Besides, we normalize all the input to the range [0,1] using the Min-Max scaler. We repeat the experiment 10 times for each model to reduce the random experimental error and take the average of the whole test results as the final result.

As for the RFR, we set $bootstrap = True, criterion = 'mse', max\_featur-es = 'auto', min\_samples\_leaf = 1, min\_samples\_split = 2, n\_estimators = 100$. As for the SVR, we set $kernel = 'rbf', gamma = 'scale', C = 1.0$. As for the Adaboost, we set $base\_estimator = None, learning\_rate = 1.0, loss = 'linear', n\_estimators = 50$. As for the GBDT, we set $n\_estimators = 100, crit-erion = 'friedman\_mse', max\_features = None, min\_samples\_leaf = 1$. As for the XGBoost, we set $booster = 'gbtree', learning\_rate = 0.1, max\_depth = 3, n\_estimators = 100$.

Table 3. The performance of each models

Model	Operator type							
	COperator				CMOperator			
	CO1		CO2		CMO1		CMO2	
	RMSE	MAE	RMSE	MAE	RMSE	MAE	RMSE	MAE
SVR	0.0924	0.0833	0.1048	0.1040	0.0833	0.0819	0.0833	0.0819
Adaboost	0.0678	0.0514	0.0469	0.0390	0.0622	0.0599	0.0598	0.0575
GBDT	0.0687	0.0536	0.0550	0.0361	0.0821	0.0646	0.0706	0.0565
XGBoost	0.0722	0.0561	0.0522	0.0431	0.0986	0.0860	0.0986	0.0860
RFR	**0.0435**	**0.0372**	**0.0220**	**0.0197**	**0.0644**	**0.0533**	**0.0477**	**0.0399**

Table 3 shows that the effectiveness of the ensemble learning is significantly better than the SVR. Because in our scenario, the size of the correlation sample set is smaller, the advantage of the ensemble learning is more prominent.

Besides, there is not much difference between the Adaboost, GBDT, and XGBoost. However, compared to the above boosting models, the RFR performs better. Because the RFR model adopts the bootstrap strategy, it can effectively prevent overfitting when the size of the sample set is small.

4   Conclusion

In this paper, we present a quantitative elastic scaling framework, named QEScalor, to allocate resources for the operator instances quantitatively based on the actual performance requirements. It contains three key modules: the OOPSer, the OPRRMer and the QECer. Firstly, we use the OOPSer to learn the correlation samples of the operator performance and resource provision online. Then we use these samples as the input of the OPRPMer to build the operator performance and resource provision model (OPRPM) by using the random forest regression model. At last, we use the QECer to adjust the scaling plan according to the real workload fluctuation. The experimental results show that, compared with the state-of-the-art methods, the QEScalor is better on the real-world datasets. And we can address the problem which ignores the correlation between the operator performance and resource provision.

Acknowledgements. This work is supported by the National Key Research and Development Plan (2018YFC0825101).

References

1. Chandrasekaran, S., et al.: TelegraphCQ: continuous dataflow processing. In: Proceedings of the 2003 ACM SIGMOD International Conference on Management of Data, p. 668 (2003)
2. Abadi, D.J., et al.: The design of the borealis stream processing engine. In: CIDR 2005, pp. 277–289 (2005)
3. Neumeyer, L., Robbins, B., Nair, A., Kesari, A.: S4: distributed stream computing platform. In: ICDMW 2010, pp. 170–177 (2010)
4. Carbone, P., Katsifodimos, A., Ewen, S., Markl, V., Haridi, S., Tzoumas, K.: Apache flink^{TM}: stream and batch processing in a single engine. IEEE Data Eng. Bull. **38**(4), 28–38 (2015)
5. Storm. http://storm.apache.org/
6. Zacheilas, N., Kalogeraki, V., Zygouras, N., Panagiotou, N., Gunopulos, D.: Elastic complex event processing exploiting prediction. In: 2015 IEEE International Conference on Big Data, Big Data 2015, pp. 213–222 (2015)
7. Hidalgo, N., Wladdimiro, D., Rosas, E.: Self-adaptive processing graph with operator fission for elastic stream processing. J. Syst. Softw. **127**, 205–216 (2017)
8. Wei, X., Li, L., Li, X., Wang, X., Gao, S., Li, H.: Pec: Proactive elastic collaborative resource scheduling in data stream processing. IEEE Trans. Parallel Distrib. Syst. **30**(7), 1628–1642 (2019)

9. Marangozova-Martin, V., Palma, N.D., El-Rheddane, A.: Multi-level elasticity for data stream processing. IEEE Trans. Parallel Distrib. Syst. **30**(10), 2326–2337 (2019)

10. Breiman, L.: Random forests. Mach. Learn. **45**(1), 5–32 (2001). https://doi.org/ 10.1023/A:1010933404324

11. Mu, W., Jin, Z., Wang, J., Zhu, W., Wang, W.: BGElasor: elastic-scaling framework for distributed streaming processing with deep neural network. In: Tang, X., Chen, Q., Bose, P., Zheng, W., Gaudiot, J.-L. (eds.) NPC 2019. LNCS, vol. 11783, pp. 120–131. Springer, Cham (2019). https://doi.org/10.1007/978-3-030-30709-7_10

12. Mu, W., Jin, Z., Liu, F., Zhu, W., Wang, W.: OMOPredictor: an online multi-step operator performance prediction framework in distributed streaming processing (2019). unpublished thesis

13. Drucker, H., Burges, C.J.C., Kaufman, L., Smola, A.J., Vapnik, V.: Support vector regression machines. In: Mozer, M., Jordan, M.I., Petsche, T. (eds.) Advances in Neural Information Processing Systems, NIPS 1996, vol. 9, pp. 155–161. MIT Press (1996)

14. Freund, Y., Schapire, R.E.: A desicion-theoretic generalization of on-line learning and an application to boosting. In: Vitányi, P. (ed.) EuroCOLT 1995. LNCS, vol. 904, pp. 23–37. Springer, Heidelberg (1995). https://doi.org/10.1007/3-540-59119-2_166

15. Friedman, J.H.: Greedy function approximation: A gradient boosting machine. Ann. Stat. **29**(5), 1189–1232 (2001)

16. Chen, T., Guestrin, C.: XGBoost: a scalable tree boosting system. In: Proceedings of the 22nd ACM SIGKDD International Conference on Knowledge Discovery and Data Mining, pp. 785–794 (2016)

From Conditional Independence to Parallel Execution in Hierarchical Models

Balazs Nemeth[1]([✉]), Tom Haber[1,2], Jori Liesenborgs[1], and Wim Lamotte[1]

[1] Hasselt University - tUL, Expertise Centre for Digital Media,
Martelarenlaan 42, 3500 Hasselt, Belgium
{balazs.nemeth,tom.haber,wim.lamotte}@uhasselt.be
[2] Exascience Lab, Imec, Kapeldreef 75, 3001 Leuven, Belgium

Abstract. Hierarchical models describe phenomena by grouping data into multiple levels. Due to the size of these models, parallel execution is required to avoid prohibitively long computing time. While it is occasionally possible to specify some of these models using parallel building blocks, this limits expressivity. Therefore, a more general generative specification is preferred. To leverage parallel computing capacity, these specifications can be annotated, but doing so effectively assumes that the modeler has expertise from computer science. This paper outlines how to identify parallel parts automatically by leveraging the conditional independence property in the graphical model extracted from the dataflow graph of model specifications. Computation related to random variables with the same depth in the graphical model are identified as candidates for parallel execution. Since subsequent proposals in the parameter space exploration of the model are clustered together, the results show that the well known longest processing time scheduling heuristic deals adequately with load imbalance. The proposed parallelization is evaluated on two pharmacometrics models, a domain where hierarchical models with load imbalance are common due to the numeric simulation of pharmacokinetics and pharmacodynamics of human subjects. The varying number of measurements taken per subject further exacerbates load imbalance.

Keywords: High performance computing · Descriptive language · Probabilistic modelling · Automatic parallelization · Dataflow · Hierarchical models

1 Introduction

In recent years, the physical limits that would otherwise prohibit Moore's Law's predicted performance increase have been circumvented by the trend towards more parallel systems [7]. The flip-side of this explicit form of parallelism is that it puts more of the burden on the software developers, or even the users. It is far from straightforward to leverage all the compute power available in parallel systems [10,13], but the complexity of the models precludes fitting a model on a single processor since it is too time intensive in practice. In the context of computational modeling, there are two prominent strategies for parallelization.

© Springer Nature Switzerland AG 2020
V. V. Krzhizhanovskaya et al. (Eds.): ICCS 2020, LNCS 12137, pp. 161–174, 2020.
https://doi.org/10.1007/978-3-030-50371-0_12

First, in some cases, as models are fit using an iterative optimization routine, multiple processors can be kept busy within each iteration. Computation fits into the Bulk Synchronous Parallel (BSP) model of parallel computing [16] with multiple candidate parameters evaluated concurrently. While this approach allows hiding the parallel constructs within the routine, improving the usability of these routines for scientists from other domains requires the optimization routine to be designed to run in parallel, which might not be feasible.

Second, depending on the model, a single candidate parameter can be evaluated in parallel. This strategy is suitable both for more sequential optimization routines as well as parallel optimization routines where it can further improve performance. Even if a task can be decomposed into smaller concurrently executable tasks, doing so manually is tedious and error-prone even when armed with the right parallel computing background. Arguably, the scientists concerned with building these models are in an even worse position; their expertise is probably not in parallel computing and a more automated approach, like the one explored here, is preferable.

The focus here is on the parallelization of hierarchical models composed of multiple interconnected levels. Computational tasks required for each model evaluation are typically spread across relatively few layers. Consequently, this brings with it the opportunity to execute each level in parallel. While it might not be the optimal parallelization, it turns out that it performs well in practice. It can even be used in conjunction with other methods that search for more fine-grained parallelism [20]. The main contribution is to show how to extract the graphical model representation from the dataflow graph of the model and how to map parallelism from the former to the latter.

When the number of tasks exceeds the number of processors in a layer, some processors will inevitably execute more than one task. Depending on the variability of execution times between these tasks and the ratio between the number of tasks and processors, neglecting the scheduling problem can result in inefficient use of the underlying hardware. The parallelization approach is augmented with the well-known Longest Processing Time (LPT) static scheduling heuristic [9], where independent jobs with varying execution time are scheduled on p identical processors.

The reachable efficiency is model-dependent; in general, the more compute-intensive tasks are available at each level of the hierarchy, the better performance will scale. Therefore, two different models are considered for evaluation: one containing only a few tasks and another with many more compute-intensive tasks. While parallelization adds overhead introduced by inter-processor communication, overall run time decreases in both cases.

The remainder of this paper is structured as follows. Section 2 references related work. Section 3 discusses hierarchical models, their structure in the dataflow graph representation and the relationship with conditional independence. Section 4 describes the parallelization approach. Section 5 discusses performance results. Section 6 provides future work directions and concludes the paper.

2   Related Work

The input to the optimization routines or sampling algorithms is a function that evaluates a model and returns a score that reflects the quality of the parameters. In this paper, the input is a model description specified similarly to the probabilistic languages used in Turing [11], Stan [5] and WinBUGS [19].

The Turing system [11] relies on explicit vectorization syntax to gain performance. The presented approach relies on the message passing model [16] for parallelism and vectorization is an extension that is left as future work.

Stan [5] is a platform for statistical modeling and high-performance statistical computation. Recently, an extension to its modeling language has been proposed for parallelization [22], but use requires changing the model description. In contrast, the parallelization outlined below does not require the user to specify additional input signifying how computation should be scheduled on the hardware, but the downside is that it can be too aggressive causing performance to degrade in some cases.

Gibbs sampling [6] draws samples from the marginal target distribution by combining samples taken from conditional distributions. The concept of a graphical model is fundamental for Bayesian inference Using Gibbs Sampling (BUGS), implemented in WinBUGS [19]. MultiBUGS [12] has added parallel execution to WinBUGS by working directly on the graphical model from which conditionally independent parts are identified and scheduled to parallel processors only when deemed beneficial by a heuristic. Execution of Gibbs Sampling requires synchronization between phases more closely resembling the BSP model. The difference with the work presented below is that the graphical model is used indirectly to detect parallel parts of the dataflow graph. Since the posterior is evaluated as a whole with less synchronization instead of being separated into smaller conditional densities, the applicability is not limited to Gibbs sampling. Another difference is that MultiBUGS ignores load imbalance by explicitly assuming that tasks have the same running time.

Even if the outlined approach is applied in a Gibbs setting, the parallelization within a single phase is different. For example, given a posterior $p(\theta|\mathcal{D})$, if $p(\theta_i|\ldots)$ and $p(\theta_j|\ldots)$ are assigned to one Gibbs phase, computation shared between these two conditional distributions can be executed only once even without blocking, a technique that affects convergence properties of Gibbs sampling [25].

Nemeth et al. [20] uses an Evolutionary Algorithm (EA) to parallelize the evaluation of probabilistic models by optimizing schedules through simulation of a parallel system with communication overhead. The downside is that searching for a schedule can become prohibitively slow, even though, at least in theory, the optimal schedule could be found. In contrast, using the graphical model is a simpler strategy as only tasks assigned to phases can be executed in parallel. However, it turns out that such an approach already yields well-performing schedules. Another difference is that the EA approach yields a static schedule in which both the execution order and the assignment of tasks to processors are fixed while the tasks that have been identified from the graphical model can be re-assigned depending on load imbalance changes.

An extensive survey for the well researched task graph scheduling problem is provided by Yu-Kwong et al. [17]. The main difference with conventional scheduling approaches is that the target domain is rather specific. The dataflow graph of a generative model specification always obeys a specific template. From this observation, a mapping can be formulated from which the parallelism is extracted directly.

3 Hierarchical Models and Conditional Independence

The main goal of this paper is to show how model descriptions can be parallelized by relying on information from the graphical model. This section introduces the notion of a model description, its dataflow graph, and its graphical model. To distinguish between the structure of the two representations, "layers" refers to candidates for parallelism in the former and "levels" refers to the depth of variables in the latter.

From a Bayesian perspective [23], a model description defines a posterior $p(\theta|\mathcal{D})$. The numeric value of the posterior determines the quality of a chosen set of parameters θ while taking into account evidence $\mathcal{D}$. In what follows, θ_i denotes a component of the θ vector and $y_i \in \mathcal{D}$ denotes a data entry.

The description consists of likelihood expressions $y_i \sim p(.|\mathrm{pa}(y_i))$ and prior expressions of the form $\theta_i \sim p(.|\mathrm{pa}(\theta_i))$ where pa(.) is the set of random variables conditioned upon. These expressions will be generalized to $\gamma_i \sim p(.|\mathrm{pa}(\gamma_i))$ for convenience. As an example, consider the model shown by Fig. 1 on the left describing both pharmacokinetics (PK) and pharmacodynamics (PD) of a drug for type-2 diabetes treatment [24].

To convert a model description into an executable function $f(\theta, \mathcal{D})$, prior and likelihood expressions are replaced by probability density function evaluations of the density $p(.|\ldots)$ at γ_i, denoted by a call to pdf() to which the distribution and the position are passed. Finally, the product of the resulting probability densities is returned while the remaining expressions are left untouched. The resulting function is then converted into a dataflow graph [1,8]. In contrast to the typical controlflow style reasoning, a dataflow graph is an alternative model of computation where instead of executing operations on data, data flows through operators. This representation of computation lends itself well to parallelization [16]. The dataflow graph $G = (V, E)$ represents the set of computational tasks V and specifies how data flows between the tasks with edges E.

In general, the dataflow graph of a function $f(\theta, \mathcal{D})$ for a hierarchical model has the structure shown in Fig. 2. The inputs θ and $\mathcal{D}$ are shown at the top and the product over densities is shown at the bottom. These are connected with the central portion of the graph, shown by dashed lines. Considering only the part with solid lines, the relationship with the graphical model is revealed. Each level depends on any of the previous levels through density evaluation nodes in V. In the example shown, the connections are less dense; for example, the

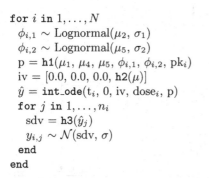
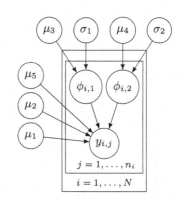

```
for i in 1,...,N
    φ_{i,1} ~ Lognormal(μ_2, σ_1)
    φ_{i,2} ~ Lognormal(μ_5, σ_2)
    p = h1(μ_1, μ_4, μ_5, φ_{i,1}, φ_{i,2}, pk_i)
    iv = [0.0, 0.0, 0.0, h2(μ)]
    ŷ = int_ode(t_i, 0, iv, dose_i, p)
    for j in 1,...,n_i
        sdv = h3(ŷ_j)
        y_{i,j} ~ N(sdv, σ)
    end
end
```

Fig. 1. The Canagliflozin model on the left and its graphical model on the right. A PK/PD model, used to describe the compound concentration over time for N individuals in a population, is numerically integrated by int_ode. The number of measurements for the ith individual is given by n_i, and h1, h2 and h3 are helper functions.

first level is only connected to the second and fourth level and not to the third level. However, it is easy to see how the structure generalizes to any hierarchical model.

The model from Fig. 1 is even less dense. Part of the first layer, μ_3, σ_1, μ_4 and σ_2 are connected with the second layer with variables $\phi_{i,1}$ and $\phi_{i,2}$ and all variables in the second layer together with the remaining part of the first layer are connected with the third layer with variables $y_{i,j}$.

One simplification made here is that an edge in Fig. 2 can represent a sequence of operations that transform random variables between layers or parts of layers like h1, h2, h3 and int_ode in Fig. 1. It is important to keep this in mind for the discussion in Sect. 4.

A graphical model $H = (R, F)$, is a representation of the conditional independence between variables. Figure 1 shows the graphical model on the right for the Canagliflozin model. For brevity, it is conventional to summarize similar variables with the plate notation by placing them into boxes with the range of iterated indices specified at the bottom [12]. For hierarchical models, H is a Directed Acyclic Graph (DAG), where the set of nodes R represents the random variables in the hierarchical model and their priors, and the edges $F \subseteq R \times R$ denote how the posterior can be factorized, i.e. $p(\theta|\mathcal{D}) \propto p(\theta, \mathcal{D}) = p(\gamma) = \prod_i p(\gamma_i|\text{pa}(\gamma_i))$. An edge from γ_j to γ_i is placed in F if $\gamma_j \in \text{pa}(\gamma_i)$.

To convert a dataflow graph G into a graphical model H, the nodes R and edges F need to be defined in terms of V and E. All nodes with input parameters in G, i.e. θ_i and y_i at the top of Fig. 2, form R. The edges F are defined by the density evaluation nodes. By traversing the edges in E in the opposite direction starting at the node that provides the density input, the variables $\text{pa}(\gamma_i)$ can be found. Similarly, following the other input, γ_i can be found. This mapping introduces a function m from R to V where $m(r)$ is either the corresponding probability evaluation node if it exists, or the input node of that variable.

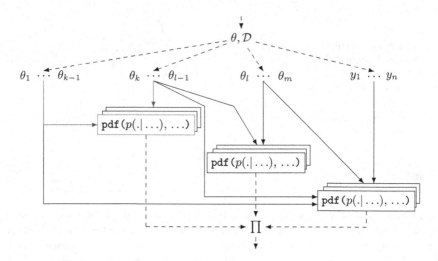

Fig. 2. Simplification of the structure of the dataflow graph of a function $f(\theta, \mathcal{D})$ built from a hierarchical model with four layers the last of which is the data layer. The structure generalizes to any generative specification of a hierarchical models with arbitrary interconnected layers. (Color figure online)

The set $\mathrm{pa}(\gamma_i)$ and γ_i for the red node in Fig. 2 can be found by following the green edges and the blue edges respectively.

4   Extracting Parallelism from the Graphical Model

Since the dataflow representation naturally exposes parallelism in a model it is possible to execute the dataflow graph by starting execution of each node when all its inputs become available. It is well known that scheduling such a computational DAG is hard to solve optimally, but many heuristics exist [3].

Scheduling each node separately is prohibitively expensive in practice due to the amount of overhead introduced on a contemporary system; not only is overhead introduced by starting a function compiled from the expression in a node, but also by tracking and storing its inputs and outputs. To reduce overhead, sets of nodes can be grouped into larger tasks and treated as a single unit at the cost of potentially reducing parallelism. It is possible to find satisfactory assignments of tasks to processors by considering the dataflow graph and the characteristics of the underlying parallel system directly [20], but this search can be slow in practice, especially with graphs that have on the order of 10^4 nodes or more. Such graphs are not uncommon for typical pharmacometrics (PMX) models. Assuming that the order of tasks is not fixed, there are n^p possible assignments to consider with p processors. Parallelization based on the graphical model is more tractable, although less detailed.

Since the posterior can be seen as a product of conditional densities as discussed in Sect. 3, the most basic approach is to create one task for each conditional density from the dataflow graph. This is accomplished by traversing the dataflow graph backwards from each node c that contains the expression $\mathtt{pdf}(p(.|\ldots), \gamma_i)$ and selecting all reachable nodes, denoted by the set $\mathrm{pred}(c)$. The procedure $\mathrm{pred}(c)$ extends naturally to sets of random variables as well.

While this leads to an embarrassingly parallel solution since each task can be computed independently, the downside is that many nodes will be recomputed due to the similarities. More formally, for two density evaluation nodes c_1 and c_2, $\mathrm{pred}(c_1) \cap \mathrm{pred}(c_2) \neq \emptyset$. For example, suppose that one of the input parameters γ_i is first transformed to $g(\gamma_i)$ and there are multiple γ_j such that $\gamma_i \in \mathrm{pa}(\gamma_j)$. Then, to compute each conditional density $p(\gamma_j | g(\gamma_i), \ldots)$, $g(\gamma_i)$ will need to be recomputed for every conditional density, a situation that is undesirable if a significant portion of computation effort is spent in g.

Therefore, this paper proposes to use the conditional independence to find similarities between conditional densities. Computationally, conditionally independence of two random variables γ_i and γ_j given γ_k means not only that $p(\gamma_i, \gamma_j | \gamma_k) = p(\gamma_i | \gamma_k) \cdot p(\gamma_j | \gamma_k)$ holds, but also that some of the tasks related to γ_i and γ_j are to be executed after some of the tasks related to γ_k. In addition, there might be some similarity between the computation related to γ_i and γ_j, but there will also be some differences. If this was not the case, then $p(\gamma_i | \gamma_k) = p(\gamma_j | \gamma_k)$.

Computational similarities can be captured by introducing deterministic variables β so that $p(\gamma_i | \beta, \gamma_k) = p(\gamma_i | \beta)$ and $p(\gamma_j | \beta, \gamma_k) = p(\gamma_j | \beta)$. Probabilistically, after marginalizing the deterministic variables β, Eq. (1) holds.

$$p(\gamma_i, \gamma_j | \gamma_k) = p(\gamma_i | \beta) \cdot p(\gamma_j | \beta) \cdot p(\beta | \gamma_k) \tag{1}$$

Once $p(\beta | \gamma_k)$ has been computed, both $p(\gamma_i | \beta)$ and $p(\gamma_j | \beta)$ can be computed sharing as little information as possible. If no information is shared, they can be computed in parallel. Algorithm 1 shows how to accomplish this by processing random variables in the graphical model.

The end goal is to assign random variables to layers and to construct tasks from the variables in these layers. The assumption is that tasks constructed from a layer are independent. In the extreme, when a deterministic variable is introduced for each node in the dataflow graph, all tasks will be independent given their predecessors. Note however that Algorithm 1 introduces only a limited number of deterministic variables. Therefore, the predecessor relationship imposed by E will still need to be respected since there might still be some dependencies. The rationale behind this is that variables in the same layer tend to share computation through their connection with previous layers, on a layer by layer basis, while little or no computation is shared within a layer.

First, following the depth definition from MultiBUGS [12], the level $d(r)$ is computed for $r \in R$. If $\mathrm{pa}(r) = \emptyset$, then $d(r) = 1$. Otherwise, $d(r) = 1 + \max_{p \in \mathrm{pa}(r)} d(p)$. In Fig. 1 the depth is 1 for all μ and σ variables, 2 for all ϕ

Algorithm 1. Extracting layers to construct parallel tasks.

---

procedure EXTRACTLAYERS(G, H, m)   ▷ G, H and m defined in Section 3
    Compute $d(r)$ for $r \in R$ as in MultiBUGS [12]
    for $i = 1, \ldots, D$ **do**
        $L_{i,i} = \{\{m(r)\} | r \in R_i\}$
        $P_i = \cup_{r \in R_i} \mathrm{pa}(r)$ ▷ All direct parents of level i
        for $j = 1, \ldots, i - 1$ **do**
            $P_{i,j} = P_i \cap R_j$ ▷ Direct parents in level $j < i$
            $L_{i,j} = \{\mathrm{lcpred}(\mathrm{ch}(p) \cap R_i) | p \in P_{i,j}\}$ ▷ Find computational similarities
        end for
    end for
    return $L_{1,1} \ldots, L_{D,D}$
end procedure

---

variables, and 3 for the data variables $y_{i,j}$. The levels of the random variables partition R into sets $R_1, \ldots, R_D$. Here, R_i contains all the random variables at level i.

It might seem that D layers can now be constructed, one for each set of variables R_i. However, this does not expose computational similarities present between layers. Instead, multiple layers will be introduced for each level i, represented by $L_{i,j}$. The elements of layer $L_{i,j}$ are sets of dataflow graph nodes.

For a level i, the deepest layer $L_{i,i}$ contains the dataflow graph node associated with the random variables $r \in R_i$ as singletons. Next, the directly reachable parents of the variables in R_i are collected in P_i. For each $j < i$, $|L_{i,j}| = |P_{i,j}|$ where $P_{i,j} \subseteq P_i$ are the direct parents on level j. Each element of $L_{i,j}$ is given by the last common predecessors of the children of $p \in P_{i,j}$ in R_j, denoted by $\mathrm{lcpred}(\mathrm{ch}(p) \cap R_j)$. For a set of nodes $S \subseteq R$, $\mathrm{lcpred}(S)$ is computed by taking the nodes in $\cap_{c \in S} \mathrm{pred}(m(c))$ for which edges lead to nodes in $\cup_{c \in S} \mathrm{pred}(m(c)) \setminus \cap_{c \in S} \mathrm{pred}(m(c))$. The expressions from the dataflow graph in each set in $L_{i,j}$ with $j < i$ constitute the computational similarities of random variables with depth i with respect to parents at depth j. These similarities correspond to deterministic variables like β.

For the model from Fig. 1, $L_{1,1}$ contains singletons for the random variables at depth 1 like $\{m(\mu_1)\}$ and $\{m(\sigma_2)\}$. Analogously, $L_{2,2}$ and $L_{3,3}$ contains singletons for the random variables ϕ and the data entries y respectively. The direct parents of the variables with depth 3 are $\mu_1, \mu_2, \mu_5, \phi_{i,1}$ and $\phi_{i,2}$. Since $d(\phi_{i,1}) = d(\phi_{i,2}) = 2$ and $d(\mu_1) = d(\mu_2) = d(\mu_5) = 1$, two additional layers $L_{3,2}$ and $L_{3,1}$ will be introduced. Here, among others $\mathrm{lcpred}(\mathrm{ch}(\phi_{i,1}) \cap R_2)$ contains calls to `int_ode`, and $\mathrm{lcpred}(\mathrm{ch}(\phi_{i,1}) \cap R_2)$ contains calls to `h2`.

Finally, to turn the constructed layers $L_{i,j}$ into a partitioning of V, they are processed from shallowest to deepest while assigning all nodes in V. Each set $S' \in L_{i,j}$ is replaced by nodes in $\cup_{s \in S'} \mathrm{pred}(s)$ except for those that have already been assigned. The resulting sets form the final tasks.

While it might seem that annotating the for-loops in the model description like the one given in Fig. 1 to specify that these should be parallelized is straightforward, the parallelization described here will not only automatically detect this, but it will also work for more arbitrarily interrelated models in which loops need not necessarily match the levels of the hierarchy.

The tasks within each layer can be scheduled to run in parallel. To maximize parallel efficiency [13], idle times need to be kept to a minimum. The only heuristic considered during performance evaluation is LPT [9] although other heuristics could be used as well. The focus is not so much on scheduling, but on presenting a mapping between two representations of a model to identify parallel parts.

Since subsequent posterior evaluations occur at similar positions in the parameter space, i.e. $\theta^t \approx \theta^{t+1}$, it turns out that the execution time for each task changes only gradually. For this reason, after running one iteration with tasks scheduled using a round-robin (RR) strategy, subsequent rounds can be scheduled with LPT using the execution time measured during evaluation of the previous candidate parameter θ.

5   Performance Evaluation

PMX models are key computational components leveraged for decision making during drug development. Here, only a limited amount of data is available [4]. The data includes the compound concentration in the blood of subjects, a costly measurement to make. In contrast to more classical models where all data is "independent and identically distributed", the data also specifies from which patient each measurement is taken creating a hierarchy as discussed above.

In this section, the performance of the proposed method is evaluated using two models from PMX. The first model, called the Nimotuzumab model, describes a humanized monoclonal antibody mAb, in patients with advanced breast cancer [21]. The second model is the Canagliflozin model used as the example in Sect. 3.

The structure is similar in both models; it consists of a population layer in which a set of patients that have taken part in the clinical trial are each modeled separately. However, it is important to note that the parallelization outlined in this paper can be applied to models with more layers assuming that there are enough computationally intensive tasks in each layer.

The data for the Nimotuzumab model contains measurements of 12 patients resulting in limited amount of parallelization. On the other hand, the data for the Canagliflozin model consists of measurements of 1144 patients. For this model, it is important to note that some patients in the placebo group are not given the compound, while others are given the compound for either a shorter or longer period. Therefore, the time required to simulate PK and PD for each patient varies drastically [14]. For example, execution time of numeric integration varies up to 100× across patients for Canagliflozin.

If all expressions are compiled separately, respectively 6643 and 46261 tasks are created for the two models. The overhead of running these tasks separately, estimated by a run on a single system, slows down execution time by a few orders of magnitude. By applying the steps outlined in Sect. 4, the number of tasks drops to 375 and 9080 reducing task management overhead.

Table 1. The number of tasks per layer and the percentage of time in each layer for the two test models. Most of the time is spent in the fifth layer, where tasks that perform the numeric integration are concentrated. The final layer, with the most tasks, contains likelihood evaluations.

Model	Metric	$L_{1,1}$	$L_{2,1}$	$L_{2,2}$	$L_{3,1}$	$L_{3,2}$	$L_{3,3}$
Nimotuzumab	Tasks (#)	1	3	36	1	12	321
	Coverage (%)	0.00%	0.09%	1.43%	0.03%	90.35%	8.10%
Canagliflozin	Tasks (#)	1	2	2694	1	1144	5237
	Coverage (%)	0.00%	0.01%	0.03%	0.00%	99.90%	0.06%

The distribution of tasks across layers is shown in Table 1. Most of the computation time, 90% and 99% respectively, is spent in the numeric integration of the PK and PD equations. The tasks that perform this integration are captured in a single layer. Both models compile to 5 layers with the most tasks in the last layer containing likelihood evaluations. Since likelihood evaluations in these models are lightweight, they also serve to demonstrate that the presented parallelization can be too aggressive as all layers are parallelized while manual parallelization would only assign more resources in the layer that captures numeric integration tasks.

The number of messages exchanged between processors depends on how tasks are scheduled, and varies at runtime for each evaluation when the scheduling step reassigns tasks. It is important to note that the LPT heuristic has a local view. Tasks in each layer are scheduled without considering the assignment of tasks in other layers.

Figure 3 compares performance when tasks in a phase are scheduled using a RR strategy or by using the LPT heuristic on a single Haswell system with 2 Xeon E5-2699 v3 @ 2.30 GHz CPUs, each with 18 physical cores for a total of 36 cores. The parallelization was implemented in the Julia programming language [2]. For the sake of stability of the results, frequency scaling was disabled. While other custom message passing implementations were also tested, the results are reported for an implementation relying on Intel MPI Version 2018 as it is widely available. Preparing and copying messages adds overhead, but note that since the results are for a single system, this could be avoided by using threads instead. Nevertheless, the mapping between the two representations with this overhead still shows promising performance scalability. It is also applicable to larger systems with a higher latency interconnect as long as the tasks are sufficiently compute intensive.

Since the outlined approach uses the message passing model for parallel execution, the more general term "processor" is used here [13]. The comparison is made in terms of the speedup achieved by running on p processors, denoted by S_p and given by the ratio between the execution time with one processor and p processors, i.e. T_1/T_p. As both T_1 and T_p are stochastic due to noise in the system [15,18], execution time is measured 200 times for each choice of p to obtain stable results. Samples for T_1 are paired with T_p to generate samples for S_p. The 5th and 95th quantiles are shown to quantify the spread of S_p.

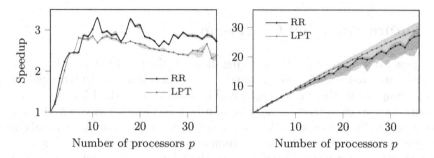

Fig. 3. Mean scalability of the Nimotuzumab model on the left and the Canagliflozin model on the right with the shaded regions showing the uncertainty range for the 5th and 95th quantile of the speedup. The efficacy of the parallelization approach is model dependent, but performance improves for both models.

The limited number of patients in the Nimotuzumab model causes execution time not to scale past approximately 10 processors. Note also that performance does not reach $10\times$ with respect to baseline. Through profiling, it became apparent that this is not only due to the varying computational requirements between tasks associated with different patients, but also due to communication overhead. With the relatively small amount of available parallelism, this cannot be neglected, and it causes performance to degrade past 10 processors.

Note that the LPT heuristic results in slightly slower performance when compared to RR. This is due to the increase in the time spent communicating between some cores in some layers, an aspect not taken into account by the heuristic while in RR communication cost is spread more evenly.

Note also that initially, there is little to no difference between the two strategies. This is due to the two strategies behaving similarly when a few processors are used. As the number of processors increases, the performance of the two strategies diverges.

The Canagliflozin model scales better since there is a much larger opportunity for parallelization. Due to the amount of imbalance between patients, the LPT scheduling heuristic further improves performance by about 8%. Around 10% is lost due to overhead introduced by communication between processors and task management. This is verified by comparing to theoretically computed execution time where this overhead is ignored. Note that efficiency, computed by comparing actual scalability with linear scalability, stays above 90%. From this, it can be concluded that most of the available parallelism is exploited.

Since multiple processors are employed in each layer of the hierarchy, it only improves performance in layers with tasks that take a sufficient amount of computation to dwarf communication overhead. For layers with small tasks, the benefits of parallel execution will be outweighed by the overhead introduced by communication. In this case, overall performance will improve only when other compute-intensive layers make the overhead for layers with many small tasks negligible.

6 Future Work and Conclusion

This paper introduces a novel way to parallelize evaluation of hierarchical models by observing that conditional independence in a graphical model representation can be mapped to the dataflow graph. The presented method has been shown to work for two characteristic models from PMX. Note that it is not limited to this domain. The efficacy of the model depends on the amount of parallelism inherent in the input models and the computational size of its tasks. The results show that by using a simple well-known scheduling heuristic within each layer, performance can further improve in case execution time varies between tasks.

One drawback of the presented method is that *all* layers are parallelized. As long as there are enough layers with many compute-intensive tasks, the presented approach results in high utilization of parallel resources. However, the communication introduced in layers with small, but numerous, tasks can degrade performance. Therefore, future work will explore how to disable parallelization selectively if communication overhead is high relative compared to the amount of computation.

The scheduling heuristic relies on the measured execution time of tasks during previous model evaluations. As long as the assumption holds that the execution time of tasks changes gradually while the encompassing sampling algorithm or optimization routine takes small steps in the parameter space, such an approach will suffice. There is additional overhead introduced by measuring and collecting the execution time of each task. Therefore, future work will study the trade-off of occasionally disabling these measurements while the scheduling heuristic uses less up-to-date measurements.

The current results were limited to a single system with communication between processors accomplished through memory. Another aspect that will be explored next is how to mitigate the latency of contemporary interconnects.

Finally, while the partitioning of nodes is used in this paper to construct tasks, using the resulting assignments for initializing more complex heuristics as those used in other work [20] to speed up convergence will be studied next.

Acknowledgments. Part of the work presented in this paper was funded by Johnson & Johnson.

References

1. Beck, M., Pingali, K.: From control flow to dataflow. Cornell University, Technical report (1989)
2. Bezanson, J., Edelman, A., Karpinski, S., Shah, V.B.: Julia: a fresh approach to numerical computing. SIAM Rev. **59**(1), 65–98 (2017). https://doi.org/10.1137/141000671
3. Błażewicz, J., Ecker, K.H., Pesch, E., Schmidt, G., Weglarz, J.: Handbook on Scheduling: From Theory to Applications. Springer, Heidelberg (2007). https://doi.org/10.1007/978-3-540-32220-7
4. Carey, V., Wang, Y.G.: Mixed-Effects Models in S and S-PLUS. Springer, New York (2001). https://doi.org/10.1007/b98882
5. Carpenter, B., et al.: Stan: a probabilistic programming language. J. Stat. Softw. **76**(1), (2017)
6. Casella, G., George, E.I.: Explaining the Gibbs sampler. Am. Stat. **46**(3), 167–174 (1992)
7. Chakravarthi, V.S.: SOC Physical Design. A Practical Approach to VLSI System on Chip (SoC) Design, pp. 173–199. Springer, Cham (2020). https://doi.org/10.1007/978-3-030-23049-4_9
8. Culler, D.E.: Dataflow architectures. Annu. Rev. Comput. Sci. **1**(1), 225–253 (1986)
9. Della Croce, F., Scatamacchia, R.: The longest processing time rule for identical parallel machines revisited. J. Sched. **23**(2), 163–176 (2018). https://doi.org/10.1007/s10951-018-0597-6
10. Eijkhout, V.: Introduction to High Performance Scientific Computing. Lulu press, Morrisville (2012)
11. Ge, H., Xu, K., Ghahramani, Z.: Turing: a language for flexible probabilistic inference. In: International Conference on Artificial Intelligence and Statistics, pp. 1682–1690 (2018)
12. Goudie, R.J., Turner, R.M., De Angelis, D., Thomas, A.: Multibugs: a parallel implementation of the bugs modelling framework for faster Bayesian inference. arXiv preprint arXiv:1704.03216 (2017)
13. Grama, A., Kumar, V., Gupta, A., Karypis, G.: Introduction to Parallel Computing. Pearson Education, London (2003)
14. Haber, T., van Reeth, F.: Improving the runtime performance of non-linear mixed-effects model estimation. In: Schwardmann, U., et al. (eds.) Euro-Par 2019: Parallel Processing Workshops, Euro-Par 2019. Lecture Notes in Computer Science, vol. 11997. Springer, Cham (2020). https://doi.org/10.1007/978-3-030-48340-1_43
15. Hoefler, T., Belli, R.: Scientific benchmarking of parallel computing systems: twelve ways to tell the masses when reporting performance results. In: Proceedings of the International Conference for High Performance Computing, Networking, Storage and Analysis, pp. 1–12 (2015)
16. Kessler, C., Keller, J.: Models for parallel computing: review and perspectives. Mitteilungen-Gesellschaft für Informatik eV, Parallel-Algorithmen und Rechnerstrukturen **24**, 13–29 (2007)
17. Kwok, Y.K., Ahmad, I.: Static scheduling algorithms for allocating directed task graphs to multiprocessors. ACM Comput. Surv. (CSUR) **31**(4), 406–471 (1999)
18. Lameter, C.: Shoot first and stop the OS noise. In: Linux Symposium, p. 159. Citeseer (2009)
19. Lunn, D.J., Thomas, A., Best, N., Spiegelhalter, D.: Winbugs-a Bayesian modelling framework: concepts, structure, and extensibility. Stat. Comput. **10**(4), 325–337 (2000). https://doi.org/10.1023/A:1008929526011

20. Nemeth, B., Haber, T., Liesenborgs, J., Lamotte, W.: Automatic parallelization of probabilistic models with varying load imbalance. In: International Symposium on Cluster, Cloud and Grid Computing (CCGRID) Workshop on High Performance Machine Learning Workshop (2020)
21. Rodríguez-Vera, L., et al.: Semimechanistic model to characterize nonlinear pharmacokinetics of nimotuzumab in patients with advanced breast cancer. J. Clin. Pharmacol. **55**(8), 888–898 (2015)
22. Saintes, F.: I-56 sebastian weber supporting drug development as a Bayesian in due time?!. In: Euro-Par, vol. 2020 (2019)
23. Sivia, D., Skilling, J.: Data Analysis: A Bayesian Tutorial. OUP Oxford, Oxford (2006)
24. de Winter, W., et al.: Dynamic population pharmacokinetic-pharmacodynamic modelling and simulation supports similar efficacy in glycosylated haemoglobin response with once or twice-daily dosing of canagliflozin. Br. J. Clin. Pharmacol. **83**(5), 1072–1081 (2017)
25. Yildirim, I.: Bayesian Inference: Gibbs Sampling. MIT Press, New York (2012)

Improving Performance of the Hypre Iterative Solver for Uintah Combustion Codes on Manycore Architectures Using MPI Endpoints and Kernel Consolidation

Damodar Sahasrabudhe$^{(\boxtimes)}$ and Martin Berzins

SCI Institute, University of Utah, Salt Lake City, UT, USA
{damodars,mb}@sci.utah.edu

Abstract. The solution of large-scale combustion problems with codes such as the Arches component of Uintah on next generation computer architectures requires the use of a many and multi-core threaded approach and/or GPUs to achieve performance. Such codes often use a low-Mach number approximation, that require the iterative solution of a large system of linear equations at every time step. While the discretization routines in such a code can be improved by the use of, say, OpenMP or Cuda Approaches, it is important that the linear solver be able to perform well too. For Uintah the Hypre iterative solver has proved to solve such systems in a scalable way. The use of Hypre with OpenMP leads to at least 2x *slowdowns* due to OpenMP overheads, however. This behavior is analyzed and a solution proposed by using the MPI Endpoints approach is implemented within Hypre, where each team of threads acts as a different MPI rank. This approach minimized OpenMP synchronization overhead, avoided slowdowns, performed as fast or (up to 1.5x) faster than Hypre's MPI only version, and allowed the rest of Uintah to be optimized using OpenMP. Profiling of the GPU version of Hypre showed the bottleneck to be the launch overhead of thousands of micro-kernels. The GPU performance was improved by fusing these micro kernels and was further optimized by using Cuda-aware MPI. The overall speedup of 1.26x to 1.44x was observed compared to the baseline GPU implementation.

Keywords: Hypre · OpenMP · GPUs · MPI End Point

The authors thank Department of Energy, National Nuclear Security Administration (under Award Number(s) DE-NA0002375) and Intel Parallel Computing Center, for funding this work. This research used resources of the Argonne Leadership Computing Facility, which is a DOE Office of Science User Facility supported under Contract DE-AC02-06CH11357 and also of Lawrence Livermore National Laboratory. J. Schmidt, J. Holmen, A. Humphrey and the Hypre team are thanked for the help.

© Springer Nature Switzerland AG 2020
V. V. Krzhizhanovskaya et al. (Eds.): ICCS 2020, LNCS 12137, pp. 175–190, 2020.
https://doi.org/10.1007/978-3-030-50371-0_13

1   Introduction

The asynchronous many task Uintah Computational Framework [3] solves complex large-scale partial differential equations (pdes) involved in multi physics problems such as combustion and fluid interactions. One of the important tasks in the solution of many such large scale pde problems is to solve a system of linear equations. Examples are the linear solvers used in the solution of low-Mach-number combustion problems or incompressible flow. Uintah-based simulations of next generation combustion problems have been successfully ported to different architectures, including heterogeneous architectures and have scaled up to 96K, 262K, and 512 K cores on the NSF Stampede, DOE Titan, and DOE Mira respectively [3]. Such simulation employs the Arches component of Uintah. Arches is a three dimensional, Large Eddy Simulation (LES) code developed at the University of Utah. Arches is used to simulate heat, mass, and momentum transport in reacting flows by using a low Mach number ($Ma < 0.3$) variable density formulation [14]. The solution of a pressure projection equation at every time sub-step is required for the low-Mach-number pressure formulation. This is done using the Hypre package [14]. Hypre supports different iterative and multigrid methods, has a long history of scaling well [2,5] and has successfully weak scaled up to 500000 cores when used with Uintah [11].

While Uintah simulations were carried out [3] on DOE Mira and Titan systems [11], the next generation of simulations will be run on many core architectures such as DOE's Theta, NSFs Frontera, Riken's Fugaku and on GPU architectures such as DOEs Lassen Summit, Frontier and Aurora. On both classes of machines, the challenge for library software is then to move away from an MPI-only approach in which one MPI process runs per core to a more efficient approach in terms of storage and execution models. For many cores a common approach is to use a combination of MPI and OpenMP to achieve this massive parallelism. In the case of GPUs an offload of the OpenMP parallel region to GPU with CUDA or OpenMP 4.5 may be used. It is also possible to use portability layers such as Kokkos [7] to automate the process of using either OpenMP or Cuda. The MPI-only configuration for Uintah is to have one single threaded rank per core and one patch per rank. In contrast, the Uintah's Unified Task Scheduler was developed to leverage multi-threading and also to support GPUs [8]. Work is in progress to implement portable multi-threaded Kokkos - OpenMP and Kokkos - Cuda [7] based schedulers and tasks to make Uintah portable for future heterogeneous architectures. These new Uintah schedulers are based on teams of threads. Each rank is assigned with multiple patches, which are distributed among teams. Teams of threads process patches in parallel (task parallelism) while threads within a team work on a single patch (data parallelism). This design has proven useful on many core systems and in conjunction with Kokkos has led to dramatic improvements in performance [7].

The challenge addressed here is to make sure that similar improvements may be seen with Uintah's use of Hypre and its Structured Grid Interface (Struct) at the very least performs as well in a threaded environment as in the MPI case. Hypre's structured multigrid solver, PFMG, [2] is designed to be used with unions of logically rectangular sub-grids and is a semi-coarsening multigrid method for solving scalar diffusion equations on logically rectangular grids discretized with up to 9-point stencils in 2D and up to 27-point stencils in 3D. Baker et al. [2] report that various version of PFMG are between 2.5 and 7 times faster than the equivalent algebraic multigrid (AMG) options inside Hypre because they are able to take account of the grid structure. When Hypre is used with Uintah the linear solver algorithm uses the Conjugate Gradient (CG) method with the PFMG preconditioner based upon a Jacobi relaxation method inside the structured multigrid approach [14].

The Eq. (1) that is solved in Uintah is derived from the numerical solution of the Navier-Stokes equations and is a Poisson equation for the pressure, p, whose solution requires the use of a solver such as Hypre for large sparse systems of equations. While the form of (1) is straightforward, the large number of variables, for example 6.4 Billion in [14], represents a challenge that requires large scale parallelism. One key challenge with Hypre is that only one thread per MPI rank can call Hypre. This forces Uintah to join all the threads and teams before Hypre can be called, after which the main thread calls Hypre. Internally Hypre uses all the OpenMP threads to process cells within a domain, while patches are processed serially. From the experiments reported here, it is this particular combination that introduces extra overhead and causes the observed performance degradation. Thus, the challenge is to achieve performance with the multi-threaded and GPU versions of Hypre but without degrading the optimized performance of the rest of the code.

$$\nabla^2 p = \nabla \cdot \mathbf{F} + \frac{\partial^2 \rho}{\partial t^2} \equiv R \qquad (1)$$

1.1 Moving Hypre to New Architectures

In moving the Hypre to manycore architectures OpenMP was introduced to support multithreading [6]. However, in contrast to the results in [6], when using Uintah with Hypre in the case of one MPI process and OpenMP with multiple cores and mesh patches, a dramatic slowdown of up to 3x to 8x slowdown was experienced when using Hypre with Uintah as in a multi-threaded environment, as compared to the MPI-only version. Similar observations were made by Baker using a test problem with PFMG solver and up to 64 patches per rank and slowdown of 8x to 10x was observed between the MPI-only and MPI+OpenMP versions [2]. The potential challenges with OpenMP and Hypre either force Uintah with Hypre to singlethreaded (MPI only) version or use OpenMP with one patch per rank. This defeats the purpose of using OpenMP.

This work will show that the root cause of the slowdown to be the use of OpenMP pragmas at the innermost level of the loop structure. However the obvious solution of moving these OpenMP pragmas to a higher loop level does not seem to offer the needed performance either. The solution adopted here is to use a variant of an alternate threading model "MPI scalable Endpoints" [4,16] to solve the problem and to achieve a speedup consistent with the observed results of [2,6]. The approach described here is referred to as "MPI Endpoints", and abbreviated as MPI Ep, requires overriding MPI calls to simulate MPI behavior, parallelizing packing and unpacking of MPI buffers.

In optimizing Hypre performance for GPUs, Hypre 2.15.0 was run as a baseline code on Nvidia V100 GPUs, to characterize performance. Profiling on GPU reveals the launch overhead of GPU kernels to be the primary bottleneck and occurs because of launching thousands of "micro" kernels. The problem was fixed by fusing these micro kernels together and using GPU's constant cache memory. Finally, Hypre was modified to leverage Cuda-aware MPI on Lassen cluster which gives extra 10% boost.

The main contributions of this work are: (i) Introduce MPI EP model in Hypre to avoid slowdowns observed in the OpenMP version, which can achieve faster overall performance in the future while running the full simulation using multi-threaded task scheduler within Uintah AMT. (ii) Optimize the Cuda version of Hypre to improve CPU to GPU speedups ranging from 2.3x to 4x in the baseline version to the range of 3x to 6x in the optimized version, which can benefit the future large-scale combustion simulations on GPU based supercomputers.

2   Analysis of and Remedies for OpenMP Slowdown

The slowdown of OpenMP was investigated by profiling of Hypre using the PFMG preconditioner and the PCG solver with a representative standalone code that solves a 3D Laplace equation on a regular mesh, using a 27 point stencil. Intel's Vtune amplifier and gprof were used to profile on a single node KNL with 64 cores. The MPI-Only version of the code was executed with 64 single threaded ranks and the MPI + OpenMP version used 1×64, 2×32, 4×16, 8×8 and 16×4 ranks and threads, respectively. The focus was on the solve step that is run at every time step rather than the setup stage that is only called once. This example mimicked the use of Hypre in Uintah in that each MPI rank derived its own patches (Hypre boxes) based on the rank and allocated the required data structures accordingly. Each rank owned from a minimum of 4 patches to a maximum of 128 patches and each patch was then initialized by its respective rank. The Struct interface of Hypre was then called - first to carry on the setup and then to solve the equations. The solve step was repeated up to 10 times to simulate timesteps in Uintah by slightly modifying cell values every time. Then each test problem used different combinations of domain and patch sizes: a 64^3 or 128^3 domain was used with 4^3 patches of sizes 16^3 or 32^3. A 128^3 or 256^3 domain was used with 8^3 patches of sizes 16^3 or 32^3. Multiple combinations of MPI ranks, number of OpenMP threads per rank and patches per rank were tried and compared against the MPI-only version. Each solve step took about 10 iterations to converge on average.

The main performance bottlenecks were noted as follows.

(a) **OpenMP fork-join overhead.** Figure 1a shows the code structure of how an application (Uintah) calls Hypre. Uintah spawns its own threads, generates patches, and executes tasks scheduled on these patches. When Uintah

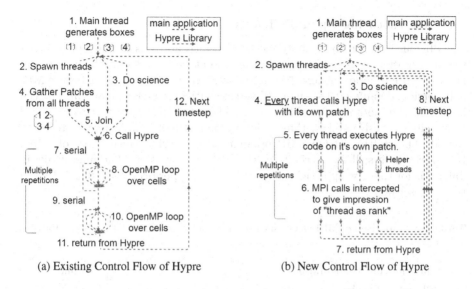

(a) Existing Control Flow of Hypre   (b) New Control Flow of Hypre

Fig. 1. Software design of hypre

encounters the Hypre task, all threads join and the main thread calls Hypre. Hypre then spawns its own OpenMP threads and continues.

With 4 MPI ranks and 16 OpenMP threads in each, Vtune shows that Hypre solve took 595 s. Of this time the OpenMP fork-join overhead was 479 s and spin time was 12 s.

The PFMG-CG algorithm calls 1000 s of "micro-kernels" during the solve step. Each micro kernel performs lightweight operations such matrix vector multiplication, scalar multiplication, relaxation, etc. and uses OpenMP to parallelize over the patch cells. However, the light workload does not offset the overhead of the OpenMP thread barrier at the end of every parallel for and results into 6x performance degradation. As a result, Hypre does not benefit from multiple threads and cores, with a performance degradation from OpenMP that grows with the number of: OpenMP threads per rank, patches per rank and points per patch.

(b) **Load imbalance due to serial sections**. Profiling detected three main serial parts - namely: 1. Packing and unpacking of buffers before and after MPI communication, 2. MPI communication and 3. Local data halo exchanges. Furthermore, the main thread has to do these tasks on behalf of worker threads while in the MPI-only version, each rank processes its own data and, of course, it does not have to wait for other threads.

(c) **Failure of auto-vectorization**. Hypre has "loop iterator" macros (e.g. BoxLoop) which expand into multidimensional for loops. These iterator macros use a dynamic stride passed as an argument. Although the dynamic stride is needed for some use cases, many use cases have a fixed unique stride. As the compiler cannot determine the dynamic stride a priori, the loop is not auto-vectorized.

2.1   Restructuring OpenMP Loops

One obvious solution to the bottlenecks identified above is to place pragmas at the outermost loop possible, namely the loop at "patch" level. This was tested for the Hypre function `hypre_PointRelax`. Table 1 shows timings for the MPI only version, default MPI + OpenMP version with OpenMP pragmas around cell loops and the modified OpenMP version where OpenMP pragmas were moved from cells to mesh patches, thus assigning one or more mesh patches to every thread. The shifting of OpenMP pragma gave a performance boost of 1.75x. However this is still 2x slower than the MPI only version. The final result in Table 1 is for the new approach suggested here that performs as well as MPI and is now described.

Table 1. Comparison of MPI vs OpenMP execution time(s) using 64 32^3 Mesh Patches

Hypre run time configuration	Runtime (s)
MPI Only 64 ranks	1.45
Default 4 ranks, each with 16 threads, OpenMP on cells loop	5.61
Modified 4 ranks, each with 16 threads, OpenMP on boxes loop	3.19
MPI Endpoints: 4 ranks each with 4 teams each with 4 threads	1.56

The MPI Endpoints approach adopted to overcome these challenges is shown in Fig. 1b. In this new approach, each Uintah "team of threads" acts: independently as if it is a separate rank (also known as MPI End Point or EP) and calls Hypre, passing its own patches. Each team processes its own patches and communicates with other real and virtual ranks. The mapping between teams and ranks is *virtual rank = real rank * number of teams + team id*. MPI wrappers are updated to convert virtual ranks to real ranks and vice versa during MPI communication. This conversion generates an impression of each team being an MPI rank and the code behaves as if it is MPI only version. The smaller team size (compared to the entire rank) minimizes overhead incurred in fork join in the existing OpenMP implementation, yet can exploit data parallelism.

The design and implementation of this approach posed the following challenges.

(a) **Race Conditions:** All global and static variables were converted to thread_local variables to avoid race conditions.
(b) **MPI Conflicts:** A potentially challenging problem was to avoid MPI conflicts due to threads. In Hypre only the main thread was designed to handle all MPI communications. With the MPI Endpoints approach, each team is required to make its own MPI calls. As Hypre already has MPI wrappers in place for all MPI functions, adding some code in every wrapper function to convert between a virtual rank and a real rank and to synchronize teams during MPI reductions was enough to avoid MPI conflicts.

(c) **Locks within MPI:** The MPICH implementation used as a base for Intel MPI and Cray MPI for the DOE Theta system uses global locks. As a result, only one thread can be inside the MPI library for most of the MPI functions. This is a potential problem for the new approach as the number of threads per rank are increased. To overcome the problem, one extra thread was spawned and all the communication funneled through the communication thread during the solve phase. This method provides a minimum thread wait time and gives the best throughput.

2.2 Optimizations in Hypre

The implementation of this approach needed following changes:

```
int g_num_teams;
__thread int tl_team_id;
int hypre_MPI_Comm_rank( MPI_Comm comm, int *rank ){
   int mpi_rank, ierr;
   ierr = MPI_Comm_rank(comm, &mpi_rank);
   *rank = mpi_rank * g_num_teams + tl_team_id;
   return ierr;
}
```

Fig. 2. Pseudo code of MPI EP wrapper for MPI_Comm_rank

(a) **MPI Endpoint:** The approach adopted a dynamic conversion mechanism between the virtual and the real rank along with encoding of source and destination team ids within the MPI message tag. Also MPI reduce and probe calls need extra processing. These changes are now described below.

 (i) **MPI_Comm_rank:** this command was mapped by using the formula above relating ranks and teams. Figure 2 shows pseudo code used to convert the real MPI rank to the virtual MPI EP rank using formula "mpi_rank * g_num_teams + tl_team_id". The global variable g_num_teams and the thread local variable tl_team_id are set to the number of teams and the team id during initialization. Thus the each end point gets an impression of a standalone MPI rank. The similar conversion is used in the subsequent wrappers.

 (ii) **MPI_Send, Isend, Recv, Irecv:** The source and destination team ids were encoded in the tag values. The real rank and the team id are easily recalculated from the virtual rank by dividing by the number of teams.

(iii) **MPI_Allreduce:** All teams within a rank carry out a local reduction first and then only the zeroth thread calls the real MPI_Allreduce and passes the locally reduced buffer as an input. Once the real MPI_Allreduce returns, all teams copy the data from the globally reduced buffer back to their own output buffers. C11 atomic operations are used for busy waiting rather than using any locks.

(iv) **MPI_Iprobe and Improbe:** Each team is assigned with a message queue
internally. Whenever a probe is executed by any team, it first checks its
internal queue for the message. If the handle is found, it is retrieved using
MPI_mecv. If the handle is not found in the queue, then the real Improbe
is issued and if the message at the head of the MPI queue is destined for
the same team, then again MPI_mecv is issued. If the incoming message is
tagged for another team, then the receiving team inserts the handle in the
destination team's queue. The method avoids the blocking of MPI queues
when the intended recipient of the MPI queue's head is busy and does not
issue probe.

(v) **MPI_GetCount:** In this case, the wrapper simply updates source and tag
values.

(vi) **MPI_Waitall:** A use of global global locks in MPICH MPI_Waitall stalls
other threads and MPI operations do not progress. Hence a MPI_Waitall
wrapper was implemented by calling MPI_Testtall and busy waiting until
MPI_Testtall returns true. This method provided about 15–20% speedup
over threaded MPI_Waitall.

(b) **Parallelizing serial code:** The bottleneck of fork - join was no longer
observed after profiling MPI Endpoints. However, this new approach exposed
a load imbalance due to serial code. The packing and unpacking of MPI
buffers and a local data transfer are executed by the main thread for all
the data. Compared to the MPI-only version, the amount of data per rank
is "number of threads" times larger, assuming the same workload per core.
Thus the serial workload of packing - unpacking for the main thread also
increases by "number of threads" times. The solution was to introduce
OpenMP pragmas to parallelize the loops associated with these buffers. Thus
each buffer could then be processed independently.

(c) **Interface for parallel_for:** A downside of explicitly using OpenMP
in Hypre is possible incompatibilities with other threading models. in the
spirit of [7] an interface was introduced that allows users to pass their own
version of "parallel_for" as a function pointer during initialization and
this user-supplied parallel for is called by simplified BoxLoop macros. Users
of Hypre can implement parallel_for in any threading model they wish
and pass on to Hypre to make flexible.

(d) **Improving auto-vectorization:** The loop iterator macros in Hypre operate using dynamic stride which prevents the compiler from vectorizing these
loops. To fix the problem, additional macros were introduced specifically for
the unit stride case. The compiler was then able to auto-vectorize some of
the loops and gave additional 10 to 20% performance boost depending on
the patch size.

3   GPU Hypre Performance Characterization and Profiling

While Hypre has had CUDA support from version 2.13.0, version 2.15.0 is used
here to characterize performance, to profile for bottlenecks and to optimize the

solver code. The GPU experiments are carried out on LLNL's Lassen cluster. Each node is equipped with two IBM Power9 CPUs with 22 cores each and four Nvidia V100 GPUs. Hypre and Uintah both were compiled using gcc 4.9.3 and cuda 10.1.243. The initial performance characterization was done on 16 GPUs of Lassen using a standalone mini-app which called Hypre to solve a simple Laplace equation and run for 20 iterations. GPU strong scaling is carried out using 16 "super-patches" of varying sizes $44^3, 64^3$ and 128^3. The observed GPU performance is evaluated against the corresponding CPU performance, which is obtained using the MPI only CPU version of Hypre. Thus, corresponding to every GPU, 10 CPU ranks are spawned and super-patches are decomposed smaller patches into smaller patches to feed each rank, keeping the total amount of work the same. Figure 3 shows the CPU performs 5x faster than the GPU for patch size 44^3. Although 64^3 patches decrease the gap, it takes the patch size of 128^3 for GPU to justify overheads of data transfers and launch overheads and deliver better performance than CPU. Based on this observation, all further work as carried out using 128^3 patches. HPCToolkit and Nvidia nvprof were used to profile CPU and GPU executions. The sum of all GPU kernel execution time shown by nvprof was around 500 ms, while the total execution time was 1.6 s. Thus the real computation work was only 30% and nearly 70% of the time was spent in the bottlenecks other than GPU kernels. Hence, tuning individual kernels would not help as much. This prompted the need for CPU profiling which revealed about 30 to 40% time consumed in for MPI wait for sparse matrix-vector multiplication and relaxation routines. Another 30 to 40% of solve time was spent in the cuda kernel launch overhead. It should be noted that although the GPU kernels are executed asynchronously, the launching itself is synchronous. Thus to justify the launching overhead, the kernel execution time should be at least $10\,\mu s$ - the launch overhead of the kernel on V100 (which was shown in the nvprof output).

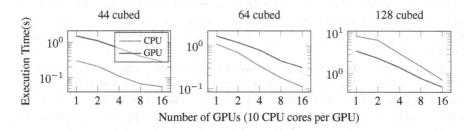

Fig. 3. GPU performance variation based on patch size

Table 2 shows the top five longest running kernels for the solve time of 128^3 patches on 16 GPUs with one patch per GPU. InitComm and FinComm kernels which are used to pack and unpack MPI buffers are fourth and fifth in the list. The combined timing of these two kernels can take them to the second position. More interestingly, together these kernels are called for 41,344 times, but the

Table 2. Top five longest running kernels before and after merging

Before merging				After merging			
Name	Calls	Tot time	Avg time	Name	Calls	Tot time	Avg time
MatVec	3808	110.69 ms	29.067 us	MatVec	3808	110.59 ms	29.040 us
elax1	2464	55.326 ms	22.453 us	Relax1	2464	55.350 ms	22.463 us
Relax0	2352	45.153 ms	19.197 us	Relax0	2352	44.987 ms	19.126 us
InitComm	20656	38.544 ms	1.8650 us	Axpy	1660	35.664 ms	21.484 us
inComm	20688	37.894 ms	1.8310 us	Memcpy-HtoD	12862	26.689 ms	2.0750 us

average execution time per kernel execution is just 1.8 μs. On the other hand the launch overhead of the kernel on V100 is 10 μs (which was revealed in the profile output). Thus the launch overhead of pack-unpack kernels consumes 0.4 s of 1.6 s (25%) of total execution time.

The existing implementation iterates over neighboring dependencies and patches and launches the kernel to copy required cells from the patch into the MPI buffer (or vice a versa). This results in thousands of kernel launches as shown in Table 2, but the work per launch remains minimal due to a simple copying of few cells. The problem can be fixed by fusing such kernel launches - at least for a single communication instance. To remedy the situation, the CPU code first iterates over all the dependencies to be processed and creates a buffer of source and destination pointers along with indexing information. At the end, all the buffers are copied into GPU's constant memory cache and the pack (or unpack) cuda kernel is launched *only once* instead of launching it for every dependency. After the fix InitComm and FinComm disappeared from the top five longest running kernels as shown in Table 2. The combined number of calls for InitComm and FinComm reduced from 41,344 to 8338. As a result, the communication routines perform 3x faster than before and the overall speedup in solve time achieved was around 20%. The modified code adds some overhead due to copying value to the GPU constant memory, which is reflected Memcpy-HtoD being called 12862 times compared to 4524 times earlier, but still the new code performs faster.

With the first major bottleneck resolved, the second round of profiling using HPCToolkit showed that the MPI wait time for matrix vector multiplication and for relaxation routines was now more than 60%. The problem is partially over-come by using cuda aware MPI supported on Lassen. The updated code directly passes GPU pointers to the MPI routines and avoids copying data between host and device. This decreased the communication wait time to 40 to 50% and resulted in an extra speedup of 10%.

4   Experiments

4.1   CPU (KNL) Experiments

Choosing the Patch Size: Initial experiments using only the Hypre solve component on a small node count showed the speedups increase with the patch size. Both MPI+OpenMP and MPI EP versions were compared against the MPI only version for different patch sizes. As shown in Table 3, MPI+OpenMP version always performs slower than the MPI Only version, although the performance improves a little as the patch size is increased. On the other hand, the MPI EP model performed nearly as well as the MPI Only version for 16^3 and 32^3 patch sizes on 2 and 4 nodes, but broke down at the end of scaling. With 64^3 patches, however, MPI EP performed up to 1.4x faster than the MPI Only version. As a result, the patch size of 64^3 was chosen for the scaling experiments on the representative problem. These results carry across to the larger node counts. Strong scaling studies with 16^3 patches show the MPI+OpenMP approach works 4x to 8x slower than the MPI Only version. In case of Hypre-MPI EP, the worst case slowdown of 1.8x was experienced for 512 nodes and the fastest execution matched the time of Hypre-MPI Only. This experience together with the results presented above straces the importance of using larger patch sizes, 64^3 and above, to achieve scalability and performance.

Table 3. Speedups of the MPI+OpenMP and MPI EP versions compared to the MPI Only version for different the patch sizes

Patch size:	16^3		32^3		64^3	
Nodes	MPI+OpenMP	MPI EP	MPI+OpenMP	MPI EP	MPI+OpenMP	MPI EP
2	0.2	0.9	0.2	1.2	0.5	1.4
4	0.2	0.8	0.2	0.9	0.4	1.4
8	0.2	0.5	0.3	0.6	0.5	1.3

As the process of converting Uintah's legacy code to Kokkos based portable code which can use either OpenMP or cuda is still in progress, not all sections of the code can be run efficiently in the multi-threaded environment. Hence a representative problem containing the two most time consuming components was chosen for the scaling studies on DOE Theta. The two main components are: (i) Reverse Monte Carlo Ray Tracing (RMCRT) which is used to solve for the radiative-flux divergence during the combustion [9] and (ii) pressure solve which uses Hypre. RMCRT has previously been converted to utilize multi-threaded approach that preforms faster than the MPI only version and also reduces memory utilization [12]. The second component, Hypre solver, is optimized as part of this work for a multi-threaded environment. The combination of these two components shows the impact of using an efficient implementation of multi-threaded Hypre code on the overall simulation of combustion. Three different mesh sizes were used for strong scaling experiments on DOE Theta: small (512^3), medium (1024^3) and large (2048^3). The coarser mesh for RMCRT was fixed to 128^3.

Each node of DOE Theta contains one Intel's Knights Landing (KNL) processor with 64 cores per node, 16 GB of the high bandwidth memory (MCDRAM) and AVX512 vector support. The MCDRAM was configured in cache-quadrant mode for the experiments. Hypre and Uintah were compiled using Intel Parallel Studio 19.0.5.281 with Cray's MPI wrappers and compiler flags "-std=c++11 -fp-model precise -g -O2 -xMIC-AVX512 -fPIC". One MPI process was launched per core (i.e., 64 ranks per node) while running the MPI only version. For the MPI+OpenMP and MPI EP version, four ranks were launched per node (one per KNL quadrant) with 16 OpenMP threads per rank. The flexibility of choosing team size in MPI EP allowed running the multiple combinations of teams x worker threads within a rank: 16x1, 8x2 and 4x4. The fastest results among these combinations were selected.

4.2   GPU Experiments

The GPU experiments were carried out on LLNL's Lassen cluster. Each node is equipped with two IBM Power9 CPUs with 22 cores each and four Nvidia V100 GPUs. Hypre and Uintah both were compiled using gcc 4.9.3 and cuda 10.1.243 with compiler flags "-fPIC -O2 -g -std=c++11 --expt-extended-lambda".

Strong and weak scaling experiments on Lassen were run by calling Hypre from Uintah (instead of mini-app) and the real equations originating from combustion simulations were passed to generate the solve for the pressure at each mesh cell. Strong scaling experiments were conducted using three different mesh sizes: small (512x256x256), medium (512^3) and large (1024^3). Each mesh is divided among patches of size 128^3 - such a way that each GPU gets one patch at the end of the strong scaling. CPU scaling was carried out by assigning one MPI rank to the every available CPU core (40 CPU cores/node) and by decomposing the mesh into smaller patches to feed each rank.

5   Results

5.1   KNL Results on Theta:

Table 4 shows the execution time per timestep in seconds for the RMCRT and Hypre solve components on DOE Theta. The multi-threaded execution of RMCRT shows improvements between 2x to 2.5x over the MPI Only version for the small problem and 1.4x to 1.9x for the medium size problem. Furthermore, the RMCRT speedups increase with the scaling. This performance boost is due to the all to all communication needed for the RMCRT algorithm is reduced by 16 times when 16 threads are used per rank. The multi-threaded version also results in up to 4x less memory allocation per node. However, the RMCRT performance improvements are hidden by poor performance of Hypre in the MPI+OpenMP version. As compared to the MPI Only version, a slowdown of 2x can be observed in Hypre MPI+OpenMP in spite of using 64^3 patches. The slowdowns observed are as bad as 8x for smaller patch sizes. Using optimized

Table 4. Theta results: The execution time per timestep in seconds for RMCRT, Hypre and total time up to 512 KNLs.

Nodes	MPI Only			MPI+OpenMP			MPI EP		
	Solve	RMCRT	Total	Solve	RMCRT	Total	Solve	RMCRT	Total
2	36	35	71	76	17	93	24	16	40
4	18	23	41	38	10	48	13	9	22
8	10	18	28	20	7	27	8	7	15
16	40	34	74	80	25	105	32	24	56
32	20	30	50	41	19	60	16	17	33
64	10	29	39	22	15	37	10	15	25
128	42	74	116	83	23	106	36	21	57
256	19	82	101	44	21	65	18	21	39
512	11	72	83	23	20	43	12	22	34

version of Hypre (MPI EP + partial vectorization) not only avoids these slow-downs, but also provides speedups from 1.16x to 1.5x over the MPI Only solve. The only exceptions are 64 nodes and 512 nodes, where there is no extra speedup for Hypre because the scaling breaks down. Because of the faster computation times (as observed in "Solve Time" of Table 4), lesser time is available for the MPI EP model to effectively hide the communication and also wait time due to locks within MPI starts dominating. Table 5 shows the percentage of solve time spent in waiting for the communication. During first two steps of scaling, the communication wait time also scales, but increases during the last step for eight and 64 nodes. The MPI wait time increases from 24% for 32 nodes to 50% for 64 nodes and the communication starts dominating the computation because there is not enough work per node.

Table 5. Theta results: communication wait time for MPI EP.

Nodes	2	4	8	16	32	64	128	512
MPI wait	2.4	1.4	1.7	6	3.9	5	11	6
Solve	24	13	8	32	16	10	36	12
% Comm	10%	11%	21%	19%	24%	50%	30%	50%

As both the components take advantage of the multi-threaded execution, the combination the overall simulation can lead to the combined performance boost of up to 2x as can be observed in the "Total" column of Table 4. It shows how the changes made to Hypre attribute to an overall speedups up to 2x.

5.2   GPU Results on Lassen

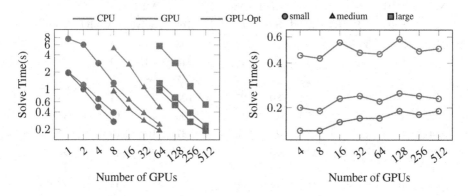

Fig. 4. Strong scaling of solve time   **Fig. 5.** Weak scaling of solve time

The strong scaling plot in Fig. 4 shows GPU version performs 4x faster than CPU version in the initial stage of strong scaling when the compute workload per GPU is more. As the GPU version performs better than the CPU version, it runs out of compute work sooner than the CPU version and the scaling breaks down with speedup reduced to 2.3x. Similarly, the optimized GPU version performs up to 6x faster than the CPU version (or 1.44x faster than the baseline GPU version) with the heavy workload. As the strong scaling progresses, the speedup by the optimized version against CPU reduces to 3x (or 1.26x against baseline GPU version). The communication wait time of both GPU versions is reduced by 4x to 5x as the number of ranks is reduced by ten times (not shown for brevity). Thanks to faster computations, the optimized GPU version spends 15 to 25% more time in waiting for MPI compared to the baseline GPU version.

The weak scaling was carried out using one 128^3 patch per GPU (or distributed among ten CPU cores) from four GPUs to 512 GPUs. Figure 5 shows good weak scaling for all three versions. The GPU version shows 2.2x to 2.8x speedup and the optimized GPU code performs 2.6x to 3.4x better than the CPU version.

Preliminary experiments with the MPI EP model on Lassen showed that the MPI EP CPU version performed as well as the MPI Only CPU version (not shown in Fig. 4 for brevity). Work is in progress to improve GPU utilization by introducing the MPI EP model for the GPU version and assigning different CUDA streams to different endpoints which may improve overall performance.

6   Conclusions and Future Work

In this paper it has been shown that the MPI-Endpoint approach makes a threaded version of Hypre as fast or faster than the MPI-only version when used with multiple patches and enough workload. Thus other multi threaded applications which use Hypre could benefit from this approach and achieve overall speedup as demonstrated on Theta. Similarly, improved GPU speedups can help in gaining overall speedups for other Hypre-cuda users.

One of the bottlenecks for the MPI EP version was locks within MPI - especially for smaller patches. This bottleneck can be improved if the lock-free MPI implementations are available or if the End Point functionality [4] is added into the MPI standard. This work used MPI EP to reduce the OpenMP synchronization overhead. However, the EP model can achieve a sweet spot between "one rank per core" and "one rank per node with all cores using OpenMP" and reduce the communication time up to 3x with the minimal OpenMP overhead, which can lead to better strong scaling as shown in [15].

On GPUs the current optimized version shows around 40 to 50% time consumed in waiting for MPI communication during sparse matrix vector multiplication and relaxation routines. If the computations and communications are overlapped, then a new kernel needs to be launched for the dependent computations after the communication is completed. As these kernels do not have enough work to justify the launch it resulted into slightly slower overall execution times during the initial experiments of overlapping communications. Similar behavior was observed by [1]. A possible solution is to collect kernels as "functors" and to launch a single kernel later, which calls these functors one after another as a function call. This is the work in progress, as is the application of the code to full-scale combustion problems. Another option for speeding up the algorithm is to use communication avoiding approaches e.g., see [10] which uses a multi-grid preconditioner and spends less than 10% of the solve time in the global MPI reductions on Summit. As this work here also used a multi-grid preconditioner [13], similar behavior was observed in our experiments and the global reduction in the CG algorithm is not a major bottleneck so far. However, these options will be revisited when applying the code to full scale combustion problems at Exascale.

References

1. Ali, Y., Onodera, N., Idomura, Y., Ina, T.: GPU acceleration of communication avoiding Chebyshev basis conjugate gradient solver for multiphase CFD simulations. In: 2019 IEEE/ACM 10th Workshop on Latest Advances in Scalable Algorithms for Large-Scale Systems (ScalA), pp. 1–8. IEEE (2019)
2. Baker, A., Falgout, R., Kolev, T., Yang, U.: Scaling hypre's multigrid solvers to 100,000 cores. In: Berry, M., et al. (eds.) High-Performance Scientific Computing, pp. 261–279. Springer, London (2012). https://doi.org/10.1007/978-1-4471-2437-5_13
3. Berzins, M., et al.: Extending the Uintah framework through the petascale modeling of detonation in arrays of high explosive devices. SIAM J. Sci. Comput. **38**, S101–S122 (2016)
4. Dinan, J., et al.: Enabling communication concurrency through flexible MPI endpoints. Int. J. High Perform. Comput. Appl. **28**(4), 390–405 (2014)

5. Falgout, R.D., Jones, J.E., Yang, U.M.: Pursuing scalability for hypre's conceptual interfaces. ACM Trans. Math. Softw. (TOMS) **31**(3), 326–350 (2005)
6. Gahvari, H., Gropp, W., Jordan, K.E., Schulz, M., Yang, U.M.: Modeling the performance of an algebraic multigrid cycle using hybrid MPI/OpenMP. In: 2012 41st International Conference on Parallel Processing, pp. 128–137, September 2012
7. Holmen, J.K., Peterson, B., Berzins, M.: An approach for indirectly adopting a performance portability layer in large legacy codes. In: 2nd International Workshop on Performance, Portability, and Productivity in HPC (P3HPC) (2019). In conjunction with SC19
8. Humphrey, A., Berzins, M.: An evaluation of an asynchronous task based dataflow approach for Uintah. In: 2019 IEEE 43rd Annual Computer Software and Applications Conference (COMPSAC), vol. 2, pp. 652–657, July 2019
9. Humphrey, A., Harman, T., Berzins, M., Smith, P.: A scalable algorithm for radiative heat transfer using reverse Monte Carlo ray tracing. In: Kunkel, J.M., Ludwig, T. (eds.) ISC High Performance 2015. LNCS, vol. 9137, pp. 212–230. Springer, Cham (2015). https://doi.org/10.1007/978-3-319-20119-1_16
10. Idomura, Y., et al.: Communication avoiding multigrid preconditioned conjugate gradient method for extreme scale multiphase CFD simulations. In: 2018 IEEE/ACM 9th Workshop on Latest Advances in Scalable Algorithms for Large-Scale Systems (scalA), pp. 17–24. IEEE (2018)
11. Kumar, S., et al.: Scalable data management of the Uintah simulation framework for next-generation engineering problems with radiation. In: Yokota, R., Wu, W. (eds.) SCFA 2018. LNCS, vol. 10776, pp. 219–240. Springer, Cham (2018). https://doi.org/10.1007/978-3-319-69953-0_13
12. Peterson, B., et al.: Demonstrating GPU code portability and scalability for radiative heat transfer computations. J. Comput. Sci. **27**, 303–319 (2018)
13. Schmidt, J., Berzins, M., Thornock, J., Saad, T., Sutherland, J.: Large scale parallel solution of incompressible flow problems using Uintah and hypre. SCI Technical report UUSCI-2012-002, SCI Institute, University of Utah (2012)
14. Schmidt, J., Berzins, M., Thornock, J., Saad, T., Sutherland, J.: Large scale parallel solution of incompressible flow problems using Uintah and hypre. In: 2013 13th IEEE/ACM International Symposium on Cluster, Cloud and Grid Computing (CCGrid), pp. 458–465 (2013)
15. Sridharan, S., Dinan, J., Kalamkar, D.: Enabling efficient multithreaded MPI communication through a library-based implementation of MPI endpoints. In: International Conference for High Performance Computing, Networking, Storage and Analysis, SC14, pp. 487–498. IEEE (2014)
16. Zambre, R., Chandramowlishwaran, A., Balaji, P.: Scalable communication endpoints for MPI+Threads applications. In: 2018 IEEE 24th International Conference on Parallel and Distributed Systems (ICPADS), pp. 803–812, December 2018

Analysis of Checkpoint I/O Behavior

Betzabeth León[✉][iD], Pilar Gomez-Sanchez[iD], Daniel Franco[iD],
Dolores Rexachs[iD], and Emilio Luque[iD]

Computer Architecture and Operating Systems Department,
Universitat Autònoma de Barcelona, 08193 Bellaterra, Barcelona, Spain
{betzabeth.leon,pilar.gomez,daniel.franco,dolores.rexachs,
emilio.luque}@uab.es

Abstract. Nowadays, checkpoints have gained some relevance, given
the increasing complexity of scientific applications for the use of many
resources over a long period of time. Thus, in fault tolerance strategies,
in addition to taking into account the impact that the application itself
has on HPC systems, we must add the impact of the checkpoint. The
checkpoint saves information about the application and the system in
order to be able to restore the application, if necessary, in stable storage.
The checkpoint can be considered as an intensive I/O application, so its
storage need can have a great impact on the application. Therefore, in
this paper, the analysis of the checkpoint's I/O behavior is presented.
The number of checkpoints to be performed in an application is often
related to the maximum overhead that you want to introduce in the
application. If we know the maximum overload the user wants to pay
for and the overhead that a checkpoint introduces, we can calculate the
number of checkpoints to be performed. This overhead depends significantly on the I/O operations. The PIOM-PX tool was used to analyze
the spatial and temporal I/O patterns of the checkpoint. Based on this
analysis, a model was designed to predict their behavior. This information is used to calculate the number of checkpoints to be performed in
an application given a maximum overhead predefined by the user. This
will allow us to understand what happens when a checkpoint is created
in an HPC system, in order to make decisions that adapt to the user's
requirements.

Keywords: Checkpoint · Fault tolerance · I/O behavior · PIOM-PX

1 Introduction

Input/Output (I/O) is an important element that greatly affects the performance
of parallel applications in High Performance Computing (HPC) systems. As
it generates a lot of readings and writes, if they are very frequent they could
impact significantly by collapsing storage and slowing down the execution of
applications. Among the elements related to I/O behavior are the design of the
same applications executed (I/O patterns), the HPC system and, especially, the
storage subsystem (workload and resource management).

© Springer Nature Switzerland AG 2020
V. V. Krzhizhanovskaya et al. (Eds.): ICCS 2020, LNCS 12137, pp. 191–205, 2020.
https://doi.org/10.1007/978-3-030-50371-0_14

In [1], the input and output is differentiated in two ways: productive I/O and defensive I/O. Productive I/O is the writing of data that the user needs for actual science, such as visualization dumps and traces of key scientific variables over time. Defensive I/O is employed to manage a large application executed over a period of time much larger than the platform's Mean-Time-Between-Failure (MTBF). Defensive I/O is used for restarting a job in the event of application failure in order to retain the state of the computation, and hence the forward progress since the last checkpoint. Thus, one would like to balance the amount of resources devoted to defensive I/O and computation lost due to platform failure, which would require restarting the application. As the time spent on defensive I/O (mechanisms for fault tolerance) is reduced, the time spent on useful computations will increase. Checkpoints are a Fault Tolerance (FT) strategy which require intensive large-scale access to the storage system, through I/O operations.

Checkpoints can be differentiated into several types depending on how the processes involved in the execution of the application work with fault tolerance. In this way, if the processes are coordinated to create and store the checkpoint, they are said to be coordinated checkpoints. If each process performs the checkpoint independently, they are non-coordinated checkpoints and if they are coordinated by process groups the checkpoints are named semi-coordinated. In this work, in order to study I/O, coordinated checkpoints will be used because, when carrying out large amounts of simultaneous writes, they intensively access the storage system, so the file system must manage all this information, which can significantly affect the execution time of the application and influence its scalability. In this way, to analyze the influence of scalability, mapping on the size and time to make a checkpoint and having all the complete information of these accesses to the file system can help you manage them better. It is required to have the detailed information of the patterns generated to be able to replicate them without performing the computation, to achieve the most appropriate configuration and management of the file system and to be able to tune our applications with fault tolerance within the system, as well as to be able to carry out an analysis of the performance. This paper will focus on the analysis of the patterns generated by the checkpoint and its impact on the use of resources and generated overheads.

This article is structured as follows: In Sect. 2 related work related to this research will be presented. As a next step, in Sect. 3 we present the tools used to obtain this information and we will justify the selection of the tool used in this work. In Sect. 4 we will make a comparison between some I/O instrumentation tools, thus selecting the most suitable for our work. In Sect. 5 the Analysis of the checkpoint generated by the Distributed MultiThreaded CheckPointing (DMTCP) will be presented. In Sect. 6, a model to estimate the size and storage time of the checkpoint will be implemented. We finish with the conclusions and future work.

2   Related Work

Different authors have studied the I/O behavior and fault tolerance and its impact generating overhead in the applications. Below are some studies related to this research.

In [2] the authors indicate that among the elements that determine the frequency of the checkpoint I/O depend on the choice of the checkpoint interval, the period between checkpoints and the number of checkpoint I/O operations performed by the application. This work is similar to the proposed objective, the authors analyze the I/O operations but they do not consider the behavior (spatial and temporal patterns of the I/O operations). The authors propose an analytical model designed for the simulation, while we propose a behavior model that allows the replication of the behavior in different systems or system configurations. In [3], a checkpointing technique is presented that significantly reduces the checkpoint overhead and is highly scalable. For this asynchronous checkpointing technique, a theoretical model is developed to estimate the checkpoint overhead. In this paper, we propose a model that permits replicating the behavior of the checkpoint in different systems or system configurations, so that with reduced resources we can obtain information that allows us to predict the overhead. In [4], the goal of this research has been to develop a strategy for optimizing parallel I/O over the wide variety of possible access patterns. They have demonstrated that, with the two-phase access strategy for parallel I/O, it is possible to obtain significant improvements in performance over previously used methods. Data distribution and storage distribution have been decoupled, enabling the most effective configuration to be used for parallel. Overlapping and parallel I/O strategies are also compatible with our proposal.

3   Previous Analysis

The coordinated checkpoint requires the coordination of the processes involved in the execution of the application with fault tolerance. For each process a file is generated relative to the work done by that process. There are several factors that influence the size of the checkpoint files that have been generated and the time they have used, among which is the number of accesses to the files and the pattern of these accesses, which is immersed in the workload handled. These patterns can be analyzed or described by their spatial behavior (related to the type of access), the mode (sequential, striped, random), the size of the access, among others, and their temporal behavior (related to the number and frequency of access). Therefore, the checkpoint storage time depends on the size and pattern, as shown below:

$$T_{storage_{ckpt}} = f(size, pattern) \qquad (1)$$

This checkpoint storage time (see expression 1) is part of the total time (see expression 2) required to perform the coordinated checkpoint, as follows:

$$T_{ckpt} = T_{coord_{ckpt}} + T_{storage_{ckpt}} \qquad (2)$$

The coordination time $(T_{coord_{ckpt}})$ consists of the preparation and management time at which the checkpoint starts, after this the storage time begins, where it carries out the storage of the content that concentrates the state of the process at a specific time. The storage stage is where the checkpoint spends more time. Another element that influences the storage time of the checkpoint is the place where it will be stored in the I/O subsystem, such as device type (SSD, HDD) and file system (ext3, NFS, PVFS). But the patterns generated are dependent on the application and therefore independent of the system. Once we have the pattern, we can analyze its behavior in different systems or different configurations. It is important to study it, as it does not only influence the size of the checkpoint, it is also important in relation to the pattern consisting of the frequency, size and amount of writes made in the time of storage, management files, mapping dependence, the number of nodes and the files generated. All these factors become important when the number of processes increases. In our case, one file is generated per process.

The characterization model of the checkpoint I/O patterns shows that it makes different writes of different sizes, both in the coordination and in the content that the checkpoint stores, and these form the total checkpoint size. This stored information is composed of the following three zones: (1) A data zone which depends on the application and where, if we increase the number of processes, the zone decreases. This is because when increasing the degree of parallelism, the same amount of data is split into smaller pieces for each process. (2) A library zone which remains fixed, because it depends on what the application needs functionally from the system. (3) A shared memory zone in which, as we increase the number of processes within the same node, the size of this zone increases. This is because, depending on the number of processes used, a variable amount of memory is reserved for inter-process communications within the same node. This memory reserve depends on the MPI implementation used.

Therefore, the storage times of each zone depend on the size of each write made. Adapting this ratio to the Pearson correlation coefficient (see Eq. 3), whilst the size of the checkpoint zones increases, so does storage time.

$$r_{xy} = \frac{\sum Z_x Z_y}{N} \tag{3}$$

In our case, "x" corresponds to the size of each write, "y" corresponds to the operation write time and "N" is the number of writes made. In this way, the larger the size, the more time it takes and this also depends on the congestion for the number of files. Likewise, for analysis, it is important to have the right tool that can implement all the necessary information to perform the required analysis of these patterns generated by fault tolerance. To get the information about the behavior, we must monitor the application's I/O. There are different trace and analysis tools that help describe the application's I/O behavior. In Sect. 4, a comparative study of three important tools capable of tracing the I/O of the applications is carried out the coordinated checkpoint will also be taken as the case of central use as an application with intensive access to the storage system.

4   Tools Used

According to the parameters obtained above and to have information about their I/O behavior, different tools have been used. Below are the details of the information provided by each and a comparison, taking into account the type of information we need.

4.1   Experimentation Environment

The experiments have been carried out on: AMD Opteron^{TM} 6200 @ CPU 1,56 GHz, Processors: 4, CPU cores: 16, Memory: 256 GiB, File system: ext3 and Disk type: HDD. The MPI implementation used was MPICH 3.2.1. For checkpoints the DMTCP-2.4.5 (Distributed MultiThreaded Checkpointing) has been used for this study [5].

4.2   Application I/O Behavior Tools

To perform the analysis of the checkpoint I/O patterns, it is necessary to have a tool that provides all the information required to study the checkpoint and its structure, how it is formed and what the information it stores is. In this way, there are a large number of tools that can trace the I/O of an application. In this work, we have used and compared three tools widely referenced in the literature used in HPC: Darshan, PIOM and Strace.

Darshan captures information about each file opened by the application. Rather than trace all operation parameters, however, Darshan captures key characteristics that can be processed and stored in a compact format. Darshan instruments POSIX, MPI-IO, Parallel netCDF, and HDF5 functions in order to collect a variety of information [6].

PIOM allows us to define an I/O behavior model based on the I/O phases of HPC applications at MPI-IO and POSIX-IO level. For every file used by the application, an I/O file is created. The spatial and temporal patterns are extracted from information contained in the I/O file [7].

Strace is a diagnostic, debugging and instructional user space utility for Linux. It is used to monitor and tamper with interactions between processes and the Linux kernel, which include system calls, signal deliveries, and changes of process state. Strace works by using the ptrace system call which causes the kernel to halt the program being traced each time it enters or exits the kernel via a system call [8].

In the following tables we will make a comparison between Darshan, PIOM and Strace; and we show some general characteristics that are important for the user. In Table 1 we can observe the characteristics related to its installation and operation. The selected criteria were the following: Table 1 shows the elements that the tool traces. -File Access Type: This corresponds to the type of operation that is identified, for example: open, close, write, read, among others.

Table 1. Elements traced by the tool

Input				
Tool	File access type	Bursts	Trace the checkpoint	Trace the restart
Darshan	Yes	Together	Yes	No
PIOM	Yes	Independent	Yes	No
Strace	Yes	Independent	Yes	Yes

-I/O Bursts: A burst is the grouping of I/O operations of the same type. -Trace the checkpoint: This indicates the type of fault tolerance strategy being implemented. -Trace the restart: This indicates whether there is a trace if the starting point is a restart.

The three tools can identify the types of file access, but Darshan summarizes the total operations of the same type (burst) while PIOM and Strace show the operations separately, thus being able to identify the size and time of each one, as well as other indicators. With respect to the checkpoint, the three tools can trace their information, but with respect to the restart, only Strace can do it. However, this aspect does not present a major problem because the restart behaves in the same way as the checkpoint but instead of being "writings" they are "readings". Table 2 shows some elements of the monitoring considered. -Overhead: Corresponding to the overhead introduced by the tool. -Administrative Privileges: Corresponding to whether it is necessary for the tool to be executed with administrator privileges. -Transparent: Not interfering in the application execution. The overhead introduced by Darshan and PIOM is smaller than the one introduced by Strace, because it implements a lot of information related to the monitoring and manipulation of interactions between processes and the Linux kernel, which includes system calls, deliveries of signals and changes in the state of the process. Regarding the administration privileges for its use, once they are already installed, the three tools can be used without having administrator privileges, as none of the three hinder the execution of the application.

Table 3 shows the execution times of the NAS parallel benchmarks with one checkpoint, the execution times with fault tolerance, as well as the instrumentation tool used in each case. The purpose of this is to be able to check the time overhead generated by each of them.

Table 2. Monitoring tool

Monitoring			
Tool	Overhead	Administrative privileges	Transparent
Darshan	Small	No	Yes
PIOM	Small	No	Yes
Strace	Medium	No	Yes

Table 3. Overhead introduced by each of the tools

App	Time (sec.)			
	APP	DARSHAN	STRACE	PIOM
BT.B.4 N:1	139.93	140.12	142.15	141.52
BT.C.4 N:4	431.12	432.16	433.81	432.71
BT.D.25 N:1	2568.30	2834.91	3508.72	2641.45
SP.D.25 N:1	3810.71	4032.89	5221.62	4011.14
LU.D.25 N:1	3947.86	4186.34	4378.81	4097.10

In Table 3 it can be seen that the executions of the application BT, SP and LU applications class B [9] with fault tolerance implemented with Strace is the longest execution. As for the other two tools, Darshan and PIOM, it remained similar, that is, in the cases studied, half the time was less with PIOM and the other half with Darshan. This indicates that these two tools introduce less overhead when they are implementing the application with checkpoint.

I/O Information: Other important information to be analyzed is that generated by the I/O when the checkpoint is being stored. In the case of the PIOM tool, spatial information (number and size of writes), type of access (sequential, random) and temporal information (frequency of accesses) are analyzed in order to identify the bursts or consecutive accesses to the disk. Darshan gives profile information with a number of writes grouped by size and does not allow us to identify bursts.

Table 4 shows the difference in the I/O behavior identified in the BT, SP and LU class B applications with 4 processes with DMTCP, identified by the PIOM and Darshan tools. It also shows the writes identified, which present a difference between both tools of less than 4% and for the size of less than 3%. In addition, it shows the information per file (each process generates a file), that is, for BT, SP and LU class B applications with 4 processes executed on a single node (N: 1), the information for each of the four files (Fx). As all four files have similar behavior, to present the information in the table, only two files are shown for each application.

Table 4. I/O information identified by PIOM and Darshan by file

App	Files	PIOM	Darshan	Difference %Num. writes	PIOM	Darshan	Difference %MiB
		No. writes			MiB		
BT.B.4	F0	209	216	3.24	155.14	156.59	0.93
N:1	F1	209	216	3.24	154.56	156.00	0.92
SP.B.4	F0	215	228	4.44	139.55	141.02	1.05
N:1	F1	221	226	1.34	138.99	140.43	1.02
LU.B.4	F0	219	225	2.67	93.16	95.57	2.52
N:1	F1	217	224	3.13	93.16	94.68	1.60

In Table 5, a comparison is made between PIOM and Strace with respect to the "writes" generated and the bursts, considering a burst as a number of continuous operations of the same type, in this case of "writes", as well as the size of each of them. Table 5 shows that, in terms of the first burst, both tools detect the same number of "writes" 44 writes and a size of 0.0078 MiB for the checkpoint. On the other hand, for burst two there is a difference between 0 to 5% between the number of writes that detect these tools and in terms of size, it is very similar, so this aspect is less than 1% difference.

Table 5. I/O information identified by PIOM and Strace by file

App	File	PIOM			Strace			Difference
		Burst 1	Burst 2	Total	Burst 1	Burst 2	Total	%Num.
		No. writes			No. writes			No.writes
BT.B.4	F0	44	165	209	44	155	199	5.03
N:1	F1	44	165	209	44	157	201	3.98
	File	MiB			MiB			%MiB
	F0	0.0078	155.13	155.14	0.0078	154.52	154.48	0.43
	F1	0.0078	154.55	154.56	0.0078	154.56	154.49	0.05
SP.B.4	File	No. writes			No. writes			%No. writes
N:1	F0	44	171	215	44	173	217	0.92
	F1	44	177	221	44	173	217	1.84
	File	MiB			MiB			%MiB
	F0	0.0078	139.54	139.55	0.0078	138.97	138.98	0.41
	F1	0.0078	138.98	138.99	0.0078	138.97	138.98	0.01
LU.B.4	File	No. writes			No. writes			%No. writes
N:1	F0	44	175	219	44	167	211	3.79
	F1	44	173	217	44	166	210	3.33
	File	MiB			MiB			%MiB
	F0	0.0078	93.22	93.23	0.0078	93.15	93.16	0.07
	F1	0.0078	93.22	93.23	0.0078	93.14	93.15	0.08

Table 6 shows some aspects related to the output report: Postprocessing (It implies what must be done after the tool is executed in order to analyze the information), Temporary Pattern (Informing on the order that the access events are carried out to the file), Space Pattern (Providing information about what file is used by the application, which process accessed a file and where (File positions) it was accessed by the process during the application execution), Generated File (Referring to the amount of trace files generated) and Size Generated File (Referring to the size of trace files generated).

With respect to the reports generated by these three tools, Darshan does not need a post-processing, because with the same utility programs that it features, several ready reports can be generated that show the information in a summary way. With respect to PIOM, post-processing is medium because it presents an orderly report with each of the operations identified, as well as generating a report for each process executed with all its information. However, certain calculations must be made to obtain the total operations, total size and time. As for Strace, post-processing must be carried out by the user, because it is only a monitoring tool and it is much more complex because it generates a lot of information and as it issues a single report, the required information must be thoroughly searched. The three tools show the temporal and spatial patterns, PIOM and Strace show them by operation, although Darshan shows them summarized. In relation to the size of the reports generated, the smallest is Darshan, on average. In the case of PIOM and Strace, as they generate so much information, the size of the report is larger.

We consider that the three tools used adequately instrument the I/O of the applications with checkpoint. But based on the results of these experiments and the qualitative comparison, we consider that PIOM is the tool of instrumentation that most adapts to the study we want to perform in relation to the analysis of the generated patterns of the I/O of the applications with fault tolerance.

5   Analysis of the Checkpoint Generated by the DMTCP

5.1   Spatial and Temporal Pattens

Figure 1 shows the information identified by PIOM of a checkpoint of a BT.D.25 application. We can observe the moment in which the checkpoint began to be stored and each of the zones that compose it, that is, the zone of data (DTAPP), the library zone (LB) and the shared memory zone (SHMEM), as well as the size of each zone. Knowing this, we can know the time it took to store each zone and the total time to store all the zones.

Likewise, PIOM also provides us with information about the total execution time of the application plus the checkpoint and the total size that will be stored. All this information has been validated with the information generated by the DMTCP, which is the following: Checkpoint time: 138.57 s. Checkpoint size: 1205.51 MiB (This size is per file, as is a BT.D.25 must be multiplied by 25 processes, to obtain the total stored). App time + checkpoint: 2641.45 s.

Table 6. Output tool

Output					
Tool	Post-processing	Show temporary pattern	Show space pattern	Generated file	Size generated file
Darshan	Easy	Summarized	Summarized	1	Small
PIOM	Medium	By operation	All	1 per process	Medium
Strace	Complex	By operation	All	1	Large

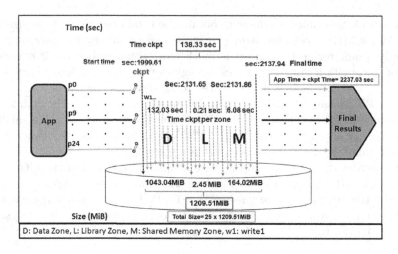

Fig. 1. Spatial and temporal patterns identified by PIOM

In the case of this example, it indicates that the time to store the checkpoint is 138.57 s, which is similar to that shown by PIOM when adding the three zones 138.33 s. In addition to the total size stored by each checkpoint file 1205.51 MiB and is also similar to the information identified by PIOM 1209.51 MiB. In this sense, the total time of the application with fault tolerance in this case is 2641.45 s and the one identified by PIOM for this example was 2237.03 s. With this information generated by PIOM, we obtain important details of the patterns, being able to identify and compare them with great congruence with what the DMTCP generates.

Table 7 shows a comparison between the times and sizes of each zone in the execution of the BT benchmark with class D, with 64 and 16 processes and with I/O. In this table we can see some elements which impact the execution time of the application with fault tolerance.

Table 7. Elements that affect the temporal and spatial pattern

App		Ckpt zones		
		Data zone	Library zone	Shared Memory zone
BT.D.64.mpi_io_full	Time	327.96	2.67	537.28
	Size per file	478.16	2.53	533.86
	Total size	30602.25	161.75	34167.04
BT.D.16.mpi_io_full	Time	135.28	0.12	3.01
	Size per file	1662.89	2.52	108.76
	Total size	26606.24	40.32	1740.16

If we analyze the results presented by each zone, we can observe the following: the data zone with 64 processes occupied a larger amount of time than in the other case with 16 processes, although the workload was equal in all two cases. When distributing it among the 64 processes, the size to handle for each process was smaller. Therefore, this data zone is impacted by the amount of information that must be stored from the application itself and the number of processes that it handles, producing congestion in the storage system.

In relation to the library zone, it can be observed that when it is handling 16 processes, the size per file is the same, regardless of the workload, but when increasing the number of processes to 64, the time also increases, as does the total to be stored in to this zone.

With respect to the shared memory zone, it can be seen that this is directly impacted by the number of processes used within the same node. In the case of 64 processes, it can be observed that the size is much larger than that of 16 processes, although the workload handled is the same. Therefore this zone is independent of the workload handled, so the time increases as the number of processes increases, because the size also grows.

In the case shown in Table 7, all the files generated by the checkpoint have been stored locally and an ext3 file has been used (Experiments have been carried out on several file systems, but we have illustrated this with ext3). In this way, the number of processes used is an element that significantly impacts the runtime of the application with fault tolerance, because a bottleneck is formed when all processes try to access the storage system at the same time because it is a coordinated checkpoint.

5.2   A Comparative analysis of the I/O Patterns of Applications with Fault Tolerance

In order to compare the I/O of an application with the I/O of the checkpoint of the same application, the first graph in Fig. 2 shows the behavior of the I/O of BT.C.16.mpi_io_full. In this case, it made a total of 440 writes in forty bursts of 10 writes, each write of 16 MiB, and the last of each burst 2.18 MiB. Therefore, the number of writes and their sizes exhibit regular behavior. In the second graph in Figure 2, the I/O behavior of a checkpoint executed in BT.C.16.mpi_io_full can be observed. In this case, it made a total of 241 writes, and we can observe that the writes have different sizes; there are a great number of very small 4 KiB ones and few large ones of up to 111.73 MiB. In the same way, the time varies if we compare when making the small writes, which can take thousandths of seconds, whereas a big write can take more than 12 s. Therefore, the I/O behavior of the checkpoint is not regular. In both cases, we can observe that the time depends on the writing size.

After observing the trace generated by PIOM, we need to know the correlation between the size of each writing with the time it took to make. In this way, we can find this if we use the Pearson correlation coefficient, which is defined as the covariance between two typified variables. By applying Eq. 3 to our traces to understand the correlation between the size and time of the checkpoint scripts,

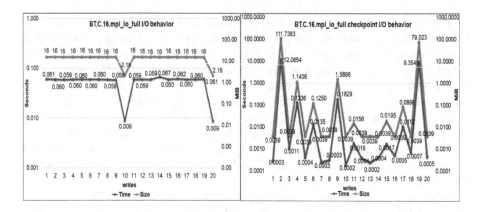

Fig. 2. I/O behavior (writes size and time)

we can see an association level of 0.992. Therefore, a positive correlation between both variables is observed whereas the write size increases the time. Therefore, the size of each write is an important element to study because it significantly affects the time of the checkpoint.

5.3   Checkpoint Storage Time Estimation

In [10] a study was conducted to predict the size of the checkpoint. In this article, equations were obtained to calculate the size of the areas that make up the checkpoint. In the case of the DTAPP Zone, it was calculated through an equation that was made from the number of processes and the size of the zone. In this way, the size of this zone could be estimated with a different number of processes. With respect to the LB zone in all cases, the size of this zone of 2.45 MiB was obtained. In this case, it was recommended to identify the area in a run and verify the size, because if it is the same application it does not change even if the number of processes and the workload change. With regards to the SHMEM zone, as the size of this zone depends on the number of processes, the equation obtained was as follows:

$$SHMEM_{Size} = 0.0617x^2 + 3.9983x + 25.47 \qquad (4)$$

The storage time of the checkpoint depends on the size and as noted above, we can already predict the size of each of the checkpoint zones. Therefore, we can estimate the storage time approximately. In order to try to get closer to the value of time, we have calculated the regression equation for each of the zones, obtaining the following equations:

$$DTAPP_{Time} = 1.46041308E - 07x - 0.00404367 \qquad (5)$$

$$LB_{Time} = 1.3205654E - 07x - 0.00014156 \qquad (6)$$

$$SHMEM_{Time} = 1.5359626E - 07x - 0.0109242 \qquad (7)$$

In these equations "x" corresponds to the size of each of the zones. So as to validate the three equations shown above 6, 7, 5, the execution of the following applications was executed and measured with a checkpoint: BT.C.16, BT.C.25, BT.C.36, BT.C.49, BT.C.64 , BT.D.25, BT.D.36, BT.D.49, BT.D.64, SP.C.64, LU.C.36 on one node and BT.D.64 on two nodes.   These applications were used

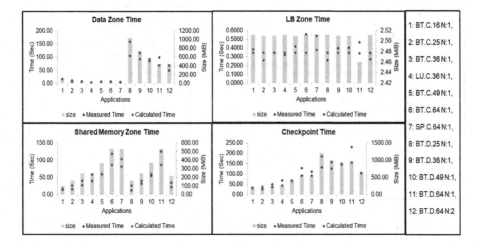

Fig. 3. Comparison between measured and calculated time

with different workloads, number of processes and nodes, in order to be able to vary the sizes of the checkpoint zones and check the validity of the equations. In Fig. 3 you can see the time in which the data zone was stored, the library zone, the shared memory zone and the total time of the checkpoint. In general, it is observed that the time calculated with the equations increases as the size grows; with respect to the measured time this also remains similar, but in some cases there is variability that occurs because it is being stored on the hard disk and this can impact on the variability of the storage time.

In this way the calculated results obtained from the equations presented an approximation with the actual measured values. The variability in the times can be considered acceptable to estimate the storage time of the checkpoint.

5.4   Approximate Model the Number of Checkpoints to Execute

This section presents an approximate model for obtaining the number of checkpoints in a given execution (Fig. 4). For this, the checkpoint size must first be calculated, identifying and calculating the zones that integrate it. The DTAPP zone, depending on the application, should look for the regression equation, the LB zone is constant and once verified it is sufficient and the SHMEM zone is calculated using the equation indicated in the model.   As a next step, the storage time of each zone must be calculated, since the time depends on the size of the

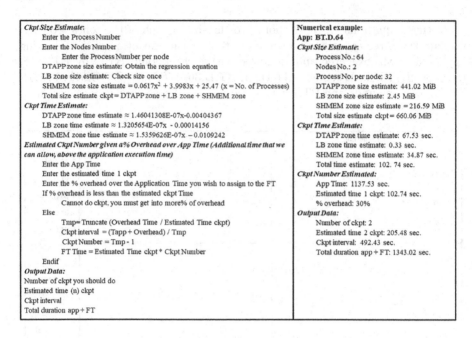

Ckpt Size Estimate:	**Numerical example:**
Enter the Process Number	**App: BT.D.64**
Enter the Nodes Number	*Ckpt Size Estimate:*
Enter the Process Number per node	Process No.: 64
DTAPP zone size estimate: Obtain the regression equation	Nodes No.: 2
LB zone size estimate: Check size once	Process No. per node: 32
SHMEM zone size estimate $= 0.0617x^2 + 3.9983x + 25.47$ (x = No. of Processes)	DTAPP zone size estimate: 441.02 MiB
Total size estimate ckpt = DTAPP zone + LB zone + SHMEM zone	LB zone size estimate: 2.45 MiB
Ckpt Time Estimate:	SHMEM zone size estimate = 216.59 MiB
DTAPP zone time estimate $\approx 1.46041308E\text{-}07x\text{-}0.00404367$	Total size estimate ckpt = 660.06 MiB
LB zone time estimate $\approx 1.3205654E\text{-}07x - 0.00014156$	*Ckpt Time Estimate:*
SHMEM zone time estimate $\approx 1.5359626E\text{-}07x - 0.0109242$	DTAPP zone time estimate: 67.53 sec.
Estimated Ckpt Number given a% Overhead over App Time (Additional time that we	LB zone time estimate: 0.33 sec.
can allow, above the application execution time)	SHMEM zone time estimate: 34.87 sec.
Enter the App Time	Total time estimate: 102. 74 sec.
Enter the estimated time 1 ckpt	*Ckpt Number Estimated:*
Enter the % overhead over the Application Time you wish to assign to the FT	App Time: 1137.53 sec.
If % overhead is less than the estimated ckpt Time	Estimated time 1 ckpt: 102.74 sec.
Cannot do ckpt. you must get into more% of overhead	% overhead: 30%
Else	*Output Data:*
Tmp= Truncate (Overhead Time / Estimated Time ckpt)	Number of ckpt: 2
Ckpt interval = (Tapp + Overhead) / Tmp	Estimated time 2 ckpt: 205.48 sec.
Ckpt Number = Tmp - 1	Ckpt interval: 492.43 sec.
FT Time = Estimated Time ckpt * Ckpt Number	Total duration app + FT: 1343.02 sec.
Endif	
Output Data:	
Number of ckpt you should do	
Estimated time (n) ckpt	
Ckpt interval	
Total duration app + FT	

Fig. 4. Approximate model of the number of checkpoints to execute

zone, three regression equations have been obtained which are approximate when there is no congestion in the node, because if there were congestion the values could vary a lot. In order to calculate the estimated checkpoint number given an overhead percentage over the application time, the application execution time, the estimated checkpoint storage time obtained with the previous equations must be known. Then you enter the percentage of overhead you want to have on the execution of the application and you get the total time of the application with fault tolerance, the number of checkpoints that must be performed in a given time interval.

6   Conclusions and Future Work

The defensive I/O that generates fault tolerance directly affects the application, increasing the execution time. We have observed in this article that the pattern generated by the checkpoint is very irregular, which makes it difficult to predict. The storage time of the checkpoint depends on the size and patterns. In this way, knowledge of these spatial and temporal patterns will allow us to predict the size and storage time of the checkpoint so that with this information you can calculate the number of approximate checkpoints that can be made in a given time, when there is no congestion in the node. In addition, it allows us to establish policies and develop tools that help to replicate their behavior in any system as well as being able to establish configuration methods and strategies to reduce the overload generated by the fault tolerance I/O. In this way, as a future work, we plan to obtain a more exact model of prediction, taking into consideration the elements that affect the congestion of the node. In addition, we will continue to research with other types of fault tolerance strategies and develop utilities that focus on the behavior of defensive I/O in order to assess its impact and reduce it.

Acknowledgment. This publication is supported under contract TIN2017-84875-P, funded by the Agencia Estatal de Investigación (AEI), Spain and the Fondo Europeo de Desarrollo Regional (FEDER) UE and partially funded by a research collaboration agreement with the Fundación Escuelas Universitarias Gimbernat (EUG).

References

1. Subramaniyan, R., Grobelny, E., Studham, S., George, A.D.: Optimization of checkpointing -related I/O for high-performance parallel and distributed computing. J. Supercomput. **46**, 150–180 (2008)
2. Arunagiri, S., Daly, J.T., Teller, P.J.: Modeling and analysis of checkpoint I/O operations. In: Al-Begain, K., Fiems, D., Horváth, G. (eds.) ASMTA 2009. LNCS, vol. 5513, pp. 386–400. Springer, Heidelberg (2009). https://doi.org/10.1007/978-3-642-02205-0_27
3. Shahzad, F., Wittmann, M., Zeiser, T., Hager, G., Wellein, G.: An evaluation of different I/O techniques for checkpoint/restart. In: 2013 IEEE International Symposium on Parallel Distributed Processing, pp. 1708–1716 (2013)
4. del Rosario, J.M., Bordawekar, R., Choudhary, A.: Modeling and analysis of checkpoint I/O operations. ACM SIGARCH Comput. Archit. News **21**, 31–38 (1993)
5. Jason, A., Arya, K., Cooperman, G.: DMTCP: transparent checkpointing for cluster computations and the desktop. In: 23rd IEEE International Parallel and Distributed Processing Symposium (2007)
6. Carns, P., et al.: Understanding and improving computational science storage access through continuous characterization. ACM Trans. Storage (TOS) **7**, 1–26 (2011)
7. Gomez-Sanchez, P., Mendez, S., Rexachs, D., Luque, E.: PIOM-PX: a framework for modeling the I/O behavior of parallel scientific applications. In: Kunkel, J.M., Yokota, R., Taufer, M., Shalf, J. (eds.) ISC High Performance 2017. LNCS, vol. 10524, pp. 160–173. Springer, Cham (2017). https://doi.org/10.1007/978-3-319-67630-2_14
8. The strace developers. Strace, 2001–2019. https://strace.io/
9. Bailey, D.H., et al.: The NAS parallel benchmarks. Int. J. High Perform. Comput. Appl. **5**, 63–73 (1991)
10. Leon, B., Franco, D., Rexachs, D., Luque, E.: Impact of the checkpoint on the scalability of the parallel applications. In: The 2019 International Conference on High Performance Computing & Simulation (HPCS 2019)(Accepted) (2019)

Enabling EASEY Deployment of Containerized Applications for Future HPC Systems

Maximilian Höb[(✉)] and Dieter Kranzlmüller

MNM-Team, Ludwig-Maximilians-Universität München,
Oettingenstraße 67, 80538 Munich, Germany
hoeb@mnm-team.org
www.mnm-team.org

Abstract. The upcoming exascale era will push the changes in computing architecture from classical CPU-based systems towards hybrid GPU-heavy systems with much higher levels of complexity. While such clusters are expected to improve the performance of certain optimized HPC applications, it will also increase the difficulties for those users who have yet to adapt their codes or are starting from scratch with new programming paradigms. Since there are still no comprehensive automatic assistance mechanisms to enhance application performance on such systems, we propose a support framework for future HPC architectures, called EASEY (Enable exASclae for EverYone). Our solution builds on a layered software architecture, which offers different mechanisms on each layer for different tasks of tuning, including a workflow management system. This enables users to adjust the parameters on each of the layers, thereby enhancing specific characteristics of their codes. We introduce the framework with a Charliecloud-based solution, showcasing the LULESH benchmark on the upper layers of our framework. Our approach can automatically deploy optimized container computations with negligible overhead and at the same time reduce the time a scientist needs to spent on manual job submission configurations.

Keywords: Auto-tuning · HPC · Container · Exascale

1 Introduction

Observation, Simulation and Verification build the pillars of most of today's HPC applications serving different goals of diverse scientific domains. Those applications have changed in the last decades and years, and they will and have to change again, driven by several factors. More applications, more scientists, more levels of detail, more data, more computing power, more of any contributing part. This more of everything needs to be satisfied by current and future computing systems, including not only computing power, storage and connectivity, but also direct application support from computing centers.

Such a support is essential to execute a demanding high performance application with huge data sets in an efficient manner, where efficiency can have several

© Springer Nature Switzerland AG 2020
V. V. Krzhizhanovskaya et al. (Eds.): ICCS 2020, LNCS 12137, pp. 206–219, 2020.
https://doi.org/10.1007/978-3-030-50371-0_15

objectives like time to solution or energy consumption. The latter will be the crucial factor to minimize to achieve an exascale system that serves the scientific community and does not stress our environment.

Computing efficiency in means of likely optimal usage of resources in an acceptable time to solution is heavily investigated from different sites. Many workflow management frameworks promise enhancements on data movement, deployment, management or reliability, like in SAGA, a grid based tool suit described in [11] or Pegasus, a scientific pipeline management system presented in [7]. The scalability of such frameworks is limited by the state-of-the-art services, where bottlenecks are found inside and across today's supercomputing facilities.

Solutions to this challenge will not be found in one single place. Instead, it will be the most optimal interaction of exascale-ready services, hardware and applications. To support todays and future application developers, automatic assistant systems will be a key enabling mechanism, also shown by Benkner et al. in [2]. Computing systems will continue to develop and change faster than applications can be adapted to build likely optimal synergies between the application and the underlying system. Heterogeneity among already connected centers will additionally increase the complexity.

Although scientist want to simulate, calculate, calibrate or compare data or observations always with the same application core, many hurdles slow down or stop those scientist to deploy on newer systems, since the effort to bring their application on new hardware or on new software stacks is too high in comparison to their actual work in their scientific domains. An astro physicist needs to focus on physical problems, not on how to deploy applications on a computing cluster.

This is supported by a survey from Geist and Reed in [9] who state, that the complexity of software needs to be reduced to reduce software development costs. This could also be solved if we can encourage application developers to include building blocks, which will adapt and optimize code (compare Sect. 6) or executions automatically, like proposed in our paper.

Therefore, new assistant systems are needed to make access to new super-computing centers easier and possible also for unexperienced scientist. The complexity of those systems requires knowledge and support, which usually only the computing centers themselves can offer. Since their time is limited also the diversity of the applications might be limited.

This work introduces a framework to enable container applications based on Docker to be transformed automatically to Charliecloud containers and executed on leading HPC systems by an integrated workflow management system. This transformation includes also the possibility to define mounting points for data. Charliecloud is considered the most secure container technology for supercomputers, since Singularity is not entirely free of breaches as reported in [6].

In this work we propose initial steps towards the first comprehensive framework, already including auto-tuning mechanisms focused on containerized applications. Using containers to encapsulate an application with all its dependencies and libraries introduces portability in general. With EASEY also specific libraries can be added to the portable container to optimize the performance on a target system automatically.

Within this paper we introduce the underlying technology and present the functionality of the framework, evaluated by a hydrodynamics stencil calculation benchmark. We show, that our approach adds automatically cluster dependent building bricks, which improve the utilization of the underlying hardware only by acting on the software layer of the container. With this auto-tuning mechanism, we reduce the necessary time domain scientists need to invest in deploying and managing their HPC jobs on different clusters and can in stead concentrate on their actual work in their domain.

The paper is ordered as follows. Section 2 introduces the architecture of our framework integrated into the layered architecture of supercomputing systems. Afterwards, Sect. 3 presents the necessary configuration needed for EASEY. Section 4 evaluates this approach with a benchmark use case. Related Work to this paper is presented in Sect. 5 and Sect. 6 closes with a summary and an outlook of future work to extend this approach to other layers of the HPC architecture.

2   EASEY Architecture

Enabling scientists to focus on their actual work in their domain and remove all deployment overhead from their shoulders is the main goal of tour approach. And while we reduce the necessary interaction between scientist and compute cluster, we also apply performance optimization on the fly. High performance systems need applications to adapt their technology to talk their language. We are able to add such optimizations while preparing the containerized application.

The architecture of the EASEY system is detailed in Fig. 1, integrated as two building bricks in the layered HPC architecture. On the upper *Applications and Users layer* the EASEY-client is mainly responsible for a functional build based on a Dockerfile and all information given by the user. The middleware on the *local resource management layer* takes care of the execution environment preparation, the data placement and the deployment to the local scheduler. The additional information service can be pulled for monitoring and status control of the execution through. The *hardware layer* of the compute cluster underneath remains not included in any optimization in this release of the framework (compare future work in Sect. 6).

2.1   EASEY Client

The client as the main service for any end-user prepares the basis of the execution environment by collecting all needed information to build a Charliecloud container. Therefore, the main information is given by the user with the Dockerfile. This file can also be pulled from an external source and needs to include all necessary steps to create the environment for the distinguished tasks. The client needs at some point root privileges to build the final container, hence, it needs to be deployed on a user's system like a workstation or a virtual machine in a cloud environment.

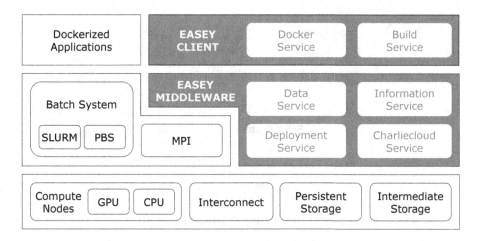

Fig. 1. EASEY integration in the layered architecture of HPC systems

As a first step, the *Docker Service* builds a docker image. Since HPC clusters can require local configurations, the *Docker Service* adds local dependencies to the Dockerfile. Therefore, the user needs to specify the target system in the build command: `easey build Dockerfile --target cluster`, e.g. *easey build Dockerfile --target "lrz:supermuc-ng"*.

In addition for mpi-based applications, the actual mpi-version needs to match the system's version. In the Dockerfile the user can specify the position where the cluster's mpi-version needs to be integrated by including the line `###includelocalmpi###`, which will be replaced by the client with the actual purge of all other mpi-versions and the compilation of the needed one. This should be done before the target application is compiled to include the right mpi libraries.

As a final step the later mounting point will be created as a folder inside the Docker image. The path was defined inside the configuration file (see Listing 1.2). Also specific requirements from the target system will be handled here, for example to include local libraries and functionalities inside the container (e.g. symlinks to special folders). Those requirements are known by the EASEY system and don't need to be provided by the user.

In the same environment as the *Docker Service* the *Build Service* will transform the before created Docker image to a Charliecloud container archive. The service will call the Charliecloud command `ch-builder2tar` and specify the Docker image and the build location.

2.2 EASEY Middleware

The second building brick is the EASEY Middleware, which connects and acts with the resource manager and scheduler. The main tasks are *job deployment*, *job management*, *data staging* and creating the *Charliecloud environment*.

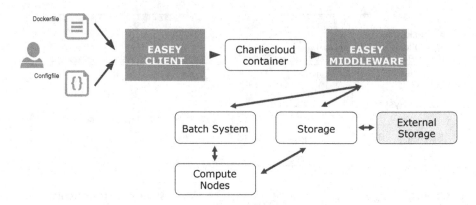

Fig. 2. EASEY workflow of the job submission on HPC systems

Thereby, a workflow is started including besides the before created Charliecloud container and configuration file, the local cluster storage and batch system as well as the user specified external storage, if needed. A schematic view of the major steps is given in Fig. 2.

Starting with the user, the Dockerfile and the filled EASEY configuration file need to be included in a build call of the EASEY client, which is running on a user system with root privileges. Within this process a Docker container is built and transformed to a Charliecloud container, which again is packed in a tar-ball.

The EASEY middleware can be placed inside the target cluster or outside on a virtual machine for example. The framework will start the preparation of the submission based on the information given in the configuration. This can also include a data stage-in from external sources. To place data and later to submit a cluster job on behalf of the user, EASEY needs an authentication or delegation possibility on each contributing component. At this time of the development the only possibility included is access grants via public keys. This means in detail, if the EASEY middleware runs outside the cluster, that the public key of the host system needs to be added to the *authorized keys* inside the cluster. Thereby, EASEY can transfer data on the user's storage inside the cluster.

Also the following job deployment needs a working communication from the middleware to the cluster submission node. The deployment based on the given configuration (see Listing 1.4) follows a well reinforced approach. The complete algorithm is shown in Algorithm 1.

Additionally to the already mentioned tasks the *data folder* is only created if at least one input or output file in specified. EASEY requires the user to place the output file after the computations inside this folder, mounted inside the Charliecloud container. For each input file, EASEY controls the transfer inside the data folder.

For the submission on the local batch system a batch file is required following also local specifications, known by EASEY. The resource allocation information is provided by the user in the configuration (*number of nodes, ram, ...*).

Algorithm 1. EASEY submission

Require: Charliecloud tar-ball
Require: EASEY configuration file
Require: User credentials
  Move tar-ball to cluster storage
  Extract tar-ball and create execution environment
  if *data* in configuration **then**
    mkdir data_folder
  end if
  while *input* in configuration **do**
    transfer *input*[*source*] to data_folder
  end while
  create batch_file
  for each *deployment* in configuration **do**
    parse to SLURM or PBS command in batch_file
  end for
  while *execution* in configuration **do**
    add command to batch_file
  end while
  submit batch_file to local scheduler and return *jobID* to EASEY

For SLURM or PBS those are parsed into a valid form, other scheduler are not supported so far.

The actual computations follow after this prolog, described as *executions* by the user. For each the corresponding bash or mpi commands are also included. If data is required as input parameters, the user has to specify them relatively to the data folder, where they are placed.

Since the local job ID is known by the middleware the user can pull for the status of the job. In addition to *pending, running, finished* or *failed*, also error log and standard output is accessible, also at an intermediate state. After the job ended EASEY will transfer output files if specified.

This workflow includes all necessary steps to configure and run an originally Docker based application on a HPC cluster. Thereby, it saves time any scientist can use for actual work in their domain and removes any human overhead especially if such computations need to be deployed regularly. In the same time, it adds optimization mechanisms for the actual computing resource. In the following section, details on the user's mandatory configuration are presented.

3 EASEY Configuration

Our approach requires a full and valid description of all essential and optional parts of the submission. Therefore we defined a json-based configuration file including all required information. This file needs to be provided by the user together with the application's Dockerfile. The configuration consists of four main parts: *job, data, deployment* and *execution*.

Job Specification. This part of the configuration can also be seen as mandatory meta data for a valid job management. The keys of the key-value pairs are presented in Listing 1.1.

An EASEY job needs to have an unique identifier, a hash which is determined by the system at the moment of submission, a user specified *name* and a *mail address* to contact the end-user if specified. Besides those, no further information is mandatory.

Listing 1.1. Job Specification

```
{"job":{"name","id","mail",
 "data":{..},
 "deployment":{..},
 "execution":{..}}
}
```

Listing 1.2. Data Specification

```
"data":{"input":[
 {"source","protocol",
  "user","auth"}],
"output":[
 {"destination","protocol",
  "user","auth"}],
"mount":{"container-path"}}
```

Data Service Specification. Our backend is able to fetch accessible data files via the protocols https, scp, ftp and gridftp. The latter is planed to be implemented in the next release. For the others already available only the path to the source and the protocol needs to be declared. If the data needs to be accessed on a different site, authentication with public-key mechanism is necessary. The *input* is declared as an array and can include several declarations of different input files.

The backend will place all input files in one folder, which will be mounted into the container on the relative position declared as *path*.

After the complete execution an *output* recommend as an archive can be also moved again to a defined destination. Also here a public key mechanism would be mandatory.

Deployment Service Specification. The deployment service offers basic description possibilities to describe necessary resources for the execution.

As shown in the next section, within one deployment only one job is allocated. Therefore, each execution commands specified in Listing 1.4 will be run on the same allocation. The specifications regarding *nodes, ram, taks-per-node* and *clocktime* will be translated into scheduler specific commands and need to be specified given in Listing 1.3.

Listing 1.3. Deployment Specification

```
"deployment":{"nodes",
 "ram","cores-per-task",
 "tasks-per-node","clocktime"
}
```

Listing 1.4. Execution Specification

```
"execution":[{
 "serial":
  {"command"},
 "mpi":
  {"command","mpi-tasks"}
}]
```

Although there exist much more possible parameters, at this state of the framework only those are implemented, since all others are optional.

Execution Service Specification. The main ingredients of HPC jobs are the actual commands. The *execution* consists of in principle unlimited *serial* or *mpi* commands. Those are executed in order of sequence given inside the *execution* array as shown in Listing 1.4. In all considered HPC jobs the only kinds of commands are bash (*serial*) or *mpi*-based commands.

A complete example is given in Listing 1.5 showing a practical description of the evaluated use case. The presented configuration will of course be adapted whenever necessary. However, the main goal is to stay as generic as possible to connect and enable as many combinations of applications on the one side and resource management systems on the other. The next section evaluates this approach regarding the computational and the human overhead.

4   Evaluation

The previously presented framework builds the basis for further development. The main goal is to introduce auto-tuning on several layers. In this paper, we presented the EASEY client and middleware to support scientists deploying a Docker-based application on a HPC cluster without interacting with the local resource themselves.

This framework was tested on one of the fastest HPC systems in the world, the SuperMUC-NG, a general purpose system at the Leibniz Supercomputing Center[1] in Garching, listed ninth in the Top500 list in November 2019 and has a peak performance of 26.87 Petaflops, computing on 305,856 Intel Xeon Platinum 8174 CPU cores, without any accelerators. All compute nodes of an island are connected with a fully non-blocking Intel Omnipath OPA network offering 100 Gbit/s, detailed in [3]. SuperMUC-NG uses SLURM as a system scheduler.

We used a Dockerimage for LULESH, the Livermore Unstructured Lagrangian Explicit Shock Hydrodynamics benchmark. It is a widely used proxy application to calculate the Sedov blast problem that highlights the performance characteristics of unstructured mesh applications. Details on the application and the physics are described by Karlin et al. in [13].

This benchmark in version 2.0.3 was ported by the MNM research team (Fürlinger et al., described in [8]) to DASH, a C++ template library for distributed data structures, supporting hierarchical locality for HPC and data-driven science. Adopting the Partitioned Global Address Space (PGAS) programming model, DASH developed a template library that provides PGAS-like abstraction for important data containers and allows a developer to control and take advantage of the hierarchical data layout of global data structures. The authors showed, that DASH offers a performance advantages of up to 9%.

As described in Sect. 3 the EASEY client requires a Dockerfile and a configuration specifying the deployment, data and execution parameters. Since LULESH

---

[1] https://www.lrz.de/english/.

does not require any data, the json configuration shown in Listing 1.5 contains only *job meta data*, *deployment* and *execution*. Values, which are determined by EASEY, or which are not defined by the user (e.g. *ram* since there are no special memory requirements) are not set.

The actual execution is given by the *command* keyword. In this case a charliecloud container is started with *ch-run* and a data volume is mounted with the -b flag, *-b source:target*. Inside the container *lulesh.dash* the command */built/lulesh.dash -i 1000 -s 13* is executed. Together with the *mpi-tasks* LULESH is ran with a cube size of 2.197 cells, a cube mesh length of 13, and in 1.000 iterations. The maximum runtime is limited to *6 hours* and passed to the SLURM scheduler.

Listing 1.5. LULESH:DASH Execution Specification

```
{"job":{
 "name":"LULESH:DASH","id":"",
 "mail":"hoeb@mnm-team.org",
 "deployment":{
  "nodes":"46","ram":"","cores-per-task":"1",
  "tasks-per-node":"48","clocktime":"06:00:00"
 },
 "execution":{
  "serial":
   {"command":"echo \"Starting LULESH:DASH\""},
  "mpi":
   {"command":"ch-run -b /lrz/sys/.:/lrz/sys -w lulesh.dash
      -- /built/lulesh -i 1000 -s 13",
    "mpi-tasks":"2197"},
  "serial":
   {"command":"echo \"Finished LULESH:DASH\""},
 }
}
```

This setup was used to run several execution of the DASH LULESH and the DASH LULESH inside the Charliecloud container, on up to 32,768 cores. As it can be seen in Fig. 3, the figure of merit (FOM) shows slightly higher values (higher is better) for native DASH than for the Charliecloud runs. The FOM values of the Charliecloud executions are lower for runs with more than 4,000 cores. With less cores they differ under 1% (compare Table 1). This can be seen in detail in Fig. 4, where the FOM value is divided through the number of cores. Ideally we would see a horizontal line, however, the difference between this line and the measurements corresponds to the application, which does not have perfect linear scaling. However, the scaling behavior of the containerized application is similar to the native one although some overhead introduced by Charliecloud is visible.

The detailed mean measurements between the dash version of native LULESH and the EASEY container execution inside the Charliecloud container can be seen in Table 1, where the number of cores (and mpi-tasks) correspond

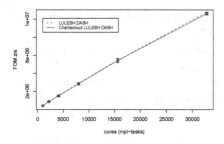
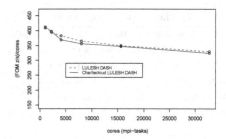

Fig. 3. Weak scaling on SuperMUC-NG   **Fig. 4.** FOM per cores SuperMUC-NG

to cubic numbers of the given input cube side length. The shown delta varies from $+0,8\%$ to $-3,6\%$ of the FOM values. This spread can be explained by the limited number of runs, that could be performed on the target system, the SuperMUC-NG of the LRZ, and statistical deviations.

However, the usage of a Charliecloud container adds some overhead to the execution shown in the measurements with more the 4,000 cores. This overhead needs to be compared to the invested effort on executing this application on the system. With EASEY it was possible to execute and measure the application without manual interaction on the system itself. The so added performance overhead is within an acceptable interval. Especially for runs with many CPUs (10,000+) this overhead does not increase significantly. This is especially important, since our framework targets later Exascale systems and can already show today its scalability.

Table 1. FOM comparison: lulesh:dash native and inside charliecloud container.

Cube length p	Cores p^3	Nodes	FOM EASEY	FOM NATIVE	Δ
10	1,000	21	412,122.1	409,204,8	0,71 %
13	2,197	46	873,366.4	866,515,2	0,78 %
16	4,096	86	1,511,665.1	1,566,899,9	−3,65 %
20	8,000	167	2,846,589.0	2,916,102,0	−2,44 %
25	15,625	326	5,423,072.1	5,461,509,5	−0,71 %
32	32,768	683	10,627,767.7	10,805,287,0	−1,67 %

The goal of these measurements was not to show an optimal scaling behavior of the application, it was to demonstrate the validity of the approach. Although there might be some additional overhead due to Charliecloud, EASEY could reproduce the scaling behavior and very closely the overall performance of the original, manually compiled application without any container framework. This shows that the approach of EASEY adds only negligible overhead to the performance. In the same time it saves the scientist time by automatically tuning some adjusting screws.

With the weak scaling shown in Fig. 3 we can show, that our approach scales as well as the manually compiled application without any container environment. Automatically enabling a container deployment on such a supercomputing cluster and in the same time applying local tuning possibilities show, that EASEY is a promising approach. It is also likely that such assistance systems will increase to number of users using those HPC systems and in the same time enabling them to include as much optimization as possible, without changing anything manually.

The time of scientists is limited and we want to enable physicists, chemists, engineers and all others to focus on their domain. They want to optimizes the application regarding the scientific outcome, while our framework takes care of the deployment. We also want to encourage more scientists not to be afraid of such huge systems. The learning curve is high, if someone wants to use a Top500 supercomputer system. However, with EASEY, there exists a solution to use a more and more common praxis: Docker container. General purpose systems like the SuperMUC-NG are made for general purpose applications. With the presented performance in this section, we can substantially offer an additional deployment approach on those systems, for everybody.

5   Related Work

The presented framework and its implementation bases on the development towards containerization and the abilities such encapsulated environments offer.

Charliecloud and Docker. Priedhorsky and Randles from the Los Alamos National Laboratory introduced in 2017 in [16] a lightweight open source implementations of a container framework: Charliecloud. The authors followed their basic assumptions that the need for user-defined software stacks (UDSS) increases. Dependencies of application's still need to be compiled on the actual target HPC systems since not all of them are available in the stack provided by the compute center. Todays and future users need particular dependencies and build requirements, and more over also portability and consistency to deploy applications on more than one system. This is offered by Charliecloud, which bases on Docker to build an UDSS image.

The advantage of Charliecloud lays in the usage of the user namespace, supporting non-privileged launch of containerized applications. Within this unprivileged user namespace also all other privileged namespaces are created without the requirement of root privileges. Therewith, any containerized application can be launched, without requiring privileged access to the host system, as described from Brayford et al. in [3]. In the same paper, the authors investigated the performance of Charliecloud scaling an AI framework up to 32 nodes. Their findings showed a similar, negligible overhead, although our performance analysis included more nodes. Concerning possible security issues the authors stated that Charliecloud is safe, since it only runs inside the non-privileged user namespace.

Docker, described by Merkel in [14] is considered an industry standard container to run an applications in an encapsulated environment. Nevertheless, since some containers require root privileges by default and others can not prevent privilege-escalation in all cases, as shown in [4], Docker is not considered a safe solution when deployed on shared host systems.

Besides Charliecloud and Docker a newer daemon less container engine attracts more and more attention. Podman, described in [15], provides functionalities for developing, managing, and running Open Container Initiative containers and container images. Future work of this paper will include a substantial analysis of Podman and its possible enhancements for EASEY.

Shifter and Singularity. Charliecloud was also compared to other approaches like Shifter and Singularity. The Shifter framework also supports Docker images (and others, e.g. vmware or squashfs), shown by Gerhardt et al. in [10]. In contrast, it is directly tied into the batch system and its scalability and security outside the initial cluster is not shown so far.

Also Singularity was developed to be encapsulated into a non-privileged namespace, security issues have been detected, for example in [6], where users could escalate the given privileges. An example is detailed in [3].

Choosing the right container technology is crucial, especially regarding the security of the host and other users. Since Charliecloud is considered secure and shows promising scaling behavior, we choose this technology for our framework, however, a Singularity extension might be added at a later stage.

Including Charliecloud in such a framework, only one related approach could be discovered so far: BEE.

BEE. The authors of *Build and Execution Environment BEE* in [5] propose an execution framework which can, besides others, also deploy Charliecloud container on a HPC infrastructure. Their approach focuses on a variety of different cloud and cluster environments managed by the same authority. This broad approach tries to unify the execution of the same container. Compared to EASEY, which aims to auto-tune the performance for Petaflop-systems, it does not include any optimization to the underlying infrastructure.

BEE also includes a submission system for deployment of jobs, but deploys each single run command as one job. Our approach focuses on complex computations which might also include several steps within one *job*. Regarding the data service, no possibility is provided in *BEE* to connect to an external resource for data stage-in or -out.

We consider *BEE* as a valid approach to deploy the same container on many different technologies. However, we focus on auto tuning of adjusting screws to gain performance advantages, and our target infrastructures are high performance systems in the range of Petaflops and later Exaflops.

6 Conclusion

The presented architecture and its components are considered the starting point for further investigations. The heterogeneous landscape of computing and storage facilities need applicable and efficient solution for the next generation computing challenges. Exascale is only the next step as Petascale was a few years ago. We need to built assistent systems which can be adapted to the actual needs of different communities and enable those to run their applications on more and more different and complex hardware systems. On the other hand, building and using entire Exascale systems will require enhancements in all pillars like fault tolerance, load-balancing and scalability of algorithms themselves. EASEY aims to enable scientists to deploy their applications today on very large systems with minimal interaction. With the ongoing research, we aim also to scale well on a full Exascale system in the future.

To close the gap between application developers and resources owners, a close collaboration is needed. In fact, the presented solution for an efficient usage of containerized applications with Charliecloud is only the first part. We need to go deep in the systems, optimize the hardware usage on a node or even CPU, accelerator and memory level and convince the scientific communities, that also their applications are able to scale up to such a level. Here, such an assistant system like EASEY, which automatically deploys optimized applications, will convince the communities and enable scientists to focus on their work.

Future Work. As mentioned throughout this paper, this version of EASEY is the first step towards a comprehensive framework to enable easy access to future Exascale systems. Those systems might have a hybrid setting with CPUs and accelerators side by side. EASEY will be extended to more layers which will also operate on the abstraction of the computing unit and introduce a code optimization which aims to optimize certain executions with more efficient ones adapted to the target system.

Porting originally CPU-based applications to such hybrid systems will require more research. Enable an efficient but easy to use approach will base on several layers, which functionalities and interfaces will be the main target of the future research question of this work.

The next direct steps focus on efficient data transfers, also including the model of data transfer nodes as investigated in the EU-funded project PROCESS and published in [12] and [1]. For such a data access also the authentication mechanism needs to be enhanced.

Acknowledgment. The research leading to this paper has been supported by the PROCESS project, which has received funding from the European Union's Horizon 2020 research and innovation programme under grant agreement No 777533.

References

1. Belloum, A., et al.: Design of data infrastructure for extreme-large data sets. In: Deliverable D5.1. PROCESS (2018)
2. Benkner, S., Franchetti, F., Gerndt, H.M., Hollingsworth, J.K.: Automatic application tuning for HPC architectures (dagstuhl seminar 13401). In: Dagstuhl Reports, vol. 3. Schloss Dagstuhl-Leibniz-Zentrum fuer Informatik (2014)
3. Brayford, D., Vallecorsa, S., Atanasov, A., Baruffa, F., Riviera, W.: Deploying AI frameworks on secure HPC systems with containers. arXiv preprint arXiv:1905.10090 (2019)
4. Bui, T.: Analysis of docker security. arXiv preprint arXiv:1501.02967 (2015)
5. Chen, J., et al.: Build and execution environment (BEE): an encapsulated environment enabling HPC applications running everywhere. In: 2018 IEEE International Conference on Big Data (Big Data), pp. 1737–1746, December 2018. https://doi.org/10.1109/BigData.2018.8622572
6. CVE-2019-11328. Available from MITRE, CVE-ID CVE-2019-11328, May 2019. http://cve.mitre.org/cgi-bin/cvename.cgi?name=CVE-2019-11328
7. Deelman, E., et al.: Pegasus, a workflow management system for science automation. Future Gener. Comput. Syst. **46**, 17–35 (2015)
8. Fürlinger, K., Fuchs, T., Kowalewski, R.: DASH: A C++ PGAS library for distributed data structures and parallel algorithms. In: Proceedings of the 18th IEEE International Conference on High Performance Computing and Communications (HPCC 2016), pp. 983–990. Sydney, Australia, December 2016. https://doi.org/10.1109/HPCC-SmartCity-DSS.2016.0140
9. Geist, A., Reed, D.A.: A survey of high-performance computing scaling challenges. Int. J. High Perform. Comput. Appl. **31**(1), 104–113 (2017)
10. Gerhardt, L., et al.: Shifter: containers for HPC. J. Phys. Conf. Ser. **898**, 082021 (2017)
11. Goodale, T., et al.: SAGA: a simple API for grid applications. High-level application programming on the grid. Comput. Meth. Sci. Technol. **12**(1), 7–20 (2006)
12. Hluchý, L., et al.: Heterogeneous exascale computing. In: Kovács, L., Haidegger, T., Szakál, A. (eds.) Recent Advances in Intelligent Engineering. TIEI, vol. 14, pp. 81–110. Springer, Cham (2020). https://doi.org/10.1007/978-3-030-14350-3_5
13. Karlin, I., Keasler, J., Neely, J.: Lulesh 2.0 updates and changes. Technical report, Lawrence Livermore National Lab. (LLNL), Livermore, CA (United States) (2013)
14. Merkel, D.: Docker: lightweight linux containers for consistent development and deployment. Linux J. **2014**(239), 2 (2014)
15. Podman. Available on GitHub, March 2020. https://podman.io
16. Priedhorsky, R., Randles, T.: Charliecloud: unprivileged containers for user-defined software stacks in HPC. In: Proceedings of the International Conference for High Performance Computing, Networking, Storage and Analysis, pp. 1–10 (2017)

Reproducibility of Computational Experiments on Kubernetes-Managed Container Clouds with HyperFlow

Michał Orzechowski[1], Bartosz Baliś[1(✉)], Renata G. Słota[1],
and Jacek Kitowski[1,2]

[1] Department of Computer Science, AGH University of Science and Technology,
Krakow, Poland
{morzech,balis,rena,kito}@agh.edu.pl
[2] AGH University of Science and Technology, ACK Cyfronet AGH, Krakow, Poland

Abstract. We propose a comprehensive solution for reproducibility of
scientific workflows. We focus particularly on Kubernetes-managed container clouds, increasingly important in scientific computing. Our solution addresses conservation of the scientific procedure, scientific data,
execution environment and experiment deployment, while using standard
tools in order to avoid maintainability issues that can obstruct reproducibility. We introduce an Experiment Digital Object (EDO), a record
published in an open science repository that contains artifacts required
to reproduce an experiment. We demonstrate a variety of reproducibility scenarios including experiment repetition (same experiment and conditions), replication (same experiment, different conditions), and propose a smart reuse scenario in which a previous experiment is partially
replayed and partially re-executed. The approach is implemented in the
HyperFlow workflow management system and experimentally evaluated
using a genomic scientific workflow. The experiment is published as an
EDO record on the Zenodo platform.

Keywords: Scientific workflows · Reproducibility · Cloud computing ·
Application containers · Container clouds · Kubernetes

1   Introduction

Reproducibility, a fundamental quality of the experimental scientific method,
requires conservation of three basic component of scientific experiments [15]: *scientific procedure* describing the steps of the experiment, *scientific data* required
as input of the experiment and produced as its results, and *scientific equipment* necessary to conduct the experiment. In the case of computational sciences, the scientific procedure can be described as a scientific workflow [6], input
data are typically files, while the equipment is the execution environment of the
experiment.

Application containers are often viewed as a means to scientific workflow reproducibility [3,17], but containers alone solve only part of the problem enabling conservation of individual software components. Some approaches

© Springer Nature Switzerland AG 2020
V. V. Krzhizhanovskaya et al. (Eds.): ICCS 2020, LNCS 12137, pp. 220–233, 2020.
https://doi.org/10.1007/978-3-030-50371-0_16

tackle reproducibility of the broader execution environment, but propose non-standard solutions that suffer from lack of maintainability [16], and do not support reproducibility of complex experiment deployments. In this paper, we discuss modern *container clouds* as a comprehensive solution for reproducibility of computational experiments. We propose a solution for reproducibility of scientific workflows, focusing on Kubernetes-managed container clouds. The main contributions of this paper are as follows:

- We provide a solution that supports conservation of *in silico* experiments at various levels: the scientific procedure, the scientific data, the execution environment, and the experiment deployment.
- The solution covers a variety of reproducibility scenarios, including repetition, replication, and *smart reuse* of in silico experiments.
- We introduce an *Experiment Digital Object*, a record that captures information required to reproduce a scientific workflow experiment.
- We study the implementation of the solution in the context of Kubernetes-managed container clouds, an increasingly important computing infrastructure for scientific computing.
- We show an evaluation of the solution by studying reproducibility scenarios of a genomic scientific workflow. The experimental part is published as an Experiment Digital Object on the Zenodo platform as a demonstration of the solution.

In the paper, we adopt the following terminology for different types of reproducibility (partially adopted from [5]):

- **repetition**: the experiment is reproduced in exactly the same conditions, including the workflow specification, input data and the execution environment (OS, libs, even machine types and cluster configuration).
- **replication**: the same experiment is reproduced but in a different execution environment.
- **(smart) reuse**: intermediate results from a previous experiment are partially reused (without re-computing them), whereas the rest is re-executed due to changes, e.g. a new version of the scientific software.
- **reproduction**: broad term denoting any of the above cases.

The paper is organized as follows. Section 2 surveys current work on reproducibility of scientific workflows. Section 3 describes the proposed reproducibility solution, while Sect. 4 contains its experimental evaluation. Section 5 concludes the paper.

2   Related Work

Container clouds are increasingly adopted as an infrastructure for computational experiments [4,9]. A container cloud adds an additional layer on top of virtual machines, in which clusters of nodes are formed from VMs and are managed by a *container management platform*. Container managers such as Mesos, Docker Swarm and Kubernetes have been studied in the context of scientific workflow management [11,18]. However, while many papers focus on application containers as a tool for achieving reproducibility of computational experiments [3,17], containers alone address only the conservation of the runtime environment of individual software components, not the entire execution environment.

Several approaches employ logical conservation of the target execution environment for scientific workflows. In [16], a custom OWL ontology for describing the computational infrastructure used in a computational experiment is proposed. In [14], TOSCA specifications of the underlying infrastructure are used to deploy workflows on different cloud platforms in a portable way. The problem with such proprietary solutions that focus on portability and abstraction [1,12] is their maintainability, a very important quality crucial for reproducibility. Given the complexity and dynamic evolution of computing infrastructures, it is difficult to maintain in-house provisioning and configuration management tools dedicated for scientific computing. In our experience, even industry-leading tools, such as Terraform, sometimes have this problem and cannot be treated as a universal solution in every situation. Finally, existing approaches do not address reproducibility of the entire complex *experiment deployment*.

Our approach supports conservation of the experiment procedure, data, execution environment, and experiment deployment. First, using Docker containers ensures conservation of the operating system and software packages. For provisioning of the computing infrastructure (virtual machines, storage resources), we rely on standard tools that fit best the particular computing environment and satisfy infrastructure reproducibility requirements, such as Terraform, Kubespray, or native tools of the cloud provider. Finally, we argue that configuration management of individual components and nodes is not enough. Complex experiment deployments – orchestration of component startup, dynamic provisioning of persistent volumes, replication and fault tolerance strategies, etc. – also need to be described in a reproducible way. That is where Kubernetes comes into play as a universal application deployment and execution platform. We chose Kubernetes [13] because it is supported by virtually all major cloud providers (Amazon EKS, Azure AKS, Google GKE) and because it is portable and universal – complex deployments can be reproduced using the same Kubernetes description files on various infrastructures. Kubernetes is also increasingly used in scientific computing [4]. The advantage of this approach is not only high reproducibility, but also portability and maintainability.

3   Experiment Reproducibility with HyperFlow

This section describes the overall methodology for reproducibility of scientific workflows on Kubernetes-managed container clouds using the HyperFlow workflow management system. Details regarding reproducibility capabilities of HyperFlow on the level of workflow description, workflow execution and workflow deployment are also described.

3.1 Experiment Digital Object Record

Figure 1 presents an overview of the *Experiment Digital Object* (EDO) record, a collection of data and executable files that contain information on all software and data artifacts involved in the experiment, execution traces, and scripts for their basic analysis. This record is created after the experiment and can be published on an Open Science Platform. We have chosen Zenodo because of its relative popularity. Zenodo allocates a unique DOI identifier to the record which can be used in references. Records published in Zenodo are immutable which is important for reproducibility. Changes can be made and published as a new version of the record.

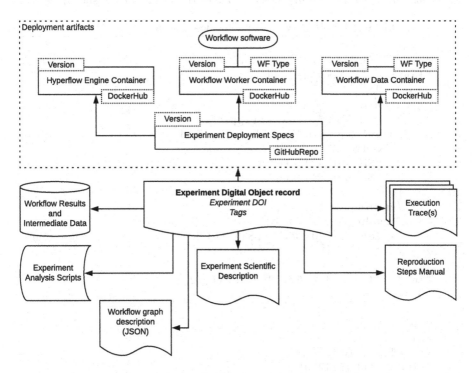

Fig. 1. Experiment Digital Object (EDO) record contains all information required to reproduce a computational experiment managed by the HyperFlow WMS.

An EDO record is an entry point to experiment reproduction. It describes specific steps required to reproduce the experiment, but also points to specific versions of all deployment artifacts involved in this particular experiment. It also contains execution traces collected during the experiment run, for convenience converted to a CSV format. Experiment analysis scripts are also provided that process the execution trace as a data frame and produce useful charts.

3.2   Workflow Description Language

HyperFlow introduces a simple workflow description language with strong semantics [2]. HyperFlow has been shown to successfully convert and run work-flows from such system as Pegasus [7]. Such workflow interoperability is an important aspect of reproducibility. Listing 1.1 shows a fragment of a Hyper-Flow workflow description. The description contains mainly two sections: an array of *processes* (workflow tasks), and an array of *signals* (inputs and outputs of tasks). The listing shows an example process entry in which the *name* identifies the type of processing performed by the task, while *config* contains information passed to the executor on a remote node in order to run the task. The signal entry describes a file which is one of the inputs of the shown task.

Listing 1.1. HyperFlow workflow description (fragment).

```
1  processes: [ {
2    "name": "alignment_to_reference",
3    "function": "k8sCommand",
4    "config": {
5      "executor": {
6        "executable": "bwa-wrapper",
7        "args": [ "mem", "-t", "2", "-M",
8          "Gmax_275_v2.0.fa",
9          "USB-001_1.fastq",
10         "USB-001_2.fastq"
11       ],
12       "cpuRequest": "1",
13       "memRequest": "500M",
14       "stdout": "20180321-083514-USB-001_aligned_reads.sam"
15     }
16   },
17   "ins": [1,2,3,4,5,6,7,8,9,10,11,12],
18   "outs": [13]
19 }, ... ],
20 signals: [ {
21   "name": "Gmax_275_v2.0.fa",
22   "type": "file",
23   "size": "990744229",
24   "md5sum": "3aa6cf1962f5260cf1405e82efb25c71"
25 }, ... ]
```

Such workflow representation is flexible and annotations can be easily (manually or automatically) added to the workflow description without interfering with other parts of the description. Several such annotations are shown in this example:

- CPU and memory requests (lines 12–13) are inserted by an execution planner (on the basis of historical execution traces or a prediction model) and are used by the underlying Kubernetes scheduler to optimize the placement of workflow jobs. During experiment replication these values can be changed to perform controlled experiments regarding observation of their effect, e.g., on the execution time.

- Detailed information about file size and its md5 hash (lines 23–24) are useful during experiment reproduction to verify if the file has not been changed or corrupted.

3.3   Workflow Execution

Workflow Execution Traces. During workflow execution, HyperFlow collects various information, such as job execution events (creation, pending, start and completion), resource usage metrics (time series) and provenance records. This information is converted to data frames and saved as CSV files which are included in the Experiment Digital Object record, along with the description of columns. Such an approach is quite versatile. Data from many experiments can be analyzed in a data frame parallel processing framework (e.g. Pyspark), or imported into a chosen database. Addition of new columns preserves backward compatibility, provided that the format and semantics of "old" columns remain unchanged. Simple data analysis tools (python scripts) are included in the EDO record for quick and simple presentation of the experiment.

Workflow Persistence and Smart Reuse. HyperFlow provides mechanisms for persistence and recovery of workflow executions using the *event sourcing* approach. In event sourcing, all changes made to the system during the execution are recorded as *events*. To recover the state of the system, events are replayed and the state changes are applied again but without repeating the side effects of the original operations. In the case of scientific workflows, the record of execution events contains information about completed job executions along with inputs consumed and outputs produced by them. The side effects not repeated during the recovery are actual running of the jobs and creation of their output files.

HyperFlow supports advanced *smart reuse* scenarios in which a previously recorded experiment is partially replayed (by reusing intermediate data from the experiment record), while other parts of it have been changed and need to be re-executed. To support smart reuse, intermediate data produced during the experiment must be persisted. Smart reuse scenarios include the following cases:

- Some input data files have been updated. Consequently, tasks consuming these inputs and all their *successors* (dependent tasks) need to be re-executed.
- Part of the workflow software has been upgraded to a new version. Tasks using this software and all their *successors* need re-execution in this case.
- Some intermediate data has been corrupted, so it cannot be reused during experiment reproduction. In this case, the intermediate data needs to be recreated, but only if it is needed by the parts of the workflow that must be re-executed. This can lead to a cascade re-execution of some or all *predecessor* tasks of the affected task that produces the intermediate data in question.

HyperFlow provides a tool called `hflow-recovery-analysis` which annotates an existing experiment execution records with information on which tasks need re-execution based on a configuration file consisting of the following entries:

```
1  changed: {
2    "selector": <spec>,
3    "value": "<value>"
4  }
```

Each entry specifies one or more workflow objects (processes or data) that have been changed. For example {"selector":"process.name","value":"foo"} selects all processes whose *name* is *foo*. Consequently all tasks invoked by such processes and their successors will be re-executed. Since the selector may apply to any attribute found in the workflow description and can address groups of objects, this mechanism is flexible and effective even for very large workflow graphs. Moreover, the tool can also be pointed to a directory where workflow intermediate data is located to check if it is not missing or corrupt. If this is the case, additional tasks can be be marked for re-execution.

3.4 Workflow Deployment and Execution Environment

The final aspect of reproducibility concerns the computing environment and the deployment of the experiment in this environment. Let us describe this aspect at different levels in the context of Kubernetes-managed container clouds.

Computing Infrastructure. The computing infrastructure level comprises Virtual Machines, storage resources, and the Kubernetes cluster. There are several options to address reproducibility at this layer, depending on the target infrastructure. In fact, we argue that choosing the right tool in the right situation is crucial.

- Terraform[1] is a widely used Infrastructure-as-Code tool designed to provision reproducible infrastructures, and supporting a large number of providers, including all major clouds.[2]
- Kubespray a tool that automates deployment of "bare-metal" Kubernetes clusters (that includes deployments on IaaS clouds).[3]
- Native clients of a particular infrastructure. This approach can be very effective for managed Kubernetes clusters offered in the PaaS model, such as Google GKE, Amazon EKS, or Azure AKS.

In the experiments shown in this paper we use the third option as it is sufficient and effective.

OS, Libs and Application Software. This layer is where application containers come into play. We use Docker containers for the software involved in the workflow execution: HyperFlow engine, Redis server, workflow workers, and – in the particular deployment used in this paper – the NFS server. Versioning of container

---

[1] https://www.terraform.io.

[2] https://www.terraform.io/docs/providers/type/major-index.html.

[3] https://github.com/kubernetes-sigs/kubespray.

images is crucial for reproducibility. We employ a Continuous Integration/Deployment (CI/CD) pipeline which automatically builds and publishes new container images whenever a tagged commit is pushed into the appropriate repository.

Workflow Data. We support conservation of workflow data by employing *data containers*. A data container contains input files of the workflow or, alternatively, provisions them before the execution. Data containers with specific input data sets can be prepared, versioned and published. They can be generic and used in a family of similar workflows, or very specific, created for a particular workflow, and even contain the workflow graph description file. The image of the workflow data container contains metadata information regarding such details. Moreover, data containers can be created to conserve output and intermediate data of a particular experiment run.

Experiment Deployment. A computational experiment is not only a collection of software and data artifacts, but rather their concrete, complex deployment which also must be reproducible. For this purpose, we use the *Kubernetes YAML manifests* which allow reproducible, declarative deployment approach. The structure of the deployment used in the experimental evaluation is depicted in Fig. 2. The *HyperFlow engine* container runs the workflow described in the `workflow.json` file and creates *Workflow worker containers* running workflow jobs. A job container runs a *HyperFlow job executor* which communicates with the HyperFlow engine through a *Redis service* running in a separate container. Workflow data is shared through a *Persistent volume* which uses an *NFS server* running in another container. The NFS server is populated with workflow data via the *Workflow data container*. A deployment contains many important configuration parameters that needs to be preserved, e.g. environment variables that influence the behavior of individual components.

Listing 1.2. Example customization of the experiment deployment configuration.

```
1   apiVersion: apps/v1
2   kind: Deployment
3   metadata:
4     name: hyperflow-engine
5   spec:
6     template:
7       spec:
8         containers:
9         - name: hyperflow
10            env:
11            - name: HF_VAR_WORKER_CONTAINER
12              value: "hyperflowwms/soykb-workflow-worker:v1.0.6"
```

To version the deployment, we simply point to a particular commit in the github repository and use Kustomize[4] to configure the deployment with specific

---

[4] https://github.com/kubernetes-sigs/kustomize.

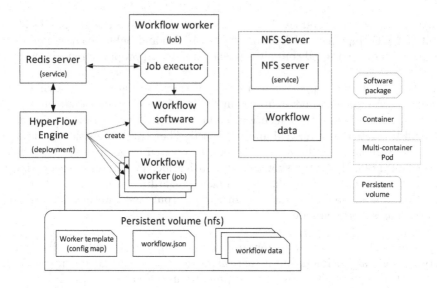

Fig. 2. Scientific workflow deployment used in the reproduction experiments.

versions of the containers. Kustomize allows one to create declarative customizations of the base Kubernetes YAML manifests. A customization is simply one or more YAML files that change (patch) specific parts of their counterparts from the chosen base. In our case, the customization captures specific parameters of the deployment of the original experiment, mainly the containers and their versions, as shown in the example on Listing 1.2. A customization is also a convenient place to modify these parameters and reproduce the experiment in different conditions, e.g. using new software versions. Finally, a customization can be published as an artifact documenting the experiment conditions and enabling its reproducibility.

4   Experimental Evaluation

4.1   Methodology, Experiment Setup and Original Run

In the experiments, we have used the SoyKB genomic workflow [10]. Four experiments have been conducted:

- The *original run* of the workflow on a Google Kubernetes Engine with four nodes, each having 2 virtual CPUs and 1.93 GB memory. This run produced the Experiment Digital Object record along with other artifacts.
- The *repetition run* in which the experiment was repeated exactly in the same conditions.
- The *replication run* wherein the experiment was replicated on our in-house Kubernetes cluster deployed at the Cyfronet Computing Centre. For this experiment, we set up a 4-node cluster, where each node had 6 vCPUs and 22 GB of memory.

– The *reuse run* wherein the smart reuse capabilities of HyperFlow were demonstrated using the GKE cluster.

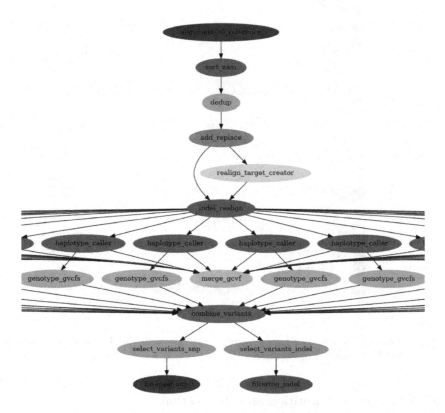

Fig. 3. Visualization of the experimental SoyKB workflow (simplified).

The experimental workflow consists of 52 tasks, as depicted in Fig. 3, and requires 2.6 GB of input data. The EDO record of the original experiment run was assigned a DOI and was published on Zenodo[5]. The execution of the original run is visualized in Fig. 4 (top). The chart, produced from the execution traces using a python script (published as part of the EDO record), shows the execution time of all workflow tasks, type of tasks (denoted by different colors), and their mapping to nodes whose names are indicated on the y axis. Note that the same node occurs multiple times, denoted as *nodeName-n*, if tasks run in parallel on this node. In this case, all tasks were configured with a *cpuRequest* of 0.5 vCPU. Since each node provided 1.93 vCPU and about 0.5 of it was reserved by the middleware, only two parallel tasks could fit in a node without exceeding the 1.93 vCPU supply, which is visible on the graph. We have tried to lower the vCPU requests so that more tasks would fit a node, but it resulted in job evictions due to out of memory errors.

---

[5] https://doi.org/10.5281/zenodo.3659211.

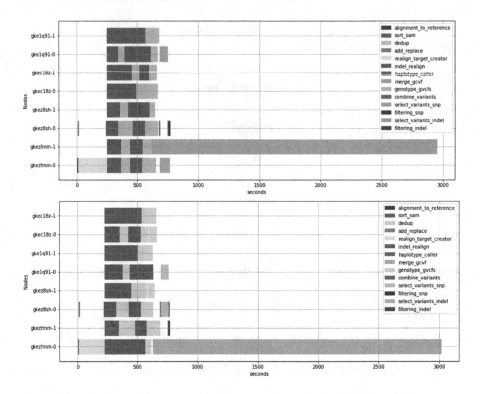

Fig. 4. Visualization of the original and repeated run of the SoyKB workflow on a 4-node Google Kubernetes Engine cluster.

4.2   Experiment Repetition and Replication

Figure 4 (bottom) shows a visualization of the repetition of the experiment in the same infrastructure, while Fig. 5 shows its replicated execution on the Kubernetes cluster in the Cyfronet Computing Centre.

The repeated run is very similar, even the general mapping of tasks to nodes is much the same, albeit with some differences. This hints that the execution environment behaves in a relatively predictable way despite its complexity. This was proven by more runs which yielded similar results. The replicated experiment runs on much faster nodes, so all concurrent workflow tasks could run in parallel in the cluster. The resulting makespan is therefore significantly better.

4.3   Smart Experiment Reuse

The final experiment demonstrates the smart reuse scenario. This scenario is made possible by HyperFlow's capability to persist its execution state. In compliance with the event sourcing model, this is done in the following steps: (1) run a workflow job to completion; here, job represents an operation that changes the workflow execution state but also causes side effects (writes files); (2) persist an

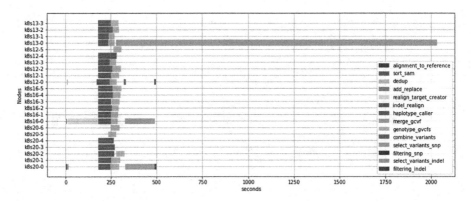

Fig. 5. Visualization of the replicated experiment run on Cyfronet Kubernetes cluster.

event `job completed` in an *execution journal* file; (3) update the internal state of the workflow engine. The execution journal file can be used to *replay* a previous workflow execution, either in part or in full. A common scenario is fault tolerance where the workflow is restarted from a point its execution failed, after the cause of the failure has been resolved. The smart reuse scenario involves a full replay of the workflow, but some its parts are marked as changed and these are forced to be re-executed. To this end, the user prepares an appropriate configuration file which in our case looks as follows:

```
1  [ changed: {
2    "selector": "process.name",
3    "value": "genotype_gvcfs"
4  } ]
```

This configuration indicates that all workflow tasks whose name is `geno-type_gvcfs` have been changed. The nature of the change can vary, e.g. it can denote the upgrade of the workflow software used by the tasks. The user runs the `hflow-recovery-analysis` tool which annotates the execution journal file marking all `genotype_gvcfs` tasks and their successors (task subtree) for re-execution.

The visualization of this execution is shown in Fig. 6. As one can see, only the last 25 tasks of the workflow have been repeated. It must be noted that the workflow can be re-executed in such a way from an arbitrary point only as long as the side effects of the original run – i.e. the intermediate data – are persisted and can be reused. The annotation tool can optionally verify if this is the case and, if some intermediate data are missing, analyze the execution graph and mark additional predecessor tasks for re-execution. In our case, we simply reused the storage volume of the original run which contained the intermediate data, but in a production setting a more robust storage solution would be required. Also, a trade-off should be considered whether it is more cost-effective to re-execute a task and generate its intermediate data again, or persist the data for future reuse. This is, however, out of scope of this paper.

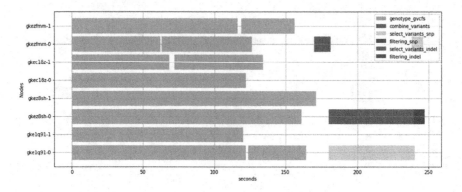

Fig. 6. Visualization of the experiment reuse scenario.

5   Conclusions and Future Work

The reproducibility of the computational experiments is important not only for the domain scientists, but also very useful for workflow research, e.g. for conducting controlled experiments in the area of distributed computing management. We have shown a solution which facilitates repetition and replication of *in silico* experiments, focusing on scientific workflows and container cloud infrastructures managed by Kubernetes. Our solution supports conservation of not only the scientific procedure, but also the execution environment and the experiment deployment. We have also proposed a smart reuse scenario which supports partial reuse of previous experimental runs while configuring which parts have changed and need re-execution. An Experiment Digital Object that we have introduced contributes to open science and the idea of so called executable papers whose results are fully reproducible. Future work involves enrichment of the EDO record with more useful data and software for presentation of the experiment, and integration with a more robust storage solution combined with transparent data access using the Onedata platform [8].

Acknowledgements. The research presented in this paper has been partially supported by the funds of Polish Ministry of Science and Higher Education assigned to the AGH University of Science and Technology.

References

1. Azarnoosh, S., et al.: Introducing precip: an API for managing repeatable experiments in the cloud. In: 2013 IEEE 5th International Conference on Cloud Computing Technology and Science, CloudCom, pp. 19–26. IEEE (2013)
2. Balis, B.: Hyperflow: a model of computation, programming approach and enactment engine for complex distributed workflows. Future Gener. Comput. Syst. **55**, 147–162 (2016)

3. Bartusch, F., Hanussek, M., Kruger, J., Kohlbacher, O.: Reproducible scientific workflows for high performance and cloud computing. In: 19th IEEE/ACM International Symposium on Cluster, Cloud and Grid Computing, CCGRID, pp. 161–164 (2019)

4. Beltre, A.M., Saha, P., Govindaraju, M., Younge, A., Grant, R.E.: Enabling HPC workloads on cloud infrastructure using Kubernetes container orchestration mechanisms. In: IEEE/ACM International Workshop on Containers and New Orchestration Paradigms for Isolated Environments in HPC, pp. 11–20. IEEE (2019)

5. Cohen-Boulakia, S., Belhajjame, K., et al.: Scientific workflows for computational reproducibility in the life sciences: status, challenges and opportunities. Future Gener. Comput. Syst. **75**, 284–298 (2017)

6. Deelman, E., Gannon, D., Shields, M., Taylor, I.: Workflows and e-science: an overview of workflow system features and capabilities. Future Gener. Comput. Syst. **25**(5), 528–540 (2009)

7. Deelman, E., et al.: Pegasus, a workflow management system for science automation. Future Gener. Comput. Syst. **45**, 17–35 (2014)

8. Dutka, Ł., et al.: Onedata - a step forward towards globalization of data access for computing infrastructures. Proc. Comput. Sci. **51**, 2843–2847 (2015)

9. Herbein, S., et al.: Resource management for running HPC applications in container clouds. In: Kunkel, J.M., Balaji, P., Dongarra, J. (eds.) ISC High Performance 2016. LNCS, vol. 9697, pp. 261–278. Springer, Cham (2016). https://doi.org/10.1007/978-3-319-41321-1_14

10. Joshi, T., Valliyodan, B., Khan, S.M., et al.: Next generation resequencing of soybean germplasm for trait discovery on XSEDE using pegasus workflows and iplant infrastructure (2014)

11. Liu, K., Aida, K., Yokoyama, S., Masatani, Y.: Flexible container-based computing platform on cloud for scientific workflows. In: 2016 International Conference on Cloud Computing Research and Innovations, ICCCRI, pp. 56–63. IEEE (2016)

12. Nguyen Minh, B., Tran, V., Hluchy, L.: Abstraction layer for development and deployment of cloud services. Comput. Sci. **13**, 79–88 (2012)

13. Orzechowski, M., Balis, B., Pawlik, K., Pawlik, M., Malawski, M.: Transparent deployment of scientific workflows across clouds-kubernetes approach. In: IEEE/ACM International Conference on Utility and Cloud Computing Companion, pp. 9–10. IEEE (2018)

14. Qasha, R., Cała, J., Watson, P.: A framework for scientific workflow reproducibility in the cloud. In: 2016 IEEE 12th International Conference on e-Science (e-Science), pp. 81–90. IEEE (2016)

15. Santana-Perez, I., Pérez-Hernández, M.S.: Towards reproducibility inscientific workflows: an infrastructure-based approach. Sci. Program. **2015**, 11 (2015)

16. Santana-Perez, I., da Silva, R.F., Rynge, M., Deelman, E., Pérez-Hernández, M.S., Corcho, O.: Reproducibility of execution environments in computational science using semantics and clouds. Future Gener. Comput. Syst. **67**, 354–367 (2017)

17. Stubbs, J., Talley, S., Moreira, W., Dooley, R., Stapleton, A.: Endofday: a container workflow engine for scalable, reproducible computation. In: Proceedings of the 8th International Workshop on Science Gateways, IWSG (2016)

18. Zheng, C., Tovar, B., Thain, D.: Deploying high throughput scientific workflows on container schedulers with makeflow and mesos. In: 17th IEEE/ACM International Symposium on Cluster, Cloud and Grid Computing, CCGRID, pp. 130–139. IEEE (2017)

GPU-Accelerated RDP Algorithm for Data Segmentation

Pau Cebrian$^{(\boxtimes)}$ and Juan Carlos Moure

Computer Architecture and Operating Systems Department,
Universitat Autonoma de Barcelona, Barcelona, Spain
{pau.cebrian,juancarlos.moure}@uab.cat

Abstract. The Ramer-Douglas-Peucker (RDP) algorithm applies a recursive split-and-merge strategy, which can generate fast, compact and precise data compression for time-critical systems. The use of GPU parallelism accelerates the execution of RDP, but the recursive behavior and the dynamic size of the generated sub-tasks, requires adapting the algorithm to use the GPU resources efficiently. While previous research approaches propose the exploitation of task-based parallelism, our research advocates a general fine-grained solution, which avoids the dynamic and recursive execution of kernels. The segmentation of depth images, a typical application used on autonomous driving, reaches speeds of almost 1000 frames per second for typical workloads using our massively parallel proposal on low-consumption, embedded GPUs. The GPU-accelerated solution is at least an order of magnitude faster than the execution of the same program on multiple CPU cores with similar energy consumption.

Keywords: GPU acceleration · Data segmentation · Segmented scan

1 Introduction

The line-simplification algorithm defined by Ramer-Douglas-Peucker (RDP) [3] transforms the complex strokes described by a set of points into a set of a few connected segments that capture the inherent structure of the data. This algorithm is defined with a *split-and-merge* strategy, which increases the details of the representation recursively. Although RDP was originally designed to simplify cartographic coastlines, nowadays it applies to a multitude of computer vision problems that need to define or simplify the perimeters of objects within a scene.

The low complexity and the high degree of underlying parallelism of the RDP algorithm make it especially suitable for time-limited segmentation tasks. One usage scenario is autonomous driving systems, which need to provide segmented data within a real-time pipeline and with limited hardware resources. In this context, and compared to a classical segmentation proposal as Stixels [6,7], the RDP algorithm reduces the execution time, generating compressed and accurate representations, as shown in Fig. 1. While most of nowadays proposals [4,16,18] follow a global maximization strategy, with quadratic complexity, the RDP proposal allows solving the segmentation problem also globally, but with linear-logarithmic complexity on the practical cases that are

© Springer Nature Switzerland AG 2020
V. V. Krzhizhanovskaya et al. (Eds.): ICCS 2020, LNCS 12137, pp. 234–247, 2020.
https://doi.org/10.1007/978-3-030-50371-0_17

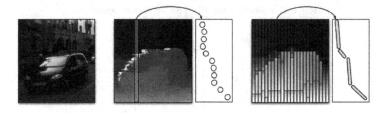

Fig. 1. Column-level segmentation of depth images for autonomous driving tasks.

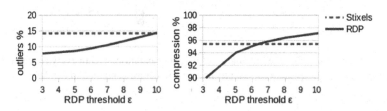

Fig. 2. Variation of the accuracy (left, lower is better) and compression (right, higher is better) of the segmented depth representations depending on the ε parameter.

addressed. Moreover, the RDP algorithm only relies on a single configuration parameter (ε), which allows adjusting the quality and compression levels of the segmentations.

Our preliminary work on this topic [2] focused mainly on the quality and compression capabilities of the RDP algorithm with respect to the Stixels proposal. We concluded that the split-and-merge strategy allows to compute representations with similar quality and compression to those obtained using expensive global maximization strategies, as shown in Fig. 2.

This paper focuses on defining an efficient implementation of the algorithm for GPU acceleration rather than on evaluating the properties of the RDP segmentation, which can be found in the research done by Heckbert *et al.* [5]. Previous approaches propose the exploitation of task-based parallelism for CPU execution [11,20], which represents a natural way to extract parallelism from a recursive algorithm. However, the dynamic generation of tasks and its irregular size do not adapt well to the GPU execution model. Instead, we propose to process the independent tasks generated in the same recursive level together, using a segmented reduction pattern. This approach matches better with the underlying SIMT (Single Instruction Multiple Threads) model of a GPU architecture, but the algorithm must be refactored and the data structures must be redesigned to be statically allocated. Our parallelization strategy transforms dynamic and irregular parallel work into homogeneous work, more appropriate for GPUs.

This paper describes in detail our massively-parallel proposal and analyzes the optimization strategies applied to minimize its execution time. Also, we present: a theoretical evaluation of the computational behavior and the number of recursion levels expected for our executions; an experimental evaluation of the performance in GPUs; and a performance comparison with the recursive algorithm executed in CPUs. We believe that the methods and the analysis presented on this work can be very useful to accelerate the execution on GPUs of similar recursive split-and-merge problems.

Ramer-Douglas-Peucker (RDP) Segmentation Algorithm:

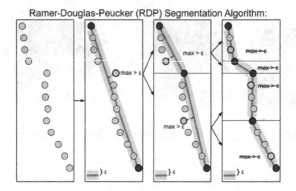

Fig. 3. Segmentation process: the granularity of the representation is recursively refined until the actual set of segments accurately represents the lineal structure contained in the data.

2   Analysis of the RDP Algorithm

This section describes the Ramer-Douglas-Peucker algorithm [3] and presents a qualitative and quantitative analysis of its computational complexity.

The algorithm approximates the one-dimensional data input, which contains n values, into a set of m connected linear segments. The cut-off points between segments are calculated with a split-and-merge strategy, refining the details of the representation recursively. A user-defined constant (ε) determines the accuracy and the compression level of the final segmented description.

Figure 3 illustrates the dynamics of the algorithm. The input data points on the left are substituted by a line drawn from the first to the last point. Then, a certain user-defined distance with respect to the line is computed for each input data point, and the point at the maximum distance is identified. If this maximum distance is greater than a given ε value, the corresponding point is selected as a cut-off point to partition the initial line into two connected segments. Then, the same process is performed recursively on the two new segments created from the cut-off point. The algorithm stops when the maximum distance in all segments is less than ε, which provides control over the level of detail of the segmentation. The higher the value of ε, the lower the number of segments required to represent the data (higher compression), but the lower the accuracy of the segmentation of the data.

Algorithm 1 presents a recursive definition of RDP, and Fig. 4 shows the binary tree representing the recursively-generated tasks of a segmentation example. The bulk of the computation work is done at lines 3 and 4, which is the calculation of a distance and the comparison with the maximum distance, and is proportional to the number of points involved in the current segment. The total computation work is the addition of the work done on each task, and depends on the input vector size n, and on the topology of the recursion tree. The number of levels in the tree and the size of each node in the tree (see Fig. 4) depends on the combination of the actual data values (i.e., the underlying structure of the data) and the tolerance ε. The total number of generated segments, m, is not enough to fully determine the algorithm's complexity.

Algorithm 1. Segmentation($data^*$, ε, $first$, $last$, $segments^*$)

1: $maxId = 0, maxDist = 0$
2: **for** $i := first \to last$ **do**
3:    $dist$ = computeDist($data^*, first, last, i$)
4:    $maxDist, maxId$ = maxArg($dist, maxDist, i, maxId$)
5: **if** $maxDist > \varepsilon$ **then**
6:    Segmentation($data^*, \varepsilon, first, maxId, segments^*$)
7:    Segmentation($data^*, \varepsilon, maxId, last, segments^*$)
8: **else**
9:    $segments^*$.add($[first, last]$)

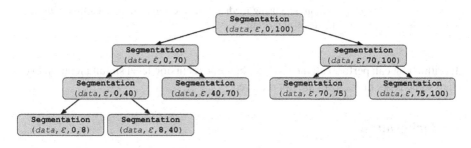

Fig. 4. Recursion tree of an example of execution of RDP.

The best-case complexity, $\Theta(n)$, occurs when the input data fits as single segment, *i.e*, $m = 1$. The worst-case complexity, $\Theta(n^2)$, occurs for very noisy input data that is converted into $m \approx n$ segments, and with a configuration that requires $m - 1$ levels of recursive calls. In the average case the recursive tree should be quite balanced and the computational complexity is $\Theta(n \log m)$.

We focus our analysis on the data sets used in the segmentation of depth image columns; like KITTI [14], which contains pairs of stereo images of autonomous-driving scenes and their corresponding depth maps, captured with stereo cameras and LIDAR. In this context, from hundreds to a thousand data columns must be processed, with vector sizes ranging from 256 to 1024 values. The raw depth images obtained from stereo matching algorithms usually contain outlier values, which are typically rectified using median filters. This preprocessing step reduces the average amount of segments generated by the RDP algorithm.

We next present an analysis of the actual execution complexity achieved when using a typical autonomous driving dataset. Figure 5 shows, for different tolerance values (ε), the average and standard deviation of the number of recursion levels required to process the data columns.

As expected, smaller values of ε generate more segments and then more recursion levels. In the practical range of the application for computer vision, with ε between 4 and 8, most of the cases require less than 5 or 6 recursion levels, generating between 20 and 50 segments, while some exceptional cases may need at most 8 recursion levels. The overall result is that the computation complexity for the data inputs considered are far from the worst case. This result is key to design an efficient massively-parallel

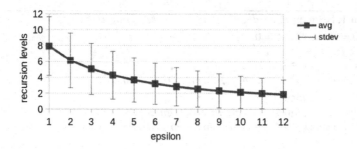

Fig. 5. Experimental analysis on the KITTI data set [14] (described later), with depth images scaled to $n = 1024$ pixels/column, and using median filters. Complexity is measured indirectly as the levels or recursion of the RDP algorithm.

algorithm that can perform all the tasks of the same recursion level simultaneously, even when some tasks in the level are not required.

3   Background

The RDP algorithm has been revised several times, either to apply the algorithm as a subroutine of some scene-understanding task, such as Romadi or Hu investigations [8, 17]; or with the aim of selecting the best points of start and end for closed segments, such as the proposals of Li or Mahmoudi [10, 12].

Researches focused on parallelization and computation efficiency, such as Jingsong *et al.* [11], proposes a task-based parallelization strategy for multicore CPU systems, where processors receive the segments on which to apply RDP from a list of tasks in which they add segmentation sub-tasks dynamically. On the other hand, Scherger *et al.* [20] transform the split-and merge or top-down strategy of the RDP algorithm to an only-merge or bottom-up strategy, initializing as many segments as pixels in the column, and defining their own multiple associative computing (MASC) model to make iterative merging of the segments.

Other approaches have been suggested for solving the split-and-merge strategy, but not directly related with the RDP algorithm.

Paravecino's research [15] studies the nested parallelism of GPU systems, and analyzes the execution of several well-known algorithms with recursive kernels.

Argüello *et al.* [1] treat the split-and-merge problem with a bottom-up strategy by calling a single GPU kernel. However, his research is more related to convolutional filters and the amount of boundary data (thus, repeated) that they require for the blocks to be independent. Applying bottom up strategies in our problem, as they did, can lead us to obtain non-global solutions, and then losing accuracy. Also, our problem benefits from needing only 2 boundary points per segment regardless of the recursion level, and from being able to store the boundary points in local registers.

Mei *et al.* [13] proposes the parallelization of the *QuickHull* algorithm, another split-and-merge strategy similar to RDP, which looks for the furthest points for a series of segments. They focus on the use of a single GPU kernel, applying the collaborative and segmented patterns defined by Shubhabrata *et al.* [19] to nest recursion. In addition, they propose rearranging the data so that the resolved points are packed at the end of the block of blocks, and intra-warp divergence is minimized.

Our work focuses on the use of collaborative strategies at the segment level to solve segmentation with a single GPU kernel. Our strategy allows to reduce the collaborative work of the computer patterns proposed by Shubhabrata *et al.* [19], and benefits from operating on a fixed number of columns containing open segments, performing dynamic and recursive work in a homogeneous and iterative manner.

4   RDP Parallelization

This section summarizes the working behaviour of GPU architectures, for those readers who are not familiar with these systems, and describes our proposal to parallelize the RDP algorithm, as well as the optimization strategies applied to minimize its execution time.

4.1   Summary of the GPU Parallel Architecture

GPU systems are composed by tens of streaming multiprocessors (SMs), where each SM is made up of shared memory areas, schedulers for massively parallel execution, and thousands of registers, to be distributed among the execution threads. These systems are specialized in performing homogeneous computing, executing the instructions in a vectorized way (SIMD).

The CUDA programming model was designed as a logical layer over the GPU hardware to ease the use of parallel resources. CUDA allows to define kernels, C/C++ functions for which you can specify the number of parallel instances of the code that will be executed as threads. In addition, CUDA allows the creation of CTAs (Cooperative Thread Arrays), blocks of threads that can perform collaborative work, and with areas of low-latency shared memory. Each thread has its own local memory space, consisting of tens of registers, a shared memory area with quick access at the CTA level, and a global memory area, public for all the threads in the kernel, but with a large latency. Finally, unlike CPU threads, GPU threads are not completely independent; CUDA executes the threads of a block in warps, sub-groups of almost-synchronous threads that share execution instructions in SIMD fashion.

The efficiency of a parallel implementation is directly related to the level of local or shared memory usage versus global memory usage, the level of computation dependencies and divergences between threads, and the amount of hardware used to do the job. The objective of this work is to define an efficient GPU version of data segmentation, adapting the RDP algorithm to massively parallel systems.

4.2   Parallelism Granularity of Our Proposal

Our proposal defines a single kernel to generate the segmentation of all the columns of the image, and assigns a single execution thread to each pixel of the image. Each thread calculates (iteratively) if a pixel is a cut-off point, and the threads corresponding to the pixels of the same data column are grouped into CTAs. This allows all the segmentation tasks on the same iteration to be performed in parallel, limiting the cooperative work at the segment level within a CTA.

The usage of a single kernel allows maximizing parallelism at the column level, and minimizing global memory transactions between kernels. The task-level parallelization alternative entails dynamic kernel creation and, therefore, the serialization of the kernels, which requires many data transactions over the long-latency global memory.

Another way to parallelize at the task level could be to assign to each CTA the calculation of a single segment. However, this requires the dynamic creation of CTAs of different sizes, which can only be done by calling new kernels. Our proposal avoids the dynamic creation of blocks. Treating the segmentation as a single function, instead of a task-level function, minimizes the amount of memory transactions and maximizes parallelism.

Our proposal redefines the recursive segmentation as a global and iterative process, detecting the boundaries between segments instead of creating new segments. Algorithm 2 shows a high-level view of our proposed segmentation strategy.

Algorithm 2. IterativeSegmentation

1: Set initial segment
2: **repeat**
3:     Compute the distance of each data value w.r.t. their segments.
4:     Find the position of the data at maximum distance on its corresponding segment
5:     Update list of segments for those segments whose maximum distance if larger than ε
6: **until** no new segments

4.3   Description of Our Parallel Proposal

We redefine the recursive algorithm iteratively, to avoid dynamic task generation. The calculation of the maximum distance performed on lines 2 to 4 of Algorithm 1, is now performed in parallel by all the threads, for each input data point. Each thread computes the distance to its corresponding segment using copies of the segment ends, stored in local registers, as shown in lines 6 to 9 of the Algorithm 3. By applying a parallel *reduce* computing pattern at the segment level, the last thread of each segment obtains the position of the data point at maximum distance in $\log n$ steps. Then, each thread reads the position of the maximum distance of its segment, and updates the local information of its ends.

The exit condition from the convergence loop is controlled at the CTA level, using a single register in shared memory, which is updated in parallel for each new cut-off point found. The loop stops when no modification is done in this convergence register.

Algorithm 3. ParallelSegmentation($data, cuts, \varepsilon$)

1: $i = threadIdx, first = 0, last = colSize - 1, localCut = 1$
2: **if** $i == first$ **or** $i == last$ **then** $localCut = 0$
3: **repeat**
4:    $cutPointsConvergence[0] = yes$
5:
6:    $idx = i - first$
7:    $slope = (data[last] - data[first])/(last - first)$
8:    $maxCost[i] = |(slope * idx + data[first]) - data[i]|$
9:    **if** $localCut == 0$ **or** $maxCost[i] < \varepsilon$ **then** $maxCost[i] = 0$
10:    $argMax[i] = i$
11:    $\_syncthreads()$
12:
13:    **for** $stride := colSize/2 \rightarrow 1; stride = stride/2$ **do**
14:       **if** $idx > stride$ **and** $maxCost[i] <= maxCost[i - stride]$ **then**
15:          $maxCost[i] = maxCost[i - stride])$
16:          $argMax[i] = argMax[i - stride]$
17:       $\_syncthreads()$
18:
19:    **if** $argMax[last] == i$ **then**
20:       $cutPointsConvergence[0] = no$
21:       $localCut = 0$
22:    **else if** $argMax[last] < i$ **then** $first = argMax[last]$
23:    **else if** $argMax[last] > i$ **then** $last = argMax[last]$
24:    $\_syncthreads()$
25: **until** $cutsConvergence[0] == yes$
26: $cuts[i] = localCut$

4.4   Optimization Strategies Applied in Our Proposal

The parallel *reduce* computing pattern is a basic subroutine for many of the tasks performed with GPUs, and has been studied in depth by the community to take advantage of hardware features efficiently. The *stride* strategy shown in Fig. 6 has become a *de facto* standard; since it allows maximizing the bandwidth of memory transactions, the warp occupancy, and the thread parallelism.

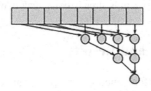

Fig. 6. Strided reduction pattern

On the one hand, every time a thread makes a memory access, the system loads a chunk of memory that contains the required data and some of its adjacent ones. When adjacent threads access adjacent memory locations, then the number of chunks loads is minimized. On the other hand, as the threads are executed in warps (adjacent thread groups), the *stride* strategy allows to maximize the use of intra-warp operations, and reduces synchronization latencies.

Our segmentation proposal applies the *reduce* computing pattern for each segment of the same block in parallel. In addition, we limit the cooperation of segments with copies of the indexes of the ends for each thread of the segment. As segments are independent between them, all reductions in the same column can be calculated in parallel, but some strategy must be defined so that the reductions do not use elements from other segments. This problem was previously discussed by Shubhabrata *et al.* [19], whose proposal uses variables with the ID of the segment to limit *segmented-scans* within a block. However, in an algorithm with recursive behavior like RDP, the ID of the segment needs to be recalculated for each level of recursion, and the calculation of the IDs requires $log\ n$ steps. Our strategy allows to limit the cooperative work with the information of the ends of the segment stored locally in each thread, in this way we can avoid the calculation of the IDs of segment to limit the cooperative work.

The main downside of our proposal is that the hardware efficiency decreases as the segmentation of Algorithm 3 progresses; as some segments are solved, they do not require more *useful* computation, as shown in Fig. 7. We have found that including specific conditions for these cases can decrease the number of floating-point operations and memory communication processes, but it represents a significant increase in control flow instructions and an increase in the execution time.

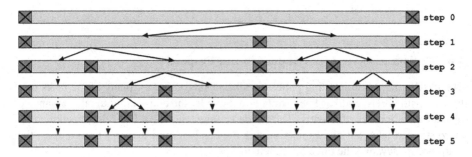

Fig. 7. Computation utility as segmentation progresses. Blue stands for useful computation, and green stands for threads whose segment is already computed. (Color figure online)

5   Experimental Study and Results

This section evaluates the performance of our proposal for RDP parallelization. Our evaluation focuses on the performance of the proposal defined in Algorithm 3 (unrolling the last 5 iterations of the loop of lines 13 to 17); and in the performance gained on the parallel implementation with respect to the recursive implementation from Algorithm 1 executed in CPU. Our GPU proposal has been evaluated using *nvprof*, which allows to capture metrics of the kernels, such as the number of instructions, memory transactions, or percentage of hardware usage. The CPU version has been evaluated using *perf* and the C++11 library *chrono*, in order to capture only the execution time of the recursive segmentation algorithm.

Since the underlying structure of the data determines the behavior of the RDP proposal, the input used for the experiments has been obtained by applying the SGM algorithm [9] to the stereo images of the KITTI [14] dataset. This dataset is composed by 200 images with 1242 columns of 375 pixels, with realistic structures on which to analyze the average performance of our proposal. Depth images generated with SGM have been scaled vertically so that each column has 1024 pixels, the maximum number of threads per block in Jetson's Xavier GPUs, in order to evaluate the maximum performance of our GPU algorithm.

The Jetson's Xavier architecture used for the experiments integrates a 512-core Volta GPU with 64 tensor cores, and a 8-core Carmel ARM CPU. This architecture allows to operate with different energy modes by specifying the clock frequencies for the internal GPU cores and CPU cores, the number of usable CPUs, and the level of power usage. Table 1 summarizes an evaluation of the execution time required for modes with maximum and minimum power consumption.

Table 1. Energy mode configurations for the GPU and CPU cores (top). Execution time in miliseconds of the average input for columns of $n = 1024$ pixels for GPU execution and for single-core and multi-core CPU execution (bottom).

	Low energy	Default energy	Max energy
GPU Max. Frequency (MHz)	520	670	1337
CPU Max. Frequency (MHz)	1200	1200	2265.5
Number of Online CPUs	2	4	8
Memory Max. Frequency (MHz)	1066	1333	2133
Total Power Budget (W)	10	15	30+
GPU Time (ms)	5.66	2.33	1.09
CPU Single-Core Time (ms)	43.9	40.10	21.15
CPU Multi-Core Time (ms)	23.3	14.31	4.97

Table 1 shows that our proposal can be executed in real time in a GPU regardless the energy mode. Assuming a target frame rate for real-time of 20 fps, a maximum slot of 50 ms per image, our proposal leaves, as average, from 28.5 to 23 ms of margin to the rest of the pipeline tasks, i.e. it occupies from 3.35% to 17.4% of the pipeline's time.

Figure 8 shows the selected input cases with the best and worst performance, and a full-noise input such as those that can be generated by unexpected SGM errors. We have found that even our proposal is able to ignore unknown depth points, unknown points are usually surrounded by errors. Smoothing depth pixels near unknown values could increase our proposal performance.

Fig. 8. Selected input cases used in our evaluations. From left to right: a corrupt input case, the worst input case from our dataset, one average input case, and the best input case.

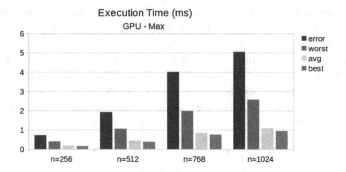

Fig. 9. Execution times of our GPU proposal for different input cases, and images with fixed width (1242 px) and variable height (n); tested with max energy mode.

Figure 9 includes the average execution time (ms) of our proposal for different data sizes, and the execution times for the best and worst case of the SGM evaluation data. The columns sizes chosen for this evaluation ranges from $n = 256$ to $n = 1024$ pixels, the expected input size for autonomous driving systems. We aware that images with column sizes of 256 px may include data smoothness given to the reduction of the original size (378 px). This figure shows that by doubling the size of the column from 256 to 512, the execution time increases $\times 2.5$; while doubling the size from 512 to 1024, increases the execution time by $\times 6$. Also, it shows that the content of the depth image that is segmented has a great influence on the execution time: the higher the quality of the input, the shorter the execution time. The most remarkable result is that the execution times required by our proposal (<5 ms) are good enough for real-time applications, even in our worst case scenarios.

Finally, Fig. 10 summarizes and ranks the execution time required by the CPU and GPU proposals for different energy modes. This figure shows that the GPU implementation is 4.6 times faster than the best CPU option, in maximum power mode and for 1242×1024 px images; and that this speedup can increase to 6.8 for 1242×256 px images. This can be a great advantage in autonomous driving systems, where images have few rows (short columns), or are horizontally cropped. The small speed-up of the GPU execution with respect to the multi-core execution is due to the fact that number of operations per second of the CPUs, and therefore, the CPU energy consumption, is considerably higher. The energy efficiency of GPU cores implies that a GPU-accelerated solution is at least an order of magnitude faster than the execution of the same

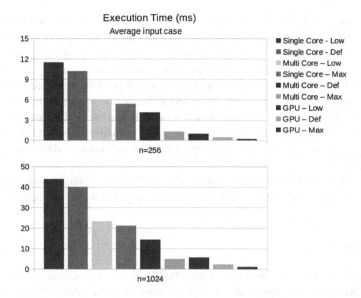

Fig. 10. Ranked execution times on CPUs and GPUs with different energy modes. At the top for 1242×256 px images; at the bottom images with 1242×1024 px.

program on multiple CPU cores with similar energy consumption. This is another great advantage for autonomous driving systems.

6   Conclusions

In this paper we have presented a fine-grained parallelization of the RDP algorithm for the segmentation the multiple data vectors, a common task on many computer vision applications. Our research adapts the dynamic and recursive execution of the RDP algorithm to massively parallel environments, minimizing the execution of kernels, and using collaborative computing strategies to reduce its execution time. The great advantage of our proposal is the capability to manage the segmented-collaboration patterns with local registers, and avoiding the computation of the segment ids.

Our GPU proposal is from $3.8\times$ to $6.8\times$ times faster than an standard RDP recursive implementation executed on a multi-core CPU, and also consumes much less energy. Its low energy consumption, and its high execution speed (up to 1000 frames per second in 1024×1242 px images), make it especially suitable for latency-limited environments such as autonomous driving systems. Using more CPU-cores allows to reach the same speed as our GPU solution, but this implies a significant increase in energy consumption, which might be critical in autonomous systems.

The experimental results also show that our proposal is sensitive to the contents of the input data, i.e. the execution time reduces as the structure of the input data is simpler and smoother. This can be taken as an advantage, by defining filters with which to pre-process the segmentation inputs. For example, we can add filters to detect undefined outliers and delete them from the input data.

Future work could also focus on extending the size of the input data to values higher than $n = 1024$. This will require assigning more than one data point to each thread and carefully redistributing the larger amount of program's data among the different memory spaces (private, shared and global). In the same line of research, we can include specific conditions for the worst-case input cases, which generate many recursive levels. In this situation, most of the threads are doing redundant work, since their associated data points have already been fitted into a segment. When the RDP algorithm reaches a deep recursion level, detecting those threads doing redundant work and setting them idle could save resources that will benefit the execution of other thread blocks executing on the same SM.

We finally hope that our research will serve as base for studies on increasing the dimensionality of parallel split-and-merge strategies, so it can also be applied over 2D and 3D data.

References

1. Argüello, F., Heras, D., Bóo, M., Lamas-Rodríguez, J.: The split-and-merge method in general purpose computation on GPUs. Parallel Comput. **38**(6), 277–288 (2012). https://doi.org/10.1016/j.parco.2012.03.003. http://www.sciencedirect.com/science/article/pii/S0167819112000208
2. Cebrian, P., Hernandez-Juarez, D., Moure, J.C.: Column-level segmentation of depth images for autonomous driving. Technical report, Autonomous University of Barcelona, Department of Computer Architecture and Operative Systems, Barcelona, Spain, February 2020
3. Douglas, D.H., Peucker, T.K.: Algorithms for the reduction of the number of points required to represent a digitized line or its caricature. Cartographica Int. J. Geogr. Inf. Geovisualization **10**(2), 112–122 (1973)
4. Dunham, J.G.: Optimum uniform piecewise linear approximation of planar curves. IEEE Trans. Pattern Anal. Mach. Intell. **8**(1), 67–75 (1986). https://doi.org/10.1109/TPAMI.1986.4767753
5. Heckbert, P.S., Garland, M.: Survey of polygonal surface simplification algorithms (1997)
6. Hernandez-Juarez, D., Espinosa, A., Moure, J.C., Vázquez, D., López, A.M.: GPU-accelerated real-time Stixel computation. In: IEEE Winter Conference on Applications of Computer Vision (WACV), pp. 1054–1062, March 2017. https://doi.org/10.1109/WACV.2017.122
7. Hernandez-Juarez, D., et al.: Slanted Stixels: representing San Francisco's steepest streets. In: British Machine Vision Conference (BMVC), 2017 (2017)
8. Hu, X., Ye, L.: A fast and simple method of building detection from lidar data based on scan line analysis. ISPRS Ann. Photogram. Remote Sens. Spat. Inf. Sci. **II–3/W1**, 7–13 (2013). https://doi.org/10.5194/isprsannals-II-3-W1-7-2013
9. Juárez, D.H., Chacón, A., Espinosa, A., Vázquez, D., Moure, J.C., Peña, A.M.L.: Embedded real-time stereo estimation via semi-global matching on the GPU. CoRR abs/1610.04121 (2016). http://arxiv.org/abs/1610.04121
10. Li, L., Jiang, W.: An improved Douglas-Peucker algorithm for fast curve approximation. In: 3rd International Congress on Image and Signal Processing, vol. 4, pp. 1797–1802, October 2010. https://doi.org/10.1109/CISP.2010.5647972
11. Ma, J., Xu, S., Pu, Y., Chen, G.: A real-time parallel implementation of Douglas-Peucker polyline simplification algorithm on shared memory multi-core processor computers. In: International Conference on Computer Application and System Modeling (ICCASM 2010), vol. 4, pp. V4–647-V4-652, October 2010. https://doi.org/10.1109/ICCASM.2010.5620612

12. Mahmoudi, S.A., Lecron, F., Manneback, P., Benjelloun, M., Mahmoudi, S.: GPU-based segmentation of cervical vertebra in x-ray images. In: IEEE International Conference On Cluster Computing Workshops and Posters (CLUSTER WORKSHOPS), pp. 1–8, September 2010. https://doi.org/10.1109/CLUSTERWKSP.2010.5613102

13. Mei, G., Zhang, J., Xu, N., Zhao, K.: A sample implementation for parallelizing divide-and-conquer algorithms on the GPU. Heliyon **4**(1), e00512 (2018). https://doi.org/10.1016/j.heliyon.2018.e00512. http://www.sciencedirect.com/science/article/pii/S2405844016326032

14. Menze, M., Geiger, A.: Object scene flow for autonomous vehicles. In: Conference on Computer Vision and Pattern Recognition (CVPR) (2015)

15. Paravecino, F.N.: Characterization and exploitation of nested parallelism and concurrent kernel execution to accelerate high performance applications. Ph.D. thesis, Northeastern University (2017)

16. Pikaz, A., Dinstein, I.: Optimal polygonal approximation of digital curves. In: Proceedings of 12th International Conference on Pattern Recognition, vol. 1, pp. 619–621, October 1994. https://doi.org/10.1109/ICPR.1994.576378

17. Romadi, M., Faizi, R., Chiheb, R., Romadi, R.: A shape-based approach for detecting and recognizing traffic signs in a video stream. In: 4th International Conference on Control, Decision and Information Technologies (CoDIT), pp. 0254–0258, April 2017. https://doi.org/10.1109/CoDIT.2017.8102600

18. Schneider, L., et al.: Semantic stixels: depth is not enough. In: IEEE Intelligent Vehicles Symposium (IV), pp. 110–117, June 2016. https://doi.org/10.1109/IVS.2016.7535373

19. Sengupta, S., Harris, M., Garland, M., et al.: Efficient parallel scan algorithms for GPUs. NVIDIA, Santa Clara, CA, Technical report. NVR-2008-003 vol. 1, no. 1, pp. 1–17 (2008)

20. Tran, H., Scherger, M.: A massively parallel algorithm for polyline simplification using an associative computing model. In: Proceedings of the International Conference on Parallel and Distributed Processing Techniques and Applications (PDPTA), p. 1. The Steering Committee of The World Congress in Computer Science (2011)

Sparse Matrix-Based HPC Tomography

Stefano Marchesini[1]([✉]), Anuradha Trivedi[2], Pablo Enfedaque[3],
Talita Perciano[3], and Dilworth Parkinson[4]

[1] Sigray, Inc., 5750 Imhoff Drive, Ste I, Concord, CA 94520, USA
smarchesini@sigray.com
http://sigray.com
[2] Virginia Polytechnic Institute and State University, Blacksburg, VA 24061, USA
[3] Computational Research Division, Lawrence Berkeley National Laboratory,
1 Cyclotron Rd., Berkeley, CA 94720, USA
[4] Advanced Light Source, Lawrence Berkeley National Laboratory,
1 Cyclotron Rd., Berkeley, CA 94720, USA

Abstract. Tomographic imaging has benefited from advances in X-ray
sources, detectors and optics to enable novel observations in science,
engineering and medicine. These advances have come with a dramatic
increase of input data in the form of faster frame rates, larger fields of
view or higher resolution, so high performance solutions are currently
widely used for analysis. Tomographic instruments can vary significantly
from one to another, including the hardware employed for reconstruction: from single CPU workstations to large scale hybrid CPU/GPU
supercomputers. Flexibility on the software interfaces and reconstruction engines are also highly valued to allow for easy development and
prototyping. This paper presents a novel software framework for tomographic analysis that tackles all aforementioned requirements. The proposed solution capitalizes on the increased performance of sparse matrix-
vector multiplication and exploits multi-CPU and GPU reconstruction
over MPI. The solution is implemented in Python and relies on CuPy
for fast GPU operators and CUDA kernel integration, and on SciPy
for CPU sparse matrix computation. As opposed to previous tomography solutions that are tailor-made for specific use cases or hardware,
the proposed software is designed to provide flexible, portable and high-
performance operators that can be used for continuous integration at
different production environments, but also for prototyping new experimental settings or for algorithmic development. The experimental results
demonstrate how our implementation can even outperform state-of-the-
art software packages used at advanced X-ray sources worldwide.

Keywords: Tomography · SpMV · X-ray imaging · HPC · GPU

1 Introduction

Ever since Wilhelm Röntgen shocked the world with a ghostly photograph of
his wife's hand in 1896, the imaging power of X-rays has been exploited to help
see the unseen. Their penetrating power allows us to view the internal structure of many objects. Because of this, X-ray sources are widely used in multiple

© Springer Nature Switzerland AG 2020
V. V. Krzhizhanovskaya et al. (Eds.): ICCS 2020, LNCS 12137, pp. 248–261, 2020.
https://doi.org/10.1007/978-3-030-50371-0_18

imaging and microscopy experiments, e.g. in Computed Tomography (CT), or simply tomography. A tomography experiment measures a transmission absorption image (called radiograph) of a sample at multiple rotation angles. From 2D absorption images we can reconstruct a stack of slices (*tomos* in Greek) perpendicular to the radiographs measured, containing the 3D volumetric structure of the sample. Tomography is used in a variety of fields such as medical imaging, semiconductor technology, biology and materials science. Modern tomography instruments using synchrotron-based light sources can achieve measurement speeds of over 200 volumes per second using 40 kHz frame rate detectors [7]. Tomography can also be combined with microscopy techniques to achieve resolutions down to a single atom using electrons [17]. Its experimental versatility has also been exploited by combining it with spectroscopic techniques, to provide chemical, magnetic or even atomic orbital information about the sample.

Nowadays, tomographic analysis software faces three main challenges. 1) The volume of the data is constantly increasing as X-ray sources become brighter and newer generation detectors increase their resolution and acquisition frame rate. 2) Instruments from different facilities (or even from the same one) present a variety of experimental settings that can be exclusive to said instrument, such as the geometry of the measurements, the data layout and format, noise levels, etc. 3) New experimental use cases and algorithms are frequently explored and tested to accommodate new science requisites. These three requirements strongly force tomography analysis software to be HPC and flexible, both in terms of modularity and interfaces, as well as in hardware portability. Currently, TomoPy [9] and ASTRA [22] are the most popular solutions for tomographic reconstruction at multiple synchrotron and tabletop instruments. TomoPy is a Python-based open source framework optimized for performance using a C backend that can process a variety of data formats and algorithms. ASTRA is a tomography toolbox accelerated using both GPU and CPU computing and it is also available through TomoPy [16]. Although both solutions are highly optimized at different levels, they do not provide the level of flexibility required to be easily extendable by third parties regarding solver modifications or accessing specific operators.

In this work we present a novel framework that focuses on providing multi-CPU and GPU acceleration with flexible operators and interfaces for both 1-step and iterative tomography reconstruction. The solution is based on Python 3 and relies on CuPy, mpi4py, SciPy and NumPy to provide transparent CPU/GPU computing and innocuous multiprocessing through MPI. The idea is to provide easy HPC support without compromising the solution lightweight so that development, integration and deployment is streamlined. The current operators are based on sparse matrix-vector multiplication (SpMV) computation which benefit from preexisting fast implementations on both CuPy and SciPy and provide faster reconstruction time than direct dense computation [10]. By minimizing code complexity, we can efficiently implement advanced iterative techniques [13,19] that are not normally implemented for production, also due to their computational complexity; prior implementations could take up to a full day of a supercomputer to reconstruct a single tomogram [20]. The high level technologies and modular design employed in this project permits the proposed solution to be particularly flexible, both for exploratory uses (algorithm development or new experimental settings), and also in terms of hardware: we can scale the reconstruction from a single CPU, to a workstation using multiple CPU and GPU processors, to large distributed memory systems. The experimental results demonstrate how the proposed solution can reconstruct datasets of 68 GB in less than 5 s, even surpassing the performance of TomoPy's fastest reconstruction engine by 2.2X. This project is open source and available at [12].

The paper is structured as follows: Sect. 2 overviews the main concepts regarding tomography reconstruction. Section 3 presents the proposed implementation with a detailed description of the challenges behind its design and the techniques employed, and Sect. 4 assesses its performance through experimental results. The last section summarizes this work.

2   Tomography

Tomography is an imaging technique based on measuring a series of 2D radiographs of an object rotated at different angles relative to the direction of an X-ray beam (Fig. 1). A radiograph of an object at a given angle is made up of line integrals (or projections). The collection of projections from different angles at the same slice of the object is called sinogram (2D); and the final reconstructed volume is called tomogram (3D), which is generally assembled from the independent reconstruction of each measured sinogram.

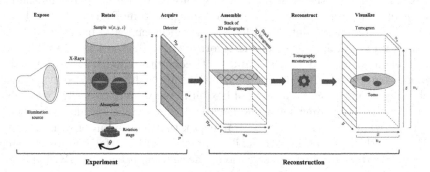

Fig. 1. Overview of a tomography experiment and reconstruction. A 3D sample is rotated at angles $\theta = 0, \ldots, 180°$ as X-rays produce 2D radiographs onto the detector. The collection of radiographs is combined to provide a sinogram for each detector row. Each sinogram is then processed to generate a 2D reconstructed slice, the entire collection of which can be assembled into a 3D tomogram.

Physically, the collected data measures attenuation, which is the loss of flux through a medium. When the X-ray beams are parallel along the optical axis, the beam intensity impinging on the detector is given by:

$$I_\theta(p, z) = I_0 e^{-\mathcal{P}_\theta(p,z)}, \quad \mathcal{P}_\theta(p, z) = \int u\left(p\cos\theta - s\sin\theta, p\sin\theta + s\cos\theta, z\right) ds,$$

where $u(x, y, z)$ is the attenuation coefficient as a function of position $\boldsymbol{x} = (x, y, z)$ in the sample, I_0 is the input intensity collected without a sample, $\mathcal{P}_\theta$ is the projection after rotating θ around the z axis, and (p, z) are the coordinates on the detector that sample the data onto $(n_p \times n_z)$ detector pixels. The negative log of the normalized data provides the projection, also known as X-ray transform:

$$\mathcal{P}_\theta(p, z) = -\ln\left(\frac{I_\theta(p, z)}{I_0(p, z)}\right),$$

with element-wise log and division. The Radon transform (by H.A. Lorentz [2]) at a fixed z is then given by the set of n_θ projections for a series of angles θ:

$$\text{Radon}_{(\theta, p) \leftarrow (x, y)} u(x, y, z) = \text{Sinogram}_z(p, \theta) = \mathcal{P}_\theta(p, z).$$

2.1 Iterative Reconstruction Techniques

The tomography inverse problem can be expressed as follows:

$$\text{To find } u \text{ s.t. Radon}(u) = -\log(I/I_0).$$

The pseudo-inverse iRadon$(-\log(I/I_0))$ described below provides the fastest solution to this problem, and it is typically known as *Filtered Back Projection* in the literature. When implemented in Fourier space, the algorithm is referred to as *non-uniform inverse FFT* or *gridrec*.

The inverse problem can be under-determined and ill-conditioned when the number of angles is small. The equivalent least squares problem is:

$$\arg\min_u \|\mathbb{P}\left(\text{Radon}(u) + \log(I/I_0)\right)\|,$$

where $\mathbb{P} = \mathcal{F}^\dagger \mathcal{D}^{1/2} \mathcal{F}$ is a preconditioning matrix, with $\mathcal{D}$ a diagonal matrix, and $\mathcal{F}$ denotes a 1D Fourier transform. Note that $\mathcal{F}$ does not need to be computed when using the Fubini-Radon operator (see below). The model-based problem is:

$$\arg\min_u \|\widehat{\mathbb{P}}\left(\text{Radon}(u) + \log(I/I_0)\right)\|_w + \mu \cdot \text{Reg}(u),$$

where $\|\cdot\|_w$ is a weighted norm to account for the noise model, $\widehat{\mathbb{P}}$ may incorporate streak noise removal [11] as well as preconditioning, Reg is a regularization term such as the Total Variation norm to account for prior knowledge about the sample, and μ is a scalar parameter to balance the noise and prior models.

Many algorithms have been proposed over the years including Filtered Back Projection (FBP), Simultaneous Iterative Reconstruction Technique (SIRT), Conjugate Gradient Least Squares (CGLS), and Total Variation (TV) [3]. FBP can be viewed as the first step of the preconditioned steepest descent when starting from 0. To solve any of these problems, one needs to compute a Radon transform and its (preconditioned) adjoint operation multiple times.

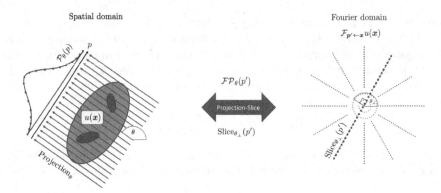

Fig. 2. Depiction of the Fubini-Radon transform (1), based on the Fourier slice theorem. The projection $\mathcal{P}_\theta(p, z)$ at a given angle θ, height z, is related to an orthogonal 1D slice (different from the tomographic slice) by a Fourier transform: $\mathcal{F}_{p' \leftarrow p} \left(\mathcal{P}_\theta(u(\boldsymbol{x})) \right) =$ $\text{Slice}_{\theta_\perp, p'} \mathcal{F}_{p' \leftarrow x}(u(\boldsymbol{x}))$. The slicing interpolation between the Cartesian and polar grid is the key step in this procedure and can be implemented with a sparse matrix operation.

2.2 The Fubini-Radon Transform

One of the most efficient ways to perform Radon(u) is to use the Fourier central slice theorem by Fubini [5]. It consists on performing first a 2D Fourier transform, denoted by $\mathcal{F}$ of u, and then interpolating the transform onto a polar grid, to finally 1D inverse Fourier transforming the points along the radial lines:

$$\text{Radon}_{(\theta,p) \leftarrow (x,y)} = \mathcal{F}^*_{p \leftarrow p'} \text{Slice}_{(\theta_\perp, p') \leftarrow (p'_x, p'_y)} \mathcal{F}_{(p'_x, p'_y) \leftarrow (x,y)} \qquad (1)$$

We will refer to this approach as the *Fubini-Radon* transform (Fig. 2). Numerically, the slicing interpolation between the Cartesian and polar grid in the Fubini-Radon transform is the key step in the procedure. It can be carried out using a *gridding* algorithm that maintains the desired accuracy with low computational complexity. The *gridding* algorithm essentially allows us to perform a *non-uniform FFT*. The projection operations require $\mathcal{O}(n^2) \cdot n_\theta$ arithmetic operations when computed directly using $n = n_p = n_x = n_y$ discretization of the line integrals, while the Fubini-Radon version requires $\mathcal{O}(n) \cdot n_\theta + \mathcal{O}(n^2 \log(n))$ operations, where the first term is due to the slicing operation and the second term is due to the two dimensional FFT. For sufficiently large n_θ, the Fubini-Radon transform requires fewer arithmetic operations than the standard Radon transform using projections. Early implementations on GPUs used ad hoc kernels to deal with atomic operations and load-balancing of the highly non-uniform distribution of the polar sampling points [11], but became obsolete with new compute architectures. In this work we implement the slicing interpolation using a sparse matrix-vector multiplication. The SpMV and SpMM operations are level 2 and level 3 BLAS functions which have been heavily optimized (see e.g. [21]) on numerous architectures for both CPUs and GPUs.

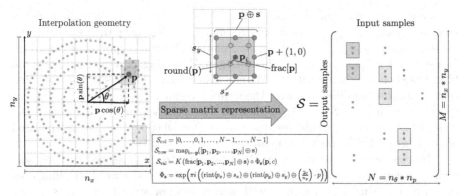

Fig. 3. Sparse matrix $\mathcal{S} \in \mathbb{C}^{M \times N}$ (right) representation of a set of convolutional kernel windows of width $k_w = 3$ with stencils $s = (s_x, s_y)$, $s_x = (-1, 0, 1)$, $s_y = s_x^T$ (top), centered around a set of coordinates $p_i = p_i(\cos\theta_i, \sin\theta_i) + \frac{1}{2}(n_x, n_y)$ of each input point of the sinogram on the output image (left).

3 Radon Transform by Sparse Matrix Multiplication

The gridding operation requires the convolution between regular samples and a kernel to be calculated at irregular sample positions, and vice versa for the inverse gridding operation. To maintain high numerical accuracy and minimize the number of arithmetic operations, we want to limit the width of the convolution kernel. Small kernel width can be achieved by exploiting the finite sample dimensions ($u(x) > 0$ in the field of view) using a pair of functions $k^\star(x), k(x)$ so that $k^\star(x)k(x) = \{1 \text{ if } u(x) > 0, 0 \text{ otherwise}\}$. By the convolution theorem:

$$u = (k^\star \circ k) \circ u = k^\star \circ \mathcal{F}^{-1}(K \circledast \mathcal{F}u), \quad K = \mathcal{F}(k),$$

where $\circledast$ is the convolution operator, $\circ$ denotes the Hadamard or elementwise product, $K = \mathcal{F}k$ is called the convolution *kernel* and $k^\star$ the *deapodization* factor. We choose K with finite width k_w, and the *deapodization* factor can be pre-computed as $k^\star = \{(\mathcal{F}^*K(x))^{-1} \text{ if } u(x) > 0, 0 \text{ otherwise}\}$. Several kernel functions have been proposed and employed in the literature, including truncated Gaussian, Kaiser-Bessel, or an interpolation kernel to minimize the worst-case approximation error over all signals of unit norm [6].

The Fubini-Radon transform operator and its pseudo-inverse iRadon can be expressed using a sparse matrix (Fig. 3) to perform the interpolation [14]:

$$\text{Radon}(u(x)) = \mathcal{F}^\dagger_{p \leftarrow p'} \mathcal{S}^\dagger_{(\theta, p') \leftarrow (p')} \mathcal{F}_{(p') \leftarrow (x)} k^\star(x) \circ u(x),$$

$$\text{iRadon}(\text{Sino}(p, \theta)) = k^\star(x) \circ \mathcal{F}^\dagger_{(x) \leftarrow (p')} \mathcal{S}_{(p') \leftarrow (\theta, p')} \mathcal{D} \mathcal{F}_{p' \leftarrow p} \text{sino}(p, \theta), \quad (2)$$

where bold indicates 2D vectors such as $p = (p_x, p_y)$, $x = (x, y)$, and $\mathcal{D}$ is a diagonal matrix to account for the density of sampling points in the polar grid.

Fourier transforms and multiplication of $\mathcal{S} \in \mathbb{C}^{M \times N}$ with a sinogram in vector form ($1 \times N$, $N = n_\theta \cdot n_p$), (sparse matrix-vector multiplication or SpMV) produces a tomogram of dimension $(1 \times M \rightarrow n_y \times n_x)$; multiplication with a stack of sinograms (sparse matrix-matrix multiplication or SpMM) produces the 3D tomogram(n_z, n_y, n_x)). The diagonal matrix $\mathcal{D}$ can incorporate standard filters such as the Ram-Lak ramp, Shepp-Logan, Hamming, or a minimum residual filter based on the data itself [15]. We can also employ the density filter solution that minimizes the difference with the impulse response (a constant $\mathbf{1}$ in Fourier space) as $\arg\min_{\mathcal{D}_v} \|\mathcal{S}\mathcal{D}_v - \mathbf{1}_{n_x \cdot n_y}\|$, with $\mathcal{D}_v$ as the vector of the diagonal elements of $\mathcal{D}$. Note that in this case, the matrix $\mathcal{D} = \text{Diag}(\mathcal{D}_v)$ can be incorporated directly into $\mathcal{S}$ for better performance.

The row indices and values of the sparse matrix are related to the coordinates where the kernel windows are added up on the output 2D image as $\boldsymbol{p}_i = p_i(\cos\theta_i, \sin\theta_i) + \frac{1}{2}(n_x, n_y)$, with kernel window stencil $\boldsymbol{s} = (s_x, s_y)$, and $s_x = (-1, 0, 1)$, $s_y = s_x^T$ (for $k_w = 3$). The column index is simply given by the consecutive sequence of natural numbers $\mathbb{N}_1^N = [0, 1, \ldots, N-1]$, repeated k_w^2 times, and the row index and value are given by:

$$\mathcal{S}_{\text{col}} = \mathbb{N}_0^{N-1} \otimes \mathbf{1}_{k_w^2} = [0, 0, \ldots, 0, 1, 1, \cdots, N-1, \ldots, N-1],$$

$$\mathcal{S}_{\text{row}} = \text{map}_{i \leftarrow p}([\boldsymbol{p}_1, \boldsymbol{p}_2, \ldots, \boldsymbol{p}_N] \oplus \boldsymbol{s}),$$

$$\mathcal{S}_{\text{val}} = K\left(\text{frac}[\boldsymbol{p}_1, \boldsymbol{p}_2, \ldots, \boldsymbol{p}_N] \oplus \boldsymbol{s}\right) \circ \Phi_s(\boldsymbol{p}, c),$$

$$\Phi_s(\boldsymbol{p}, c) = \exp\left(\pi i \left((\text{rint}(p_x) \oplus s_x) \oplus (\text{rint}(p_y) \oplus s_y) \oplus \left(\frac{2c}{n_p}\right) \cdot \boldsymbol{p}\right)\right),$$

where $\oplus \boldsymbol{s}$ is the broadcasting sum with the window stencil reshaped to dimensions $(1, 1, k_w, k_w)$, $\text{map}_{i \leftarrow p}(\boldsymbol{p}) = \text{rint}(p_x) * n_y + \text{rint}(p_y)$ is the lexicographical mapping from 2D to 1D index, K is the kernel function and $\text{frac}[\boldsymbol{p}] = \boldsymbol{p} - \text{round}[\boldsymbol{p}]$ is the decimal part, and $\otimes \mathbf{1}$ represents the Kronecker product with the unit vector $\mathbf{1}_{(k_w)^2} = [1, 1, \ldots, 1]$, for a window of width k_w (see Fig. 3). $\mathcal{S}$ has at most $\text{nnz} = n_\theta \cdot n_p \cdot k_w^2$ non-zero elements, and the sparsity ratio is at most $\frac{N \cdot k_w^2}{N \cdot M} = \frac{k_w^2}{n_x \cdot n_y}$, or $\frac{(k_w - 1)^2}{n_x \cdot n_y}$ when the kernel is set to 0 at the borders. We account for a possible shift c of the rotation axis and avoid FFTshifts in the tomogram and radon spaces by applying a phase ramp as $\Phi_s(\boldsymbol{p}, c)$, with $c = \frac{n_p}{2}$ when the projected rotation axis matches the central column of the detector. For better performance, FFTshifts in Fourier space are incorporated in the sparse matrix by applying an FFTshift of the p coordinate, and by using a $\left(\begin{smallmatrix} +1 & -1 \ldots \\ -1 & +1 \ldots \end{smallmatrix}\right)$ checkerboard pattern in the deapodization factor $k^\star$.

3.1   Parallel Workflow

The Fubini-Radon transform operates independently on each tomo and sinogram, so we can aggregate sinograms into chunks and distribute them over multiple processes operating in parallel. Denoising methods that operate across multiple slices can be handled using halos with negligible reduction in the final Signal-to-Noise-Ratio (SNR), while reducing or avoiding MPI neighborhood communication.

Pairs of sinograms are combined into a complex sinogram which is processed simultaneously, by means of complex arithmetic operations, and is split back at the end of the reconstruction. We can limit the amount of chunks assigned to each process in order to avoid memory constraints. Then, when the data has more slices than what can be handled by all processes, it is divided up ensuring that each process operates on similar size chunks of data and all processes loop through the data. When the number of slices cannot be distributed equally to all processes, only the last loop chunk is split unequally with the last MPI ranks receiving one less slice than the first ones.

The setup stage uses the experimental parameters of the data (number of pixels, slices and angles) and the choice of filters and kernels to compute the sparse matrix entries, deapodization factors and slice distribution across MPI ranks. During the setup stage, the output tomogram is initialized as either a memory mapped file, a shared memory window (whenever possible) or an array in rank-0 to gather all the results.

Several matrix formats and conversion routines exist to perform the SpMV operation efficiently. In our implementation, the sparse matrix entries are first computed in Coordinate list (COO) format which contains a list of (row, column, value) tuples. Zero-valued and out-of-bound entries are removed, and then the sparse matrix is converted to compressed sparse row (CSR) format, where the entries are sorted by column and row, and the row index is replaced by a compressed pointer. The sparse matrix and its transpose are stored separately and incorporate preconditioning filters and phase ramps to avoid all FFTshifts. The CSR matrix entries are saved in a cache file for reuse, with a hash function derived from the experimental parameters and filters to identify the corresponding sparse matrix from file. The FFT plans are computed at the first application and stored in memory until the reconstruction is restarted. When the data is loaded from file and/or the results are saved to disk, parallel processes pre-load the input to memory or flush the output from a double buffer as the next section of the data is processed.

Our implementation uses cuSPARSE and MKL libraries for the SpMV and FFT operations, MPI for distributed parallelism through shared memory when available, or scatterv/gatherv and non-blocking double buffers for I/O and MPI operations. All these libraries are accessed through Python, NumPy, CuPy, SciPy and mpi4py; we also rely on h5py or tifffile modules to interface with data files. This framework also provides the capability to call TomoPy and ASTRA solvers on distributed architectures using MPI.

We used this framework to implement the most popular algorithms described in Sect. 2.1, namely FBP, SIRT, CGLS, and TV. To achieve high throughput, our implementation of SIRT uses the Hamming preconditioning and the BB-step acceleration [1], which provides 10-fold convergence rate speedup and makes it comparable to the conjugate gradient method but with fewer reductions and lower memory footprint. The CGLS implementation is based on the conjugate gradient squared method [18], and the TV denoising employs the split-Bregman [8] technique.

4   Experiments and Results

The experimental evaluation presented herein is two-fold. We assess the performance of our implementation on both shared and distributed memory systems and on CPU and GPU architectures, and we also study how it compares to TomoPy, the state-of-the-art solution on X-ray sources, in terms of run time and quality of reconstruction.

We employ two different datasets for this analysis. The first one is a simulated Shepp-Logan phantom generated using TomoPy, with varying sizes to analyze the performance and scalability of the solution. The second one is an experimental dataset generated at Lawrence Berkeley National Laboratory's Advanced Light Source during an outreach program with local schools out of a bread-crumb inserted at the micro-tomography beamline 8.3.2. The specifics of the experiments were: 25 keV X-rays, pixel size $0.65\,\mu$, 200 ms per image and 1313 angles over $180°$. The detector consisted of $20\,\mu$ LuAG:Ce scintillator and Optique Peter lens system with Olympus 10x lens, and PCO.edge sCMOS detector. The total experiment time, including camera readout/overhead, was around 6 min, generating a sinogram stack of dimension $(n_z, n_\theta, n_p) = (2160, 1313, 3620)$.

We use two different systems for this evaluation. The first is the Cori supercomputer (`Cori.nersc.gov`), a Cray XC40 system comprised of 2,388 nodes containing two 2.3 GHz 16-core Intel Haswell processors and 128 GB DDR4 2133 MHz memory, and 9,688 nodes containing a single 68-core 1.4 GHz Intel Xeon Phi 7250 (Knights Landing) processor and 96 GB DDR4 2400 GHz memory. Cori also provides 18 GPU nodes, where each node contains two sockets of 20-core Intel Xeon Gold 6148 2.40 GHz, 384 GB DDR4 memory, 8 NVIDIA V100 GPUs (each with 16 GB HBM2 memory). For our experiments, we use the Haswell processor and the GPU nodes.[1] The second system employed is CAM, a single node dual socket Intel Xeon CPU E5-2683 v4 @ 2.10 GHz with 16 cores 32 threads each, 128 GB DDR4 and 4 NVIDIA K80 (dual GPU with 12 GB of GDDR5 memory each).

The first experiment reports the performance results and scaling studies of our iRadon implementation and of TomoPy-Gridrec, when executed on both Cori and CAM, over the simulated dataset. The primary objective is to compare their scalability using both CPUs and GPUs. We executed both algorithms at varying levels of concurrency using a simulation size of $(2048, 2048, 2048)$. On Cori, we used up to 8 Haswell nodes in a distributed fashion, only using physical cores in each node. On CAM, we ran all the experiments on a single node, dual socket. The speedup plots are shown in Fig. 4. The reported speedup is defined as $S(n,p) = \frac{T^*(n)}{T(n,p)}$ where $T(n,p)$ is the time it takes to run the parallel algorithm on p processes with an input size of n, and $T^*(n)$ is the time for the best serial algorithm on the same input.

First, we notice that the iRadon algorithm running on GPU has a super-linear speedup on both platforms. This is unusual in general, however possible in some cases. One known reason is the cache effect, i.e. the number of GPU changes,

---

[1] Cori configuration page: https://docs.nersc.gov/systems/cori/.

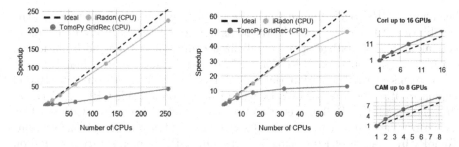

Fig. 4. Speedup of iRadon and TomoPy-Gridrec algorithms on CPU Cori (left), CPU CAM (center) and GPU Cori and CAM (right) for a $(n_z, n_\theta, n_p) = (2048, 2048, 2048)$ simulation. The horizontal axis is the concurrency level and the vertical axis measures the speedup.

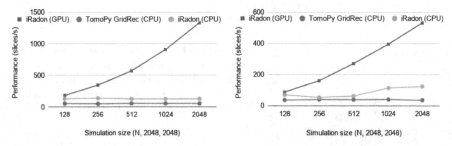

Fig. 5. Performance on Cori (left) and CAM (right), for varying sizes of simulated datasets as $(n_z, n_\theta, n_p) = (N, 2048, 2048)$, running both the iRadon and TomoPy-Gridrec algorithms. The horizontal axis is the number of slices (sinograms) of the input data, and the vertical axis measures performance as slices reconstructed per second. CPU experiments employ 64 processes and GPU experiments use 8 on CAM and 16 on Cori.

and so does the size of accumulated caches from different GPUs. Specifically, in a multi-GPU implementation, super-linear speedup can happen due to configurable cache memory. In the CPU case, we see a close to linear speedup. On CAM, the performance decreases because of MPI oversubscribe, i.e. when the number of processes is higher than the actual number of processors available.

Finally, there is a clear difference in speedup results compared to the TomoPy-Gridrec implementation. We believe that the main difference here is due to the fact that TomoPy only uses a multithreaded implementation with OpenMP, while our implementation relies on MPI. For the purpose of comparison with our implementation, we use MPI to run TomoPy across nodes.

We also evaluate our implementation by running multiple simulations with a fixed number of angles and rays (2048) and varying number of slices (128–2048) on 64 CPUs and 8 GPUs. Performance results in slices per second are shown in Fig. 5. One can notice that the GPU implementation of iRadon presents an increase in performance when the number of GPU increases. This is a known behavior of GPU performance when the problem is too small compared to the capabilities of the GPU, and the device is not completely saturated with data, not taking full advantage of the parallelized computations. For both platforms, our CPU implementation of iRadon performs significantly better than TomoPy.

In terms of raw execution time, TomoPy-Gridrec outperforms our iRadon implementation by a factor of 2.3× when running on a single CPU on Cori. On the other hand, the iRadon execution time using 256 CPU cores on Cori is 4.11 s, outperforming TomoPy by a factor of 2.2×. Our iRadon version also ourperforms TomoPy by a factor of 1.9× using 32 cores. Our GPU implementation of iRadon runs in 1.55 s using 16 V100 GPUs, which improves the CPU implementation (1 core) by a factor of 600×, and runs 2.6× faster compared with 256 CPU cores. Finally, our GPU version of iRadon runs 7.5× faster (using 2 GPUs) than TomoPy (using 32 CPUs), which could be considered the level of hardware resources accessible to average users.

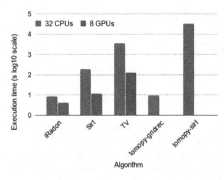

Fig. 6. Comparison of execution time (seconds in log10 scale) for different algorithms, reconstructing 128 slices of the bread-crumb dataset on CAM. SIRT and TV run for 10 iterations.

Table 1. Execution times for CPU and GPU (minutes) and SNR values for each reconstruction algorithm implemented. SNR is computed for a simulation of size (256, 1024, 1024).

Alg.	CPU	GPU	SNR
iRadon	**0.14**	**0.07**	3.51
SIRT	3.13	0.19	17.11
TV	57.8	2.07	**17.78**

The last experiment focuses on the analysis of the different algorithms implemented in this work, in terms of execution time and reconstruction quality. Figure 6 shows the reconstruction of 128 slices of the bread-crumb experimental dataset on CAM (32 CPUs and 8 GPUs), for 3 different implemented algorithms: iRadon, SIRT, and TV, and also for TomoPy-Gridrec and TomoPy-SIRT. All iterative implementations (SIRT and TV) run for 10 iterations. Our iRadon implementation presents the best execution time for CPU (9 s), while on GPU, it runs in 4 s. Our SIRT implementation outperforms TomoPy's by a factor of 175×. We report the SNR values (and corresponding execution times) of our implemented algorithms in Table 1, using a simulation dataset of size (256, 1024, 1024). We can observe how both SIRT and TV present the best results in terms of reconstruction quality.

Fig. 7. Example of reconstructed slice from the bread-crumb dataset using the iRadon and the TV algorithms. This visual result shows a better quality of reconstruction obtained using iRadon.

Figure 7 shows a reconstructed slice of the bread-crumb data using the iRadon and the TV algorithms, along with a zoomed-in region of the same slice. The difference in reconstruction quality is minor in this case due to the dataset presenting high contrast and a large number of angles. Still, in the zoomed-out image we can appreciate higher contrast fine features on the TV reconstruction. Sparser datasets would be analyzed in the future to assess the performance of TV and iterative solutions on more challenging scenarios.

It is important to remark that all the execution times presented in this section refer to the solver portion of the calculations. When running the TV algorithm on the complete bread-crumb data using 8 GPUs on CAM, for example, the solver time takes approximately 78% of the total execution time (44.82 min). Most of the remaining time is taken by I/O (18%) and gather (2%).

5   Conclusions

This paper presents a novel solution for tomography analysis based on fast SpMV operators. The proposed software is implemented in Python relying on CuPy, SciPy and MPI for high performance and flexible CPU and GPU reconstruction. As opposed to existing solutions, the software presented tackles the main requirements existing in tomography analysis: it can run over most hardware setups and can be easily adapted and extended into new solvers and techniques, while greatly simplifying deployment at new beamlines. The experimental results of this work demonstrate the remarkable performance of the solution, being able to iteratively reconstruct datasets of 68 GB in less than 5 s using 256 cores and in less than 2 s using 16 GPUs. For the simulated datasets analyzed, the proposed software outperforms the reference tomography solution by a factor of up to 2.7×, while running on CPU. When reconstructing the experimental data, our implementation of the SIRT algorithm outperforms TomoPy by a factor of

175× running on CPU. The code of this project is also open source and available at [12].

As future work, we will employ CPU and GPU co-processing, Block Compressed Row (BSR) format and sparse matrix-dense matrix multiplication (SpMM) to enhance the throughout of the solution. We will also explore the Toeplitz approach [14], which permits combining the Radon transform with its adjoint into a single operation, while also avoiding the forward and backward 1D FFTs. Half-precision arithmetic is also probably sufficient to deal with experimental data with photon counting noise obtained with 16 bits detectors and can further improve performance by up to an order of magnitude using tensor cores. Generalization to cone-beam, fan beam or helical scan geometries using generalized Fourier slice methods [23] will also be subject of future work. We will also explore the implementation of advanced denoising schemes based on wavelets or BM3D [4], combining the operators presented in this work.

Acknowledgments. Work by S. M. was supported by Sigray, Inc. A. T. work was in part sponsored by Sustainable Research Pathways of the Sustainable Horizons Institute. P. E. was funded through the Center for Applied Mathematics for Energy Research Applications. T. P. is supported by the grant "Scalable Data-Computing Convergence and Scientific Knowledge Discovery", program manager Dr. Laura Biven. D. P. is supported by the Advanced Light Source. This research used resources of the National Energy Research Scientific Computing Center (NERSC) and the Advanced Light Source, U.S. DOE Office of Science User Facility operated under Contract No. DE-AC02-05CH11231.

References

1. Barzilai, J., Borwein, J.M.: Two-point step size gradient methods. IMA J. Numer. Anal. **8**(1), 141–148 (1988)
2. Bockwinkel, H.: On the propagation of light in a biaxial crystal about a midpoint of oscillation, Verh. Konink Acad. V. Wet. Wissen. Natur **14**(636), 20 (1906)
3. Bouman, C., Sauer, K.: A generalized Gaussian image model for edge-preserving map estimation. IEEE Trans. Image Process. **2**(3), 296–310 (1993)
4. Dabov, K., Foi, A., Katkovnik, V., Egiazarian, K.: Image denoising by sparse 3-D transform-domain collaborative filtering. IEEE Trans. Image Process. **16**(8), 2080–2095 (2007)
5. Davison, M., Grunbaum, F.: Tomographic reconstruction with arbitrary directions. Commun. Pure Appl. Math. **34**(1), 77–119 (1981)
6. Fessler, J.A., Sutton, B.P.: Nonuniform fast Fourier transforms using min-max interpolation. IEEE Trans. Signal Process. **51**(2), 560–574 (2003)
7. García-Moreno, F., et al.: Using x-ray tomoscopy to explore the dynamics of foaming metal. Nature Commun. **10**(1), 1–9 (2019)
8. Goldstein, T., Osher, S.: The split Bregman method for L1-regularized problems. SIAM J. Imaging Sci. **2**(2), 323–343 (2009)
9. Gürsoy, D., De Carlo, F., Xiao, X., Jacobsen, C.: TomoPy: a framework for the analysis of synchrotron tomographic data. J. Synchrotron Radiat. **21**(5), 1188–1193 (2014)

10. Jackson, J.I., Meyer, C.H., Nishimura, D.G., Macovski, A.: Selection of a convolution function for Fourier inversion using gridding (computerised tomography application). IEEE Trans. Med. Imaging **10**(3), 473–478 (1991)
11. Maia, F., et al.: Compressive phase contrast tomography. In: Image Reconstruction from Incomplete Data VI, vol. 7800, p. 78000F. SPIE, International Society for Optics and Photonics (2010)
12. Marchesini, S., Trivedi, A., Enfedaque, P.: https://bitbucket.org/smarchesilni/xpack/
13. Mohan, K.A., Venkatakrishnan, S., Drummy, L.F., Simmons, J., Parkinson, D.Y., Bouman, C.A.: Model-based iterative reconstruction for synchrotron x-ray tomography. In: IEEE International Conference on Acoustics, Speech and Signal Processing (ICASSP), pp. 6909–6913. IEEE (2014)
14. Ou, T.: gNUFFTW: auto-tuning for high-performance GPU-accelerated nonuniform fast Fourier transforms. Technical report. UCB/EECS-2017-90, University of California, Berkeley, May 2017
15. Pelt, D.M., Batenburg, K.J.: Improving filtered backprojection reconstruction by data-dependent filtering. IEEE Trans. Image Process. **23**(11), 4750–4762 (2014)
16. Pelt, D.M., Gürsoy, D., Palenstijn, W.J., Sijbers, J., De Carlo, F., Batenburg, K.J.: Integration of tomopy and the astra toolbox for advanced processing and reconstruction of tomographic synchrotron data. J. Synchrotron Radiat. **23**(3), 842–849 (2016)
17. Scott, M., et al.: Electron tomography at 2.4-ångström resolution. Nature **483**(7390), 444 (2012)
18. Sonneveld, P.: CGS, a fast Lanczos-type solver for nonsymmetric linear systems. SIAM J. Sci. Stat. Comput. **10**(1), 36–52 (1989)
19. Venkatakrishnan, S.V., Bouman, C.A., Wohlberg, B.: Plug-and-play priors for model based reconstruction. In: IEEE Global Conference on Signal and Information Processing, pp. 945–948. IEEE (2013)
20. Venkatakrishnan, S., et al.: Making advanced scientific algorithms and big scientific data management more accessible. Electron. Imaging **2016**(19), 1–7 (2016)
21. Williams, S., Oliker, L., Vuduc, R., Shalf, J., Yelick, K., Demmel, J.: Optimization of sparse matrix-vector multiplication on emerging multicore platforms. In: SC 2007: Proceedings of the 2007 ACM/IEEE Conference on Supercomputing, pp. 1–12. IEEE (2007)
22. van Aarle, W., et al.: Fast and flexible X-ray tomography using the ASTRA toolbox. Opt. Express **24**(22), 25129–25147 (2016)
23. Zhao, S.R., Jiang, D., Yang, K., Yang, K.: Generalized Fourier slice theorem for cone-beam image reconstruction. J. X-ray Sci. Technol. **23**(2), 157–188 (2015)

heFFTe: Highly Efficient FFT for Exascale

Alan Ayala[1]([✉]), Stanimire Tomov[1], Azzam Haidar[2], and Jack Dongarra[1,3,4]

[1] Innovative Computing Laboratory, The University of Tennessee,
Knoxville, TN, USA
{aayala,tomov,dongarra}@icl.utk.edu
[2] Nvidia Corporation, Santa Clara, CA, USA
azzamhaidar@nvidia.com
[3] Oak Ridge National Laboratory, Oak Ridge, USA
[4] University of Manchester, Manchester, UK

Abstract. Exascale computing aspires to meet the increasing demands from large scientific applications. Software targeting exascale is typically designed for heterogeneous architectures; henceforth, it is not only important to develop well-designed software, but also make it aware of the hardware architecture and efficiently exploit its power. Currently, several and diverse applications, such as those part of the Exascale Computing Project (ECP) in the United States, rely on efficient computation of the Fast Fourier Transform (FFT). In this context, we present the design and implementation of *heFFTe* (Highly Efficient FFT for Exascale) library, which targets the upcoming exascale supercomputers. We provide highly (linearly) scalable GPU kernels that achieve more than $40\times$ speedup with respect to local kernels from CPU state-of-the-art libraries, and over $2\times$ speedup for the whole FFT computation. A communication model for parallel FFTs is also provided to analyze the bottleneck for large-scale problems. We show experiments obtained on Summit supercomputer at Oak Ridge National Laboratory, using up to 24,576 IBM Power9 cores and 6,144 NVIDIA V-100 GPUs.

Keywords: Exascale · FFT · Scalable algorithm · GPUs

1 Introduction

Considered one of the top 10 algorithms of the 20th century, the Fast Fourier transform (FFT) is widely used by applications in science and engineering. Such is the case of applications targeting exascale, e.g. LAMMPS (EXAALT-ECP) [14], and diverse software ranging from particle applications [21] and molecular dynamics, e.g. HACC [7], to applications in machine learning, e.g. [16]. For all these applications, it is critical to have access to an heterogeneous, fast and scalable parallel FFT library, with an implementation that can take advantage of novel hardware components, and efficiently exploit their benefits.

Highly efficient implementations to compute FFT on a single node have been developed for a long time. One of the most widely used libraries is FFTW [10],

© Springer Nature Switzerland AG 2020
V. V. Krzhizhanovskaya et al. (Eds.): ICCS 2020, LNCS 12137, pp. 262–275, 2020.
https://doi.org/10.1007/978-3-030-50371-0_19

which has been tuned to optimally perform in several architectures. Vendor libraries for this purpose have also been highly optimized, such is the case of MKL (Intel) [13], ESSL (IBM) [8], clFFT (AMD) [1] and CUFFT (NVIDIA) [18]. Novel libraries are also being developed to further optimize single node FFT computation, e.g. FFTX [9] and Spiral [22]. Most of the previous libraries have been extended to distributed memory versions, some by the original developers, and others by different authors.

1.1 Related Work

In the realm of *distributed-CPU* libraries, FFTW supports MPI via slab decomposition, however it has limited scalability and hence it is limited to a small number of nodes. P3DFFT [19] extends FFTW functionalities and supports both pencil and slab decompositions. Large scale applications have built their own FFT library, such as FFTMPI [20] (built-in on LAMMPS [14]) and SWFFT [23] (built-in on HACC [7]). These libraries are currently being used by several molecular-dynamics applications.

Concerning *distributed-GPU* libraries, the slab-approach introduced in [17] is one of the first heterogeneous codes for large FFT computation on GPUs. Its optimization approach is limited to small number of nodes and focus on reducing tensor transposition cost (known to be the bottleneck) by exploiting infiniband-interconnection using the IBverbs library, which makes it not portable. Further improvements to scalablity have been presented in FFTE library [26] which supports pencil decompositions and includes several optimizations, although with limited features and limited improvements on communication. Also, FFTE relies on the commercial PGI compiler, which may limit its usage. Finally, one of the most recent libraries is AccFFT [11], its approach consists in overlapping computation and blocking collective communication by reducing the PCIe overhead, they provide good (sublinear) scalability results for large real-to-complex transforms using NVIDIA K20 GPUs.

Even though the fast development of GPUs has enabled great speedup on local computations, the cost of communication between CPUs/GPUs on large-scale computations remains as the bottleneck, and this is a major challenge supercomputing has been facing over the last decade [6]. Large parallel FFT is well-known to be communication bounded, experiments and models have shown that for large node counts the impact on communication needs to be efficiently managed to properly target exascale systems [5,15].

In this context, we introduce *heFFTe* (pronounced "*hefty*"), which provides very good (linear) scalability for large node count, it is open-source and consists of C++ and CUDA kernels with (CUDA-aware) MPI and OpenMP interface for communication. It has a user-friendly interface and does not require any commercial compiler. Wrappers to interface with C, Fortran and Python are available. It is publicly available in [2] and documented in [24,27–29]. Its main objective is to become the standard for large FFT computations on the upcoming exascale systems. Figure 1 shows how *heFFTe* is positioned on the ECP software stack, and some of its target exascale applications (gray boxes).

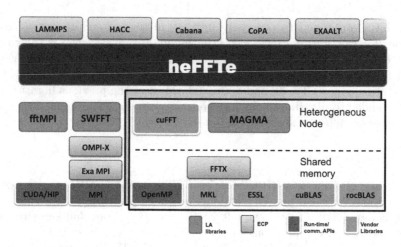

Fig. 1. *heFFTe* in the Exascale Computing Project (ECP) software stack. (Color figure online)

This paper is organized as follows, Sect. 2 describes the classical FFT multidimensional algorithm and its implementation phases within *heFFTe*. We then present *heFFTe*'s main features and functionalities. Next, Sect. 3 presents a multi-node communication model for parallel FFTs and an approach to reduce its computational complexity for a small number of nodes. Section 4 presents numerical experiments on Summit supercomputer, and evaluates the multi-GPU communication impact on performance. Finally, Sect. 5 concludes our paper.

2 Methodology and Algorithmic Design

Multidimensional FFTs can be performed by a sequence of low-dimensional FFTs (see e.g. [12]). Typical approaches used by parallel libraries are the *pencil* and *slab* decompositions. Algorithm 1 presents the pencil decomposition approach, which computes 3D FFTs by means of three 1D FFTs. This approach is schematically shown in Fig. 2. On the other hand, slab decomposition relies on computing sets of 2D and 1D FFTs.

Figure 2 schematically shows the steps during a 3D FFT computation in parallel, using a 3D partition of processors. On the top part of this figure, we present the pencil methodology, as described in Algorithm 1, in which $\hat{N}_i$ denotes output data obtained from applying 1D FFT of size N_i on the i-th direction. This approach can be summarized as follows, the input data of size $N_0 \times N_1 \times N_2$ is initially distributed into a grid processors, $P_{i0} \times P_{i1} \times P_{i2}$, in what is known as *brick* decomposition. Then, a reshape (transposition) puts data into pencils on the first direction where the first set of 1D FFTs are performed. These two steps are repeated for the second and third direction. Observe that intermediate reshaped data is handled in new processor grids which must be appropriately created to ensure load-balancing, for simplicity a single $Q_0 \times Q_1$ grid is used

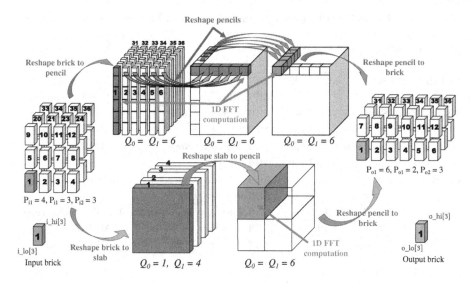

Fig. 2. 3D FFT computation steps via pencil decomposition approach (top) and via slab-pencil decomposition (bottom), c.f., Algorithm 1.

Algorithm 1. 3D FFT algorithm via pencil decomposition approach

---

Require: Initial and final processor grids, $P_{i0} \times P_{i1} \times P_{i2} - P_{o0} \times P_{o1} \times P_{o2}$.
  Data in spatial domain, $N_0/P_{i0} \times N_1/P_{i1} \times N_2/P_{i2}$
Ensure: FFT transform in frequency domain, $\widehat{N_0}/P_{i0} \times \widehat{N_1}/P_{i1} \times \widehat{N_2}/P_{i2}$.
  Calculate a 2D grid Q_0 and Q_1 s.t. $Q_0 \times Q_1 = P_{i0} \times P_{i1} \times P_{i2}$.

$$N_0/P_{i0} \times N_1/P_{i1} \times N_2/P_{i2} \xrightarrow{\text{Reshape}} N_0 \times N_1/Q_0 \times N_2/Q_1$$

$$N_0 \times N_1/Q_0 \times N_2/Q_1 \xrightarrow{\text{First Dimension 1D FFTs}} \widehat{N_0} \times N_1/Q_0 \times N_2/Q_1$$

$$\widehat{N_0} \times N_1/Q_0 \times N_2/Q_1 \xrightarrow{\text{Reshape}} \widehat{N_0}/Q_0 \times N_1 \times N_2/Q_1$$

$$\widehat{N_0}/Q_0 \times N_1 \times N_2/Q_1 \xrightarrow{\text{Second Dimension 1D FFTs}} \widehat{N_0}/Q_0 \times \widehat{N_1} \times N_2/Q_1$$

$$\widehat{N_0}/Q_0 \times \widehat{N_1} \times N_2/Q_1 \xrightarrow{\text{Reshape}} \widehat{N_0}/Q_0 \times \widehat{N_1}/Q_1 \times N_2$$

$$\widehat{N_0}/Q_0 \times \widehat{N_1}/Q_1 \times N_2 \xrightarrow{\text{Third Dimension 1D FFTs}} \widehat{N_0}/Q_0 \times \widehat{N_1}/Q_1 \times \widehat{N_2}$$

$$\widehat{N_0}/Q_0 \times \widehat{N_1}/Q_1 \times \widehat{N_2} \xrightarrow{\text{Reshape}} \widehat{N_0}/P_{o0} \times \widehat{N_1}/P_{o1} \times \widehat{N_2}/P_{o2}$$

---

in Algorithm 1. Finally, a last data-reshape takes pencils on the third direction into the output brick decomposition.

Several applications provide input data already on pencil distribution on the first direction and require the output written as pencils on the third direction. In this case, only two data-reshapes are required, this is the default for FFTE [26] and AccFFT [11] libraries. On the other hand, *heFFTe* can treat input and output shapes with high flexibility, generalizing features of modern libraries, and with a friendly interface as presented in Sect. 2.3.

Table 1. MPI routines required by parallel FFT libraries.

Libraries	Point-to-point routines		Collective routines		Process topology
	Blocking	Non-blocking	Blocking	Non-blocking	
FFTMPI	MPI_Send	MPI_Irecv	MPI_Allreduce MPI_Alltoallv	None	MPI_Group MPI_Comm_create
SWFFT	MPI_Sendrecv	MPI_Isend MPI_Irecv	MPI_Allreduce MPI_Barrier	None	MPI_Cart_create MPI_Cart_sub
AccFFT	MPI_Sendrecv	MPI_Isend MPI_Irecv	MPI_Alltoallv MPI_Bcast	None	MPI_Cart_create
FFTE	None	None	MPI_Alltoallv MPI_Bcast	None	None
heFFTe	MPI_Send MPI_Recv MPI_Sendrecv	MPI_Isend MPI_Irecv	MPI_Alltoallv MPI_Allreduce MPI_Barrier	heFFTe_Alltoall	MPI_Comm_create MPI_Group MPI_Cart_sub

Finally, in the bottom part of Fig. 2, we show the slab approach which saves one step of data reshape by performing 2D FFTs, this has a considerable impact in performance for a small number of nodes [25].

2.1 Kernels Implementation

Two main sets of kernels intervene into a parallel FFT computation:

1. *Computation of low dimensional FFTs*, which can be obtained by optimized libraries for single node FFT, as those described in Sect. 1.
2. *Data reshape*, which essentially consists on a tensor transposition, and takes a great part of the computation time.

To compute low-dimensional FFTs, *heFFTe* supports several open-source and vendor libraries for single node, as those described in Sect. 1. And it also provides templates to select types and precision of data. For direct integration to applications, *heFFTe* provides example wrappers and templates to help users to easily link with their libraries.

Data reshape is essentially built with two sets of routines, the first one consists in *packing* and *unpacking* kernels which, respectively, manage data to be sent and to be received among processors. Generally, these set of kernels account for less than 10% of the reshaping time. Several options for packing and unpacking data are available in *heFFTe*, and there is an option to tune and find the best one for a given problem on a given architecture. The last set of routines correspond to communication kernels, *heFFTe* supports binary and collective communications as presented in Table 1.

The main impact on performance obtained by *heFFTe* in comparison to standard libraries, comes from the kernels optimization ($> 40\times$ speedup w.r.t CPU libraries, c.f. Fig. 6), and also by a novel efficient asynchronously-management

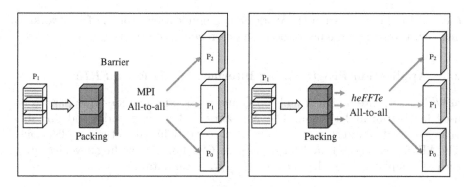

Fig. 3. Standard approach of packing and synchronous data transfer (left), and its asynchronous approach coupling packing and communication (right).

of packing and communication, as shown in Fig. 3, where we can observe the classical packing process supported by most libraries and a novel approach via routine *heFFTe_alltoallv*, introduced in [4], which overlaps packing/unpacking with MPI communication.

2.2 Communication Design and Optimization

Parallel FFT libraries typically handle communication by moving data structures on the shape of pencils, bricks, and slabs of data. For each of these options the total amount of data communicated is always the same. Hence, decreasing the number of messages between processors yields to increasing the size of the messages they send. On the other hand, for modern hardware architectures, it is well-known that latency and bandwidth improvements do not grow as quickly as the arithmetic computation power [6]. Therefore, it is important to choose the appropriate communication scheme. For instance, reshaping brick to pencil data requires $O(P^{1/3})$ messages, this can be verified by overlapping both grids. Analogously, the number of messages for reshaping pencil to pencil is $O(P^{1/2})$, while $O(P^{2/3})$ for brick to slab, and $O(P)$ for slab to pencil.

Choosing the right communication scheme highly depends on the problem size and hardware features, *heFFTe* supports standard MPI_Alltoallv within subgroups, which generally yields better performance compared to poin-to-point communication. However, optimizations of all-to-all routines on heterogeneous clusters are still not available (e.g. on the NVIDIA Collective Communications Library [3]), even though, as can be regarded in Fig. 6, improvements to all-to-all communication are critical. For this reason, we developed a routine called *heFFTe_alltoallv* [4] which includes several all-to-all communication kernels and can be used for tuning and selecting the best one for a given architecture. This routines aimed to provide a better management of multi-rail communication. The asynchronous approach of *heFFTe_alltoallv* was proved efficient for up to 32 nodes, and the multi-rail management, although promising, negatively impacts the performance when increasing node count, degrading the potential benefit of the multi-rail optimization [4]. Work on communication avoiding frameworks is ongoing, targeting large node count on heterogeneous clusters.

2.3   Application Programming Interface (API) for *heFFTe*

heFFTe's software design is built on C and C++, and provides wrappers to be used in Fortran and Python. The API aims to be user-friendly and portable among several architectures, allowing users to easily link it to their applications. The API follows styles from FFTW3 and CUFFT libraries, adding novel features, such as templates for multitype and multiprecision data.

Define FFT Parameters. Distributed FFT requires data split on a processors grid. In 3D, input/output arrays are typically defined by the six vertices of a *brick*, e.g. (i_{lo}, i_{hi}), as shown in Fig. 2; *heFFTe* allows this definition using any MPI sub-communicator, fft_comm, which has to be provided by user together with data definition.

```
#include <heffte.h>

int main(int argc, char *argv[]) {

MPI_Init(&argc, &argv);
MPI_Comm fft_comm = MPI_COMM_WORLD;

heffte_init(); /* heFFTe initialization */
float *work; /* Single precision input */
FFT3d <float> *fft = new FFT3d <float> (fft_comm);
```

FFT Plan Definition. Once data and processors grids are locally defined, user can create an FFT plan by simply providing the following parameters,

- dim: Problem dimension, e.g., dim = 3
- N: Array size, e.g., N = [nx,ny,nz]
- permute: Permutation storage of output array.

Next, we show how a *heFFTe* plan is created; memory requirements are returned to the user as a workspace array.

```
...
/* Create FFT plan */
heffte_plan_create(dim, work, fft, N, i_lo, i_hi, o_lo, o_hi, permute,
                   workspace);
```

Note that a single plan can be used for several FFT computations, which is typical for several applications where grids are fixed.

FFT Execution. One of *heFFTe*'s most important kernels is the one in charge of the FFT computation. This kernel has the same syntax for any type of data and its usage follows APIs from CUFFT and FFTW3.

```
...
/* Compute an in-place complex-to-complex (C2C) forward FFT */
heffte_execute(fft, work, work);
```

Similar execution function is available for the case of real-to-complex (R2C) transforms, `heffte_execute_r2c`.

3 Multi-node Communication Model

To analyze the bottleneck of exascale FFT computation, communication models can be deduced and experimentally verified for different types of computer clusters [5]. These models can be built for specific frameworks, as for pencil and slab data exchanges [25]; or they could be oriented to the hardware architecture [11].

In this section, we propose an inter-node communication model for large FFTs. We focus on inter-node effects since fast interconnection is typically available intra-node, e.g. NVLINK. And properly scheduling intra-node communications can overlap their cost with the inter-node communications. In Table 2, we summarize the parameters to be used for the communication model.

Table 2. Parameters for communication model

Symbol	Description
N	Size of FFT
P	Number of nodes
r	Number of reshapes (tensor transpose)
α	Size of datatype (Bytes)
M	Message size per node (Bytes)
W	Inter-node bandwidth (GB/s)

To create a communication model, we analyze the *computational intensity* (φ) in Flops/Byte. For the case of FFT, we have that the number of FLOPS is $5N\log(N)$ and the volume of data moved at each reshape is αN, then for the total FFT computation using P nodes, we get,

$$\varphi := P\frac{C}{M} = \frac{5P\log(N)}{\alpha r}, \tag{1}$$

and the peak performance (in GFlops) is defined as,

$$\Psi := \varphi B = \frac{5P\log(N)B}{\alpha r}. \tag{2}$$

For the case of Summit supercomputer, we have a node interconnection of $B = 25$ GB/s, considering $r = 4$ (c.f. Fig. 2) and data-type as double-precision complex (i.e. $\alpha = 16$). Then,

$$\Psi_{\text{Summit}} = \frac{5P \log(N) * 25}{16 * 4} = 1.953 P \log(N). \tag{3}$$

Figure 8, shows *heFFTe*'s performance for a typical FFT of size $N = 1024^3$, and compares it to the roofline peak for increasing number of nodes, getting about to 90% close to peak value.

4   Numerical Experiments

In this section we present numerical experiments on Summit supercomputer, which has 4,608 nodes, each composed by 2 IBM Power9 CPUs and 6 Nvidia V100 GPUs. For our experiments, we use the pencil decomposition approach, which is commonly available in classical libraries and can be shown to be faster than the slab approach for large node count [25]. In Fig. 4, we first show strong scalability comparison between *heFFTe* GPU and CPU implementations, being the former ∼2× faster than the latter. We observe very good linear scalability in both curves. Also, since *heFFTe* CPU version was based on improved versions of kernels from FFTMPI and SWFFT libraries [29], then its performance is at least as good as them. Therefore, *heFFTe* GPU is also ∼2× faster than FFTMPI and SWFFT libraries. Drop in performance for the CPU implementation after 512 nodes (12,288 cores) is due to latency impact, which we verified with several experiments. This is because for a 1024^3 size, the number of messages become very large while their size becomes very small. When increasing the problem size, the CPU version keeps scaling very well as shown in Fig. 5.

Next, Fig. 5 shows weak scalability comparison of *heFFTe* GPU and CPU implementations for different 3D FFT sizes, showing over 2× speedup and very good scaling.

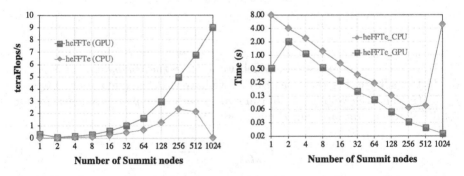

Fig. 4. Strong scalability on 3D FFTs of size 1024^3, using 24 MPI processes (1 MPI per Power9 core) per node (blue), and 24 MPI processes (4 MPI per GPU-V100) per node (red). (Color figure online)

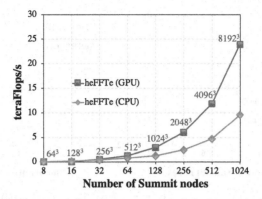

Fig. 5. Weak scalability for 3D FFTs of increasing size, using 24 MPI processes (1 MPI per Power9 core) per node (blue), and 24 MPI processes (4 MPI per GPU-V100) per node (red). (Color figure online)

In order to show the impact of local kernels acceleration, Fig. 6 shows a profile of a single 3D FFT using both, CPU and GPU, versions of *heFFTe*, where over 40× speedup of local kernels and the great impact of communication are clearly displayed.

Next, in Fig. 7, we compare strong and weak scalability of *heFFTe* and FFTE libraries. Concluding that *heFFTe* overcomes FFTE in performance (by a factor >2) and having better scalability. We do not include results with AccFFT library, since its GPU version did not verify correctness on several experiments performed in Summit. However, AccFFT reported a fairly constant speedup of ~1.5 compared with FFTE, while having very similar scalability [11].

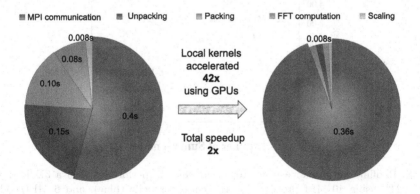

Fig. 6. Profile of a 3D FFT of size 1024^3 on 32 Summit nodes with all-to-all communication – using 40 MPI processes per node (left) and 6 MPIs per node with 1 GPU per MPI (right)

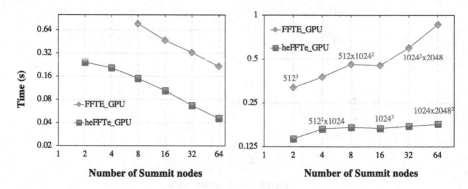

Fig. 7. Strong scalability for a 1024^3 FFT (left), and weak scalability comparison (right). Using 40 MPI processes, 1MPI/core, per node (blue), and 6 MPI processes with 1MPI/GPU-Volta100 per node (red). (Color figure online)

4.1 Multi-node Communication Model

In Fig. 8, we numerically analyze how we approach to the roofline peak performance as described in Sect. 3. We observe that by appropriately choosing the transform size and the number of nodes, we approach to the proposed peak, and hence a correlation could be established between these two parameters to ensure that maximum resources are being used, while still leaving GPU resources to simultaneously run other computations needed by applications.

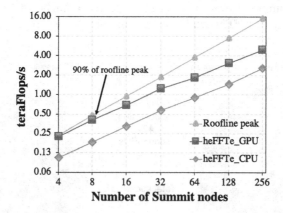

Fig. 8. Roofline performance from Eq. 3 and *heFFTe* performance on a 3D FFT of size 1024^3; using 40 MPI processes, 1MPI/core, per node (blue), and 6 MPI/node, 1MPI/1GPU-Volta100, per node (red). (Color figure online)

4.2 Using *heFFTe* with Applications

Diverse applications targeting exascale make use of FFT within their models. In this section, we consider LAMMPS [14], part of the EXAALT ECP project. Its KSPACE package provides a variety of long-range Coulombic solvers, as well as pair styles which compute the corresponding pairwise Coulombic interactions. This package heavily rely on efficient FFT computations, with the purpose to compute the energy of a molecular system.

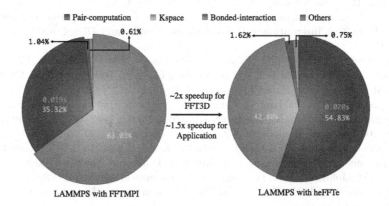

Fig. 9. LAMMPS Rhodopsin protein benchmark on a 128^3 FFT grid, using 2 nodes, 4 MPI processes per node. For FFTMPI we use 1 MPI per core plus 16 OpenMP threads, and for *heFFTe* we use 1 MPI per GPU.

In Fig. 9 we present an experiment obtained using a LAMMPS benchmark experiment, where we compare the performance when using its built-in FFTMPI library, and then using the GPU version of *heFFTe* library. As shown in Fig. 4, it is expected that even for large runs, using LAMMPS with *heFFTe* would provide a 2× speedup of its KSPACE routine.

5   Conclusions

In this article, we presented the methodology, implementation and performance results of *heFFTe* library, which performs FFT computation on heterogeneous systems targeting exascale. We have provided experiments showing considerable speedups compared to state-of-the-art libraries, and that linear scalability is achievable. We have greatly speedup local kernels getting very close to the experimental roofline peak on Summit (a large heterogeneous cluster which ranks first on the top 500 supercomputers). Our results show that further optimizations would require better hardware interconnection and/or new communication-avoiding algorithmic approaches.

References

1. https://github.com/clMathLibraries/clFFT
2. https://bitbucket.org/icl/heffte
3. https://github.com/NVIDIA/nccl
4. Ayala, A., et al.: Impacts of Multi-GPU MPI collective communications on large FFT computation. In: IEEE/ACM Workshop on Exascale MPI (ExaMPI) (2019)
5. Czechowski, K., McClanahan, C., Battaglino, C., Iyer, K., Yeung, P.K., Vuduc, R.: On the communication complexity of 3D FFTs and its implications for exascale, June 2012. https://doi.org/10.1145/2304576.2304604
6. Demmel, J.: Communication-avoiding algorithms for linear algebra and beyond. In: IEEE 27th International Symposium on Parallel and Distributed Processing (2013)
7. Emberson, J., Frontiere, N., Habib, S., Heitmann, K., Pope, A., Rangel, E.: Arrival of first summit nodes: HACC testing on phase I system. Technical report. MS ECP-ADSE01-40/ExaSky, Exascale Computing Project (ECP) (2018)
8. Filippone, S.: The IBM parallel engineering and scientific subroutine library. In: Dongarra, J., Madsen, K., Waśniewski, J. (eds.) PARA 1995. LNCS, vol. 1041, pp. 199–206. Springer, Heidelberg (1996). https://doi.org/10.1007/3-540-60902-4_23
9. Franchetti, F., et al.: FFTX and SpectralPack: a first look. In: IEEE International Conference on High Performance Computing, Data, and Analytics (2018)
10. Frigo, M., Johnson, S.G.: The design and implementation of FFTW3. Proc. IEEE **93**(2), 216–231 (2005). Special issue on "Program Generation, Optimization, and Platform Adaptation"
11. Gholami, A., Hill, J., Malhotra, D., Biros, G.: Accfft: a library for distributed-memory FFT on CPU and GPU architectures. CoRR abs/1506.07933 (2015). http://arxiv.org/abs/1506.07933
12. Grama, A., Gupta, A., Karypis, G., Kumar, V.: Accuracy and Stability of Numerical Algorithms, 2nd edn. Addison Wesley, Boston (2003)
13. Intel: Intel Math Kernel Library. http://software.intel.com/en-us/articles/intel-mkl/, https://software.intel.com/en-us/mkl/features/fft
14. Large-scale atomic/molecular massively parallel simulator (2018). https://lammps.sandia.gov/
15. Lee, M., Malaya, N., Moser, R.D.: Petascale direct numerical simulation of turbulent channel flow on up to 786k cores. In: Proceedings of the International Conference on High Performance Computing, Networking, Storage and Analysis (2013)
16. Lin, S., et al.: FFT-based deep learning deployment in embedded systems. In: Design, Automation Test in Europe Conference Exhibition (DATE), pp. 1045–1050 (2018)
17. Nukada, A., Sato, K., Matsuoka, S.: Scalable multi-GPU 3-D FFT for TSUBAME 2.0 supercomputer. In: High Performance Computing, Networking, Storage and Analysis (2012)
18. CUDA NVIDIA: cuFFT library (2018). http://docs.nvidia.com/cuda/cufft
19. Pekurovsky, D.: P3DFFT: a framework for parallel computations of Fourier transforms in three dimensions. SIAM J. Sci. Comput. **34**(4), C192–C209 (2012). https://doi.org/10.1137/11082748X
20. Plimpton, S., Kohlmeyer, A., Coffman, P., Blood, P.: fftMPI, a library for performing 2d and 3d FFTs in parallel. Technical report, Sandia National Lab. (SNL-NM), Albuquerque, NM, United States (2018)

21. Plimpton, S.J.: FFTs for (mostly) particle codes within the DOE exascale computing project (2017)
22. Popovici, T., Low, T.M., Franchetti, F.: Large bandwidth-efficient FFTs on multicore and multi-socket systems. In: IEEE International Parallel and Distributed Processing Symposium (IPDPS). IEEE (2018)
23. Richards, D., Aziz, O., Cook, J., Finkel, H., et al.: Quantitative performance assessment of proxy apps and parents. Technical report, Lawrence Livermore National Lab. (LLNL), Livermore, CA, United States (2018)
24. Shaiek, H., Tomov, S., Ayala, A., Haidar, A., Dongarra, J.: GPUDirect MPI Communications and Optimizations to Accelerate FFTs on Exascale Systems. Extended Abstract icl-ut-19-06, September 2019
25. Takahashi, D.: Implementation of parallel 3-D real FFT with 2-D decomposition on Intel Xeon Phi clusters. In: 13th International Conference on Parallel Processing and Applied Mathematics (2019)
26. Takahashi, D.: FFTE: a fast Fourier transform package (2005). http://www.ffte.jp/
27. Tomov, S., Haidar, A., Ayala, A., Schultz, D., Dongarra, J.: Design and implementation for FFT-ECP on distributed accelerated systems. ECP WBS 2.3.3.09 Milestone report FFT-ECP ST-MS-10-1410, Innovative Computing Laboratory, University of Tennessee, revision, April 2019
28. Tomov, S., Haidar, A., Ayala, A., Shaiek, H., Dongarra, J.: FFT-ECP Implementation optimizations and features phase. Technical report, ICL-UT-19-12, October 2019
29. Tomov, S., Haidar, A., Schultz, D., Dongarra, J.: Evaluation and design of FFT for distributed accelerated systems. ECP WBS 2.3.3.09 Milestone report FFT-ECP ST-MS-10-1216, Innovative Computing Laboratory, University of Tennessee, revision, October 2018

Scalable Workflow-Driven Hydrologic Analysis in HydroFrame

Shweta Purawat[1]([✉]), Cathie Olschanowsky[2], Laura E. Condon[3], Reed Maxwell[4], and Ilkay Altintas[1]

[1] San Diego Supercomputer Center, UC San Diego, San Diego, USA
spurawat@gmail.com, shpurawat@sdsc.edu
[2] Boise State University, Boise, USA
[3] University of Arizona, Tucson, USA
[4] Colorado School of Mines, Golden, USA

Abstract. The HydroFrame project is a community platform designed to facilitate integrated hydrologic modeling across the US. As a part of HydroFrame, we seek to design innovative workflow solutions that create pathways to enable hydrologic analysis for three target user groups: the modeler, the analyzer, and the domain science educator. We present the initial progress on the HydroFrame community platform using an automated Kepler workflow. This workflow performs end-to-end hydrology simulations involving data ingestion, preprocessing, analysis, modeling, and visualization. We demonstrate how different modules of the workflow can be reused and repurposed for the three target user groups. The Kepler workflow ensures complete reproducibility through a built-in provenance framework that collects workflow specific parameters, software versions, and hardware system configuration. In addition, we aim to optimize the utilization of large-scale computational resources to adjust to the needs of all three user groups. Towards this goal, we present a design that leverages provenance data and machine learning techniques to predict performance and forecast failures using an automatic performance collection component of the pipeline.

Keywords: Computational hydrology · Scientific Workflow · Reproducibility · Machine learning

1 Introduction

Hydrology aims to model freshwater on earth with the aim of facilitating large-scale and long-term impact on humanity. Resource contention and rapid growth of the human population demand a precise understanding of physics behind the flow of large water reserves and their interactions with interfacing geological systems. Computational Hydrology focuses on leveraging large-scale computational infrastructure to build hydrologic models. Hydrologic researchers are beginning to exploit the full potential of computational resources by converting hydrologic equation solvers into scalable simulations. However, to make a substantial

© Springer Nature Switzerland AG 2020
V. V. Krzhizhanovskaya et al. (Eds.): ICCS 2020, LNCS 12137, pp. 276–289, 2020.
https://doi.org/10.1007/978-3-030-50371-0_20

improvement, scientists need a frictionless mechanism to perform continental-scale hydrologic experiments.

In the recent developments, platforms such as National Water Model(NWM) [1] and ParFlow-CONUS [2–4] have become critical toolsets on which hydrologic community depends. These platforms extend the reach of scientists in performing hydrologic simulations. However, to make a continental scale impact in the field of hydrology, there is a need for an efficient mechanism that matches the needs of hydrology with advancements in Computer Science. In recent years, the field of Computer Science has grown tremendously delivering capabilities such as cloud computing, parallelization mechanisms such as map-reduce, ultra-fast I/O hardware that facilitate efficient data movement, and Graphical Processing Units to perform large scale model development. Isolated developments in hydrology have not exploited these computational developments to the fullest, mainly due to the rapid rate of change in computer science techniques. Hydrologic researchers' main problem-solving abilities should focus on their domain expertise. To bridge this gap between hydrology and computer science, we are building a framework that provides abstractions to allow domain scientists to unleash the underlying capabilities of computer science, with the same ease as driving a car without knowing the internal workings of an internal combustion engine. We challenge ourselves to develop abstractions that make hydrologic research transparent to computational complexity, while still exploiting the benefits of computational developments.

The key challenges that hinder hydrologic research at the continental scale are experiment reproducibility especially due to the advent of distributed computing, code-portability, efficient data storage and movement, fault tolerance, and functionality to auto-scale hydrologic simulations. Our framework provides a graphical user interface oriented approach based on Kepler Scientific Workflow's modular code-encapsulation and execution structure. In addition, we leverage SmartFlows Suite to provide hydrology experts the benefit of using Machine Learning for automated failure prediction, resource allocation, dynamic fault tolerance, and 24×7 anytime anywhere control of simulations [5].

Kepler: The Kepler Scientific Workflow System provides domain scientists a simplified graphical front end, and a sophisticated workflow execution engine for backend computation [6,7]. Hydrologic scientists can design large-scale simulations by connecting blocks of code on a canvas, and execute their experiments on local or remote hardware. Kepler provides two critical components for building new workflows: (a) Actor and (b) Director. An *Actor* encapsulates a piece of code and can be connected to another actor through a directed edge. The directed edge signifies the dataflow in the graph. A *Director* orchestrates the execution of the entire graph, based on parameters set by the user. A hydrologic researcher can formulate a complex experiment as a directed acyclic graph in Kepler, by building new actors or by re-using existing actors that encapsulate hydrologic code. Such an actor-oriented approach promotes code-reusability and rapid prototyping while the director performs heavy-lifting tasks of executing the instructions on distributed hybrid computational architecture.

SmartFlows: The SmartFlows Suite couples tightly with the Kepler system. It utilizes Machine Learning algorithms to predict failures, assign resources for execution, re-run workflows when faults occur, and provide an interface for anytime-anywhere control of workflow by scientists [5,8,9].

The SmartFlows Suite architecture is designed to provide hydrology scientists a range of services to choose from and create a customized simulation. Once implemented, the SmartFlows Suite will provide the functionality to monitor workflow execution continuously and dynamically make intelligent decisions, as per the initial parameterization of the experiment. Specifically, through the proposed SmartFlows framework, we seek to create a new paradigm where computer scientists and domain science experts work as a unified team. The unified Kepler-SmartFlows framework will act as an automated data assimilator and decision-maker, which will allow domain experts to focus their problem solving capabilities on hydrology specific challenges. Our Machine Learning driven framework will provide assurance to hydrology experts by intelligent handling of computational challenges such as parallelization and fault tolerance with minimal supervision.

In this paper, we present an architecture for an intuitive scientific toolbox that encapsulates computational complexity needed to execute large-scale hydrology codes on distributed hardware platforms. The framework modularizes hydrology algorithms. We have wired the modules of the framework to provide (a) native reproducibility support, (b) 24×7 monitoring and anytime-anywhere control feature, (c) real-time fault prediction by machine learning, (d) dynamic resource allocation based on hardware cost and system behavior, (e) ability to transparently deploy hydrology codes in containers, which facilitate code-portability, and (f) a shared platform for community engagement to upload hydrology workflows to cloud for solving the grand challenges of hydrology at scale.

We envision the proposed framework to become a critical partner that facilitates the hydrology community to unleash an era of continental-scale simulations: a vital step for solving future hydrology challenges.

The key contributions of this paper include:

- A unified architecture that proposes (i) modular hydrologic workflow development with multiple execution modes customized for hydrologic experimentation and (ii) a Machine Learning toolset that is pre-trained on hydrologic workflows' performance monitoring data.
- In-depth illustration of specific benefits the proposed framework brings to each target stakeholder group: the modeler, the analyzer, and the domain science educator.
- A scientific workflow tool that contains the widely-utilized ParFlow-CLM Model.
- An open-source containerization toolset that enables scientists to bootstrap the Kepler-ParFlow framework on their machine within minutes.

2   Related Work

Hydrologic researchers are actively involved in addressing global problems through innovation and continuous improvement. However, the challenges associated with water are not always bound by geographic partitioning. Most solutions that form isolated hydrologic models are approximations that deviate significantly from reality. Large-scale water bodies are essentially continental scale, and demand computational tools that operate at this scale. [10] have specifically highlighted continental scale simulations in the field of hydrology as a "grand challenge." Their work emphasizes the need to exploit massively parallel computational systems to solve hyperresolution hydrologic equations, which contain a large number of unknowns. [11] demonstrates the benefits gained by the use of parallel execution capabilities of ParFlow and the urge for the development of parallel system platforms in hydrology. [11–16] remark the requirement of adopting best practices in parallel computation from other fields, with the goal of improving infrastructure utilization when solving large-scale hydrologic problems.

There has been growing interest in the hydrology community in adapting the scientific workflow approach due to its modularity, reproducibility, code shareability, execution choices, and ease of use. [18] demonstrates the effect of automating the pre-processing step in hydrologic systems using Rule Oriented Data Management System (iRODS). However, this work in limited in scope to enhancing only a portion of the hydrology pipeline. [19–21] utilizes the Microsoft Trident Workflow System based solutions to demonstrate the benefits of using workflow design and execution paradigm in hydrology studies. [22] examines the challenges in hydrology related to big data, data flow, and model management, and build a preliminary adaptation of the Kepler Scientific Workflow System for hydrology applications. In contrast to these developments, the solution we present is first of its kind framework that will not only provide modularity, choice of execution, reproducibility but also dynamically lookout for predicting failures and course correct the execution trail to improve chances of successful workflow execution.

The framework presented in this paper utilizes the Kepler Scientific Workflow System [6] as a core pillar. Kepler has evolved over the last decade to serve the requirements of domain scientists from multiple research areas, due to its extensive repository of features, and a robust execution engine that can orchestrate complex workflows on distributed infrastructure. bioKepler [7], a workflow system for Biological Scientists, is a prominent example of Kepler's flexibility to adapt its features to a domain orthogonal to Computer Science. There are multiple works [24–27,29] that leverage Kepler's Workflow Engine to amplify and accelerate scientific discovery rate, and broaden the impact of research with improved shareability and provenance. Specifically, [26] deploy Kepler to perform single-cell data analysis using the Kepler System. [30,31] leverage the Kepler System to improve disaster management of Wild Fires by conducting real-time data-driven simulations that forecast a wildfire's trajectory, and visualizing wildfires to design mitigation response strategies. In addition, [28] provides an

open-source platform, BBDTC, that allows domain science educators to build
and share knowledge with a larger community. [32] develop fundamental techniques that allow precise prediction of workflow performance using Machine
Learning techniques.

In contrast to isolated and partially focused development efforts, this paper
presents an end-to-end solution that couples the Kepler System and the Smart
Flows toolset to provide Hydrology researchers a framework critical for tackling the grand challenges of hydrology. This framework aims to bridge the
gap between Hydrology Research and Computer Science. In coordination with
Hydrology experts, we plan to build a solution that meets the needs of all three
hydrology user categories: the modeler, the analyzer, and the educator. The
framework will promote code-shareability, enhance data preparation and model
development pipeline, provide multiple execution choices, automatically capture
performance metrics, and leverage Machine Learning to dynamically forecast
failures and adapt execution trails to reduce the impact of hardware faults.

3   HydroFrame Target User Groups

The design of the presented framework is driven by the quest to provide meaningful answers to every individual interested in conducting a hydrology study.
To ensure large scale adoption of the framework, we've partitioned our potential
consumers into three orthogonal categories shown in Fig. 1(i) HydroFrame Target User groups. This organization enables us to write user-specific expectations
and map the framework's capabilities to the demands of our users.

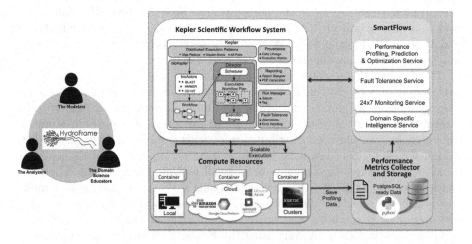

Fig. 1. (i) HydroFrame Target User groups (ii) End-To-End-Architecture of Scalable
Workflow-Driven Hydrologic Analysis

The first category represents 'the modeler.' A modeler is a researcher interested in the mechanics of a hydrology model and needs to control the inputs to a model for research. The technical bottleneck faced by the modeler is the investment of time required to gain an understanding of computational concepts that enable large scale distributed processing. Our proposed framework obviates such requirements by providing intuitive abstractions that act as powerful levers. Using our framework's graphical user interface, domain experts can leverage distributed data-parallel execution patterns to solve complex hydrologic equations. The modelers can design the flow of hydrologic data through an array of modular operations, choose from multiple execution patterns, and execute hydrologic instructions remotely on compute resources (e.g., a Kubernetes cluster), without necessarily writing lengthy programs with thousands of lines of code.

The second category represents the direct consumer of model outputs - 'the analyzer.' An analyzer is an applied researcher or scientist, who consumes a model's output, but not necessarily interested in changing the input parameters. The analyzer wants to visualize and derive a meaningful understanding of the complex hydrologic phenomenon, and impact the output of a project or create new downstream solutions based on the insights extracted. To benefit this community, our portal can facilitate model and output shareability and ensure reproducibility.

The last critical consumers are the educators who play a pivotal role in building the next generation hydrology researchers and scientists. These users are passionate about creating educational material that sparks community interest in hydrology research and enables students to understand the grand challenges. The educators want simplified visual outputs that effectively communicate the importance of solving the challenges of today's hydrologists. Our framework will enable these educators to export the model output, share model design, and input data in a streamlined manner with other educators and students. Our framework will provide K-12 educators the ability to give feedback and drive continuous improvement of the framework's usability.

4   A Reference Architecture for Measurement-Driven Scalable Hydrologic Workflows

In the architecture illustrated in Fig. 1(ii) End-To-End-Architecture of Scalable Workflow-Driven Hydrologic Analysis, we build four tightly coupled systems, which will orchestrate a system that continues to operate efficiently, in the face of faults, and completes hydrology workflow through the cost-effective utilization of computational hardware. In addition, the architecture is designed with ease-of-use as a central theme, in order to facilitate community engagement among Earth System Science communities.

The core component of the architecture is the Kepler Scientific Workflow System, which leverages modularity and parallel execution strategies to transparently execute instructions from hydrology equation solvers. Continental-scale hydrology can generate massive amounts of data in a short duration, presenting opportunities for parallel execution. Earth Science Researchers can delegate parallelization and containerization to Kepler's execution engine. Kepler Workflow Engine provides native support to measure and log critical execution steps of hydrology workflows via the 'Provenance Module.' Kepler workflow continuously monitors and records performance statistics into a relational database. This data can be consumed by Machine learning algorithms for preemptive resource allocation decisions. Automated provenance collection by Kepler ensures real-time measurement of performance data and dynamic course correction.

The SmartFlows component plugs into the Kepler Scientific Workflow System and consumes critical data streams to perform real-time fault prediction. The performance prediction module of SmartFlows consumes hardware utilization data of hydrology workflows and provides real-time resource allocation strategies to meet performance goals in a cost-effective manner. In addition, the Smart-Flows enhances the Earth Science Researchers' capability to monitor and change the course of a real-time workflow by providing anytime anywhere access to workflow execution on multiple devices. Through it 24×7 monitoring services, the SmartFlows component can inform the modeler of any critical change in expected behavior, or notify the analyzer when results become available. The domain-specific intelligence service can be used to record hydrology-specific execution patterns, which allow the SmartFlows module to make predictions with higher accuracy. For example, Earth Science Research Community can share execution histories of multiple hydrology workflows into the cloud and have a dedicated Machine Learning model that specializes in predicting hydrology specific workflows.

To enhance the ability of modelers and analyzers to repurpose existing workflow modules, and design complex workflows with sophisticated debugging capabilities, we aim to provide 'containerization' as a built-in option for each actor. Containerized actor execution can enhance fault tolerance by leveraging the Kubernetes framework [33]. Containerization promotes standardization, rapid code-portability, and scalability. Kubernetes containers will ensure replication of software stack, hence providing the same initial conditions for equation solvers. Once the software stack variation has been eliminated, any deviation in outcome can be traced to reasons such as algorithmic randomization or data distribution.

The framework design keeps automation at the core of each component. The seamless provenance collection feeds the machine learning algorithms and continuously makes them more accurate with each workflow execution. The containerization and modularization of code increase code portability and further facilitates performance metrics distribution among hydrologic experts, thus enabling performance prediction machine learning algorithms to learn at a much faster pace. This shared-training of domain-specific models can lead to the rapid maturity of the hydrology domain-specific intelligence service module in the Smart-Flows toolbox.

The modular nature of code, and transparent execution of containers on the hardware of scientist's choice, will enable domain science educators to engage and build scientific community interest in hydrology at large scale. Hence, our architecture keeps not just enhances the computational scalability of hydrologic workflows, but also enables knowledge dissemination of domain to a wider audience in a seamless manner.

Keeping up with the rate of innovation in Computer Science: In the future, as the computer science community continues to innovate and build new hardware and software systems, the hydrologic domain experts will not have to go through the training cycles of new computational techniques. The Kepler Workflow Engine abstracts the computational complexity and provides a simplified interface for domain experts. In the future, the mechanism to orchestrate containers will continue to evolve; Kepler will leverage the advancements in this area while maintaining the simplicity of the interface to end-users. We envision the Kepler-SmartFlows architecture to become an auto-upgrading abstraction for hydrology research. The architecture will ensure that hydrology scientists continue to innovate by leveraging advances in parallelization, resource allocation, fault tolerance, reproducibility, provenance, and performance forecasting, without the need to catch up with ever-evolving computer science vocabulary.

5   Example: A Scalable Workflow Tool for ParFlow Hydrologic Model Simulations

As progression on the above-described architecture, in this section, we present a Kepler workflow tool that performs end-to-end hydrology simulations by integrating data ingestion, preprocessing, modeling, analysis, and performance profile collection steps in one automated tool. The workflow uses the Parflow CONUS model for hydrologic simulations. The workflow employs an in-built provenance feature of Kepler to record execution history. The workflow is built to support three execution choices – i) Local Execution, ii) Remote Cluster Execution, and iii) Kubernetes Execution. In this section, we will illustrate the workflow tool and different execution styles in detail; describe different features of the tool and how it can be repurposed to the requirements of the three target HydroFrame user groups – the Modelers, the Analyzers, and the Domain Science Educators.

5.1   End-to-end Hydrologic Simulation Tool

The workflow Fig. 2 combines five major stages of ParFlow-CLM(Common Land Model) hydrologic simulation in one automated tool: data ingestion, data preprocessing, Parflow-CLM model simulation, data analysis, and performance profile collection steps. The stages are further broken down into modules. The modular design of the workflow supports repurpose and customization to create workflows for specific user groups. Each of the modules is autonomous and can be reused to create custom workflows with steps relevant to that user group's study.

Data Ingestion Stage: The Data Ingestion stage setups input files, dependent on execution style selected by a user, for subsequent processing and simulation steps. For local execution, it will specify the path of input directory containing

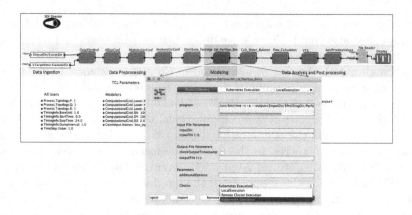

Fig. 2. A Scalable Workflow tool for ParFlow Hydrologic Model Simulations.

CLM (Community Land Model) input files, ParFlow input files, and TCL scripts. For Remote execution, the Kepler actor will copy files securely to the remote cluster.

Data Pre-processing Stage: The Data Pre-processing stage includes three different UserConf actors and a Distribute_Forcing actor. The UserConf actor facilitates users to configure frequently used hydrological model parameters through a graphical user interface (GUI), command-line, or web interfaces. The design ports model parameters as top-level workflow parameters. The UserConf modules programmatically update model parameters and other dependent files with user-specified values. The workflow contains three different actors 'ModelerUsrConf', 'AnalystsUsrConf,' and 'AllUsrConf,' focused on the needs of the three user groups – the Modelers, the Analyzers, and Common to all users, respectively. We ported the most frequently changed parameters from the TCL scripts to the top-level of the workflow, so users do not need to open TCL scripts every time to perform hydrologic simulation or analysis. The 'Distribute_Forcings' actor executes Dist_Forcings.tcl script that distributes the 3D meteorological forcings used by the ParFlow model.

Model Simulation Stage: This module simulates the Little Washita real domain (41 km × 41 km) using ParFlow (Parallel Flow) coupled with the Common Land Model (CLM) plus 3D meteorological forcing. Parflow is an open-source simulation platform that solves the Richards solver equation in 3D. This Kepler module runs the Tcl/TK script that provides data to the ParFlow model and runs the models.

Data Analysis Stage: This step performs an analysis of the simulation outputs obtained in the previous step. This stage performs water balance calculation, flow calculation, VTK output files generation. The 'Calc_Water_Blance' actor estimates changes in subsurface water storage, surface water storage, and total surface runoff over the simulation time. The 'Flow_Calculation' actor calculates the pressure and flow at the outlet over the simulation time. The 'VTK' actor runs VTK_example.tcl script that loads pfb or silo output and generates 3D vtk files. The vtk files are stored in the project subdirectory and are used for visualization using VisIt.

Containerization: We built a Kepler-ParFlow Docker image. It is a lightweight package that includes ParFlow, Kepler software, dependencies, and everything needed to execute the Kepler-ParFlow workflow tool. The workflow tool plus docker container provides a standardized, reproducible, and shareable unit to quickly set up the environment required to perform hydrologic simulations and analysis. The combination ensures quick and reliable code-portability from one computing resource to another. The Modeler and the Analyzer user communities can use the Kepler-ParFlow docker to shorten the environment set up time and concentrate on advancing hydrology science. The ease-of-use of the workflow and docker container will empower Domain Science Educators to effortlessly use in education and training of students and budding hydrologists.

5.2   Features of the Kepler-ParFlow Workflow Tool

Multiple Execution Choices: Kepler "Build once – run many times" policy allows users to operate the same workflow tool in three different execution ways depending on the requirement, input data size, or available compute resources. We used the Kepler execution choice actor and built multiple sub-workflows to offer different alternatives to execute the same workflow, as shown in Fig. 3. The Kepler workflow tool operates in three modes:

- Local Execution: This is the default mode of operation; it uses External execution Kepler actor to execute TCL/TK scripts on the local machine. To execute the workflow, users can install ParFlow and Kepler on their machine and follow the README guide to execute the workflow. Alternatively, users can download and use the Kepler-ParFlow Docker image from the Docker hub to speed up the development activity. The docker image includes ParFlow, Kepler software, and all dependencies needed to execute the workflow tool. This mode of execution is well suited for exploration, limited-scale hydrologic simulations, education, and training.
- Remote Cluster Execution: This mode of operation enables execution of ParFlow tools on a remote cluster, such as BlueM at Colorado School of Mines, an HPC resource. The subworkflow copies input files to a remote cluster using Kepler SCP (secure copy) actor; creates, submits and manages job on a remote cluster using Kepler SSH Session and GenericJobLauncher actors; and copies back the results to the local machine.
- Kubernetes Execution: This execution choice facilitates to run and manage different steps in the workflow tool in docker containers on a remote cluster managed by Kubernetes. Figure 3 illustrates Containerized scalable execution with Kubernetes Execution Choice. The Kepler abstraction layer provides an easy-to-use interface on top of Kubernetes hiding all the underlying complexities. Specifically, Kepler actors create all the YAML files and implement all the steps required by Kubernetes to create persistent volume, transfer data from your machine the remote cluster, create pods, submit and manage the

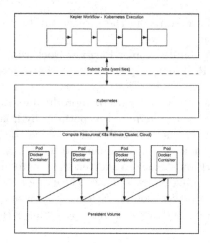

Fig. 3. Containerized Scalable Execution with Kubernetes Execution Choice.

job on the remote cluster and scalability. Each actor in the Kepler-ParFlow tool can be executed within a docker container on a pod. The docker containers are created using the Kepler-ParFlow Docker image. Kepler-ParFlow docker image packages all software and dependencies needed to run the application in a standard unit.

Provenance, Reproducibility, and Reporting: The workflow employs an inbuilt provenance feature of Kepler to record execution history. We advanced the provenance to collect and save the System and Software information for the execution for complete reproducibility. The system and software version information are gathered and recorded in the provenance database using the 'AddProvValues' actor. The builtin reporting suite in Kepler assists users to generate reports from workflow runs. Users can execute Kepler workflow within Jupyter Notebook using Kepler Magic functions https://github.com/words-sdsc/Jupyter_Kepler_Integration and create reports in Jupyter Notebook.

Automated Performance Metrics Collection: We integrated the Linux time application profiler in the hydrologic workflow that enables systematic collection of application performance diagnostic features: execution time, memory, and io measurements. The collected data will be utilized by machine learning algorithms in the SmartFlow services to predict Performance, in Fault Tolerance, and to optimize utilization of large-scale computational resources dynamically, as the experiment progresses.

6   Conclusion

In this paper, we present the initial developments of our long-term vision to produce an integrated framework that enables the Hydrology research community to leverage advancements in Computer Science, such as distributed computing, fault-tolerance, reproducibility. We developed a Kepler Scientific Workflow that allows scientists to execute the Parflow-CLM model on large-scale hybrid infrastructure. The workflow consists of five modules that involve data preparation, ParFlow model simulation, analysis, and visualization. To facilitate performance measurements, we have added functionality to measure and collect performance metrics such as CPU usage, memory usage, I/O rates within the Parflow-CLM Workflow. This will enable scientists to record both hardware and software configurations automatically, enhancing the reproducibility of the workflow. In addition to providing the core functionality, the workflow will enable the Hydrology community users such as modelers, data analyzers, and educators to configure modeling parameters through both command line and graphical user interface, without the need to edit code. Through continuous development, we aim for the presented framework to become a critical workbench for Hydrologists for building modeling simulations and sharing knowledge.

In the future, we plan to add Machine Learning modules that leverage the collected configurations and performance data to dynamically predict hardware usage. We plan to add an auto-deploy feature, which will initiate a new instance of the workflow, once the probability of failure reaches a threshold, preset by the modeler. These functionalities will be encapsulated in the SmartFlows Toolset, which will allow modelers, analyzers, and domain science educators to control and monitor workflow progress from any connected device, through SmartFlow's 24×7 monitoring module.

Software Availability for Community Outreach and Education:

- https://cloud.docker.com/repository/docker/spurawat/kepler-parflow
- https://github.com/hydroframe/Workflows.git

Acknowledgements:. This work is supported by NSF OAC CSSI 1835855, and DOE DE-SC0012630 for IPPD.

References

1. Gochis, D.J., Yu, W., Yates, D.N.: The WRF-Hydro model technical description and user's guide, version 1.0. NCAR Technical Document. Boulder, CO, National Center for Atmospheric Research, p. 120 (2013)
2. Ashby, S.F., Falgout, R.D.: A parallel multigrid preconditioned conjugate gradient algorithm for groundwater flow simulations. Nucl. Sci. Eng. **124**(1), 145–159 (1996)
3. Jones, J.E., Woodward, C.S.: Newton-Krylov-multigrid solvers for large-scale, highly heterogeneous, variably saturated flow problems. Adv. Water Resour. **24**(7), 763–774 (2001). https://doi.org/10.1016/S0309-1708(00)00075-0
4. Kollet, S.J., Maxwell, R.M.: Integrated surface-groundwater flow modeling: a free-surface overland flow boundary condition in a parallel groundwater flow model. Adv. Water Resour. **29**(7), 945–958 (2006). https://doi.org/10.1016/j.advwatres.2005.08.006

5. Altintas, I., Purawat, S., Crawl, D., Singh, A., Marcus, K.: Toward a methodology and framework for workflow-driven team science. Comput. Sci. Eng. **21**(04), 37–48 (2019). https://doi.org/10.1109/MCSE.2019.2919688
6. Ludaescher, B., et al.: Scientific workflow management and the Kepler system. Concurr. Comput. Pract. Exp. **18**, 1039–1065 (2006)
7. Altintas, I., Wang, J., Crawl, D., Li, W.: Challenges and approaches for distributed workflow-driven analysis of large-scale biological data: vision paper. In: Proceedings of the 2012 Joint EDBT/ICDT Workshops. ACM (2012)
8. Singh, A., Rao, A., Purawat, S., Altintas, I.: A Machine learning approach for modular workflow performance prediction. In: Proceedings of the 12th Workshop on Workflows in Support of Large-Scale Science, New York, NY, USA, pp. 7:1–7:11 (2017). https://doi.org/10.1145/3150994.3150998
9. Singh, A., Schram, M., Tallent, N., Altintas, I.: Deep learning for enhancing fault tolerant capabilities of scientific workflows. In: IEEE International Workshop on Benchmarking, Performance Tuning and Optimization for Big Data Applications, at the IEEE Big Data: Conference, Seattle, WA (2018)
10. Wood, E.F., et al.: Hyperresolution global land surface modeling: meeting a grand challenge for monitoring Earth's terrestrial water. Water Resour. Res. **47**, W05301 (2011). https://doi.org/10.1029/2010WR010090
11. Kollet, S.J., et al.: Proof of concept of regional scale hydrologic simulations at hydrologic resolution utilizing massively parallel computer resources. Water Resour. Res. **46**, W04201 (2010). https://doi.org/10.1029/2009WR008730
12. Bierkens, M.F.P.: Global hydrology 2015: state, trends, and directions. Water Resour. Res. **51**, 4923–4947 (2015). https://doi.org/10.1002/2015WR017173
13. Clark, M.P., et al.: A unified approach for process-based hydrologic modeling: 1. modeling concept. Water Resour. Res. **51**, 2498–2514 (2015). https://doi.org/10.1002/2015WR017198
14. Maxwell, R.M.: A terrain-following grid transform and preconditioner for parallel, largescale, integrated hydrologic modeling. Adv. Water Resour. **53**, 109–117 (2013)
15. Maxwell, R.M., Condon, L.E., Kollet, S.J.: A high-resolution simulation of groundwater and surface water over most of the continental US with the integrated hydrologic model ParFlow v3. Geosci. Model Dev. **8**, 1–15 (2015)
16. Maxwell, R.M., Condon, L.E.: Connections between groundwater flow and transpiration partitioning. Science **353**(6297), 377 (2016)
17. Hutton, C., Wagener, T., Freer, J., Han, D., Duffy, C., Arheimer, B.: Most computational hydrology is not reproducible, so is it really science? Water Resour. Res. **52**, 7548–7555 (2016). https://doi.org/10.1002/2016WR019285
18. Billah, M.M.: Using a data grid to automate data preparation pipelines required for regional-scale hydrologic modeling. Environ. Model. Softw. **78**, 31–39 (2016). https://doi.org/10.1016/j.envsoft.2015.12.010. ISSN 1364–8152
19. Fitch, P., Perraud, J.M., Cuddy, S., Seaton, S., Bai, Q., Hehir, D.: The Hydrologists Workbench: more than a scientific workflow tool. In: Sims, J., Merrin, L., Ackland, R., Herron, N., (eds.) Water Information Research and Development Alliance: Science Symposium Proceedings, 1–5 August 2011, Melbourne, Australia, CSIRO, pp. 61–69 (2012). http://hdl.handle.net/102.100.100/100717?index=1
20. Cuddy, S.M., Fitch, P.: Hydrologists Workbench - a hydrological domain workflow toolkit. In: International Congress on Environmental Modelling and Software, vol. 246 (2010)
21. Piasecki, M., Lu, B.: Using the workflow engine TRIDENT as a hydrologic modeling platform, vol. 12, p. 3680 (2010)

22. Guru, S.M., Kearney, M., Fitch, P., Peters, C.: Challenges in using scientific workflow tools in the hydrology domain. In: 18th IMACS World Congress; MODSIM 2009 International Congress on Modelling and Simulation, Cairns, Qld, pp. 3514–3520 (2009). http://hdl.handle.net/102.100.100/111463?index=1
23. Perraud, J., Fitch, P.G., Bai, Q.: Challenges and Solutions in Implementing Hydrological Models within Scientific Workflow Software. AGU Fall Meet. Abstr. (2010)
24. Chen, R., et al.: EPiK - a workflow for electron tomography in Kepler. Proc. Comput. Sci. **29**, 2295–2305 (2014)
25. Gan, Z., et al.: MAAMD: a workflow to standardize meta-analyses and comparison of affymetrix microarray data. BMC Bioinform. **15**(1), 69 (2014). https://doi.org/10.1186/1471-2105-15-69
26. Qian, Y., et al.: FlowGate: towards extensible and scalable web-based flow cytometry data analysis. In: Proceedings of the 2015 XSEDE Conference: Scientific Advancements Enabled by Enhanced Cyberinfrastructure. ACM (2015)
27. Purawat, S.: A Kepler workflow tool for reproducible molecular dynamics. Biophys. J. **112**(12), 2469–2474 (2017). https://doi.org/10.1016/j.bpj.2017.04.055
28. Purawat, S., Cowart, C., Amaro, R.E., Altintas, I.: Biomedical big data training collaborative (BBDTC): an effort to bridge the talent gap in biomedical science and research. J. Comput. Sci. **20**, 205–214 (2017)
29. Wang, J., Tang, Y., Nguyen, M., Altintas, I.: A scalable data science workflow approach for big data bayesian network learning. In: Proceedings of the 2014 IEEE/ACM International Symposium on Big Data Computing. IEEE Computer Society (2014)
30. Altintas, I., et al.: Towards an integrated cyberinfrastructure for scalable data-driven monitoring, dynamic prediction and resilience of wildfires. In: Proceedings of the International Conference on Computational Science, ICCS 2015 (2015)
31. Nguyen, M.H., Uys, D., Crawl, D., Cowart, C., Altintas, I.: A scalable approach for location-specific detection of Santa Ana conditions. In: Proceedings of the 2016 IEEE International Conference on Big Data (2016)
32. Singh, A., Nguyen, M., Purawat, S., Crawl, D., Altintas, I.: Modular resource centric learning for workflow performance prediction. In: The 6th Workshop on Big Data Analytics: Challenges, and Opportunities (BDAC) at the 27th IEEE/ACM International Conference for High Performance Computing, Networking, Storage, and Analysis, SC15 (2015). http://arxiv.org/abs/1711.05429
33. Burns, B., Grant, B., Oppenheimer, D., Brewer, E., Wilkes, J.: Borg, Omega, and Kubernetes. Queue **14**(1), 2016 (2016). https://doi.org/10.1145/2898442.2898444

Patient-Specific Cardiac Parametrization from Eikonal Simulations

Daniel Ganellari[1], Gundolf Haase[1], Gerhard Zumbusch[2], Johannes Lotz[3],
Patrick Peltzer[3], Klaus Leppkes[3], and Uwe Naumann[3(✉)]

[1] Institute for Mathematics and Scientific Computing,
University of Graz, 8010 Graz, Austria
[2] Institut für Angewandte Mathematik,
Friedrich-Schiller-Universität Jena, 07743 Jena, Germany
[3] Software and Tools for Computational Engineering, RWTH Aachen University,
52062 Aachen, Germany
naumann@stce.rwth-aachen.de

Abstract. Simulations in cardiac electrophysiology use the bidomain
equations to describe the electrical potential in the heart. If only the
electrical activation sequence in the heart is needed, then the full bidomain equations can be substituted by the Eikonal equation which allows
much faster responses w.r.t. the changed material parameters in the
equation. We use our Eikonal solver optimized for memory usage and
parallelization. Patient-specific simulations in cardiac electrophysiology
require patient-specific conductivity parameters which are not accurately
available in vivo. One chance to improve the given conductivity parameters consists in comparing the computed activation sequence on the
heart surface with the measured ECG on the torso mapped onto this surface. By minimizing the squared distance between the measured solution
and the Eikonal computed solution we are able to determine the material parameters more accurately. To reduce the number of optimization
parameters in this process, we group the material parameters and introduce a specific scaling parameter γ_k for each group. The minimization
takes place w.r.t. the scaling γ. We solve the minimization problem by
the BFGS method and adaptive step size control. The required gradient
$\nabla_\gamma f(\gamma)$ is computed either via finite differences or algorithmic differentiation using dco/c++ in tangent as well as in adjoint mode. We present
convergence behavior as well as runtime and scaling results.

Keywords: Eikonal equation · Domain decomposition · Tetrahedral
mesh · Parallel algorithm · Shared memory · Optimization ·
Algorithmic differentiation · Adjoints

Support from FWF project F32-N18, Erasmus Mundus JoinEUsee PENTA scholarship
and Horizon 2020 Project-Nr.: 671697 MONT-BLANC 3. The computational results
presented have been achieved in part using the Vienna Scientific Cluster (VSC).

V. V. Krzhizhanovskaya et al. (Eds.): ICCS 2020, LNCS 12137, pp. 290–303, 2020.
https://doi.org/10.1007/978-3-030-50371-0_21

1   Introduction

Simulations in cardiac electrophysiology (EP) use the bidomain equations consisting of two partial differential equations (PDEs) coupled nonlinearly by a set of ordinary differential equations which describe the intercellular and the extracellular electrical potential. Its difference, the transmembrane potential, is responsible for the excitation of the heart and its steepest gradients form an excitation wavefront propagating in time. The arrival time $\varphi(x)$ of this excitation wavefront at some point $x \in \Omega$ can be approximated by the simpler Eikonal equation with given heterogeneous, anisotropic velocity information $M(x)$. The domain $\Omega \subset \mathbb{R}^3$ is discretized by planar-sided tetrahedrons with a piecewise linear approximation of the solution $\varphi(x)$ inside each of them.

It is almost impossible to consider the bidomain equation for inverse problems such as determining the material parameters of a cardiac model. The Eikonal equation reduces the computational intensity of the bidomain equation significantly. It is much faster, and it can provide an activation sequence in seconds [18].

Recent work in [2,5,17] has shown that building an efficient 3D tetrahedral Eikonal solver for multi-core and SIMD (single instruction multiple data) architectures poses many challenges. It is important to keep the memory footprint low to reduce the costly memory accesses and achieve a good computational density on GPUs and other SIMD architectures with limited memory and register capabilities. In this paper we briefly address our algorithms for shared memory parallelization and global solution algorithms for a fast many-core Eikonal solver with a low memory footprint [3,4] which makes it suitable for the inverse problem we are considering.

Accurate patient specific conductivity measurements are still not possible. Current clinical EP models lack patient-specificity as they rely on generic data mostly obtained by generalized measurements done on dead tissues. One way to accurately determine these parameters is to compare the computed solution w.r.t. the Eikonal computed activation sequence on the heart surface with the measured ECG. By minimizing the squared distance between the measured solution ϕ^* and the Eikonal computed solution $\phi(\gamma, x)$ with material domains scaled by the parameter $\gamma \in \mathbb{R}^m$, we are able to determine the scaling parameters which may identify heart tissues with a low conductivity that might indicate a dead or ischemic tissue.

The minimization problem is solved by the steepest descent and the BFGS method. To compute accurate gradients up to machine accuracy, we use algorithmic differentiation (AD) [16] based on the AD tool dco/c++ [12]. This paper provides mathematical and implementational details of the method followed by comparative results for the shared memory parallelization for the tangent as well as for the adjoint AD approach.

2 Eikonal Solver

We used the following variational formulation of the Eikonal equation [5]

$$\sqrt{(\nabla\varphi(x))^T M(x)\nabla\varphi(x)} = 1 \qquad x \in \Omega \qquad (1)$$

to model an activation sequence on the heart mesh. It is an elliptic non-linear PDE that does not depend on time and does not describe the shape of the wavefront but only its position in time. The solution of the Eikonal equation presents the arrival time of the wavefront at each point x in the discretized domain. In this context it is less accurate in comparison to the bidomain equations which describe both the shape and the position of the wavefront [18,20]. On the other hand, Eikonal equation provides a much lower computational intensity making it a good candidate for the inverse problem that we are considering in this paper.

Our Eikonal solver [3,4] builds on the fast iterative method (FIM) [1,10] by Fu, Kirby, and Whitaker [2] which is the state of the art for solving the Eikonal Eq. 1 on fully unstructured tetrahedral meshes with given heterogeneous, anisotropic velocity information M. We further improved the solver w.r.t. the algorithm, memory footprint and parallelization. A task based parallel algorithm for the Eikonal solver [5] is shown in Algorithm 1. The wavefront active list L is dynamically partitioned into sublists which are assigned for further processing to a number of processors. This strategy is particularly suitable for the OpenMP shared memory parallelization. On top of that we build our parallel automatic differentiation approach for computing the gradients needed to solve the inverse problem. The solution to Eikonal equation representing a wave front propagation is depicted in Fig. 1. Here we start with a single excitation point at the bottom of the domain and the isosurfaces of the solution of the Eikonal equation travel from the bottom to the top of the domain where are thereafter cut off.

Algorithm 1. Task based parallelism example

1: partition L dynamically into sub-sets L_i
2: launch kernel on GPU processors or start threads on thread i
3: **for all** vertices x in L_i **do**
4:     $\Phi_x = min(\Phi_x, Solver(t, \Phi))$
5: **end for**
6: wait for threads or processors to terminate

Detailed numerical results can be found in our previous work [5], showing that our code scales very well and provides an activation sequence in seconds. The low memory footprint [4] of the solver makes it well suitable for the inverse problem especially when considering an algorithmic differentiation approach where one of the main bottlenecks in the adjoint implementation is the increased memory footprint.

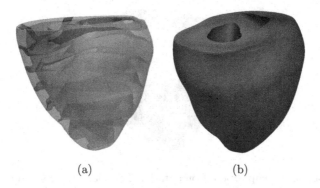

(a)   (b)

Fig. 1. Arrival time $\varphi(x)$ ranging from 0 (bottom (a), blue (b)) to 1 (top (a), red (b)). (Color figure online)

3 Cardiac Parametrization

3.1 Eikonal Equation with Material Domains

According to physiology, the human heart consists of different tissues, such as the heart chamber, with different conductivity parameters. Let us denote these different material domains by $\overline{\Omega}_k$. Then our discretized domain can be expressed as

$$\Omega \xrightarrow{tets} \overline{\Omega}_h = \bigcup_{k=1}^{m} \overline{\Omega}_k, \qquad \Omega_l \bigcap_{k \neq l} \Omega_k = \emptyset \ . \tag{2}$$

The velocity information $M(x)$ is specific but constant in each tetrahedron $\tau \in \Omega_h$. An additional tag indicates to which material domain $\overline{\Omega}_k$ that element is assigned. It allows to scale the velocity information $M(x)$ in each material domain $\overline{\Omega}_k$ by some $\gamma_k \in \mathbb{R}$, i.e., the tetrahedron specific information is preserved but all tetrahedrons in the same material class have the same scaling parameter.

The decomposition of the domain into material domains leads to an Eikonal equation with material domains

$$\sqrt{(\nabla \phi)^T \gamma_k \cdot M \cdot (\nabla \phi)} = 1 \quad \forall x \in \Omega_k. \tag{3}$$

Now the activation time ϕ depends on the scaling parameters $\gamma \in \mathbb{R}^m$. Let us emphasize that this change does not affect the runtime of the Eikonal solver.

The domain Ω is statically partitioned into a number of non-overlapping sub-domains Ω_k, see Fig. 2, and a scaling parameter γ_k is assigned to each of them. We use two different ways to partition the domains into subdomains. The first approach uses ParMETIS to achieve an equal size subdomain partitioning and the second approach follows the physiology of the heart strictly. The latter produces currently worse optimization results, and therefore we focus on the first approach.

Fig. 2. Domain decomposition. Computational domain Ω and sub-domains Ω_i.

3.2   Optimization

Accurate patient specific conductivity measurements are still not possible. Current clinical EP models lack patient-specificity as they rely on generic data mostly obtained by generalized measurements done on dead tissues. One way to accurately determine these parameters consists in comparing the computed solution w.r.t. the Eikonal computed activation sequence on the heart surface with the measured ECG on the torso which is mapped onto the heart surface. Doing so one needs to solve a minimization problem with the objective functional as follows

$$f(\gamma) := \| \; \phi^*(x) - \phi(\gamma, x) \; \|_{\ell_2(\omega_h)}^2$$

with ω_h denoting the vertices in the discretization of Ω_h.

By minimizing the squared distance between the measured solution ϕ^* and the Eikonal computed solution $\phi(\gamma, x)$ with material domains scaled by the parameter $\gamma \in \mathbb{R}^m$, we are able to determine the scaling parameters which identify heart tissues with low conductivity that might indicate a dead or ischemic tissue. The scaling parameters γ do not only determine low conductivity tissues but also scale the generalized conductivity measurements to the accurate patient-specific conductivity parameters.

The mapping $\gamma \rightarrow \phi$ in (3) is nonlinear and nonconvex which results in a nonconvex optimization problem. It requires a good initial guess γ_0 and equal sized material domains as regularization. We consider generalized measurements done on dead tissues as an initial guess for the scaling.

The minimization problem is solved either by the steepest descent or the BFGS method. Both methods include an adaptive step size control and require the calculation of the gradient $\nabla_\gamma f(\gamma)$. The gradient calculation is done in two ways. First, we use finite differences as a verification method. The second approach uses AD that delivers gradients up to machine accuracy. We implemented a shared memory parallelization for both the tangent and the adjoint model. Adjoint shared memory optimizations are still ongoing work.

4 Implementation Details Using Algorithmic Differentiation

Algorithmic differentiation [7, 16] is a semantic program transformation technique that yields robust and efficient derivative code. For a given implementation of a k-times continuously differentiable function $f : \mathbb{R}^n \to \mathbb{R}, y = f(\gamma)$ for $\gamma \in \mathbb{R}^n$ and $y \in \mathbb{R}$, AD generates implementations of corresponding *tangent* and *adjoint* models automatically. The tangent model computes directional derivatives $\dot{y} = \nabla_\gamma f \cdot \dot{\gamma}$, whereas the adjoint model computes gradients $\overline{\gamma} = [\nabla_\gamma f(\gamma)]^T \cdot \overline{y}$. The vectors $\dot{\gamma}, \overline{\gamma} \in \mathbb{R}^n$ and the scalars $\dot{y}, \overline{y} \in \mathbb{R}$ are first-order tangents and adjoints of input and output variables, respectively, see [16] for more details. The computational cost of evaluating either of both models is a constant multiple of the cost of the original function evaluation. This observation becomes particularly advantageous when computing the gradient of f for $n \gg 1$. In this case, instead of n evaluations of the tangent model to compute the gradient element by element, only one evaluation of the adjoint model is required. Nonetheless, it needs to be mentioned that the adjoint mode of AD usually requires a lot more memory, which makes further techniques like checkpointing [6], preaccumulation, or hybrid algorithmic/symbolic approaches necessary [15]. Higher derivatives can be obtained by recursive instantiations of the tangent and adjoint models. In the optimization problem defined in Sect. 3.2, only first derivatives are required.

In addition to the *scalar* modes presented above, *vector* modes can be used. The tangent scalar mode evaluates the tangent model once, i.e., performs one matrix-vector multiplication. The tangent vector mode, on the other hand, is able to perform multiple matrix-vector multiplications in one go, being more efficient than doing multiple scalar tangent evaluations by avoiding recomputation of temporary results. More details on scalar and vector mode can also be found in [16].

AD can be implemented either via source code transformation or operator overloading techniques. In this project, we use the AD overloading tool dco/c++. It has been successfully applied to a number of problems in, for example, computational fluid dynamics [19, 23], optimal control [13], or computational finance [22]. dco/c++ supports a wide range of features, two of which are used in this project: Checkpointing as well as advanced preaccumulation techniques on different levels (automatically on assignment-level, user-driven on higher level). Both techniques help to bring down memory requirements for the adjoint necessary to get a feasible code. The performance of AD tools is usually measured by the *run time factor*, which is the ratio of the runtime of one gradient computation w.r.t. one primal function evaluation, i.e.

$$\mathcal{F} = \frac{\text{Cost}(\nabla_\gamma f(\gamma))}{\text{Cost}(f(\gamma))} . \tag{4}$$

This factor and its scaling behavior is shown in Sect. 5.

As mentioned earlier, the Eikonal solver is OpenMP parallelized. Assuming that the original code is correctly parallelized, i.e., no data races, the (vector) tangent code can safely be run in parallel as well. Producing an adjoint of OpenMP parallel code, on the other hand, is a major challenge since computing the adjoint requires a data flow reversal of the original code. Reversing the flow of data of parallel code introduces additional race conditions. Handling those automatically is not trivial and still actively researched [9,14]. We resolved this issue by splitting the code into active and passive segments, where only the active segment influences the derivative. As shown in Algorithm 1, in each parallel section, a minimum over many elements is computed. A simplified representation is given by $v = \min_i (S_i)$, where each S_i is calculated by an expensive solver call. Decomposing this into

$$k = \arg\min_i (S_i) \qquad (5)$$

$$v = S_k \qquad (6)$$

clearly shows, that we can compute Eq. 5 passively and only calculate Eq. 6 actively. This has the drawback that S_k is calculated twice. However, since only the data flow of the active segment needs to be reversed in adjoint mode AD, we can run the passive segment safely in parallel and keep the active segments sequential. This code version is referred to by the term *most-passive* in Sect. 5 where runtime results and scaling factors are analyzed. In contrast to this, the basic approach of running all segments actively is called *all-active*. The *all-active* version needs to run sequentially if computing adjoints. In case of running the tangent (vector) code, both, the *most-passive* and the *all-active* versions are executed fully in parallel.

5   Numerical Tests and Performance Analysis

5.1   Steepest Descent vs. BFGS

This section presents numerical results related to the dco/c++ implementation for both, tangent and adjoint models, tested on Intel(R) Xeon(R) CPU E5-2690 v4 @ 2.60 GHz, 512 GB RAM. Initially we test with a small number of material domains and the coarsest mesh, Tbunny_C2, with 266,846 tetrahedrons in order to analyze the different behaviors of the steepest descent and BFGS. We test for $\gamma \in \mathbb{R}^6$, $\gamma_0 = (2\ 3\ 1\ 4\ 2\ 3)$ and convergence criteria $\epsilon = 0.5$ with an exact solution of $\gamma^* = (1\ 1\ 1\ 1\ 1\ 1)$. The run time and the resulting γ are presented in Table 1. Tangent and adjoint are based on *all-active* version described in the previous section. It can be seen that the BFGS method is faster and converges in less iterations. Please note that the tangent (vector) model and the adjoint model have a similar run time for the small number of material domains. The adjoint model makes use of its advantage only for a larger number of parameters as shown later.

Table 1. Single-threaded run times in seconds and results.

Method	FD	Tangent	Adjoint	Result γ
Steepest descent	623	1570	1469	(1.003 0.980 1.003 0.958 1.002 1.227)
BFGS	194	684	696	(1.005 1.004 0.997 0.984 0.993 1.031)

Based on the actual physiology, the larger case has 21 material domains. The decomposition itself is currently not based on the physiology but using ParMETIS for an equally sized subdomain decomposition. Later on, we discuss what happens if we use a decomposition based on the physiology. The initial scaling was set to $\gamma_0 = \mathbf{1} \in \mathbb{R}^{21}$, i.e. the velocity information computed from the measurements done on dead tissues is not changing. Then we set the measured solution to be: $\gamma^* = (1\,1\,1\,1\,1\,1\,0.2\,0.2\,0.2\,0.2\,0.2\,0.2\,1\,1\,1\,1\,1\,1\,1\,1\,1)$. The scaling parameters differ only in 5 subdomains in which we assume lower conductivity measured from the electrocardiogram.

The simulation in Fig. 3 shows in four steps how the material scaling changes throughout the optimization process. The whole optimization takes 19 iterations in total, but we only present four: the first, the last and two iterations in between to give an idea of the optimization process. Each subfigure represents the material scaling for each material domain of the Tbunny_C2 mesh. Here one can identify low conductivity tissue in the heart and in this case these material domains are exactly the ones in blue color as shown in the last step.

5.2 Gradient Verification and BFGS Convergence

To verify the AD implementation, we compare the gradient with finite differences results. The derivatives using tangent and adjoint AD are expected to be accurate up to machine precision. Since finite differences suffer from truncation and round-off errors a 'V'-shape is expected as error for varying perturbations h. As error, we compute the difference between finite difference and AD gradient, since we assume the AD gradient to be correct. This is indeed supported by observing the 'V'-shape as shown in Fig. 4a.

Results of running BFGS using the three different methods for derivative calculation is shown in Fig. 4. Though the gradients are expected to match up to machine precision for the two AD versions, BFGS shows different convergence behavior. Embedded into an iterative optimizer, the small differences in the gradients can lead to reaching a local stopping criterion earlier. This is the case here for iteration five. Nonetheless, both AD modes then fulfill the global convergence criteria of $\epsilon = 0.5$ after 16 and 18 iterations respectively. In contrast to that, the finite difference version fails to converge and requires a less strict convergence criteria.

Fig. 3. Four steps from the simulation of the optimization using the BFGS algorithm with 21 material domains on the Tbunny_C2 mesh.

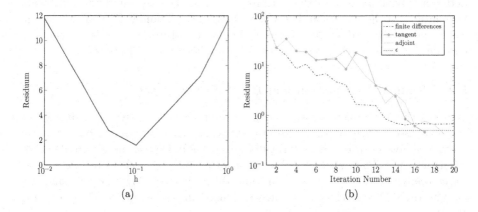

Fig. 4. Error of finite difference gradient (a) and convergence behavior of BFGS (b).

5.3   Sequential Timings for Gradient Computation

In the following, we measure the run time and the factor $\mathcal{F}$ (see Eq. 4) of a single gradient computation of the objective function. As introduced in Sect. 4, two different variants are implemented: *most-passive* and *all-active*. As shown in Fig. 5 tangent and adjoint codes give better run times for the *most-passive* version. In fact, the factor of the adjoint mode w.r.t. the primal run time was successfully improved from 23.8 to 2.7 in the *most-passive* variant which can be considered as very good. Besides this, the adjoint mode clearly wins over the tangent mode.

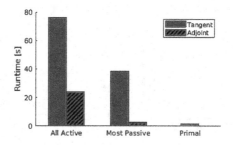

Fig. 5. Run time comparison for sequential gradient computations.

Table 2. Sequential run times for different tangent vector sizes.

Vector size	*all-active* [s]	*most-passive* [s]
7	50.9	9.6
11	54.0	8.6
21	50.6	6.5

While the above-shown run times correspond to the scalar tangent mode, measurements have also been made for the tangent vector mode. The gradient has been computed sequentially for both code variants for a vector size of 7, 11 and 21. To compute the full gradient, three tangent model evaluations are needed for a vector size of 7, two for 11, and only one for a vector size of 21. The runtime results are shown in Table 2. As the tangent vector mode by principle avoids some passive evaluations, a better runtime compared to the scalar mode can be expected. Therefore, the speedup compared to the previously shown tangent scalar run time of 76 s is reasonable. Since the problem size is a multiple of 7 and equals exactly 21, running with these two vector sizes perform the minimum amount of required tangent direction computations for the full gradient. Using a vector size of 11 ends up calculating one extra tangent direction, which results in slightly worse runtime. The *most-passive* variants on the other hand only run a small segment of the code actively. The slow-down due to additional directions only affect the active segment, which results in the run times shown in Table 2.

5.4   Scaling

For analyzing the OpenMP scaling behavior, we compare the AD versions to the scaling of the primal code, see Fig. 6a. Ideally, the AD code should show a similar scaling behavior as the primal. Since the *most-passive* implementation is the more efficient one, the following scaling analysis uses this version. Figure 6b contains scaling information for tangent and adjoint mode. One interesting aspect is the runtime factor for increasing number of threads. As one can see, the tangent and adjoint factors increase for more threads. This indicates a worse scaling behavior, which is reflected in the scaling factors. Recalling the introduction

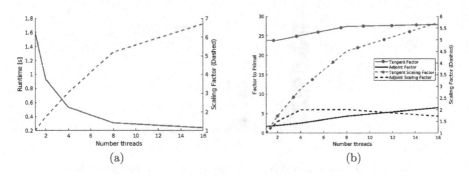

Fig. 6. Scaling behavior of the primal code (a) and scaling factors for the *most-passive* AD variant (b).

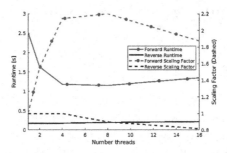

Fig. 7. Forward and reverse run time and scaling factor for the *most-passive* variant.

given in Sect. 4, the tangent (vector) mode can be fully parallelized. However, although the scaling is similar to the primal scaling, it is a little worse (6.7 compared to 5.7 with 16 threads). The adjoint code on the other hand can only be parallelized for the passive segment. For a further discussion, detailed adjoint measurements are shown in Fig. 7.   An AD adjoint code by overloading consists of two sweeps, the forward and the reverse sweep. During the forward sweep, a data structure is build up (the *tape*) while executing the overloaded function. This data structure is then used during the reverse sweep to actually propagate adjoints. The reverse sweep can not be parallelized at all, i.e. the scaling factor is expected to be one. However, due to data allocated on different physical memory banks (first-touch policy during forward sweep), multithreading eventually slows down the reverse section. Since the run time of the reverse sweep is shorter than of the forward sweep, the impact on the overall scaling behavior is so as well. As stated before, the forward sweep timings include the execution of the passive segment, which explains a descent scaling. However, at some point (in this case already with 8 threads), the sequential active segment, running in an OpenMP critical section, leads to stalling of threads, which results in a decreasing scaling factor. Combining this insight for forward and reverse mode explains the adjoint scaling factor displayed in Fig. 6b.

6 Conclusions and Future Work

Patient-specific simulations in cardiac electrophysiology require patient-specific conductivity parameters which are not accurately available in vivo. In this article, we have shown algorithms and a proper software stack for improving the given conductivity parameters by minimizing the squared distance between the measured solution and the Eikonal computed solution. An efficient parallelized Eikonal solver has successfully been coupled with the AD tool dco/c++ to not only speed up the optimization, but also to get more accurate answers. As demonstrated by carrying out numerical run time tests, the scaling behavior of the tangent model implementation is similar to the original scaling behavior. The adjoint model on the other hand only scales up to a small amount of cores with the current approach. Nonetheless, small to medium sized problems can be processed much faster using the adjoint model implementation.

Further work needs to be done towards a scalable adjoint model implementation not only for use with OpenMP but special care needs to be taken for getting a working GPU adjoint. Usually, adjoint GPU code cannot be generated automatically but has to be written by hand. Nonetheless, there exist tools that support the developer in writing adjoint GPU code, e.g., using source transformation [8] or overloading [11]. None of those tools are as automatic as state-of-the-art tools for CPU code currently are. In addition, we are considering an analytic approach to compute the gradient using the adjoint-state method [21]. The analytic approach shall not suffer from excessive memory requirements as does the adjoint model using AD tool [16]. From the problem definition perspective, the decomposition with equally sized subdomains where the regularization holds is not only much more accurate, but it also converges in fewer iterations. It is preferable to the pure physiological decomposition. Mapping the equal sizes subdomains to the physiology is ongoing work. Possible solution strategies could be:

- Refine the physiologically based material domain decomposition.
- Interpret the information $\frac{dy}{dM}$ returned without additional costs from the adjoint model.

The first idea relates to the equally sized subdomains. It means that we split the larger subdomains into smaller subdomains similar to the smallest one and preserve the same scaling parameter therein to end up with equally sized subdomains in which case the regularization holds. The second idea tells us what the problem is and leads the way to a more global regularization which shall apply in all cases. The information returned without additional costs from the adjoint model, $\frac{dy}{dM}$, shows how each component of the velocity information changes for each tetrahedron. This sensitivity analysis from the detailed element-wise information shall give us hints for an improved regularization.

References

1. Fu, Z., Jeong, W.K., Pan, Y., Kirby, R.M., Whitaker, R.T.: A fast iterative method for solving the Eikonal equation on triangulated surfaces. SIAM J. Sci. Comput. **33**, 2468–2488 (2011)
2. Fu, Z., Kirby, R.M., Whitaker, R.T.: Fast iterative method for solving the Eikonal equation on tetrahedral domains. SIAM J. Sci. Comput. **35**(5), C473–C494 (2013)
3. Ganellari, D., Haase, G.: Fast many-core solvers for the Eikonal equations in cardiovascular simulations. In: International Conference on High Performance Computing Simulation (HPCS), pp. 278–285. IEEE (2016). https://doi.org/10.1109/HPCSim.2016.7568347. peer-reviewed
4. Ganellari, D., Haase, G.: Reducing the memory footprint of an Eikonal solver. In: International Conference on High Performance Computing Simulation (HPCS). IEEE (2017, accepted)
5. Ganellari, D., Haase, G., Zumbusch, G.: A massively parallel Eikonal solver on unstructured meshes. Comput. Visual. Sci. **19**(5), 3–18 (2018). https://doi.org/10.1007/s00791-018-0288-z
6. Griewank, A., Walther, A.: Algorithm 799: revolve: an implementation of checkpointing for the reverse or adjoint mode of computational differentiation. ACM Trans. Math. Softw. **26**(1), 19–45 (2000). https://doi.org/10.1145/347837.347846
7. Griewank, A., Walther, A.: Evaluating Derivatives: Principles and Techniques of Algorithmic Differentiation. Other Titles in Applied Mathematics, vol. 105, 2nd edn. SIAM, Philadelphia (2008)
8. Hascoët, L., Pascual, V.: The Tapenade automatic differentiation tool: principles, model, and specification. ACM Trans. Math. Softw. **39**(3), 20:1–20:43 (2013)
9. Hückelheim, J., Hovland, P., Strout, M.M., Müller, J.D.: Reverse-mode algorithmic differentiation of an OpenMPparallel compressible flow solver. Int. J. High Performan. Comput. Appl. **33**(1), 140–154 (2019). https://doi.org/10.1177/1094342017712060
10. Jeong, W.K., Whitaker, R.T.: A fast iterative method for Eikonal equations. SIAM J. Sci. Comput. **30**, 2512–2534 (2008)
11. Leppkes, K., Lotz, J., Naumann, U., du Toit, J.: Meta adjoint programming in C++. Technical report, AIB-2017-07, Department of Computer Science, RWTH Aachen University (2017)
12. Lotz, J.: Hybrid Approaches to Adjoint Code Generation with dco/C++. Dissertation, RWTH Aachen University (2016)
13. Lotz, J., Naumann, U., Hannemann-Tamas, R., Ploch, T., Mitsos, A.: Higher-order discrete adjoint ODE solver in C++ for dynamic optimization. Procedia Comput. Sci. **51**, 256–265 (2015)
14. Lotz, J., Naumann, U., Sagebaum, M., Schanen, M.: Discrete adjoints of PETSc through dco/c++ and adjoint MPI. In: Wolf, F., Mohr, B., an Mey, D. (eds.) Euro-Par 2013. LNCS, vol. 8097, pp. 497–507. Springer, Heidelberg (2013). https://doi.org/10.1007/978-3-642-40047-6_51
15. Naumann, U., Lotz, J., Leppkes, K., Towara, M.: Algorithmic differentiation of numerical methods: Tangent and adjoint solvers for parameterized systems of nonlinear equations. ACM Trans. Math. Softw. **41**(4), 1–21 (2015)
16. Naumann, U.: The Art of Differentiating Computer Programs: An Introduction to Algorithmic Differentiation. Society for Industrial and Applied Mathematics, Philadelphia (2012)

17. Noack, M.: A two-scale method using a list of active sub-domains for a fully par-allelized solution of wave equations. J. Comput. Sci. **11**, 91–101 (2015)
18. Pezzuto, S., Kal'avský, P., Potse, M., Prinzen, F.W., Auricchio, A., Krause, R.: Evaluation of a rapid anisotropic model for ECG simulation. Frontiers Physiol. **8**, 265 (2017). https://doi.org/10.3389/fphys.2017.00265
19. Sagebaum, M., Gauger, N.R., Naumann, U., Lotz, J., Leppkes, K.: Algorithmic differentiation of a complex C++ code with underlying libraries. Procedia Comput. Sci. **18**, 208–217 (2013)
20. Sali, A.: Coupling of monodomain and Eikonal models for cardiac electrophysi-ology. Master's thesis (2016). https://opus4.kobv.de/opus4-zib/frontdoor/index/index/docId/6051
21. Taillandier, C., Noble, M., Chauris, H., Calandra, H.: First-arrival travel time tomography based on the adjoint-state method. Geophysics 74(6)(2009)
22. du Toit, J., Lotz, J., Naumann, U.: Adjoint algorithmic differentiation of a GPU accelerated application. http://www.nag.co.uk/Market/articles/adjoint-algorithmic-differentiation-of-gpu-accelerated-app.pdf
23. Towara, M., Naumann, U.: A discrete adjoint model for OpenFOAM. Procedia Comput. Sci. **18**, 429–438 (2013)

An Empirical Analysis of Predictors for Workload Estimation in Healthcare

Roberto Gatta[1] ⓘ, Mauro Vallati[2](✉) ⓘ, Ilenia Pirola[3], Jacopo Lenkowicz[1],
Luca Tagliaferri[4], Carlo Cappelli[3], and Maurizio Castellano[3]

[1] Universitá Cattolica del Sacro Cuore, Istituto di Radiologia, Rome, Italy
`roberto.gatta.bs@gmail.com`
[2] School of Computing and Engineering, University of Huddersfield, Huddersfield, UK
`m.vallati@hud.ac.uk`
[3] Universitá degli Studi di Brescia, Spedali Civili di Brescia, Brescia, Italy
[4] Fondazione Policlinico Universitario A. Gemelli IRCCS, Rome, Italy

Abstract. The limited availability of resources makes the resource allocation strategy a pivotal aspect for every clinical department. Allocation is usually done on the basis of a workload estimation, which is performed by human experts. Experts have to dedicate a significant amount of time to the workload estimation, and the usefulness of estimations depends on the expert's ability to understand very different conditions and situations. Machine learning-based predictors can help in reduce the burden on human experts, and can provide some guarantees at least in terms of repeatability of the delivered performance. However, it is unclear how good their estimations would be, compared to those of experts.

In this paper we address this question by exploiting 6 algorithms for estimating the workload of future activities of a real-world department. Results suggest that this is a promising avenue for future investigations aimed to optimising the use of resources of clinical departments.

Keywords: Workload estimation · Machine learning · Predictors

1 Introduction

Global spending on health is consistently growing worldwide [7], and it is expected to grow in the near future. This is the result of two main driving forces:

- in countries where the economy is developing, the increase is due to the improvement of services overtime.
- In the so-called first-world countries, the growing life expectancy and the low birth rate are already increasing the pressure on the healthcare (see for instance [10]).

Remarkably, the problem faced in developed countries envisages a scenario where optimising the available resources will be a mandatory way to increase the efficiency of the healthcare system, and to optimise delivered services.

© Springer Nature Switzerland AG 2020
V. V. Krzhizhanovskaya et al. (Eds.): ICCS 2020, LNCS 12137, pp. 304–311, 2020.
https://doi.org/10.1007/978-3-030-50371-0_22

There are many different aspects and perspectives that can be subject to resource optimisation: optimisation can focus on different levels of the organisational charts, can focus on geographical clusters, can be tuned for the type of delivered services or clinical domains, and can address both administrative and clinical issues. Examples of approaches aimed at optimising the use of resources include a dynamic appointment scheduling system to cope with no-show patients and appointments deletion [5]; a scheduler for Radiology Departments [9]; and a chemotherapy appointment scheduling model under uncertainty [2]. There is a growing interest in optimisation approaches, thanks to the potentially large benefits that their application would result in for an hospital or a clinical department.

Notably, most of the existing optimisation approaches deal with the allocation of resources, as soon as appointment requests are received or an estimation of future workload has been performed. In a sense, this is a kind of *reactive* optimisation. Intuitively, optimising resources on the basis of estimated workload can lead to better resource optimisation, due to the fact that there is no need to wait for actual appointments to be made. This would allow a shift from *reactive* to *pro-active* otpimisation. However, this kind of pro-active optimisation is very sensitive to the quality of predictions that are provided. Despite being pivotal for the allocation and exploitation of available resources, the workload estimation is still mostly performed manually by human experts, that have to devote a usually significant amount of their time to perform such task. Moreover, the usefulness of estimations depends on the expert's ability to understand very different conditions and situations, and is very hard to verify. In fact, the same expert can provide both very accurate and very inaccurate estimations, undermining the subsequent allocation processes.

Machine learning-based predictors may help to overcome some of the aforementioned issues. In particular, their use can reduce the burden on experts, and provide some general guarantees on the quality of the predictions. Furthermore, machine learning can be used to quickly generate multiple scenarios, that can then be compared by experts to select the most appropriate. However, in order to understand the usefulness of well-known machine learning approaches for this task, it is mandatory to assess their ability in estimating future workloads in real-world circumstances.

In order to address the above issue, in this paper we present the results of a large empirical analysis aimed at comparing the performance in workload estimation of a number of algorithms on real-world data obtained from a Centre of study on Thyroid. To minimise the risk of providing results that are only specific for the case taken into account, we trained the considered algorithms on a restricted set of information, commonly available on the vast majority of Electronic Health Records (EHR) or appointment booking systems.

The remainder of this paper is organised as follows. First, we describe material and methods of the performed analysis. Then, in Sect. 3 we present results and a discussion. Next, we provide the conclusion of this paper and we envisage future steps.

2   Materials and Methods

From the EHR of a Centre of study on Thyroid we extracted the complete clinical pathways of patients treated for acute and chronic thyroid diseases (e.g., cancer, age related, genetic hyper-hypo thyroidism, etc.). The investigated events were:

- (i) oncological examinations: ambulatory visits aimed at staging the Thyroid neoplasm, to assess the progression during the treatment or follow-up visits;
- (ii) Non-oncological examinations: ambulatory visits for generic consultations for specific non-oncologic diseases such as hypo or hyper-thyroidism (e.g. due to physiological ageing or more specific reasons, such as the Basedow diseases).
- (iii) Free triiodothyronine (fT3), a thyroid hormone. This analysis only requires a blood sample; for this reason it tends to be relatively cheap to perform and it is commonly prescribed.
- (iv) Free thyroxine (fT4), a thyroid hormone similar to fT3. It can be analysed as the fT3 and, together, they are primarily responsible for regulation of metabolism.
- (v) Parathyroid hormone (PTH), an hormone secreted by the parathyroid glands with a relevant role in the regulation of the serum calcium.
- (vi) Thyroglobulin (Tg), a protein produced and consumed within the Thyroid.
- (vii) Other common laboratory exams, such as complete blood count, cholesterol, etc.
- (viii) Thyroid ultrasound investigation,
- (ix) Fine Needle Aspiration Cytology (FNAC): the aspiration of some thyroid cells with a fine needle guided via ultrasound. Due to the invasive nature of the procedure, it require specific clinical skills and can be considered the most demanding event.

We decided to focus on this level of granularity because, also as a result of discussions with human experts of the considered medical field, these are key events with regards to human resources of a department (e.g. FNAC, Ultrasound, Medical examinations) or with regards to lab time and costs (e.g. fT3, fT4, Tg, PTH). Furthermore, those events are commonly recorded in EHRs, and would therefore provide a general ground to exploit workload estimation predictors in different units or departments.

Other clinical variables, such as co-morbidities, drugs or biomarkers was not considered: such kind of data are not always present in the EHR and when present are often represented without any specific reference to a shared ontology. For this reason, even if their inclusion had increased the performances of the predictions, it would also had reduced the reproducibility.

We considered a total of 5,941 patients treated by the thyroid centre, which lead to 42,839 events. The available data has been processed as follows. For each of the 9 clinical events analysed in this study, and considering all the patients involved in the event, we divided the logs in two parts, corresponding to an observation time window of at least 18 months before and 18 months after. The predicting task is to estimate the number of events that will occur in the 18 months after the event, given information about the 18 months before. It should be noted that a different predictor is built for each of the 9 events, and such predictor is only used to predict the number of future occurrences of such event. We then trained and tested the predictors exploiting a cross-validation jackknife approach, where 90% of the available data is used for training purposes, and the remaining 10% for testing predictors.

The 18-months time window reflects, to some extent, the nature of the treatments performed in the considered centre. This represents the common follow-up time, and includes a prudential margin to allow enough informative content for the prediction of the following 18. Of course, for different departments, this value can be straightforwardly adapted.

2.1 Algorithms

For the sake of this experimental analysis, we considered six well-known algorithms for building predictors, spreading from naive approaches –exploited as baselines– to widely-exploited Machine Learning techniques.

- **Mean**: considering the entire training set it calculates the density of each kind of event during the time (how many, on average, per month) and use this density to predict how many events are expected in the future.
- **TipOver**: each prediction is simply made by replicating the past recorded events. More specifically, for each patient, the kind and number of events of the next x months are exactly the same of the previous x months.
- k-nearest neighbours algorithm (**kNN**) [4]: uses the neighbourhood of the 8 most similar clinical cases and uses them to estimate the future, exploiting the mean of events occurred in the past 18 months. The metric is built on an n dimensional space where n is the number of kind of events. In this way, any patient can be seen as a point and the euclidean distance is used to select the neighbourhood. The axes are normalised between 0 and 1 to avoid overweighting the most frequent events.
- Generalised linear model (**lm**) [11]: uses generalised linear regression to estimates the next 18 months, adopting all the entire training data set.
- Random Forest (**rf**) [3]: Random forests are a combination of predictors such that each predictor is randomly generated, and all the predictorsa have the same weight. We built a Random Forest-based models using 500 random trees;
- Support-vector machine (**svm**) [1]: Support Vector Machines-based models the exploits a Gaussian kernel to perform the prediction.

2.2 Domain Expert

The director of the Centre of study on Thyroid is the human expert that is in charge of estimating the workload of the unit. She has some 20+ years experience in the specific domain. In order to make the comparison as fair as possible, she was asked to make estimations on the basis of the same data that is made available to the considered algorithms. Notably, this is not the usual amount of data that is provided to human experts. In most of the cases, they are required to estimate future workloads by relying on a significantly smaller amount of explicit knowledge; however, they can leverage on their extensive experience in the field. For this reason, we believe we put the human expert in the best possible condition to perform her work: a large amount of available data that can provide a good and compelling overview of the past months.

3   Results

Results of the performed comparison are shown in Fig. 1. For each considered event, we provide a box-and-whisker plot showing the distribution of the error percentage, measured as the percentage of events as follows:

$$\frac{(predicted - actual)}{actual} \tag{1}$$

An average percentage error value of 0.0 indicates that the predictor has always provided the perfect estimation. In each box, the mid-line is the median of the performance, with the upper and lower limits of the box being the third and first quartile respectively. The whiskers will extend up to 1.5 times the interquartile range from the top (bottom) of the box to the furthest datum within that distance. In predicting the expected numbers of Other Lab exams, for example, the algorithm tipOver has a median error close to 100% with the 50% of the measured performances included approximately between and 80% and 110%. Admittedly, tipOver is not a good approach to estimate the workload for that type of clinical events.

Dispersion is also to take into account, as a high value indicates that the corresponding predictor's performance can vary greatly according to the considered circumstances. The solid horizontal (red) line represents the performance of the human expert. In this case, we could not show any dispersion value, as the expert made only a limited number of estimations, due to the complexity of the task when performed manually.

3.1   Discussion

The results presented in Fig. 1 indicate that the machine learning-based predictors tend to estimate better than the very basics *mean* and *tipOver* approaches. However, even such naive approaches can deliver good performance in a couple of cases, indicating that the corresponding events are trivially easy to predict, given a suitable amount of available information. Notably, in some cases the *mean* approach is able to deliver prediction that outperform human experts: it can indeed be the case that even such naive approaches can be useful in supporting humans, by clearly highlighting regular patterns that would otherwise be hard to identify. On the other hand, more sophisticated ML approaches tend to consistently deliver better performance also on more complex cases.

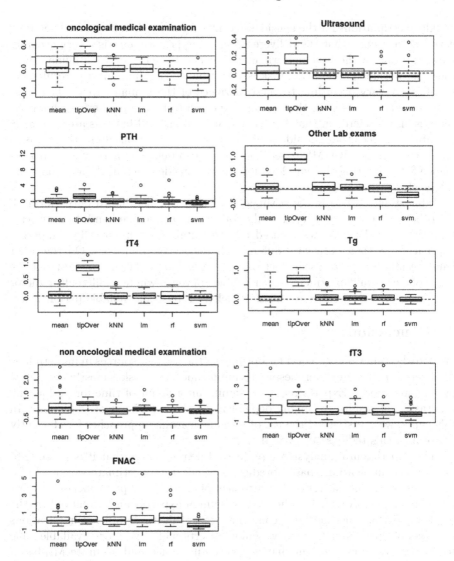

Fig. 1. Performance, in terms of average estimation error percentage (y-axis) of the considered algorithms when predicting the number of occurrences of the 9 clinical events. In each box, the mid-line is the median of the performance, with the upper and lower limits of the box being the third and first quartile respectively. The whiskers will extend up to 1.5 times the interquartile range from the top (bottom) of the box to the furthest datum within that distance. The solid horizontal indicates the performance of the human expert. (Color figure online)

In most of the considered cases, the performance of the human expert are impressive, even though ML-based techniques can still help in reducing mistakes and improving predictions. Noteworthy, the human expert has been making workload estimations for the considered centre for more than 20 years. Therefore, it is safe to assume that the delivered predicting performance is a very accurate representation of the best performance that can be achieved by a human. Further, the workload estimation task is very time consuming, and the results can significantly vary according to the experience of the human expert. The more experienced the expert is, the best are expected to be the predictions: however, there is also to factor in that fact that more experienced humans are extremely valuable resources that should spend their precious time on more critical tasks. Given this perspective, ML-based approaches can deliver generally good performance for estimating the workload for all the considered clinical events, and are extremely quick.

Interestingly, there is not a single algorithm that is able to outperform all the others in all the considered prediction tasks. On the one hand, this suggests that the clinical events we focused on are suitable for empirically comparing approaches as they pose very different challenges to predictors. On the other hand, results also point to the fact that an ensemble predictor may best suit the needs of a clinical department. An ensemble approach where a different predictor is trained for each event may therefore deliver robust and reliable performance.

4   Conclusions

Workload estimation is pivotal for optimising the use of resources in modern hospital departments. However, despite its importance, this task is mostly performed by human experts. Experts require a significant amount of time for performing this task, and results are highly dependent on the experience of the human. In this paper, we investigated the use of machine learning approaches for efficiently performing this tedious yet pivotal task.

The experimental analysis we performed demonstrates that it is possible to exploit machine learning-based predictors to accurately estimate workload of a clinical department, in terms of occurrences of a number of personnel or lab/cost intensive clinical events. In other words, human experts can be relieved by the burden of performing such time-consuming task: this has significant implications in terms of optimisation. Firstly, senior experts will have more available time to dedicate to more relevant matters. Secondly, quick and accurate ML-based predictions can be used as input to schedule-optimiser, in order to optimise the allocation of resources via more robust and better informed scheduling.

We see several avenues for future work. Firstly, we are interested in investigating the use of ensemble-based approaches for maximising the predicting performance of a wide range of clinical events. Secondly, we plan to extend our analysis to different departments, in order to evaluate how general the presented results are. Thirdly, we are interested in evaluating whether sharing information between departments of different hospitals can help improving the performance of predictors, by leveraging on privacy-preserving approaches [6].Finally, we will focus on approaches aimed at integrating the strengths of machine learning with the capabilities of human experts, possibly using an overarching framework that encompasses all the relevant steps of the process [8].

References

1. Aizerman, M.A.: Theoretical foundations of the potential function method in pattern recognition learning. Autom. Remote Control **25**, 821–837 (1964)
2. Alvarado, M., Ntaimo, L.: Chemotherapy appointment scheduling under uncertainty using mean-risk stochastic integer programming. Health Care Manage. Sci. **21**(1), 87–104 (2016). https://doi.org/10.1007/s10729-016-9380-4
3. Breiman, L.: Random forests. Mach. Learn. **45**(1), 5–32 (2001)
4. Cover, T., Hart, P.: Nearest neighbor pattern classification. IEEE Trans. Inf. Theory **13**, 21–27 (1967)
5. Creps, J.R., Lotfi, V.: A dynamic approach for outpatient scheduling. J. Med. Econ. **20**, 786–798 (2017)
6. Damiani, A., et al.: Distributed learning to protect privacy in multi-centric clinical studies. In: Holmes, J.H., Bellazzi, R., Sacchi, L., Peek, N. (eds.) AIME 2015. LNCS (LNAI), vol. 9105, pp. 65–75. Springer, Cham (2015). https://doi.org/10.1007/978-3-319-19551-3_8
7. Dieleman, J.L., et al.: National spending on health by source for 184 countries between 2013 and 2040. Lancet **387**(10037), 2521–2535 (2016)
8. Gatta, R., et al.: Towards a modular decision support system for radiomics: a case study on rectal cancer. Artif. Intell. Med. **96**, 145–153 (2019)
9. Gatta, R., et al.: On the efficient allocation of diagnostic activities in modern imaging departments. In: Pereira, F., Machado, P., Costa, E., Cardoso, A. (eds.) EPIA 2015. LNCS (LNAI), vol. 9273, pp. 103–109. Springer, Cham (2015). https://doi.org/10.1007/978-3-319-23485-4_10
10. Guzman-Castillo, M., et al.: Forecasted trends in disability and life expectancy in england and wales up to 2025: a modelling study. Lancet **2**(1), e307–e313 (2017)
11. Nelder, J.A., Wedderburn, R.W.: Generalized linear models. J. Roy. Stat. Soc. Ser. A (Gen.) **135**(3), 370–384 (1972)

How You Say or What You Say? Neural Activity in Message Credibility Evaluation

Łukasz Kwaśniewicz[1], Grzegorz M. Wójcik[1(✉)], Andrzej Kawiak[1], Piotr Schneider[1], and Adam Wierzbicki[2]

[1] Chair of Neuroinformatics and Biomedical Engineering, Maria Curie-Sklodowska University, ul. Akademicka 9, 20-033 Lublin, Poland
gmwojcik@live.umcs.edu.pl
[2] Polish-Japanese Academy of Information Technology, ul. Koszykowa 86, 02-008 Warsaw, Poland

Abstract. In the Internet era, understanding why humans find messages from unknown receivers credible (such as fake news) is an important research topic. Message credibility is an important theoretical aspect of credibility evaluation that relies only on message contents and design. For the first time in the field, we study message credibility by directly measuring brain activity of humans who make credibility evaluations in an experiment that controls message design. Brain activity as measured using EEG is used to investigate areas of the brain involved in message credibility evaluation. We also model and predict human message credibility evaluations using EEG brain activity measurements.

1 Introduction

"It is not what you say, but how you say it." This proverb summarizes our common sense knowledge about persuasive communication. Philosophy, eristics, as well as modern marketing studies contain a wealth of knowledge on how to design a persuasive message. Unfortunately, such knowledge can be misused, as in the case of a phenomenon that has a huge detrimental effect on modern society, i.e. "fake news". Proliferation of fake news today exploits the Web, which has been designed for a fast, cheap and easy publishing of information, but does not have adequate mechanisms that would support evaluation of information credibility. Fake news are, unfortunately, widely accepted by Web users who are not sufficiently critical or able to evaluate credibility of Web content [2]. Fake news are a subject of active research, but are still poorly understood [13].

The situation of a communication of (potentially) fake news can be abstracted into a model based on information theory. Consider Bob (the source of the message) who sends a message to Alice (the receiver): "If you have a headache, put some aromatic oil on your temples. Do not take aspirin, because it has bad side effects." For simplicity, let us consider that the message only contains textual content. While Bob could communicate the message face-to-face (in this case,

V. V. Krzhizhanovskaya et al. (Eds.): ICCS 2020, LNCS 12137, pp. 312–326, 2020.
https://doi.org/10.1007/978-3-030-50371-0_23

Bob's looks and demeanor would probably strongly influence Alice's reception of the message), but he could also use an e-mail, tweet, or Messenger message.

Upon receiving the messages, Alice must make the decision whether she believes the message or not. This decision is based on the credibility of the communicated message (which can be affected by several message properties, such as its persuasiveness) and on Alice's knowledge concerning the subject. Researchers could simply ask Alice whether she believes the message. This was the approach used in social psychology [9,20] that has studied factors that affect credibility evaluations, such as confirmation biases, group polarization, overconfidence and statistical illiteracy [9]. However, Alice may not want to answer such a question (for example, if the message concerns a sensitive subject), or be unable to give reasons of her credibility evaluation. Understanding these reasons could have important practical implications; for example, how to best design messages and information campaigns that correct and counteract fake news?

We hypothesize that credibility evaluation creates a mental state in Alice's brain: Alice decides whether or not Bob (in her opinion) told her the truth. This mental state can be observed and measured using electroencephalography (EEG). Observing Alice's brain activity could enable researchers to create an EEG-based measure of credibility and to better understand the factors (even unconscious ones) that influence Alice's credibility evaluation.

However, most research that used EEG or fMRI in the context of credibility focused on lie detection [15,26]. In other words, it studied Bob's brain signals, not Alice's. No attempts have been described in the research literature to investigate message credibility using EEG, which is the main subject of this article.

1.1 Research Problem and Contributions

The goal of this article is to address the following research questions:

- What brain areas are active when a receiver is evaluating message credibility?
- Does brain activity during credibility evaluation depend on message design?
- Can we model and predict human message credibility evaluations using EEG brain activity measurements?

One of the difficulties in addressing these question lies in the fact that experiment participants can have different levels of knowledge about the message. (In the above example: does Alice know the side effects of aspirin, or not?) An experiment for studying message credibility must ensure that participants have the same knowledge, and control the factors that may influence message credibility evaluation. In this article, we describe a pilot experiment that mimics the situation when the receiver has very little knowledge about the message subject, and can be influenced by irrelevant factors of message design. This situation reflects the reality of many Web users who encounter fake news on various subjects.

While the full answer to the questions listed above would require additional studies, this article makes significant contributions to the subject. We propose an experiment for studying message credibility evaluations using EEG. For the first time in literature, areas of the brain (Brodmann areas - BA) involved in the decision making based on message credibility are described in this article. Based on this knowledge, the article describes an operational model of decision making based on message credibility that uses EEG measurements as an input. Not only does this model provide basic knowledge in the area of neuroinformatics, but it can be seen as a first step towards a practical EEG-based measurement method of message credibility.

In the next section, we introduce a definition of message credibility and discuss theoretical research that can guide the design of empirical experiments for studying credibility. We also discuss related work that studied brain activity related to credibility evaluation. In Sect. 3, we describe the design of our experiment. Section 4 discusses the experiment results. Section 5 concludes the article and introduces our plans for future work.

2   Related Work

2.1   Source, Message, Media Credibility

The concept of credibility, similarly to the concept of trust, is grounded in science and in common sense. Credibility has been subject to research by scientists, especially in the field of psychology and media science. One of the earliest theoretical works on credibility dates back to the 1950s. This influential work of the psychologist Carl Hovland [10] introduced the distinction between *source, message, and media credibility*. Out of these three, two are a good starting point for a top-down study of the complex concept of credibility: source credibility and message credibility. These two concepts are closely related to the natural-language definitions of the term "credibility". In the English language dictionary (Oxford Advanced Learner's Dictionary), credibility is defined as "the quality that somebody/something has that makes people believe or trust them". When this definition is applied to a person ("somebody"), it closely approximates source credibility – an essential concept in real-life, face-to-face communication. However, notice that the dictionary definition of credibility can also be applied to "something" - the message itself. In many online environments, message credibility must be evaluated without knowledge about the source.

Information scientists have studied credibility evaluations with the goal of designing systems that could evaluate Web content credibility automatically or support human experts in making credibility evaluations [14,27]. Credibility evaluations, especially of source credibility, are significant in online collaboration, for example on Wikipedia [25,29]. However, human credibility evaluations are often subjective, biased or otherwise unreliable [12,18], making it necessary to search for new methods of credibility evaluation, such as the EEG-based methods proposed in this article.

2.2   Message Credibility

A search for the term "message credibility" on Google Scholar returns over 1000 results (for an overview of recent publications, especially on the subject of Web content credibility, see [28]). Researchers from the media sciences have attempted to create scale for declarative measurements of message credibility [3]. The importance of message credibility on social media has been recognized in many studies [28], for example in the area of healthcare [6].

As defined by Hovland, message credibility is the aspect of credibility that depends on the communicated message, not on its source or the communication medium. As such, message credibility depends on all information contained in the message itself. Consider a Web page that includes an article. The entire Web page is (in the information-theoretic sense) a message communicated to a receiver. Message credibility can depend on the article's textual content, on images or videos embedded in the article, on Web page design and style, or even on advertisements embedded in the Web page.

This simple example shows that message credibility can be affected by many factors, or features of the message. Even if we limit ourselves to just the textual content of the message, message credibility is affected by both the semantic content of the message (its "meaning") and by the pragmatic content of the message (its style, persuasiveness, sentiment, etc.) This is especially important since message credibility is usually evaluated rapidly. The work of Tseng and Fogg [24] introduced the two concepts of "surface credibility" and "earned credibility", both of which can be applied to message credibility. Surface credibility is the result of a fast and superficial examination of the message. Earned credibility is the result of a slower and more deliberative reasoning about the message. The two concepts are similar to Kahneman's distinction about the fast, heuristic-based System I and the slower, deliberative System II [11]. Surface credibility is message credibility based on System I reasoning, while earned credibility is message credibility based on System II reasoning. Research results [28] have established that most users evaluate Web page credibility quickly, in a matter of minutes (three minutes are enough for most Web page credibility evaluations). These results are relevant for our experiment design. In order to begin to understand brain activity during message credibility evaluation, we shall limit message design to a single aspect that can be rapidly evaluated.

2.3   Research on Brain Activity Related to Message Credibility Evaluation

To our knowledge, message credibility has not been investigated to date using neuroimaging methods. There are, however, some studies that may be in some way associated with our interests. Research conducted on patients with Alzheimers disease and controls showed engagement of hyperactivity of Brodmann Area 38 in Positron Emission Tomography (PET) experiment discussed in [19]. Similarly BA38 was hyperactive in functional Magnetic Resonance Imaging fMRI studies reported in [8] where prospective customers were obliged to choose between similar goods making economic decisions. The activation of BA38 in language related tasks is also reported in [5]. Language and visual perception associations are postulated in [4] where some meta-analyses on BrainMap were conducted. BA10 and BA47 play important role in decision-making related fMRI studies classified by neural networks in [1]. Another fMRI experiment shows that

Fig. 1. Typical screen shown to participant during the experiment with the long note in the top and short in the bottom.

BA10, BA46 and through connectivity BA38 are involved in high- and low-risk decisions [7] as well as BA47 reported to be engaged in lexical decisions concerning nonwords or pseudowords in PET/fMRI studies in [22]. Activation of selected BA related to language, decision making and in some way trustworthiness is often investigated using competition or game element like in [21], but in our experiments one should remember there is no game component.

3   Experiment Design

We have designed and conducted a pilot experiment to study message credibility evaluations using EEG. The pilot experiment was carried out at Marie Curie Skłodowska University in Lublin, Poland, from June, 15th till July, 14th, 2019 (MCSU Bioethical Commission permission 13.06.2019).

The goal of the pilot experiment was to observe the electrical activity and the most active areas of the participant's brain cortex during tasks involving message credibility evaluation, as well as the influence of the message design on this process. In order to ensure that participants could only rely on message design during the experiment, the experiment was designed so that the participants would not be familiar with the topic of the messages. The chosen topic of the messages concerned the meaning of Japanese kanji signs.

The experiment was designed to create a condition when participants assess truth or falsehood with practically no knowledge of the message subject. Participants were completely unsure about the correct answer. This situation resembles the case when a person who has no knowledge of the subject receives fake news.

The participants to the experiment were right-handed male students without any knowledge of Japanese. A total of 62 participants took part in the pilot experiment.

3.1   Message Credibility Evaluation Task

Participants were requested to choose the meaning of a Japanese Kanji sign, based on a provided explanations (the message). There were 128 different Kanji signs to be assessed and their meaning was described by a single word or by a longer description (consisting of at most 20 words) in participant's native language (see Fig. 1). The longer descriptions were designed to logically explain the relationship between the shape and meaning of the Kanji sign, for example: "gutter, curves symbolize pipes" In the remainder of this article, we shall refer to the single word and longer descriptions as "short note" and "long note", respectively.

In half (64) of the cases, the true meaning was given by a single word, and in the remaining 64 cases the true meaning was described by a longer note. In half of the screens the long note was at the top of the screen and the single-word note was at the bottom, while in the other half, the notes were placed in the opposite way.

The participants' task was formulated in the form of a question: "Does the Japanese character ... mean: ..." (see Fig. 1). The meaning contained in the question was always the meaning that was displayed at the top. The participants could answer the question by selecting "Yes" or "No". The actual answer was not important for the analysis; what actually mattered was whether the answer matched the single-word note or the longer note (it could only be one of the two). The agreement of the answer with the note is equivalent to a positive evaluation of the note's message credibility.

Such a setup allowed us to register EEG measurements in four cases:

1. ST-SC: Short Top-Short Chosen. The short note was on top and the long note was at the bottom. The short note was chosen as the proper meaning of the kanji sign,
2. ST-LC: Short Top-Long Chosen. The short note was on top and the long note was at the bottom. The long note was chosen as the proper meaning of the kanji sign,
3. SB-SC: Short Bottom-Short Chosen. The short note was at the bottom and the long note was on top. The short note was chosen as the proper meaning of the kanji sign,
4. SB-LC: Short Bottom-Long Chosen. The short note was at the bottom and the long note was on top. The long note was chosen as the proper meaning of the kanji sign.

This also could indicate whether participants prefer the length of the note or their position in the interpretation of Kanji signs.

3.2 EEG Measurements

In the empirical experiments we were using top EEG devices. We were equipped with a dense array amplifier recording cortical activity with up to 500 Hz frequency through 256 channels HydroCel GSN 130 Geodesic Sensor Nets provided by Electrical Geodesic Systems (EGI). In addition, in the EEG Laboratory the Geodesic Photogrammetry System (GPS) was used. The position of the electrodes is precisely defined in EGI Sensor Nets documentation.

The responses of the cohort of 62 participants were examined. Showing the screen with two notes (single-word and long) was treated as the stimulus (event)

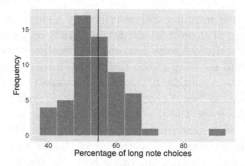

Fig. 2. Histogram of frequency of evaluating long note as credible among all participants.

evoking the ERP. For each of 256 electrodes the mean electric charge was calculated in the way as follows: first the ERP was estimated in the interval of 900 ms (beginning 100 ms before showing the stimulus and ending 800 ms after showing that), next the source localisation algorithm sLORETA (GeoSource Parameters set as follow: Dipole Set: 2 mm Atlas Man, Dense: 2447 dipoles Source Montages: BAs) was applied to each signal from each electrode and the average electric current varying in time and flowing through each BA was tabularized. Finally, the Mean Electric Charge (MEC) flowing through each BA was calculated by integration of electric current in 10 ms intervals of time. The procedure of calculating MEC has been described in detail in [31] and using the MEC as one more method of quantitative EEG analysis has been verified in different ways and discussed in [30, 32].

3.3   Experiment Hypotheses

We formulated the following hypotheses:

1. The length of the note has a significant positive influence on the participant's decision about message credibility.
2. The length of the note has a significant influence on Brodmann areas' activity during making decisions about message credibility.
3. The decision of participants about message credibility can be predicted based on measurements of mean electric charges in participant's brains.
4. There are significant differences in the models that predict decisions of participants who frequently choose the long note as compared to models of other participants.

   Hypotheses 1 is not directly related to participants' brain activities. Rather, it is a test of our experiment's internal validity. A positive validation of hypothesis 1 would confirm that there exists a relationship between the main independent variable of our experiment and the participant's decision. Such a relationship would confirm the internal validity of our experiment. Hypothesis 2 is related to

Table 1. The Brodmann Areas manifested statistically significant differences in MEC measured during choice of long or short note during the experiment, and were selected by the model trained on experimental data to predict long or short note choices. Presented with their corresponding anatomical structure of the brain and known functions as listed in [17]. The prefix of particular BAs stands for Left and Right hemispheres.

BA	Anatomical structure	Known functions
L-BA19	Associative Visual Cortex (V3)	Processing phonological properties of words, confrontation naming
L-BA38	Temporopolar area	Semantic processing, naming of items learned in early life, word retrieval for specific entities, lexico-semantic ambiguity processing, narrative comprehension
L-BA41	Primary Auditory Cortex	Working memory
L-BA34	Entorhinal area, superior temporal gyrus	Working memory, memory encoding, experiencing/processing emotions
R-BA38	Temporopolar area	Irony processing
L-BA40	Supramarginal Gyrus	Semantic processing, verbal creativity, deductive reasoning
R-BA39	Somatosensory Association Cortex	Language processing, literal sentence comprehension, word comprehension (imageability)

the first and second research questions. To validate this hypotheses, we need to study brain activity during the experiment and compare this activity in cases when the short note is evaluated as credible (ST-SC or SB-SC) and when the long note is chosen (ST-LC or SB-LC).

Hypothesis 3 can be validated by constructing a classifier that will predict the (binary) decision of participants with sufficiently high accuracy. However, the validation of 4 requires training two such classifiers, one based on the set of participants who tend to evaluate long messages as credible, and another on the remaining set of participants. The comparison of these two classifiers is only possible if the two classifiers are explainable, which excludes the use of black-box classifiers such as neural networks.

4   Experiment Results

After the recording phase of the experiment the EEG signal of 57 participants was analysed. Each participant answered 128 questions requiring making decision by choosing the short or the long note as a correct answer. That gave $57 \times 128 = 7296$ responses. In fact, 7296 responses were collected.

The long note was chosen 1782 times when it was on top of the screen (SB-LC), while the short one was chosen 1442 times when it was on top (ST-SC). On the other hand, the long note was chosen 2206 times when it was at the bottom (ST-LC), as compared to 1866 times when the short note was chosen while at the bottom (SB-SC). This means that no matter whether the long note

Table 2. Five best models found by the brute force method to classify participants' decisions using selected BAs from the set of BAs manifesting statistically significant differences in activity. For more details see text.

No.	BAs used by the model	Accuracy	Precision	Recall	F1 score
1	LBA19, LBA38, LBA41, LBA34, RBA38, LBA40, RBA39	0.793	0.786	0.786	0.786
2	LBA19, RBA13, LBA38, LBA41, LBA34, RBA38, LBA40, RBA39	0.793	0.714	0.833	0.769
3	LBA13, LBA19, RBA13, LBA38, LBA41, LBA34, RBA38, LBA40, RBA39	0.793	0.786	0.786	0.786
4	LBA13, LBA19, LBA38, LBA41, RBA41, LBA34, RBA38, LBA40, RBA39	0.793	0.714	0.833	0.769
5	LBA13, LBA34, RBA38, RBA47, RBA39	0.793	0.786	0.786	0.786

was on top or at the bottom of the screen it was chosen more frequently and the total difference is 680 (or 9.3%) towards the long note. The length of the note influences the participants' decision regardless of the note's position. This positively verifies hypothesis 1.

We can also test whether or not the message credibility evaluation was random. Participants could evaluate the long or short note as credible. If the choice is be random, the choices of the long or short note should form a binomial distribution with probability 0.5. We used the binomial test and calculated the p-value, which was less than 0.00001 (we observed 3988 choices of the long note out of 7296 message credibility evaluations). Therefore, we concluded that we could reject the possibility that the choices of notes in the experiment where binomially random.

We also found that a subset of participants who chose the long note more frequently than other participants. For each participant, we calculated how frequently they evaluated the longer note as credible. This frequency is shown on a histogram in Fig. 2. The median of this distribution, which is shown as red vertical line in Fig. 2, is approximately 54% because overall the long note was evaluated as credible more frequently than the single word note. 25 of 57 participants chose the long note in more than 55% of all possible choices, and 17 participants evaluated the long note as credible in more than 60% of cases.

Statistically significant differences in MEC of all participants were observed using the Mann-Whitney-Wilcoxon test for choosing long/short note in the time between 150 ms to 650 ms from stimulus, but only in the case when the short note was on top of the screen. The spontaneous decisions in ERP experiments are usually made is such interval [16,23,30–32]. If the short note was at the bottom, there were no statistical differences in MEC observed in the cognitive processing time interval. This partially verified the hypothesis 2. Note that in every scenario we asked the participant about the meaning of the note that was explained at top of the screen. This questions most likely affected participants' focus. For this reason, in the analysis we shall focus on the comparison of the cases ST-SC and ST-LC.

The BAs for which there was a statistically significant difference in Mann-Whitney-Wilcoxon Test in MEC during the interval [150 ms, 650 ms] from the stimulus are: L-BA19, L-BA38, L-BA41, L-BA34, R-BA38, L-BA40, R-BA39, R-BA10, L-BA13, R-BA13, R-BA20, L-BA29, R-BA29, L-BA30, R-BA37, R-BA41, L-BA42 and R-BA47. However, only a subset of them was chosen by the model of participants credibility evaluations of the long or short notes, as described in the next section. This subset of 7 most significant BAs is shown on Table 1.

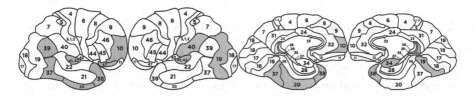

Fig. 3. Brodmann areas most significant for predicting message credibility evaluations

4.1 Regression Model of Message Credibility

We used a generalized logistic regression classifier to predict message credibility evaluations. The classifier was trained on evaluations of a subset of 42 (75%) participants and validated on evaluations of 15 participants. This means that the classifier had 42 observations of each class available for training. Only the BAs manifesting statistically significant differences in their activity were taken into consideration as independent variables. These were BA13, BA19, BA29, BA30, BA34, BA38, BA40, BA41, BA42 in the left hemisphere and BA10, BA13, BA20, BA29, BA37, BA38, BA39, BA41, BA47 in the right hemisphere see. Table 1.

A set of independent variables for the logistic regression classifier was constructed in a special manner, using the following reasoning:

1. 18 different BAs had significant differences during the experiment.
2. It is not obvious that all of them should be taken as features of classifiers.
3. If so, then each subset of these 18 BAs could be considered as independent variables for classifiers.
4. Why not to check all possible combinations of all subsets and then find the ideal model for the whole population?

There are 262143 statistically significant BAs possible subsets as a sum of combinations as in the following equation: $N = \sum_{i=1}^{Q} \binom{Q}{i}$ where $Q = 18$.

The Logistic-regression classifiers were built for all subsets in the above-mentioned manner. Results achieved by 5 best classifiers are presented in Table 2. Note that there are only 7 BAs engaged in the best classifier and also used in the following 3 next-best classifiers: LBA19, LBA38, LBA41, LBA34, RBA38,

LBA40, RBA39. This confirms that the proposed methodology of selecting variables produces models that have consistently similar sets of independent variables. The 7 BAs used in the best classifier are clearly the most useful independent variables overall.

For the best combination of independent variables, we have used 10-fold cross-validation to confirm our results. The average efficiency achieved by classifiers in this cross-validation is: Accuracy 0.68, Precision 0.6582, Recall 0.7184 and F1 Score 0.7184 which proves its stability.

Such classification results positively confirm the hypothesis 3.

4.2 Modeling Brain Activity of Participants that Prefer Long Notes

The simulations described above were repeated in order to find the best classification models for the participants who found long notes credible more frequently. The goal was to compare the best classifiers for this subset of participants (25 out of 57, or 43% of all participants who evaluated the long note as credible in more than 55% of the cases) to the best classifiers for the remaining participants. Note that the classifier for participants who preferred long notes had 25 observations of each class (single-word, or long note selection) in the training set and in the validation set.

For participants who were choosing long notes more frequently, the following BAs manifest statistically significant differences in activity: LBA13, RBA25, LBA19 LBA30, RBA29, LBA29, LBA46, RBA13, RBA19, RBA37, LBA34, RBA47, RBA39. The best classifier for the long-note-choosers involved: LBA19, RBA29, LBA29, LBA46 and reached the characteristics of accuracy: 0.923, precision: 0.833 and recall: 1.000. For the best combination of independent variables, we have used 10-fold cross-validation to confirm our results. The average efficiency achieved by classifiers in this cross-validation is: Accuracy 0.6615, Precision 0.6285, Recall 0.6562 and F1 Score 0.909 which also proves its stability. The BAs that were most significant for classification of participant's decisions are visualized on Fig. 3.

The BA29 (granular retrosplenial cortex) associated with memory retrieval and emotion related to language appeared in the best models of participants who preferred long notes in contrast to the set of best models generated for the entire population. This positively verified hypothesis 4, as it was a significant difference in the models of message credibility decision for the two groups.

4.3 Discussion and Limitations

The aim of this article was to address three research questions.

Firstly, we wondered if it is possible to indicate the brain regions that are engaged in receiver's cortex while evaluating the message credibility? Indeed, we found 18 BAs that manifested statistically significant differences during tasks of message credibility evaluation.

Some of the BAs: BA38, BA10, BA46, BA47 confirmed previous findings from the experiments where decision-making was involved. Our models, however, chose the best candidates for future research of brain functional anatomy related to credibility evaluation.

It was also interesting to check whether the design of the message had the influence on brain activity? The experiment was designed so that participants had no knowledge of the message subject and could only be influenced by structure of the note that described the Kanji sign. Participants had a tendency to choose a longer note, and we found statistically significant differences in brain activity when a longer note was chosen, as compared to the situation of choosing a single-word note.

The most interesting results were observed while addressing the third research question, i.e. whether it is possible to model and predict brains' message credibility evaluation?

Only 18 BAs that were statistically significantly different in the spectrum of MEC were taken into consideration, and, therefore, it was possible to find the best classifier for the entire investigated population. The chosen method, consisting of checking all possible combinations of BA subsets, was computationally demanding and required parallel computations.

The analysis of independent variables used by the best classifiers of message credibility revealed a set of 7 BAs that are the best independent variables to predict message credibility evaluations based on EEG measurements. The classifiers achieve a satisfactory accuracy of 79% (against the baseline of 50%). This result is a step forward in the study of brain functional anatomy, as this is a first finding concerning areas in the brain involved in message credibility evaluation. It is also an achievement in the field of neuroinformatics, as our results give a firm basis for the development of an EEG-based measurement method of message credibility. This method could have many practical applications, as for example in evaluating messages aimed at combating and debunking fake news. Note that our models used generalized logistic regression, as our goal was to have an explanatory model. The use of other classifiers in the future could most likely further improve the accuracy of predicting message credibility evaluation.

We attempted to investigate incorrect message credibility evaluations as well. In our experiment, participants had no access to ground truth. However, the experiment was designed to create a situation when participants could choose to be influenced by message design in their credibility evaluation. The only factor of message design used in the experiment was message length. We found that a subset of participants tended to more frequently evaluate long notes as credible, as compared to other participants. By modeling the decisions of this subset of participants separately, we identified the area in the brain (BA29, the granular retrosplenial cortex) associated with memory retrieval and emotion related to language that has impact on message credibility evaluations of participants who preferred long notes. While this finding needs to be verified on larger samples, it demonstrates the possibility of studying other factors that impact message credibility evaluation using EEG measurements.

5   Conclusion and Future Work

Although the subject of credibility has received significant attention from psychologists, media scientists, and computer scientists aiming at detecting and combating fake news, it had not been studied before using brain activity measurements. This is probably due to the fact that it is difficult to design and carry out brain measurement experimentation. As a result, no method of message credibility measurement (or any credibility measurement) based on brain activity had been proposed before.

We have described in this article the first experiment designed to study message credibility evaluation. The experiment controlled the influence of participants' knowledge and beliefs on message credibility evaluation by using a subject with which the participants where not familiar (Kanji signs). Furthermore, the experiment controlled message design factors that could influence message credibility evaluation and limited them to one factor, i.e. message length.

We have identified and described 7 BAs that have the greatest impact on message credibility evaluations. The selection of these brain functional areas was done by a statistical analysis of EEG signals and based on generalized logistic regression classifiers of participants' decisions. These classifiers achieved an accuracy of 79%, demonstrating the feasibility of creating an EEG-based message credibility measurement method.

Our experiment has several limitations due to its design. One limitation is the small amount of users who are influenced by the message design. We intend to continue investigating the impact of message design using other design factors, such as message persuasiveness.

In the future, we plan to run similar experiments that require basic knowledge of Japanese by participants, in order to observe the impact of participant's own knowledge on message credibility evaluation.

References

1. Ahmad, F., Ahmad, I., Dar, W.M.: Identification and classification of voxels of human brain for rewardless-related decision making using ANN technique. Neural Comput. Appl. **28**(1), 1035–1041 (2017). https://doi.org/10.1007/s00521-016-2413-6
2. Allcott, H., Gentzkow, M.: Social media and fake news in the 2016 election. J. Econ. Perspect. **31**(2), 211–36 (2017)
3. Appelman, A., Sundar, S.S.: Measuring message credibility: construction and validation of an exclusive scale. Journal. Mass Commun. Q. **93**(1), 59–79 (2016)
4. Ardila, A., Bernal, B., Rosselli, M.: Language and visual perception associations: meta-analytic connectivity modeling of brodmann area 37. Behav. Neurol. **2015** (2015). 14 pages
5. Ardila, A., Bernal, B., Rosselli, M.: The elusive role of the left temporal pole (BA38) in language: a preliminary meta-analytic connectivity study. Int. J. Brain Sci. **2014** (2014). 7 pages

6. Borah, P., Xiao, X.: The importance of 'likes': the interplay of message framing, source, and social endorsement on credibility perceptions of health information on facebook. J. Health Commun. **23**(4), 399–411 (2018)
7. Cohen, M., Heller, A., Ranganath, C.: Functional connectivity with anterior cingulate and orbitofrontal cortices during decision-making. Cogn. Brain. Res. **23**(1), 61–70 (2005)
8. Deppe, M., et al.: Nonlinear responses within the medial prefrontal cortex reveal when specific implicit information influences economic decision making. J. Neuroimaging **15**(2), 171–182 (2005)
9. Forgas, J.P., Baumeister, R.: The Social Psychology of Gullibility: Conspiracy Theories, Fake News and Irrational Beliefs. Routledge, London (2019)
10. Hovland, C.I., Weiss, W.: The influence of source credibility on communication effectiveness. Public Opin. Q. **15**(4), 635–650 (1951)
11. Kahneman, D.: Thinking, Fast and Slow. Macmillan, New York (2011)
12. Kakol, M., Jankowski-Lorek, M., Abramczuk, K., Wierzbicki, A., Catasta, M.: On the subjectivity and bias of web content credibility evaluations. In: Proceedings of the 22nd International Conference on World Wide Web, pp. 1131–1136. ACM (2013)
13. Lazer, D.M., et al.: The science of fake news. Science **359**(6380), 1094–1096 (2018)
14. Liu, X., Nielek, R., Adamska, P., Wierzbicki, A., Aberer, K.: Towards a highly effective and robust web credibility evaluation system. Decis. Support Syst. **79**, 99–108 (2015)
15. Meijer, E.H., Verschuere, B.: Deception detection based on neuroimaging: better than the polygraph? J. Forensic Radiol. Imaging **8**, 17–21 (2017)
16. Mikołajewska, E., Mikołajewski, D.: The prospects of brain–computer interface applications in children. Open Med. **9**(1), 74–79 (2014). https://doi.org/10.2478/s11536-013-0249-3
17. Pascual-Leone, A., Davey, N.J., Rothwell, J., Wasserman, E.M., Puri, B.K.: Handbook of Transcranial Magnetic Stimulation, vol. 15. Arnold, London (2002)
18. Rafalak, M., Abramczuk, K., Wierzbicki, A.: Incredible: is (almost) all web content trustworthy? Analysis of psychological factors related to website credibility evaluation. In: Proceedings of the 23rd International Conference on World Wide Web, pp. 1117–1122. ACM (2014)
19. Rinne, J., et al.: Semantic decision making in early probable ad: a pet activation study. Cogn. Brain. Res. **18**(1), 89–96 (2003)
20. Rutjens, B.D., Brandt, M.D.: Belief Systems and the Perception of Reality. Routledge, London (2018)
21. Sailer, U., Robinson, S., Fischmeister, F.P.S., Moser, E., Kryspin-Exner, I., Bauer, H.: Imaging the changing role of feedback during learning in decision-making. Neuroimage **37**(4), 1474–1486 (2007)
22. Specht, K., et al.: Lexical decision of nonwords and pseudowords in humans: a positron emission tomography study. Neurosci. Lett. **345**(3), 177–181 (2003)
23. Tadeusiewicz, R., et al.: Neurocybernetyka teoretyczna. Wydawnictwa Uniwersytetu Warszawskiego (2009)
24. Tseng, S., Fogg, B.: Credibility and computing technology. Commun. ACM **42**(5), 39–44 (1999)
25. Turek, P., Wierzbicki, A., Nielek, R., Datta, A.: WikiTeams: how do they achieve success? IEEE Potentials **30**(5), 15–20 (2011)
26. Wang, H., Chang, W., Zhang, C.: Functional brain network and multichannel analysis for the p300-based brain computer interface system of lying detection. Expert Syst. Appl. **53**, 117–128 (2016)

27. Wawer, A., Nielek, R., Wierzbicki, A.: Predicting webpage credibility using linguistic features. In: Proceedings of the 23rd International Conference on World Wide Web, pp. 1135–1140. ACM (2014)
28. Wierzbicki, A.: Web Content Credibility. Springer, Cham (2018). https://doi.org/10.1007/978-3-319-77794-8
29. Wierzbicki, A., Turek, P., Nielek, R.: Learning about team collaboration from Wikipedia edit history. In: Proceedings of the 6th International Symposium on Wikis and Open Collaboration, pp. 1–2 (2010)
30. Wojcik, G.M., et al.: Mapping the human brain in frequency band analysis of brain cortex electroencephalographic activity for selected psychiatric disorders. Front. Neuroinform. **12**, 73 (2018)
31. Wojcik, G.M., et al.: New protocol for quantitative analysis of brain cortex electroencephalographic activity in patients with psychiatric disorders. Front. Neuroinform. **12**, 27 (2018)
32. Wojcik, G.M., et al.: Analysis of decision-making process using methods of quantitative electroencephalography and machine learning tools. Front. Neuroinform. **13**, 73 (2019)

Look Who's Talking: Modeling Decision Making Based on Source Credibility

Andrzej Kawiak[1], Grzegorz M. Wójcik[1(✉)], Lukasz Kwasniewicz[1], Piotr Schneider[1], and Adam Wierzbicki[2]

[1] Chair of Neuroinformatics and Biomedical Engineering,
Maria Curie-Sklodowska University, ul. Akademicka 9, 20-033 Lublin, Poland
`gmwojcik@live.umcs.edu.pl`
[2] Polish-Japanese Academy of Information Technology,
ul. Koszykowa 86, 02-008 Warsaw, Poland

Abstract. Understanding how humans evaluate credibility is an important scientific question in the era of fake news. Source credibility is among the most important aspects of credibility evaluations. One of the most direct ways to understand source credibility is to use measurements of brain activity of humans who make credibility evaluations. Nevertheless, source credibility has never been investigated using such a method before. This article reports the results of an experiment during which we have measured brain activity during source credibility evaluation using EEG. The experiment allowed for identification of brain areas that were active when a participant made positive or negative source credibility evaluations. Based on experimental data, we modelled and predicted human source credibility evaluations using EEG brain activity measurements with F1 score exceeding 0.7 (using 10-fold cross-validation).

1 Introduction

Fake news, or false news, has become a buzz-word to describe a general problem that emerged when the Web achieved critical mass in developed societies. In 2016, Web-based social media became a source of news for over 60% of adult Americans [1]. This situation coincided with the American presidential election campaign, which revealed the Web's weakness: lack of mechanisms for Web content credibility evaluation. Google trends show that since then (October, 2016), the term fake news (which has been in use since 19th century) has rapidly grown in popularity[1]. Fake news are a subject of active research, but at the same time they are still poorly understood [14]. Web-based social media, such as Facebook or Twitter, are especially vulnerable to fake news proliferation [1], because messages on social media are forwarded based on the trust that receivers have in their virtual friends (or followers in Twitter sources). It is, therefore, crucial to research the evaluation of source credibility.

---

[1] Leetaru, Kalev. "Did Facebook's Mark Zuckerberg Coin The Phrase 'Fake News'?". Forbes.

© Springer Nature Switzerland AG 2020
V. V. Krzhizhanovskaya et al. (Eds.): ICCS 2020, LNCS 12137, pp. 327–341, 2020.
https://doi.org/10.1007/978-3-030-50371-0_24

Social psychology has contributed to our understanding of social and psychological factors that affect credibility evaluations made by a information receiver [7,21]. However, all these findings are based on behavioral studies that rely on participant declarations or on indirect inferences from observations of participant behavior. Simply asking experiment participants whether they believe fake news (or inferring this information from observed behavior) may not reveal the real reasons for such a decision (some of these reasons may not even be consciously known by the experiment participants). By directly observing brain activity, researchers can understand basic processes that occur in the brain during credibility evaluations.

In turn, knowledge about brain activity during credibility evaluation could be applied by neuroinformatics in order to create an EEG-based measure of credibility. However, most research that used EEG or fMRI in the context of credibility has focused on lie detection [16,24], which is based on the investigation of the brain activity of the author, and not the receiver of the message.

1.1   Research Problem and Contributions

In order to study credibility empirically, it is useful to deconstruct this complex concept into simpler ones. This is the approach adopted in our research. In this article, we deal with one of such simpler concepts: source credibility. Because source credibility evaluation has not been studied using brain activity analysis before, our research goals are largely exploratory and we do not have preconceived expectations based on literature. The goal of this article is to address the following research questions:

- What brain areas, and in which time intervals since the stimulus, are active when a receiver makes positive or negative source credibility evaluations?
- How does brain activity depend on the level of source credibility?
- Can we model and predict human source credibility evaluations using EEG brain activity measurements?

While a comprehensive answer to the questions listed above would require additional studies, this article makes significant contributions to this matter. For the first time in literature, in this article, there are described areas of the brain (Brodmann areas) involved in the decision making process based on source credibility. Using this knowledge, this article describes an operational model of decision making based on source credibility that uses EEG measurements as an input. Not only does this model provide basic knowledge in the field of neuroinformatics, but it can be seen as a first step towards a practical EEG-based measurement method of source credibility.

In the next section, we introduce a definition of source credibility and discuss theoretical research that can guide the design of empirical experiments for studying credibility. We also discuss related work that studied brain activity related to source credibility evaluation. In Sect. 3, we describe the design of our experiment. Section 4 discusses the experiment results. Section 5 concludes the article and introduces our plans for future work.

2   Related Work

2.1   Source, Message, Media Credibility

The concept of credibility, similarly to the concept of trust, is grounded in science and in common sense. Credibility has been subject to research by scientists, especially in the field of psychology and media science. One of the earliest theoretical works on credibility dates back to the 1950s. This influential work of the psychologist Carl Hovland [8] introduced the distinction between *source, message, and media credibility*. Out of these three, two are a good starting point for a top-down study of the complex concept of credibility: source credibility and message credibility. These two concepts are closely related to the natural-language definitions of the term "credibility". In the English language dictionary (Oxford Advanced Learner's Dictionary), credibility is defined as "the quality that somebody/something has that makes people believe or trust them". When this definition is applied to a person ("somebody"), it closely approximates source credibility – an essential concept in real-life, face-to-face communication. However, notice that the dictionary definition of credibility can also be applied to "something" - the message itself. In many online environments, message credibility must be evaluated without knowledge about the source.

Information scientists have studied credibility evaluations with the goal of designing systems that could evaluate Web content credibility automatically or support human experts in making credibility evaluations [15,25]. Credibility evaluations, especially of source credibility, are significant in online collaboration, for example on Wikipedia [23,27]. However, human credibility evaluations are often subjective, biased or otherwise unreliable [11,19], making it necessary to search for new methods of credibility evaluation, such as the EEG-based methods proposed in this article.

2.2   Source Credibility

A search for the term "source credibility" on Google Scholar returns an excess of 12,000 results (for an overview of recent publications, especially on the subject of Web content credibility, see [26]). Research on this subject has ranged from investigating impact of source credibility on politics [6] to healthcare [12].

Previous theoretical research hypothesized that *source credibility is closely related to credibility trust* [26]. Credibility trust is an expectation that the source will observe the social norm of not lying (not communicating a false message). Following the analogy to trust, source credibility can also be based on the trustworthiness of the source in the context of veracity; it is difficult, however, to reliably observe, measure or predict this property. Most observations or valuations concerning credibility are done in a relational setting: communication of a *message* from a *source* to a *receiver*. A proxy for credibility trustworthiness may be *source reputation* in the context of veracity, estimated based on the past performance of the source. Therefore, it can be concluded that *source credibility is a combination (or multiple criteria evaluation) of two kinds of trust: credibility trust and the trust in the expertise of the source.* These two types of trust are independent and complementary; a source may, after all, usually tell the truth, but not be able to do so because of lack of expertise in a given subject. On the other hand, an expert in the subject may not be trustworthy due to the fact of being a habitual liar.

2.3 Research on Brain Activity Related to Source Credibility Evaluation

Not much has been done in the field of source credibility research as far as neuroimaging methods are concerned. Source credibility is associated with trustworthiness which was discussed, for example, by [20]. They state that usually the amygdala is involved in trusting others [20]. However, some findings [3,10] prove that insular cortex has function similar to orbitofrontal cortex and plays an important role in decision making process [20]. The Brodmann area BA47 is anatomically located in orbitofrontal cortex and its activity during the process of making risky decisions was observed in Positron Emission Tomography (PET) study reported in [5]. The BA46 was found to be involved in decision making process in the presence of fatiguing factors in magnetoencephalographic research (MEG) [9] as well as in the [4] where comparison of moral and cognitive judgments was conducted. Together with BA47 where patients with reduction of ventrolateral prefrontal cortex gray matter played an economic game with some degree of irrationality implemented [2]. All above-mentioned papers show (using various techniques like PET, MEG and fMRI) that decision making, game theory, trustworthiness and judgment tasks are related and involve the BA46 and BA47 Brodmann areas. However, no research so far has identified areas of the brain involved in source credibility evaluation.

To sum up the Dorsolateral Prefrontal Cortex (DLPFC) BA46 is reported to be engaged in working memory oriented tasks requiring cognitive effort while BA47 is involved in decision-making process, especially in morally difficult problems [2–5,9,10,20]. Those are potential candidates to observation with greatest attention in our experiment.

3 Experiment Design

We have designed and conducted a pilot experiment to study source credibility evaluations using EEG. The pilot experiment was carried out at Marie Curie Skłodowska University in Lublin, Poland, from June, 15th till July, 14th, 2019 (MCSU Bioethical Commission permission 13.06.2019).

The aim of the pilot experiment was to observe activity of the participant's brain cortex during performance of a task involving source credibility evaluation. In order to ensure that the participants could rely only on source credibility during the experiment, the experiment was designed so that the participants would not be familiar with the topics of the messages. The selected message topics concerned a Japanese language test. All experiment participants had no knowledge of Japanese.

In order to simplify EEG measurement, all participants selected for the experiment were right-handed males. A total of 57 participants took part in the pilot experiment.

Fig. 1. Typical screen shown to participant during the experiment. The student's hint, accuracy during the test, and avatar are in the bottom section of the screen. The participant is asked to agree or disagree with the student's answer (top section) based on their trust or distrust in the hint that are influenced by the student's accuracy.

3.1   Source Credibility Evaluation Task

In the introduction to the experiment, participants were informed that students of another university had solved a test of their knowledge of Japanese Kanji signs (after one semester of learning completed) and that we know the results achieved by all students.

To the end of our experiment, randomly generated names of students were chosen together with avatars representing their faces. Avatars were generated by means of repository available in www.makeavatar.com and a simple Python script. The neutral emotions were provided by the smile-option turned off. The 'generated' students fell into three groups – those who received 50%, 70% and 90% of the maximum score to be gained during the test.

Participants were shown 180 screens with one Kanji sign on each of them and the question if the translation of that sign was correct or not. As a hint, participants received information about students (represented by name and avatar) who had an overall accuracy of 50%, 70% or 90% during the test. The hint was the student's answer ('Yes' or 'No') to the question posed to the participant (see Fig. 1). The same accuracy of the hinting student was shown on 60 screens (with values 50%, 70% or 90%).

Note that the participants did not know whether the student's response was correct or incorrect. The only thing a participant knew was the student's result in the entire test. In this way, we have created a situation in which the participant had to make a decision whether to accept a message (the student's hint) based on source credibility (the student's overall score in the test).

In the experiment, the participants faced a binary decision: they were asked to press a 'YES' or 'NO' button. This decision could comply with the student's hint, in which case we shall refer to this decision as "trusting". The participants could also disagree with the student's hint, in which case we shall speak of "distrust". Note that both trusting and distrusting decisions can be 'YES' or

'NO', but this is not relevant for the experiment. The only relevant aspect of the participant's decision is whether it is trusting or distrusting, corresponding to a positive or negative source credibility evaluation, respectively.

Recall that source credibility can be measured as the source's reputation in the context of expertise or veracity. In our experiment, the only information that participants had about the students' reputation was the test score. If the participants were informed that the suggesting student's test score was 90%, they would probably make a trusting decision. If the test score was 50%, we could expect that the participant would respond randomly. The most interesting situation was when the test score was 70%. We shall refer to the hinting student's test score as the Source Credibility Level (SCL).

Moreover, the participants were not in any case given the correct meaning of the current and previous signs presented to them. Thus, the participants were not rewarded for a good answer and were not punished for a bad one. This experiment design ensured that the participant made decisions in a non-competitive setting and without consideration for a reward.

Repeating similar screens 60 times for each source credibility level made it possible to observe the so-called Event-Related Potentials (ERPs) in the electroencephalographic activity registered by the amplifier in our lab. The methodology of ERP is probably most often used in experimental psychology and observations made using source localisation methods allowed us to measure brain cortex activity quantitatively.

3.2 Experimental Cases and Data

All decisions made by experiment participants can be classified into the six following cases that allow us to compare brain activity for trusting and distrusting decisions under stimulus of various source credibility levels:

- T50: Source credibility was 50% and the subject trusted the message.
- D50: Source credibility was 50% and the subject did not trust the message.
- T70: Source credibility was 70% and the subject trusted the message.
- D70: Source credibility was 70% and the subject did not trust the message.
- T90: Source credibility was 90% and the subject trusted the message.
- D90: Source credibility was 90% and the subject did not trust the message.

Additionally, let us introduce three larger sets of all decisions made while the participant was shown a particular source credibility level: $A50 = T50 \cup D50$, $A70 = T70 \cup D70$ and $A90 = T90 \cup D90$.

3.3 Experiment Hypotheses

The experiment was designed to study the three research questions described in the introduction. We observed and analysed participants' brain activity during the source credibility evaluation task. The experiment's design had limited the stimulus received by participants to source credibility, and the participants had to make a binary decision. Our pre-hypothesis was, therefore, that there existed a positive relationship between source credibility and average number of trusting decisions.

The next hypotheses concern participants' brain activity: specifically, the average amplitudes of ERP signals measured by cognitive processing electrodes in cognitive processing time interval (450–580 ms from stimuli) that will be referred as CPTI. For short, we shall refer to these ERP signals as cognitive ERP signals. We make the following hypotheses related to the first two research questions concerning brain activity for trusting and distrusting decisions on the basis of source credibility:

1. the ERP signals from all 26 cognitive electrodes in a certain time interval have statistically significant differences for different source credibility levels of 50%, 70% and 90% (in cases A50, A70, A90);
2. the ERP signals from all 26 cognitive electrodes in a certain time interval have statistically significant differences for pairs of cases: T50 and T90, D50 and D90, T70 and T90, D70 and D90;
3. the ERP signals from all 26 cognitive electrodes in a certain time interval have statistically significant differences for cases: T90 and D90, T70 and D70, T50 and D50;

When verifying hypotheses 1, 2, and 3, we will investigate time intervals within the cognitive decision making time interval and select the longest time interval during which a hypothesis holds. The comparison of these time intervals for the various hypotheses brings additional insight into the analysis.

The next hypothesis concerns the mean electric charge (MEC) flowing through all Brodmann Areas (BAs). We have used these measurements to consider the third research question: whether it is possible to model and predict source credibility evaluations using EEG measurements.

5. the mean electric charge flowing through estimated Brodmann Areas (BA) is sufficient to predict the decision to trust or distrust during the experiment, with an accuracy that significantly exceeds the baseline.

Note that hypothesis 5 concerns the possibility of creating an EEG-based method of source credibility measurement. While this is only a first step, a positive validation of hypothesis 5 would open an avenue of investigating EEG-based source credibility measurement in other, more complex and realistic scenarios. Note that the baseline accuracy for hypothesis 5 is 50% (experiment participants make binary decisions).

3.4 EEG Measurements

Our empirical experiments involved top EEG devices. We were equipped with a dense array amplifier recording the cortical activity with up to 500 Hz frequency

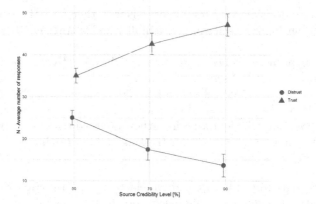

Fig. 2. Average number of trusting and distrusting responses given by participants when evaluating different source credibility levels 50%, 70% and 90%. The increase in trust and decrease in distrust can be observed with source credibility level increase.

through 256 channels HydroCel GSN 130 Geodesic Sensor Nets provided by EGI[2]. In addition, in the EEG Laboratory the Geodesic Photogrammetry System (GPS) was used.

Estimating ERP for each of the 256 electrodes is not necessary for ERP observation, as in general standards there are just a few electrodes (in our case 26) playing an important role in cognitive tasks.[3] However, for the sLORETA source localisation analyses (used for verification of the next hypotheses) the ERP for all 256 electrodes had to be in fact calculated on the fly. Therefore, in the beginning the raw EEG time series were post-processed, averaged and ERPs were estimated from 26 cognitive electrodes [13,17,22]. In the following discussion, when we refer to differences in the cognitive ERP signal, it means that in a certain time interval the average ERP signal from all 26 cognitive electrodes was different.

Having the ERP signal estimated for each electrode out of 256, it was possible to calculate the mean electric charge (MEC) flowing through the BA situated under these electrodes on the brain cortex in CPTR. Moreover, it was also possible to conduct the full source localisation analysis of the signal originating from all 256 electrodes using sLORETA algorithm (GeoSourse parameters set as follow: Dipole Set: 2 mm Atlas Man, Dense: 2447 dipoles Source Montages: BAs). Mean electric current flowing through each BA and varying in time was given as

---

[2] Electrical Geodesic Systems, Inc., 500 East 4th Ave. Suite 200, Eugene, OR 97401, USA.

[3] The electrodes are described in EGI 256-channel cap specification as best for cognitive ERP observations, covering the scalp regularly and numbered as follows: E98, E99, E100, E101, E108, E109, E110, E116, E117, E118, E119, E124, E125, E126, E127, E128, E129, E137, E138, E139, E140, E141, E149, E150, E151, E152. Those electrodes are automatically chosen for observing P-300 ERP signal by NetStation software.

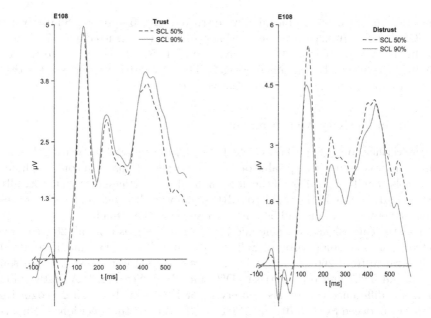

Fig. 3. The ERP plots averaged for all participants from -100 ms before stimuli to 600 ms after that for source credibility levels 50% and 90%. Results obtained for one electrode No. 108. The statistical difference in cortical activity is present both for the T50 & T90 (in the left) and distrust D50 & D90 (in the right) responses.

an output. Having those values calculated, it was possible to integrate that current in time and then get the MEC. The mean electric charge calculated for each electrode using source localisation techniques could, as we intended, indicate the hyperactivity of some BAs that are not necessary precisely situated under the cognitive electrodes. For all calculations of MEC the CPTR was divided into 10 ms time intervals. The procedure of calculating MEC has been described in detail in [29].

We shall denote the MEC by $\mu_b^{(t_1,t_2)}$, where b is the index of the Brodmann area, while (t_1,t_2) is the time interval. Note that $t_2 - t_1 \geq 10$ ms, but we can also calculate the MEC in longer time intervals. Note also that we calculate the MEC based on ERP signals calculated from a subset of participant decisions - usually from a subset of identical decisions (to trust or to distrust the source). Therefore, the variables $\mu_b^{(t_1,t_2)}$ can be used as independent variables related to a single participant's decision.

4   Experiment Results

As far as the pre-hypothesis is concerned, it was indeed possible to observe a relationship between subjects' responses and Source Credibility Level (SCL) during the experiment. When the SCL was set to 50%, on average 35 out of 60

336   A. Kawiak et al.

suggestions of correct word meaning were trusted and 25 were not. Similarly, for the SCL equal to 70%, on average 42.64 suggestions were trusted and 17.47 were not. In the case of SCL = 90%, the average of 44.52 suggestions were trusted and 10.98 were not. For details see Fig. 2. This observation demonstrates that the pilot experiment has a high internal validity.

4.1 Brain Activity Measurements

For the verification of hypotheses 1 the Pearson's chi-squared test with 2 degrees of freedom was used with a p value of 0.05. Hypothesis 1 has been confirmed. For all 3 levels of SCL (regardless of the decision made by participants), a statistically significant difference in the cognitive ERP signal was observed. These differences manifest between 340 ms and 540 ms after showing the stimulus.

A statistically significant difference in the ERP signals from all 26 cognitive electrodes for the comparison of SCL = 50% and SCL = 90% manifests itself in time range from 260–600 ms in both cases when subjects trust (comparison of T50 and T90) and do not trust (D50 and D90). See Fig. 3. A statistically significant difference can be also observed for ERP signals on all 26 cognitive electrodes between SCL = 70% and SCL = 90% for trusting decisions (comparison of T70 and T90) cover the entire CPTI. It is interesting that there is no statistically significant difference in case of distrusting decisions (comparison of D70 and D90). This means that the hypothesis 2 is partially confirmed.

As far as the hypothesis 3 is concerned, it was also possible to find statistically significant differences for certain time intervals when comparing trusting and distrusting ERPs for all three SCLs. The ERP signal collected from cognitive electrodes was significantly different between 360 ms and 600 ms when the subject was responding at SCL = 70% (comparison of T70 & D70). For SCL = 50%, statistically significant differences could be observed in 390–460 ms and 500–560 ms (comparison of T50 & D50), while for SCL = 90% the ERP was significantly different in the interval of 480–530 ms and 550–600 ms since the stimulus (comparison of T90 & D90).

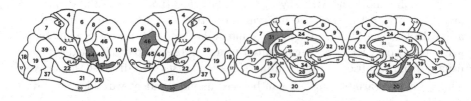

Fig. 4. Brodmann areas most significant for predicting source credibility evaluations

Table 1. Brodmann Areas that manifest statistically significant differences in the spectrum of MEC for trust and distrust decisions during source credibility evaluation. Presented with their corresponding anatomical structure of the brain and known functions as listed in [18]. The prefix of particular BAs stands for L-eft and R-ight hemispheres.

BA	Anatomical structure	Known functions
L-BA46	Dorsolateral Prefrontal Cortex	Memory encoding and recognition internal mental calculation processing emotions self-reflections in decision making
R-BA29	Granular retrosplenial cortex	Related to language memory retrieval
L-BA20	Inferior Temporal Gyrus	Lexico-semantic processing language comprehension and production
L-BA35	Perirhinal area	Memory encoding
L-BA43	Primary Gustatory Cortex	Language comprehension
R-BA47	Inferior Prefrontal Gyrus	Decision making working memory semantic encoding active semantic retrieval phonological expression single word reading
R-BA44	Pars Opercularis	Working memory expression of emotional information sentence comprehension word and face encoding Solving arithmetical tasks
R-BA31	Dorsal posterior cingulate	Emotion related to language attention to speech

4.2   Regression Model of Source Credibility Evaluations

We divided participant responses into two equal parts. The first part consisted of cognitive ERPs calculated on the basis of first 90 questions (30 for every level of SCL), and the second part consisted of cognitive ERPs calculated on the questions 91–180. Next, the MECs were calculated for both parts. The first part was used to create the training set, while the second part to create a validation set for the models of the participants' brain processes and decision making.

The explanatory (independent) variables of the model were MECs based on the ERPs of participants. Recall that we have denoted MEC by $\mu_b^{(t_1,t_2)}$, where b is the index of the Brodmann area, while (t_1, t_2) is the time interval. The model is based on MECs of all participants. These MECs are calculated in the time interval that had the highest differences of ERP for all Brodmann areas. The independent variables of the model are $x_b^p = \mu_b^{(150\,\text{ms},600\,\text{ms})}$. The training set of the model is 68 observations (the number of participants) of each class (136 observations in total), and the validations set has the same size.

The number of explanatory variables was reduced by considering a subset of variables that had the largest Wald statistic in the first model with all variables.

We created a universal Backward Stepwise Logistic Regression with Wald criterion classifier in SPSS. We choose generalized logistic regression in order to gain higher insight into the brain activities during source credibility evaluation (using other classifiers could increase classification accuracy).

The classifier achieved satisfactory characteristics in 14 steps with Nagelkerke's $R^2 = 0.38$. The Brodmann areas L-BA46, R-BA29, L-BA20 and L-BA35 (see first four rows of Table 1) had the highest impact on the classification with regression parameters β equal to 180.24, 146.17, 126.30, 90.87 for each mentioned above BA respectively. The most significant Brodmann areas selected by the classifier are also shown on anatomical maps of the brain in Fig. 4.

We have used 10-fold cross-validation to confirm our results and the average efficiency achieved by classifiers in this cross-validation are as follows: $Accuracy = 0.695$, $Precision = 0.667$, $Recall = 0.781$ and $F1score = 0.719$ which proves the stability of our results.

An F1 accuracy of 70% with a baseline of 50% is sufficiently high to consider that hypothesis 5 holds.

4.3 Discussion and Limitations

The results of our analysis concerning the first two research questions: "What activities occur in the brain when a receiver is evaluating source credibility?" and "How does brain activity depend on the level of source credibility?" revealed differences in brain activity on all cognitive electrodes during time intervals of at least 100 ms. The statistically significant difference in the EEG activity for different source credibility levels manifests both in the ERP curves' shapes generated for the cognitive electrodes, and in the mean electric charge flowing through particular Brodmann Areas. The mean electric charge approach was proposed in [29] and seems to work in the full spectrum of qEEG analysis [28,30].

Differences are observed in the interval of time in which the cognitive processing takes place. As the SCL had a strong influence on the decisions made by participants we can believe that it was methodologically correct. In most cases, the ERP signals significantly differ both for the trust and distrust decisions and for the particular SCLs.

When analysing the mean electric charge, the most significant differences in brain activity were found in the Dorsolateral Prefrontal Cortex, granular retrosplenial cortex, Inferior Temporal Gyrus and perirhinal area. These brain modules are responsible for a variety of functions (see Table 1), including working memory, decision making and lexical-semantic processing. The observations reported in this article are new results and point out new functions of these brain areas: the function of source credibility evaluation.

It is worth noting, however, that the word and face encoding areas, as well as the unit responsible for solving mathematical tasks (Pars Opercularis), were also engaged (although they had weaker impact on decisions). This can lead to the conclusion that during the source credibility tasks, some probabilistic calculations are conducted and the face of the information sender may also play an important role in the decision making process. However, these observations may also be a consequence of our experiment design.

Note, that BA47 and BA46 reported by us as manifesting statistically significant difference in activity and playing such an important role in our model were also reported in similar tasks involving decision-making about trustworthiness and judgements [2–5,9,10,20].

Our results concerning the other two research questions: "Can we model and predict human source credibility evaluations using EEG brain activity measurements?" and "Can we create a source credibility measurement method based on EEG brain activity measurements?" are promising. We showed that using EEG methods, we can model and predict human source credibility evaluations.

A model based on logistic regression was proposed for classifying the trusting or distrusting decisions of experiment participants. The accuracy of the models is high compared to the baseline. This shows that our model can be considered a first step towards source credibility measurement method based on EEG. Such a method would have wide applications, for example the measurement of source credibility of election candidates or journalists. However, these applications would have to be tested using follow-up experiments in the future.

5   Conclusion and Future Work

In future work, we intend to test other classifiers to check their accuracy in the discussed classification tasks. Some parameter search simulations will be performed as well, and we hope that it is possible to find the set of Brodmann Areas characteristic not only for the entire cohort, but also for individual brains.

On the other hand, increasing the number of participants should let us build a universal model classifying brain cortical dynamics with a high level of accuracy. It is justified to hypothesise that while individual brains are different, when we evaluate more participants, there is a higher chance of finding similar brains.

Our investigation studied source credibility, which is a single aspect of credibility evaluation. Our research will continue with the investigation of message credibility, as well as a holistic credibility evaluation. We hope that our research will one day lead to proposing a comprehensive, quantitative model of the credibility phenomenon, based on electrical brain activity.

References

1. Allcott, H., Gentzkow, M.: Social media and fake news in the 2016 election. J. Econ. Perspect. **31**(2), 211–236 (2017)
2. Chung, H.K., Tymula, A., Glimcher, P.: The reduction of ventrolateral prefrontal cortex gray matter volume correlates with loss of economic rationality in aging. J. Neurosci. **37**(49), 12068–12077 (2017)
3. Clark, L., Bechara, A., Damasio, H., Aitken, M., Sahakian, B., Robbins, T.: Differential effects of insular and ventromedial prefrontal cortex lesions on risky decision-making. Brain **131**(5), 1311–1322 (2008)
4. Deppe, M., et al.: Nonlinear responses within the medial prefrontal cortex reveal when specific implicit information influences economic decision making. J. Neuroimaging **15**(2), 171–182 (2015)
5. Ernst, M., et al.: Decision-making in a risk-taking task: a pet study. Neuropsychopharmacology **26**(5), 682–691 (2002)
6. Flanagin, A.J., Metzger, M.J.: Digital media and perceptions of sourcecredibility in political communication. Oxf. Handb. Polit. Commun. **417**, 1 (2017)

7. Forgas, J.P., Baumeister, R.: The Social Psychology of Gullibility: Conspiracy Theories, Fake News and Irrational Beliefs. Routledge, Abingdon (2019)
8. Hovland, C.I., Weiss, W.: The influence of source credibility on communication effectiveness. Public Opin. Q. **15**(4), 635–650 (1951)
9. Ishii, A., Tanaka, M., Watanabe, Y.: The neural mechanisms underlying the decision to rest in the presence of fatigue: a magnetoencephalography study. PLoS ONE **9**(10), e109740 (2014)
10. Jones, C.L., Ward, J., Critchley, H.D.: The neuropsychological impact of insular cortex lesions. J. Neurol. Neurosurg. Psychiatry **81**(6), 611–618 (2010)
11. Kakol, M., Jankowski-Lorek, M., Abramczuk, K., Wierzbicki, A., Catasta, M.: On the subjectivity and bias of web content credibility evaluations. In: Proceedings of the 22nd International Conference on World Wide Web, pp. 1131–1136. ACM (2013)
12. Kareklas, I., Muehling, D.D., Weber, T.: Reexamining health messages in the digital age: a fresh look at source credibility effects. J. Advert. **44**(2), 88–104 (2015)
13. Kawala-Sterniuk, A., et al.: Comparison of smoothing filters in analysis of EEG data for the medical diagnostics purposes. Sensors **20**(3), 807 (2020)
14. Lazer, D.M., et al.: The science of fake news. Science **359**(6380), 1094–1096 (2018)
15. Liu, X., Nielek, R., Adamska, P., Wierzbicki, A., Aberer, K.: Towards a highly effective and robust web credibility evaluation system. Decis. Support. Syst. **79**, 99–108 (2015)
16. Meijer, E.H., Verschuere, B.: Deception detection based on neuroimaging: better than the polygraph? J. Forensic Radiol. Imaging **8**, 17–21 (2017)
17. Mikołajewska, E., Mikołajewski, D.: Ethical considerations in the use of brain-computer interfaces. Open Med. **8**(6), 720–724 (2013)
18. Pascual-Leone, A., Davey, N.J., Rothwell, J., Wasserman, E.M., Puri, B.K.: Handbook of Transcranial Magnetic Stimulation, vol. 15. Arnold, London (2002)
19. Rafalak, M., Abramczuk, K., Wierzbicki, A.: Incredible: is (almost) all web content trustworthy? Analysis of psychological factors related to website credibility evaluation. In: Proceedings of the 23rd International Conference on World Wide Web, pp. 1117–1122. ACM (2014)
20. Rosenbloom, M.H., Schmahmann, J.D., Price, B.H.: The functional neuroanatomy of decision-making. J. Neuropsychiatry Clin. Neurosci. **24**(3), 266–277 (2012)
21. Rutjens, B.D., Brandt, B.D.: Belief Systems and the Perception of Reality. Routledge, Amsterdam (2018)
22. Tadeusiewicz, R., et al.: Neurocybernetyka teoretyczna. Wydawnictwa Uniwersytetu Warszawskiego, Warsaw (2009)
23. Turek, P., Wierzbicki, A., Nielek, R., Datta, A.: Wikiteams: how do they achieve success? IEEE Potentials **30**(5), 15–20 (2011)
24. Wang, H., Chang, W., Zhang, C.: Functional brain network and multichannel analysis for the P300-based brain computer interface system of lying detection. Expert. Syst. Appl. **53**, 117–128 (2016)
25. Wawer, A., Nielek, R., Wierzbicki, A.: Predicting webpage credibility using linguistic features. In: Proceedings of the 23rd International Conference on World Wide Web, pp. 1135–1140. ACM (2014)
26. Wierzbicki, A.: Web Content Credibility. Springer, Cham (2018). https://doi.org/10.1007/978-3-319-77794-8
27. Wierzbicki, A., Turek, P., Nielek, R.: Learning about team collaboration from Wikipedia edit history. In: Proceedings of the 6th International Symposium on Wikis and Open Collaboration, pp. 1–2 (2010)

28. Wojcik, G.M., et al.: Mapping the human brain in frequency band analysis ofbrain cortex electroencephalographic activity for selected psychiatricdisorders. Front. Neuroinformatics **12**, 73 (2018)
29. Wojcik, G.M., et al.: New protocol for quantitative analysis of braincortex electroencephalographic activity in patients with psychiatricdisorders. Front. Neuroinformatics **12**, 27 (2018)
30. Wojcik, G.M., et al.: Analysis of decision-making process using methods of quantitative electroencephalography and machine learning tools. Front. Neuroinformatics **13**, 73 (2019)

An Adaptive Network Model for Burnout and Dreaming

Mathijs Maijer[1], Esra Solak[1], and Jan Treur[2]($\boxtimes$)

[1] Computational Science, University of Amsterdam, Amsterdam, The Netherlands
m.f.maijer@gmail.com, s.3.solak@gmail.com
[2] Social AI Group, Vrije Universiteit Amsterdam, Amsterdam, The Netherlands
j.treur@vu.nl

Abstract. As burnouts grow increasingly common, the necessity for a model describing burnout dynamics becomes increasingly apparent. The model discussed in this paper builds on previous research by adding dreams, a component that has been shown to have an adaptive regulating effect on emotions. The proposed model is a first-order adaptive temporal-causal network model, incorporating emotions, exercise, sleep, and dreams. The model was validated against given patterns found in empirical literature and it may be used to gain a better understanding of burnout dynamics.

1 Introduction

A burnout is seen as a buildup of long-term unresolved work-related stress. Recently, the World Health Organization (WHO) has classified an occupational burnout as a syndrome [1]. There may be multiple causes for the development of this syndrome, but according to the WHO, it can arise from a failure to manage chronic work-related stress. This can cause feelings of energy depletion, exhaustion, job detachment, negative feelings and cynicism towards the job, and reduced professional efficacy as put forward in [1]. The number of people complaining about burnout related symptoms has increased over time and it is rising fast. In 2015, 13% of the Dutch employees mentioned burnout symptoms, this figure has risen to 16% by 2017 as shown in [2, 3]. Around 20% of employees aged between 25 and 35 have mentioned burnout symptoms in the Netherlands.

Due to the increase in burnout complaints and the severity of the issue, it has become more important to not only analyse the causes of burnouts, but also how to prevent, avoid, and if possible, cure them. A few possibilities to fight against burnouts may be lifestyle or habit changes. This paper uses previous research on modeling burnout phenomena [4, 5], to build forth on the already created temporal-causal network models that describe the burnout syndrome as a dynamic interplay between symptoms in line with [6]. In [4], the initial model was created and the effects of physical exercise were analysed, in [5] sleep factors were added to improve the model. This paper adds relevant dream components as described in [7] and also makes the model adaptive, resulting in a first-order adaptive temporal-causal network model. The main adaptive network-oriented modelling approach was adopted from [8].

© Springer Nature Switzerland AG 2020
V. V. Krzhizhanovskaya et al. (Eds.): ICCS 2020, LNCS 12137, pp. 342–356, 2020.
https://doi.org/10.1007/978-3-030-50371-0_25

The aim is to gain a better understanding of the development of burnouts and to create a more realistic model that can be used in real-life scenarios. In Sect. 2, a brief overview of the relevant background knowledge will be provided; in Sect. 3, the network-oriented modelling approach will be explained in some detail. In Sect. 4, the designed adaptive network model will be described. Multiple simulations will be given in Sect. 5. Section 6 and Sect. 7 will respectively address empirical and mathematical validation of the model. Finally, a discussion concludes the paper.

2   Theoretical Background

This paper assumes the classic definition of the burnout by Maslach and Jackson [9]. It describes the symptoms of emotional exhaustion, a decline in experiencing personal accomplishment, and a sense of depersonalization. These symptoms are also described in the WHO classification of a burnout provided in [1]. The main idea behind the definition of Maslach and Jackson, was to invent a measurement instrument for burnouts [9]. Various components are mentioned that interact together to form a burnout. The most important components are described as risk factors and suppressive factors, also called protective factors. Among the risk factors are subjective stress, job ambiguity, work pressure, thoughts of work during leisure time, the amount of sleep, and the quality of sleep. Protective factors are, for example, confidence, or the amount of physical exercise, as modeled in [4] and sleep modeled in [5]. Factors that influence the progression or development of burnout symptoms are influenced by personal characteristics as described in [10]. For example, neuroticism has shown a strong correlation with experiencing stress. Next to this, openness has been shown to be negatively correlated with depersonalization and emotional exhaustion as discussed in [11]. Because of the link between openness and physical exercise addressed, for example, in [12], previous recomputational modeling search has analyzed the effects of physical exercise on the dynamics of a burnout [4]. In [5], the relation between sleep and burnout dynamics is analysed. This is partially done by analysing the results of a questionnaire which showed that insufficient sleep can be used as a predictive factor for clinical burnout as put forward in [13]. A noteworthy finding is that the amount of sleep is a better predictor for a clinical burnout than the amount of stress someone experiences at work.

Next to sleep, dreams have been found to have a regulating effect on some emotions as described by [7, 15–17] which is considered a form of internal simulation. The internal simulation consists of activation of memory elements, which are sensory representations in relation to emotions. Dream episodes occur after competing, which will activate different sensory representations (e.g., images) during dreams. The level of how much the feeling states and sensory representations are activated is controlled by different control states. In [7], Ch 5, an adaptive temporal-causal network model is introduced to model dream dynamics that shows a form of adaptiveness called fear extinction as described in [17]. Here emotion regulation is included of which the connections become stronger as they are used more according to the principle of Hebbian learning as discussed in [18]. Figure 1 shows a conceptual representation of part of the adaptive temporal-causal network model that is presented in [7], Ch. 5, Fig. 5.1. Five states are shown, a sensory representation state, an emotion regulation control state, a dream episode state, a feeling state and a preparation for a bodily response state. The red arrows indicate inhibition, or a negative impact, and the black and green arrows indicate a positive impact, where the green arrows are adaptive.

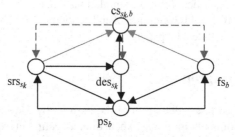

Fig. 1. Conceptual representation of part of an adaptive temporal-causal network model for dream dynamics. Here, (1) the arrows between srs_{s_k} and ps_b indicate the bidirectional associations between sensory representations srs_{s_k} and emotional fear responses ps_b, (2) the arrows between ps_b and fs_b indicate an as-if body loop generating fear feelings fs_b, (3) the arrow from sensory representation srs_{s_k} to dream episode des_{s_k} indicates the triggering of a (pseudo-conscious) dream episode, (4) the arrow from des_{s_k} to ps_b indicates the (amplifying) impact of the dream episode on the emotional fear response, (5) the upward arrows from des_{s_k} to emotion regulation control state $cs_{s_k,b}$ indicate monitoring of the (emotional) state of the person, and (6) the red dotted downward arrows indicate suppression of the target states as a form of emotion regulation. (Color figure online)

3   Network-Oriented Modelling

The temporal-causal network model presented here was designed using a network-oriented modelling approach that is described in [7] and [8]. Network-Oriented Modelling uses nodes and edges, which are the connections between nodes. Nodes are states with values that vary over time, while the connections can be seen as the causal relationships between these nodes. For an adaptive network model, besides the states also the causal relationships can change over time. Table 1 summarises the main concepts of network oriented modelling. The connections indicate the impact that states have on each other. Every connection has a connection weight, which is a numerical value indicating the connection strength. The connections and their weights define the network's *connectivity characteristics*.

Next to this, every state has a combination functions that describes the manner in which the incoming impacts per connection are combined to form an aggregated impact. This defines the network's *aggregation characteristics*. A combination function can be a basic combination function from the available Combination Function Library or a weighted average of a number of such basic combination functions. Which combination function is used depends on the application and can also be node-specific. To define the network's *timing characteristics*, every state has a speed factor that determines how fast a state changes because of its received causal impact. The numerical representation derived from the network characteristics is summarised in Table 2.

Table 1. An overview of the concepts in the conceptual component of temporal-causal networks.

Concepts	Notation	Explanation
States and connections	X, Y $X \to Y$	Denotes the nodes and edges in the conceptual representation of a network
Connection weights	$\omega_{X,Y}$	A connection between states X and Y has a corresponding *connection weight*. In most cases $\omega_{X,Y} \in [-1, 1]$
Aggregating multiple impacts on a state	$\mathbf{c}_Y(..)$	Each state has a *combination function* and is responsible for combining causal impacts of all states connected to Y on that same state
Timing of the effect of causal impact	η_Y	The *speed factor* determines how fast a state is changed by any causal impact. In most cases: $\eta_Y \in [0, 1]$

Table 2. Numerical representations of temporal-causal networks.

Concepts	Notation	Explanation
State value at time t	$Y(t)$	For every time t a state Y has a value in $[0,1]$
Single causal impact	$\mathbf{impact}_{X,Y}(t)$ $= \omega_{X,Y}X(t)$	At any time t a state X (if connected to Y) impacts Y through a connection weight $\omega_{X,Y}$
Aggregating multiple impacts on a state	$\mathbf{aggimpact}_Y(t)$ $= \mathbf{c}_Y(\mathbf{impact}_{X_1,Y}(t),...,$ $\mathbf{impact}_{X_k,Y}(t))$ $= \mathbf{c}_Y(\omega_{X_1,Y}X_1(t), ..., \omega_{X_k,Y}X_k(t))$	The combination function $\mathbf{c}_Y$ determines the aggregated causal impact of states X_i on Y
Timing of the effect of causal impact	$Y(t + \Delta t) = Y(t) +$ $\eta_Y [\mathbf{aggimpact}_Y(t) - Y(t)] \Delta t$ $= Y(t) +$ $\eta_Y [\mathbf{c}_Y(\omega_{X_1,Y}X_1(t), ..., \omega_{X_k,Y}X_k(t)) - Y(t)] \Delta t$	The speed factor Y determines how fast a state Y is changed by the aggregated causal impact of states X_i

The last row of this Table 2, shows the difference equation. Adaptive networks are networks for which some of the characteristics $\omega_{X,Y}$, $\mathbf{c}_Y(..)$, η_Y change over time. To model this, extra states are added that represent the adaptive characteristics. For example, for an adaptive connection weight $\omega_{X,Y}$ a new state $\mathbf{W}_{X,Y}$ is added (called a *reification state* or *adaptation state* for $\omega_{X,Y}$) representing the dynamic value of $\omega_{X,Y}$.

Table 3 shows an overview of the combination functions used in the designed model. The first is the *identity* function **id(.)**, which is commonly used when a state only has one incoming connection.

The *advanced logistic sum* function **alogistic$_{\sigma,\tau}$(..)** is used to aggregate impact for each state that has multiple incoming connections; it has as parameters steepness σ and threshold τ. The combination function **hebb$_\mu$(..)** is used for adaptation states $\mathbf{W}_{X,Y}$, representing the adaptive value of a connection weight. It has one parameter μ, which is the persistence of the state. In all formula of Table 3 the variables $V_1,\ldots,V_k$ are used for incoming single impacts, and W for the value of the connection weight reification state.

Table 3. Overview of the combination functions used.

Combination function	Description	Formula $\mathbf{c}_Y(V_1,\ldots,V_k) =$
id(.)	Identity	V
alogistic$_{\sigma,\tau}$(..)	Advanced logistic sum	$\left[\dfrac{1}{1+e^{-\sigma(V_1+\ldots+V_k-\tau)}} - \dfrac{1}{1+e^{\sigma\tau}}\right](1+e^{-\sigma\tau})$
hebb$_\mu$(..)	Hebbian learning	$V_1 V_2(1-W) + \mu W$

4   Modeling Adaptive Burnout Dynamics with Dreams

This section describes the details of the designed adaptive network model. The states shown in the model are mainly based on the literature mentioned in Sect. 2, specifically [4] and [5]. Table 4 shows the different states that are used in the model, as well as their respective types. There are 5 different state types:

- **Protective:** Protective states are states that protect a person against a clinical burnout; if they have high values, then the chance of developing a burnout is lower.
- **Risk:** Risk states are states that increase the chance of developing a burnout.
- **Burnout Element:** Burnout elements are affected by the protective factors and the risk factors, they are the states that will grow in value when a burnout is developing. Looking at these states is the best way to identify the level of burnout progression.
- **Consequent:** Consequent states are states that are affected by the burnout elements. By introducing a feedback loop from protective or risk states to burnout elements and then from burnout elements to the protective or risk factors, realistic positive or negative feedback becomes possible. Thus, some states may have a consequent type, as well as another type.
- **Dream:** The dream type states are newly introduced in this model compared to previous literature on burnout modeling. These dream states regulate the emotions in an adaptive manner as new dream episodes occur, as described in Sect. 2.

Table 4. The states used in the model and their respective types.

State	Abbr.	Description	Type
X_1	CO	Confidence	Protective
X_2	OP	Openness	Protective
X_3	PE	Physical exercise	Protective
X_4	PA	Personal accomplishment	Protective / Consequent
X_5	NR	Night rest	Protective / Consequent
X_6	CW	Charged work	Risk
X_7	JA	Job ambiguity	Risk
X_8	JS	Job satisfaction	Protective / Consequent
X_9	NE	Neuroticism	Risk / Consequent
X_{10}	SC	Social contact	Protective
X_{11}	EE	Emotional exhaustion	Burnout element
X_{12}	CY	Cynicism	Burnout element
X_{13}	JP	Job performance	Consequent
X_{14}	JD	Job detachment	Consequent
X_{15}	DU	Drugs	Consequent
X_{16}	ST	Stress	Combination
X_{17}	fs_b	Feeling state for b	Dream
X_{18}	srs_{ST}	Sensory representation state for ST	Dream
X_{19}	$cs_{ST,b}$	Control state for regulation of sensory representation of ST and feeling b	Dream
X_{20}	des_{ST}	Dream episode state for ST	Dream
X_{21}	$\mathbf{W}_{srs,cs}$	Reification state for connection weight $\omega_{srs_{ST},cs_{ST,b}}$	Dream
X_{22}	$\mathbf{W}_{fs,cs}$	Reification state for connection weight $\omega_{fs_b,cs_{ST,b}}$	Dream

There are two special states shown in Table 4, namely state X_{21} and X_{22}. These are the reification states that introduce the adaptivity in the dream component of the model, as described in [7]. These states use a Hebbian combination function, whereas the other states all use a logistic function. Furthermore, state X_{16} has a type 'combination', which was not mentioned above. According to [1], the main symptom of a clinical burnout is the high stress level. Stress is an abstract concept that can be approached using multiple emotions mentioned in Sect. 2. The stress state was added to the model as a kind of aggregate state that represents an overall combination of some other important states in the model. The stress state is not directly regulated; instead states X_{17} and X_{18} corresponding to the feeling state of stress and the sensory representation of stress are regulated by the Control state X_{19}. A conceptual representation of the introduced adaptive network model is shown in Fig. 2. This Fig. 2 shows two planes, of which the second plane (blue) represents the adaptation states in the network: state X_{21} and X_{22} represent the values for the weights of the connections from state X_{17} to state X_{19} and from state X_{18} to state X_{19}, which allows the connection weight values to change over time. In contrast to the simplified representation shown in Fig. 2, the actual model contains many more causal relations between all states, as most emotions slightly affect each other, as shown in literature and mentioned in Sect. 2. The network characteristics for *connectivity* (the connections and their weights), *aggregation* (the combination functions and their parameters), and *timing* (the speed factors) have been specified in the form of *role matrices,* which provides a compact specification format for (adaptive) temporal-causal network models. For two of them, **mb** (*base matrix*) and **mcw** (*matrix for connection weights*) specifying the *connectivity characteristics,* see Box 1.

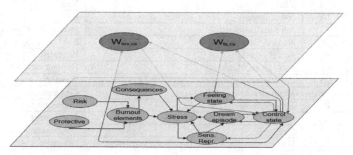

Fig. 2. A simplified conceptual representation of the designed first-order adaptive temporal-causal network. The upper plane indicates the adaptiveness of the model. (Color figure online)

Most of the values for these network characteristics were selected based on empirical data as well as previous works like [4] and [5]. Role matrices have rows for all the states and at each row indicate the elements that for the specific role have impact on that state. For example, in Box 1 in **mb** it is indicated which other states have basic impact, and in **mcw** it is indicated what is the connection weight impact for that state. Note that the cells with nonadaptive values are green and the cells with adaptive values are red. In the latter cells, not a value but the name of the reification state is specified which represents the adaptive value. This can be seen in the row for the control state X_{19}, where X_{21} and X_{22} are indicated as the states representing the adaptive values.

mb / State Abbr		1	2	3	4	5	6	7	8	9	10	11	12
X_1	CO	PE	PA	JA	JS	NE	JP	JD					
X_2	OP	CO	PE	JS	NE	SC	JD						
X_3	PE	OP											
X_4	PA	CO	PA	CW	JA	JS	NE	EE	JP	JD			
X_5	NR	PE	PA	CW	JS	NE	EE	DU					
X_6	CW	CO	PE	PA	NR	CW	JA	JS	NE	SC	EE	JD	DU
X_7	JA	CO	PA	NR	JA	JS	NE	SC	EE	JD	DU		
X_8	JS	PA	CW	JA	NE	EE	CY	JP	JD				
X_9	NE	CO	OP	PE	PA	NR	CW	JA	JS	SC	EE	JP	JD
X_{10}	SC	CO	OP	PE	NR	NE							
X_{11}	EE	CO	OP	NR	CW	JA	JS	NE	JP	JD	DU		
X_{12}	CY	CO	OP	NE	EE	JD							
X_{13}	JP	CO	PA	NR	CW	JA	JS	NE	SC	EE	JD	DU	
X_{14}	JD	CO	PA	CW	JA	JS	NE	EE	CY	JP	DU		
X_{15}	DU	OP	PE	CW	NE	SC	EE	DU					
X_{16}	ST	EE	CY	JP	JD								
X_{17}	fs_b	ST	$cs_{st,b}$										
X_{18}	srs_{ST}	ST	$cs_{st,b}$										
X_{19}	$cs_{st,b}$	fs_b	srs_{st}	des_{st}									
X_{20}	des_{ST}	srs_{st}	$cs_{st,b}$										
X_{21}	$\mathbf{W}_{srs,cs}$	srs_{st}	$cs_{st,b}$	$\mathbf{W}_{srs,cs}$									
X_{22}	$\mathbf{W}_{fs,cs}$	fs_b	$cs_{st,b}$	$\mathbf{W}_{fs,cs}$									

mcw / State Abbr		1	2	3	4	5	6	7	8	9	10	11	12
X_1	CO	0.5	1	-1	1	-1	1	-0.5					
X_2	OP	1	0.5	1	-1	1	-0.5						
X_3	PE	1											
X_4	PA	1	1	-0.5	-1	1	-1	-0.5	1	-1			
X_5	NR	0.5	0.25	-1	1	-0.5	-0.5	1					
X_6	CW	-1	-0.5	-0.5	-1	0.5	1	-1	1	-0.5	1	-1	1
X_7	JA	-1	-1	-1	1	-1	1	-0.5	1	-1	-1		
X_8	JS	1	-1	-1	-1	-1	-0.5	1	-1				
X_9	NE	-0.25	-0.25	-0.25	-0.5	-0.5	1	1	-0.5	-0.5	1	-0.5	1
X_{10}	SC	1	1	0.5	0.5	-1							
X_{11}	EE	-0.5	-1	-1	1	1	-1	1	-0.25	-1	-0.5		
X_{12}	CY	0.25	-1	0.5	0.25	1							
X_{13}	JP	1	1	1	-0.5	-0.75	1	-1	0.25	-0.5	-1	-1	
X_{14}	JD	-0.5	-1	-1	1	-1	1	1	1	-0.5	0.25		
X_{15}	DU	0.25	-0.5	0.25	1	0.25	0.25	0.5					
X_{16}	ST	1	1	-1	1								
X_{17}	fs_b	1	-1										
X_{18}	srs_{ST}	0.5	-1										
X_{19}	$cs_{ST,b}$	X_{22}	X_{21}	0.3									
X_{20}	des_{ST}	1	-1										
X_{21}	$\mathbf{W}_{srs,cs}$	1	1	1									
X_{22}	$\mathbf{W}_{fs,cs}$	1	1	1									

Box 1 Role matrices for the *connectivity characteristics*: role matrix **mb** (base connectivity) and role matrix **mcw** (connection weights).

Similarly, in Box 2 a role matrix **ms** (*matrix for speed factors*) specifies what speed factor value has impact on the state, role matrix **mcfw** (*matrix for combination function weights*) specifies what combination function weights have impact and **mcfp** (*matrix for combination function parameters*) what parameter values of the combination function. As can be seen in these role matrices, no further adaptive characteristics were considered.

	mcfw			mcfp		hebb		alogistic		ms		
		hebb	alogistic			1	2	1	2			1
State	Abbr			State	Abbr	μ		σ	τ	State	Abbr	
X_1	CO		1	X_1	CO			50	0.5	X_1	CO	0.1
X_2	OP		1	X_2	OP			50	0.5	X_2	OP	0.1
X_3	PE		1	X_3	PE			50	0.5	X_3	PE	0.1
X_4	PA		1	X_4	PA			50	0.5	X_4	PA	0.1
X_5	NR		1	X_5	NR			50	0.5	X_5	NR	0.1
X_6	CW		1	X_6	CW			50	0.5	X_6	CW	0.1
X_7	JA		1	X_7	JA			50	0.5	X_7	JA	0.1
X_8	JS		1	X_8	JS			50	0.5	X_8	JS	0.1
X_9	NE		1	X_9	NE			50	0.5	X_9	NE	0.1
X_{10}	SC		1	X_{10}	SC			50	0.5	X_{10}	SC	0.1
X_{11}	EE		1	X_{11}	EE			50	0.5	X_{11}	EE	0.1
X_{12}	CY		1	X_{12}	CY			50	0.5	X_{12}	CY	0.1
X_{13}	JP		1	X_{13}	JP			50	0.5	X_{13}	JP	0.1
X_{14}	JD		1	X_{14}	JD			50	0.5	X_{14}	JD	0.1
X_{15}	DU		1	X_{15}	DU			50	0.5	X_{15}	DU	0.1
X_{16}	ST		1	X_{16}	ST			50	0.5	X_{16}	ST	0.1
X_{17}	fs_b		1	X_{17}	fs_b			50	0.5	X_{17}	fs_b	0.1
X_{18}	srs_{ST}		1	X_{18}	srs_{ST}			50	0.1	X_{18}	srs_{ST}	0.1
X_{19}	$cs_{ST,b}$		1	X_{19}	$cs_{ST,b}$			50	0.5	X_{19}	$cs_{ST,b}$	0.1
X_{20}	des_{ST}		1	X_{20}	des_{ST}			60	0.25	X_{20}	des_{ST}	0.1
X_{21}	$W_{srs,cs}$	1		X_{21}	$W_{srs,cs}$	0.99				X_{21}	$W_{srs,cs}$	1
X_{22}	$W_{fs,cs}$	1		X_{22}	$W_{fs,cs}$	0.99				X_{22}	$W_{fs,cs}$	1

Box 2 Role matrices for the *aggregation characteristics*: role matrix **mcfw** (combination function weights) and role matrix **mcfp** (combination function parameter values); and role matrix for the *timing characteristics*: role matrix **ms** (speed factors).

5   Simulation Results

This section shows the results obtained by running the model described in Sect. 4. The model was simulated using the modeling environment described in [14] and [8], Ch 9, using different initial values to create different scenarios. This modeling environment uses the above role matrices and initial values (and the step size Δt and end time of the simulation) as input and then runs the simulations. These scenarios were tested to gain a better understanding of the model and of clinical burnout progression. The main scenarios tested are non-burnout versus burnout scenarios, which are obtained by proper setting of the initial values of the relevant protective and risk states. Due to the many causal relationships, initial values can severely impact the outcome of the simulation. The scenarios were obtained by using the initial values for the states shown in Table 5 and $\Delta t = 0.01$. Figure 3 shows the progression of the states for a non-burnout scenario (left) and a burnout scenario (right). For the non-burnout scenario the initial values of the states that form high risks are low and the protective factors are high, as can be seen in Table 5. The lines in the figure are colored based on their type: protective factors in green, risk factors yellow, burnout elements red, consequences blue, dream factors magenta, and finally, the stress state in black. The difference between states that share the same type is shown using different line styles.

The pattern can be explained by looking at where the states are converging to. The protective factors are converging to one (all had high initial values), while the risk factors converge to zero and (they had low initial values).

Table 5. The initial values used for the non-burnout and burnout scenarios.

State	X_1	X_2	X_3	X_4	X_5	X_6	X_7	X_8	X_9	X_{10}
Non-burnout	0.85	0.80	0.70	0.75	0.72	0.21	0.15	0.75	0.10	0.82
Burnout	0.25	0.22	0.23	0.20	0.35	0.90	0.95	0.30	0.98	0.17
State	X_{12}	X_{13}	X_{14}	X_{15}	X_{16}	X_{17}	X_{18}	X_{19}	X_{20}	X_{21}
Non-burnout	0.20	0.78	0.15	0.22	0.00	0.00	0.00	0.00	0.00	0.00
Burnout	0.15	0.90	0.05	0.075	0.00	0.00	0.00	0.00	0.00	0.50

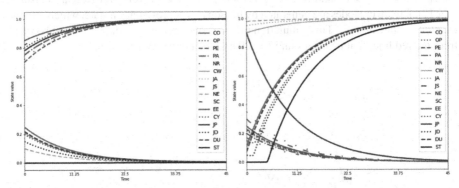

Fig. 3. Non-burnout scenario (left) and a burnout scenario (right) simulated, all states except the dream states are shown. States are colored by their type and can be distinguished by looking at their line types. (Color figure online)

When the initial values are changed to the values noted in the burnout scenario rows in Table 5, the plot shown in the right hand side of Fig. 3 is acquired. This graph is a bit more complex compared to the left graph, as more dynamics are shown. The figure shows how the protective states, which have low initial values, converge to zero this time. In contrast, the risk states, with high initial values, converge to one. This has multiple repercussions for the model, as can be seen from the burnout element states (red), EE and CY. As the burnout element states start to increase in value, the consequent states (blue) are affected. For example, the job performance state, which had an initial value of 0.90, starts to converge to zero, even though it had a high initial value. Next to this, the consequent states job detachment and drugs, start to converge to one, which starts to indicate the condition of the simulated person. Finally, the black line shows the stress factor, which is a combination of the most important stress-related states. At first, the stress level is not changing, until a tipping point is reached where it starts to converge to one, as the consequent states are starting to affect the stress level too much compared to the protective factors.

Figure 4 shows the progression of the dream states during the burnout scenario simulation. The dream states are not shown in Fig. 3 to prevent the figure from becoming unreadable. The brown and purple lines in Fig. 4 indicate the reification states that show the progression of the adaptive weights from the sensory representation state of stress to the control state and the feeling state to the control state. The reification states are affected by the control state, their own state, and respectively the sensory representation state and feeling state. This can also be seen in Fig. 4, as the reification state $\mathbf{W}_{srs,cs}$ starts to increase when the control state as well as the sensory representation are increasing and then starts to slowly decrease when the sensory representation state converges to zero. The reification states increase due to the fear extinction learning cycle [7], which means that when the connections are used more, they are strengthened over time. Furthermore, the sensory representation state starts to increase rapidly, but when the control state starts to increase, the sensory representation state starts to decrease, until it converges all the way to zero. This is due to the negative emotion regulation cycle, where the control state affects the feeling state and sensory representation state as well as the dream episode state [7]. Dream episodes are generated by the sensory representation state, which affects the control state as well. This simulation includes one dream episode, which can be seen from the red line, that peaks around $t = 20$ and then converges to zero.

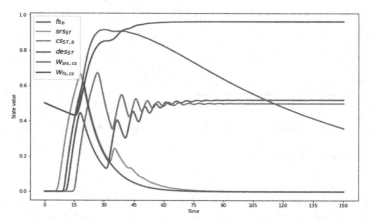

Fig. 4. A display of the change in the dream states values during the burnout scenario simulation. (Color figure online)

The interesting part of the figure, is where in contrast to the sensory representation of stress, the feeling state starts to increase and then decrease, which is in line with literature and the expected behavior, but then starts to increase again due to the resonance with the control state. At that point, both the control state and the feeling state of stress are resonating with one another and are creating a cycle where they are both enforcing the behavior of each other. This continues until the feeling state fs_b reaches a point where its value is higher than the value of the control state, after this they converge to their respective values.

6   Empirical Validation

Although no numerical empirical data is available that outlines the exact influence that emotions have on one another in a quantified manner, certain patterns can still be found in the literature, which can be used to validate the model in an empirical manner. This section will describe how the network model's characteristics were tuned in accordance with the patterns found in literature. In [4] and [5], the models were tuned in accordance with respectively physical exercise and sleep components, which allowed for a more realistic selection of parameter values.

This paper tunes the most important dream states, the feeling state and the control state, in accordance with the patterns found in [7, 15–17]. This was done by first creating data points in a manner that corresponds with the noted literature. The data points used can be found in Table 6. The pattern that would be acquired by tuning the parameters to be in accordance with the data points, would be more in line with the emotion regulation cycle, instead of the resonance pattern that was shown in Fig. 4. The network characteristics that were selected to be tuned as parameters were the connection weights for the incoming connections to the feeling state X_{17} and the control state X_{19}, making 5 parameters in total.

To tune the parameters, a simulated annealing algorithm was used with the default settings of Matlab's Optimization Toolkit. A final Root Mean Squared Error (RMSE) of $9.87 * 10^{-2}$ was acquired using 10^4 iterations. Table 7 shows the optimal values that were found to achieve the RMSE in accordance with Table 6. After simulating the model, using the initial values for the burnout scenario shown in Table 5, it can be seen that the results are more in accordance with literature, as shown in Fig. 5.

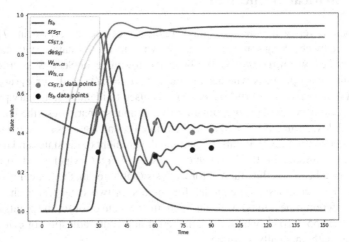

Fig. 5. Burnout scenario simulation using the optimal parameters found in Table 7 with the initial values shown in Table 5 compared to the empirical data as dots.

Extinction learning and the reduction in feeling level can now be properly shown in accordance with [5]. The resonance is still present, but in contrast to Fig. 4, the feeling state does not surpass the control state, as they both converge before intersecting a second time. The Appendix (see https://www.researchgate.net/publication/340162256) shows the development of the RMSE over iterations during the tuning process.

Table 6. Data points created for the feeling state and control state in accordance with patterns found in literature.

Time	fs_b	$cs_{ST,b}$
30	0.50	0.30
60	0.45	0.28
80	0.40	0.31
99	0.38	0.30

Table 7. Optimal parameter values found for network characteristics ω_{ST,fs_b}, $\omega_{cs_{ST,b},fs_b}$ and $\omega_{des_{ST},cs_{ST,b}}$ (indicated as parameters P1, P2, and P3, resp.) using simulated annealing to minimize the RMSE in accordance with the points in Table 6.

ω_{ST,fs_b}	$\omega_{cs_{ST,b},fs_b}$	$\omega_{des_{ST},cs_{ST,b}}$
P1	P2	P3
0.632	−0.333	−0.037

7 Mathematical Verification

The methods to verify if a model is mathematically correct described in [7] and [8] were followed by checking some of the stationary points for the states. Stationary points can be identified when $dY(t)/dt = 0$. Given the formulae in Sect. 3, a criterion for finding a stationary point, is whether **aggimpact**$_Y(t) = Y(t)$ holds, or $c_Y(\omega_{X_1,Y}X_1(t),$ ..., $\omega_{X_k,Y}X_k(t)) = Y(t)$. This criterion can thus be used to identify stationary points in a temporal-causal network. This was done for the burnout scenario described in Sect. 5.

The model was run until $t = 100$ and then state fs_b and state srs_{ST} were analysed to see if they reached stationary points, by plotting the gradient of the states and finding the points where the gradient is 0. The result can be seen in Fig. 6, which yields some of the points that have been noted in Table 8 for analysis usable for mathematical verification. To estimate the correctness of the model, four points for two states (for each of the two states two different time points) were analysed; the average error for the points as shown in Table 8 is $2.855 * 10^{-4}$, which is a small error and is an indication of evidence that the model is mathematically accurate.

The errors were acquired by calculating the difference between the state values $X_i(t)$ and **aggimpact**$_{X_i}(t)$, which is based on the logistic combination function with as input the incoming state values with their corresponding weights, with $\sigma = 50$ and $\tau = 0.5$ for state X_{17} and $\tau = 0.1$ for state X_{18}.

Table 8. Stationary point identification to verify the model.

State X_i	$fs_b = X_{17}$	$fs_b = X_{17}$	$srs_{ST} = X_{18}$	$srs_{ST} = X_{18}$
Time point t	17.94	31.19	18.14	32.73
$X_i(t)$	0.4449	0.1303	0.6643	0.1665
aggimpact$_{X_i(t)}$	0.4446	0.1305	0.6648	0.1665
deviation	$3 * 10^{-4}$	$2.5 * 10^{-4}$	$5.1 * 10^{-4}$	$8.2 * 10^{-5}$

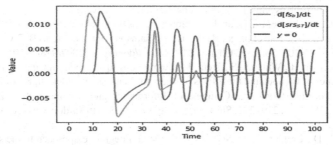

Fig. 6. Plot of the gradient (derivative over time) of states fs_b and srs_{ST} to identify where intersections are with $y = 0$, which indicate stationary points.

8 Discussion

The goal of this study was to design an adaptive temporal-causal network model incorporating dream components to create adaptive and more realistic burnout dynamics than in earlier models [4, 5]. Not only dream states were added, the model was also turned into a first-order adaptive model using a hebbian learning approach for adaptive weights between states involved in dreaming. Using the methodology described in [7, 8] and the environment described in [14], a model was created that can be simulated as well as optimised. The results acquired by introducing dream states, do substantially differ from previous work that only introduced states corresponding to sleep [5], as dreams are powerful regulators of emotions such as fear [17].

Further application of the model may address a portrayal of how a clinical burnout might develop, as this might give more insights into how they can be prevented. This could be done for example, by creating an agent-based model, that keeps track of the emotional wellbeing of a person and then scheduling them in manners where they gain enough sleep which allows for enough dreams to take place, to prevent them from developing burnouts. One issue is still that there is no numerical data available, which means that the model had to be validated based on qualitative empiric information using a simulated annealing algorithm to tune parameters to find the behavior of the model that is in accordance with the literature.

This paper serves as a first step to create an adaptive temporal-causal network describing burnout dynamics, which can still be expanded in the future, by adding more real-life states. If more empirical data becomes available in regard to burnout, it will also become possible to optimise the relationships between states, that were now based on qualitative literature. When these components are optimised, a foundation can be created to prevent, treat, or identify burnouts as well as gain a better understanding of the underlying dynamics.

References

1. Burn-out an "occupational phenomenon": International Classification of Diseases. WHO
2. CBS, TNO: Psychosociale arbeidsbelasting (PSA) werknemers (2017). http://statline.cbs.nl/StatWeb/publiction/?VW=T&DM=SLNL&PA=83049NED&LA=NL
3. CBS: Meer psychische vermoeidheid ervaren door werk (2018). https://www.cbs.nl/nl-nl/nieuws/2018/46/meer-psychische-vermoeidheid-ervaren-door-werk
4. Dujmić, Z., Machielse, E., Treur, J.: A temporal-causal modeling approach to the dynamics of a burnout and the role of physical exercise. In: Samsonovich, A.V. (ed.) BICA 2018. AISC, vol. 848, pp. 88–100. Springer, Cham (2019). https://doi.org/10.1007/978-3-319-99316-4_12
5. von Kentzinsky, H., Wijtsma, S., Treur, J.: A temporal-causal modelling approach to analyse the dynamics of burnout and the effects of sleep. In: Yang, X.-S., Sherratt, S., Dey, N., Joshi, A. (eds.) Fourth International Congress on Information and Communication Technology. AISC, vol. 1027, pp. 219–232. Springer, Singapore (2020). https://doi.org/10.1007/978-981-32-9343-4_18
6. Borsboom, D., Cramer, A.O.: Network analysis: an integrative approach to the structure of psychopathology. Ann. Rev. Clin. Psychol. **9**, 91–121 (2013)
7. Treur, J.: Network-Oriented Modeling: Addressing Complexity of Cognitive, Affective and Social Interactions. UCS. Springer, Cham (2016). https://doi.org/10.1007/978-3-319-45213-5
8. Treur, J.: Network-Oriented Modeling for Adaptive Networks: Designing Higher-Order Adaptive Biological, Mental and Social Network Models. SSDC, vol. 251. Springer, Cham (2020). https://doi.org/10.1007/978-3-030-31445-3
9. Maslach, C., Jackson, S.E.: The measurement of experienced burnout. J. Organ. Behav. **2**(2), 99–113 (1981)
10. Huang, L., Zhou, D., Yao, Y., Lan, Y.: Relationship of personality with job burnout and psychological stress risk in clinicians. Chin. J. Ind. Hyg. Occup. Dis. **33**(2), 84–87 (2015)
11. Emilia, I., Gómez-Urquiza, J.L., Cañadas, G.R., Albendín-García, L., Ortega-Campos, E., Cañadas-De la Fuente, G.A.: Burnout and its relationship with personality factors in oncology nurses. Eur. J. Oncol. Nurs. **30**, 91–96 (2017)
12. Sutin, A.R., Stephan, Y., Luchetti, M., Artese, A., Oshio, A., Terracciano, A.: The five-factor model of personality and physical inactivity: a meta-analysis of 16 samples. J. Res. Pers. **63**, 22–28 (2016)
13. Söderström, M., Jeding, K., Ekstedt, M., Perski, A., Åkerstedt, T.: Insufficient sleep predicts clinical burnout. J. Occup. Health Psychol. **17**(2), 175–183 (2012)
14. Treur, J.: Modeling higher-order adaptivity of a network by multilevel network reification. Netw. Sci. J. (2020, in press)
15. Levin, R., Nielsen, T.: Nightmares, bad dreams, and emotion dysregulation: a review and new neurocognitive model of dreaming. Curr. Dir. Psychol. Sci. **18**(2), 84–88 (2009)
16. Walker, M.P., van der Helm, E.: Overnight therapy? The role of sleep in emotional brain processing. Psychol. Bull. **135**(5), 731–748 (2009)
17. Pace-Schott, E.F., Germain, A., Milad, M.R.: Effects of sleep on memory for conditioned fear and fear extinction. Psychol. Bull. **141**(4), 835–857 (2015)
18. Kempter, R., Gerstner, W., Van Hemmen, J.L.: Hebbian learning and spiking neurons. Phys. Rev. E **59**(4), 4498 (1999)

Computational Analysis of the Adaptive Causal Relationships Between Cannabis, Anxiety and Sleep

Merijn van Leeuwen, Kirsten Wolthuis, and Jan Treur[✉]

Social AI Group, Vrije Universiteit Amsterdam, Amsterdam, The Netherlands
merijnvnl@gmail.com, k.wolthuis@student.vu.nl, j.treur@vu.nl

Abstract. In this paper an adaptive computational temporal-causal network model is presented to analyse the dynamic and adaptive relationships between cannabis usage, anxiety and sleep. The model has been used to simulate different well-known scenarios varying from intermittent usage to longer periods of usage interrupted by attempts to quit and to constant usage based on full addiction. It is described how the model has been verified and validated by empirical information from the literature.

1 Introduction

In this paper an adaptive network model is used to show how an adolescent turns to cannabis (a synonym to marijuana) in an attempt to ease his or her anxiety and associated negative emotions. These processes involve emotional regulation, but also habit formation and addiction. The following short scenario depicts how an individual may find temporary relief in cannabis or marijuana usage, but ends up with higher levels of anxiety than at baseline:

An adolescent is experiencing anxiety in his daily life and struggles to cope with the difficulties it brings, leading to negative emotions which further fuel his anxiety. He remembers cannabis advocates promoting the stress-relieving effects of cannabis and decides to give it a try. Cannabis seems to instantly relieve his stress, reducing his negative emotions. This positive effect on his emotions causes him to develop a usage habit over time. However, the impact cannabis has on his sleep quality gradually leads to a further increase in his anxiety and consequently his negative emotions. Besides, his friends and those around him start to express their discontent with his new habit and his reduced interest in social activities, while the negative effects cannabis has on his cognition lead to forgetfulness and reduced performance. Together, these two factors cause stressful events, increasing his anxiety.

The literature shows that cannabis usage may provide an instant relief of anxiety symptoms, making many individuals seek this desirable effect [2, 6]. Despite the initial benefits of self-treating with cannabis, its usage poses two major implications for sleep

© Springer Nature Switzerland AG 2020
V. V. Krzhizhanovskaya et al. (Eds.): ICCS 2020, LNCS 12137, pp. 357–370, 2020.
https://doi.org/10.1007/978-3-030-50371-0_26

architecture. This refers to the different phases of sleep an individual goes through. The relative proportions of sleep phases within the sleep architecture determine the sleep quality an individual is getting [4].

First, cannabis has been shown as bringing about reductions in Rapid-Eye Movement (REM) sleep, both acutely and long-term [8, 16]. Secondly, acute usage causes a temporary increase in the restorative Slow Wave Sleep (SWS) during the first 4 days of usage [1]. After 4 days, time spent in SWS phase gradually reduces until it is significantly below baseline levels after about 8 days [1]. These disruptions of normal sleep architecture may negatively affect the anxiety for which cannabis use was started in the first place. Disrupted sleep has been associated with a wide range of mental and physical health problems and impairments in daily life [5, 17]. Particularly the gradual buildup of sleep deprivation over a longer period of time has been associated with an increase in anxiety [9, 13]. Thus, the alterations of sleep architecture induced by cannabis usage may indirectly lead to higher levels in anxiety. Besides this mediation of the relationship between cannabis usage and anxiety by sleep deprivation, other factors associated with cannabis usage may also cause an increase in anxiety.

The first factor could be fear of being caught with illegal possession of cannabis, as a consequence of cannabis being illegal in many parts over the world. Second, its daily usage is often associated with a certain stigma, even in the Netherlands. In turn, being stigmatized is a known stressor [11]. Satterlund et al. [15] found this specific stigma and the associated stress in a study among Californian medical cannabis users. Third, cannabis usage reduces motivation among users by, among other factors, blunting the dopamine system [21]. This reduced motivation may translate into an unwillingness to pursue previously valued goals or (social) activities, making the person seem uninterested. Finally, marijuana usage is known to negatively affect cognitive performance [3, 14, 22], which may be noticed by colleagues or reflected in academic performance. Stress over academic performance is known to induce anxiety among college students [12]. Combined, these factors may increase a cannabis users' stress levels which, in turn, increases anxiety [10]. Finally, the anxiety response to stressful events increases over time as the frequency increases, termed priming [20].

In the next section, scenarios like the one depicted above are modeled by an adaptive temporal-causal network model [18, 19]. Section 3 shows simulations for three variants of this scenario corresponding to patterns known from the empirical literature, while Sects. 4 and 5 show verification by analysis and validation using parameter tuning, respectively.

2   The Adaptive Computational Network Model

In this section, first the Network-Oriented Modeling approach used is described, after which the specific introduced network model for dynamics and adaptation of the interaction between marijuana usage, anxiety and sleep quality is presented.

2.1   Network-Oriented Modeling for the Design of Adaptive Networks

The adaptive computational network model is based on the Network-Oriented Modelling approach based on reified temporal-causal networks described [19]. The *network* structure characteristics used are as follows. A full specification of a network model provides a complete overview of their values in so-called role matrix format.

- **Connectivity:** The strength of a connection from state X to Y is represented by weight $\omega_{X,Y}$
- **Aggregation:** The aggregation of multiple impacts on state Y by combination function $c_Y(..)$.
- **Timing:** The timing of the effect of the impact on state Y by speed factor η_Y

Given initial values for the states, these network characteristics fully define the dynamics of the network. For each state Y, its (real number) value at time point t is denoted by $Y(t)$. Each of the network structure characteristics can be made adaptive by adding extra states for them to the network, called *reification states* [19]: states $\mathbf{W}_{X,Y}$ for $\omega_{X,Y}$, states $\mathbf{C}_Y$ for $c_Y(..)$, and states $\mathbf{H}_Y$ for η_Y; for specific models such reification states can be given names that are more informative in the context of the particular application. Such reification states get their own network structure characteristics to define their (adaptive) dynamics and are depicted in a higher level plane, as shown by the upper plane in blue in Fig. 1. For example, using this, the adaptation principle called Hebbian learning [7], considered as a form of plasticity of the brain in cognitive neuroscience ("neurons that fire together, wire together") can be modeled.

A dedicated software environment (implemented in Matlab) is available by which the conceptual design of an adaptive network model is automatically transformed into a numerical representation of the model that can be used for simulation; this is based on the following type of (hidden) difference of differential equation defined in terms of the above network characteristics:

$$Y(t + \Delta t) = Y(t) + \eta_Y [\mathbf{aggimpact}_Y(t) - Y(t)] \Delta t \text{ or } dY(t)/dt = \eta_Y [\mathbf{aggimpact}_Y(t) - Y(t)]$$
$$\text{with } \mathbf{aggimpact}_Y(t) = c_Y(\omega_{X_1,Y} X_1(t), \dots, \omega_{X_k,Y} X_k(t)) \tag{1}$$

where the X_i are all states from which state Y has incoming connections. Different combination functions are available in a library that can be used to specify the effect of the impact on a state (see [18, 19]). The following three of them are used here:

- the *advanced logistic sum* combination function with steepness σ and threshold τ

$$\mathbf{alogistic}_{\sigma,\tau}(V_1, \dots, V_k) = \left[\frac{1}{1 + e^{-\sigma(V_1 + \dots + V_k - \tau)}} - \frac{1}{1 + e^{\sigma\tau}} \right] (1 + e^{-\sigma\tau}) \tag{2}$$

- the *Hebbian learning combination function* $\mathbf{hebb}_\mu(..)$

$$\mathbf{hebb}_\mu(V_1, V_2, W) = V_1 V_2 (1 - W) + \mu W \tag{3}$$

with μ the persistence parameter, where V_1 stands for $X(t)$, V_2 for $Y(t)$ and W for $\mathbf{W}_{X,Y}(t)$, where X and Y are the two connected states
- the *complementary identity combination function* $\mathbf{compid}(..)$

$$\mathbf{compid}(V) = 1 - V \tag{4}$$

2.2   The Introduced Adaptive Computational Network Model

The adaptive temporal-causal network model introduced here is depicted graphically in Fig. 1. Table 1 provides an overview of the states. A state X_8 with complementary identity function was used to model the relaxing effect that marijuana has on anxiety. Thus, marijuana usage V leads to an effect of $1 - V$ which is then propagated to lower anxiety state X_1. A similar state with a complementary identity function was used for SWS, so that there is a delay in onset of decreased SWS as described in the literature.

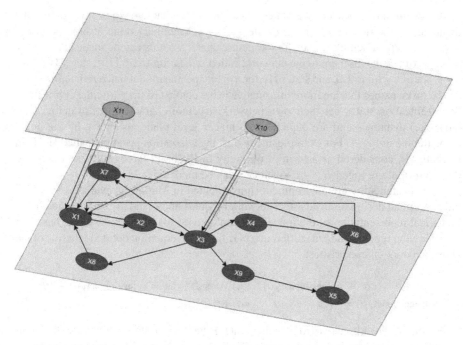

Fig. 1. Graphical conceptual representation of the introduced adaptive network model

Table 1. Overview of states within the network model

Name	Nr	Explanation
A	X_1	Anxiety
NE	X_2	Negative emotions
MU	X_3	Marijuana usage
BRS	X_4	Bad REM Sleep
BSS	X_5	Bad SWS sleep
BSQ	X_6	Bad sleep quality
SE	X_7	Stressful events
MRE	X_8	Marijuana Relaxation Effect
SSD	X_9	SWS dummy
HF	X_{10}	Habit Formation
P	X_{11}	Priming

The reification states represent the habit formation of cannabis usage (by Hebbian learning) HF (state X_{10} or $\mathbf{W}_{X_7, X_1}$) and the priming of anxiety P (state X_{11} or $\mathbf{W}_{X_2, X_3}$). This priming refers to developing a more sensitive and increased anxiety response to stressful events over time.

As can be seen in Fig. 1, affected by Stressful events X_7, Anxiety X_1 leads to Negative emotions X_2 which trigger Marijuana usage X_3. This leads to a relaxation effect X_8 which decreases Anxiety X_1 and in turn Negative emotions X_2. However, Marijuana usage X_3 after a while also leads to Bad SWS sleep X_5 and Bad REM sleep X_4, together making Bad sleeping quality X_6, which in turn increases Anxiety.

Moreover, Marijuana usage X_3 and Bad sleep quality also contribute to Stressful events X_7. So, there are a number of cycles in this causal model, which gives it non-trivial basic dynamics. In addition, also adaptive dynamics occurs in the form of Habit formation X_{10} and Priming X_{11} that model forms of Hebbian learning by which the causal connections to Marijuana usage X_3 and Anxiety X_1 are strengthened. For simulations the chosen combination functions for the different states were as follows:

Logistic function	$\mathbf{alogistic}_{\sigma,\tau}(V_1, \ldots, V_k)$	X_1 to X_4, X_6, X_7, X_9
Complementary identity function	$\mathbf{compid}(V)$	X_5, X_8
Hebbian learning function	$\mathbf{hebb}_\mu(V_1, V_2, W)$	X_{10}, X_{11}

The specification of the adaptive network model by role matrices [19] can be found at https://www.researchgate.net/publication/340162147. In Sect. 3 the behaviour for three different scenarios (describing patterns well-known from empirical literature) will be explored.

3 Simulation Experiments

Three scenarios were simulated using the dedicated modeling environment. The scenarios are cyclical use (Scenario A); more substantial usage with occasionally a failed quit attempt (Scenario B); and constant usage in an equilibrium state where the person does not stop using marijuana (Scenario C).

3.1 Scenario A: Cyclic Usage

This first scenario shows cyclical behaviour that does not reach an equilibrium. The cyclical behaviour simulates a person who decides to use marijuana, but then soon afterwards again decides to stop using it; however, due to high anxiety that occurs after stopping, the person starts using again, and this pattern repeats itself indefinitely. Settings for most of the network characteristics by role matrices [19] can be found in Table 2, in particular for role matrices **mb** (*matrix for base connections*), **mcw** (*matrix for connection weights*) and **ms** (*matrix for speed factors*). Each row in such a role matrix indicates for a given state the states (in red cells) or constant values (in green cells) that affect this state for the given role. For example, in **mb** the row for the anxiety state X_1 indicates that this state is affected from the base role by states X_2, X_6, X_7 and X_8. In role matrix **mcw** it is indicated that X_1 is affected from the connection weight role by 0.1,

0.1, 0.1, and 1, respectively, and in matrix **ms** it is indicated it is affected from the speed factor role by 0.4. For the adaptive cases, in role matrix **mcw** it is indicated that X_1 is affected from the connection weight role for its third incoming connection by X_{11}, and similarly X_3 is affected from the connection weight role for its incoming connection by X_{10}.

Table 2. Role matrices **mb**, **mcw** and **ms** for Scenario A

mb		1	2	3	4
A	X_1	X_2	X_6	X_7	X_8
NE	X_2	X_1			
MU	X_3	X_2			
BRS	X_4	X_3			
BSS	X_5	X_9			
BSQ	X_6	X_4	X_5		
SE	X_7	X_3	X_6		
MRE	X_8	X_3			
SSD	X_9	X_3			
HF	X_{10}	X_1	X_3	X_{10}	
P	X_{11}	X_1	X_7	X_{11}	

mcw		1	2	3	4
A	X_1	0.1	0.1	X_{11}	1
NE	X_2	1			
MU	X_3	X_{10}			
BRS	X_4	1			
BSS	X_5	1			
BSQ	X_6	1	1		
SE	X_7	1	1		
MRE	X_8	1			
SSD	X_9	0.1			
HF	X_{10}	1	1	1	
P	X_{11}	1	1	1	

ms		1
A	X_1	0.4
NE	X_2	0.1
MU	X_3	0.1
BRS	X_4	0.1
BSS	X_5	0.1
BSQ	X_6	0.1
SE	X_7	0.1
MRE	X_8	0.1
SSD	X_9	0.1
HF	X_{10}	0.1
P	X_{11}	0.1

The simulation of this scenario is shown in the graphs in Figs. 2 and 3. In Fig. 2, it is visible that the states X_5, X_6, and X_7 for Bad sleep and Stressful events reach an equilibrium state. State X_5 (bad SWS sleep) takes a longer time to reach an equilibrium because the slow wave sleep becomes worse the longer marijuana is used. State X_6 and X_7 also reached an equilibrium, since they were influenced by the bad SWS sleep.

The other four states do not reach an equilibrium, but showed the cyclical behaviour that was expected. It is interesting to see that the first use led to a lower increase in anxiety in comparison to the use of marijuana after the first time. The behaviour of the habit formation (state X_{10}) and primig (state X_{11}) are shown in Fig. 3. It is visible that the Habit formation fluctuates around 0.6, while the Priming fluctuates around 0.85. These adaptive states both did not reach an equilibrium, since their incoming states showed the fluctuating behaviour.

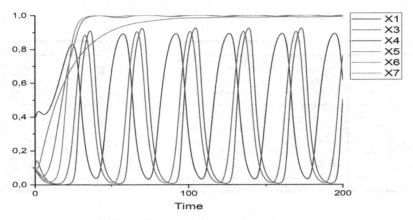

Fig. 2. Cyclic usage behaviour for the base states

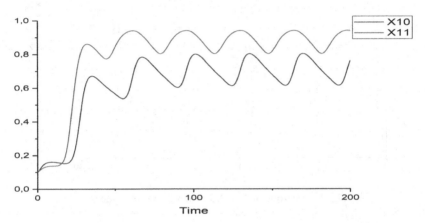

Fig. 3. Cyclic usage behaviour for the reification states modeling adaptation.

3.2 Scenario B: Occasional Failed Quit Attempts

In this scenario, a user starts using marijuana and uses it for some time, which reduces his/her anxiety. However, after some time the user decides to stop using because his anxiety levels are higher than at baseline. The user makes a quit attempt that does not succeed because his/her anxiety increases too much. The user then again decides to start using again until a new (failed) quit attempt. Table 3 specifies the role matrices **mcw** and **ms** for this scenario.

Figures 4 and 5 show the results from the simulation. In Fig. 4 it is visible that again the states X_5, X_6, and X_7 reach an equilibrium because of the behaviour of state X_5 (bad SWS sleep). The other states, in comparison to their behaviour in Scenario A, have longer stationary time periods, which only decrease during the quit attempt. It is also noticeable that the quit attempts are short and that the user starts using again within a couple of days. Figure 5 shows that the habit formation and priming increase fast in the

Table 3. Role matrices **mcw** and **ms** for Scenario B

mcw		1	2	3	4
A	X_1	0.33	0.33	X_{11}	1
NE	X_2	1			
MU	X_3	X_{10}			
BRS	X_4	1			
BSS	X_5	1			
BSQ	X_6	1	1		
SE	X_7	1	1		
MRE	X_8	1			
SSD	X_9	0.1			
HF	X_{10}	1	1	1	
P	X_{11}	1	1	1	

ms		1
A	X_1	0.6
NE	X_2	0.6
MU	X_3	0.6
BRS	X_4	0.6
BSS	X_5	0.1
BSQ	X_6	0.6
SE	X_7	0.6
MRE	X_8	0.2
SSD	X_9	0.6
HF	X_{10}	0.6
P	X_{11}	0.6

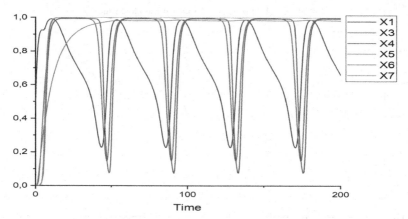

Fig. 4. Recurring quitting behaviour for the base states.

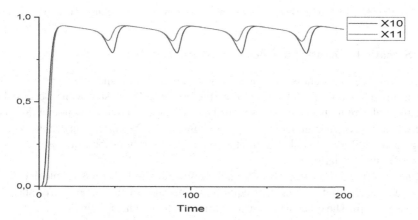

Fig. 5. Recurring quitting behaviour for the reification states modeling adaptation.

beginning. After they reached their highest value, they start to fluctuate because of the failed quit attempts made by the user.

3.3 Scenario C: Equilibrium Addiction State

Scenario C depicts a constant increase in usage. After a time, all states reach an equilibrium, meaning that the habit of using cannabis has solidified, the individual is constantly having bad sleep quality and higher anxiety levels than at baseline. This represents a cannabis user who might fear quitting his habit, and craves the initial relaxing effects it gave him. Table 4 specifies the role matrices **mcw** and **ms** for this scenario.

Table 4. Role matrices **mcw** and **ms** for Scenario C

mcw		1	2	3	4
A	X_1	0.4	0.4	X_{11}	1
NE	X_2	1			
MU	X_3	X_{10}			
BRS	X_4	1			
BSS	X_5	1			
BSQ	X_6	1	1		
SE	X_7	1	1		
MRE	X_8	1			
SSD	X_9	1			
HF	X_{10}	1	1	1	
P	X_{11}	1	1	1	

ms		1
A	X_1	0.6
NE	X_2	0.6
MU	X_3	0.6
BRS	X_4	0.6
BSS	X_5	0.1
BSQ	X_6	0.6
SE	X_7	0.6
MRE	X_8	0.2
SSD	X_9	0.6
HF	X_{10}	0.6
P	X_{11}	0.6

Figures 6 and 7 show the behaviour of the states over time.

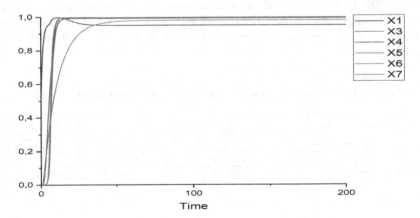

Fig. 6. Equilibrium addiction for the base states.

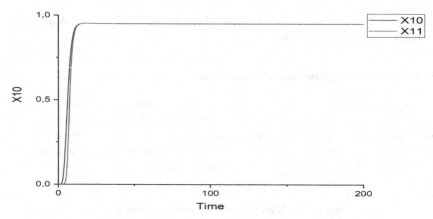

Fig. 7. Equilibrium addiction for the reification states modeling adaptation

4   Verification and Validation of the Computational Model

In this section it is shown how the model was verified by mathematical analysis and how it was validated using empirically based data and parameter tuning.

4.1   Verification of the Model by Mathematical Analysis

To verify the model, the above Scenario B was analysed, representing a user who uses fairly constantly but also occasionally attempts to quit, albeit unsuccessfully. For a number of selected stationary points, the difference between the aggregated impact and the state value assigned was calculated, following [18], Ch 12. According to Eq. (1), theoretically this difference should be 0 in a stationary point, and in practice it should be close to 0. The formulas used were dependent upon the combination function utilized for the concerning state. The deviations found are significantly small (see Table 5, last row), meaning that there were no notable errors within the network model which would require attention.

Table 5. Verification of the model by mathematical analysis of stationary points

state X_i	X_2	X_4	X_5	X_6	X_7	X_8	X_9	X_{10}	X_{11}
time point t	13	60	200	200	200	200	120.5	120.5	118
$X_i(t)$	0.988955	0.992882	0.982017	1	1	0.007488	0.961621	0.831518	0.879718
aggimpact$_{X_i}(t)$	0.985927	0.992923	0.975546	0.993935	0.993396	0.007982	0.963867	0.831452	0.879498
deviation	0.003028	-0.000041	0.006470	0.006065	0.006604	0.000494	-0.002246	0.000066	0.000220

4.2   Validation of the Model by Parameter Tuning

For the following five states empirically based values were used for the tuning: Marijuana usage (X_3), Bad REM sleep (X_4), Bad Slow-Wave Sleep (X_5), Stressful events (X_7), Habit formation (X_{10}). Table 6 shows the time points and the corresponding empirical values for which the model was tuned for these five states. These empirical data simulate a case in which the user increasingly uses more cannabis over time, reflected by the gradual increase among all values (similar to Scenario B in Sect. 3). Bad Slow-Wave-Sleep sets in after 4 days, as described in the literature. As this literature only provides qualitative indications which had to be hand-mapped onto numbers, the precision of these empirical data cannot be expected to be perfect.

Table 6. Empirically based values used for parameter tuning

Time points states	1	4	6	12	20	29
X_3	0.1	0.5	0.7	0.8	0.9	1
X_4	0.15	0.5	0.75	0.85	0.95	1
X_5	0	0.2	0.3	0.5	0.8	1
X_7	0	0.1	0.15	0.6	0.8	0.9
X_{10}	0	0.4	0.85	0.85	0.75	1

The parameter tuning by Simulated Annealing was applied to 56 parameters (for all nonadaptive network characteristics and initial values, represented by the values in the role matrices), with minimum value 0 and a maximum value of [50, 1] for all logistic function parameter values σ and τ; for the rest of the parameters the value interval was [0, 1]. The five lowest RMSE values found are 0.202517, 0.235468, 0.236067, 0.252116, 0.253504. Given the imprecision of the data, this may not be worse than to be expected. The first option was used for the parameters for the final simulation. This final simulation is compared to the empirical data per state in Fig. 8.

In the graphs, the red line represents the simulated data and the black points represent the empirical data. The first state that is compared is state X_3. It is visible in this graph that the simulated data lies a bit higher than the empirical data.

It is visible that for X_4 the simulated data has got a high initial value, which does not correspond with the empirical data. Except for the first two points of the empirical data, the tuned model and empirical data have the same shape, so that part of the tuning did work out. For X_5, the initial value of the simulation is again set higher than is expected from the empirical data. Furthermore, the form of the simulated data does not follow the empirical data well. So, for the tuning for this state there is room for improvement.

The comparison between the empirical data and the tuned model for state X_7 is good, since the tuned data follows the empirical data and the empirical data lies around the simulated data. For state X_{10}, the empirical data was not chosen that well and therefore the simulated tuned data does not fit the empirical data. The tuned model shows a habit formation that does not rise above 0.7, which was not what was expected. This is an aspect that needs to be improved for future models.

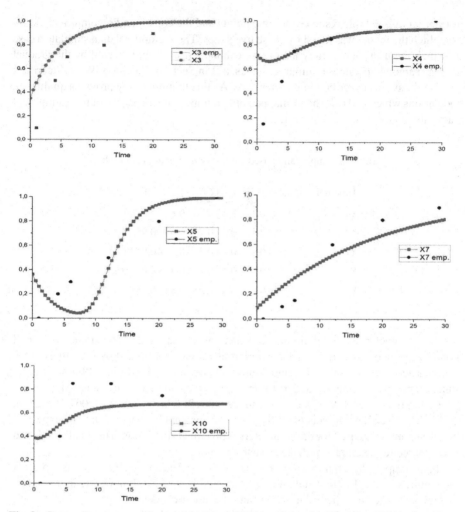

Fig. 8. Comparison between the empirical data (black) and the simulated data (black) for states X_3 (Marijuana usage), X_4 (Bad REM sleep), X_5 (Bad Slow-Wave Sleep), X_7 (Stressful events), X_{10} (Habit formation)

5   Discussion

In conclusion, the introduced adaptive network model was able to show behaviour that well suits three different nontrivial dynamic and adaptive patterns that are well-known from the empirical literature. Only qualitative empirical information was available, which had to be approximated by numerical data in order to apply parameter tuning by Simulated Annealing. The obtained empirical data is approximated by the model somewhat different than the best hand-tuned simulation. Improving these empirical data could improve the tuned simulation. This could be attained by doing more research into the relationships between the different states and also quantifying these relationships. Furthermore, not tuning the initial values may also already lead to a more realistic simulation. The initial values the tuning came up with are not realistic; for example, the sleep quality improves first before it becomes worse because the initial value was set by the tuning at a high value.

References

1. Barratt, E.S., Beaver, W., White, R.: The effects of marijuana on human sleep patterns. Biol. Psychiatry **8**, 47–54 (1974)
2. Buckner, J.D., Schmidt, N.B.: Social anxiety disorder and marijuana use problems: the mediating role of marijuana effect expectancies. Depress. Anxiety **26**(9), 864–870 (2009). https://doi.org/10.1002/da.20567
3. Dahlgren, M.K., Sagar, K.A., Racine, M.T., Dreman, M.W., Gruber, S.A.: Marijuana use predicts cognitive performance on tasks of executive function. J. Stud. Alcohol Drugs **77**(2), 298–308 (2016). https://doi.org/10.15288/jsad.2016.77.298
4. Deatherage, J.R., Roden, R.D., Zouhary, K.: Normal sleep architecture. Seminar. Orthodont. **15**(2), 86–87 (2009). https://doi.org/10.1053/j.sodo.2009.01.002
5. Fairholme, C.P., Manber, R.: Sleep, emotions, and emotion regulation: an overview. In: Babson, K.A., Feldner, A. (eds.) Sleep and Affect: Assessment, Theory, and Clinical Implications, Chap. 3, pp. 45–61. Academic Press, San Diego (2015). https://doi.org/10.1016/B978-0-12-417188-6.00003-7
6. Glodosky, N.C., Cuttler, C.: Motives matter: cannabis use motives moderate the associations between stress and negative affect. Addict. Behav. **102**, 106188 (2019). https://doi.org/10.1016/j.addbeh.2019.106188
7. Hebb, D.O.: The Organization of Behavior: A Neuropsychological Theory. Wiley, London (1949)
8. Jacobus, J., Bava, S., Cohen-Zion, M., Mahmood, O., Tapert, S.F.: Functional consequences of marijuana use in adolescents. Pharmacol. Biochem. Behav. **92**(4), 559–565 (2009). https://doi.org/10.1016/j.pbb.2009.04.001
9. Kahn-Greene, E.T., Killgore, D.B., Kamimori, G.H., Balkin, T.J., Killgore, W.D.S.: The effects of sleep deprivation on symptoms of psychopathology in healthy adults. Sleep Med. **8**(3), 215–221 (2007). https://doi.org/10.1016/j.sleep.2006.08.007
10. Kurebayashi, L.F.S., Do Prado, J.M., Da Silva, M.J.P.: Correlations between stress and anxiety levels in nursing students. J. Nurs. Educ. Pract. **2**(3), 128 (2012)
11. Link, B.G., Phelan, J.C.: Stigma and its public health implications. Lancet **367**(9509), 528–529 (2006)
12. Misra, R., Mckean, M.: College students' academic stress and its relation to their anxiety, time management, and leisure satisfaction. Am. J. Health Stud. **16**, 41–51 (2000)
13. Pires, G.N., Bezerra, A.G., Tufik, S., Andersen, M.L.: Effects of acute sleep deprivation on state anxiety levels: a systematic review and meta-analysis. Sleep Med. **24**, 109–118 (2016). https://doi.org/10.1016/j.sleep.2016.07.019
14. Pope Jr., H.G., Gruber, A.J., Hudson, J.I., Huestis, M.A., Yurgelun-Todd, D.: Neuropsychological performance in long-term cannabis users. Arch. Gen. Psychiatry **58**(10), 909–915 (2001). https://doi.org/10.1001/archpsyc.58.10.909
15. Satterlund, T.D., Lee, J.P., Moore, R.S.: Stigma among California's medical marijuana patients. J. Psychoact. Drugs **47**(1), 10–17 (2015). https://doi.org/10.1080/02791072.2014.991858

16. Schierenbeck, T., Riemann, D., Berger, M., Hornyak, M.: Effect of illicit recreational drugs upon sleep: cocaine, ecstasy and marijuana. Sleep Med. Rev. **12**(5), 381–389 (2008). https://doi.org/10.1016/j.smrv.2007.12.004
17. Strine, T.W., Chapman, D.P.: Associations of frequent sleep insufficiency with health-related quality of life and health behaviors. Sleep Med. **6**(1), 23–27 (2005). https://doi.org/10.1016/j.sleep.2004.06.003
18. Treur, J.: Network-Oriented Modeling: Addressing Complexity of Cognitive, Affective and Social Interactions. Springer Publishers, Heidelberg (2016). https://doi.org/10.1007/978-3-319-45213-5
19. Treur, J.: Network-Oriented Modeling for Adaptive Networks: Designing Higher-Order Adaptive Biological, Mental and Social Network Models. Springer Publishers, Heidelberg (2020). https://doi.org/10.1007/978-3-030-31445-3
20. Vytal, K.E., Overstreet, C., Charney, D.R., Robinson, O.J., Grillon, C.: Sustained anxiety increases amygdala-dorsomedial prefrontal coupling: a mechanism for maintaining an anxious state in healthy adults. J. Psychiatry Neurosci. **39**(5), 321–329 (2014). https://doi.org/10.1503/jpn.130145
21. Volkow, N.D., et al.: Effects of cannabis use on human behavior, including cognition, motivation, and psychosis: a review. JAMA Psychiatry **73**(3), 292–297 (2016). https://doi.org/10.1001/jamapsychiatry.2015.3278
22. Wadsworth, E.J.K., Moss, S.C., Simpson, S.A., Smith, A.P.: Cannabis use, cognitive performance and mood in a sample of workers. J. Psychopharmacol. **20**(1), 14–23 (2005). https://doi.org/10.1177/0269881105056644

Detecting Critical Transitions in the Human Innate Immune System Post-cardiac Surgery

Alva Presbitero[1(✉)], Rick Quax[2], Valeria V. Krzhizhanovskaya[2,3], and Peter M. A. Sloot[2,3,4]

[1] Asian Institute of Management, Makati, Philippines
avpresbitero@gmail.com
[2] University of Amsterdam, Amsterdam, The Netherlands
[3] ITMO University, Saint Petersburg, Russia
[4] Nanyang Technological University, Singapore, Singapore

Abstract. Coronary artery bypass grafting with cardiopulmonary bypass activates the human innate immune system (HIIS) and invokes a vigorous inflammatory response that is systemic. This massive inflammatory reaction can contribute to the development of postoperative complications that could topple the state of the system from health to disease, or even to some extent, death. The body, after all, is in a state where majority of its immune cell populations have been depleted, and sometimes needs days or even longer to recuperate. To obtain a deeper understanding on how HIIS responds to complications after cardiac surgery, we perturb the immune system model that we have developed in an earlier work *in-silico* by adding another source of inflammation triggering moieties (ITMs) hours after surgery in various regimes. A critical transition occurs upon the addition of a critical concentration of ITMs when the insult is sustained for approximately 3 h – a total concentration that corresponds to the fatal concentration of ITMs documented in literature. By perturbing HIIS *in-silico* with additional sources of ITMs to mimic persistent and recurring episodes of post-surgery complications, we are able to specify under which conditions critical transitions occur in HIIS, as well as pinpoint important blood parameters that exhibit critical transitions in our model. More importantly, by applying early warning signals on the clinical trial data used to calibrate and validate HIIS model, we are able to detect blood parameters that exhibit critical transitions in patients who died post-surgery, where pro-inflammatory cytokines are deemed potential markers for critical transitions.

Keywords: Human innate immune response · Post-surgery complications · Critical transitions · Early warning signals

1 Introduction

Coronary artery bypass grafting (CABG) with cardiopulmonary bypass (CPB) invokes a systemic inflammatory response that activates HIIS. Contact of blood components with the artificial surface of the bypass circuit induces sheer stress on blood cells. Ischemia-reperfusion injury due to accumulated ITMs that have crossed the gut-barrier during hypo-perfusion [1], endotoxemia or the presence of endotoxins such as ITMs in

© Springer Nature Switzerland AG 2020
V. V. Krzhizhanovskaya et al. (Eds.): ICCS 2020, LNCS 12137, pp. 371–384, 2020.
https://doi.org/10.1007/978-3-030-50371-0_27

the blood, as well as tissue damage caused by the surgical wound are all possible causes of systemic inflammatory response syndrome (SIRS). This massive inflammatory reaction may contribute to the development of postoperative complications such as myocardial dysfunction, respiratory failure, renal and neurologic dysfunction, bleeding disorders, altered liver function, and sequentially, multiple organ failure [2]. Taking into account that more than 800,000 patients per year undergo coronary artery bypass grafting (CABG) surgery worldwide while approximately 150,000 patients undergo valve surgery [3, 4], postoperative respiratory failure has a mortality rate of 80% in patients undergoing cardiac surgery [5, 6]. Myocardial dysfunction that escalates to symptomatic heart failure accounts for 50% of medical admissions to hospitals, and is associated with in-hospital mortality of 12% and a 1-year mortality of 20–35% [7, 8]. The Society of Thoracic Surgeons National Database reported that 20% (22,000 patients) of "low-risk" patients developed postoperative complications.

Using the HIIS model that we have developed in an earlier work [9], we show how HIIS reacts to complications after surgery by adding a source of ITMs *in-silico* hours post-surgery. The developed model is an ordinary differential equations model of that of HIIS in response to systemic inflammation. The model has been calibrated and validated against clinical trials data of patients undergoing cardiac surgery. ITMs may refer to any cell or enzyme that triggers the innate immune response, such as bacterial lipopolysaccharides (LPS) and extracellular nucleotides [10, 11]. In case of a massive insult, HIIS' response becomes amplified and dysregulated [12], which leads to the imbalance between pro-inflammatory and anti-inflammatory cytokines [13]. By perturbing the *in-silico* system with different intensities of ITMs, we aim to test the resilience of HIIS and assess at which point the system shifts between alternative regimes: from state of health to disease.

Various and diverse complex dynamical systems have been shown to exhibit transitions or so-called tipping points, where there occurs an abrupt shift in stable states. In biological systems, such as the human body, this tipping point can occur as a rapid shift from state of health to disease in various manners [14, 15]. In depression, fluctuations of emotions serve as indicators for tipping points from normal to the onset of a depressive state [16]. Other examples also include systemic market crashes observed in financial systems [17, 18], the slowing down of fluctuations before a climate shift [19, 20], trends of a declining population prior to extinction [21, 22], blood parameters as indicators of tipping points in patients undergoing cardiac surgery [23], and early warning systems in floods [24] and dams.

Early warning signals (EWS) are hypothesized to serve as indicators of loss of system resilience prior to transitions between regimes. Subtle statistical properties of measurements in the system are assessed to indicate presence of critical transitions [25]. Sometimes, these transitions are observed in changes in correlations, standard deviation, and skewness of system measurements through time [26].

We define critical transition occurring in the *in-silico* model when blood parameter concentrations exhibit either a saturation to a maximum value, as in the case of increasing concentrations of ITMs, accompanied by the depletion of other immune cell populations. These serve as strong indicators that the body is no longer able to neutralize the ongoing inflammation. We show that the system shifts abruptly and irreversibly from the state of health to disease given a critical threshold of ITMs in our model.

This startling transition in HIIS poses an urgent and crucial concern as it might be difficult, or even impossible for medical practitioners to act upon beforehand due to the abrupt nature of the transition. Due to the urgency of the situation, it calls for a deeper understanding on the nature of the instances that contribute to the occurrence of these transitions. More importantly, there is a need to investigate the possibility of detecting these transitions at a considerable time before the event happens. We define a *healthy* state when HIIS can resolve or neutralize all ITMs, while *disease* when ITMs are not resolved within incubation time. Consequently, critical transition is the point when the state of the system shifts from health to disease. Finally, we assess the capability of EWS in detecting critical transitions in clinical trials data of patients undergoing cardiac surgery that was used to calibrate and validate HIIS model in [9]. In the clinical trials data, 3 out of 52 patients died post-surgery. In the context of our model, we define the 3 patients who died as *critical* patients who exhibited critical transitions in their blood parameters, while the remaining patients we refer to as *non-critical*.

2 Methods

2.1 Metric and Model-Based Indicators

EWS for detecting critical transitions in systems can be divided into two categories: metric and model-based. Both methods aim to quantify the variations in correlation structure, and changes in variability in measurements prior to the system's transition between alternate regimes [27]. Metric-based indicators aim to quantify changes in statistical properties of measurements without attempting to fit the measurements onto a model. We use *variance*, *skewness*, and *kurtosis* as metric-based indicators for transition from state of health to disease, which are explained each in turn next.

The most important hints of whether a system is close to a critical transition is referred to in dynamical systems theory as "critical slowing down" [28]. It's most straightforward implication is when the rate of recovery after tiny perturbations can be used as an indicator on whether a system is close to a bifurcation point [29]. That is, the time it takes to return to equilibrium even after tiny perturbations strongly increases as the system approaches the threshold of bifurcation. Hence, referring to how the system "slows down" going back to equilibrium [30, 31].

Variance. An increase in variance in fluctuation patterns could be another consequence of critical slowing down. As a system approaches a tipping point it could exhibit increasingly strong variations at measurements around the equilibrium as the impacts of perturbations do not decay, and only accumulates. *Skewness.* Perturbations drive the state of the system to shift between alternate regimes. Critical slowing down, which refers to a decreasing return rate of the system towards equilibrium results in distribution asymmetry [32]. Hence skewness either increases or decreases depending on the direction of transition. *Kurtosis.* Strong perturbations provokes the system to take on extreme values close to transition, increasing the occurrence of rare values in the measurements [33]. Therefore, an increase in *kurtosis*, or "bulging" is observed in the measurements leading to a tipping point.

Model-based indicators quantify variations in measurements by fitting the data to a model. Autocorrelation is a simple method used to quantitatively describe slowing down in a system nearing tipping point. *Autocorrelation* is one of the simplest ways in measuring slowing down. Increasing autocorrelation implies that consecutive points in the time series have become increasingly similar [34]. *Time-varying Autoregressive models (AR)* at time lag p is also one of the numerous methods used to estimate the local dynamics in measurements of a system [35]. The first step is calculating the inverse of the characteristic root (λ), by estimating the autoregressive function. Values for λ that approaches 0 imply that the system quickly returns or stabilizes towards the mean. This is because we used a time lag equal to one, which indicates that the current value is based on the value immediately preceding it. Hence, λ would simply be the slope of change between two time points, $y(t)$ and $y(t-1)$. See equation for time-varying AR(1) model in Eq. (1). The smaller this slope is, the more similar the measurements are at time $t-1$ with t. Hence, it would be quicker for the system to go back to equilibrium. On the other hand, when values for λ approach 1, measurements become increasingly varied hence implying instability.

$$y(t) = a(t)y(t-1) + \varepsilon(t), \tag{1}$$

where $a(t)$ corresponds to the autoregressive coefficient, and $\varepsilon(t)$ corresponds to the environmental variability [27].

2.2 Trend Detection

Any presence of statistically significant increasing trends captured by early warning indicators are evaluated using the Mann-Kendall trend test. The Mann-Kendall trend test is a non-parametric test that analyzes consistent increasing or decreasing patterns in data series. The null hypothesis being a monotonic trend does not exist, while the alternate hypothesis assumes the existence of a trend. These trends are tested to a significance level of 5%. We used a one-tailed test. This means that we only look at positive trends in values of EWS to be able to fully understand the system.

3 Results and Discussion

3.1 Effects of Adding Inflammation Triggering Moieties *In-Silico* to the Human Innate Immune System 2 Days After Surgery

Cardiac surgery with CABG activates HIIS, which invokes a vigorous response that most likely depletes the body's reservoir of immune cells, proteins, and enzymes, such as macrophages, neutrophils. Depending on the patient's conditions, it may take days, weeks or even months for immune cell levels to fully recuperate to normal levels. Nguyen et al. have shown that the activity of immune cells in cardiac surgery patients was impaired on the 3rd day post-surgery. These levels, however, returned to normal after a week after surgery [36]. The occurrence of complications post-operation becomes a serious threat as the body has not yet fully recovered. Complications sometimes happen from 2 to 9 days after surgery [37]. In a study conducted by

Hashemzadeh et al., the majority of the complications, more specifically postoperative atrial fibrillation, develop within the first 2 days after surgery [38]. Hence, in all our experiments, we add a source of ITMs that starts at 48 h after cardiac surgery.

Below we explore the effects of adding various concentrations of ITMs *in-silico* 48 h after surgery for a duration of 3 h. These ITMs may come from complications from inflicted wound due to surgery, oxidative stress coming from various sources in the body, or external factors that invoke further production of ITMs. 3 h is the duration of insult that is typically observed in patients undergoing cardiac surgery before they stabilize back to normal values, often 7 days after surgery [36, 39]. We show that this duration of adding ITMs is able to tip the balance, pushing the state of the system from health to disease, which we will later show numerically in Sect. 3.2. We summarize our results in Fig. 1.

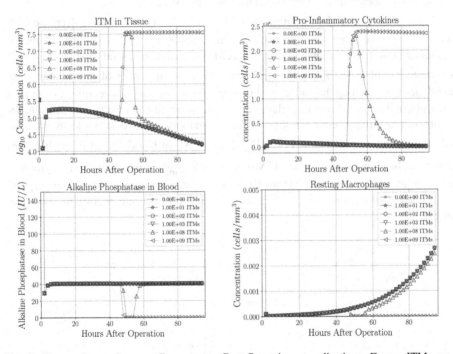

Fig. 1. Human Innate Immune Response to Post-Operative complications. Excess ITMs are continuously added for 3 h in-silico at exactly 48 h (2 days) after surgery. Our results show that at an ITM concentration of $1 \times 10^9 \frac{cells}{mm^3}$, the concentration of ITMs in the tissue remains unneutralized even after 96 h of surgery. Compared to $1 \times 10^8 \frac{cells}{mm^3}$, this concentration HIIS can completely neutralize the inflammation at 60 h post-surgery. Pro-inflammatory cytokines, proteins responsible for opening the endothelial barrier to allow recruitment of more neutrophils from the bloodstream into the tissue, exhibit a saturation of concentration at added ITMs of $1 \times 10^9 \frac{cells}{mm^3}$. AP, enzymes known to neutralize ITMs, are depleted both in blood and tissue at added ITM concentration of $1 \times 10^9 \frac{cells}{mm^3}$. Resting macrophages show slight differences for various ITM concentration regimes due to the slow replenishment rate from the bone marrow. Nonetheless, we still see a depletion of concentration of resting macrophages at a critical ITM concentration of $1 \times 10^9 \frac{cells}{mm^3}$. Cells, in the context of our work, also refer to proteins, enzymes, and molecules as a unifying unit in our system. (Color figure online)

The abrupt change in blood parameter concentrations shown in our results imply that there seems to be a critical concentration of ITMs where HIIS is no longer able to neutralize the inflammation. We highlight this in red as shown in Fig. 1. With overwhelming concentration of ITMs, activated neutrophils that are at the site of inflammation go into necrosis, as an attempt, paradoxically, to aggravate the inflammation, which results in the recruitment of more neutrophils into the site of inflammation. This peculiar choice in death pathway (apoptosis or necrosis) is explained and modeled in [40, 41]. Necrosis, a violent death pathway that involves the rupture of the neutrophil's cytoplasmic content into its surroundings, releases an additional source of ITMs that invokes a series of immune cell responses, which fuels, and further aggravates the ongoing inflammatory response. One could imagine the effect of a considerable amount of ITMs on HIIS. More specifically, how it induces a magnified and continuous production of concentrations of pro-inflammatory cytokines.

With additional ITM concentrations of $1 \times 10^8 \frac{cells}{mm^3}$, ITMs in tissue decrease 60 h after surgery, implying that the body is still capable of neutralizing the additional amount of insult. On the other hand, this ITM concentration saturates when the added concentration of ITMs is $1 \times 10^9 \frac{cells}{mm^3}$. One of the key functions of pro-inflammatory cytokines is to open up the endothelial barrier, which consequently recruits a fresh fleet of neutrophils into the site of inflammation. We show in our results that the concentration of pro-inflammatory cytokines increases and saturates to a steady level when the added concentration of ITMs $1 \times 10^9 \frac{cells}{mm^3}$ while AP in blood and in tissue becomes depleted. In contrast, for added ITMs of $1 \times 10^8 \frac{cells}{mm^3}$, pro-inflammatory cytokines level slides back to zero and AP stabilizes back to normal roughly 60 h after surgery.

Added ITMs do not seem to affect activated macrophages and neutrophils as shown in Fig. 2.

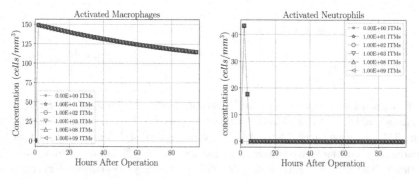

Fig. 2. Resting and Activated Macrophages and Neutrophils' Response to Added ITMs. Even without additional ITMs, our model predicts the activation of all resting macrophages and neutrophils due to the scale of insult cardiac surgery with CABG invokes on HIIS. Therefore, additional source of ITMs, especially when the immune cells, proteins, and enzymes are already depleted, will still invoke the maximum effect on macrophages and neutrophils. Cells, in the context of our work, also refer to proteins, enzymes, and molecules as a unifying unit in our system.

This is because even without a new source of ITMs, resting macrophages and neutrophils have already been fully activated. Hence, *additional* source of ITMs will not significantly change the profiles of these immune cells, proteins, and enzymes. During systemic insult, the bone marrow releases both mature and immature neutrophils into the bloodstream. This is the so-called "left shift," which refers to the increase in the number of immature neutrophils in the bloodstream [42]. After which, it takes roughly a week for the bone marrow to release a new set of *mature* neutrophils into the bloodstream [43, 44].

3.2   How Does the Human Innate Immune System Respond to Persistent and Recurrent Episodes of Post-surgery Complications?

In this section, we further explore how HIIS responds to complications that are either recurring or persistent by adding ITMs in various regimes: 1) changing intervals and 2) changing durations.

Effects of Adding Inflammation Triggering Moieties In-Silico at Different Time Intervals?
Here we introduce an additional source of ITMs at various intervals: 8 h, 16 h, and 24 h intervals. The concentration of ITMs is continuously added for 30 min to mimic those complications that are persistent. Our results are summarized in Fig. 3.

Our results show that recurrent episodes of post-surgery complications that are sustained for 30 min only exhibit critical transitions when the intervals between episodes are 8 h. Our initial results show a proof-of-concept that there exists a critical interval between episodes that drives the state of the system to shift from one regime to another, which could possibly make interventions by medical practitioners feasible.

Effects of Adding Inflammation Triggering Moieties In-Silico at Different Time Range?
In order to mimic post-surgery complications that are persistent, we added ITMs in various durations starting from 30 min of continuous infusion, to 1 h, 2 h and 3 h. Our results are summarized in Fig. 4.

Our results show that the system can no longer neutralize the inflammation when the added insult is sustained for 3 h. This can be deduced based on the profiles of ITMs in the tissue as well as pro-inflammatory cytokines, which portray high values. AP in blood and tissue, however, are depleted.

Intuitively, we are able to show numerically that the duration of added ITMs in the system has prominent effects on ITMs in tissue, pro-inflammatory cytokines, and AP concentrations in blood and in tissue. As the body recuperates after cardiac surgery, there comes a point when the system can no longer neutralize the inflammation. We have shown in the previous section that recurrent episodes of post-surgery complications could tip the balance between health and disease when the time interval reaches 8 h apart. In this section, we show that this critical transition happens when the post-surgery complication is persistent and lasts for 3 h. This is in fact consistent with the findings of Damas et al., where the overall concentration of ITMs within this 3-h duration corresponds to the fatal concentration of ITMs in humans [45].

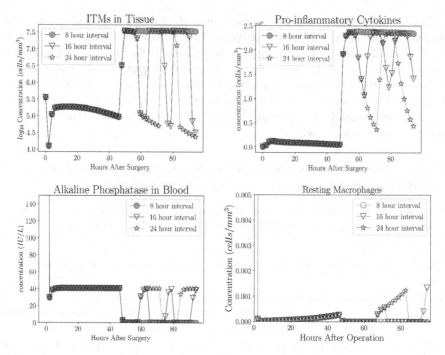

Fig. 3. Human Innate Immune Response to Additional Sources of ITMs at Varying Time Intervals. A non-fatal concentration of $1 \times 10^9 \frac{\text{cells}}{\text{mm}^3}$ ITMs [45] was added at different time intervals starting at 2 days (48 h) after surgery continuously for 30 min to mimic a persistent and recurring post-surgery complication. Our results show that when the interval between each episode decreases to 8 h, the system undergoes a transition where it is no longer able to neutralize the ITMs effectively. Hence, we see that the ITMs in the tissue remain at a stable concentration because the remaining population of immune cells, proteins and enzymes are no longer able to neutralize the ITMs. Moreover, more pro-inflammatory cytokines are induced due to the intense scale of insult. Cells, in the context of our work, also refer to proteins, enzymes, and molecules as a unifying unit in our system.

Critical Transitions in Blood Parameter Timeseries of Patients Undergoing Cardiac Surgery?

The clinical trials data is composed of concentrations of 43 various blood parameters sampled from 52 patients who have undergone cardiac surgery with bypass filter. Time stamps at which the samples were taken were also recorded and indicated in the data. The data was collected from two separate hospitals: Catharina Hospital Eindhoven (The Netherlands), and Zuid Oost-Limburg Hospital (Belgium). The conditions at which the patients have undergone, methods used to obtain the blood parameter samples, as well as time intervals for the data collection were standardized between the hospitals. A more detailed description of the population of patients can be found in [9].

The raw data contains a huge amount of missing data points (58.7%) because not all blood parameters are sampled. Missing values are inevitable in clinical trial data, so it is necessary that the methods are able to deal with this type of data. Numerous

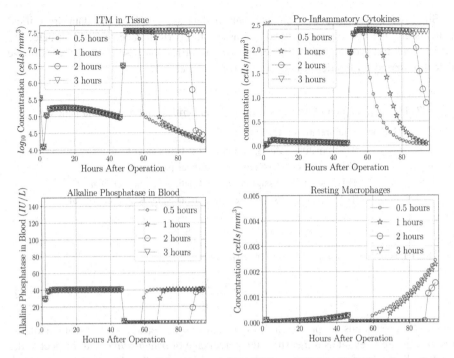

Fig. 4. Human Innate Immune Response to Added ITMs at Different Time Durations. A non-fatal concentration of $1 \times 10^9 \frac{cells}{mm^3}$ ITMs [45] was added at increasing durations starting at 2 days (48 h) after surgery continuously for 30 min, 1, 2, and 3 h to model persistent post-surgery complications. Our results show that when the infusion of ITMs is 3 h, the system undergoes a transition where it is no longer able to neutralize the ITMs effectively

techniques are able to handle missing values. But what is important is that, these techniques should not significantly increase the rate at which *false positives* are being detected or labeling critical patients as non-critical; labeling critical patients as healthy. Otherwise, it makes the signal noisy as well as impractical for medical practitioners to act upon.

Missing values are dealt with by using a simple technique called *bootstrapping*. The basic idea behind bootstrapping involves a repeated random sampling with replacement from the original data to come up with random samples (or bootstrap samples) that have the same size as the original data. Each measurement can be sampled more than once and only within the distribution of the type of patients involved. That is, bootstrapping of non-critical patient is only resampled within the distribution of non-critical patients. The same goes with critical patients, where missing data points are resampled within the distribution of critical patients. In this way we limit the possibility of increasing false negatives in our bootstrapped data. We resampled 100 times to ensure variability in the bootstrap samples.

Since we are dealing with an imbalanced data set – 6% of the data are critical patients and the rest are non-critical, we assess the performance of EWS in detecting critical and non-critical patients by calculating the F1 score based on outcomes of the detection based on the definitions summarized in Table 1.

Table 1. Definition of terms used for assigning critical and non-critical patients.

Symbol	Interpretation	Definition
T_P	True positive	Assigning critical patients as critical
F_P	False positive	Assigning non-critical patients as critical
F_N	False negative	Assigning critical patients as non-critical
T_N	True negative	Assigning non-critical patients as non-critical

The F1 score is calculated based on Eq. (2):

$$F_1 = 2\frac{P \cdot R}{P + R}, \tag{2}$$

where P corresponds to precision, which provides a measure or percentage of the results that are relevant as it measures the percentage of *true positive* with respect to the total predicted positive (*true positive* + *false positive*). R is Recall, which measures the fraction of relevant instances retrieved or what percentage of the actual number of critical patients are correctly identified by the methods. Precision provides a good measure when the cost of *false positive* is high. On the other hand, Recall is a good measure when the cost of *false negative* is high. F1 score provides a good measure that seeks the balance between precision and recall especially when the dataset exhibits an imbalanced class distribution. We correct this from a previously published work, where we used Recall and Precision as measures of our model [23].

Using Early Warning Signals to Pinpoint Blood Parameter Markers of Death
Each time series corresponding to a timely record of a patient's concentrations of blood parameter is assessed on whether a critical transition is detected or not using EWS. This is done by using a rolling window of half the size of the time series data for each methodology for EWS. The Mann Kendall trend test is then used to test the presence of a significant increasing trend. The results are evaluated by calculating for the F1 scores per blood parameter. The motivation here is to pinpoint blood parameters that may be the best option for medical practitioners to focus on, as opposed to doing an extensive scan on all blood parameters that in fact do not reveal signs of critical transitions in patients at all. In this way, resources as well as time are wisely conserved and patients, who are prone to criticalities, can readily be given the immediate treatment they need. We processed both bootstrapped and original data, but the results of our simulations are similar for both data sets. These results are summarized in Fig. 5.

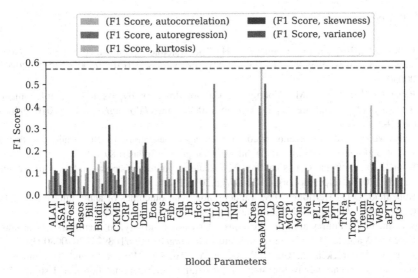

Fig. 5. F1 Score of model output after using early warning signals in detecting critical and non-critical patients. The highest F1 score corresponds to KreaMDRD, which corresponds to the level of creatinine in blood calculated using the MDRD (Modification of Diet in Renal Disease Study) equation with Kurtosis as EWS. This is followed by IL6 (pro-inflammatory cytokine) and LD (Lactate Dehydrogenase) with autoregression and variance as EWS respectively.

4   Summary and Conclusion

Using our model of the human innate immune response for patients undergoing cardiac surgery, we show how HIIS reacts to complications that occur post-surgery. We did this by adding *in-silico ITMs* at 48 h (2 days) after surgery. We showed that an additional concentration of $1 \times 10^9 \frac{\text{cells}}{\text{mm}^3}$ ITMs continuously added for 3 h lead to a rapid and irreversible critical transition from health to disease. In fact, this concentration of ITMs corresponds to the fatal concentration of ITMs documented in literature. We used EWS to detect the presence or absence of critical transitions in clinical trials data of patients undergoing cardiac surgery. Our initial findings show that by using EWS, blood parameter markers such as Creatinine, IL6 and Lactate Dehydrogenase reveal significant presence of critical transitions. IL6, a pro-inflammatory cytokine, was also pinpointed in the *in-silico* model as one of the blood parameters that exhibit critical transitions. However, more experiments need to be done to carefully assess the strength of positive trends that we have detected using EWS.

We have provided a proof-of-concept on the existence of critical transitions in HIIS model, with ITMs as the driving force for this bifurcation. Our initial findings call for a thorough investigation on the conditions at which critical transitions occur in HIIS. More importantly, to explore if the onset of this bifurcation can be detected using known methods in EWS, which we perceive as potentially interesting and helpful to medical practitioners as these might serve as indicators to warn, or better yet prevent the onset of disease leading to fatalities.

References

1. Laffey, J.G., Boylan, J.F., Cheng, D.C.H.: The systemic inflammatory response to cardiac surgery. Anesthesiology **97**, 215–252 (2002). https://doi.org/10.1097/00000542-200207000-00030
2. Paparella, D., Yau, T.M., Young, E.: Cardiopulmonary bypass induced inflammation: pathophysiology and treatment. Update (2002). https://doi.org/10.1016/S1010-7940(01)01099-5
3. Nalysnyk, L.: Adverse events in coronary artery bypass graft (CABG) trials: a systematic review and analysis. Heart (2003). https://doi.org/10.1136/heart.89.7.767
4. Rong, L.Q., Di Franco, A., Gaudino, M.: Acute respiratory distress syndrome after cardiac surgery (2016). https://doi.org/10.21037/jtd.2016.10.74
5. Rubenfeld, G.D., Herridge, M.S.: Epidemiology and outcomes of acute lung injury. Chest (2007). https://doi.org/10.1378/chest.06-1976
6. Weissman, C.: Pulmonary complications after cardiac surgery. In: Seminars in Cardiothoracic and Vascular Anesthesia (2004). https://doi.org/10.1177/108925320400800303
7. Jong, P., Vowinckel, E., Liu, P.P., Gong, Y., Tu, J.V.: Prognosis and determinants of survival in patients newly hospitalized for heart failure: a population-based study. Arch. Intern. Med. (2002). https://doi.org/10.1001/archinte.162.15.1689
8. Lloyd-Jones, D., et al.: Heart disease and stroke statistics - 2010 update: A report from the American heart association (2010). https://doi.org/10.1161/CIRCULATIONAHA.109.192666
9. Presbitero, A., Mancini, E., Brands, R., Krzhizhanovskaya, V.V., Sloot, P.M.A.: Supplemented alkaline phosphatase supports the immune response in patients undergoing cardiac surgery: clinical and computational evidence. Front. Immunol. **9**, 2342 (2018). https://doi.org/10.3389/fimmu.2018.02342
10. Poelstra, K., Bakker, W.W., Klok, P.A., Hardonk, M.J., Meijer, D.K.: A physiologic function for alkaline phosphatase: endotoxin detoxification. Lab. Invest. **76**, 319–327 (1997)
11. Kats, S., et al.: Anti-inflammatory effects of alkaline phosphatase in coronary artery bypass surgery with cardiopulmonary bypass. Recent Pat. Inflamm. Allergy Drug Discov. **3**, 214–220 (2009). IADD-01 [pii]
12. Cohen, J.: The immunopathogenesis of sepsis (2002). https://doi.org/10.1038/nature01326
13. Schulte, W., Bernhagen, J., Bucala, R.: Cytokines in sepsis: potent immunoregulators and potential therapeutic targets—an updated view. Mediat. Inflamm. (2013). https://doi.org/10.1155/2013/165974
14. Trefois, C., Antony, P.M.A., Goncalves, J., Skupin, A., Balling, R.: Critical transitions in chronic disease: transferring concepts from ecology to systems medicine (2015). https://doi.org/10.1016/j.copbio.2014.11.020
15. Liu, R., Yu, X., Liu, X., Xu, D., Aihara, K., Chen, L.: Identifying critical transitions of complex diseases based on a single sample. Bioinformatics **30**, 1579–1586 (2014). https://doi.org/10.1093/bioinformatics/btu084
16. van de Leemput, I.A., et al.: Critical slowing down as early warning for the onset and termination of depression. Proc. Natl. Acad. Sci. U. S. A. **111**, 87–92 (2014). https://doi.org/10.1073/pnas.1312114110
17. May, R.M., Levin, S.A., Sugihara, G.: Complex systems: ecology for bankers. Nature **451**, 893–895 (2008). https://doi.org/10.1038/451893a
18. Quax, R., Kandhai, D., Sloot, P.M.A.: Information dissipation as an early-warning signal for the Lehman Brothers collapse in financial time series. Sci. Rep. **3**, 1898 (2013). https://doi.org/10.1038/srep01898

19. Dakos, V., Scheffer, M., van Nes, E.H., Brovkin, V., Petoukhov, V., Held, H.: Slowing down as an early warning signal for abrupt climate change. Proc. Natl. Acad. Sci. U. S. A. **105**, 14308–14312 (2008). https://doi.org/10.1073/pnas.0802430105

20. Lenton, T.M., Livina, V.N., Dakos, V., van Nes, E.H., Scheffer, M.: Early warning of climate tipping points from critical slowing down: comparing methods to improve robustness. Philos. Trans. R. Soc. A Math. Phys. Eng. Sci. **370**, 1185–1204 (2012). https://doi.org/10.1098/rsta.2011.0304

21. Clements, C.F., Ozgul, A.: Including trait-based early warning signals helps predict population collapse. Nat. Commun. **7**, 10984 (2016). https://doi.org/10.1038/ncomms10984

22. Drake, J.M., Griffen, B.D.: Early warning signals of extinction in deteriorating environments. Nature **467**, 456–459 (2010). https://doi.org/10.1038/nature09389

23. Presbitero, A., Quax, R., Krzhizhanovskaya, V., Sloot, P.: Anomaly detection in clinical data of patients undergoing heart surgery. Procedia Comput. Sci. (2017). https://doi.org/10.1016/j.procs.2017.05.002

24. Pyayt, A.L.: Combining data-driven methods with finite element analysis for flood early warning systems. Procedia Comput. Sci. **51**, 2347–2356 (2015). https://doi.org/10.1016/j.procs.2015.05.404

25. Fisher, W.D., Camp, T.K., Krzhizhanovskaya, V.V.: Crack detection in earth dam and levee passive seismic data using support vector machines. Procedia Comput. Sci. **80**, 577–586 (2016). https://doi.org/10.1016/j.procs.2016.05.339

26. DeAngelis, D.L.: Energy flow, nutrient cycling, and ecosystem resilience. Ecology **61**, 764–771 (1980). https://doi.org/10.2307/1936746

27. Dakos, V., et al.: Methods for detecting early warnings of critical transitions in time series illustrated using simulated ecological data. PLoS One **7** (2012). https://doi.org/10.1371/journal.pone.0041010

28. Wissel, C.: A universal law of the characteristic return time near thresholds. Oecologia (1984). https://doi.org/10.1007/BF00384470

29. Van Nes, E.H., Scheffer, M.: Slow recovery from perturbations as a generic indicator of a nearby catastrophic shift. Am. Nat. (2007). https://doi.org/10.1086/516845

30. Scheffer, M., et al.: Early-warning signals for critical transitions. Nature **461**, 53–59 (2009). https://doi.org/10.1038/nature08227

31. Dakos, V., van Nes, E.H., D'Odorico, P., Scheffer, M.: Robustness of variance and autocorrelation as indicators of critical slowing down. Ecology **93**, 264–271 (2012)

32. Guttal, V., Jayaprakash, C.: Changing skewness: an early warning signal of regime shifts in ecosystems. Ecol. Lett. **11**, 450–460 (2008). https://doi.org/10.1111/j.1461-0248.2008.01160.x

33. Biggs, R., Carpenter, S.R., Brock, W.A.: Turning back from the brink: detecting an impending regime shift in time to avert it. Proc. Natl. Acad. Sci. U. S. A. **106**, 826–831 (2009). https://doi.org/10.1073/pnas.0811729106

34. Held, H., Kleinen, T.: Detection of climate system bifurcations by degenerate fingerprinting. Geophys. Res. Lett. (2004). https://doi.org/10.1029/2004GL020972

35. Ives, A.R., Dakos, V.: Detecting dynamical changes in nonlinear time series using locally linear state-space models. Ecosphere **3** (2012). https://doi.org/10.1890/ES11-00347.1. art58

36. Nguyen, D.M., Mulder, D.S., Shennib, H.: Effect of cardiopulmonary bypass on circulating lymphocyte function. Ann. Thorac. Surg. (1992). https://doi.org/10.1016/0003-4975(92)90319-Y

37. Peretto, G., Durante, A., Limite, L.R., Cianflone, D.: Postoperative arrhythmias after cardiac surgery: incidence, risk factors, and therapeutic management. Cardiol. Res. Pract. (2014). https://doi.org/10.1155/2014/615987

38. Hashemzadeh, K., Dehdilani, M., Dehdilani, M.: Postoperative atrial fibrillation following open cardiac surgery: predisposing factors and complications. J. Cardiovasc. Thorac. Res. (2013). https://doi.org/10.5681/jcvtr.2013.022
39. Diegeler, A., et al.: Humoral immune response during coronary artery bypass grafting : a comparison of limited approach, "Off-Pump" technique, and conventional cardiopulmonary bypass. Circulation (2000). https://doi.org/10.1161/01.cir.102.suppl_3.iii-95
40. Presbitero, A., Mancini, E., Castiglione, F., Krzhizhanovskaya, V.V., Quax, R.: Evolutionary game theory can explain the choice between apoptotic and necrotic pathways in neutrophils. In: 2018 IEEE International Conference on Bioinformatics and Biomedicine (BIBM), pp. 1401–1405. IEEE (2018). https://doi.org/10.1109/BIBM.2018.8621127
41. Presbitero, A., Mancini, E., Castiglione, F., Krzhizhanovskaya, V.V., Quax, R.: Game of neutrophils: modeling the balance between apoptosis and necrosis. BMC Bioinformatics **20**, 475 (2019). https://doi.org/10.1186/s12859-019-3044-6
42. Honda, T., Uehara, T., Matsumoto, G., Arai, S., Sugano, M.: Neutrophil left shift and white blood cell count as markers of bacterial infection (2016). https://doi.org/10.1016/j.cca.2016.03.017
43. Athens, J.W.: Blood: leukocytes. Annu. Rev. Physiol. (2003). https://doi.org/10.1146/annurev.ph.25.030163.001211
44. Summers, C., Rankin, S.M., Condliffe, A.M., Singh, N., Peters, A.M., Chilvers, E.R.: Neutrophil kinetics in health and disease (2010). https://doi.org/10.1016/j.it.2010.05.006
45. Damas, P., et al.: Cytokine serum level during severe sepsis in human IL-6 as a marker of severity. Ann. Surg. **215**, 356–362 (1992). https://doi.org/10.1097/00000658-199204000-00009

Using Individual-Based Models to Look Beyond the Horizon: The Changing Effects of Household-Based Clustering of Susceptibility to Measles in the Next 20 Years

Elise Kuylen[1,2]([✉]), Jori Liesenborgs[3], Jan Broeckhove[4], and Niel Hens[1,5]

[1] Centre for Health Economics Research and Modelling Infectious Diseases (CHERMID), Vaccine and Infectious Disease Institute, University of Antwerp, Antwerp, Belgium
`elise.kuylen@uantwerpen.be`
[2] Discipline Group Computer Sciences, Hasselt University, Hasselt, Belgium
[3] Expertise Center for Digital Media, Hasselt University - Transnational University Limburg, Hasselt, Belgium
[4] IDLab, Department of Mathematics and Computer Science, University of Antwerp, Antwerp, Belgium
[5] I-BioStat, Data Science Institute, Hasselt University, Hasselt, Belgium

Abstract. Recent measles outbreaks in regions with a high overall vaccination coverage have drawn attention to other factors - aside from the overall immunity level - determining the spread of measles in a population, such as heterogeneous social mixing behavior and vaccination behavior. As households are an important context for measles transmission, the clustering of susceptible individuals within households can have a decisive effect on the risk for measles outbreaks. However, as the population ages and household constitutions change over the next 20 years, that effect may change as well. To adequately plan for the control and eventual elimination of measles, we need to understand how the effect of within-household susceptibility clustering will evolve. Individual-based models enable us to represent the different levels of heterogeneity in a population that are necessary to understand the spread of a disease in a highly immunized population. In this paper, we use such an individual-based model to investigate how the effect of household-based susceptibility clustering is expected to change over the next two decades in Flanders, Belgium. We compare different scenarios regarding the level of within-household susceptibility clustering for three different calendar years between 2020 and 2040, using projections for the age distribution of the population, the constitution of households and age-specific immunity levels. We find that a higher level of susceptibility clustering within households increases the risk for measles outbreaks and their potential to spread through the population, in current as well as in future populations.

Keywords: Individual-based model · Vaccination · Measles

© Springer Nature Switzerland AG 2020
V. V. Krzhizhanovskaya et al. (Eds.): ICCS 2020, LNCS 12137, pp. 385–398, 2020.
https://doi.org/10.1007/978-3-030-50371-0_28

1   Introduction

The WHO recommends that 95% of children born in a country should be vaccinated against measles to ensure herd immunity [25]. However, during recent years, measles outbreaks have frequently occurred in regions with a high overall vaccination coverage [7,8]. To control and eventually eliminate measles, we need to understand the mechanisms that enable such outbreaks to happen. The threshold proposed by the WHO is based on the assumption that susceptibility to measles is always homogeneously distributed in a population. This, however, might not be a realistic assumption.

When measles outbreaks occur in a highly vaccinated population, it is often observed that pockets of un-vaccinated individuals are important drivers of these outbreaks [10,20]. Susceptible individuals can be clustered in a variety of ways: geographically, in schools, or in households. Some attention has already been devoted to modeling the effect of the geographical clustering of measles susceptibility [21]. Since households are an important place for disease transmission [17], the presence of multiple susceptible individuals within the same household may have an important impact on the risk for measles outbreaks. This is the reason why we investigated the effect of household-based susceptibility clustering on the risk and persistence of measles outbreaks in a previous study [14]. Using data on the current population of Flanders (Belgium), we found that a higher level of household-based susceptibility clustering leads to an increased risk for measles outbreaks and increases the size of those outbreaks.

However, to adequately plan for the control and elimination of measles, we need to validate if these results will still apply in the future. As time progresses, the age distribution of a population changes, as does the way in which households are constituted [6]. Furthermore, different age groups will be at risk for measles infection in the future [11,15]. To estimate the effects of household-based susceptibility clustering in the future, we need to take these changes in the population into account.

Individual-based models, in which each individual is treated as a unique entity, are very well suited to model these different levels of heterogeneity in a population [24]. They allow us to take into account age- and context-dependent social mixing behavior, as well as age-specific immunity levels and heterogeneous vaccination behavior.

In this paper we use Stride [13], an individual-based model for the transmission of infectious diseases, to examine how the effects of household-based susceptibility clustering on measles outbreaks are expected to change over the next 20 years. To do this, we simulate different scenarios regarding within-household clustering of susceptibility, using projections of the Flemish population and age-specific immunity levels for 2020, 2030 and 2040. To estimate how the effects of household-based clustering evolve over the next two decades, we compare simulation results regarding the risk and persistence of measles outbreaks between the different scenarios and calendar years.

2   Methods

Stride was previously developed by our research group [13]. By supplying different input files to the simulator, it can be used to model a wide variety of populations and (air-borne) infectious diseases. It is an open-source project: the source code can be found in a public Github repository [3]. We will briefly describe the input we supplied to Stride to conduct our study and the way in which certain relevant aspects of the simulator were implemented.

2.1   Population

We used projections for the age distribution of the population of Flanders and the constitution of households for 2020, 2030, and 2040. These projections are based on currently unpublished work, and were provided to us after personal communication with the authors [16]. For each calendar year we examined, we used a population of about 300,000 individuals, 5% of the total population size of Flanders. Aside from in their households, there are three other social contexts in which individuals can contact each other in our model: schools, workplaces, and more general communities.

The distribution of school group sizes by student age is based on registration data collected by the Flemish government in 2019 [23]. We assumed that all children aged between 3 and 18 years of age attend school during weekdays. The distribution of workplace sizes was based on data extracted from Eurostat [5]. We assumed an employment rate of 70% for individuals aged between 18 and 65 years old, based on data for Belgium in 2018 obtained from the Eurostat database [5]. Finally, each individual was also assigned to a separate week- and weekend community, to represent more general contacts made during, respectively, week- and weekend days. These communities consist of about 1,000 individuals each, in line with earlier work by Chao et al. [4].

Age-specific contact rates in each of these contexts - households, schools, workplaces, and communities - are based on a social contact study conducted in Flanders in 2010 and 2011 [12].

2.2   Age-Specific Immunity Levels

To inform our model, we used projections for age-specific immunity levels for Belgium in 2020, 2030 and 2040. These projections are based on a recent study by Hens et al. [11], and were obtained in the same way as described in a recent paper [15]. In Fig. 1, the projected percentage of immune individuals by age can be observed for 2020 (solid blue line), 2030 (dashed orange line) and 2040 (dotted green line).

In Belgium, uptake of a measles-containing vaccine took off on a large scale from 1985 on - although a vaccine had been available on the Belgian market since 1975 [2]. Therefore, we will assume that most individuals born in Belgium before 1985 have acquired natural immunity after surviving a measles infection.

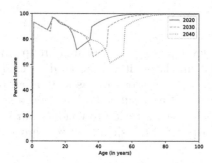

Fig. 1. Projected percentages of immune individuals by age for 2020 (solid blue line), 2030 (dashed orange line) and 2040 (dotted green line). (Color figure online)

For the generation born between 1985 and 1995, the situation is different. As measles no longer circulated, fewer persons were infected and, consequently, natural immunity became less common. However, due to the introduction period of the vaccine, many individuals born during this period were not vaccinated, or were incompletely vaccinated - receiving only one dose instead of the recommend two doses of a measles-containing vaccine. This is reflected in the immunity level of this age group, as can be seen in Fig. 1: for 2020, a dip in the immunity level for individuals aged 25 to 35 years can be observed. As this generation ages, their immunity level is expected to decrease further due to the waning of vaccine-induced immunity: the dip in immunity level that can be observed for 2020 becomes deeper for 2030 and 2040. As a consequence of this, the overall immunity level of the population is also projected to decline in the future (from about 92% in 2020 to a little over 86% in 2040).

Since 1995, vaccination coverage in Flanders has been fairly stable. In the projections we used, it is assumed that this will also remain the case in the future. The recommended age for infants to receive the first dose of a measles-containing vaccine in Flanders is at 12 months [1]. As such, children younger than 12 months of age constitute another group at risk for measles infection. Furthermore, in Flanders, the recommended age to receive the second dose of a measles-containing vaccine is at 10 years of age [1]. Until they receive this second dose, children might not be fully protected against measles.

2.3 Distribution of Immunity: Implementation

To reflect both the age-specific immunity levels discussed above, and the target level of household-based susceptibility clustering that is supplied to the simulator as an input parameter, the following procedure was used to distribute immunity in the population before the start of each simulation.

We assume that individuals born before 1985 have acquired natural immunity, and that the household-based clustering of susceptibility is the result of decisions made about vaccination. As such, we do not take into account household-based susceptibility clustering for individuals born before 1985. To distribute immunity among individuals in this age group, we follow the algorithm described below.

First, we calculate for each age the target number of immune individuals, based on the age-specific immunity level and the total number of individuals of this age present in the simulated population. Next, we draw a random individual from the population. If this individual is susceptible and there are not yet enough immune individuals in the age category that this individual belongs to, we make this person immune. We repeat this process until all age-dependent immunity quota for individuals born before 1985 have been fulfilled.

For individuals born since 1985, we do take into account the target clustering level. The target clustering level is an input parameter between 0 and 1. It represents the probability that if an individual (born since 1985) is immune, all other individuals (also born since 1985) that belong to the same household are also immune. To immunize this part of the population, we use the following procedure.

As before, we calculate the target number of immune individuals for each age category. Next, we select a random individual from a random household. If the selected individual is still susceptible, and if there are not yet enough immune individuals in their age category, we make this person immune. Then, we compare a random draw to the target clustering level. If the random draw is lower than the target clustering level, we immunize all other individuals born since 1985 that belong to the same household. We do this within the limits posed by age-specific immunity quota: we do not allow the actual immunity level for each age to exceed the target immunity level by more than 5%. We repeat this procedure until the total target number of immune individuals born since 1985 has been reached.

2.4 Contact and Transmission Events

Once the population has been initialized, we simulate the actual contact and transmission events that occur in the population. This procedure is explained in more detail in a previous publication [13], but we will briefly repeat relevant elements here. The simulator moves forward in discrete time-steps of one day. During each time-step, we update the presence of individuals in their respective social contact pools (households, schools, workplaces and communities), based on the day of the week and each individual's health status (symptomatic individuals only make contacts within their own household).

Next, we simulate contact and transmission events for each social contact pool. Based on the age- and context-specific contact rates discussed above, we check whether contact occurs between infectious and susceptible members of each contact pool. If an infectious and a susceptible individual contact each other, we compare a random draw to the transmission probability. This transmission probability - $P_{transmission}$ - is supplied as an input parameter to the simulator and represents the probability that if an infectious and a susceptible individual have contact, a transmission of the disease occurs. If the random draw is lower than the transmission probability, we set the health status of the susceptible individual to *exposed*.

Individuals that are not marked as immune from the beginning of the simulation can be in one of 5 health states: susceptible, exposed, infectious, infectious and symptomatic, and recovered. We modeled the natural history of measles as described in previous work [14].

2.5 Scenarios

A common measure used in epidemiology to estimate the transmission potential of a disease is R_0, the basic reproduction number. R_0 represents the average number of new cases one infected individual would cause in a completely susceptible population. However, as R_0 not only depends on the pathogen itself, but also the structure of the population and on social contact behavior, we used $P_{transmission}$ as an input parameter to represent the transmission potential of the disease. Despite this, we want to be able to express our results in terms of R_0, and so we established a relationship between $P_{transmission}$ and R_0 for the different populations that we used in our experiments.

We estimated this relationship for the populations projected for 2020, 2030 and 2040. For each calendar year, we ran 1000 simulations each for 21 values of $P_{transmission}$ between 0 and 1. At the beginning of each simulation, we introduced one infected individual into the population. We kept track of the number of secondary cases caused by this index case in a completely susceptible population, and used this to estimate R_0. We ran each simulation for 30 days: by this time the index case has recovered and can no longer infect any new cases.

We used the *optimize.curve_fit* function in the *scipy* Python package [22] to fit a function through the 21,000 data points we collected for each calendar year, allowing us to estimate a corresponding R_0 value for each $P_{transmission}$ that we used as an input parameter. The *optimize.curve_fit* function uses a non-linear least squares method to fit the data obtained from our simulation runs to a function of the form shown in Eq. (1).

$$a + b \times \log(1 + P_{transmission}) \tag{1}$$

After we established a relationship between $P_{transmission}$ and R_0, we investigated different scenarios regarding household-based clustering of susceptibility for different calendar years. We tested 9 values for $P_{transmission}$ between 0.40 and 0.80 - corresponding to a basic reproduction number of about 11.16 to 19.71 for 2020, 10.95 to 19.36 for 2030 and 10.80 to 19.12 for 2040. We also tested 5 values for the target clustering level between 0 and 1. We compared these 45 scenarios between 3 different calendar years: 2020, 2030 and 2040. For each of these 135 scenarios, we ran 200 stochastic simulations.

At the beginning of each simulation, we introduced one infectious individual into the population. Next, we ran every simulation for 730 days. We assumed that after this period the outbreak had run its full course - as no more new infections were recorded after day 730 in previous, exploratory simulations.

3   Results

3.1   Relationship $P_{transmission} \sim R_0$

As discussed above, we established a relationship between the input parameter $P_{transmission}$ and R_0, the basic reproduction number. As R_0 depends on both the transmission potential of a disease as well as on the structure and social mixing behavior of a population, we estimated this relationship separately for each different population projection we used (2020, 2030, and 2040).

The functions of the form shown in Eq. (1) that we fit for 2020, 2030, and 2040 can bee seen in Table 1. Even though the populations used for 2020, 2030 and 2040 differ from each other in terms of age distribution and household constitution, the relationship between $P_{transmission}$ and R_0 does not appear to change a lot.

Table 1. Coefficients for fitted functions of the form shown in Eq. (1) to estimate the relationship between $P_{transmission}$ and R_0.

Calendar year	a	b
2020	−0.29	34.03
2030	−0.31	33.46
2040	−0.33	33.09

An overview of the simulation results we used to fit these functions can be seen in Fig. 2. For all three fits, we observe that a value for $P_{transmission}$ of 0 corresponds to a value of 0 for R_0. Furthermore, as $P_{transmission}$ increases, we see that both the mean number of secondary cases caused by our index case (solid blue line) and the median number of secondary cases (dashed pink line) increase. These values both roughly follow the shape of a logarithmic function, which is the reason why we chose to fit them to a function of the form shown in Eq. (1).

We also added the fitted function which estimates the relationship between $P_{transmission}$ and R_0 for each calendar year to the plot seen in Fig. 2 (dotted brown line). For all calendar years we tested, the fitted function neatly follows both the mean and median number of secondary cases we observed for each value of $P_{transmission}$.

The basic reproduction number of measles is commonly estimated to be between 12 and 18. In our model, this would thus correspond to a value of $P_{transmission}$ of about 0.44 ($\hat{R}_0 = 12.12$) to 0.72 ($\hat{R}_0 = 18.17$) for 2020, 0.45 ($\hat{R}_0 = 12.12$) to 0.73 ($\hat{R}_0 = 18.03$) for 2030, and 0.46 ($\hat{R}_0 = 12.19$) to 0.75 ($\hat{R}_0 = 18.19$) for 2040. As we ran simulations for values of $P_{transmission}$ from 0.40 to 0.80, we are certain to have included a relevant range of transmission probabilities for measles.

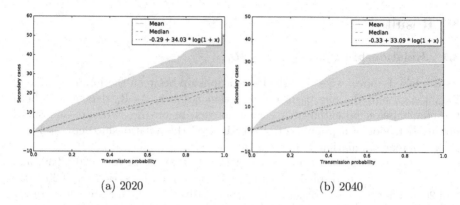

(a) 2020   (b) 2040

Fig. 2. Estimated relationship between $P_{transmission}$ and R_0 in our simulated populations for 2020 (a) and 2040 (b) (results for 2030 not shown here). Besides the mean number of secondary cases for each value of $P_{transmission}$ (solid blue line), the median number of secondary cases (dashed pink line), and the 95% percentile interval of secondary cases observed (gray shape), the fitted function (dotted brown line) is also shown for each tested calendar year. (Color figure online)

3.2   Household Assortativity Coefficient

Our goal in this study was to investigate how the effect of the clustering of measles susceptibility within households evolves as the population ages. We used a target clustering level as an input parameter to inform how immunity is to be distributed in the simulated population. To check in how far changes in this input parameter actually led to more clustering of susceptibility in the simulated population, and to obtain a measure that could be used to measure clustering in reality, we constructed a measure to estimate the actual level of household-based susceptibility clustering in a population: the household assortativity coefficient [14,18].

To calculate this household assortativity coefficient, we first build a network. The nodes in this network correspond to individuals in the population. Each node has a single attribute: an individual is either susceptible to measles at the beginning of the simulation or they are not. An edge connects two nodes if the two individuals represented by the nodes belong to the same household.

Once this network has been constructed, we can calculate the attribute assortativity coefficient, based on the susceptibility attribute. This coefficient describes in how far similar nodes - here in respect to their immunity status - in the network are connected to each other. To construct the networks and calculate the attribute assortativity coefficient, we used the *networkX* Python package [9].

In Fig. 3, the distribution of household assortativity coefficients by target clustering level can be seen for simulations for 2020 (red), 2030 (yellow), and 2040 (green). For all calendar years the same trend can be observed: as the target clustering level is increased, the household assortativity coefficient also increases. Furthermore, there seems to be a consistent relationship between the target clustering level and the household assortativity coefficient for each calendar year.

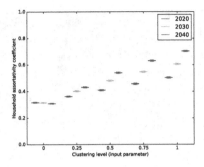

Fig. 3. Distribution of household assortativity coefficients by input clustering level for simulations for 2020 (red), 2030 (yellow) and 2040 (green). (Color figure online)

When we compare the different calendar years, we observe that, in later years, the household assortativity coefficient increases more sharply as the clustering level is increased. This can be expected when we consider that we only took the target clustering level into account for individuals born since 1985. In 2020, this age group constitutes a smaller part of the population than it does in 2030 and in 2040. As such, clustering is applied to a larger part of the population in later calendar years, which is reflected in the corresponding household assortativity coefficients.

3.3 Risk and Persistence of Measles Outbreaks

Effective R. To estimate the impact of household-based susceptibility clustering on the risk for measles outbreaks, we calculated the Effective R for each scenario that we tested. We defined the Effective R as the average number of secondary cases an infected individual causes in a partially immune population. The method we used to calculate the Effective R is similar to how we calculated R_0. For each scenario we tested, we calculated the average number of secondary cases caused by the index case over the 200 stochastic simulations.

In Fig. 4, the Effective R by $P_{transmission}$ and clustering level is shown for 2020 (a), 2030 (b), and 2040 (c). For all calendar years, the same trend can be observed. As expected, increasing the transmission probability leads to an increase in the Effective R. However, when the clustering level is increased while $P_{transmission}$ remains the same, the Effective R also increases.

When we compare the results for the different calendar years to each other, we observe that overall, the Effective R is higher in later calendar years - even for the lowest values of $P_{transmission}$ and a clustering level of 0. This can be explained by the fact that the overall immunity level of the population is also decreasing as time progresses.

Escape Probability. To estimate the risk for measles outbreaks, it is important to know whether it is likely that an outbreak will be contained to a few secondary

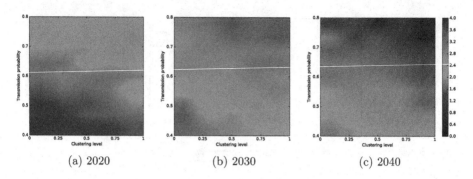

Fig. 4. Heat-maps of the Effective R by $P_{transmission}$ and clustering level for 2020 (a), 2030 (b), and 2040 (c).

cases, or has the potential to spread to a large part of the susceptible population. For this reason, we calculated, for each simulation run, the escape probability. We defined the escape probability as the chance that an individual who is susceptible at the beginning of the simulation will remain uninfected over the entire course of the simulation (730 days). We estimated this probability as shown in Eq. (2), with $N_{susceptible}$ the number of susceptible individuals in the population at the beginning of the simulation and N_{cases} the total number of cases infected over the course of the entire simulation (730 days).

$$\hat{P}_{escape} = \frac{N_{susceptible} - N_{cases}}{N_{susceptible}} \tag{2}$$

In Fig. 5, the average escape probability over 200 stochastic runs for each scenario is shown for 2020 (a), 2030 (b), and 2040 (c). Again, a clear relationship can be observed between $P_{transmission}$ and the escape probability: as the transmission probability is increased, the escape probability decreases. Increasing the clustering level has the same effect: a higher clustering level corresponds to a lower escape probability.

In 2020 (see Fig. 5 (a)), when $P_{transmission}$ is 0.50 (corresponding to an R_0 of about 13.5) and the clustering level is set at 0, the average escape probability is 0.79. However, when $P_{transmission}$ remains the same, but the clustering level is increased to 1, the average escape probability decreases to 0.68. The same trend can be observed for 2030 (b) and 2040 (c).

Outbreak Size. Finally, we looked at the average size of persistent outbreaks. We defined a persistent outbreak as an outbreak that spreads throughout the population, instead of being contained to only a few secondary cases. To determine what we would consider a persistent outbreak, we looked at the frequency of outbreak sizes over all scenarios. We observed that an outbreak either dies out after only a few infections, or spreads through a large part of the susceptible population, with very few cases in between. As such, we chose a threshold of 5,000 infected cases (about 1.5% of the population), in between those two

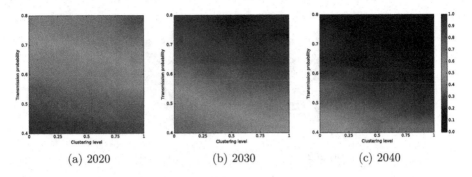

(a) 2020   (b) 2030   (c) 2040

Fig. 5. Average escape probability by $P_{transmission}$ and clustering level for 2020 (a), 2030 (b), and 2040 (c).

extremes. Any outbreak that leads to more than 5,000 infected cases, we thus regard as a persistent outbreak.

In Fig. 6, the average outbreak size of persistent outbreaks by $P_{transmission}$ and by clustering level is shown for 2020 (a), 2030 (b), and 2040 (c). Again, an increase in $P_{transmission}$ leads to an increase in the average size of persistent outbreaks. Furthermore, as the clustering level is increased, the average size of persistent outbreaks also increases.

However, we observe that the increase in average outbreak size as the clustering level is raised from 0 to 1 is smaller for later calendar years, and conversely, is also smaller for larger values of $P_{transmission}$. Indeed, when we consider that the overall population immunity decreases from 2020 to 2040, these are two sides of the same coin: an increase in the transmission potential of a disease can be expected to have the same effects as a decrease in the overall population immunity to a disease.

4 Discussion

We used an individual-based model to investigate how the effect of household-based susceptibility clustering on the risk for measles outbreaks is expected to change over the next 20 years. We compared different scenarios regarding the clustering of susceptibility over three calendar years: 2020, 2030 and 2040. For each of these calendar years we used projections of age distribution, household constitution, and age-specific immunity levels.

We compared the results of these simulations regarding the risk, persistence and size of measles outbreaks. We found that for all tested calendar years, an increase in the level of household-based susceptibility clustering leads to an increase in the risk for measles outbreaks, their potential to spread throughout the population, and the eventual size of these outbreaks. However, for later calendar years, and for higher values of $P_{transmission}$, the effect of susceptibility clustering on the size of persistent outbreaks is less pronounced.

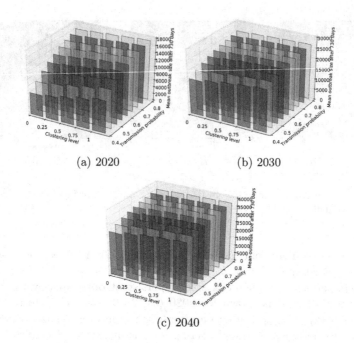

(a) 2020   (b) 2030

(c) 2040

Fig. 6. Average sizes of persistent outbreaks (threshold = 5,000 cases) by $P_{transmission}$ and clustering level for 2020 (a), 2030 (b), and 2040 (c).

As such, we need to take household-based susceptibility clustering into account when modeling the spread of measles both in current and in future populations. Not doing so could lead to an under-estimation of the herd immunity threshold and the efforts needed to control measles. Furthermore, the clustering of susceptibility within households becomes especially important when taking into account the projections for age-specific immunity levels that we used. The generation born between 1985 and 1995, who have a lower immunity level, are now becoming parents, meaning they will share a roof with susceptible infants, who are still too young to be vaccinated. Increasing the immunity level of this generation - for example by organizing a catch-up campaign - should thus be a priority.

There are some limitations that should be taken into account when interpreting the results of this study. First, the populations we used are closed populations: no individuals were born or died over the course of each simulation. As such, there were no new susceptible individuals entering the population. In reality, newborn infants are an important group of susceptible individuals, that may thus have been underrepresented in our study [1].

Furthermore, the projected age-specific immunity levels that we used for 2020, 2030 and 2040 were based on a serological survey [11]. A serological survey measures antibody titres in surveyed individuals, which provide an estimate of humoral immune response. However, cellular immune response mechanisms may offer protection against a disease even when antibody levels are low [19]. As such, the projections we used may underestimate the level of immunity against measles in Flanders.

Finally, this study is a case-study for the population of Flanders, Belgium. It should be verified whether the effect of household-based susceptibility clustering evolves in the same manner for other populations. We also recommend that, in the future, data on the level of susceptibility clustering in different contexts and for different populations should be collected and used to update estimations of the herd immunity threshold for measles.

Acknowledgments. We thank Signe Møgelmose for providing us with projections for the age distribution of the population of Flanders, Belgium and the constitution of households in 2020, 2030, and 2040.

References

1. Agentschap Zorg & Gezondheid: Richtlijn infectieziektebestrijding Vlaanderen - Mazelen (Morbilli), March 2019. https://www.zorg-en-gezondheid.be/sites/default/files/atoms/files/Mazelen%20Final.pdf. Accessed 6 Dec 2018
2. Beutels, P., Van Damme, P., Van Casteren, V., Gay, N., De Schrijver, K., Meheus, A.: The difficult quest for data on "vanishing" vaccine-preventable infections in Europe: the case of measles in Flanders (Belgium). Vaccine 20(29–30), 3551–3559 (2002)
3. Broeckhove, J., Kuylen, E., Willem, L.: Stride Github repository. https://github.com/broeckho/stride
4. Chao, D.L., Halloran, M.E., Obenchain, V.J., Longini, I.M.: FluTE, a publicly available stochastic influenza epidemic simulation model. PLoS Comput. Biol. 6(1), e1000656 (2010)
5. European Commission: Eurostat. https://ec.europa.eu/eurostat/. Accessed 10 Jan 2020
6. Geard, N., et al.: The effects of demographic change on disease transmission and vaccine impact in a household structured population. Epidemics 13, 56–64 (2015)
7. George, F., et al.: Measles outbreak after 12 years without endemic transmission, Portugal, February to May 2017. Eurosurveillance 22(23), 30548 (2017)
8. Grammens, T., et al.: Ongoing measles outbreak in Wallonia, Belgium, December 2016 to March 2017: characteristics and challenges. Eurosurveillance 22(17), 30524 (2017)
9. Hagberg, A., Schult, D., Swart, P.: NetworkX. http://networkx.github.io/. Accessed 10 Oct 2019
10. Hanratty, B., et al.: UK measles outbreak in non-immune anthroposophic communities: the implications for the elimination of measles from Europe. Epidemiol. Infect. 125(2), 377–383 (2001)
11. Hens, N., et al.: Assessing the risk of measles resurgence in a highly vaccinated population: Belgium anno 2013. Eurosurveillance 20(1), 20998 (2015)
12. Hens, N., Goeyvaerts, N., Aerts, M., Shkedy, Z., Van Damme, P., Beutels, P.: Mining social mixing patterns for infectious disease models based on a two-day population survey in Belgium. BMC Infect. Dis. 9(1), 5 (2009)

13. Kuylen, E., Stijven, S., Broeckhove, J., Willem, L.: Social contact patterns in an individual-based simulator for the transmission of infectious diseases (Stride). Procedia Comput. Sci. **108**, 2438–2442 (2017)
14. Kuylen, E., Willem, L., Broeckhove, J., Beutels, P., Hens, N.: Clustering of susceptible individuals within households can drive measles outbreaks: an individual-based model exploration. MedRxiv (2019)
15. Kuylen, E., Willem, L., Hens, N., Broeckhove, J.: Future ramifications of age-dependent immunity levels for measles: explorations in an individual-based model. In: Rodrigues, J.M.F., et al. (eds.) ICCS 2019. LNCS, vol. 11536, pp. 456–467. Springer, Cham (2019). https://doi.org/10.1007/978-3-030-22734-0_33
16. Møgelmose, S., et al.: FORTHCOMING (2020)
17. Muscat, M.: Who gets measles in Europe? JID **204**(Suppl 1), S353–S365 (2011)
18. Newman, M.E.J.: Mixing patterns in networks. Phys. Rev. E **67**, 026126 (2003)
19. Plotkin, S.A.: Complex correlates of protection after vaccination. Clin. Infect. Dis. **56**(10), 1458–1465 (2013)
20. Sugerman, D.E., et al.: Measles outbreak in a highly vaccinated population, San Diego, 2008: role of the intentionally undervaccinated. Pediatrics **125**(4), 747–755 (2010)
21. Truelove, S.A., Graham, M., Moss, W.J., Metcalf, C.J.E., Ferrari, M.J., Lessler, J.: Characterizing the impact of spatial clustering of susceptibility for measles elimination. Vaccine **37**(5), 732–741 (2019)
22. Virtanen, P., Gommers, R., Oliphant, T.E., Haberland, M., Reddy, T., Cournapeau, D., et al.: SciPy 1.0 - fundamental algorithms for scientific computing in Python. arXiv e-prints arXiv:1907.10121 (2019)
23. Vlaams Ministerie van Onderwijs en Vorming: Leerlingenaantallen basis- en secundair onderwijs en hbo5 (2019). https://onderwijs.vlaanderen.be/nl/leerlingenaantallen-basis-en-secundair-onderwijs-en-hbo5. Accessed 8 Jan 2020
24. Willem, L., Verelst, F., Bilcke, J., Hens, N., Beutels, P.: Lessons from a decade of individual-based models for infectious disease transmission: a systematic review (2006–2015). BMC Infect. Dis. **17**(1), 612 (2017). https://doi.org/10.1186/s12879-017-2699-8
25. World Health Organization: Measles vaccines: WHO position paper. Wkly Epidemiol. Rec. **92**(17), 205–228 (2017)

Modelling the Effects of Antibiotics on Gut Flora Using a Nonlinear Compartment Model with Uncertain Parameters

Thulasi Jegatheesan and Hermann J. Eberl[✉]

Department of Mathematics and Statistics, Biophysics Interdepartmental Program,
University of Guelph, Guelph, Canada
{tjegathe,heberl}@uoguelph.ca

Abstract. We present a mathematical model of a microbial community in the human colon, including transport, exchange and metabolic processes. The colon is represented as a bioreactor with separate lumen and mucus microhabitats. The microbial community in the colon is grouped into four biomass functional groups based on their metabolic activity. While computational models present a challenge in the selection of model parameters, they also provide a more systematic approach to addressing uncertainties when compared to *in vivo* and *in vitro* experiments. We conduct an exploratory study on the uncertainty of our input parameters and a simulation study of the perturbation and recovery of gut microbiota from antibiotic exposure. We consider our parameters as random variables drawn from a uniform random distribution to reflect the diversity of gut microbial composition and variability between individuals.

Keywords: Mathematical model · Colon · Simulation experiments · Parameter uncertainty · ADM1

1 Introduction

The human intestine, or colon, is inhabited by a diverse and dynamic microbial community that harbours over 1000 types of microbes [8]. This microbial community plays an important role in human health and has a mutualistic relationship with the host [16]. The intestinal microbiota are responsible for many functions, such as maintaining the gastrointestinal epithelial barrier, preventing pathogens from adhering to intestinal surfaces, and the development of the immune system [17]. A core function of the gut microbiota is carbon fermentation and the breakdown of complex carbohydrates into short chain fatty acids (SCFAs) and gas. The conversion of these carbohydrates into SCFAs contributes to human energy requirements, regulation of intestinal physiology, and immune function [9].

Supported by Natural Sciences and Engineering Research Council of Canada (NSERC): RGPIN-2019-05003, RTI-2016-00080.

© Springer Nature Switzerland AG 2020
V. V. Krzhizhanovskaya et al. (Eds.): ICCS 2020, LNCS 12137, pp. 399–412, 2020.
https://doi.org/10.1007/978-3-030-50371-0_29

The composition of gut microbial communities varies between individuals and may be altered by many factors such as diet, genetics, antibiotic treatment, and environmental factors [3,6,8]. Many *in vivo* studies have shown that an imbalance in gut microbiota composition through loss of microbial and metabolic diversity can result in the development of disease [15]. Due to host-specific gut community composition, as well as confounding environmental factors and lack of easy access to the gut, it is difficult to draw conclusions about underlying function and behaviour of the gut microbiota [19]. *In vitro* experiments aim to provide insight by experimentally simulating the gut using a bioreactor. These experiments are more economical and allow for greater experimental control when compared to human studies, however repeated experiments can be expensive and it is difficult to exactly replicate the physiology of the colon, particularly the mucus layer. Additionally, microbial community composition can vary between bioreactor models due to the source of the inocula [20], making it difficult to draw population-level conclusions. *In silico*, or computational models can be used as a predictive and explanatory tool for biological processes and phenomena. They can allow for the mechanistic investigation of the gut microbiota while overcoming several limitations of *in vivo* and *in vitro* studies. They eliminate the ethical considerations of *in vivo* studies, can incorporate interactions and processes that are not possible *in vitro*, and repeated *in silico* simulations are more economical than *in vivo* trials and *in vitro* experiments.

Presented in this paper is a mechanistic model of the microbiota in the human colon that considers the lumen and mucus environments of the colon, microbial interactions, and transport and exchange processes within the colon. Despite the large inter-individual variation between gut microbial communities, core metabolic pathways are conserved across individuals due to functional redundancy in microbial groups [8]. Rather than representing individual species of microbes, gut microbiota can be organized based on their metabolic activity into biomass functional groups (BFGs). Each BFG is responsible for a particular metabolic pathway based on substrate preference and fermentation products. This is modelled using mass balance equations for biomass growth, substrate consumption and product formation. There are a number of studies that model the gut microbiota using mass balance equations with varying degrees of complexity. The review by Williams et al. [19] includes models that range from a single-species in a simple chemostat reactor to multi-species models with complex physical representations of the colon. Several models are based on the Anaerobic Digestion Model No. 1 (ADM1) [1], which is a generalised model of anaerobic digestion, typically applied to wastewater. The ADM1 was modified by [14] to consider only the degradation of carbohydrates to glucose, lactate, hydrogen, carbon dioxide, water, methane and SCFAs (acetate, propionate, butyrate). Since the majority of substrates available for fermentation in the human colon consist of carbohydrates, [14] excludes the digestion of proteins and lipids, acid-base reactions in ADM1. Many current mechanistic models of anaerobic digestion in the gut are adaptions and extensions of the carbohydrate model in [14], modified to include more realistic physical representations of the colon [7,11], digestion of proteins and lipids [4,5] and further subdivision of BFGs [4,11]. Our model uses the metabolic processes in [11] with a simplified physical representation of the colon. A bacteriostatic antibiotic treatment which inhibits the growth of the microbiota is included. The subsequent recovery of microbiota after perturbation is investigated with probiotics and prebiotics.

Complex models lead to more realistic representations of biological systems but increased complexity often limits possible theoretical analyses. Complex mathematical models typically require computational simulations with large parameter spaces. A limitation of mechanism-based and predictive modeling often stems from the difficulties in model parameter estimation, since experimental data is needed to validate and calibrate model behaviour. These models require clear estimates of stoichiometric and kinetic properties of model processes in order to make meaningful predictions. However, bioreactor operating conditions such as temperature, pH and source of inocula determine the composition of bioreactor microbial communities and can result in variability in the measured parameters. Due to the diversity of gut microbial composition, there is an inherent variability of parameters between individuals. A specific set of parameters could be taken to represent the microbial composition and physiology of a particular individual. As such, the uncertainty of the parameter space must be considered when drawing conclusions from simulation results of computational models of the gut, to account for both variability in measured experimental parameters and inter-individual variability in gut microbiota composition. In [4], each BFG includes multiple strains with kinetic parameters drawn randomly from a uniform distribution. We consider all model parameters, including kinetic and exchange parameters, as a uniform random variables to account for the variability in gut microbiota diversity and physiological variability between individuals.

2 Model Formulation

2.1 Reactor Representation

The human colon can be described as an anaerobic bioreactor. Previous work has described the human colon as a single-stage reactor [4,10,13], a three-stage reactor [10,14] and a plug-flow reactor [7,11,12]. In this study, to capture the lumen and mucus microhabitats, the colon is modelled as a continuous stirred-tank reactor (CSTR) with a lateral diffusion compartment (see Fig. 1). The main (lumen) compartment, has inflow from the upper gastrointestinal tract (GIT) and outflow out of the colon as well as removal of water and metabolites from the lumen into the host. There is no inflow and outflow from the lateral diffusion (mucus) compartment but there is exchange between the lumen and mucus compartments. In a typical system, fiber is the only inflow into the lumen compartment and mucins are endogenously produced in the mucus compartment. The flow rate through the lumen compartment is assumed to be continuous and is chosen such that washout of biomass does not occur.

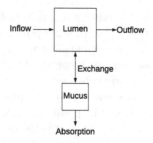

Fig. 1. Reactor representation of the gut with inflow from the upper GIT, outflow out of the colon and exchange between the mucus and lumen compartments

2.2 Reaction Kinetics

In both the lumen and mucus compartment, anaerobic digestion of carbohydrates is carried out by intestinal microbiota. Rather than characterising individual species, microbiota are categorised into four biomass functional groups (BFGs) based on their metabolic activity: sugar degrading biomass (SD), lactate degrading biomass (LD), acetogenic biomass (HDA), and methanogenic biomass (HDM). Due to the redundancy in metabolic pathways in the colon, five main reaction pathways are considered (Fig. 2), each facilitated by a biomass functional group (BFG). Fiber and mucin are assumed to be the only fermentable carbohydrates available to the microbiota. Fiber enters the lumen compartment at a constant rate from the upper GIT and mucin is produced endogenously by the host. The SD are the only BFG that are able to degrade the fiber and mucin into consumable monomer sugars. The hydrolysis of fiber and mucin into monomer sugars by SD is based on Contois kinetics ($\phi_1(c)$ in Table 1). Monod kinetics is applied to the fermentation of sugar by SD, lactate by LD, acetogenesis by HDA and methanogenesis by HDM ($\phi_2(c)$–$\phi_5(c)$ in Table 1).

The entirety of the reaction processes are shown in Table 1. The rows correspond to the processes and the columns correspond to the components. The components consist of the biomass and metabolites and are associated with the lumen of mucus microhabitat that they occupy. The reaction rates are given by $\phi_j(c)$ and include growth kinetics and decay of biomass. Endogenous mucin production is modelled as in [11] and is a source of fiber in the mucus compartment.

2.3 Exchange

The exchange of materials between the lumen and mucus compartment can be categorised into passive and active exchange. Sugar is the only component to passively diffuse between compartments. In general, all other components are exchanged through attachment (lumen to mucus), detachment (mucus to lumen), and absorption (mucus to host).

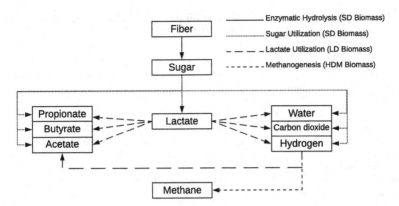

Fig. 2. Metabolic reaction pathways showing the degradation of fiber into short chain fatty acids and gas by the four biomass functional groups

2.4   Complete Model

The matrix representation of the complete model including transport, exchange and reaction kinetics is given by:

$$\frac{d}{dt}\begin{pmatrix} C_L \\ C_M \end{pmatrix} = \begin{pmatrix} S^T & 0 \\ 0 & S^T \end{pmatrix}\begin{pmatrix} \phi(C_L) \\ \phi(C_M) \end{pmatrix} - \begin{pmatrix} -D_{LL} & D_{LM} \\ D_{ML} & -D_{MM} \end{pmatrix}\begin{pmatrix} C_L \\ C_M \end{pmatrix} + \begin{pmatrix} u_L \\ u_M \end{pmatrix}. \quad (1)$$

The variables C_L and C_M are each vectors of length 14 and contain the concentrations of biomass, substrates and products in the lumen and mucus, respectively. S^T is the transpose of the stoichiometric matrix in Table 1 and is of size 14×9. $\phi(C_L)$ and $\phi(C_M)$ are the reaction rates given in Table 1 and are each vectors of length 9. The matrices $D_{..}$ include all transport and exchange terms, are non-negative and diagonal and are of dimension 14×14. The vectors u_L and u_M are each of length 14 and contain the inflow of substrate, biomass of products into the system. In a typical simulation, the only external input into the system is a fiber inflow to the lumen compartment. Since fiber is the only input into the system, u_L contains only one non-zero entry. u_M is a vector of zeros as there is no external input into the mucus compartment.

2.5   Antibiotics Mechanism of Action

A bacteriostatic antibiotic is considered in the simulation studies. The antibiotic mechanism is designed to elicit a nonlethal response from the microbiota by preventing the proliferation of bacteria. Antibiotic action is modelled as a growth inhibition term such that biomass growth is inhibited as antibiotic concentration increases. The growth inhibition term is written as $\frac{K_A}{K_A+A}$, where A is the concentration of antibiotic. In the antibiotic simulation experiments, the input concentration of the antibiotic is constant and held at $A = 1$. As a result, the parameter K_A term solely controls the strength of the antibiotic. Only single continuous doses of finite duration of antibiotics are considered in the following simulation study.

Table 1. Stoichiometric matrix, S, of reaction processes. This matrix S and the reaction rates ϕ correspond to Eq. (1) in Sect. 2.4

For soluble components

Component i	1	2	3	4	5	6	7	8	9	Rate
j Process	S_1	S_2	S_3	S_4	S_5	S_6	S_7	S_8	S_9	
1 Hydrolysis	$Y_{1,1}$									$\phi_1(c)$
2 Glucose utilization	-1	$Y_{2,2}$	$Y_{3,2}$	$Y_{4,2}$	$Y_{5,2}$	$Y_{6,2}$		$Y_{8,2}$	$Y_{9,2}$	$\phi_2(c)$
3 Lactate utilization		-1	$Y_{3,3}$	$Y_{4,3}$	$Y_{5,3}$	$Y_{6,3}$		$Y_{8,3}$	$Y_{9,3}$	$\phi_3(c)$
4 Homoacetogenesis			-1	$Y_{4,4}$				$Y_{8,4}$	$Y_{9,4}$	$\phi_4(c)$
5 Methanogenesis			-1				$Y_{7,5}$	$Y_{8,5}$	$Y_{9,5}$	$\phi_5(c)$

For particulate components

Component i	10	11	12	13	14	Kinetic Rate
j Process	I_1	X_1	X_2	X_3	X_4	
1 Hydrolysis	-1					$\phi_1(c) = \kappa_1 \dfrac{I_1 X_1}{K_1 X_1 + I_1}$
2 Glucose utilization		$Y_{11,2}$				$\phi_2(c) = \kappa_2 \dfrac{S_1 X_1}{K_2 + S_1}$
3 Lactate utilization			$Y_{12,3}$			$\phi_3(c) = \kappa_3 \dfrac{S_2 X_2}{K_3 + s_2}$
4 Homoacetogenesis				$Y_{13,4}$		$\phi_4(c) = \kappa_4 \dfrac{S_3 X_3}{K_3 + S_3}$
5 Methanogenesis					$Y_{14,5}$	$\phi_5(c) = \kappa_5 \dfrac{S_3 X_4}{K_5 + S_3} I_{pH}$ $I_{pH} := \begin{cases} \exp(-3(\frac{pH - pH_U}{pH_U - pH_L})^2) & \text{if } pH < pH_U, \\ 1 & \text{if } pH \geq pH_U \end{cases}$
6 Decay of X_1		-1				$\phi_6(c) = \kappa_{6,1} X_1$
7 Decay of X_2			-1			$\phi_7(c) = \kappa_{7,1} X_2$
8 Decay of X_3				-1		$\phi_8(c) = \kappa_{8,1} X_3$
9 Decay of X_4					-1	$\phi_9(c) = \kappa_{9,1} X_4$

3   Results of Simulation Experiments

3.1   Exploration of Parameter Space

The presented model involves a large multi-dimensional parameter space. In order to increase our confidence in the qualitative predictions of the model, we used a brute force approach to investigating the sensitivity of potentially influential parameter sets. Considering the parameters as random variables, simulations were conducted in replicate, drawing parameters from a uniform random distribution prior to the start of the simulation. The parameters were held constant over the course of a simulation. Default parameter values from the literature are taken as mean values. This approach is used to ensure that the observed qualitative behaviour is typical for a reasonable parameter range. If any parameter sets significantly impact the longterm behaviour of the system, they may require extra consideration in simulation studies. Simulations were performed in replicates of 100 with a parameter upper and lower bound of $\pm 10\%$ the mean value unless otherwise specified. With limited information about the distributions of model parameters, a uniform random distribution was chosen to reflect species heterogeneity with BFGs, host heterogeneity in species composition and to account for variability in bioreactor setups. The uniform distribution allows for parameters to be selected in only a biologically relevant (nonzero) range. The equilibrium concentration of the biomass species, particularly the sugar degrading biomass, are compared to the default equilibrium values to determine the sensitivity of the system to a parameter set.

The default parameters used for simulations were obtained from [11], which were in turn adapted from [1,14]. In general, the parameters that most influence the behaviour of the system can be classified as reaction parameters, which control substrate consumption and biomass growth, and exchange parameters, which control the exchange of materials between the lumen and mucus compartments. Both sets can be determined empirically, with the variability of reaction parameters being more well-studied experimentally. The yield coefficients (Y in Table 1) are derived from the balanced chemical equations describing the conversion of reactants to products and are not varied in this study.

Four preliminary studies were conducted to investigate the sensitivity of the system to the reaction and exchange parameters. The parameter sets are as follows: (1) 10% uncertainty on all reaction parameters performed in 200 replicates, (2) 50% uncertainty on all reaction parameters performed in 200 replicates, (3) 10% uncertainty on exchange parameters performed in 100 replicates, and (4) 10% uncertainty on exchange parameters responsible for the movement of mucins and sugar between compartments in 100 replicates. The sensitivity of the output was determined using the variance of the steady state values of the SD since they are at the top of the metabolic hierarchy and highly influence the behaviour of the other three BFGs. For the reaction parameters, the distribution of steady state values of the lumen SD biomass had a standard deviation of 0.219 and 4.965 for an uncertainty of 10% and 50% respectively. The distribution of steady state values of the mucus SD biomass had a standard deviation of 1.264 and 27.899. Figures 3 and 4 shows the distribution of the steady state values for the sugar degrading biomass with an uncertainty of 10% and 50%. The sensitivity of the output to exchange parameters similarly shows a small variance in SD steady state values for an input uncertainty of 10% on the exchange parameter set. The distributions of steady state values for the full set of exchange parameters and the subset of top level exchange parameters are shown in Figs. 5 and 6.

3.2   Calibrating Antibiotic Strength

Model output was found to be relatively stable in response to uncertainty in reaction and exchange parameters. Since these parameters did not have a significant effect on model output, they were held at their default values for the following simulation studies. As a result, simulating the effects of antibiotics on the gut microbiota only requires the consideration of the parameter K_A and the duration of antibiotic administration. Simulations were conducted by selecting a K_A value within a biologically relevant range. Small values of K_A result in

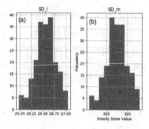

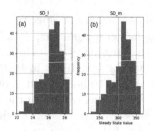

Fig. 3. Distribution of sugar degrading biomass steady state values in lumen (a) and mucus (b) with uncertainty of 10% for reaction and exchange parameters

Fig. 4. Distribution of sugar degrading biomass steady state values in lumen (a) and mucus (b) with uncertainty of 50% for reaction and exchange parameters

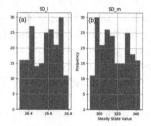

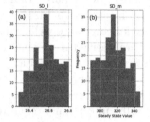

Fig. 5. Distribution of sugar degrading biomass steady state values in lumen (a) and mucus (b) with uncertainty of 10% for top level exchange parameters

Fig. 6. Distribution of sugar degrading biomass steady state values in lumen (a) and mucus (b) with uncertainty of 10% for all exchange parameters

eradication of the population (Figs. 7, 8, 9 and 10) or too drastic of an effect (Figs. 11, 12, 13 and 14) with an antibiotic administered for 14 days.

A K_A value of 60 with a duration of 14 days results in a moderate perturbation of the biomass with a recovery to equilibrium values (Figs. 15, 16, 17 and 18). This value of K_A with a dose period of 14 days is used in the following simulation study as the baseline antibiotic strength. At this strength, the antibiotic effect in the lumen and mucus are similar with the SD biomass being affected the most. This is followed by the LD biomass, with little effect on the methanogens and acetogens. The SCFAs and substrates are not heavily impacted.

3.3 Mitigation Strategies and Recovery Time

Antibiotic use can shift gut microbiota composition from a healthy to an unhealthy equilibrium that can persist for years [8]. Prebiotics and probiotics are often used to alleviate the side effects of antibiotics that result from the loss of gut microbiota. A simulation study was carried out to investigate the effect of different antibiotic mitigation strategies in the form of prebiotics and probiotics. Prebiotics, which are a 'dietary component that fosters the growth of beneficial bacteria' [18] were considered in the form of a single dose of fiber

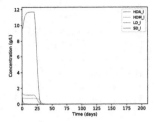

Fig. 7. Lumen concentrations of biomass with 14 day dose of lethal antibiotic, K_A = 1.35

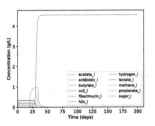

Fig. 8. Lumen concentrations of metabolites with 14 day dose of lethal antibiotic, K_A = 1.35

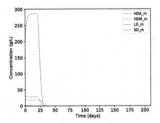

Fig. 9. Mucus concentrations of biomass with 14 day dose of lethal antibiotic, K_A = 1.35

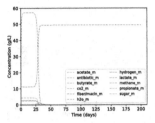

Fig. 10. Mucus concentrations of metabolites with 14 day dose of lethal antibiotic, K_A = 1.35

or sugar before or after antibiotic treatment. Probiotics, which are live microorganisms that provide a health benefit to the host [2,18], were in the form of a discrete input of either all BFGs or only sugar degrading biomass before or after antibiotic treatment. A 14 day dose with a weak antibiotic is investigated, as well as shorter doses of stronger antibiotics. The antibiotic strength K_A and the antibiotic dose duration are adjusted such that the total amount of antibiotic administered remains the same for all dosing regimes. The prebiotics and probiotics are administered at a constant input into the system for 7 days. Recovery time of each simulation is measured as the time from the end of the antibiotic dose to the time when the biomass has reached 99% of the equilibrium values prior to antibiotic perturbation. A treatment is classified as effective if the percent difference in recovery time decreases for at least one biomass species when compared to the recovery time with no treatment. Tables 2 and 3 summarize the efficacy of the different treatments and dosing regimes. For both prebiotic and probiotic treatments, administration during the antibiotic dosing reduced recovery time most consistently. Sugar was a more effective prebiotic treatment when compared to fiber, and a probiotic with only sugar degrading biomass was more effective than a complete probiotic. It was also consistently easier to mitigate the effects of an antibiotic administered at moderate strength for a moderate amount of time when compared to a long course weak antibiotic. It was however not possible to decrease recovery time for strong antibiotics that were

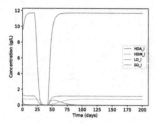

Fig. 11. Lumen concentrations of biomass with 14 day dose of strong antibiotic, $K_A = 1.45$

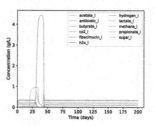

Fig. 12. Lumen concentrations of metabolites with 14 day dose of strong antibiotic, $K_A = 1.45$

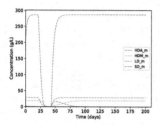

Fig. 13. Mucus concentrations of biomass with 14 day dose of strong antibiotic, $K_A = 1.45$

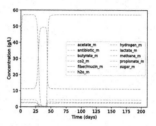

Fig. 14. Mucus concentrations of metabolites with 14 day dose of strong antibiotic, $K_A = 1.45$

administered over less than a day. In all cases, prebiotic and probiotic treatment had little effect on substrate and product concentrations. Figures 19 and 20 show the percent difference in recovery time for the most effective treatment.

Table 2. Summary of effectiveness of prebiotics for recovery of the communities in lumen and mucus compartments for different regimes. Each row represents one simulation, with the X's indicating the treatment regime

Antibiotic short dose	Antibiotic moderate dose	Antibiotic long dose	Before antibiotic	During antibiotic	Sugar prebiotic	Fiber prebiotic	Efficacy
X			X			X	None
	X		X			X	None
		X	X			X	None
X				X		X	None
	X			X		X	Both
		X		X		X	None
X			X		X		None
	X		X		X		Both
		X	X		X		Mucus
X				X	X		None
	X			X	X		Both
		X		X	X		Mucus

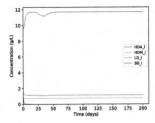

Fig. 15. Lumen concentrations of biomass with 14 day dose of moderate strength antibiotic, $K_A = 60$

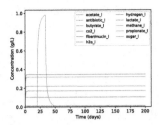

Fig. 16. Lumen concentrations of metabolites with 14 day dose of moderate strength antibiotic, $K_A = 60$

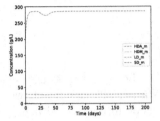

Fig. 17. Mucus concentrations of biomass with 14 day dose of moderate strength antibiotic, $K_A = 60$

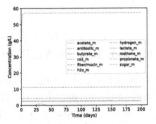

Fig. 18. Mucus concentrations of metabolites with 14 day dose of moderate strength antibiotic, $K_A = 60$

Table 3. Summary of effectiveness of probiotics for recovery of the communities in lumen and mucus compartments for different regimes. Each row represents one simulation, with the X's indicating the treatment regime

Antibiotic short dose	Antibiotic moderate dose	Antibiotic long dose	Before antibiotic	During antibiotic	SD probiotic	Complete probiotic	Efficacy
X			X			X	None
	X		X			X	Lumen
		X	X			X	None
X				X		X	None
	X			X		X	Both
		X		X		X	Lumen
X			X		X		None
	X		X		X		Both
		X	X		X		None
X				X	X		None
	X			X	X		Both
		X		X	X		Both

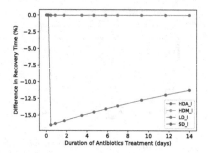
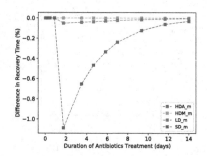

Fig. 19. Lumen concentrations of biomass after treatment with sugar degrading probiotic during antibiotic dosing

Fig. 20. Mucus concentrations of biomass after treatment with sugar degrading probiotic during antibiotic dosing

3.4 Uncertainty in Antibiotic Parameter K_A

To confirm whether the difference in recovery time between prebiotic/probiotic treatment and untreated antibiotic exposure is statistically significant, we conducted a simulation study to compare two regimes of antibiotics, with $K_A = 60$ and $K_A = 30$ as default values and an antibiotic course of 14 days. A subset of prebiotic and probiotic treatments are repeated with randomly generated K_A and growth parameters using the method described in Sect. 3.1. See Table 4 for a summary of the results. The recovery time with no prebiotic or probiotic treatment is taken as the base value for comparisons. A t-test is used to determine whether the recovery time with prebiotic/probiotic treatment with uncertainty on K_A and growth parameters is statistically different when compared to the recovery time with no treatment. Treatment recovery times with p-values less than 0.05 are considered significantly different from recovery times with no treatment. Treatment regimes where there is no improvement in recovery time using default parameter values also show no improvement in recovery time in the sample set.

Table 4. Summary of Effectiveness of Probiotics for Different Regimes

Simulation	K_A	Base	Population mean	Sample mean	Std. dev	p-value
Pre. before	60	11.23	5.409	5.465	0.1815	0.00263
Pre. before	30	14.58	8.677	8.754	0.2697	0.00523
Pre. during	60	11.22	2.075	2.088	0.1056	0.228
Pre. during	30	14.53	4.820	4.842	0.09125	0.0229
Pro. before	60	11.23	0.0	0.0	0.0	N/A
Pro. before	30	14.58	0.0	0.0	0.0	N/A
Pro. during	60	11.23	0.0	0.0	0.0	N/A
Pro. during	30	14.53	4.753	4.748	0.09334	0.573

4   Conclusions

Uncertainty of parameters is an inherent part of computational models, particularly in physiological models with heterogeneity across individuals. Including parameter uncertainty as an intrinsic part of computational experimental design allows for model outcomes to be used in a predictive or explanatory capacity on a population level. For many computational models there is an underlying idea that there is a true set of parameter values. However, for many biological models, parameter values may exist within a range. For the model presented in this study, treating the parameters as random variables helps to address the diversity of microbial gut composition and variability within and between individuals. In this preliminary computational study, we find that variability in parameters do not have a significant impact on long-term model outcomes. A more thorough analysis of the parameter distribution and uncertainty of the input parameters is necessary to draw further conclusions.

Theoretical analyses of mathematical models can provide insight into system behaviour. It is often difficult or not possible to carry out this type of analysis for large and complex models. Due to the hierarchical structure of the metabolic processes in our system, the long-term survival of the gut microbiota are dependent on the survival of the sugar degrading biomass group, which is responsible for the degradation of carbohydrates into usable monomer sugars. Antibiotic mitigation strategies were found to be most effective if they targeted this group by supplementing their population with a probiotic or their diet with a prebiotic. Other applications of this model to human health and disease, such as the role of SCFAs may require the consideration of the full system. However, in the context of recovery after antibiotic treatment, the top level of our system can be studied independently. A reduced system would consist of the equations for fiber, mucin and SD biomass, corresponding to rows 1, 2 and 6 in Table 1. This reduced system allows for theoretical analysis and results in a reduced parameter space. Studying this system could aid in the design of computational studies of the full system, particularly in the scope of microbial recovery after antibiotic perturbation.

References

1. Batstone, D.J., et al.: The IWA anaerobic digestion model no 1 (ADM1). Water Sci. Technol. **45**(10), 65–73 (2002)
2. Hickson, M.: Probiotics in the prevention of antibiotic-associated diarrhoea and Clostridium difficile infection. Ther. Adv. Gastroenterol. **4**(3), 185–197 (2011)
3. Huttenhower, C., et al.: Structure, function and diversity of the healthy human microbiome. Nature **486**(7402), 207 (2012)
4. Kettle, H., Petra, L., Holtrop, G., Duncan, S.H., Flint, H.J.: Modelling the emergent dynamics and major metabolites of the human colonic microbiota. Environ. Microbiol. **17**(5), 1615–1630 (2015)
5. Kettle, H., Holtrop, G., Louis, P., Flint, H.J.: microPop: modelling microbial populations and communities in R. Methods Ecol. Evol. **9**(2), 399–409 (2018)

6. Kolde, R., et al.: Host genetic variation and its microbiome interactions within the Human Microbiome Project. Genome Med. **10**(1), 6 (2018)
7. Labarthe, S., Polizzi, B., Phan, T., Goudon, T., Ribot, M., Laroche, B.: A mathematical model to investigate the key drivers of the biogeography of the colon microbiota. J. Theor. Biol. **462**, 552–581 (2019)
8. Lozupone, C., Stomabaugh, J., Gordon, J., Jansson, J., Knight, R.: Diversity, stability and resilience of the human gut microbiota. Nature **489**(7415), 220–230 (2012)
9. Marchesi, J.R., et al.: The gut microbiota and host health: a new clinical frontier. Gut **65**(2), 330–339 (2016)
10. Moorthy, A.S., Eberl, H.J.: Assessing the influence of reactor system design criteria on the performance of model colon fermentation units. J. Biosci. Bioeng. **117**(4), 478–484 (2014)
11. Moorthy, A.S., Brooks, S.P.J., Kalmokoff, M., Eberl, H.J.: A spatially continuous model of carbohydrate digestion and transport processes in the colon. PLoS One **10**(12), 1–17 (2015)
12. Moorthy, A.S., Eberl, H.J.: compuGUT: an in silico platform for simulating intestinal fermentation. SoftwareX **6**, 237–242 (2017)
13. Motelica-Wagenaar, A.M., Nauta, A., van den Heuvel, E.G., Kleerebezem, R.: Flux analysis of the human proximal colon using anaerobic digestion model 1. Anaerobe **28**, 137–148 (2014)
14. Muñoz-Tamayo, R., Laroche, B., Walter, E., Doré, J., Leclerc, M.: Mathematical modelling of carbohydrate degradation by human colon microbiota. J. Theor. Biol. **266**(1), 189–201 (2010)
15. Reid, G., Younes, J.A., Van Der Mei, H.C., Gloor, G.B., Knight, R., Busscher, H.J.: Microbiota restoration: natural and supplemented recovery of human microbial communities. Nat. Rev. Microbiol. **9**(1), 27–38 (2011)
16. Rowland, I., et al.: Gut microbiota functions: metabolism of nutrients and other food components. Eur. J. Nutr. **57**(1), 1–24 (2017). https://doi.org/10.1007/s00394-017-1445-8
17. Sánchez, B., Delgado, S., Blanco-Míguez, A., Lourenço, A., Gueimonde, M., Margolles, A.: Probiotics, gut microbiota, and their influence on host health and disease. Molec. Nutr. Food Res. **61**(1), 1–15 (2017)
18. Sartor, B.R.: Therapeutic manipulation of the enteric microflora in inflammatory bowel diseases: antibiotics, probiotics, and prebiotics. Gastroenterology **126**(6), 1620–1633 (2004)
19. Williams, C.F., Walton, G.E., Jiang, L., Plummer, S., Garaiova, I., Gibson, G.R.: Comparative analysis of intestinal tract models. Ann. Rev. Food Sci. Technol. **6**, 329–350 (2015)
20. Yen, S., et al.: Metabolomic analysis of human fecal microbiota: a comparison of feces-derived communities and defined mixed communities. J. Proteome Res. **14**(3), 1472–1482 (2015)

Stochastic Volatility and Early Warning Indicator

Guseon Ji[1] , Hyeongwoo Kong[2] , Woo Chang Kim[2] ,
and Kwangwon Ahn[3]([✉])

[1] Graduate School of Future Strategy, KAIST, Daejeon 34141, South Korea
[2] Department of Industrial and Systems Engineering, KAIST,
Daejeon 34141, South Korea
[3] Department of Industrial Engineering, Yonsei University, Seoul 03722, South Korea
k.ahn@yonsei.ac.kr

Abstract. We extend Merton's framework by adopting stochastic volatility to propose an early warning indicator for banks' credit risk. Bayesian inference is employed to estimate the parameters of Heston model. We provide empirical evidence and demonstrate the comparative strength of our risk measure over others.

Keywords: Early warning indicator · Credit risk · Heston model · Bayesian inference

1 Introduction

Monitoring the risk borne by the banking sector and detecting early warning signals were propelled to the forefront of regulation after the catastrophic financial crisis of 2007–08 much of which was attributable to banks such as Lehman Brothers. In finance, Merton's probability of default (PoD) is regarded as an informative and reliable credit risk measure [1].[1] Under this model, the firm's equity is considered as a call option with a strike price equal to the face value of the firm's debt [1–4]. Merton model assumes that the value of the firm's assets follows a lognormal diffusion process that has a constant volatility; however, it is restricted in terms of being able to adequately describe the real world [5].

Hence, we adopt the concept of time-varying volatility, specifically Heston model [6–10] in which volatility is driven by its own mean-reverting stochastic process where log-returns of an asset exhibit heavy tails. Bu and Liao employed stochastic volatility and jumps to explain the time variation in credit default swaps, a proxy for credit risk [11]. Fulop and Li suggested a simulation-based

---

[1] Network approach is also applied to assess the credit risk of banks. Angelini et al. and Khashman employed neural network using the real-world credit approval data of Italy and Germany to evaluate banks' credit risk [13,14]. González-Avella et al. adopted network topology, i.e., loans are interpreted as links between banks (nodes), to examine the interbank credit risk with financial contagion [15].

© Springer Nature Switzerland AG 2020
V. V. Krzhizhanovskaya et al. (Eds.): ICCS 2020, LNCS 12137, pp. 413–421, 2020.
https://doi.org/10.1007/978-3-030-50371-0_30

parameter learning methodology to estimate parameters, and applied their approach to stochastic volatility and jump models [12]. Based on these studies, we
propose an indicator that delivers early warning signals for banks' credit risk,
and compare the performance of our measure with others.

This paper is organized as follows. In the second section, we adopt stochastic
volatility to probability of undercapitalization (PoU) and propose an early warning indicator. The third section explains the parameter estimation strategy, and
we discuss the application of our risk indicator for two US banks in the fourth
section. Finally, the fifth section concludes.

2   Stochastic Volatility and the Effect of Capital Buffer

Among the pool of stochastic volatility models, we pick out Heston model due to
its semi-closed form solution and realistic assumptions such as mean-reversion
of variance and statistical dependence between an asset and its volatility. The
value of a firm at time t, V_t, is assumed to evolve with a stochastic variance, σ_t^2,
that follows a Cox, Ingersoll and Ross process [16,17]

$$dV_t = \mu V_t dt + \sigma_t V_t dW_{1,t} \quad \text{and} \quad d\sigma_t^2 = \kappa(\theta - \sigma_t^2)dt + \sigma_v \sigma_t^2 dW_{2,t},$$

where μ is the growth rate of firm value, κ is the mean reversion speed for the
variance, θ is the mean reversion level for the variance, σ_v is the volatility of
the variance, and $W_{i,t}$ (for $i = 1, 2$) is a standard Brownian motion. The Feller
condition, $2\kappa\theta > \sigma_v^2$, is imposed to ensure that the variance is strictly positive
[18]. It is further assumed that the asset value and its variance are driven by a
correlated stochastic component of $d\langle W_{1,t}, W_{2,t} \rangle = \rho dt$. When the asset return
and the variance are positively correlated ($\rho > 0$), the distribution of return has
a fat right tail [7].

A firm's asset consists of equity and debt. In particular, a bank's equity is
considered as an European call option with a strike price equal to the obligated
debt payment L at the maturity T as $E_T = \max\{V_T - L, 0\}$. Thus, the calculation of PoD with Heston model is as follows. For the simplicity of notation, all
subscripts are suppressed, and the proof is provided in the Appendix.

Proposition: *Let* $x_t = \log V_t$ *and* $v_t = \sigma_t^2$. *PoD admits a semi-analytical
expression*

$$PoD = P(V_T \leq L) = \frac{1}{2} - \frac{1}{\pi} \int_0^\infty \text{Re} \left[\frac{e^{-iu \log L} \varphi(u; 0, x(0), v(0))}{iu} \right] du,$$

where $\varphi(u; t, x, v)$ *takes an exponential linear form as*

$$\varphi(u; t, x, v) = \exp(A(T - t, u) + B(T - t, u)v + iux) \quad \text{for} \quad 0 \leq t \leq T,$$

$$A(t, u) = \left(i\mu u - \frac{\kappa\theta\lambda_2}{a}\right)t + \frac{\kappa\theta}{a} \log \frac{1 - l}{1 - le^{dt}} \quad \text{and} \quad B(t, u) = -\frac{\lambda_2}{a} \cdot \frac{1 - e^{dt}}{1 - le^{dt}}.$$

The terms are defined as

$$a = \frac{1}{2}\sigma_v^2, \qquad b = iu\sigma_v\rho - \kappa, \qquad c = -\frac{1}{2}(u^2 + iu),$$

$$d = \sqrt{b^2 - 4ac}, \qquad \lambda_2 = \frac{b - d}{2} \qquad and \qquad l = \frac{b - d}{b + d}.$$

Unlike general firms, banks are subject to capital adequacy requirement, however PoD simply focuses on a firm's debt-paying ability. Hence, Chan-Lau and Sy proposed distance-to-capital (DC) to address banks' undercapitalization risk [19]. PoU and DC are considered as more conservative measures than PoD and distance-to-default [20,21]. A bank is regarded as undercapitalized once $V_T - L < c \cdot V_T$ holds at time T after debt payment and PoU of a bank can be computed as

$$\text{PoU} = P\left(V_T < \frac{L}{1 - c}\right) = \text{PoD} + P\left(L \leq V_T < \frac{L}{1 - c}\right).$$

We further propose an early warning indicator, namely the effect of capital buffer (ECB). When bank failure looms, the elevated possibility of insolvency risk eats up the capital buffer, and the regulation on a bank's capital plays a lesser role in governing risk. Hence, the ECB drops to small numbers, which can be interpreted as warning signals,

$$\text{ECB} = \frac{\text{PoU} - \text{PoD}}{\text{PoU}}.$$

3   Estimation Strategy

A firm's value and its variance are not directly observable, thus, we need to estimate these variables from equity prices. However, the observed equity prices may be contaminated by the microstructure of noise [22]

$$\log S_t = \log \hat{S}_t(V_t, \sigma_t^2) + \delta\nu_t, \tag{1}$$

where ν_t is *i.i.d.* standard normal random variable. Thus, the fundamental component of equity price is a function of V_t and σ_t^2 [7]

$$\hat{S}_t(V_t, \sigma_t^2) = V_t P_1 - Le^{-r(T-t)}P_2,$$

$$P_j = \frac{1}{2} + \frac{1}{\pi}\int_0^\infty Re\left[\frac{\exp(-iuL)}{iu} \times \exp(C_j + D_j\sigma_t^2 + iu\ln V_t)\right]du,$$

where $u \in R$ is the characteristic index, and C_j and D_j are known functions of the model parameters for $j = 1, 2$.

The estimation can be simplified as the input of observed equity prices $y_{1:t} = \{\log S_1, \cdots, \log S_t\}$, the output of a parameter set $\Theta = \{\mu, \theta, \kappa, \sigma_v, \rho, \delta\}$, and the latent states $x_{1:t} = \{(V_1, \sigma_1^2), \cdots, (V_t, \sigma_t^2)\}$. Then, we apply the sequential Bayesian inference to estimate the parameters and hidden states [12], hence, our objective is to find the joint posterior distribution $p(x_t, \Theta|y_{1:t})$ of states and parameters at each time t. Since there is no analytical solution of the joint posterior distribution, we need to draw samples from this distribution. The underlying idea of sampling is to break up the interdependence of hidden states and fixed parameters

$$p(x_t, \Theta|y_{1:t}) = p(x_t|y_{1:t}, \Theta)p(\Theta|y_{1:t}).$$

Thus, the procedure of sampling from the posterior distribution can be divided into: (i) state filtering $p(x_t|y_{1:t}, \Theta)$; and (ii) parameter learning $p(\Theta|y_{1:t})$. State filtering estimates the probability of latent state variables for a given static parameter set, and we can derive the recursion of the filtering density (the parameter set Θ is suppressed in this step)

$$p(x_t|y_{1:t}) \propto p(y_t|x_t) \int p(x_t|x_{t-1})p(x_{t-1}|y_{1:t-1})dx_{t-1},$$

where $p(x_t|y_{1:t-1}) = \int p(x_t|x_{t-1})p(x_{t-1}|y_{1:t-1})dx_{t-1}$. Suppose that we have a weighted sample to represent the target distribution $p(x_{t-1}|y_{1:t-1})$ at time $t-1$, i.e., $\{(x_{t-1}^{(i)}, \omega_{t-1}^{(i)}), i = 1, \cdots, M\}$, where $\omega_{t-1}^{(i)} \doteq p(y_{t-1}|x_{t-1}^{(i)})$. Then, a new weighted sample $\{(x_t^{(i)}, \omega_t^{(i)}), i = 1, \cdots, M\} \sim p(x_t|y_{1:t})$ can be drawn by a recursive approach: (i) obtain a new sample $x_t^{(i)}$ from $p(x_t|x_{t-1}^{(i)})$; and (ii) assign a weight $\omega_t^{(i)} = p(y_t|x_t^{(i)})$ for each $x_t^{(i)}$.

Parameter learning evaluates the probability of parameter set Θ for the given observed equity prices. For each parameter particle $\{\Theta^{(j)}, j = 1, \cdots, N\}$, the likelihood is

$$\hat{p}(y_{1:t}|\Theta) = \prod_{l=2}^{t} p(y_l|y_{1:l-1}, \Theta)p(y_1|\Theta), \tag{2}$$

where $p(y_l|y_{1:l-1}, \Theta) = \int p(y_l|x_l, \Theta)p(x_l|y_{1:l-1}, \Theta)dx_l$. From state filtering, we already have a sample of $\{x_l^{(i)}, i = 1, \cdots, M\} \sim p(x_l|y_{1:l-1}, \Theta)$, so

$$\hat{p}(y_l|y_{1:l-1}, \Theta) = \frac{1}{M} \sum_{i=1}^{M} p(y_l|x_l^{(i)}, \Theta), \tag{3}$$

then, we can calculate the posterior distribution of Θ

$$p(\Theta|y_{1:t}) = p(y_{1:t}|\Theta)p(\Theta). \tag{4}$$

The transition density $p(x_t|x_{t-1})$ and the likelihood of measurement $p(y_t|x_t)$ are necessary for both state filtering and parameter learning. The transition law is determined by

$$\log V_{t+\tau} = \log V_t + (\mu - \frac{1}{2}\sigma_t^2)\tau + \sigma_t\sqrt{\tau}\varepsilon_{1,t}, \tag{5}$$

$$\sigma_{t+\tau}^2 = \sigma_t^2 + \kappa(\theta - \sigma_t^2)\tau + \sigma_v\sigma_t\sqrt{\tau}\varepsilon_{2,t}, \tag{6}$$

where $\varepsilon_{1,t}$ and $\varepsilon_{2,t}$ follow $N(0,1)$ with correlation ρ, and τ is the time interval of one period. The likelihood of measurement y_t is determined by Eq. (1).

The process of obtaining the posterior distribution from the real data is as follows. Assume that we have $\{(V_{t-1}^{(i)}, \sigma_{t-1}^{2(i)}); i = 1, \cdots, M\} \sim p(x_{t-1}|y_{t-1})$, then we can obtain $p(x_t|y_t)$ from the following steps: (i) draw the next volatility $\sigma_t^{2(i)}$ from $p(\sigma_t^2|\sigma_{t-1}^{2(i)})$ in Eq. (6) and error term $\nu_t^{(i)} \sim N(0,1)$; and (ii) solve the equation $S_t e^{-\delta\nu_t^{(i)}} = \hat{S}_t(V_t^{(i)}, \sigma_t^{2(i)}|P_{1,t-1}, P_{2,t-1})$ for $V_t^{(i)}$, which is a rearrangement of Eq. (1). To solve V_t, Duan and Fulop found an approximate inversion, which can estimate the solution without solving nonlinear equations [22].

Using the above sample, we can obtain the posterior density of measurement from the following steps: (i) calculate $P_{1,t}$ and $P_{2,t}$ based on sample $\sigma_t^{2(i)}$ and $V_t^{(i)}$; (ii) evaluate $\hat{S}_t(V_t^{(i)}, \sigma_t^{2(i)}|P_{1,t}, P_{2,t})$; (iii) calculate the probability density $p(y_t|x_t)$, which follows a normal distribution, i.e., $\log S_t \sim N(\log \hat{S}_t, \delta^2)$, taken from Eq. (1); and (iv) calculate the posterior density of measurement based on Eqs. (2)–(4).

4   Application to Banks

To demonstrate the performance of the ECB, we apply it to Lehman Brothers and Bank of America for the period between 1 April 2006 and 29 August 2008. The capital adequacy ratio c is set as 6.25% for investment banks following the capital rules applied by the Securities and Exchange Commission, and 4% for commercial banks, which is the tier 1 capital adequacy ratio in the Basel Accords. We assume that banks' capital completely consists of equity. Both PoD and PoU show similar movements as displayed in Figs. 1 and 2, and these measures deliver warnings prior to the bankruptcy of Lehman Brothers and the bailout to Bank of America. In reality, the US government provided 25 and 20 billion USD on October 2008 and January 2009, through Troubled Asset Relief Program (preferred stock purchase) to Bank of America. The gap between PoU and PoD indicates the capital buffer (effect of capital adequacy requirement).

Moreover, the shareholders of a bank are considered to be offered put options on the bank's assets through the bank safety net since the depositors' repayment is guaranteed in case of bank run through deposit insurance scheme, which is provided by the Federal Deposit Insurance Corporation in the US. Thus, it is suggested that shareholders had exploited the bank safety net prior to crises through various risk-taking activities [23,24], leading to increases in put value, which can be another early warning indicator. The put value can be calculated from the contingent claim model

$$\hat{S}_t'(V_t, \sigma_t^2) = V_t(P_1 - 1) - L_t e^{-r(T-t)}(P_2 - 1).$$

In the case of Lehman Brothers, the ECB gave an early warning signal in mid–2007, approximately a year earlier than the put value. For Bank of America, the ECB started to decline from the end of 2007, delivering a warning signal in mid–2008, however the put value failed to deliver any warnings. Put differently, the put value of bank safety net is insufficient as an early warning indicator unlike the ECB.

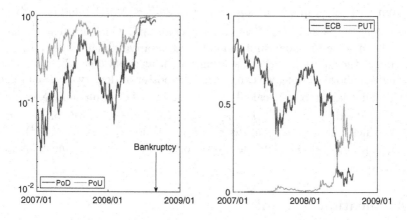

Fig. 1. Credit risk and early warning indicator (Lehman Brothers)

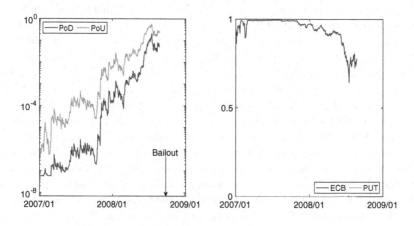

Fig. 2. Credit risk and early warning indicator (Bank of America)

5   Concluding Remarks

We extend the Merton model by incorporating stochastic volatility and the concept of undercapitalization to evaluate credit risk of banks in a more realistic manner. We elect Heston model, in which asset return distribution exhibits non-lognormal properties such as heavy tails. We employ Bayesian inference to estimate parameters. Then, capital adequacy requirement is adopted to better illustrate banks' credit risk, and we further propose an early warning indicator, namely the ECB. The application of the ECB to Lehman Brothers and Bank of America demonstrates the comparative strength of our early warning indicator compared to the put option value of the bank safety net.

Acknowledgements. This research was supported by the Future-leading Research Initiative at Yonsei University (Grant Number: 2019-22-0200; K.A.).

Appendix. Proof of the Proposition

According to Ito's formula, the dynamics of $x(t)$ are given by

$$dx(t) = \left(\mu - \frac{1}{2}v(t)\right) dt + \sqrt{v(t)}dW_V(t).$$

Let $\varphi(u;t,x,v) = E[e^{iux(T)}|x(t) = x, v(t) = v]$. Then, by Gil-Pelaez inversion formula [25], we have

$$P(V(T) < L) = P(x(T) < \log L)$$
$$= \frac{1}{2} - \frac{1}{\pi}\int_0^\infty \text{Re}\left[\frac{e^{-iu\log L}\varphi(u;t,x(0),v(0))}{iu}\right] du.$$

By Feynman-Kač theorem, φ solves the following boundary value problem

$$\frac{\partial\varphi}{\partial t} + \frac{\partial\varphi}{\partial x}\left(\mu - \frac{1}{2}v\right) + \frac{\partial\varphi}{\partial v}\kappa(\theta - v) + \frac{1}{2}\frac{\partial^2\varphi}{\partial x^2}v + \frac{1}{2}\frac{\partial^2\varphi}{\partial v^2}\sigma_v^2 v + \sigma_v v\rho\frac{\partial^2\varphi}{\partial x\partial v} = 0,$$

$$\varphi(u;T,x,v) = e^{iux}.$$

Following the guess by Heston [7], we assume that φ takes an exponential linear form

$$\varphi(u;t,x,v) = \exp(A(T-t,u) + B(T-t,u)v + iux).$$

Because of $\varphi(u;T,x,v) = e^{iux}$ for any x and v, we have boundary conditions for A and B as

$$A(0,u) = B(0,u) = 0.$$

Denoting $\tau = T - t$ and plugging the "guessed" form into a partial differential equation, we get

$$-\left(\frac{\partial A}{\partial\tau} + \frac{\partial B}{\partial\tau}v\right) + iu\left(\mu - \frac{1}{2}v\right) + B\kappa(\theta - v) - \frac{1}{2}u^2 v + \frac{1}{2}B^2\sigma_v^2 v + iu\sigma_v v\rho B = 0.$$

As this holds for any v, we get the following two ODEs

$$\frac{\partial A}{\partial\tau} = i\mu u + B\kappa\theta,$$

$$\frac{\partial B}{\partial \tau} = -\frac{1}{2}iu - B\kappa - \frac{1}{2}u^2 + \frac{1}{2}B^2\sigma_v^2 + iu\sigma_v\rho B.$$

The ODE for B takes the form of Riccati equation:

$$\frac{\partial B}{\partial \tau} = \frac{1}{2}\sigma_v^2 B^2 + (iu\sigma_v\rho - \kappa)B - \frac{1}{2}(iu + u^2) \equiv aB^2 + bB + c,$$

where

$$a = \frac{1}{2}\sigma_v^2, \quad b = iu\sigma_v\rho - \kappa \quad \text{and} \quad c = -\frac{1}{2}(u^2 + iu).$$

According to the solution of Riccati equation, the solution to the ODE for B is given by $B = -\frac{h'}{ah}$, where $h(\tau)$ solves the following ODE

$$h'' - bh' + ach = 0.$$

Denote $d = \sqrt{b^2 - 4ac}$, then h takes the form

$$h(\tau) = D_1 e^{\lambda_1\tau} + D_2 e^{\lambda_2\tau},$$

where $\lambda_1 = \frac{b+d}{2}$ and $\lambda_2 = \frac{b-d}{2}$. Letting $H = \frac{D_1}{D_2}$ and plugging h into B, we get

$$B(\tau, u) = -\frac{\lambda_1 D_1 e^{\lambda_1\tau} + \lambda_2 D_2 e^{\lambda_2\tau}}{a(D_1 e^{\lambda_1\tau} + D_2 e^{\lambda_2\tau})} = -\frac{\lambda_1 H e^{\lambda_1\tau} + \lambda_2 e^{\lambda_2\tau}}{a(H e^{\lambda_1\tau} + e^{\lambda_2\tau})}.$$

Recall the boundary condition, $B(0, u) = 0$. It follows immediately $-H = \frac{\lambda_2}{\lambda_1} \stackrel{\Delta}{=} l$. Thus, we further have

$$B(\tau, u) = -\frac{-\lambda_2 e^{\lambda_1\tau} + \lambda_2 e^{\lambda_2\tau}}{a\left(-\frac{\lambda_2}{\lambda_1}e^{\lambda_1\tau} + e^{\lambda_2\tau}\right)} = -\frac{\lambda_2}{a} \cdot \frac{1 - e^{d\tau}}{1 - le^{d\tau}}.$$

To solve A, note that the indefinite integral is

$$\int B(\tau, u)d\tau = -\frac{\lambda_2}{a}\int\frac{1 - e^{d\tau}}{1 - le^{d\tau}}d\tau = -\frac{\lambda_2\tau}{a} + \frac{\lambda_2}{ad}\left(1 - \frac{1}{l}\right)\log(1 - le^{d\tau}) + \text{const.}$$

Hence,

$$A(\tau, u) = i\mu u\tau - \frac{\kappa\theta\lambda_2}{a}\tau + \frac{\kappa\theta\lambda_2}{ad}\left(1 - \frac{1}{l}\right)\log(1 - le^{d\tau}) + \text{const.}$$

Recall the boundary condition $A(0, u) = 0$. Thus we can solve for the constant term and further simplify the expression to

$$A(\tau, u) = \left(i\mu u - \frac{\kappa\theta\lambda_2}{a}\right)\tau + \frac{\kappa\theta}{a}\log\left(\frac{1 - l}{1 - le^{d\tau}}\right),$$

and the proof is complete.

References

1. Merton, R.: On the pricing of corporate debt. The risk of structure of interest rates. J. Financ. **29**(2), 449–470 (1974)
2. Afik, Z., Arad, O., Galil, K.: Using Merton model for default prediction: an empirical assessment of selected alternatives. J. Empir. Financ. **35**, 43–67 (2016)
3. Bharath, S., Shumway, T.: Forecasting default with the Merton distance to default model. Rev. Financ. Stud. **21**(3), 1339–1369 (2008)
4. Hull, J., Nelken, I., White, A.: Merton's model, credit risk and volatility skews. J. Credit Risk **1**(1), 3–27 (2004)
5. Harada, K., Ito, T., Takahashi, S.: Is the distance to default a good measure in predicting bank failures? A case study of Japanese major banks. Jpn World Econ. **27**, 70–82 (2013)
6. Drăgulescu, A., Yakovenko, V.: Probability distribution of returns in the Heston model with stochastic volatility. Quant. Financ. **2**(6), 443–453 (2002)
7. Heston, S.: A closed-form solution for options with stochastic volatility with applications to bond and currency options. Rev. Financ. Stud. **6**(2), 327–343 (1993)
8. Hull, J., White, A.: The pricing of options on assets with stochastic volatilities. J. Financ. **42**(2), 281–300 (1987)
9. Scott, L.: Option pricing when the variance changes randomly: theory, estimation, and an application. J. Financ. Quant. Anal. **22**(4), 419–438 (1987)
10. Stein, E., Stein, J.: Stock price distributions with stochastic volatility: an analytic approach. Rev. Financ. Stud. **4**(4), 727–752 (1991)
11. Bu, D., Liao, Y.: Corporate credit risk prediction under stochastic volatility and jumps. J. Econ. Dyn. Control **47**, 263–281 (2014)
12. Fulop, A., Li, J.: Efficient learning via simulation: a marginalized resample-move approach. J. Econom. **176**(2), 146–161 (2013)
13. Angelini, E., di Tollo, G., Roli, A.: A neural network approach for credit risk evaluation. Q. Rev. Econom. Financ. **48**(4), 733–755 (2008)
14. Khashman, A.: Neural networks for credit risk evaluation: investigation of different neural models and learning schemes. Expert Syst. Appl. **37**(9), 6233–6239 (2010)
15. González-Avella, J., de Quadros, V., Iglesias, J.: Network topology and interbank credit risk. Chaos Solitons Fractals **88**, 235–243 (2016)
16. Cox, J., Ingersoll Jr., J., Ross, S.: A theory of the term structure of interest rates. Econometrica **53**(2), 385–407 (1985)
17. Dereich, S., Neuenkirch, A., Szpruch, L.: An Euler-type method for the strong approximation of the Cox-Ingersoll-Ross process. Proc. R. Soc. A: Math. Phys. Eng. Sci. **468**(2140), 1105–1115 (2012)
18. Feller, W.: Two singular diffusion problems. Ann. Math. **54**(1), 173–182 (1951)
19. Chan-Lau, J., Sy, A.: Distance-to-default in banking: a bridge too far? J. Bank. Regul. **9**(1), 14–24 (2007)
20. Ahn, K., Dai, B., Kim, C., Tsomocos, D.: Measuring financial fragility in China. Saïd Business School Research Paper No. 2015-23 (2015)
21. Ji, G., Kim, D.S., Ahn, K.: Financial structure and systemic risk of banks: evidence from Chinese reform. Sustainability **11**(13), 1–22 (2019). https://doi.org/10.3390/su11133721
22. Duan, J., Fulop, A.: Estimating the structural credit risk model when equity prices are contaminated by trading noises. J. Econom. **150**(2), 288–296 (2009)
23. Anginer, D., Demirgüç-Kunt, A., Zhu, M.: How does deposit insurance affect bank risk? Evidence from the recent crisis. Workd Bank (2012)
24. Sinn, H.: Casino Capitalism: How the Financial Crisis Came About and What Needs to be Done Now. Oxford University Press, Oxford (2010)
25. Gil-Pelaez, J.: Note on the inversion theorem. Biometrika **38**(3–4), 481–482 (1951)

Boost and Burst: Bubbles in the Bitcoin Market

Nam-Kyoung Lee[1], Eojin Yi[2], and Kwangwon Ahn[3]($\boxtimes$)

[1] KEPCO E&C, Gimcheon-si, Gyeongsangnam-do 39660, South Korea
[2] Korea Advanced Institute of Science and Technology (KAIST), Daejeon 34141, South Korea
[3] Yonsei University, Seoul 03722, South Korea
k.ahn@yonsei.ac.kr

Abstract. This study investigates bubbles and crashes in the cryptocurrency market. In particular, using the log-periodic power law, we estimate the critical time of bubbles in the Bitcoin market. The results indicate that Bitcoin bubbles clearly exist, and our forecast of critical times can be verified with high accuracy. We further claim that bubbles could originate from the mining process, investor sentiment, global economic trend, and even regulation. For policy makers, the findings suggest the necessity of monitoring the signatures of bubbles and their progress in the market place.

Keywords: Cryptocurrency · Bubble · Crash · Log-periodic power law

1 Introduction

Cryptocurrency is a digital asset that relies on blockchain technology[1] and has attracted much attention from the public, investors, and policy makers. Due to its rapid growth with extreme market volatility, concerns and warnings of bubbles in the cryptocurrency market have continued. As per the dot-com bubble of the 1990s, bubbles might occur during the introduction of new technology [1, 2]. Though it is uncertain whether the post-bubble effect on society is good or not [2], bubbles could create disastrous harm and danger as a consequence. Accordingly, understanding bubbles in the cryptocurrency market and implementing effective policies are vital to prevent such disruptive consequences.

As the market has grown, more than 3,000 cryptocurrencies have emerged. However, Bitcoin still holds leadership: Bitcoin consistently dominates others, the so-called altcoins,[2] in terms of market capitalization, number of transactions, network effects, and price discovery role [3–5]. Therefore, the literature largely discusses the turmoil in the cryptocurrency market through the experiences of Bitcoin [4, 6]. Moreover, there have

---

This paper was formerly circulated under the title "The mother of all bubbles: Episode from cryptocurrency market", Master Thesis of Nam-Kyung Lee at KAIST.

[1] "Cryptocurrency" is a medium of exchange designed as a digital currency (and/or asset) which uses cryptography, i.e., blockchain technology, to control the transactions and creation of new units. "Blockchain" is a growing list of blocks that are linked records of data using cryptography.

[2] "Altcoins" refers to all cryptocurrencies other than Bitcoin: the other cryptocurrencies launched after Bitcoin.

© Springer Nature Switzerland AG 2020
V. V. Krzhizhanovskaya et al. (Eds.): ICCS 2020, LNCS 12137, pp. 422–431, 2020.
https://doi.org/10.1007/978-3-030-50371-0_31

been several well-known episodes of bubbles and crashes in the Bitcoin market that seem to have similar patterns at first glance, but the origins and consequences are clearly different because of the internal structure of price formation (e.g., over-/undervaluation of price, market efficiency, and investor maturity) and/or environmental changes (e.g., governmental policy, public sentiment, and global economic status) [7]. In this context, we attempt to evaluate two well-known episodes of Bitcoin bubbles and the crashes that followed.

A bubble indicates excessive asset value compared to market equilibrium or that price is driven by stories and not by fundamentals [8, 9]. A market bubble includes its own limit and can "burn" itself out or experience "explosion" associated with several endogenous processes [10]. Much of the relevant studies attempted to test a speculative bubble with unit root tests [11–13] using the present value model [14, 15]. Besides, the log-periodic power law (LPPL) model has gained much attention with several successful predictions made on well-known episodes about bubbles and crashes [13, 16–22], so it has recently been applied to the Bitcoin market as well [23, 24]. In this study, we aim to address the following questions: Is there a clear signature for bubbles in the Bitcoin market? Can we precisely predict a critical time at which the bubble in the Bitcoin market will burst? What can be the possible inducers that contribute to the emergence of bubbles?

2 Method and Data

2.1 Log-Periodic Power Law (LPPL)

The LPPL combines both the power law and endogenous feedback mechanisms [10, 25, 26]. The former indicates the existence of a short head that occurs rarely but has enormous effects and a long tail that occurs frequently but with much less impact. The latter, which implies underlying self-organizing dynamics with positive feedback, describes herding behavior in the market place such as purchases in a boom and sales in a slump. Therefore, we can predict the critical time through a signature of faster-than-exponential growth and its decoration by log-periodic oscillations [17, 19]:

$$Y_t = A + B(t_c - t)^\beta \{1 + C \cos[\omega \log(t_c - t) + \phi]\}, \tag{1}$$

where $Y_t > 0$ is the log price at time t; $A > 0$ is the log price at critical time t_c; $B < 0$ is the increase in Y_t over time before the crash when C is close to 0; $C \in [-1, 1]$ restricts the magnitude of oscillations around exponential trend; $\beta \in [0, 1]$ is the exponent of the power law growth; $\omega > 0$ is the frequency of fluctuations during a bubble; and $\phi \in [0, 2\pi]$ is a phase parameter.

In this study, we estimate the critical time by mainly following Dai et al. [19]: In the first step, we produce the initial value for seven parameters using a price gyration method [27–29]; and in the second step, we optimize these parameters using a genetic algorithm [30, 31].

2.2 Data

The Bitcoin market operates for 365 days with a 24-h trading system. Thus, we retrieve daily closing prices, i.e., Bitcoin Price Index, from Coindesk[3] at 23:00 GMT. The data span for two periods: from July 2010 to December 2013 and from January 2015 to December 2017. We choose two well-known episodes of bubbles and crashes[4]: Period 1 is from July 18, 2010, when the Bitcoin market was initiated, to December 4, 2013, when the biggest peak reached at 230 USD in 2013; and Period 2 is from January 14, 2015, when the lowest point was reached after the crash in Period 1, to December 16, 2017, when the historical price run reached nearly 20,000 USD. Then we convert the data into log returns:

$$x_t \equiv \ln\left(\frac{p_{t+\Delta t}}{p_t}\right),$$

where p_t represents Bitcoin price at time t.

Table 1 summarizes the descriptive statistics for each period. The data from Period 1 are more volatile, skewed, and leptokurtic[5] than those from Period 2. The high volatility is due to decentralization and speculative demands [27]. Bitcoin exhibits positive skewness, implying more frequent drastic rise in price and investor risk-loving attitude. Lastly, excess kurtosis is obvious, indicating that a high proportion of returns are at the extreme ends of distribution.

Table 1. Summary statistics of the log returns.

	Obs.	Mean	Max.	Min.	Std.	Skewness	Kurtosis
All	1,859	8.86×10^{-3}	6.48×10^{-1}	3.75×10^{-1}	6.76×10^{-2}	1.10	11.23
Period 1	849	1.48×10^{-2}	6.48×10^{-1}	3.75×10^{-1}	8.48×10^{-2}	0.97	8.11
Period 2	734	6.88×10^{-3}	2.54×10^{-1}	-2.19×10^{-1}	4.36×10^{-2}	0.63	6.46

3 Results and Discussion

Table 2 reports parameter estimates from the LPPL including critical time t_c, and clearly shows the proximity of critical time, model implied, to the actual crash. All estimated

---

[3] www.coindesk.com.

[4] In technical perspective, the identification of a peak of the bubble is based on the following two conditions: (i) prior to the peak, there is no higher price than the peak from 262 days before; and (ii) after the peak, there is more than 25% decreased ongoing prices by following 60 days [19, 26]. In the economic context, the bursting of a bubble, for example, dramatic collapse of the market, could bring the economy into an even worse situation and dysfunction in the financial system.

[5] A leptokurtic distribution exhibits excess positive kurtosis: kurtosis has a value greater than 3. In the financial market, a leptokurtic return distribution means that there are more risks coming from extreme events.

parameters are well within the boundaries reported in the literature. Two conditions such as (i) $B < 0$ and (ii) $0.1 \leq \beta \leq 0.9$ ensure faster-than-exponential acceleration of log prices [28]. In particular, the exponent of power law β during Period 2 is 0.2, which corresponds to many crashes on major financial markets $\beta \approx 0.33 \pm 0.18$ [29]. Both angular log-frequencies ω are higher than the range of 6.36 ± 1.56, which is observed in major financial markets [30, 31]. Moreover, the value of ω during Period 2 exhibits the presence of second harmonics at around $\omega \approx 11.5$ [29], which is associated with strong amplitude and hides the existence of fundamental ω, which is common in emerging markets [31].

Table 2. LPPL parameters of the best fit.

Time span[a]	t_c	A	B	β	C	ω	ϕ
Jul 18, 2010–Nov 04, 2013	1,249 Dec 16, 2013	5.44	−0.01	0.90	−0.25	9.70	3.46
Jan 14, 2015–Nov 16, 2017	1,079 Dec 27, 2017	13.19	−1.92	0.20	−0.01	11.28	3.02

[a] We also opt for the data period as follows: (i) the time window starts from the end of the previous collapse (the lowest point since the last crash); (ii) the day with the peak value is the point of the actual bubble burst; and (iii) the endpoint is from one month before the critical point [25, 26].

Figure 1 displays the data and prediction results of the LPPL model for the two periods. Each curve represents the best fit among estimates.[6] A strong upward trend is observed, indicating fast-exponential growth of Bitcoin price and providing clear evidence of a bubble in the market. Moreover, the prediction of critical times, namely corresponding estimate, exhibits the typical hallmark of the critical time of the bubble in 2013 and 2017 (vertical arrows with red color) with high accuracy around the actual crash. The actual crashes of each term date are Dec 4, 2013 and Dec 16, 2017 (see Appendix A).

We hypothesize the plausible origins of the Bitcoin bubbles. First, the decline of a newly mined volume, along with an increase in mining[7] difficulty, generated a supply-driven impact on the market (Appendix B: Fig. 3). Moreover, changes in investor sentiment affected the internal structure of price formation on the market and made the price turbulent, namely boosting and bursting the bubbles (Appendix B: Fig. 4). A negative surprise in global markets, such as consecutive devaluations of the Chinese Yuan

---

[6] To reduce the possibility of false alarms, we conduct two diagnostics to demonstrate the robustness of our prediction. (i) Firstly, using unit root tests (augmented Dickey-Fuller (ADF) and Phillips-Perron (PP) tests) with 0 to 4 lags for each term, we conclude that the residuals do not have a unit root but are stationary at the 1% significance level. (ii) In addition, the crash lock-in plot (CLIP) further confirms that our results, in particular for the value of the predicted t_c, are robust and stable.

[7] Mining is a metaphor for the extraction of valuable things or materials from various deposits. In cryptocurrency, when computers solve complex math problems on the Bitcoin network, they produce new Bitcoins or make the Bitcoin payment network trustworthy and secure by verifying its transaction information.

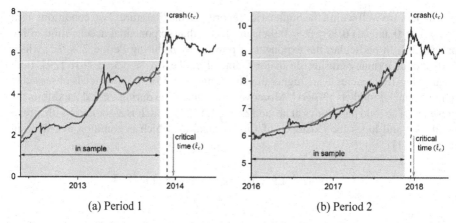

Fig. 1. Logarithm of Bitcoin prices and corresponding alarm.

(CNY) in 2015–2016, also functioned as a catalyst to rebalance the portfolios of Chinese investors (Appendix B: Fig. 5). Furthermore, the Chinese government's banning of cryptocurrency trading on major exchanges in early 2017 provoked the bubble. The sudden prohibition policy merely accomplished a quick transition of the trading currency from CNY to other key currencies, specifically the US dollar (USD) (Appendix B: Fig. 6).

4   Conclusion

In most countries, the regulatory environment appears largely opposed to Bitcoin in the early stages, and one of the key concerns has been the risk of bubbles and the consequent crashes. There is distinct evidence of multiple bubbles in the Bitcoin market, and we have successfully estimated crashes, showing the typical hallmark of the critical times in 2013 and 2017. We attribute the emergence of bubbles to the mining process, investor sentiment, global economic trend, and the regulatory action. The findings strongly suggest the necessity of ex-ante monitoring, and policy makers should be aware that technology, society, and even regulation could induce bubbles.

Acknowledgments. This research was supported by the Future-leading Research Initiative at Yonsei University (Grant Number: 2019-22-0200; K.A.).

Appendix

A. Diagnostic Tests

We further demonstrate the accuracy of our predictions using unit root test for the LPPL residuals of the best fit. As shown in Table 3, for both the ADF and PP tests with 0 to 4 lags, we reject the null hypothesis, meaning that the residuals do not have a unit root but are stationary.

Table 3. Unit root test for the residuals of the best fit.

Time span	ADF	PP
Jul 18, 2010–Nov 04, 2013	0.02**	0.02**
Jan 14, 2015–Nov 16, 2017	0.00***	0.00***

Note: ** and *** indicate that we can reject the null hypothesis that a unit root is present in a time series at the 5% and 1% significance levels, respectively.

Moreover, we implement the CLIPs with the rolling window, tracking the progress of a bubble, and examining whether a probable crash is imminent [32]. The two CLIPs for both periods shown in Fig. 2 indicate that the results of recursive estimations converge upon the actual crash dates. These further demonstrate that the closer the crash is, the more robust and precise result the LPPL model proposes.

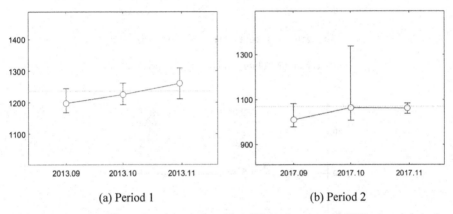

(a) Period 1   (b) Period 2

Fig. 2. CLIPs with rolling estimation window. *Note*: We implement a CLIP by changing the last observation of our sample from one to three months before the actual crash. We can see that our results are stable and robust. Error bars represent the 95% confidence interval for the forecast of critical time.

B. Plausible Origins of the Bitcoin Bubbles

The amount of newly mined volume is one of the distinctive causes of the bubbles. As mining difficulty increased, the supply of new Bitcoins became less, and it, and it further increased the price. As presented in Fig. 3(b), the net spillover from the newly mined volume to Bitcoin price increased sharply around the two standard deviations for the two bubble periods.

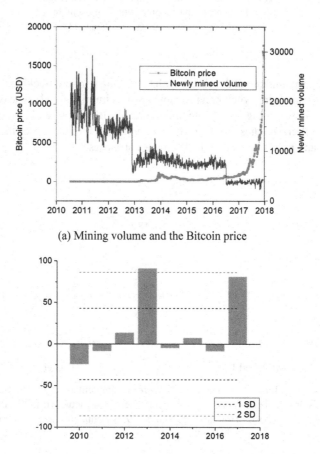

(a) Mining volume and the Bitcoin price

(b) Net spillover effects from newly mined volume to the Bitcoin price

Fig. 3. Newly mined volume and spillover effects.

We use Google Trend as a proxy for investor sentiment [33]. As shown in Fig. 4, the result exhibits a positive linear relationship between the investor sentiment and Bitcoin price during the two periods. Specifically, the Bitcoin price is more sensitive to sentiment proxy during Period 2 than during Period 1. We conjecture that this sentiment could be a substituting factor, further explaining the change in the price.

There were two distinguishable phases in the development of the Bitcoin market in Period 2. The first phase originated from the devaluation of CNY since beginning

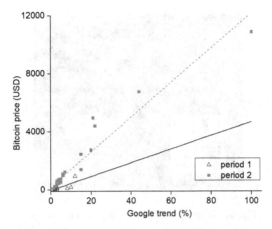

Fig. 4. Investor sentiment and the Bitcoin price.

of 2015. We can recognize the opposite direction in the price of Bitcoin and CNY: appreciating Bitcoin and depreciating CNY from early 2015.

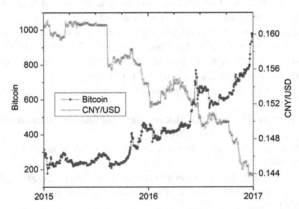

Fig. 5. Global economic trend and the Bitcoin price.

The second phase, in the development of the Bitcoin market in Period 2, was triggered by the Chinese regulatory policy introduced in early 2017. The People's Bank of China implemented policies against three major cryptocurrency exchanges (BTC China, OKCoin, and Huobi) in January 2017, and announced that the Bitcoin trading platform was running outside its business scope and provided shadow financing to investors. Accordingly, this policy resulted in a quick transition of the trading currency from CNY to other key currencies, mainly USD. Since then, the upward trend of the Bitcoin price has continued.

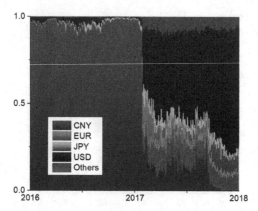

Fig. 6. Regulatory action and trading currencies.

References

1. Johnson, T.C.: Optimal learning and new technology bubbles. J. Monet. Econ. **54**(8), 2486–2511 (2007)
2. Goldfarb, B., Kirsch, D.A.: Bubbles and Crashes: The Boom and Bust of Technological Innovation, 1st edn. Stanford University Press, Stanford (2019)
3. Ciaian, P., Rajcaniova, M.: Virtual relationships: short-and long-run evidence from Bitcoin and altcoin markets. J. Int. Financ. Mark. Inst. Money **52**, 173–195 (2018)
4. Gandal, N., Halaburda, H.: Can we predict the winner in a market with network effects? Competition cryptocurrency market. Games **7**(3), 16 (2016)
5. Brauneis, A., Mestel, R.: Price discovery of cryptocurrencies: Bitcoin and beyond. Econ. Lett. **165**, 58–61 (2018)
6. Osterrieder, J., Lorenz, J.: A statistical risk assessment of Bitcoin and its extreme tail behavior. Ann. Financ. Econ. **12**(1), 1750003 (2017)
7. Sornette, D., Cauwels, P.: Financial bubbles: mechanisms and diagnostics. Rev. Behav. Econ. **2**(3), 279–305 (2015)
8. Shiller, R.J.: Market volatility and investor behavior. Am. Econ. Rev. **80**(2), 58 (1990)
9. Garber, P.M.: Famous First Bubbles: The Fundamentals of Early Manias. MIT Press, Cambridge (2001)
10. Sornette, D.: Why Stock Markets Crash: Critical Events in Complex Financial Systems. Princeton University Press, Princeton (2017)
11. Cheung, A., Roca, E., Su, J.J.: Crypto-currency bubbles: an application of the Phillips–Shi–Yu (2013) methodology on Mt. Gox bitcoin prices. Appl. Econ. **47**(23), 2348–2358 (2015)
12. Corbet, S., Lucey, B., Yarovaya, L.: Datestamping the Bitcoin and Ethereum bubbles. Financ. Res. Lett. **26**, 81–88 (2018)
13. Geuder, J., Kinateder, H., Wagner, N.F.: Cryptocurrencies as financial bubbles: the case of Bitcoin. Financ. Res. Lett. **31**, 179–184 (2019)
14. Campbell, J.Y., Shiller, R.J.: Cointegration and tests of present value models. J. Polit. Econ. **95**(5), 1062–1088 (1987)
15. Moosa, I.A.: The bitcoin: a sparkling bubble or price discovery? J. Ind. Bus. Econ. **5**, 1–21 (2019)
16. Johansen, A., Ledoit, O., Sornette, D.: Crashes as critical points. Int. J. Theor. Appl. Financ. **3**(2), 219–255 (2000)

17. Zhou, W.X., Sornette, D.: Evidence of a worldwide stock market log-periodic anti-bubble since mid-2000. Phys. A **330**(3–4), 543–583 (2003)
18. Filimonov, V., Bicchetti, D., Maystre, N., Sornette, D.: Quantification of the high level of endogeneity and of structural regime shifts in commodity markets. J. Int. Money Financ. **42**, 174–192 (2014)
19. Dai, B., Zhang, F., Tarzia, D., Ahn, K.: Forecasting financial crashes: revisit to log-periodic power law. Complexity **2018**, 4237471 (2018). https://doi.org/10.1155/2018/4237471
20. Jang, H., Ahn, K., Kim, D., Song, Y.: Detection and prediction of house price bubbles: evidence from a new city. In: Shi, Y., et al. (eds.) ICCS 2018. LNCS, vol. 10862, pp. 782–795. Springer, Cham (2018). https://doi.org/10.1007/978-3-319-93713-7_76
21. Jang, H., Song, Y., Sohn, S., Ahn, K.: Real estate soars and financial crises: recent stories. Sustainability **10**(12), 4559 (2018). https://doi.org/10.3390/su10124559
22. Jang, H., Song, Y., Ahn, K.: Can government stabilize the housing market? The evidence from South Korea. Phys. A **550**(15), 124114 (2020). https://doi.org/10.1016/j.physa.2019.124114
23. Wheatley, S., Sornette, D., Huber, T., Reppen, M., Gantner, R.N.: Are Bitcoin bubbles predictable? Combining a generalized Metcalfe's law and the log-periodic power law singularity model. R. Soc. Open Sci. **6**(6), 180538 (2019)
24. Gerlach, J.C., Demos, G., Sornette, D.: Dissection of Bitcoin's multiscale bubble history from January 2012 to February 2018. R. Soc. Open Sci. **6**(7), 180643 (2019)
25. Sornette, D., Johansen, A.: Significance of log-periodic precursors to financial crashes. Quant. Financ. **1**(4), 452–471 (2001)
26. Brée, D.S., Joseph, N.L.: Testing for financial crashes using the log periodic power law model. Int. Rev. Financ. Anal. **30**, 287–297 (2013)
27. Sapuric, S., Kokkinaki, A.: Bitcoin is volatile! Isn't that right? In: Abramowicz, W., Kokkinaki, A. (eds.) BIS 2014. LNBIP, vol. 183, pp. 255–265. Springer, Cham (2014). https://doi.org/10.1007/978-3-319-11460-6_22
28. Lin, L., Ren, R.E., Sornette, D.: The volatility-confined LPPL model: a consistent model of 'explosive' financial bubbles with mean-reverting residuals. Int. Rev. Finan. Anal. **33**, 210–225 (2014)
29. Johansen, A.: Characterization of large price variations in financial markets. Phys. A **324**(1–2), 157–166 (2003)
30. Johansen, A., Sornette, D.: Shocks, crashes and bubbles in financial markets. Brussels Econ. Rev. **53**(2), 201–253 (2010)
31. Zhou, W.X., Sornette, D.: A case study of speculative financial bubbles in the South African stock market 2003–2006. Phys. A **388**(6), 869–880 (2009)
32. Fantazzini, D.: Modelling bubbles and anti-bubbles in bear markets: a medium-term trading analysis. In: The Handbook of Trading, 1st edn, pp. 365–388. McGraw-Hill Finance and Investing (2010)
33. Kristoufek, L.: BitCoin meets Google Trends and Wikipedia: quantifying the relationship between phenomena of the Internet era. Sci. Rep. **3**, 3415 (2013)

Estimation of Tipping Points for Critical and Transitional Regimes in the Evolution of Complex Interbank Network

Valentina Y. Guleva[(⊠)]

ITMO University, Kronverkski 49, 199034 Saint-Petersburg, Russian Federation
guleva@itmo.ru

Abstract. We consider an agent-based model of dynamic interbank network, evolving under several influential factors. Co-evolution is formally performed by the connection between node states and topology of interaction, and vice versa. During the simulation, network evolves to critical regime, corresponding to cascading behaviour in the system, through transitional one. Results show these global regimes correspond to dynamics at micro-level with three types of node states. On the base of formal model of system evolution and regimes formal definitions we estimate the starting point of cascading behaviour and determine number of iterations before its early warning signal – the start of transitional regime. Experiment is made for the interbank market model, nevertheless, possible applications are not restricted by the case. We show, that the obtained estimations allow for appropriate prediction of starting points of critical and transitional regimes (which correspond to cascading behaviour and its early warning signal) and explanation of observed dynamics in the evolution of banking system model under fund infusion scenario.

Keywords: Complex dynamic network · Interbank market · Analytical estimation · Regimes · Criticality · Tipping point

1 Introduction

Complexity of systems, emerging from element interactions, their inner structure, and dynamic processes affecting network evolution, is hard to explore and predict due to wide number of affecting factors. Analytical methods are usually differential models, which assume fully connected graph for interactions between system elements. Effects of topology of nodes interaction on further system evolution is considered in simulation models. They allow for reproduction of local dynamics by means of agent-based approach in the combination with graph models. Agent-based models involve small changes and dynamics at node-level and

This research is financially supported by The Russian Science Foundation, Agreement #19–71–00153.

V. V. Krzhizhanovskaya et al. (Eds.): ICCS 2020, LNCS 12137, pp. 432–444, 2020.
https://doi.org/10.1007/978-3-030-50371-0_32

consider their effect on network structure and further changes in node states arising from their interactions with other nodes. Nevertheless, the number of details and heterogeneities in structure and behaviour obstruct analytical prediction and analysis of complex systems.

Analytical estimations are useful for models verification and validation. The validation process of agent-based models is restricted by observed data segment. Large number of model parameters puts trajectory of system evolution to a multidimensional surface, containing numerous possible trajectories [6]. Therefore, the observed data segment, used for validation, can be in several different possible trajectories, leaving a system still unpredictable, despite model validity. Other usage of analytical estimations is early-warning signals. They are usually explored to predict catastrophe bifurcations and critical failures in time-evolving systems, and are associated with systemic characteristic, served as a predictor [18]. Banking systems stability is one of the applied issues related to emergent criticality in systems, which were studied from the points of network effects, default prediction [15], single bank stress-tests [14], early-warning signals.

Here, we make an attempt to predict critical failure of interbank market and corresponding early-warning signals analytically on the base of system organisation and local dynamics, resulting in interbank network evolution. We decompose initial agent-based model of interbank market into basic local actions, formalise them, and introduce structure of interactions between agents. Node states are real-valued, which allows for consideration of individual bank effects on systemic stability. Generality of system performance allows for application of the suggested approach to wide variety of systems and to generalise some models widely used in epidemiology, ecology, etc.

For the evolution process we consider three regimes, referred to as normal, transitional, and critical. Normal regime demonstrates no special warnings, but it can be unsteady and lead to criticality. Transitional regime is related to changes in dynamics of system state variable, before a cascade, when dramatical changes are not available for observation, but they are able to be. In literature this regime is associated with early warning signals. Critical regime is associate with cascading behaviour. We build mathematical model and estimate points of transitions between regimes analytically, in contrast to early-warning signals, based on observed state variable changes [21] (like "critical slowing down"). The suggested approach application is shown for the case of interbank network with Poison structure of connections and for interbank market simulation model with varied counterparty choice, resulting in various network structures. Transition to cascading behaviour of an interbank network is taken as a case.

2 Literature

Natural systems usually evolve under internal and external drivers. The combinational effects of these drivers may shift a system to qualitatively new states, so called phase transitions, while the corresponding points are called tipping points [16]. In biological systems these transitional processes are associated with homeostasis, when positive and negative feedback links lead a system back to its stable state (negative feedback) or amplify external effects allowing for the transition to a qualitatively new state or phase [12]. In the combination with interaction patterns between elements of a system this phenomena results in self-organisation [23], and, in special cases, to self-organised criticality [20]. This can be observed as default cascades in banking systems, as synchronisation in the case of oscillators, or as avalanche in a sandpile model. In general case, there are system elements characterised by states, some results of interactions affecting elements states, and a system state, related to the combination of element states, which we call critical and want to estimate.

Transitional behaviour (preceding critical one), as it is explored in this paper, is well-presented by the avalanche model of a sandpile. Learning sandpile behavior [5] shows the systemic stability question may be reduced to the criticality of individual grains. Bouchaud et al. considers two types of grains, stable and rolling, and find a critical angle when grain turn to instable state. Systemic properties of human-made, economical and socio-technical systems can also be reduced to the local dynamics of their elements. Nevertheless, their criticality may be human made which makes the dynamics and corresponding transitions dependent on verbal definitions, notions, and regulation mechanisms. For example, bank defaults in banking systems may initialize default cascades, and in this case banks correspond to grains and a default cascade to an avalanche. Nevertheless, regulatory restrictions, applied to banks, can be considered as a critical state vector, like the critical angle of a grain, and may affect systemic evolution. This interdependence and systemic variability give rise not only to a problem of systemic phase transitions, emerging from systemic elements states, but also to the inverse problem of regulatory restrictions optimisation. But in the second case functional properties of a system and corresponding formulations of criticality can also play an important influence on resulting requirements to systemic elements.

In this section, we observe existing methods of analytical estimation of agent-based systems and network dynamics, and results related to criticality in their evolution and early-warning signals.

2.1   Agent-Based Models Estimation

Agent-based models tend to present accurate copies of observed reality and usually take many parameters and model combinations to reach the required accuracy. Nevertheless, such multidimensionality results in difficulties in model estimation, analysis, and calibration. Problem of formal description of agent-based models is not new in literature [11]. Hinkelman et al. argue, agent-based models are usually poorly formalised and suggest field theory application for their formal representation. Nevertheless, the suggested framework is aimed at providing a uniform way of agent-based system description and does not imply predictability. Laubenbacher [17] suggest formal methods like sequential dynamical systems over Boolean set of node states. Both studies propose frameworks for agent-based systems formalisation, but there are no means for state estimation, and the set of element state variables is discrete. Formalisation and estimation of agent-based models are not so developed for the best of our knowledge.

2.2 Early-Warning Signals

Early-warning signals are strongly related to questions of predictability and criticality, which are of great importance for wide range of applications. In particular, financial systems suffer from repeated crises, so authors claim, early warning signals are of great interest [13,25]. Others show how dynamics of a system at micro-level results in critical behaviour at system-level, showing synchronisation [1] as mathematically explored phenomena, homeostasis as a complex of positive and negative feedback mechanisms, resulting in regime shifts [12], and applications to other areas, involving fractal nature of river and its evolution [24]. An attempt to estimate long-term influence of node state on criticality is performed in [8], where authors introduce early-warning signal on the base of change in real-valued node state.

2.3 Complex Networks Dynamics Prediction/analysis

The most developed literature area, related to our research field, is prediction and analysis of the network evolution. Lambert and Vanni [16] explore changes of topological state variable in the dynamic graph model with edges addition and deletion at micro-level[1], and show that its fluctuations given by the derived master-equation are more significant than for the case of mean-field approximation. Nevertheless, the model does not capture influence of node state effects on micro-level dynamics.

Sole et al. [22] provide a dynamical model of node state change with consideration to neighbourhood influence according to connectivity matrix. Element states are of two values $s_i \in \{0; 1\}$, for extinct and for alive; when $t = 0 \Rightarrow \forall i \; s_i = 1$. Hernandez-Urbina and Herrmann [10] provide evolution of state variable in the iterative form with consideration of external forces impact and the influence of node states in the combination with adjacent edge weights. Nevertheless, dynamics of a network and its relation to node states are not captured.

In contrast to these methods, we consider graduate change of real-valued node states and imply three kinds of dynamics of links creation at micro-level, depending on node state value. This results in three patterns of network formation affecting changes of node states and resulting in three regimes of systemic evolution.

3 Method

3.1 Interbank Market Model

We consider an interbank network as a directed graph where nodes correspond to banks and links correspond to interbank exposures. Each bank is characterised by its balance sheet, having interbank and external assets and liabilities, s. t. assets

---

[1] The Generators–Destroyers model [19,26].

correspond to output edges, liabilities correspond to input edges. External assets and liabilities reflect bank interactions with customers and firms, and interbank exposures are related to weighted sums of output and input edges, adjacent to a considered bank. The difference between assets and liabilities determines bank state and is referred to as net worth [2, 3, 7, 9]. Changes of bank state (resulting from balance sheet changes) may initiate activity on the interbank market, which is aimed at enhancing individual bank state. In this way, dynamics on node-level in the combination with counterparty choice strategy results in patterns of network topology evolution [4]. Lack of assets initiates output links creation, and lack of liabilities results in the creation of input links, adjacent to a node. Changes in network structure in turn affect bank states, resulting in further changes. In combination with initial parameters and network configuration, the dynamics of interbank network is affected by different factors, as it is illustrated in Fig. 1.

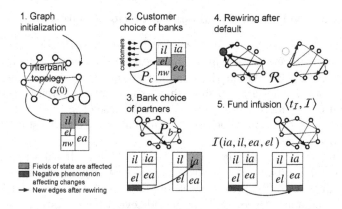

Fig. 1. Factors affecting the evolution of interbank network [9]

Therefore, the above mentioned dynamics can be decomposed to the following components: i) the bank model determines node state and is related to network structure (by definition) and activity triggers; ii) the network model connects banks and is related to the changes in their states; iii) the model of counterparty choice, resulting in emerging network topology; iv) triggers of link formation; v) bank default condition. These components provide two ways of evolution, driven by basic triggers of link formation and default condition. Prevalence of one basic mechanism over others provide transitions between normal, transitional, and critical regimes, associated with closeness to cascading behaviour. This will be formally shown further in the paper.

3.2   Formal Dynamics of Interbank Network Evolution

Let $S(t) = \{s_i(t)\}, i \in \{1, \ldots, N\}$ be a state of an evolving system of N elements with their states $s_i \in \mathbb{R}$. States are related to bank balance sheet in the interbank market model, where banks represent system elements. They interact with each other, which is formally reflected by a directed evolving graph $\Gamma(t) = \{\gamma_{ij}\}$, where nodes correspond to elements and are attributed by states. Since the structure of interactions is connected to node states, addition of output edges positively contributes to a state, while input edges negatively contribute to it. In a static case, that means $s_i = \sum_{k=1}^{N} (\gamma_{ik} - \gamma_{ki}) + C$, where C is summarised contribution of factors not related to topology. Coming to dynamics, there are triggers related to node states and resulting in different patterns of topological changes.

Types of Node States and Corresponding Triggers. Type of node states described in this section are met in literature related to interbank markets as stressed and defaulted banks [2]. Here, node states are taken real-valued to reflect the possibility of their graduate decrease. Nevertheless, number of possible reactions at micro-level, resulting in corresponding macro-level dynamic patterns, is restricted by three and fixed for each type of node state.

Let us take a node i having state s_i, fix $a, b \in \mathbb{R}$ dividing $\mathbb{R}$ into 3 semi-intervals. Without loss of generality let $a \leqslant b$.

- $s_i < a \Rightarrow$ the state is *critical* and node is removed from the network with the edges adjacent to it;
- $s_i \in [a; b) \Rightarrow$ the state is *transitional* and the node tends to enhance its state by creating new edges in the network, to make the state $s_i \geq b$;
- $s_i \in [b; \infty) \Rightarrow$ the state is *stable*, and the node does not create new edges actively, nevertheless, it can interact with other nodes if they need it.

Therefore,

$$s_i(t) < b \Rightarrow \frac{ds_i}{dt} = b - s_i(t) = \frac{\sum_{j=1}^{N} \gamma_{ij}}{dt} \tag{1}$$

$$s_i(t) < a \Rightarrow \forall s_j \in \mathcal{N}(s_i) : \frac{ds_j}{dt} = \gamma_{ij} - \gamma_{ji}; \forall j \gamma_{ij} = 0, \gamma_{ji} = 0, \tag{2}$$

where $\mathcal{N}(s_i)$ is a neighbourhood of s_i node in terms of connection graph.

Therefore, each kind of node states has a corresponding type of related dynamics at different scales (Table 1).

Table 1. Correspondence between dynamics and events at different scales. The rightest column show observable system states in the evolution trajectories, represented by number of nodes in a system and by entropy of graph Laplacian spectrum

	Node-level	Local formation pattern	Network-level
Critical	$(-\infty; a)$	del. node	
Transitional	$[a; b)$	add edge	
Stable	$[b; +\infty)$	—	

Network Formation process, initiated by Eq. (1), is determined by the considered node strategy. In random case, edges are distributed equiprobably. Counterparty choice strategies, corresponding to preferencial patterns, can be taken into account by implementing probability distribution over the all nodes for adjacency matrices of interaction. Here we suppose random connections, resulting in a network with Poisson degree distribution, as initial consideration and random choice strategy during simulation. Equation 2 can also be modified to consider rewiring process related to market clearing algorithm, nevertheless, this detail is out of this study consideration.

Co-evolution–Feedback Mechanisms. On one hand, changes in node states affect micro-scale dynamics, resulting in changes in network topology, on other, network topology contributes node states. In the system considered, each node state depends on adjacent edges and their attributes (related to $\Gamma(t)$), neighbouring nodes (system state $S(t)$), and external factors $g_i(t)$ affecting node i at time t:

$$s_i(t) = f(\Gamma(t), \ g_i(t)), \tag{3}$$

where $\Gamma(t) = \langle\{\gamma_{ij}(t)\}, V(t) = S(t)\rangle$ is a dynamic graph, reflecting the interactions between agents at the moment t. Then, following the components interplay (Eq. 1), the change in node state per iteration can be rephrased as

$$\frac{ds_i}{dt} = \sum_{k=1}^{N} f\left(t, \frac{d\gamma_{ik}}{dt} - \frac{d\gamma_{ki}}{dt}, \frac{dS}{dt}\right) + g_i(t), \tag{4}$$

where f is the rule setting dependence of a node state on its neighborhood, and $g_i(t)$ is the aggregation of external effects on the system.

Let fix parameters: N, a, b, $\{g_k(t)\}$; $s_i \in \mathbb{R}$, a and b determines 3 types of states. Initial conditions are denoted as: $S(0) = \{s_k(0)\}$; $\Gamma(0) = \{\gamma_{ij}(0)\}$. Then, coming to iterative form and using conditions (1)–(4), we obtain Eq. (5) and (6), determining dynamics in the system:

$$s_i(t+1) = s_i(t) \cdot \chi_{(-\infty;a)}(s_i(t)) + (b - s_i(t)) \cdot \chi_{[a;b)}(s_i(t)) \tag{5}$$

$$- \chi_{[a;\infty)}(s_i(t)) \cdot \left[\sum_{s_j \in [a;b)} \frac{b - s_j(t)}{N} + \sum_{s_j < a} (\gamma_{ij}(t) - \gamma_{ji}(t)) - g_i(t+1) \right]$$

$$\gamma_{ij}(t+1) = \gamma_{ij}(t) + \frac{b - s_i(t)}{N} \cdot \chi_{[a;b)}(s_i(t))\chi_{[a;+\infty)}(s_j(t)) \tag{6}$$

$$- \gamma_{ij}(t)\left(1 - \chi_{[a;+\infty)}(s_i(t))\chi_{[a;+\infty)}(s_j(t))\right)$$

Since algorithms used for simulation are discrete, formulae contain indicator function $\chi_{set}(var)$, selecting addends depending on node state, which allows for consideration of node-level dynamics variations (from Table 1).

3.3 Regime Durability Estimation

Since the transitional phase is associated with the existence of node having $s_i \in [a; b)$, while others are $\geq b$, the length of first phase, i. e. expected time before start of transitional regime is estimated as minimal number of iterations before one of nodes reach b. That is $|\Phi I| = \left|\left\{ \min t : \exists i \in \mathbb{N} \ s_i(t) \in [a; b) \ \forall j \neq i \ s_j(t) \geq b \right\}\right|$. Similarly, $|\Phi II| = \left|\left\{ \min t : \exists i \in \mathbb{N} \ s_i(t) \in (-\infty; a) \ \forall j \neq i \ s_j(t) \geq a \right\}\right|$.

Consider a set of nodes $\mathbf{B} = \{b\}$. $\forall b \in \mathbf{B}$ with the corresponding state s_b fix $g_b(t)$. Say ea_b and el_b are external assets and liabilities, therefore they are related to node state, on one hand, and to external impact – on other. Let $\forall t > 0$ $g_b(t) = const > 0 \Rightarrow$

$$\forall t > 0 \ \Delta|ea_b - el_b| = -g_b(t) \Leftrightarrow \sum_k \left[\Delta\gamma_{ik} - \Delta\gamma_{ki}\right] = -g_b(t) \tag{7}$$

Then, summarizing Eq. 7 for the whole system:

$$\sum_{i=1}^{N} \sum_{k=1}^{N} \left[\Delta\gamma_{ik} - \Delta\gamma_{ki}\right] = \sum_{b \in \mathbf{B}} -g_b \tag{8}$$

$$\Leftrightarrow \quad 0 = \sum_{b \in \mathbf{B}} -g_b, \tag{9}$$

which is obvious. Then we simplify indicator function and further equations for the cases of ΦI and ΦII, summarise equation for all nodes, and modify sums in the consideration of Erdos-Renyi graph:

$$s_i(t+1) = s_i(0) + \sum_{k=1}^{t+1} g_i(t) - \sum_{k=1}^{t} \sum_{s_j(t) \in [a;b)} \frac{b - s_j(k)}{N} \tag{10}$$

$$\sum_{i=1}^{N} N s_i(t+1) = \sum_{i=1}^{N} s_i(0) + \sum_{i=1}^{N} \sum_{k=1}^{t+1} g_i(k) - \sum_{i=1}^{N} \sum_{k=1}^{t} \sum_{s_j(t) \in [a;b)} \frac{b - s_j(k)}{N} \tag{11}$$

$$\left[\sum_{i=1}^{N} \sum_{k=1}^{t+1} g_i(k) \approx N \cdot (t+1) \cdot g\right] \Rightarrow \tag{12}$$

If we fix the number of initially stressed nodes as $\alpha_0 = \left|\left\{ i | s_i \in [a; b) \right\}\right|$, then

$$|\Phi I| = \frac{s_i(0) - b}{g} \tag{13}$$

$$|\Phi II| \geq \frac{g + b - a}{g - \frac{\alpha_0}{N}(b - s(\bar{k}))} \tag{14}$$

4 Experiment Results

4.1 Case Study: Failures Prediction

For this experiment we made simulations of implemented system and check if our estimations of tipping points correspond to simulation results. We suppose random structure of interactions, s. t. initial network configuration is provided by Erdos-Renyi model, parametrised by 1000 nodes and 0.2 connection probability. Tipping points of node states are $a = -0.5$ and $b = 2.5$, and $\alpha = 0.5$. (Parameter values were chosen arbitrary and does not affect predictability.) Taking formulae (13) and (14) we obtain estimated number of iterations before starting point of an avalanche and its early warning signal, which is the point before transitional regime. In Fig. 2 predicted values, evaluated by formulae (13) and (14) are displayed by vertical pink lines, which corresponds to dives in dynamics of node and edge counts. Right panel of Fig. 2 is aimed at demonstrating inner processes, explaining the dynamics at left panel, and shows dynamics of nodes count being in transitional and critical states.

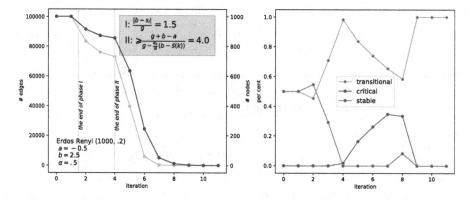

Fig. 2. Analytical estimations of transitional points and simulation results (on the left); dynamics of node count inside each category (on the **right**)

Results demonstrate satisfactory prediction ability in the case of poisson network structure. The consideration of other topologies, formally, will require other approximation techniques for final estimation formulae. Nevertheless, for scenarios, when network structure plays minor role in contrast to external and node effects, this method will work.

4.2 Case Study: Funds Infusion

The approach, provided in current paper, allows for explanation of dynamics of interbank network evolution, observed in simulations under several driving factors, like choice models, market clearing methods, and external effects [9]. Lines of different colours correspond to different combinations of counterparty choice models with choice models for external impacts (Fig. 3), and display the influence of funds infusion at different time moments.

The simulation scenario shows system evolution coming to cascading dynamics under different parameters and showing effects of funds infusion to the system. System state is observed with the number of removed nodes and with the entropy of Laplacian spectrum [27]. The left panel shows the begin of cascade when number of removed nodes increases sharply. Infusion stops this process temporally. At the same time, the right panel shows the critical regime, corresponding to decrease in entropy, has an interval of increasing entropy before it – this interval corresponds to transitional regime and show how number of edges change due to increasing number of stressed nodes. For this reason, fund infusion return nodes to prevailing stressed state from critical, and we see increase in entropy until some critical point.

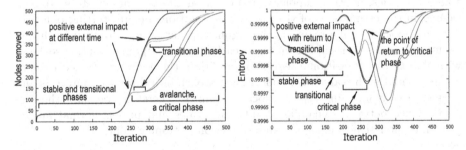

Fig. 3. Functional properties are often evaluated as a number of nodes. Transitions observed in functional properties (**left**) do not reflect hidden dynamics in topological properties (**right**)

These observations, in the combination with results of current paper, say that system has its capacity against external impacts. In this case, this combination is prevailing, so topology is not so significant. Infusion brings additional capacity to the system, allowing to avoid failures for a number of iterations when summarised external impact $\sum_{i=1}^{N} g_i(t)$ is fixed $\forall t$. In practice, this time can be appropriate for changes in managerial approaches or strategies, nevertheless, it is obvious, that in current case external impact must be balanced by other resources to provide stable system evolution.

5   Conclusion and Future Work

This study formally demonstrates, how micro-level dynamics of complex agent networks results in global patterns at system-level, in particular in the case of default cascades in interbank networks. This allows to see, how local effects are accumulated, and how this affects times between regimes of evolution. In addition, this gives a base for the exploration of which factors will have more influence and how to control it.

Cascading behaviour, observed via system state variables, is preceded by changes in inner dynamics, related to co-evolution of node states and structure of their interaction. This can be detected aforehand by means of topological features, like entropy of Laplacian spectrum, in the case of correspondence between node states and micro-level dynamics resulting in structural changes. In this way, tipping points are related to the share of nodes in each category. The consideration of real-valued state set for nodes, instead of discrete states, allows for consideration of system capacity against external impacts, which has a connection with lower levels. In the context of homogeneous structure of interactions, with no weak-connected components, the most effect is due to relation between external impact and overall systemic capacity, opposing to it.

The considered agent-based network evolves under factors, comprising node states dynamics, network topology, and external effects. These factors are considered in the estimations of the number of iterations before starting points of critical and transitional regimes. The regimes are introduced to distinguish cascading dynamics (critical regime) and early-warning tipping point, associated with the start of transitional regime. In addition, the above mentioned regimes are associated with real-valued node states, broken into three semi-intervals and triggering corresponding types of local dynamics. Therefore, we show the correspondence between dynamics at different scales and present formal model, providing inter-scale connection and prediction of tipping points.

References

1. Aleksiejuk, A., Hołyst, J.A., Kossinets, G.: Self-organized criticality in a model of collective bank bankruptcies. Int. J. Mod. Phys. C **13**(03), 333–341 (2002)
2. Bardoscia, M., Caccioli, F., Perotti, J.I., Vivaldo, G., Caldarelli, G.: Distress propagation in complex networks: the case of non-linear debtrank. PloS ONE **11**(10) (2016)
3. Battiston, S., Caldarelli, G., May, R.M., Roukny, T., Stiglitz, J.E.: The price of complexity in financial networks. Proc. Natl. Acad. Sci. **113**(36), 10031–10036 (2016)
4. Berardi, S., Tedeschi, G.: From banks' strategies to financial (in) stability. Int. Rev. Econ. Finan. **47**, 255–272 (2017)
5. Bouchaud, J.P., Cates, M., Prakash, J.R., Edwards, S.: Hysteresis and metastability in a continuum sandpile model. Phys. Rev. Lett. **74**(11), 1982 (1995)
6. Fagiolo, G., Guerini, M., Lamperti, F., Moneta, A., Roventini, A.: Validation of agent-based models in economics and finance. In: Beisbart, C., Saam, N.J. (eds.) Computer Simulation Validation. SFMA, pp. 763–787. Springer, Cham (2019). https://doi.org/10.1007/978-3-319-70766-2_31

7. Gai, P., Haldane, A., Kapadia, S.: Complexity, concentration and contagion. J. Monetary Econ. **58**(5), 453–470 (2011). https://doi.org/10.1016/j.jmoneco.2011.05.005. http://www.sciencedirect.com/science/article/pii/S0304393211000481

8. Guleva, V.Y.: The combination of topology and nodes' states dynamics as an early-warning signal of critical transition in a banking network model. Procedia Comput. Sci. **80**, 1755–1764 (2016). http://www.sciencedirect.com/science/article/pii/S1877050916309164, International Conference on Computational Science, ICCS: 6–8 June 2016, San Diego, California, USA (2016)

9. Guleva, V.Y., Bochenina, K.O., Skvorcova, M.V., Boukhanovsky, A.V.: A simulation tool for exploring the evolution of temporal interbank networks. J. Artif. Soc. Soc. Simul. **20**(4) (2017)

10. Hernandez-Urbina, V., Michael Herrmann, J.: Neuronal avalanches in complex networks. Cogent Phys. **3**(1), 1150408 (2016)

11. Hinkelmann, F., Murrugarra, D., Jarrah, A.S., Laubenbacher, R.: A mathematical framework for agent based models of complex biological networks. Bull. Math. Biol. **73**(7), 1583–1602 (2011)

12. Hsu, D., Beggs, J.M.: Neuronal avalanches and criticality: a dynamical model for homeostasis. Neurocomputing **69**(10–12), 1134–1136 (2006)

13. Iori, G., Jafarey, S.: Criticality in a model of banking crises. Physica A **299**(1–2), 205–212 (2001)

14. Kanas, A., Molyneux, P.: Macro stress testing the US banking system. J. Int. Finan. Markets Inst. Money **54**, 204–227 (2018)

15. Kreis, Y., Leisen, D.P.: Systemic risk in a structural model of bank default linkages. J. Finan. Stab. **39**, 221–236 (2018)

16. Lambert, D., Vanni, F.: Complexity and heterogeneity in a dynamic network. Chaos, Solitons Fractals **108**, 94–103 (2018). https://doi.org/10.1016/j.chaos.2018.01.024. http://www.sciencedirect.com/science/article/pii/S0960077918300249

17. Laubenbacher, R., Jarrah, A.S., Mortveit, H., Ravi, S.: A mathematical formalism for agent-based modeling. arXiv preprint arXiv:0801.0249 (2007)

18. Liu, R., Chen, P., Aihara, K., Chen, L.: Identifying early-warning signals of critical transitions with strong noise by dynamical network markers. Sci. Rep. **5**, 17501 (2015)

19. Liu, W., Schmittmann, B., Zia, R.: Extraordinary variability and sharp transitions in a maximally frustrated dynamic network. EPL (Europhysics Letters) **100**(6), 66007 (2013)

20. Pruessner, G.: Self-Organised Criticality: Theory Models and Characterisation. Cambridge University Press, New York (2012)

21. Scheffer, M., et al.: Early-warning signals for critical transitions. Nature **461**(7260), 53–59 (2009)

22. Solé, R.V., Manrubia, S.C.: Extinction and self-organized criticality in a model of large-scale evolution. Phys. Rev. E **54**(1), R42 (1996)

23. Steels, L.: Cooperation between distributed agents through self-organisation. In: IEEE International Workshop on Intelligent Robots and Systems, Towards a New Frontier of Applications, pp. 8–14. IEEE (1990)

24. Takayasu, H., Inaoka, H.: New type of self-organized criticality in a model of erosion. Phys. Rev. Lett. **68**(7), 966 (1992)

25. Tedeschi, G., Caccioli, F., Recchioni, M.C.: Taming financial systemic risk: models, instruments and early warning indicators. J. Econ. Interact. Coord. **15**(1), 1–7 (2019). https://doi.org/10.1007/s11403-019-00278-x

26. Vanni, F., Barucca, P., et al.: Time evolution of an agent-driven network model. Technical report, Laboratory of Economics and Management (LEM), Sant'Anna School of Advanced Studies, Pisa, Italy (2017)
27. Ye, C., Torsello, A., Wilson, R.C., Hancock, E.R.: Thermodynamics of time evolving networks. In: Liu, C.-L., Luo, B., Kropatsch, W.G., Cheng, J. (eds.) GbRPR 2015. LNCS, vol. 9069, pp. 315–324. Springer, Cham (2015). https://doi.org/10.1007/978-3-319-18224-7_31

Modeling of Fire Spread Including Different Heat Transfer Mechanisms Using Cellular Automata

Jarosław Wąs(✉)(iD), Artur Karp, Szymon Łukasik, and Dariusz Pałka(iD)

Faculty of Electrical Engineering, Automatics, IT and Biomedical Engineering,
AGH University of Science and Technology,
Mickiewicza 30, 30-059 Krakow, Poland
jaroslaw.was@agh.edu.pl

Abstract. The article presents a new method of modeling the spread of fire using Cellular Automata based on 3D mesh. The models based on different heat transport mechanisms were used in the study, including conductivity, convection and thermal radiation. General mechanisms of fire spread are exemplified by a fire inside a building. The tests of the created simulator for fire in rooms were conducted, and the reference point was the known simulator based on CFD, namely FDS Fire Dynamics Simulator.

Keywords: Fire spread · Cellular Automata · Heat transfer mechanisms

1 Introduction

Fire is a complex phenomenon which includes thermal degradation or pyrolysis of solid fuel into gas (of a volatile character) and combustion of the volatile fuel. It should be noted here that the generated heat influences the solid. Heat transport is carried out through thermal conductivity, convection (natural and forced) and thermal radiation. Modern computational sciences provide various methods and IT tools that allow for modeling this complex process [8] in a simplified way.

The most common models are high-fidelity CFD models, also known as *Field models*, like e.g. Fire Dynamics Simulator (FDS). based on the physical principles of conservation of: mass, energy, and momentum. In the field model, the motion of heat and smoke is calculated in particular cells in consecutive time intervals. In such models we assume that in each cell the properties like temperature, density, flow, etc. are constant. Another popular approach are *Zone models*, in which the whole space (for instance, a room) is divided into areas with different properties (like a 'combustion zone' or a 'convection zone'), and for each of them a set of algebraic equations is solved in order to calculate relevant flow properties. The most recent approach assumes looking for fire models that are realistic and computationally efficient.

The paradigm of Cellular Automata (CA) can be considered an effective and efficient solution. In the literature, CA are used to model simplified fire spreading

© Springer Nature Switzerland AG 2020
V. V. Krzhizhanovskaya et al. (Eds.): ICCS 2020, LNCS 12137, pp. 445–458, 2020.
https://doi.org/10.1007/978-3-030-50371-0_33

schemes [1], often in relation to evacuation [13] or specific scenarios like forest
fires [6,14] etc.

We propose a new model of fire spread based on Cellular Automata in which
different mechanisms of heat transfer are taken into account, namely: conductivity, convection and thermal radiation. We propose the application of the cuboid
structure of space using three-dimensional CA.

We developed a theoretical model, implemented it, and performed preliminary validation based on widely used CFD model FDS, created by NIST
(National Institute of Standards and Technology), which is often applied as a
reference model [4,10]

2   Heat Transfer Mechanisms

2.1   Thermal Conduction

The amount of energy which is transferred in thermal conduction can be computed using the law of heat conduction (Fourier's law). The differential form of
this law is as follows:

$$q = -k\nabla T \tag{1}$$

where: q - is local heat flux density (SI unit $[\frac{W}{m^2}]$), k - is material conductivity
(SI unit $[\frac{W}{mK}]$), ∇T - is the temperature gradient (SI unit $[\frac{K}{m}]$).

2.2   Thermal Radiation

The heat flux that can be emitted by radiation coming from the black body
surface is given by Stefan–Boltzmann law:

$$q = \sigma T^4 \tag{2}$$

where: q – is radiative heat flux (SI unit $[\frac{W}{m^2}]$), σ – is the Stefan–Boltzmann
constant ($\approx 5.670 \cdot 10^{-8}$ $[\frac{W}{m^2 K^4}]$), T – is the temperature of the body surface (SI
unit $[K]$)

The flux emitted by the real surface is lower than the flux emitted by the
black body surface:

$$q = \epsilon \sigma T^4 \tag{3}$$

where: ϵ – a factor called emissivity, which is a property characterising a surface
($\epsilon < 1$)

2.3   Convection

Convective heat transfer is transfer of heat from one place to another caused
by the movement of fluid. The rate equation for this form of heat transfer is
described by Newton's law of cooling:

$$q = h(T - T_f) \tag{4}$$

where: q - is convective heat flux (SI unit $[\frac{W}{m^2}]$), h - is convection heat transfer
coefficient (SI unit $[\frac{W}{m^2 K}]$), $(T_w - T_f)$ – is the temperature difference between
the surface of an object and fluid (SI unit $[K]$).

3   CAFire

The assumptions of the model used in the implementation of the CAFire programme which simulates fire spread are presented below.

The whole space under consideration can be divided into three-dimensional elements (e.g. cubes) small enough to be treated as homogeneous - i.e. all physical and chemical properties (e.g. temperature, specific heat capacity) in a given element are the same at a given time.

3.1   Neighborhood

Three basic types of neighborhoods exist in a three-dimensional array of a cubic cell:

- **6** cells sharing faces with the reference cell
- **18** cells sharing edges with the reference cell
- **26** cells sharing vertices with the reference cell

Because the simulation results obtained with the use of all the above types of neighborhoods are comparable and the first variant has the lowest calculation costs, it is used in the model (Fig. 1).

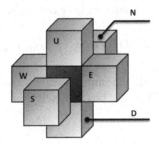

Fig. 1. Neighborhood. Red - the reference cell, U - the cell above, D - the cell below, W - the cell on the left, E - the cell on the right, S - the cell in front, N - the cell behind. (Color figure online)

3.2   Simulation of Heat Transfer Using the CA Model

Thermal Conduction. At a given temperature gradient, the heat transfer rate is proportional to the surface through which the heat flows. Generally, the amount of heat dQ that will flow L over dt with the temperature difference dT across the transverse surface S is as follows:

$$dQ = \lambda \cdot \frac{S}{L} \cdot dt \cdot dT \tag{5}$$

where: Q – heat, λ – thermal conductivity coefficient, S – transverse surface of the cell, L – distance between cells, dt – time, dT – temperature difference.

In the proposed model, the cells have the same surface area. Consider the following one-dimensional situation shown in Fig. 2.

■ – warmer cell
■ – cooler cell

Fig. 2. Heat flow between two neighbouring cells due to thermal conductivity in the model.

In this type of analyses, it is assumed that the heat exchange occurs between the cells, and so the λ factor coefficient indicates how the heat passes from one cell to another.

With two cells only one factor is determined. However, if a whole cell chain is considered, it may happen that for a given cell n the conduction coefficient from/to cell $(n - 1)$ is different than from/to $(n + 1)$.

In the discrete one-dimensional case, the formula for temperature T of the cell with index n and of a unit size can be written as:

$$T_n(t+\Delta t) = T_n(t) + \left(\lambda'_n \cdot \Big(T_n(t) - T_{n-1}(t) \Big) - \lambda'_{n+1} \cdot \Big(T_{n+1}(t) - T_n(t) \Big) \right) \Delta t \quad (6)$$

where: $\lambda' = \frac{\lambda}{m \cdot c}$, $T_n(t)$ – Cell n temperature at a given time t, λ – thermal conductivity coefficient, c – specific heat capacity, m – cell mass.

3.3   Convection

The phenomenon of convection is characteristic of liquids and gases. The model includes natural convection.

$$T_u(t + \Delta t) = T_u(t) + \alpha' \cdot \Big(T_u(t) - T_d(t) \Big) \cdot \Delta t$$
$$T_d(t + \Delta t) = T_d(t) - \alpha' \cdot \Big(T_u(t) - T_d(t) \Big) \cdot \Delta t$$
$$(7)$$

where: $\alpha' = \frac{\alpha}{m \cdot c}$, $T_u(t)$ – the temperature of the upper cell at a given time t, $T_d(t)$ – the temperature of the lower cell at a given time t, α – convection heat transfer coefficient, m – cell mass, c – specific heat capacity (Fig. 3).

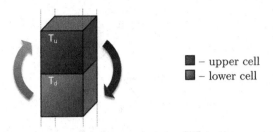

Fig. 3. Convection in the model. Heat exchange from the cell below to the cell above. T_d - the temperature of the lower cell, T_u - the temperature of the upper cell

3.4   Thermal Radiation

Thermal radiation is a process in which energy is emitted by a heated surface in all directions in the form of electromagnetic waves (electromagnetic radiation).

In the fire safety literature, thermal radiation has been identified as important and even dominant in terms of heat transfer in medium to large fires. Its mechanisms are analysed at various levels of accuracy ranging from simplified analyses to sophisticated methods of calculating radiation [9].

Stefan-Boltzmann law refers to radiation in all directions, whereas in the model only some of the energy radiated by one surface will reach another. The heat exchange from a warmer (1) to a cooler (2) surface depends on the following factors [5]:

- temperatures T_1 and T_2
- surface areas A_1 and A_2
- spatial configurations between surfaces
- surface radiation characteristics
- additional surfaces in the environment
- medium between surfaces (if it is air, it can be assumed as non-invasive)

To calculate the amount of energy that will reach from one surface to another, we need to calculate the integral, taking into account the density of the radiation at an angle under which fragments of the radiating surface are visible for the other surface. Then, an additional factor *view factor* appears in Stefan-Boltzmann law.

$$Q_{net} = F_{1-2}A_1\sigma\left(T_1^4 - T_2^4\right) \qquad (8)$$

where: Q_{net} – total radiation energy, F_{1-2} – view factor, A_1 – surface area, σ – Stefan - Boltzmann constant, T_1, T_2 – temperature in Kelvins.

The building is divided into radiation areas (for example, walls or tables) with a direction from which they can receive heat (Figs. 4 and 5).

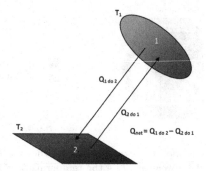

Fig. 4. General radiation pattern. Q_{net} - total energy exchange, $Q_{x \, do \, y}$ - energy transferred from the surface x to y, T_x - temperature of surface x.

View Factors for Several Three-Dimensional Configurations [5]

– Parallel surfaces

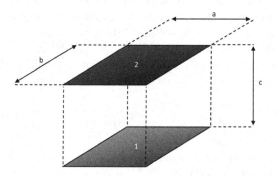

Fig. 5. Diagram of radiation for parallel surfaces in the model. a/b - length/width of surfaces, c - distance between surfaces

let:

$$X = \frac{a}{c} \text{ i } Y = \frac{b}{c} \tag{9}$$

then:

$$F_{1-2} = \frac{2}{\pi XY} \left(\ln \left(\frac{(1+X^2)(1+Y^2)}{1+X^2+Y^2} \right)^{\frac{1}{2}} - X \tan^{-1} X - Y \tan^{-1} Y \right.$$
$$\left. + X\sqrt{1+Y^2} \tan^{-1} \frac{X}{\sqrt{1+Y^2}} + Y\sqrt{1+X^2} \tan^{-1} \frac{Y}{\sqrt{1+X^2}} \right) \tag{10}$$

where: F_{1-2} – view factor, a – length of surface, b – width of surface, c – distance between surfaces (Fig. 6).

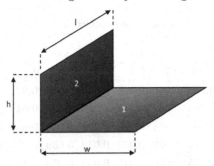

Fig. 6. Scheme of radiation for perpendicular surfaces in the model. h - width of the first surface, w - width of the second surface, l length of the surface.

– Perpendicular surfaces
  let:

$$H = \frac{h}{l} \text{ i } W = \frac{w}{l} \tag{11}$$

then:

$$F_{1-2} = \frac{1}{\pi W} \left(W \tan^{-1} \frac{1}{W} - \sqrt{H^2 + W^2} \tan^{-1} \left(H^2 + W^2 \right)^{-1/2} \right.$$

$$+ H \tan^{-1} \frac{1}{H} + \frac{1}{4} \ln \left(\left(\frac{(1+W^2)(1+H^2)}{1+W^2+H^2} \right) \right. \tag{12}$$

$$\left. \left. \left(\frac{W^2(1+W^2+H^2)}{(1+W^2)(W^2+H^2)} \right)^{W^2} \left(\frac{H^2(1+H^2+W^2)}{(1+H^2)(H^2+W^2)} \right)^{H^2} \right) \right)$$

where: F_{1-2} – view factor, h – width of the first surface, w – width of the second surface, l – length of the surface

For each wall of a fire cell parallel radiation surfaces are considered where no fire cells or solid body cells are located between them in a straight line.

$T_{surface}$ was calculated as the average temperature of the cells contained in this surface. Of course, the following condition must be met: $T_{cell} > T_{surface}$.

As an obstacle, we took into account a cell which was a solid body or a fire cell. Knowing surface dimensions a i b and the distance from the cell c coefficient, the *view factor* was calculated on the basis of Table 1.

The last step was to calculate the transferred energy Q_{net} according to Eq. 8. Energy is transferred equally to all cells belonging to the radiation surface.

In case of radiation *surface* ⇒ *surface*, the main radiation surfaces were determined at the beginning of the simulation; they can be generally treated as walls (in the model for each room, 6 such surfaces were designated for the floor,

Table 1. View factor values applied in CAFire simulation

$\frac{a}{c}$	$\frac{b}{c}$						
	<0.1	0,2	0,3	0,4	0,6	1	2<
<0,1	0,003	0,006	0,008	0,012	0,016	0,021	0,028
0,2	0,005	0,012	0,018	0,022	0,031	0,040	0,051
0,3	0,009	0,018	0,026	0,033	0,045	0,059	0,074
0,4	0,012	0,022	0,033	0,042	0,059	0,075	0,091
0,6	0,015	0,030	0,045	0,059	0,078	0,100	0,145
1	0,020	0,039	0,057	0,075	0,105	0,150	0,170
2<	0,025	0,048	0,071	0,092	0,125	0,175	0,225

ceiling and each wall). For each surface other surfaces in its "field of visions" and its *view factors* were determined on the basis of Eq. 10 or 12. The same procedure was repeated in case of *cell ⇒ surface*.

3.5 Simulation of Fire Propagation Using the CA Model

The proposed model takes into account the properties of relevant materials, because each of them has a different autoignition and ignition temperature.

Ignition is possible if a cell receives temperature proper to a given kind of material (e.g. wood 250 °C, carpet 100 °C, fabrics 170 °C) and one of the neighboring cells remains on fire long enough (Fig. 7).

The only condition for autoignition is to reach a sufficiently high temperature (e.g. wood 2000 °C, carpet 500 °C, fabrics 650 °C).

The linear fire spread rate was calculated for each material. By transforming the formula to take into account the speed, we calculated how long the cell must burn (to be able to ignite its neighbor) (Fig. 8).

If ($T > T_{ignition}$ and
burning time > ignition time)
set in the cell "fire" (red) state.

Fig. 7. Interpretation of cell ignition. White cell - a cell with temperature that ignites, red cell - an adjacent cell burning for an appropriate period of time. (Color figure online)

If ($T > T_{autoignition}$)
set in the cell "fire" (red) state.

Fig. 8. Interpretation of cell autoignition. Red cell – a cell that is autoignited (without burning neighbors). (Color figure online)

3.6   Simulation of Smoke Propagation Using the CA Model

In the presented model smoke is generated in the cell located directly above the flame, which is shown in Fig. 9. The amount of smoke emitted depends on the type of material that is burning. For example, less smoke will be released from a wooden cell than from a textile cell.

■ – cell with smoke
■ – cell with fire

Fig. 9. Smoke release mechanism (I). Smoke is generated in the cells located directly above the cells with fire.

The principle of mass conservation and the convection mechanism were retained in smoke modeling.

It was assumed that every cell tends to get rid of all its smoke. To ensure that the cell is not filled above its maximum capacity, the amount of smoke that it can take from one neighbor was limited to $\frac{1}{6}$ of the remaining free space [12].

In order to ensure that the cell does not transmit more smoke than it has, the amount of smoke that a given cell can transmit was limited to $\frac{1}{6}$ of the amount of smoke.

$$D_{0 \to n} = d_n \cdot min\left(\frac{1}{6}D_0, \frac{1}{6}\left(100\% - D_n\right)\right) \tag{13}$$

The amount of smoke in the cell will be reduced by the amount the cell will pass on to its neighbors and increased by the amount it will receive from its neighbors (Fig. 10):

$$D_0^{i+1} = D_0^i - \sum_n D_{0 \to n}^i + \sum_n D_{n \to 0}^i \tag{14}$$

where:

$D_{0 \to n}$ – the amount of smoke passed to a neighbouring cell
$D_{n \to 0}$ – the amount of smoke received from a neighbouring cell
d_n – quantitative factor, according to Table 2
n – neighbor index

Table 2. Smoke propagation factor d_n (1)

n	d_n
U	100%
N, S, E, W	50%
D	25%

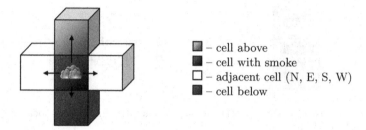

■ – cell above
■ – cell with smoke
□ – adjacent cell (N, E, S, W)
■ – cell below

Fig. 10. Smoke generation mechanism. Propagation of smoke from a given cell ('the cell with smoke') to neighboring cells.

When the cell above C_U is not air, or the amount of smoke in it D_U is close to 100% (that is, it cannot accept more smoke), the d_n factor will be modified (see Table 3).

This modification adequately demonstrates pushing smoke out of the ceiling zones.

4   Calibration and Validation

While calibrating fire scenarios, we took into account bibliographical data: standard PN-EN ISO 6946 and [3], as well as some data from reference fires [2]. The simulation was based on materials that are usually found in a typical household – Fig. 11 (Table 4).

Table 3. Smoke propagation factor d_n (2)

n	d_n
U	–
N, S, E, W	50%
D	100%

Table 4. Material properties table - simulation.

	Air	Wall ext.	Glass	Wood	Carpet	Fabrics
Solid state t = 21 °C	No	Yes	Yes	Yes	Yes	Yes
Combustibility	–	No	No	Yes	Yes	Yes
Density (kg/m^3)	1.29**	2200	2500	400	300	400
Specific heat ($J/(kg \cdot K)$)	1005	900	8400	2390	5200	2390
Conduction factor (λ)	0.026	0.250 (0.550)	0.500	0.300	0.174	0.300
Flash point (°C)	–	–	300*	250	120	350
Temp. of autoignition (°C)	–	–	–	2000	750	1400
Linear speed (m/s)	–	–	–	3.0	0.8	3.0
Smoke emission	–	–	–	0.3	0.8	1.0
Generated energy (J/s)	–	–	–	1500	500	1500

* - Since glass does not have a flashpoint, this value indicates the temperature at which the glass breaks.
** - At 1013.25 hPa and 0 degrees Celsius

4.1 Validation of the Model

The proposed model was validated in two ways: on the basis of movies recorded for fires taking place under controlled conditions, and on the basis of the results obtained with the Fire Dynamics Simulator tool. In case of movies, the spread

Movie 1   Movie 2

Model CAFire

Fig. 11. Simulation based on the data obtained from the reference fire ('Movie 1' and 'Movie 2') - time 1:00 after ignition.

of fire and smoke was visually compared, while in case of FDS, the distribution of the temperatures in selected fire areas was additionally validated.

Fire Dynamics Simulator [3, 4, 7, 11] is an application that, according to many experts in fire engineering, best reflects the course of fire in buildings. In order to compare our CA-based model with the same model in FDS, we compared two identical scenarios of a building fire. The size of the building: 7.5 m × 2.9 m × 7.5 m, including a living-room, a bathroom, a wardrobe, a study, a kitchen and a hall. Ignition temperature was 150 °C.

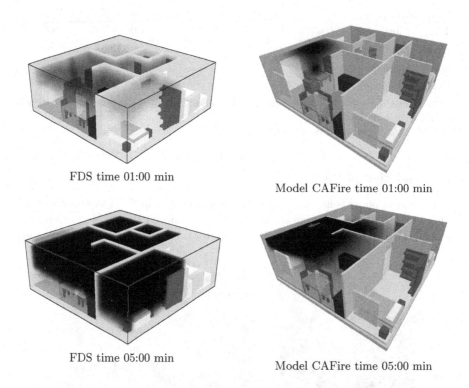

FDS time 01:00 min

Model CAFire time 01:00 min

FDS time 05:00 min

Model CAFire time 05:00 min

Fig. 12. Validation of fire and smoke spreading on the basis of CFD Fire Dynamics Simulator. The same configuration (scene) evaluated using FDS and Model CAFire.

It should be noted that other data sets were used for the calibration and validation procedures. Both, fire development scenario and basic parameters are similar in our model and the reference FDS model. Sample comparison results are presented in Fig. 12 and 13.

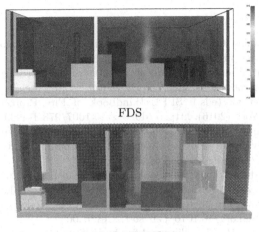

FDS

Model CAFire

Fig. 13. Validation on the basis of CFD Fire Dynamics Simulator - temperature distribution.

5   Conclusions

The main purpose of this study was to create an effective and efficient method for simulating the spread of fire. The model and simulations were based on non-homogeneous Cellular Automata to present the distribution of fire and smoke in a residential building, taking into account the temperature fields in the rooms. The current model of fire called *CAFire* includes three main phenomena accompanying the spread of heat during a fire, i.e. conductivity, convection and thermal radiation. The simulations showed relatively reliable convergence of the results compared to the reputable FDS tool.

In the presented model, only in a very simplified form, the fluid flow phenomena (air flow, smoke flow) are simulated. The interactions between the air streamlines (due to Bernoulli's principle) and the influence of pressure caused by the difference of temperatures on the formation of laminar and turbulent flows etc. are not taken into account.

The idea presented in the paper requires further development and tests - and additional emphasis should be placed on the chemical aspects of fire and exhaustive validation of the final results.

However, it should be emphasized that we consider the proposed direction of work very promising. Creating simulations in the FDS program requires time, even up to several dozen minutes, while calculations obtained using the proposed Cellular Automata method take seconds to be completed.

References

1. Curiac, D.I., Banias, O., Volosencu, C., Dan, P.: Cellular automata based simulation for smoke and fire spreading in large buildings. In: International Conference on Development, Energy, Environment, Economics - Proceedings, November 2010
2. Fires, R.: Living room fires. http://www.youtube.com/watch?v=TB42Ib3A4mg
3. Hurley, M.J., et al. (eds.): SFPE Handbook of Fire Protection Engineering. Springer, New York (2016). https://doi.org/10.1007/978-1-4939-2565-0
4. Kang, D.I., Kim, K., Jang, S.C., Yoo, S.Y.: Risk assessment of main control board fire using fire dynamics simulator. Nucl. Eng. Des. **289**, 195–207 (2015). https://doi.org/10.1016/j.nucengdes.2015.04.030
5. Lienhard, J.: A heat transfer textbook. J. Heat Transf. **108** (2013)
6. Liu, Y., Liu, H., Zhou, Y., Sun, C.: Spread vector induced cellular automata model for real-time crown fire behavior simulation. Environ. Modell. Softw. **108**, 14–39 (2018). https://doi.org/10.1016/j.envsoft.2018.07.005
7. Molkens, T., Rossi, B.: Modelling real fire by FDS and 2-zone model for structural post-fire assessment. In: Piloto, P.A.G., Rodrigues, J.P., Silva, V.P. (eds.) CILASCI 2019. LNCE, vol. 1, pp. 48–60. Springer, Cham (2020). https://doi.org/10.1007/978-3-030-36240-9_4
8. Rein, G., Bar-Ilan, A., Fernandez-Pello, C., Alvares, N.: A comparison of three fire models in the simulation of accidental fires. J. Fire Protect. Eng. **17** (2004)
9. Sacadura, J.: Radiative heat transfer in fire safety science, radiation IV proc. J. Quant. Spectrosc. Radiat. Transf. **93**, 5–24 (2005). https://doi.org/10.1016/j.jqsrt.2004.08.011
10. Walton, W.D., Carpenter, D.J., Wood, C.B.: Zone computer fire models for enclosures. In: Hurley, M.J., et al. (eds.) SFPE Handbook of Fire Protection Engineering, pp. 1024–1033. Springer, New York (2016). https://doi.org/10.1007/978-1-4939-2565-0_31
11. Wang, X., Fleischmann, C., Spearpoint, M.: Assessing the influence of fuel geometrical shape on fire dynamics simulator (FDS) predictions for a large-scale heavy goods vehicle tunnel fire experiment. Case Stud. Fire Saf. **5**, 34–41 (2016). https://doi.org/10.1016/j.csfs.2016.04.001
12. Yuan, W., Tan, K.H.: Cellular automata model for simulation of effect of guiders and visibility range. Curr. Appl. Phys. **9**(5), 1014–1023 (2009). https://doi.org/10.1016/j.cap.2008.10.007. http://www.sciencedirect.com/science/article/pii/S1567173908002733
13. Zheng, Y., Jia, B., Li, X.G., Zhu, N.: Evacuation dynamics with fire spreading based on cellular automaton. Phys. A **390**(18), 3147–3156 (2011). https://doi.org/10.1016/j.physa.2011.04.011
14. Zheng, Z., Huang, W., Li, S., Zeng, Y.: Forest fire spread simulating model using cellular automaton with extreme learning machine. Ecol. Model. **348**, 33–43 (2017). https://doi.org/10.1016/j.ecolmodel.2016.12.022

Narrow Passage Problem Solution for Motion Planning

Jakub Szkandera[1]([✉]), Ivana Kolingerová[1,2], and Martin Maňák[2]

[1] Department of Computer Science and Engineering, Faculty of Applied Sciences,
University of West Bohemia, Univerzitni 8, 30614 Plzen, CZ, Czech Republic
{szkander,kolinger}@kiv.zcu.cz
[2] New Technologies for the Information Society, Univerzitni 8, 30614 Plzen, CZ,
Czech Republic
manak@ntis.zcu.cz

Abstract. The paper introduces a new randomized sampling-based method of motion planning suitable for the problem of narrow passages. The proposed method was inspired by the method of exit points for cavities in protein models and is based on the Rapidly Exploring Random Tree (RRT). Unlike other methods, it can also provide locations of the exact positions of narrow passages. This information is extremely important as it helps to solve this part of space in more detail and even to decide whether a path through this bottleneck exists or not. For data with narrow passages, the proposed method finds more paths in a shorter time, for data without narrow passages, the proposed method is slower but still provides correct paths.

Keywords: Motion planning · Sample based algorithms · Rapidly exploring random tree · Narrow passage · Bottleneck

1 Introduction

A fast and reliable solution of the motion planning problem - to find a collision-free path for an agent (an abstraction of a moving object) between at least two spots in an environment filled with obstacles is needed in many areas (e.g. robotics, autonomous vehicle navigation, computational biology, etc.). For simple-shaped agents it is possible to use geometrical methods (e.g., Voronoi diagrams to compute centerlines). However, when the navigation of more complex or even a flexible agent is necessary, the geometric methods on themselves are not strong enough any more.

The concept of configuration space is used to interpret motion planning. The configuration space is a set of all existing configurations, where one configuration represents the specific position and rotation of the agent. These properties together form degrees of freedom. As the number of degrees of freedom increases, the dimension of the problem to be solved as well as its complexity increase. For example, the agent configuration in 3D space may be a six-dimensional vector

© Springer Nature Switzerland AG 2020
V. V. Krzhizhanovskaya et al. (Eds.): ICCS 2020, LNCS 12137, pp. 459–470, 2020.
https://doi.org/10.1007/978-3-030-50371-0_34

describing its position (3 vector components) and rotation (3 vector components) in the configuration space. A configuration space then contains a huge number of configurations that cannot be processed in a reasonable time, and, therefore, randomized sampling-based methods are used.

Randomized sampling-based methods [7,10] randomly select configurations to subsequently test for collisions. If the tested configuration is collision-free, it is added to a path-finding structure (roadmap). Otherwise, the configuration is rejected and the method creates a new random configuration. The roadmap approximates the free regions of the configuration space and enables to search a path with graph-based path planning methods. In many cases, this is a very effective way to find a passage through the environment in a reasonable time. However, the randomized sampling-based algorithms have problems with narrow passages; since the methods randomly sample space, it is very difficult to hit a sample inside a narrow passage.

This article proposes a solution to the narrow passage problem, based on the combination of Voronoi diagrams and randomization, using the idea of so-called exit areas [12]. Exit areas (exit regions) were originally proposed in the context of protein molecular models. They show the exits from deeply buried empty cavities. However, the same idea can be used in other motion planning applications as well. In general, exit areas capture the exact positions of the narrow passages, which greatly contributes to eliminating the biggest weakness of randomized sampling-based algorithms. Thanks to this knowledge, it is possible to sample the position of the narrow passage in detail and it is even possible to decide whether there is a passage through the narrow passage or not.

The paper has the following structure. Section 2 contains a description of existing motion planning methods that are useful for navigating an agent through the configuration space. Section 3 focuses on a detailed description of the proposed solution for motion planning in narrow passages. It also includes an algorithm description and improvement for the sample based algorithms to increase its acceptance of samples. Section 4 presents experiments and results on the real biomolecular data and artificially generated data. Section 5 concludes the paper.

2   Related Work

The widely used randomized sampling-based algorithms can be divided into two groups - algorithms based on Probabilistic Roadmaps (PRM) [7] and Rapidly Exploring Random Tree (RRT) [10]. The original PRM algorithm [8] builds a graph over the explored parts of the environment. This approach has two phases. First the random samples are generated and tested for collisions. The second phase tries to connect the close samples with an edge if possible. The possibilities of implementing these procedures are stated in [5] which also compares these procedures in detail. A sufficient input for the PRM algorithm is a set of obstacles. The knowledge of the start and the goal configuration is not required by the algorithm itself, but their knowledge can be used in some sampling heuristics.

There are two problems in the PRM algorithm. The first one is called boundary value problem, when it is necessary to solve whether the movement of the agent from the first state to another is possible. It rises up when connecting the two given configurations. This problem can be difficult to solve under motion constraints, so PRM is primarily used in motion planning without motion constraints. The second problem is the already mentioned narrow passage problem. It can be solved (or at least approximated) by generating random samples close to obstacles or around medial axis of the environment [9] for low-dimensional configuration spaces.

sPRM [7] is a simplified version of the Probabilistic Roadmaps algorithm. Rather than for practical use it is used for the analysis of follow-up algorithms. On the other hand, unlike the previous method, the sPRM finds the path asymptotically optimal. PRM* [6] is another possible variant which uses a heuristic function to minimise roadmap lengths. This is an algorithm based on sPRM with the only difference that potential samples for interconnection are selected from the neighborhood with radii $r > 0$.

The Rapidly-exploring Random Tree (RRT) [10] belongs to the second mentioned group. RRT has been designed for use in models with a number of complex physical constraints. A tree is generated instead of the graph, which simplifies the path planning part. Next it incrementally grows towards unexplored regions of the configuration space. In addition, it also needs the start configuration. The main RRT computation is as follows. First of all, the tree structure t_{main} is initialized and then the algorithm repeats three main steps in cycle. The first step is to randomly generate a new sample in the configuration space. Steering the new sample close to the nearest tree list of the tree t_{main} is the second step. The last step is to check the collision of the new sample. If the sample is collision-free, it is added to the tree t_{main}, otherwise it is rejected. As RRT is the base of our proposed solution, it will be explained in more detail in the next section.

In the case of RRT algorithms, there are a number of modifications that solve motion planning in general or for specialized problems. RRT* [6] is an algorithm that uses a heuristic function to find the optimal solution. The extension for dynamic environment is solved by RRT^X [13]. There are plenty of other modifications but all of them suffer from the narrow passage problem like PRM algorithms. Guiding the tree along a precomputed path by geometry-based methods [14] is a possible way how to solve the narrow passage problem.

The motion planning methods can also be applied to other areas than to the navigation of mechanical objects. The motion planning in molecular simulations, where we have found inspiration for our proposed solution, is also a very important topic of research. The problem is, e.g., a navigation of the so-called ligand in a protein. Probabilistic Roadmaps can be used to sample the configuration space of the protein [1] in order to speed up molecular dynamics simulations but atoms bounds of the ligand lead to sampling in the high-dimensional configuration space.

The RRT algorithm is an appropriate planner also for a flexible ligand [3]. The ability to generate new configurations greatly affects the performance of the RRT. The high-dimensional space problem can be time consuming and the ML-RRT (Manhattan-like RRT) copes with this problem [4]. The method was

462   J. Szkandera et al.

further extended for flexible ligands [3]. Moreover, the high-dimensional space roadmap can be projected back to 3D space [2].

Exit areas (exit regions) were originally developed for cavities in protein models [12]. Cavities and their exits are computed from a Voronoi diagram. The graph of Voronoi vertices and edges captures possible trajectories of collision avoiding spherical probes among spherical obstacles (the atoms of a protein model). When the probe is located in a cavity, it cannot get to the exterior space without a collision unless the probe radius is reduced. The exact value to which the probe radius must be reduced and the exact position where the probe will be located (the primary exit location) can be computed by analyzing the graph of Voronoi vertices and edges. The edge on which the probe could escape is then disabled and the process is repeated to discover remaining exits. Exit areas (exit regions) are then constructed as the groups of intersecting probes in the exit locations.

3   Proposed Solution

The proposed solution idea is based on incorporation of exit areas [11] into a randomized sampling-based algorithm where it helps to detect narrow passages in the data. The exit area in this context is the area that contains the nearest collision-free surrounding of the narrow passage. The position inside the exit area (more precisely in the middle of the narrowest passage) is called the exit point. Now let us first recall the original RRT algorithm in detail, see [10], and then the proposed modifications.

Each RRT algorithm contains three identical steps that have been used since the introduction of the original RRT algorithm [10], only the techniques used to solve these individual steps differ. After the tree structure t_{main} is initialized, three steps, which are repeated until the computation is finished (exceeding the maximum iteration, finding the path, etc.), are started. Generating the new random sample in the configuration space is the first step. The second step is steering the new sample close to the nearest tree list of the tree t_{main}. The third and final step is to check if the new sample is colliding with surroundings. If the sample is collision-free, it is added to the tree t_{main}, otherwise it is rejected.

Two modifications of the RRT algorithm are needed in the proposed solution. First, the exit points are computed before the main RRT cycle is started. Exit points tell us the exact position of the most problematic places (narrow passages) in the data, which can be then focused on (e.g., more detailed sampling of narrow passages can be done). In any case, the knowledge of the exact position of the narrow passage is absolutely crucial. The second change is the correction of the rejected samples. If the agent has a small collision with the surroundings and the sample would be rejected, we try to move the agent into the free space. This modification is included in the third step of the RRT algorithm (the collision test of the sample).

Let us illustrate the main idea of the proposed solution. First, the proposed solution finds exit points $v_{exit}^i, i = 1, 2, 3$, whose surrounding is then sampled

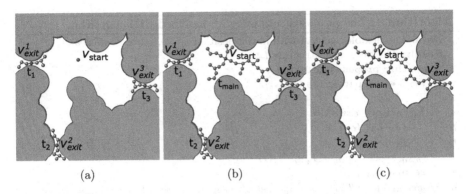

Fig. 1. (a) Sampling around exit points, (b) Sampling around the starting position v_{start}, (c) Merging exit point tree t_3 into the main tree t_{main}

by a randomized sampling method (Fig. 1a). When all narrow passages are processed, sampling from the starting position v_{start} is initiated (Fig. 1b). This sampling from the starting position v_{start} repeats until we find the pass through the sampled data. During this sampling, the tree t_{main} can reach some smaller tree t_i that has been created when the method was sampling the exit point surroundings. In this case, we connect the smaller tree t_i to the main tree t_{main} (Fig. 1c). This approach has two obvious advantages. The main tree t_{main} grows by already created samples to cover more space, and above all, it easily overcomes the narrow passage.

The whole process is shown in detail as Algorithm 1. Exit points are computed at (Algorithm 1, line 1). Each exit point v_{exit}^i is basically a new starting point for calculating the RRT algorithm, so we run it for each v_{exit}^i (Algorithm 1, line 5). It is important to note that we will let the RRT algorithm run only a limited number of steps c_{max}, because we only need to sample surroundings of a narrow passage to find if there is a possibility to get the agent through. Finally, the main RRT algorithm computation is run from the input starting point v_{start} (Algorithm 1, line 7), this run is not limited by the number of samples.

Now let us focus on the modification of the RRT algorithm itself (Algorithm 1, lines 9–19). For a given starting point v_{start}, which is the root of our main tree t_{main}, we will try to find a collision-free position (Algorithm 1, line 11). Then the main loop of the algorithm, which contains the above mentioned steps (sample, steer, connect), runs. The most important is the modification at the end of this loop where the merging of the existing trees t_{main} and t_i is done. If the new sample is collision-free and is added to the tree t_i, a check is done whether the new sample is also close to any other existing sampled tree $t_j, i \neq j$. If there is such a tree t_j, both trees are joined (i.e., the currently sampled tree t_i is extended).

We also modified the rejection of samples in collision with environment. An often case is that the sample is rejected, although only its small part is in collision with the surroundings. Therefore, in case of a collision, its 'size' is checked. If there is a 'big' collision (e.g., more than 20% of the agent is in an obstacle),

---

Algorithm 1: The proposed solution with modified RRT algorithm

---

Data: The flexible agent A with initial position v_{start}, set of obstacles O

Result: The main tree structure t_{main}

1 **Algorithm** proposed_solution
2      $exits \leftarrow$ computeExits (v_{start})
3      $T \leftarrow \emptyset$
4   **foreach** $e \in exits$ **do**
5          $t \leftarrow$ RRT (e, T, n, O)
6   Add t into T
7      $t_{main} \leftarrow$ RRT $(v_{start}, T, n_{max}, O)$
8   **return** t_{main}
9 **Procedure** RRT(*Tree root v_{root}, set of trees T, number of iterations n, set of obstacles O*)
10   Create the tree t with the root v_{root}
11   Collision-free rotation of agent in the root position v_{root}
12   **repeat** n **times**
13   Create a new sample s
14   Steer s to the tree t
15   **if** *s is collision-free* **then**
16   Add s to the tree t
17   **if** *s is close to some tree $t_j \in T$* **then**
18   Merge the tree t_j into the tree t
19   **return** t

---

the sample is rejected. Otherwise, we try to push the agent out of the obstacle to the free space following the shortest trajectory (Fig. 2a). There are multiple directions where to push the agent but we are using the direction with the shortest shift of the agent to the free space. Subsequently, it is necessary to check whether this shift was accessible or not. If the agent is collision-free (Fig. 2b), the sample is accepted and added into the tree structure. However, it may happen that the push-out of one obstacle results in a collision of the agent with another obstacle (Fig. 2c). In this case, the sample is rejected.

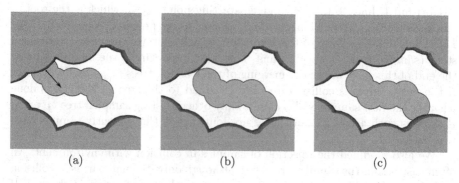

        (a)   (b)   (c)

Fig. 2. (a) Small collision with obstacles, (b) Sample pushed correctly to the free space, (c) Sample pushed to another collision

4   Experiments and Results

All experiments of the proposed solution, which was implemented in C#, were performed on a computer with the CPU Intel® Core™ i7-7700K (4.2 GHz) and 64 GB 2400 MHz RAM. Three types of environments were used for the testing of the proposed solution – two artificial environment data sets and real biomolecule *dcp* (proteins are freely available from data bank). Moreover, each of the environments was tested with a different flexible agent.

In the first column of figures there are the first artificial data (Fig. 3a) that are compounded from two hollow cubes connected by tunnels. There is also a cross-section (Fig. 3d) of these data, the location of the exit point (green circle with red cross) and the starting position (green rhombus with red cross) are shown, too. Figure 3g contains an example of one state of a flexible agent. In total, we have 15 different states of this flexible agent with different positions and rotations of individual spheres towards each other. Similarly, the second column contains data, also artificially created, that resemble a shell (Fig. 3b). To be more specific, it represents a hollow sphere with a crack. The position of the start (green rhombus with red cross), which is exactly in the middle, and the exit point (green circle with red cross) is shown in Fig. 3e. Using this data, we navigated the agent shown in Fig. 3h, which also has 15 different states. The last column contains real biomolecular data (Fig. 3c). The cross-section of these data with location of some exit points (green circles with red cross) and starting position (green rhombus with red cross) are shown in Fig. 3f. Figure 3i contains an example of one state of a flexible agent (ligand). In total, we have 100 different states of this flexible agent with different positions and rotations of individual spheres towards each other.

Now let us look at the difference in the tree structure, which is an algorithm output, using the RRT algorithm with and without exit points. The biggest visual difference between the results of RRT and its modification is at the beginning of the computation. Differences remain also in the further course of the algorithms, but it is difficult to distinguish them visually, because the environment is filled with a large number of samples. Note that the starting position v_{start} is colored green and each exit point v_{exit}^i has red color (Fig. 4). Due to the good visibility of individual trees t_i, the rendering of obstacles is disabled. In the case of a modified RRT algorithm, we may notice that we have more than one tree (Fig. 4a). A small tree t_i is created around each exit point, and then the main sampling starts from the start point v_{start}, which creates the main tree t_{main}. The isolated vertices in Fig. 4a are those exit points which did not lead to a collision-free configuration, so were not subsequently used in the calculation. Figure 4b then shows the behaviour of the standard RRT algorithm, which will subsequently have a problem with narrow passages, as there is very little probability that the algorithm will hit the right place with the correct configuration.

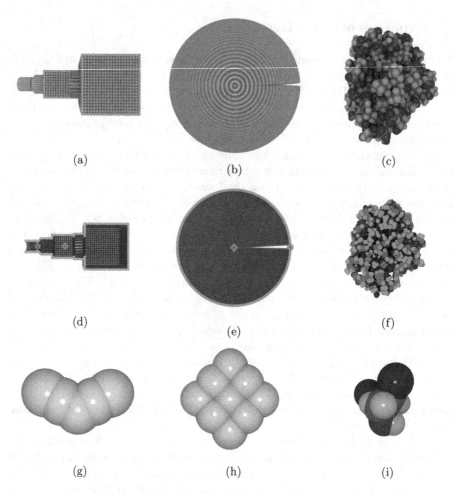

Fig. 3. (a–c) Tested data sets, (d–f) Cross-section through tested data sets, (g–h) Examples of navigated flexible agents (Color figure online)

Results of the standard RRT method and of our proposed solution (let us call it mRRT for short), applied on the first artificial data (Fig. 3a) are shown in Table 1. There is only one possible way how to get out through the obstacles and it was found in all cases by both tested algorithms with 100% success. However, mRRT has a clear superiority over the RRT algorithm as to the time of computation. As mentioned there is only one path through data but Table 1 contains 1000 found paths. This column means that we have ran the computation 1000 times with different seed of random generator.

This time difference was bigger on the other artificial data (Fig. 3b), where there is also one possible pass but the passage through the data now contains more possible space around the narrow passage, but at the same time a large amount of free space where the sampling algorithm may become congested.

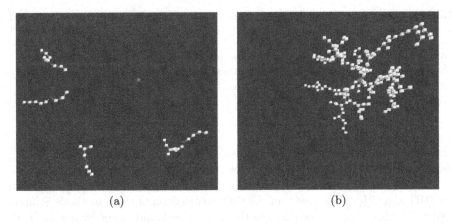

Fig. 4. Start of (a) The modified RRT algorithm with exit points, (b) Standard RRT algorithm

Table 1. Comparison of the RRT algorithm and our modified version (mRRT) - first artificial data

Algorithm	Number of found paths	Time [s]			
		Lowest	Average	Median	Biggest
RRT	1000	28.31509	281.595	259.9069	861.9697
mRRT	1000	0.145702	14.79853	0.23456	59.82809

In this case, the RRT algorithm did not find its passage through data at all (Table 2). The calculation even ended up with a low memory error, after having checked one million samples in approximately two hours of each run. Similarly to the previous table, the column with the number of paths means how many times was the path found from 5000 runs with different seeds. The mRRT found its way in every run in a very good time.

Table 3 compares the results of the real biomolecular data. The tested algorithms were run 500 times for the time of two minutes, and each run had a different seed of random sampling generator (but the same seed was used for both algorithms) Table 3 contains information on how many times the RRT algorithm was better in the evaluated property than mRRT and vice versa where the

Table 2. Comparison of the RRT algorithm and our modified version (mRRT) - second artificial data

Algorithm	Number of found paths	Time [s]			
		Lowest	Average	Median	Biggest
RRT	0	Out of memory	Out of memory	Out of memory	Out of memory
mRRT	5000	0.235554	11.2778	9.891314	53.59965

Table 3. Comparison of the RRT algorithm and our modified version (mRRT) - tested on the real biomolecular data

Algorithm	Number of paths	Time of the found path					Count of samples		
		First	Second	Third	Fourth	Fifth	Accepted	Rejected	Total
RRT	0	37	206	315	289	177	476	61	425
mRRT	500	463	294	185	211	323	24	439	75

evaluated properties were higher number of found paths, shorter path finding times, more samples accepted, fewer rejected samples, and more total samples, respectively. In all cases mRRT algorithm found a higher number of paths than the RRT algorithm. The result of the time comparison of these methods is interesting. The first two passes through the data were found faster by our method, the third and fourth by standard RRT algorithm, and since the fifth pass, mRRT began to lead again.

The last mentioned modification in the proposed solution - pushing the agent out of the obstacle if there is only a small collision with the environment - is in Table 4. There is a comparison of both RRT and mRRT algorithms with and without this modification on the real biomolecular data. Table 4 provides information on how many times the RRT algorithm was better without sample correction than with sample correction and vice versa (same for mRRT). Focusing on the RRT algorithm, we can notice that the first two passes through the data are found faster without correcting the samples. However, the other three passes are found faster when using this approach. At the same time, it is essential that using this approach, the number of samples received is higher and the number of rejected samples is lesser than the standard RRT algorithm has. The total number of samples is lower, but this is because the RRT algorithm without this approach has a much larger number of rejected samples. In the case of our proposed solution (mRRT), it is better to use the sample correction, because in almost all cases we get better results. Only in the total number of samples is the result worse, but the reason is the same as in the case of the standard RRT algorithm.

We should point out that results are highly dependent on the data type: if the data contain only narrow passages, mRRT is better. If the data contains large passes, mRRT is slower due to the calculation of exit points (useless for

Table 4. Comparison of the RRT algorithm and our modified version (mRRT) - push with and without

Algorithm	Time [s]					Samples		
	First	Second	Third	Fourth	Fifth	Accepted	Rejected	All
RRT without push	1931	1484	960	704	609	0	35	2187
RRT with push	333	780	1304	1560	1655	2264	2229	76
mRRT without push	813	727	710	696	732	0	28	1774
mRRT with push	1023	1109	1126	1140	1104	1836	1808	62

large passes). In the current data there are both types of passes (large passes and narrow passages) where our modified solution finds the narrow passages first and than the large passes. On the other hand, the standard RRT algorithm first finds the large passes and then the narrow passages (very low probability).

5   Conclusion

In this paper we introduced a modification of the RRT algorithm with improved ability to find a collision-free path for a flexible agent in an environment represented by configuration space. The proposed modification has been described for RRT algorithm but can be used in any sampling-based algorithm. Besides finding the path, the method provides knowledge where exactly the narrow passages are. This is extremely useful information as with proper tools it is possible to decide whether the narrow passage is passable or not. The RRT algorithm gives us two possible answers - there is a path through data or the algorithm cannot find any (the path may exist or not). The modified algorithm is able to give us also two answers - there is a path through the data or there is none. The proposed solution is most suitable for the data with narrow passages, where it is multiple times better than the original algorithm. For data without narrow passages the proposed method is slower than the standard sampling-based algorithm due to the extra computation of exit points and sampling their surrounding. However, a correct path will be found even for such unfavourable data.

Acknowledgement. This work was supported by the Ministry of Education, Youth and Sports of the Czech Republic, the project SGS-2019-016 Synthesis and Analysis of Geometric and Computing Models, and funded by Czech Science Foundation, the project No. 17-07690S Methods of Identification and Visualization of Tunnels for Flexible Ligands in Dynamic Proteins.

References

1. Amato, N.M., Dill, K.A., Song, G.: Using motion planning to map protein folding landscapes and analyze folding kinetics of known native structures. J. Comput. Biol. **10**(3–4), 239–255 (2003)
2. Cortés, J., Barbe, S., Erard, M., Siméon, T.: Encoding molecular motions in voxel maps. IEEE/ACM Trans. Comput. Biol. Bioinform. (TCBB) **8**(2), 557–563 (2011)
3. Cortés, J., Le, D.T., Iehl, R., Siméon, T.: Simulating ligand-induced conformational changes in proteins using a mechanical disassembly method. Phys. Chem. Chem. Phys. **12**(29), 8268–8276 (2010)
4. Ferré, E., Laumond, J.-P.: An iterative diffusion algorithm for part disassembly. In: Proceedings of the 2004 IEEE International Conference on Robotics and Automation, vol. 3, pp. 3149–3154. IEEE (2004)
5. Geraerts, R., Overmars, M.H.: A comparative study of probabilistic roadmap planners. In: Boissonnat, J.-D., Burdick, J., Goldberg, K., Hutchinson, S. (eds.) Algorithmic Foundations of Robotics V. STAR, vol. 7, pp. 43–57. Springer, Heidelberg (2004). https://doi.org/10.1007/978-3-540-45058-0_4

6. Karaman, S., Frazzoli, E.: Sampling-based algorithms for optimal motion planning. Int. J. Robot. Res. **30**(7), 846–894 (2011)
7. Kavraki, L.E., Kolountzakis, M.N., Latombe, J.-C.: Analysis of probabilistic roadmaps for path planning. IEEE Trans. Robot. Autom. **14**(1), 166–171 (1998)
8. Kavraki, L.E., Svestka, P., Latombe, J.-C., Overmars, M.H.: Probabilistic roadmaps for path planning in high-dimensional configuration spaces. IEEE Trans. Robot. Autom. **12**(4), 566–580 (1996)
9. Kurniawati, H., Hsu, D.: Workspace-based connectivity oracle: an adaptive sampling strategy for PRM planning. In: Akella, S., Amato, N.M., Huang, W.H., Mishra, B. (eds.) Algorithmic Foundation of Robotics VII. Springer Tracts in Advanced Robotics, vol. 47, pp. 35–51. Springer, Heidelberg (2008). https://doi.org/10.1007/978-3-540-68405-3_3
10. LaValle, S.M.: Planning Algorithms. Cambridge University Press, Cambridge (2006)
11. Manak, M.: Voronoi-based detection of pockets in proteins defined by large and small probes. J. Comput. Chem. **40**(19), 1758–1771 (2019)
12. Manak, M., Anikeenko, A., Kolingerova, I.: Exit regions of cavities in proteins. In: 2019 IEEE 19th International Conference on Bioinformatics and Bioengineering, BIBE, pp. 1–6. IEEE Computer Society (2019)
13. Otte, M., Frazzoli, E.: RRTX: asymptotically optimal single-query sampling-based motion planning with quick replanning. Int. J. Robot. Res. **35**(7), 797–822 (2016)
14. Vonásek, V., Faigl, J., Krajník, T., Přeučil, L.: A sampling schema for rapidly exploring random trees using a guiding path. In: Proceedings of the 5th European Conference on Mobile Robots, vol. 1, pp. 201–206 (2011)

Fault Injection, Detection and Treatment in Simulated Autonomous Vehicles

Daniel Garrido[(✉)], Leonardo Ferreira, João Jacob, and Daniel Castro Silva

Faculty of Engineering of the University of Porto, Portugal Artificial Intelligence
and Computer Science Laboratory (LIACC), Rua Dr. Roberto Frias s/n,
4200-465 Porto, Portugal
{up201403060,up201305980,joajac,dcs}@fe.up.pt

Abstract. In the last few years autonomous vehicles have been on the
rise. This increase in popularity lead by new technology advancements
and availability to the regular consumer has put them in a position where
safety must now be a top priority. With the objective of increasing the
reliability and safety of these vehicles, fault detection and treatment
modules for autonomous vehicles were developed for an existing multi-
agent platform that coordinates them to perform high-level missions.
Additionally, a fault injection tool was also developed to facilitate the
study of said modules alongside a fault categorization system to help the
treatment module select the best course of action. The results obtained
show the potential of the developed work, with it being able to detect
all the injected faults during the tests in a small enough time frame to
be able to adequately treat these faults.

Keywords: Autonomous vehicles · Unmanned aerial vehicles · Fault
injection · Fault detection · Fault treatment · Simulation · Safety

1 Introduction

Autonomous vehicles (AVs) have received a lot of attention in the last years
thanks to their ability to perform tasks in places humans can't reach or are
too dangerous [12]. This increase in popularity drives the need to guarantee that
these systems are safe to operate both for operators and surrounding population.
To assure safety of operation, AVs must be resilient to failures that create dangerous situations. Since an AV can't rely on the judgement of a human, it must
detect and handle faults internally. The simplest way to achieve this is through
redundant systems that compare each other's outputs and can take over in case
of a failure. However, this approach's disadvantages are exacerbated in small
AVs as they can't always accommodate the additional weight and space. The
alternative is to analyse the data generated from the vehicle's sensors to detect
fault-related patterns and alter its behaviour to handle the fault [2].

Because research with real vehicles can be cumbersome and expensive, the solution to this problem is going to be developed inside a simulation platform capable of
coordinating AVs to perform high-level missions, which uses FSX (Flight Simulator

© Springer Nature Switzerland AG 2020
V. V. Krzhizhanovskaya et al. (Eds.): ICCS 2020, LNCS 12137, pp. 471–485, 2020.
https://doi.org/10.1007/978-3-030-50371-0_35

X) as the simulation engine [13]. This research is a continuation of the development of this platform as it currently does not have a fault handling system, which is crucial when dealing with this kind of vehicles. While the platform and the concept of the project can be applied to any AV, it was primarily developed and tested for large fixed-wing UAVs (Unnamed Aerial Vehicles).

The goal of this project is to develop and incorporate a fault diagnosis system to the platform. This system must be easy to use and cover the most common failures in UAVs. In the end, the vehicle should be able to detect and correct fault scenarios on its own, while minimizing computational resources overhead.

To achieve this objective, several new modules were built to integrate in the existing platform. The first is a fault injection tool that allows the user to control fault injections during missions. Then, two modules were added to the vehicle agent: one for fault detection and the other for treatment. In the end, tests to these modules were conducted to assess fault detection rates and times, as well as the quality of the treatment and computational impact on the platform.

The rest of this article is structured as follows. Section 2 quickly reviews the state of the art and previous related work. Section 3 details the implementation process, starting with fault-related tests made to FSX, and the fault injection, detection and treatment modules. In Sect. 4, a description of the performed experiments is presented alongside the results, with their discussion presented in Sect. 5. Finally, Sect. 6 concludes the article and elaborates on future work.

2 State of the Art

In this section a literature review is presented in two parts. First a more general view on fault detection methods is given, before exploring some related work where these methods are applied to AVs.

2.1 Fault Detection Methods

There is a large amount of relevant literature on fault detection, which has been a serious research topic at least since the 1970s. Throughout the years, several surveys have been published which detail the advancements in fault diagnosis.

Usually, these surveys divide fault detection methods in categories to simplify their classification. Different authors propose different but similar classifications. The simplest one was proposed by Gertler, with methods divided in those that make use of a model and those that don't [6]. Miljković used three groups: data methods and signal models; process model-based methods; and knowledge-based methods [10]. The first two groups are identical to Gertler's, with a new group for the recently developed machine learning methods. Isermann's classification is the most complete and detailed, with several groups that relate to each other [9]. The studied classification methods were labeled using Gertler's approach.

Table 1. Summary of reviewed Fault Detection Methods

	Complexity	Computational cost
Model-free methods		
Limit/trend checking	Very low	Very low
Change detection	Low/Medium	Low
Neural networks/clustering	Medium/High	Medium
Model-based methods		
Parity equations	High	Medium/High
Parameter estimation	Very high	Very High
State observer	High	High
Output observer	High	High

Model-free methods, also called data-driven methods, use the input and output data from the system under diagnosis to search for fault patterns. These are usually less accurate than model-based methods but use less computational resources as they don't need to make model-related calculations. On the other hand, model-based methods use a model of the system in conjunction with a combination of inputs and outputs, depending on the method, resulting in more accurate detection, but with a computing performance penalty [8].

For more information on the other studied methods, refer to the previous work [5]. A collection and comparison of these methods regarding group, complexity and computational cost can be seen in Table 1.

Since the developed work focused on creating the whole system, the simplest method was used for fault detection, the limit and trend check methods. These are similar methods that monitor the values of specific variables while comparing them to predetermined upper and lowers bounds. When that variable is out of these bounds, a fault trigger can be activated. In limit checking, only the current value is taken into consideration, while in trend checking the rate of change of said variable is used [9].

2.2 Fault Detection in Autonomous Vehicles

In existing literature detailing the implementation of fault detection methods in autonomous vehicles, these can be either real or simulated, with some using both. In this literature review only those that study UAVs and present significant results are discussed.

Cork et al. applied the data collected from nominal flights to train Neural Networks to predict the output of a specific sensor and compare it with the measured values [3]. When a high difference between the two was detected, the system knew something was not right. For a data-driven system it obtained good results and could even train while being used.

Table 2. Summary of most relevant fault detection literature in AVs

Work	System	FD method	Results
[3]	Angular rate sensors	Neural networks	• Avg. detection rate: 84% • False-positives rate: 10% • Avg. detection time: 36 s
[7]	Positioning Sensors	Model based observer	• Avg. detection time: 0.55 s • #False-negatives: 6 • #False-positives: 9
[4]	Aileron actuators	Model-based observer and change detection with Z-test	• Model Based TIC avg.: 0.143 • Change Detection avg. detection time: 0.8 s
[11]	Pitot-static systems	Clustering (K-means and EM)	• Detection rate: 96% • False-positive rate: 1.5% • Detection time: "Almost Instant"

While not as popular as fixed wing UAVs, single rotor UAVs also exist. One of this kind of UAVs was used as a platform to create a model-based observer system to detect faults in positioning sensors. This work concluded that detection was possible but was more difficult in the case of additive and multiplicative faults, when compared to faults that made the sensors reading freeze [7].

Freeman et al. monitored the aileron actuators of a light UAV by two distinct approaches: change detection (data-driven) and observers (model-based) [4]. Both systems were tested with real flight data. It was found that the model-based approach was better at detecting faults, but it was also noted that the process of modelling the UAV was time-consuming. Meanwhile, the data-driven method was easier to implement and could also detect most of the faults.

As for a fault detection system that utilizes a game/simulation engine like the one used in this project, only one such case was found. Purvis et al. used the open-source flight simulator FlightGear to create a system that could inject, detect and treat faults related to the pitot-static system of a simulated commercial airliner. Their solution used clustering methods to label the flight data as faulty and not faulty, with very good results [11].

Table 2 summarizes the results of this small literature review, showing for each work the type of faulty system, fault detection method and experimental results. As expected, model-based methods worked better than data-driven ones, with Clustering being the better method when no model is used.

3   Implementation

The implementation process was divided in three parts. First a classification system for UAV faults was created. Next, a fault injection system was implemented in the multi-agent platform; and lastly, the agent responsible for controlling the vehicles was extended to include both a fault detection and treatment modules.

3.1   Fault Classification System

A classification system for UAV faults was created to categorize faults by severity according to the affected system and extent of the fault, while also providing recommended actions that the UAV should take in case of a fault. This system will prove helpful when there is a need to assess the impact of a detected fault to the UAV and what actions it can take to handle the situation. Table 3 presents a summary of the system and Table 4 explains the severity scale used [1]. The same failures were divided in several entries to accommodate different extents that progressively increase in severity. The failure influence on the aircraft was also included to help classify faults that are not included but cause similar problems.

Table 3. UAV Fault Classification Table

Failure	Influence	Severity	Reaction
Engine (partial)	Reduced lift and speed	Medium	Return to airport and emergency landing
Engine (complete)	Complete loss of lift	High/Extreme	Emergency landing/crash where possible
Communications	Loss of comms with ATC and potential flyaway	Medium	Return to airport and emergency landing with visual indication of communications fault
Control surfaces (single, free float)	Extra effort and care in controlling aircraft required	Medium	Return to airport and emergency landing
Control surfaces (single, stuck)	Difficulty in controlling aircraft	High	Return to airport and emergency landing
Control surfaces (multiple)	Total loss of control	Extreme	Imminent crash
Sensors (single)	None, remaining sensors should be able to compensate faulty one	Low	Procced mission
Sensors (multiple)	Loss of spatial awareness	Medium/High	Return to airport if possible, emergency landing/crash where possible otherwise
Sensors (complete)	Complete loss of spatial awareness	Extreme	Imminent crash
Electrical	Complete loss of sensors, control surfaces and electrical propulsion	Extreme	Imminent crash
Landing gear	Harsh landing	Low	Nothing
Brakes	Prolonged landing distance	Low	Abort landing and retry in longest runway, using all of the runway

3.2   Injecting Faults in Flight Simulator X

Working with FSX as a simulation platform is facilitated by using the integrated SimConnect SDK[1] which allows an external program to read and modify simulation variables through a client-server interface. Additionally, FSX includes fault injection in aircraft natively, but when the development began it was found that this was mainly supported for the player aircraft, and support for the AI-controlled aircraft that the platform uses was very limited. In spite of these limitations, some faults were able to be reliably injected in the platform's vehicles, including engines, brakes and communications.

Table 4. Fault severity scale

Severity	Flight control impact
Low	No or very subtle alterations in control; could easily reach landing site and have no problems touching down in the designated area
Medium	Significant alterations in control; can reach landing site but might have difficulty landing in the designated area
High	Very compromised control; difficulty in reaching landing site
Extreme	Very limited or no control at all

Before a fault can be injected, it first must be described. The user can create several faults that can affect any number of aircraft at any given time or during a number of special conditions. The fault itself is defined by a number of variables that determine when it should be triggered, when it ends, how strong the fault effect should be and what behaviour it should follow. Each fault contains a list of vehicles it can affect and a list of faults that can be injected to these vehicles. Different vehicles can be injected with different faults. The user can also define to great detail what conditions will trigger the fault, which can be based on the aircraft speed, altitude or location, elapsed time, weather conditions, ground surface type, etc. The value of the fault determines how severe the impact of the fault is or, in the case of control surfaces, the position at which they should be kept for the duration of the fault. The user can also choose the time behaviour that governs the fault injection, which can be set to permanent, intermittent, transient or noise. To simulate drift-like faults a ramping variable was also added that specifies how much time the fault should take to reach the desired strength. To facilitate the creation and modification of faults to be injected in a mission, a graphical interface was created to intuitively and quickly allow a user to specify changes. Figure 1 shows an example of this interface during use.

Engine faults can be injected to individual engines or to all engines. Due to the limitations of FSX, only the "all engines" fault can make use of the strength

---

[1] More information available online at https://docs.microsoft.com/en-us/previous-versions/microsoft-esp/cc526983(v=msdn.10).

value, with the single engine faults being restricted to being toggled, setting the engine on or off. Brakes fault is another toggle-type fault that affects the aircraft when it is trying to slow down after landing. The communications fault was handled entirely through the platform messaging system and effectively blocks all messages from reaching or leaving the affected vehicle.

3.3   Fault Detection and Treatment

Since no aircraft model was accessible from FSX, model-based methods could not be used. Instead, data-driven methods were used to detect faults in the three systems mentioned above. Due to FSX limitations on AI-controlled aircraft, not much data was available to use in the detectors, which limited the available methods to the simpler ones that don't require much data to be effective.

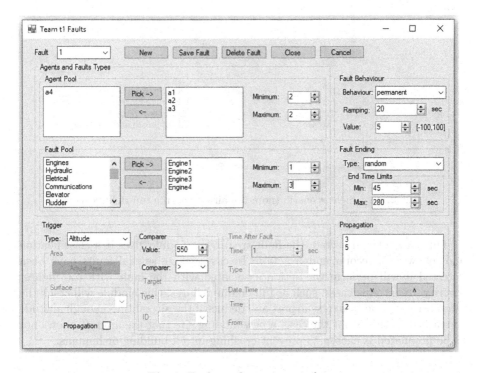

Fig. 1. Fault configuration window

The engine fault detector uses a combination of limit and trend checkers on the available engine variable: the propeller speed. The trend checker constantly analyses the propeller speed rate of change and triggers when this value is higher than a predefined value. For situations when the engine thrust descended slowly over a long period of time, also called ramping, a limit checker was also implemented that simply verifies if the engine RPM is too low (300 RPM in this case).

These two methods only trigger if the aircraft's current altitude is lower than the desired one, to prevent falsely detecting a fault when the aircraft is descending.

Faults related to brakes are detected with another trend checker. When the aircraft touches down to land, it immediately starts analysing the rate at which the aircraft slows. If this rate stays low for too long (above $-2\,\mathrm{m/s^2}$ for over 5 s in this case), a brake failure is detected.

The communications fault detector uses a very simple method to verify if the communications are working. Every 10 s the vehicle pings the closest ATC (Air Traffic Controller), who replies with an acknowledge. If the vehicle doesn't receive a response after 10 s of sending the ping, it knows the communications are not working properly. This means that in a best case scenario a fault can be detected in just 10 s, but in the worst it will take up to 20 s. The waiting time between messages could be reduced, but this could present problems when an ATC is responsible for several aircraft and can't handle all messages in a timely fashion.

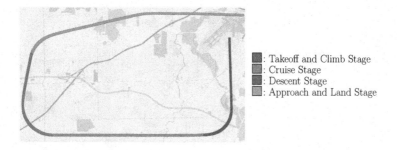

: Takeoff and Climb Stage
: Cruise Stage
: Descent Stage
: Approach and Land Stage

Fig. 2. Test flight scenario (note the airport on the top-right corner)

Once a fault is detected, the fault treatment module gives it a classification and follows the recommended action. In cases where several faults have been detected it will perform the action associated with the fault with the highest severity. In extreme cases, such as full engine failures, this module will track the aircraft return course to the airport and deduce if the aircraft has enough altitude to reach it. If this is not the case, a new landing site that the aircraft knows not to be populated is chosen to prevent crashing into a building or humans.

4   Experimental Setup and Results

This section is organized in two main parts: first, an explanation of the tests is given, followed by the presentation and analysis of the results.

4.1   Test Configuration and Scenarios

The tests to the developed work were conducted in the proximity of an airport previously modelled in detail in the platform. It was chosen because it has an interesting layout of two long and one short runway. The model of the aircraft used in the simulation was the Beechcraft Baron 58. It was picked for its relatively small size and engine configuration as it is the smallest and lightest aircraft with a twin prop engine. The small size makes it comparable to the bigger UAVs like the United States Air Force Predator, in terms of wingspan and weight, while the dual engine configuration allows for more flexibility when testing.

For every test the aircraft was given a simple mission to perform, as seen in Fig. 2, which includes taking off, making a right bank turn while ascending, holding altitude for a few miles, performing another right bank turn while now descending, approaching the smallest runway at the airport and finally landing. The different colours represent the different flight phases. The tests were all conducted with FSX running at a simulation rate of 4x to reduce test times.

The tests were separated in two phases: in the first phase only the fault detectors are active and in the second phase both the fault detection and treatment modules are operational. This way a benchmark of the outcomes of the faults can first be recorded to then compare to the outcomes when the same test is run with the fault treatment module enabled. Table 5 shows a summary of the test with all settings used.

Test #0 is a control test, with no faults active. It serves as a baseline to compare to the behaviour of the actual tests when faults are injected. Since tests #1 and #2 are not dependant of the flight stage, ramping and fault value, one test is enough to test if the module can correctly detect these faults. Test #2 is run with intermittent time behaviour to effectively allow the test to run several times to make sure the detection times don't surpass the theoretical maximum of 20 s. This is the only scenario where having an intermittent fault type is advantageous, as this type of time behaviour uses random injection times which are not ideal when the behaviour of the plane is being tested. This can result in the fault being injected for too little time to be detected or even have a meaningful effect on the aircraft.

Table 5. Tests to be performed to the fault detection module.

Test	Fault	Fault value	Stage	Ramping	Time behaviour	Duration
#0	–	–	–	–	–	–
#1	Total Brakes	–	–	–	Permanent	Unspecified
#2	Communications	–	Cruise	–	Intermittent	180 s
#3	Engine 1	–	Cruise	–	Permanent	Unspecified
#4	Engine 1	–	Climb	–	Permanent	Unspecified
#5	Engine 1	–	Descent	–	Permanent	Unspecified
#6	Engines	0	Cruise	–	Permanent	Unspecified
#7	Engines	0	Climb	–	Permanent	Unspecified
#8	Engines	0	Descent	–	Permanent	Unspecified
#9, #10	Engines	0	Cruise	30, 60	Permanent	Unspecified
#11, #12	Engines	0	Climb	30, 60	Permanent	Unspecified
#13, #14	Engines	0	Descent	30, 60	Permanent	Unspecified

Tests #3, #4 and #5 cover single engine full failure in all 3 flight phases, while tests #6, #7 and #8 do the same but with 2 failing engines. Finally, tests #9 to #14 test the effects of different ramping values in the different flight stages. This effect will only be tested with engine faults since this is the only one that supports continuous analog injection.

Finally, the fault treatment module is enabled, and tests #1, #2, #4 and #6 are ran again to test the ability to treat the faults in the expected way and comparing the outcome of the tests with the previous non-treated tests.

4.2 Fault Detection Test Results

In test #0 the fault detectors did not pick up any fault and as expected the aircraft performed the complete mission without problems.

For test #1, the brake fault detector successfully activated after the aircraft failed to slow down after landing, taking 12 s after touchdown to do so. However, when comparing the aircraft speed over time during the landing it is revealed that the aircraft only starts to slow down after 7 s in the control test, as can be seen in Fig. 3. This means that the actual fault detection time for this test was around 5 s. Because of the brake failure the aircraft ran out of asphalt and only stopped in the grassy area surrounding the runways.

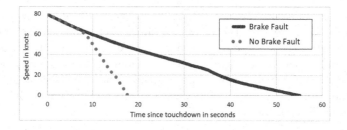

Fig. 3. Speed comparison after touchdown with and without brake fault

Since test #2 ran in an intermittent configuration where the fault was being toggled on and off repeatedly for 3 min, the detector had to correctly determine when the communications were off 3 times, as each on/off cycle takes about a minute. The results of this test can be consulted in Table 6. It achieved an average detection time of 15 s, with all detection times below the 20 s mark, as expected. In this case the aircraft completed the mission normally since communications don't affect the physical behaviour of the aircraft.

The results of the engine faults can be seen in Table 7. All failures were detected and no false positives were recorded. In general, failures that occurred during takeoff were the fastest ones to be detected, followed by the ones during cruising, the descending ones being the slowest overall. Regarding the outcome, the only tests where the aircraft was able to complete the test flight were the

Table 6. Results of intermittent communications fault

Injection timestamp (s)	Pause timestamp (s)	Injection delta (s)	Detection timestamp (s)	Detection delta (s)
96	125	29	114	18
152	187	35	165	13
235	270	35	250	12

ones with single engine failure. In the others the aircraft slowly descended until it hit the ground, without first deploying the landing gear.

While conducting the tests a strange behaviour was detected in the engine faults with ramping. It seemed that the thrust of the engines was not reducing at the expected rate, only starting to decrease after the ramping time was past the half point. This was then confirmed in the collected data when analysing the propeller speed after injecting the fault in tests #11 and #12, as seen in Fig. 4. As can be seen, the fault only starts taking effect after 2/3 of the ramping time and from there it linearly decreases to zero. This is another limitation of FSX that other tests confirm is only present in the AI-controlled vehicles and not in the user-controlled one. This means that the detection times recorded for tests that incorporate ramping are not accurate and the real detection times were included between parentheses for these tests in Table 7.

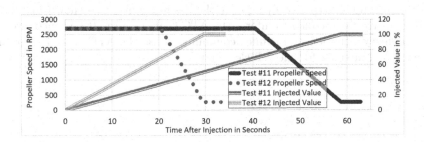

Fig. 4. Propeller speed after ramping fault injection in tests #11 and #12

4.3   Fault Treatment Test Results

With the treatment module enabled, the outcomes of the tests should vary to accommodate the injected faults. Starting with test #0, no changes were detected to mission execution and again no faults were detected.

In test #1 the brake failure was correctly identified once more on landing, but this time the aircraft aborts it, again taking off and making the necessary manoeuvres to approach the longest runway in the airport and land, as suggested in the categorization system. Even with the brake failure, the aircraft was able to stop within the length of the runway. The influence of the treatment module in this test can be seen in Fig. 5.

Table 7. Results of the various engine faults

Test	Injection timestamp (s)	Detection timestamp (s)	Detection delta (s)	Outcome
#3	35061.672	35160.782	99.11	Aircraft able to complete test flight
#4	36666.778	36667.445	0.667	Aircraft able to complete test flight
#5	38242.778	38361.663	118.885	Aircraft able to complete test flight
#6	42241.875	42252.986	11.111	Aircraft crashed
#7	41275.433	41276.099	0.666	Aircraft crashed
#8	40097.436	40148.547	51.111	Aircraft crashed
#9	45742.754	45774.976	32.222 (12.222)	Aircraft crashed
#10	46874.307	46929.862	55.555 (15.555)	Aircraft crashed
#11	50975.629	50996.296	20.667 (0.667)	Aircraft crashed
#12	49585.188	49625.855	40.667 (0.667)	Aircraft crashed
#13	54417.398	54468.509	51.111 (31.111)	Aircraft crashed
#14	56887.613	56944.058	56.445 (16.445)	Aircraft crashed

For test #2 the fault was detected the first time it was triggered, just like in the first test, and immediately the aircraft started changing its course to perform the recommended action of flying over the desired runway, as shown in Fig. 6. This maneuver is intended to inform the ATC that the aircraft has encountered an emergency situation and cannot communicate, so the ATC should clear the runways and airspace for the vehicle to land.

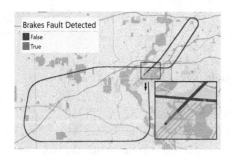

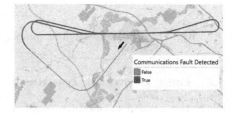

Fig. 5. Test #1 fault treatment path   Fig. 6. Test #2 fault treatment path

The fault injected in test #4 was also detected just like in the first test. The aircraft started the emergency landing protocol immediately by redirecting to the closest runway available to land as depicted in Fig. 7. Compared to the first test, where the aircraft was able to finish the mission in a safe manner, diverting to the airport immediately decreases the chances of an accident in case the fault propagates to the other engine.

Finally, in test #6 the fault was correctly identified, and the same emergency landing protocol was activated as in test #4. However, this time with both engines producing no thrust, the aircraft had no way of making it back to the airport. This was quickly detected and as a consequence the aircraft landed in a close field it knew was uninhabited, as can be seen in Fig. 8. In a real-world scenario this behaviour has the potential to decrease the number of accidents involving bystanders and decrease the probability of losing the aircraft in a crash.

Performance benchmarks were also conducted to test the impact of the new modules on the platform. The test measured the CPU (Central Processing Unit) load, memory allocated and CPU time for the platform in three scenarios: Just the Control Panel open; The Control Panel and Vehicle Agent running without the detection module; and all the modules active. Table 8 displays the results.

Since Flight Simulator is the one that controls the autonomous vehicles, a change in the performance of the platform is not detected from test #1 to test #2. Contrarily, when the detection module is being used, a small increase in CPU load and CPU time is detected but is very small to be significant to affect the overall performance of the platform.

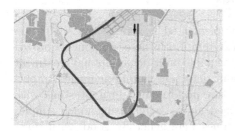
Fig. 7. Test #4 fault treatment path

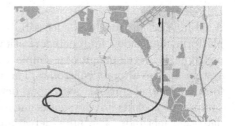
Fig. 8. Test #6 fault treatment path

Table 8. Resources used by the platform with different active modules (test were performed on a Laptop with an Intel Core i7-4710HQ processor @3.30 GHz)

Active modules	Max CPU load (%)	Max. memory (MB)	CPU time per minute (s)
Control Panel (CP)	0.4	35.5	~ 0
CP + Vehicle Agent (VA)	0.4	45.6	~ 0
CP + VA + Detection Module	0.9	46.1	~ 0.7

5 Discussion

The achieved results are promising, with all faults being detected, and no false positives. This shows that the current implementation is robust, accurate and resilient to false triggers. On the other hand, detection times were overall good but not great. This was to be expected since simple fault detection methods were used, while other authors use more advanced ones. This could be improved by using more advanced methods, such as those used in the literature mentioned in Sect. 2. Despite the slow reaction time, it was fast enough to allow the treatment module to intervene in a positive way in otherwise dangerous scenarios.

With some detection times below one second, this simple approach managed to match the detection times in other works that used model-based approaches such as Freeman et al. [4] and Heredia et al. [7], but can't keep up in more demanding scenarios. On another note, this solution managed to achieve an average detection time similar to that of Cork et al. [3]. The work of Purvis et al. [11] is the most similar to this one due to also using a flight simulator as a testbed and using a data-driven method. The use of clustering methods allowed for better results in reaction time with similar detection performance.

6 Conclusion and Future Work

A fault injection tool was successfully implemented in an existing simulation platform, alongside a fault categorization system. Both these components proved useful in the development of a simple but capable fault detection and treatment system for the aircraft controller. The fault detection module managed to perform above expectations, with good detection performance during testing, with comparable results to the works mentioned above, while using much simpler detection methods. The fault detection times were generally good, with time-sensitive faults like brakes and engines being detected quickly enough for the fault treatment module to act. This module also proved to perform well, being able to determine the best action to take when a fault occurred and maintaining the safety of bystanders always in first place by taking into consideration the surroundings of the vehicle. All of this was achieved while keeping the CPU and memory loads very minimal.

The developed work sets a solid base to continue fault-related research in this platform. The fault injection tool in particular is very useful for this kind of research as it helps create detailed fault scenarios for the detection and treatment algorithms that while being tested only with one aircraft, can handle concurrent fault injection in teams of multiple vehicles. The implementation of all the stages of a fault diagnosis system with a modular architecture also facilitates future development of new algorithms without having to redesign the system.

While the results were satisfactory, they could be improved in the future by increasing the number of failures to detect, and using different and/or more sophisticated data-driven methods that analyse more data. To do this it would likely be necessary to base the platform in another similar but more advanced simulator that can offer more data for AI-controlled vehicles and supports more fault injection options than FSX. Detection times could also be improved by using the mission details to know what should be normal and abnormal behaviour for the aircraft at a certain location or time.

References

1. Belcastro, C.M., et al.: Preliminary risk assessment for small unmanned aircraft. In: Proceedings of the 17th AIAA Aviation Technology, Integration, and Operations Conference, June 2017, Denver, Colorado, USA (2017)
2. Chen, J., Patton, R.J.: Robust Model-Based Fault Diagnosis for Dynamic Systems, 1st edn. Springer, New York (1999). https://doi.org/10.1007/978-1-4615-5149-2
3. Cork, L.R., Walker, R., Dunn, S.: Fault detection, identification and accommodation techniques for unmanned airborne vehicle. In: Proceedings of the 11th Australian International Aerospace Congress (AIAC 2005), 14–17 March 2005, Melbourne, Australia (2005)
4. Freeman, P., Pandita, R., Srivastava, N., Balas, G.J.: Model-based and data-driven fault detection performance for a small UAV. IEEE/ASME Trans. Mechatron. **18**(4), 1300–1309 (2013)
5. Garrido, D.: Fault injection, detection and handling in autonomous vehicles. Mathesis, Faculty of Engineering of the University of Porto (2019)
6. Gertler, J.J.: Survey of model-based failure detection and isolation in complex plants. IEEE Control Syst. Mag. **8**(6), 3–11 (1988)
7. Heredia, G., Ollero, A., Bejar, M., Mahtani, R.: Sensor and actuator fault detection in small autonomous helicopters. Mechatronics **18**(2), 90–99 (2008)
8. Isermann, R.: Model-based fault-detection and diagnosis - status and applications. Ann. Rev. Control **29**(1), 71–85 (2005)
9. Isermann, R.: Fault-Diagnosis Systems. Springer, Heidelberg (2006). https://doi.org/10.1007/3-540-30368-5
10. Miljković, D.: Fault detection methods: a literature survey. In: Proceedings of the 34th International Convention on Information and Communication Technology, Electronics and Microelectronics (MIPRO 2011), 23–27 May 2011, Opatija, Croatia, pp. 750–755 (2011)
11. Purvis, A., Morris, B., McWilliam, R.: FlightGear as a tool for real time fault-injection, detection self-repair. Proc. CIRP **38**, 283–288 (2015)
12. Schoenwald, D.A.: AUVs: in space, air, water, and on the ground. IEEE Control Syst. Mag. **20**(6), 15–18 (2000)
13. Silva, D.C.: Cooperative multi-robot missions: development of a platform and a specification language. Ph.D. thesis, Faculty of Engineering, University of Porto (2011)

Using Cellular Automata to Model High Density Pedestrian Dynamics

Grzegorz Bazior[ID], Dariusz Pałka[✉][ID], and Jarosław Wąs[ID]

Faculty of Electrical Engineering, Automatics, IT and Biomedical Engineering,
AGH University of Science and Technology, Mickiewicza 30, 30-059 Krakow, Poland
{bazior,dpalka,jaroslaw.was}@agh.edu.pl

Abstract. The article presents a model of pedestrian dynamics based on Cellular Automata, dedicated especially for high density conditions. Using the proposed representation, it is possible to map different behaviors of agents in various conditions. The article describes mechanisms of a pedestrian's movement for different pedestrian's shapes when a pedestrian occupies multiple cells, compared to the classical representation where a pedestrian occupies a single square cell. A greater number of cells per pedestrian can describe a pedestrian with e.g. a backpack, a suitcase on wheels, or on a bicycle more precisely than when s/he is represented by one cell. More accurate discretizations are particularly useful when high density of pedestrians occurs. The article focuses on the following aspects of a pedestrian's movement: changing direction, movement speed, a passing scheme, and an overtaking scheme.

Keywords: Crowd dynamics · Cellular automata · High density crowd

1 Introduction

Dense crowd modeling is currently perceived as a significant challenge. On the one hand, continuous methods allow for precise calculations of the superposition of the acting forces, and, on the other hand, agent-based models operating within the Cellular Automata framework [7,8] are highly efficient. The classical approach [2], in which a pedestrian is represented as a single cell, is often insufficient in modeling high pedestrian densities in a given area using two-dimensional cellular automaton. An interesting solution was proposed in [3], when sub-mesh configuration was proposed as a complement of standard mesh (lattice) in order to handle higher values of maximum density of simulation of dense crowds. Simultaneously, different concepts of finer discretization were presented: namely in [4] an idea of placing two pedestrians in one cell or the application of a dense triangular grid presented in [5].

In the classic approach, the size of a cell is usually equal to $40\,\mathrm{cm} \times 40\,\mathrm{cm}$ [2]. However, in situations when more pedestrians should be placed in a singular cell (even until 10 pers/m^2), as in Predtechensky & Milinski experimental

© Springer Nature Switzerland AG 2020
V. V. Krzhizhanovskaya et al. (Eds.): ICCS 2020, LNCS 12137, pp. 486–498, 2020.
https://doi.org/10.1007/978-3-030-50371-0_36

research [9], such granulation is insufficient. For example, the width of typical door openings differs by 10 cm, and in a situation of the high pedestrian flow (e.g. fire evacuation) the difference between 60 cm, 70 cm and 80 cm door throughput is important, as was confirmed by empirical results of competitive and non-competitive evacuation in [6]. This article presents an approach aimed at increasing the accuracy of modeling pedestrian dynamics in case of high crowd density, as a continuation of a previous study by the authors [1]. This approach is based on the density of nesting grids and the introduction of additional mechanisms of pedestrians' behavior, which can be described using CA lattice density. The article presents the methods of adaptation of the original mechanisms for dense grids proposed in the study [2] and the consequences of these adaptations.

2   Proposed Model

2.1   Reducing the Size of the Cell

The presented model is developed on the reduced size of individual cells. In order to maintain symmetry at rotations, it is recommended that densifying is made an odd number of times (3, 5, 7, ...) in the direction of each axis of the reference system (Fig. 1).

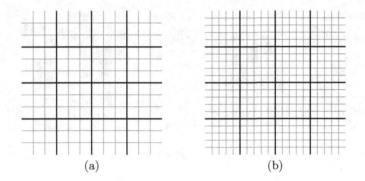

(a)   (b)

Fig. 1. Densifying the basic grid (marked by thick black lines): (a) 3 times, (b) 5 times in the direction of each axis. A denser grid is marked with thin blue lines. (Color figure online)

In case of a grid 3 times denser, the cell size is about 13 cm × 13 cm, while in case of a grid 5 times denser, the cell size is about 8 cm × 8 cm, which is usually sufficient for accurate modeling of space and pedestrian dynamics even for dense crowds.

2.2   Pedestrian's Shape

In case of the basic model [2], a pedestrian occupies one cell, so the shape of a pedestrian (in the projection to a floor plane) is a square. In reality, however, this shape is similar to an ellipse [11], where the longer half axis coincides with the axis of the arms (left-right), while the shorter half axis coincides with the axis of the front-back. Thanks to the densifying of the grid, it is possible to represent a pedestrian using a more complex shape (e.g. a rectangle instead of a square) – Fig. 2.

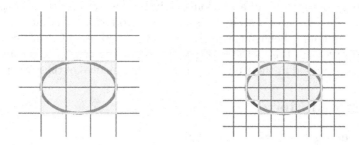

Fig. 2. Examples of shapes representing a pedestrian (yellow areas) for various grid densities. (Color figure online)

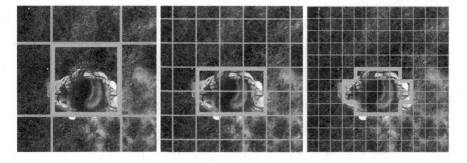

Fig. 3. A pedestrian without any objects: width (arm to arm) - 40 cm, height (forehead to back) - 25 cm.

A dense grid also allows for representing more complex shapes, such as a pedestrian with a backpack, a suitcase on wheels, a bag, etc., which can be seen in Figs. 3, 4, 5. In case of the standard pedestrian flow, these additional elements have a little influence on the pedestrian movement, however, in case of high density of pedestrians (e.g. during competitive evacuation), they become important.

2.3 Orientation of a Pedestrian

To maintain compatibility with the basic model, it is assumed that a pedestrian, regardless of his/her orientation, always occupies the same cell in the basic grid (marked in the drawings with thick black lines). The second assumption is that

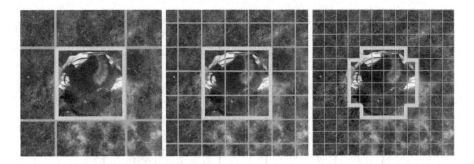

Fig. 4. A pedestrian with a small backpack.

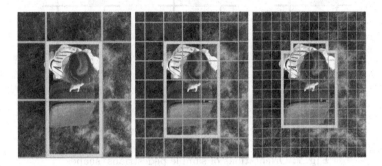

Fig. 5. A pedestrian with a suitcase on wheels.

the area occupied by a pedestrian in the projection to a floor plane (i.e. the number of occupied cells) should not depend on the pedestrian's orientation (rotation inside the basic cell).

The proposed scheme of transitions of the individual cells included in the basic cell as a result of rotation is shown in Fig. 6. The remaining orientations are created in the same manner.

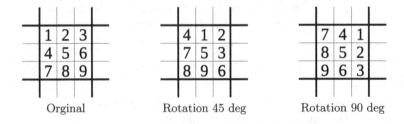

Fig. 6. Pedestrian's orientations - clockwise rotation inside the basic cell.

The orientations of sample pedestrian's shapes are shown in Fig. 7.

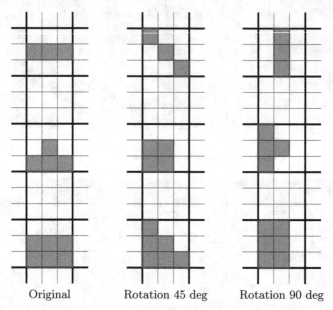

Original   Rotation 45 deg   Rotation 90 deg

Fig. 7. Orientations of sample pedestrians' shapes.

2.4   Pedestrians' Movements and Their Walking Speed

Thanks to the density of the grid, it is possible to model smaller movements
(e.g. 13 cm, 8 cm) than in the basic model for which the minimum distance
of movement is 40 cm. Such a reduction of movement granulation allows for
reproducing phenomena occurring at high and very high crowd densities - for
example, in the bottleneck area. To be able to maintain the maximum speed
(which is distance divided by time) as in the basic model, the n-times reduction
of the cell size must be accompanied by the n-times reduction of the time step.

Compared to the basic model, where only two speeds are possible within a
basic cell: $v_{max} = cell\ size/time\ step$ and 0, densifying the grid n-times gives
$n + 1$ possible speeds with which pedestrians can move:

$$v_k = \frac{k \cdot cell\ size}{n \cdot time\ step}, k = 0 \dots n \qquad (1)$$

where:

   cell size – the size of a cell in the basic model (before densifying)
   time step – the time step in the basic model

This allows for a better representation of the distribution of pedestrians'
walking speed [11], which depends on many factors such as, for example, the age
of a pedestrian, his BMI, his personal preferences, etc. Also, in case of modeling
crowd dynamics, the density of the crowd is a very important factor influencing
pedestrians' speed. Generally, for high density only low speed motion is allowed.

2.5 Passing Scheme

This scheme is applicable to the situations in which two pedestrians are walking from the opposite directions in a narrow passage, e.g. on a train, bus, tram, theater hall, cinema, etc. In such scenarios, due to limited space, it is impossible for pedestrians to pass with their normal orientation (i.e the shorter half-axis of the ellipse describing the pedestrian is in line with the direction of his movement). However, after the rotation of 90°, passing is possible – see Fig. 8.

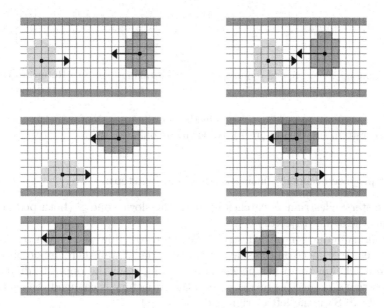

Fig. 8. A passing scheme for two pedestrians (marked yellow and green). The arrows show the pedestrians' velocity vectors. (Color figure online)

Such a phenomenon cannot be described in the basic model, in which a pedestrian is symmetrical (square), but it is possible in the proposed model thanks to densifying the grid and a more accurate representation of geometry.

2.6 Overtaking Scheme

An overtaking situation occurs when pedestrians move in the same direction at different speeds. If the trajectories of both pedestrians overlap (e.g. due to movement in the same floor field), the faster pedestrian must change his trajectory in order to overtake the slower one.

The scenario involves two pedestrians marked accordingly: S – slower (marked in green in the figure), F – faster (marked in yellow in the figure) – see Fig. 9.

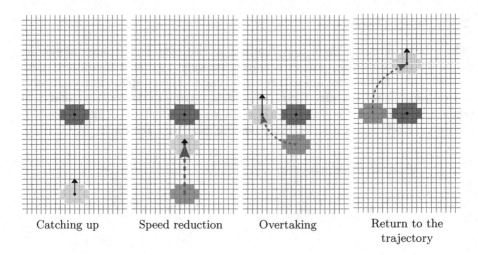

| Catching up | Speed reduction | Overtaking | Return to the trajectory |

Fig. 9. Steps in an overtaking scheme in the coordinate system associated with the slower pedestrian (marked in green). (Color figure online)

The proposed scheme consists of the following steps:

- the faster pedestrian F catches up with the slower one S (both pedestrians follow the same trajectory)
- temporary reduction of pedestrian F's speed to follow pedestrian S
- if there is free space on the side of the pedestrian S and there are no obstacles in front of him (or other pedestrians walking from the opposite direction), pedestrian's F speed is increased and the trajectory is changed (he overtakes pedestrian S on the left or right side)
- in the last step pedestrian F returns to the original trajectory and reduces his speed to the initial value (from the first step)

Figure 9 shows the flow of the overtaking scheme in the coordinate system associated with the slower pedestrian (S). The speed of the slower pedestrian (marked as a black dot) is zero in this coordinate system. The relative speed of the faster pedestrian (F) in this coordinate system is marked as a black arrow.

3 Implementation of the Proposed Model

The solutions included in the proposed model and described above were tested by the pedestrian dynamic simulator we are developing (Fig. 10). One of the main objectives of our system is to simulate the behaviour of the crowd in situations such as evacuation during a fire, storming stores by shoppers, etc. (Fig. 11).

The comparison of the simulation results obtained using the basic model [2] and the proposed model with different cell densities is presented below.

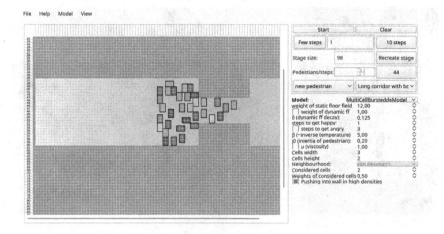

Fig. 10. Pedestrian dynamics simulator which implements the proposed model.

3.1 Crowd-Forming Scenario

In this scenario, the pedestrians initially located on the left side of the scene move towards the exit on the right side of the scene. Because the exit is closed, a crowd is formed in front of it. The initial distribution of pedestrians is the same in all variants. The dashed rectangle represents the area occupied by the crowd at the end of the simulation.

The figures present the following:

- Figure 12: the initial and final situation in the crowd-forming scenario for the basic model (cell size 40 × 40 cm, a pedestrian's shape is represented as a square 1 × 1 cell)
- Figure 13: as above for the model with 3-times densified grid and a pedestrian's shape - 2 × 3 cells rectangle
- Figure 14: as above for the model with 5-times densified grid and a pedestrian's shape - 3 × 5 cells rectangle

One hundred executions were performed for each variant, and the results are reported in Table 1.

The column 'final pedestrian density' presents the density of the crowd after simulation of its formation. As can be seen from the results, the densified grid of the cellular automata grid, which yields a more accurate description of the pedestrian's shape, allows for a higher density of people per square meter in the simulation. Such densities correspond to those found in real experiments (e.g. Fig. 11).

The column 'average execution time' shows the average time of execution of the simulation - as can be seen, increasing the density of the grid significantly increases the duration of the simulation. The theoretical influence of increasing the grid density on the time of execution of individual stages of the simulation is presented below.

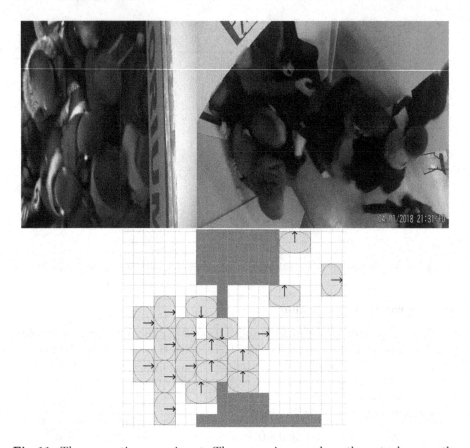

Fig. 11. The evacuation experiment. The upper images show the actual scene, the lower picture shows the simulation using the pedestrian dynamic simulator.

Shortening the Time Step. The reduction of the cell size by n-times also reduces the simulation time step n-times. Thus, in order to obtain the same pedestrian's movement (e.g. 1 m) it is necessary to perform n-times more time steps (simulation steps).

Initialisation. The most complex part of the scene preparation is static floor field initialization. During static floor field initialization for each cell in the scene all neighbours of each cell need to be checked in order to minimize the static floor field value. Thus, computational complexity of static floor field for the entire scene (C_s) is:

$$C_s = M * n \tag{2}$$

where:

- M - the number of neighbour cells for each cell in the scene
- n - the number of cells per scene

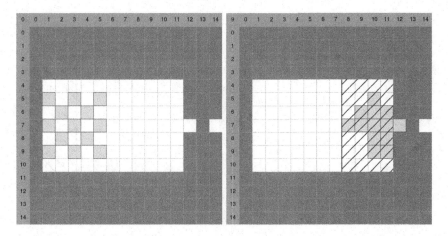

Fig. 12. The screenshot of the pedestrian dynamics simulator for the basic model (cell size 40 × 40 cm, a pedestrian's shape is represented as a square 1 × 1 cell). On the left - the initial distribution, on the right - the crowd.

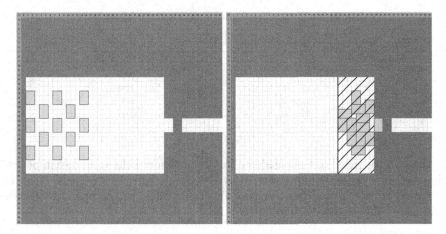

Fig. 13. The screenshot of the pedestrian dynamics simulator for the 3-times densified grid (a pedestrian's shape is represented as a rectangle 2 × 3 cell). On the left - the initial distribution, on the right - the crowd.

For the k-times densified grid (in the direction of each axis) the number of operations would increase k^2-times.

Pedestrian's Movements. According to [2], the three following factors need to be considered in calculating movement:

- static floor field
- dynamic floor field
- cell occupation by another pedestrian or obstacle

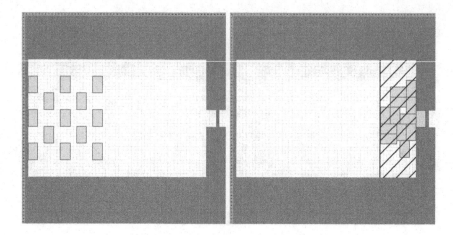

Fig. 14. The screenshot of the pedestrian dynamics simulator for the 5-times densified grid (a pedestrian's shape is represented as a rectangle 3 × 5 cell). On the left - the initial distribution, on the right - the crowd.

Table 1. Simulation results for individual model variants

Cell size (cm)	Pedestrian size (in cells)	Scene size (in cells)	Simulation time step (ms)	average execution time (ms)	Standard deviation of execution time (ms)	Final pedestrian density (pers/m²)
40 × 40	1 × 1	15 × 15	300	19.94	2.83	2.9
13 × 13	2 × 3	45 × 45	100	117.94	9.4	3.48
8 × 8	3 × 5	62 × 62	60	375.54	35.39	5.28

All these factors need to be considered for all neighbour cells around each pedestrian. So, for a pedestrian represented as a rectangle with $N_1 \times N_2$ cells $N_1 \times N_2$ cells need to be checked into each m direction. Thus, the complexity of each pedestrian's movement calculation (C_p) is:

$$C_p = f * m * N_1 * N_2 \qquad (3)$$

where:

- f - the number of factors considered in each cell (in our case they are: static floor field, dynamic floor field and cell occupation - so f=3)
- m - the number of directions considered for a pedestrian's movement
- $N_1 \times N_2$ - cells per pedestrian

So, for the k-times increasing side length of a pedestrian (due to grid densifying), the number of operations would increase to k-times.

After Movement Calculations. The most costly operation after each step of all pedestrians is recalculation of dynamic floor field, which must be done for each cell. The process of diffusion and decay of dynamic floor field is described in [2]. Thus, the complexity of dynamic floor field calculation (C_d) is:

$$C_d = m * n \tag{4}$$

where:

- m - the number of neighbour cells for each cell
- n - the number of all cells per scene

For k-times densified grid (in the direction of each axis) the number of operations would increase k^2-times.

4 Further Work

The model presented in the article allows for much more accurate representation of high density pedestrian dynamics than the basic model. However, the application of the model, as discussed earlier, is connected with increased computational complexity and thus with the increased time of the simulation. The use of the presented model is generally justified in situations with high pedestrian densities, as is the case, for example, during mass evacuation. One of the ways to improve the efficiency of the simulation may be to automatically switch the simulator working mode between the model (if the pedestrian density is low) and the model proposed in our article (if pedestrian density is medium or high). Such a model change may concern the whole scene that is being simulated or only the part of it with higher pedestrian density, leading to the concept of using hybrid models (combining different models within the same scene). This will be the subject of our further work related to the pedestrian dynamic simulator.

5 Conclusions

This work is a continuation of previous studies [1, 8] on CA-based agent models of the crowd. We wanted to create a model based on cellular automata which would allow for more accurate representations of pedestrian dynamics occurring at high crowd densities. It was developed for our pedestrian dynamics simulator (Fig. 10), one of the objectives of which is to simulate the behavior of the crowd during situations such as evacuation from a building during a fire.

The consequence of the proposed n-times densifying of the grid of the cellular automaton is an increase of space computational complexity n^2 times. In addition, the n-times reduction of a cell size requires the n-times reduction of the time step (in order to maintain the speed of the pedestrian's movement, which is distance divided by time) which leads to an additional n-times increase of time computational complexity. The hybrid model described in [10] can be a partial solution to the computation complexity problems. In the described model floor fields of some cells are grouped in order to decrease computation complexity, which can be helpful when simulation needs to be done for greater scenes.

However, thanks to the model proposed in this article, it is possible to map many situations in pedestrian dynamics in conditions of high density: a scheme of passing in a narrow corridor or an overtaking scheme. It should be stressed that thanks to a fine representation of pedestrians in a lattice it is possible to represent different shapes of pedestrians, their speed, etc. with greater precision.

References

1. Bazior, G., Pałka, D., Wąs, J.: Cellular automata based modeling of competitive evacuation. In: Mauri, G., El Yacoubi, S., Dennunzio, A., Nishinari, K., Manzoni, L. (eds.) ACRI 2018. LNCS, vol. 11115, pp. 451–459. Springer, Cham (2018). https://doi.org/10.1007/978-3-319-99813-8_41
2. Burstedde, C., Klauck, K., Schadschneider, A., Zittartz, J.: Simulation of pedestrian dynamics using a two-dimensional cellular automaton. Phys. A: Stat. Mech. Appl. **295**(3–4), 507–525 (2001). https://doi.org/10.1016/S0378-4371(01)00141-8
3. Feliciani, C., Nishinari, K.: An improved cellular automata model to simulate the behavior of high density crowd and validation by experimental data. Phys. A: Stat. Mech. Appl. **451**, 135–148 (2016). https://doi.org/10.1016/j.physa.2016.01.057. http://www.sciencedirect.com/science/article/pii/S0378437116001047
4. Gwizdałła, T.M.: The evacuation process study with the cellular automaton floor field on fine grid. In: El Yacoubi, S., Wąs, J., Bandini, S. (eds.) ACRI 2016. LNCS, vol. 9863, pp. 248–257. Springer, Cham (2016). https://doi.org/10.1007/978-3-319-44365-2_25
5. Ji, J., Lu, L., Jin, Z., Wei, S., Ni, L.: A cellular automata model for high-density crowd evacuation using triangle grids. Phys. A: Stat. Mech. Appl. **509**, 1034–1045 (2018). https://doi.org/10.1016/j.physa.2018.06.055
6. Kirchner, A., Klüpfel, H., Nishinari, K., Schadschneider, A., Schreckenberg, M.: Simulation of competitive egress behavior: comparison with aircraft evacuation data. Phys. A: Stat. Mech. Appl. **324**(3), 689–697 (2003). https://doi.org/10.1016/S0378-4371(03)00076-1. http://www.sciencedirect.com/science/article/pii/S0378437103000761
7. Li, Y., Chen, M., Dou, Z., Zheng, X., Cheng, Y., Mebarki, A.: A review of cellular automata models for crowd evacuation. Phys. A **526**, 120752 (2019). https://doi.org/10.1016/j.physa.2019.03.117
8. Lubaś, R., Wąs, J., Porzycki, J.: Cellular automata as the basis of effective and realistic agent-based models of crowd behavior. J. Supercomput. **72**(6), 2170–2196 (2016). https://doi.org/10.1007/s11227-016-1718-7
9. Predtechenskii, V.M., Milinskii, A.I.: Planning for foot traffic flow in buildings. Amerind, New Delhi (1978). Translation of: Proektirovanie zhdanii s uchetom organizatsii dvizheniya lyudskikh potokov
10. Shi, M., Lee, E.W.M., Ma, Y.: A novel grid-based mesoscopic model for evacuation dynamics. Phys. A: Stat. Mech. Appl. **497**, 198–210 (2018). https://doi.org/10.1016/j.physa.2017.12.139. http://www.sciencedirect.com/science/article/pii/S0378437117313882
11. Trans Res Board: Highway Capacity Manual. Transportation Research Board, National Research Council, Washington, D.C. (2000)

Autonomous Vehicles as Local Traffic Optimizers

Ashna Bhatia[1], Jordan Ivanchev[1,2(✉)], David Eckhoff[1,2], and Alois Knoll[2,3]

[1] TUMCREATE, 1 Create Way, Singapore 138602, Singapore
{ashna.bhatia,jordan.ivanchev,david.eckhoff}@tum-create.edu.sg
[2] Technical University of Munich, 3 Boltzmannstr., 85747 Munich, Germany
knoll@in.tum.de
[3] Nanyang Technological University, 50 Nanyang Avenue,
Singapore 639798, Singapore

Abstract. This paper explores the interaction between autonomous and human-driven cars on a microscopic level using an agent-based traffic simulator. More specifically, it deals with the design of driving logic models of "socially-aware" autonomous vehicles that can improve the performance of surrounding vehicles on the road. Congestion waves, which are created as a result of an abrupt stopping or a car joining a highway, are a known phenomenon in current traffic systems. Experiments performed, demonstrate how the presence of intelligent social vehicles on the road can reduce such effects by acting as a flexible medium between human-driven cars. Metrics to evaluate benefits ot our AV behaviour models under various states of traffic conditions/congestion are also proposed. Finally, results showing the effectiveness of these models are presented.

Keywords: Autonomous vehicles · Mixed traffic agent-based simulation · Driver models

1 Introduction

The current trends of Intelligent Transport Systems (ITS) and their applications concerning intelligent cars include advancements in Adaptive Cruise Control (ACC), Obstacle Warning, Avoidance Mechanism, Lane Detection and Collision Notification, which all contribute towards making travel more comfortable and safe [1]. In [2,3] it is reiterated that sustainable ITS application will enable the reduction in carbon dioxide and heat emission as traffic volume is reduced and managed. Research and development in the sector of Autonomous Vehicles (AVs) is being done worldwide. Benefits from them, such as more independent mobility for affluent non-drivers, may begin sooner rather than later. However, most impacts concerning improvements in safety, sustainability and comfort will only be significant when AVs become common and affordable, which might take another 20 to 30 years [4].

Consequential to these technologies, expectations and predicted trends, copious evaluations have been made on the impact of penetrating AVs into the road

© Springer Nature Switzerland AG 2020
V. V. Krzhizhanovskaya et al. (Eds.): ICCS 2020, LNCS 12137, pp. 499–512, 2020.
https://doi.org/10.1007/978-3-030-50371-0_37

transport sector. The research done in [5] shows how vehicle-to-vehicle communication, can make a difference in factors such as fuel consumption, driver safety and convenience. A mixed simulation scenario with 3 types of car models; manual, ACC, and Cooperative Adaptive Cruise Control (CACC) is analysed in [6]. The experiments carried out, analyse optimal platoon sizes feasible due to the presence of CACC, that enhance the traffic flow.

Promising improvements using V2I communications can be seen in the work done on Variable Speed Limits in [7] wherein a control algorithm is presented, which simultaneously maximizes the mobility, safety and environmental impact by finding a balanced trade-off. Another multi-objective approach considering efficiency, comfort, throughput, and safety is presented in [8] where a simplistic mixed traffic scenario is considered and optimal parametrizations of known car-following models are found.

An alternative approach to maximizing the benefits of autonomous mobility is to separate the AVs from the rest of the traffic. A macroscopic evaluation of a dedicated AV lane policy on all highways for the city of Singapore has been presented in [9]. Additionally, a microscopic analysis of replacing High Occupancy Vehicle lanes with dedicated AV lanes has been performed in [10]. Those studies, however, model expected AV behaviour rather than trying to design its logic so that it benefits traffic conditions, which is the focus of our work.

It is clear that, mixed traffic scenarios will be unavoidable with the onset of AV acceptance. Various questions arise in this scenario; 1) how should an AV behave in order to optimize the impact of its presence? 2) To what extent will the behaviour impact surrounding human-driven vehicles? 3) What traffic parameters can it improve (safety, comfort, throughput)? 4) How safe is the interaction between human controlled and automated vehicles? 5) Will additional information regarding surrounding vehicles in an autonomous vehicles' neighbourhood, help to improve traffic conditions (as opposed to only considering the vehicles in front and behind)?

In order to begin addressing those questions in an efficient and safe manner, we suggest to use a modelling and simulation approach. The main challenge when it comes to simulating mixed traffic is capturing the interactions between vehicles. This requires the usage of an agent-based simulation approach where behavioural models are responsible for the decisions that agents take on the road, given the environment around them. In our work we use a multi-agent traffic simulation environment for designing and testing the control logic of AV driver models that aim to improve traffic conditions for all traffic participants in terms of safety, fuel consumption and throughput. This allows us to fully determine and control the traffic environment, and to run all different scenarios we need, in order to find beneficial autonomous vehicle control parameters.

The contributions of this paper can be summarized as follows:

- Design of socially aware autonomous vehicle behaviour models
- Design of metrics to evaluate the performance of these models in terms of efficiency, safety, and traffic robustness

– Analysis of agent population subgroup behaviour and performance to understand the mechanism behind the improvement of traffic conditions.

2   Simulation Framework

Since mixed traffic conditions do not currently exist in the real world, using simulation seems as a viable approach to study mixed traffic. For the purposes of this research we have chosen the CityMoS simulation framework [11], which is microscopic and agent-based in nature. The behaviour of these agents (vehicles) is programmed using driver models which are introduced in the subsequent subsection. More specifically, we use the BEHAVE (Behaviour Evaluation of Human-driven and Autonomous VEhicles) tool [12], powered by the discrete-time based CityMoS engine, which is designed specifically for studying the interactions between AVs and human-driven vehicles.

Driver Models. Various car-following and lane-changing driver models have already been proposed in order to mimic the behaviour and dynamics of vehicles on the road in a simulation environment. Few of the well-known ones are the Gipps' Model [13], the Intelligent Driver Model(IDM) [14], the MOBIL Lane Changing Model [15], etc.

Driver models like IDM take as input the information about the vehicle in front and compute as output the forward acceleration, which will be applied. The models are parametrized by the personal preferences of the driver such as: time-headway, acceleration and deceleration components, minimum safety distance gaps etc. These parameters can be calibrated in order to closely match a given real-world data set.

The aspects of driving that make AVs different from humans that have been assumed in this work include almost no reaction time and perfect perception which actually is a fair description of the IDM. For this reason, we use it to model naive non-social AVs, and benchmark their performance against more elaborate models that we design later in the paper. In order to study the vehicle interactions in mixed-traffic conditions (human-driven and autonomous vehicles), an adequate model that represents human-driven vehicles should be identified as well. We use the Human Driver Model (HDM) [16], which is an extension of the work carried out for the IDM.

The IDM stands in the core of all model design in this work since it: 1) is the benchmarking model we use for non-social AVs, 2) is the model which is extended to yield the human driver model (HDM), and 3) is the model we extend to build our social autonomous vehicle models. It is a well established car-following model that is simplistic and realistic in terms of formulation and simulation, respectively. The acceleration (Eq. 1) of a vehicle α at a given time t is calculated using the continuous acceleration function $\dot{v}_\alpha(s_\alpha, v_\alpha, \Delta v_\alpha)$ depending on the vehicle's actual velocity $v_\alpha(t)$, the net distance s_α and the approaching speed $\Delta v_\alpha(t)$.

$$\dot{v}_\alpha = a \left[1 - \left(\frac{v_\alpha}{v_0} \right)^\delta - \left(\frac{s^*(v_\alpha, \Delta v_\alpha)}{s_\alpha} \right)^2 \right] \qquad (1)$$

with

$$s^*(v_\alpha, \Delta v_\alpha) = s_0 + v_\alpha T + \frac{v_\alpha \Delta v_\alpha}{2\sqrt{ab}} \qquad (2)$$

where s_0 corresponds to the desired minimum distance gap to the preceding vehicle, T to the time gap to safely come to a halt without crashing into the vehicle in front, and b to the vehicle's comfortable deceleration. Here, v_0 corresponds to the vehicles desired velocity, a corresponds to the vehicles maximum acceleration , and δ determines how the acceleration decreases once the desired velocity i.e v_0, is reached.

To incorporate more realistic behaviour into the IDM, it was extended to derive the Human Driver Model (HDM) in [16] . Human-like behaviour is integrated by taking into consideration the following destabilizing and stabilizing factors.

1. Finite reaction time: Human drivers have a certain finite reaction time T' to events occurring on the road. By computing the acceleration at time $t - T'$ instead of at time t, an effect equivalent to delayed response to stimuli on the road can be achieved.
2. Estimation errors: As shown in Eq. 1 acceleration is a function of the distance to the preceding vehicle s and the approach speed Δv. Those inputs to the model typically are the perfect measurements provided by the simulator, however, as a human driver is certainly plausible that some perception errors will be introduced. The HDM models this, using stochastic noise introduced by a Weiner process that leads to time-correlated fluctuations of the acceleration.
3. Temporal anticipation: Human drivers can be intuitive enough to anticipate their reaction time and the evolution of traffic within small time windows. To represent this awareness, of future net distance and future velocity, HDM uses *constant acceleration and velocity projections* to determine their current values, respectively.
4. Spatial anticipation: The receptivity range of the human driver can be extended to more than one preceding vehicle on accord of spatial perceptiveness. The IDM considers relative measures to only one vehicle ahead, as can be seen in its deceleration determining component. The HDM expands the interaction term by taking into account multiple nearest preceding vehicles.

The first two factors are responsible for destabilizing the traffic conditions and the last two being anticipatory in nature, contribute towards the stability. In our studies we use parameter values specified by the authors of the HDM for all additional parameters introduced in the model which achieve a balance between the 4 additional aspects of the model.

3   Social AV Models

One of the reasons IDM is widely used is because it has succeeded in showing robustness and aptness by qualitatively reproducing results from empirical data in simulation environments. This was achieved through varying mostly one parameter that the model is quite sensitive to, namely, the time-headway [14], or the distance, in terms of time, that an agent considers to be safe. Hence, in our models, which are based on IDM, we modulate this parameter as well to steer the behaviour of our agents. Consequently, the goal is to formulate models that control AVs and improve on the following measures:

1. Performance: in terms of total traffic capacity and average fuel consumption
2. Safety: in terms of accidents caused in the traffic scenario.
3. Resilience: in terms of response and recovery from perturbations that cause congestion waves.

Modelling the First Social AV Model - SAV. By exploring the interaction between mixed (IDM and HDM) agent populations using BEHAVE, the effect of the time-headway parameter was analysed in details. It was observed that as the time-headway value was increased, the overall traffic conditions became more stable and safe. This comes at the cost of reduced throughput and overall average velocity.

The central idea behind the first design of a social AV model is to provide information about more vehicles in its vicinity. Adhering to the single-lane car-following set-up, the idea is to incorporate in an AVs decision making logic, information about not just the preceding vehicle (as the IDM does), but also of the following one. The hypothesis being; an AV that is aware of the distance to its following and preceding vehicles, can make decisions regarding its own dynamics which can benefit the entire traffic situation.

The intended behaviour of this AV is designed such that, the AV adjusts its own time-headway parameter value so that it is equidistant from the vehicle ahead and the vehicle behind. The maximum and minimum permissible value of the time-headway is set to be 2.5 s and 0.5 s, respectively.

The AV is spatially aware and knows the position of the vehicle in front (x_{i-1}) and the vehicle behind (x_{i+1}), along with its own position (x_i). The modulation factor μ as shown in Eq. 3 is the difference of the current position of the vehicle from the midpoint of the leading and following vehicles. The vehicle in front has a greater position than the vehicle behind. These positions are relative to the coordinates of a highway beginning at position 0.

$$\mu = x_i - \left(\frac{x_{i-1} + x_{i+1}}{2} \right) \tag{3}$$

If the AV is closer to the vehicle ahead, μ will be positive and vice versa. Taking the logistic sigmoid of this metric, (Eq. 4) and bounding it in the range of the set limit values for the time-headway, results in a greater value for T as the AV gets closer to the vehicle ahead. This means it will now aim at increasing the distance ahead by increasing its time-headway proportionally, as per the following equation.

$$T = (T_{max} - T_{min}) * S(\mu) + T_{min} \tag{4}$$

With this underlying model logic, the AV continuously modulates its own time-headway based on its difference from the midpoint and the general equation of the IDM (Eq. 1). It always aims to be in the centre of its remote neighbourhood to reduce the overall deviation in spacing between vehicles. From this point forward, the model obeying this design will be referred to as the SAV i.e. Social Autonomous Vehicle.

Extending the SAV - SAVE. Given the advancements in V2V and V2I communications, it is plausible that an AV is capable of having more information regarding its neighbourhood, than what is provided to the SAV. Thus, in our second approach, the Social Autonomous Vehicle- Extended (SAVE) model, incorporates additional information regarding the traffic dynamics by expanding the neighbourhood awareness of the AV with a congestion predictive control parameter.

Similar to the logic of the SAV model, the SAVE model aims to achieve improvement of throughput, safety and traffic resilience by modulating the time-headway parameter. Instead of relying solely on the relative positions, this AV model was intended to understand the traffic dynamics better, for which it uses a congestion predictive control parameter called *time-to-next (TTN)*. These values facilitate a prediction into the remote future by stating the time it will take for a vehicle to reach or catch-up with the next vehicle, given its current speed and the current distance to the vehicle ahead, as shown in this equation:

$$TTN = \frac{s_{\alpha\beta}}{v_\alpha} \tag{5}$$

The AV is provided with two proximity-based sorted lists containing this TTN information, for the vehicles ahead and behind, respectively. The length N of each list denotes the size of the neighbourhood under consideration.

Before responding intelligently, an AV needs to evaluate which traffic condition it is currently in: entering a congestion wave, or exiting it. One way to do that is by using weighted means of TTN values calculated from the two sorted lists. They give a look-ahead and look-back in traffic distribution, while giving greater priority to closer vehicles.

W_1 represents the relative weighted proximity (in terms of TTN values) to the clusters. It is set according to the difference between two exponentially weighted means(one concerning the cluster of vehicles ahead, and the other concerning the cluster behind), as can be seen in Eq. 6. A positive value of W_1 means that the cluster ahead,as a whole is closer to the vehicle in question, as compared to the cluster of vehicles behind it.

$$W_1 = \sum_{i=1}^{N} \frac{2^{N-i}}{\sum_{i=1}^{N} 2^{N-i}} (TTN^{behind} - TTN^{ahead}) \tag{6}$$

W_2 is represents compactness of clusters using a relative congestion factor. First, we compute the deviation of the vehicles from the weighted mean of the cluster they belong to. Second, we subtract this value of the cluster ahead, from the cluster behind (Eq. 7). A low deviation from weighted mean implies that the vehicles in the cluster are similar in TTN values to the first vehicle in the cluster. It is to be noted that the first vehicle considered here, is the one closest to the SAVE in question. Hence, if the deviation in weighted means in the cluster ahead is lower than that of the cluster behind, W_2 is positive and the cars ahead are more closely packed compared to the ones behind.

$$W_2 = D(TTN_{behind}) - D(TTN_{ahead}) \tag{7}$$

where,

$$D(TTN) = \sqrt{\frac{\sum_{i=1}^{N}(\bar{x}_w - TTN_i)^2}{N}} \tag{8}$$

W_3 is used to make the SAVE more sensitive to its immediate neighbourhood. It uses the TTN values of the vehicle immediately ahead, and immediately behind. It tries to help the SAVE model achieve an average of these TTN values. If W_3 is positive, the current TTN value is smaller than average, and closer in proximity to the vehicle ahead. Thus the SAVE uses this weight to equalize spacing in its close proximity region. This means, when the weight is positive the current SAVE will increase its time-headway, and thus try falling back in the relative proximity centre.

$$W_3 = \left(\frac{TTN_{i-1} + TTN_{i+1}}{2}\right) - TTN_i \tag{9}$$

All three weights when positive intend to increase the time-headway, and vice-versa when negative. Experiments show that a non-linear combination of the three weights, in this case multiplying them (Eq. 10), results in better modulation of time-headway than a simple addition.

$$\mu = W_1 * W_2 * W_3 \tag{10}$$

– Increase in time-headway: By increasing the time-headway of a vehicle we increases its distance from the vehicle ahead. This is ideal when we all the weights are positive. When the cluster ahead is more compact.

4   Experiments

In our experiments, we assume that half of the agent population is human driven and modelled by HDM. The other half consists of AVs that we model with either IDM, SAV or SAVE. The BEHAVE simulator was used to simulate a 50 km single-lane highway with vehicle density of 0.067 vehicles per meter. Each simulation run lasted 45 min in real-time, which is significantly faster in the multi-core enabled simulator having high speed processing capabilities. Results derived from the experiments were averaged over 20 simulation runs, each.

Experiment 1 - Congested Traffic Flow. The aim of this experiment is to analyse the interactions of AVs with HDMs in slow moving traffic and test the hypothesis that their presence can improve overall traffic conditions. This experiment is performed on 4 group settings. The first 3 are under mixed traffic conditions with 50% HDMs and 50% of each of the 3 AVs . The last group represents the evaluation of only HDMs by themselves, i.e without the presence of any AVs.

The performance of the models is evaluated in terms of the achieved throughput, indicative of the capacity of the road, and the average absolute acceleration, indicative of the smoothness of traffic flow and the fuel consumption.

1. Throughput: The throughput of our traffic system is collected as the average number of cars passing through a chosen set point on the highway in a time period of 10 min.
2. Average Absolute Acceleration: The average fuel consumption of a vehicle, increases with increase in levels of acceleration among vehicles. The more the vehicles on the road accelerate the more fuel they will use. The stability of traffic and consequently ride comfort is also related to the acceleration profile, however, also includes the amount of deceleration that is performed. In order to capture both those aspects the average of the absolute value of the vehicles' accelerations is used as a metric. The absolute value of acceleration of each vehicle in a simulation run is summed up and averaged at every time-stamp.

Results from Fig. 1 show that the SAV models achieved the highest throughput. The scenario of a pure HDM population produces less than half the throughput.

The results from Fig. 2, indicate that the setting comprising only HDMs results in oscillatory traffic conditions that exude larger amount of fuel on average per vehicle while the SAVE vehicles provide the least oscillatory, and thus efficient, traffic conditions.

Experiment 2 - Perturbed Congested Traffic. The aim of this experiment is to analyse the traffic conditions under the presence of stop-and-go congestion waves. The extent to which these waves can affect the vehicles in terms of safety, and the traffic dynamics in terms of stability.

Similar to the previous experiment, 4 group settings were used i.e 3 with mixed dense traffic conditions and the last one representing only human drivers. Each simulation run had a duration of 45 min. After one-third of the run was done, a trigger event was invoked that caused the leading few vehicles to stop completely for the average duration of a traffic signal (1 min in real-time), post which they began to accelerate as per their own preferred velocity and, as permissible by the road density and traffic conditions. This trigger causes the phenomena of stop-and-go congestion waves which are observed to propagate downstream (away from the source of perturbation).

A congested traffic with perturbations gives rise to the following important considerations:

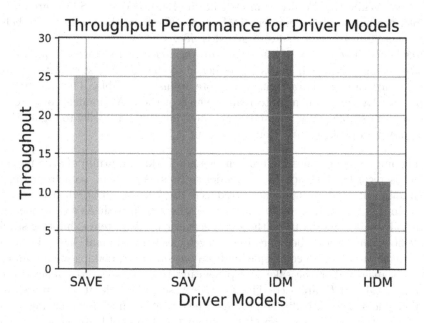

Fig. 1. Throughput

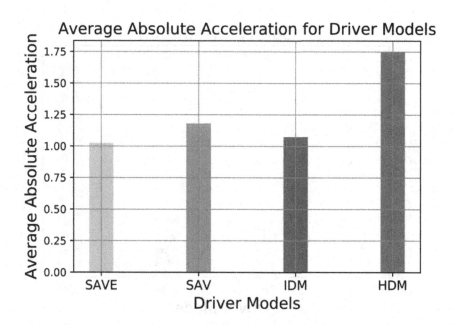

Fig. 2. Average absolute acceleration

1. Safety: While the AV driver models i.e the IDM, SAV and SAVE are crash-free, the human driver model HDM is not. The perturbation in traffic helps to magnify the safety issues in mixed traffic.
2. Resilience: The ability of a system to revert back to its state of equilibrium after being perturbed, shows its resilience to change. A system with greater resilience can be characterized as more robust or stable. The aim of this comparative experiment is to realize which of the 3 AV models could result in a more stable traffic dynamics, and how long would it take to regain their equilibrium post perturbation.

The number of crashes for 100 vehicles per 1.5 km in 45 min of slow moving traffic, is recorded for each of the model set-ups. As can be seen from Fig. 3, the number of crashes significantly reduced (by 70% to 80%) when AVs were mixed in. On an average of 20 simulations, the set-up with AVs controlled by the SAVE model showed to be the safest scenarios. A system exhibiting stable traffic flow, can be described pertaining to vehicles with minimal change in values of acceleration, and also equal spacing from one-an-other, making maximum use of road capacity. For the sake of representing these two criteria, macroscopic parameters A and D, are used. The Average Absolute Acceleration A represents the non-smoothness of traffic conditions. The deviation in average spacing of the vehicles on the road D gives an idea of about the degree of heterogeneity in the distances between the vehicles. This is an indication of the magnitude of the stop-and-go wave created by the disturbance.

Fig. 3. Crashes km/min

The instability values plotted on the $y - axis$ of Fig. 4 are the product of A and D. The higher this value, the more unstable is the system. The $x - axis$ represents the time post perturbation in minutes. Each coloured polygon shows how the system, comprising 50% of the respective AV models, reacts to the perturbation and recovers from it. The magenta coloured polygon with the greater area, corresponds to the simulation runs with HDMs only. The black dots at time 0, are called the equilibrium offset that represent the initial equilibrium value for each model set-up. As can be seen, the equilibrium offset for SAVE models are the best and most stable. The perturbation trigger is also applied at this instant causing the peak instabilities.

As the time progresses, the system is expected to converge back to its equilibrium offset. The area under the instability vs. time curve shows how efficiently the system mitigates the effect of congestion waves. The lesser the area, the better is the recovery. To determine which model is most stable, we find the product of the equilibrium offset and the area under the curve, and term it as the instability index. The HDMs alone, as can be seen, don't seem to converge back or close to their own system equilibrium. The penetration by 50% AVs has much of an alleviating effect. The instability index of SAVEs is the best, followed by SAV .

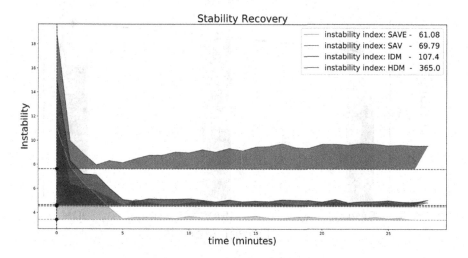

Fig. 4. Stability recovery of driver models with HDMs

The already presented evaluations and comparisons measure the overall macroscopic properties of mixed-traffic scenarios. In order to see how different vehicle populations perform in these runs, we extend the analysis done in Experiment 2 by examining the performance metrics of the two driver categories (i.e. human and autonomous) in every scenario separately. This evaluation is important as it sheds light on how the presence of AVs in a traffic scenario, could help bring about improvements in the performance of human drivers. The comparative approach also demonstrates which AV model can be more advantageous for this purpose and to what extent does it sacrifice its own gaining, in order to benefit the other population.

Figure 5, shows the values of the overall macroscopic parameter called Average Absolute Acceleration A that is computed separately for the HDM and the AV models and normalized to produce a metric that we call *Disturbance*. By comparing the difference between these two values in every group, we observe that: 1) The overall *Disturbance* for the SAVE + HDM scenario results in the best collective evaluation (as shown by the black dots), and 2) The SAVs are more "altruistic" as the difference between them and the HDMs is smaller. In other words, it seems that the SAV perform active acceleration and deceleration manoeuvres in order to minimize the oscillation of the HDMs. This means that, the human drivers in the presence of SAVs will have a smoother driving experience and more efficient fuel consumption, than with SAVEs or IDMs.

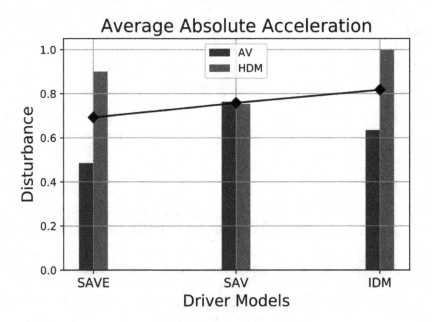

Fig. 5. Rate of deviation from equilibrium value of average absolute acceleration for experiments with 50% HDM and 50% AV driver models

5   Conclusions and Future Work

This paper analyses the interaction between AVs and human-driven vehicles in traffic states involving congestion and stop-and-go waves. We design AV driver models that aim at increasing the beneficial effect AVs can have on overall traffic properties as throughput, safety, and resilience of traffic flow against perturbations. We compare the performance of the two models that we have designed to a naive implementation of an AV that does not try to improve traffic conditions around it, modelled as an IDM, and to a scenario with only human-drivers.

Having AVs in the system improves all discussed traffic parameters. Furthermore, the SAV and SAVE models lead to improvements compared to the IDM model. The main idea behind the SAV model is to stay equidistant to the vehicles in front and behind it. This allows for quick stabilization to normal flow conditions after perturbations are introduced to the system and thus the SAV has the highest resilience score of all tested models. The SAVE model, tries to improve on traffic throughput by utilizing information about groups of vehicles in front and behind to determine whether it is approaching a local congestion zone. It achieves a better throughput than the other examined models without compromising the level of safety on the road in terms of number of accidents.

As research on AV technology advances, mixed traffic conditions will inevitably occur sooner rather than later. There are numerous concerns regarding the safe interaction between AVs and human drivers, mostly due to the unpredictability of humans. It is therefore, important to study those interactions and to design solutions which mitigate the safety risks while not compromising traffic performance. Our results demonstrate that by utilizing more information about a vehicle's surroundings, high-level control models can be designed in order to improve both safety and efficiency of the studied traffic system. This work can be considered as a first step to utilizing AVs as local traffic optimizers rather than just means of transportation.

After we have shown that dedicated strategies might improve the studied traffic characteristics, we see plenty of opportunities to extend this line of research. Reinforcement learning approaches can be used to come up with a continuously learning and adapting driver behaviour model. Model predictive control can be used to guarantee safety and continuously optimize chosen traffic characteristics. Apart from extending the variety of tools for optimizing control of car following models, we can extend those tools into lane changing decision making and overall vehicle coordination with safety and efficiency in mind. We would further like to explore a wider variety of scenarios including on and off ramps of highways, traffic junctions etc. thus also covering urban intelligent transport systems.

Acknowledgement. This work was financially supported by the Singapore National Research Foundation under its Campus for Research Excellence And Technological Enterprise (CREATE) program.

References

1. Nkoro, A., Vershinin, Y.A.: Current and future trends in applications of intelligent transport systems on cars and infrastructure. In: 2014 IEEE 17th International Conference on Intelligent Transportation Systems (ITSC), IEEE, pp. 514–519 (2014)

2. McDonald, M., et al.: Intelligent Transport Systems in Europe: Opportunities for Future Research. World Scientific, Singapore (2006)
3. Ivanchev, J., Fonseca, J.A.: Anthropogenic heat due to road transport: a mesoscopic assessment and mitigation potential of electric vehicles and autonomous vehicles in Singapore. Technical Report, ETH Zurich (2020)
4. Litman, T.: Autonomous Vehicle Implementation Predictions. Victoria Transport Policy Institute Victoria, Canada (2017)
5. Bergenhem, C., Hedin, E., Skarin, D., et al.: Vehicle-to-vehicle communication for a platooning system. Procedia-Soc. Behav. Sci. **48**(2012), 1222–1233 (2012)
6. Zhao, L., Sun, J.: Simulation framework for vehicle platooning and car-following behaviors under connected-vehicle environment. Procedia-Soc. Behav. Sci. **96**, 914–924 (2013)
7. Khondaker, B., Kattan, L.: Variable speed limit: a microscopic analysis in a connected vehicle environment. Transport. Res. Part C Emerg. Technol. **58**, 146–159 (2015)
8. Ivanchev, J., Eckhoff, D., Knoll, A.: System-level optimization of longitudinal acceleration of autonomous vehicles in mixed traffic. In: IEEE Intelligent Transportation Systems Conference (ITSC). IEEE 2019, pp. 1968–1974 (2019)
9. Ivanchev, J., Knoll, A., Zehe, D., Nair, S., Eckhoff, D.: A macroscopic study on dedicated highway lanes for autonomous vehicles. In: Rodrigues, J.M.F., et al. (eds.) ICCS 2019. LNCS, vol. 11536, pp. 520–533. Springer, Cham (2019). https://doi.org/10.1007/978-3-030-22734-0_38
10. Xiao, L., Wang, M., van Arem, B.: Traffic flow impacts of converting an HOV lane into a dedicated CACC lane on a freeway corridor. IEEE Intelligent Transportation Systems Magazine (2020)
11. Zehe, D., Nair, S., Knoll, A., Eckhoff, D.: Towards citymos: a coupled city-scale mobility simulation framework. In: 5th GI/ITG KuVS Fachgespräch Inter-Vehicle Communication, p. 26 (2017)
12. Ivanchev, J., Braud, T., Eckhoff, D., Zehe, D., Knoll, A., Sangiovanni-Vincentelli, A.: On the need for novel tools and models for mixed traffic analysis. In: 26th ITS World Congress, Singapore, October 2019, to appear
13. Gipps, P.G.: A behavioural car-following model for computer simulation. Transport. Res. Part B Methodol. **15**(2), 105–111 (1981)
14. Treiber, M., Hennecke, A., Helbing, D.: Congested traffic states in empirical observations and microscopic simulations. Phys. Rev. E **62**(2), 1805 (2000)
15. Kesting, A., Treiber, M., Helbing, D.: General lane-changing model mobil for car-following models. Transport. Res. Record **1999**(1), 86–94 (2007)
16. Treiber, M., Kesting, A., Helbing, D.: Delays, inaccuracies and anticipation in microscopic traffic models. Phys. A Stat. Mech. Appl. **360**(1), 71–88 (2006)

Modeling Helping Behavior in Emergency Evacuations Using Volunteer's Dilemma Game

Jaeyoung Kwak[1]([✉]), Michael H. Lees[2], Wentong Cai[1],
and Marcus E. H. Ong[3,4]

[1] Nanyang Technological University, Singapore 639798, Singapore
{jaeyoung.kwak,aswtcai}@ntu.edu.sg
[2] University of Amsterdam, Amsterdam 1098XH, The Netherlands
m.h.lees@uva.nl
[3] Singapore General Hospital, Singapore 169608, Singapore
marcus.ong.e.h@singhealth.com.sg
[4] Duke-NUS Medical School, Singapore 169857, Singapore

Abstract. People often help others who are in trouble, especially in emergency evacuation situations. For instance, during the 2005 London bombings, it was reported that evacuees helped injured persons to escape the place of danger. In terms of game theory, it can be understood that such helping behavior provides a collective good while it is a costly behavior because the volunteers spend extra time to assist the injured persons in case of emergency evacuations. In order to study the collective effects of helping behavior in emergency evacuations, we have performed numerical simulations of helping behavior among evacuees in a room evacuation scenario. Our simulation model is based on the volunteer's dilemma game reflecting volunteering cost. The game theoretic model is coupled with a social force model to understand the relationship between the spatial and social dynamics of evacuation scenarios. By systematically changing the cost parameter of helping behavior, we observed different patterns of collective helping behaviors and these collective patterns are summarized with a phase diagram.

Keywords: Emergency evacuation · Helping behavior · Game theory · Volunteer's Dilemma game · Social force model

1 Introduction

Pedestrian emergency evacuation is a movement of people from a place of danger to a safer place in case of life-threatening incidents such as fire and terrorist attacks. Numerical simulation has been a popular approach to perform pedestrian emergency evacuation studies, for instance, predicting total evacuation time in a class room [1] and preparing an optimal evacuation plan for a large scale pedestrian facility [2].

© Springer Nature Switzerland AG 2020
V. V. Krzhizhanovskaya et al. (Eds.): ICCS 2020, LNCS 12137, pp. 513–523, 2020.
https://doi.org/10.1007/978-3-030-50371-0_38

Based on numerical simulations, it has been identified that evacuees are often in conflict with others when more than two evacuees try to move to the same position [3]. Game theory has been used to model strategic interactions among evacuees in such a conflict. Under game theoretic assumptions, each evacuee has his own strategies and selects a strategy in a way to maximize his own payoff. Various emergency evacuation simulations have been performed based on different game theory models including evolutionary game [4], snowdrift game [5], and spatial game [6,7].

Although those game theory models successfully modeled evacuees' egress, especially from a room, other aspects of evacuees' behavior such as helping behavior have not been sufficiently studied. In the context of emergency evacuations, it has been reported that evacuees help injured evacuees to evacuate from the place of danger, for instance the WHO concert disaster occurred on December 3, 1979 in Cincinnati, Ohio, United States [8] and 2005 London bombings in United Kingdom [9].

A few studies have investigated helping behavior in emergency evacuation by means of pedestrian simulation. Von Sivers *et al.* [10,11] applied social identity and self-categorization theories to pedestrian simulation in order to simulate helping behavior observed in 2005 London bombings. In their studies, they assumed that all the evacuees share the same social identify which makes them be willing to help others rather than be selfish. Lin and Wong [12] applied the volunteer's dilemma game [13,14] to model the behavior of volunteers who removed obstacles from the exit. Their work can be considered as a helping behavior modeling study in that some evacuees were voluntarily removing the obstacles so they helped others in the same room to evacuate faster.

One can observe that such a helping behavior provides a collective good in case of emergency evacuations. This is especially true when there are not enough rescuers, more injured persons can be rescued with the help of other evacuees than by only the rescuers. In order to study the collective effects of helping behavior in emergency evacuations, we have developed an agent-based model simulating such helping behaviors among evacuees. Based on the agent based model, we represent individual behaviors with a set of behavioral rules and then systematically study collective dynamics of interacting individuals. In our agent based model, we assumed that helping an injured person can be a costly behavior because the volunteer spends extra time and take a risk to assist the injured person in the evacuation. If individuals feel that helping behavior is a costly behavior for them, they might not turn into volunteers. Thus we implemented the volunteer's dilemma game model [13,14] to reflect the cost of helping behavior. Pedestrian movement is simulated based on social force model [15].

The remainder of this paper is organized as follows. The simulation model and its setup are explained in Sect. 2. We then present its numerical simulation results with a phase diagram in Sect. 3. Finally, we discuss the findings of this study in Sect. 4.

2 Method

2.1 Volunteer's Dilemma Game

We employ the volunteer's dilemma game model to study helping behavior of passersby in a room evacuation [13,14]. A passerby is an evacuee who is not injured and can play the volunteer's dilemma game model. According to the volunteer's dilemma game, two types of players are considered. Passerby i can be either a volunteer (C) who helps an injured person to evacuate or a bystander (D) who does not help the injured person. Once a passerby decides to be a volunteer, he approaches to and then rescues the injured person. We can express the payoff of player i in terms of collective good U and volunteering cost $K < U$, see Table 1. The payoff of a bystander (D) is U if there is at least one volunteer, 0 otherwise. It can be understood that, bystanders are benefited by the volunteer. However, if nobody volunteers, the collective good U cannot be produced because all the players are bystanders. The collective good U can be produced by volunteers when they rescue injured persons. For simplicity, we assume that the value of U is constant if there is at least one volunteer. The payoff of a volunteer (C) is always $U - K$, indicating that his payoff is constant regardless other players' choice.

Table 1. Payoff of a volunteer (C) and a bystander (D) for the different number of other players choosing C (based on Refs. [13,14]). Here, U is the collective good, $K < U$ is the volunteering cost, and $N \geq 2$ is the number of players.

Player i's choice	The number of other players choosing C				
	0	1	2	...	$N - 1$
Volunteer (C)	$U - K$	$U - K$	$U - K$	$U - K$	$U - K$
Bystander (D)	0	U	U	U	U

Actor i's expected payoff E_i is given as:

$$E_i = q_i \left(1 - \prod_{j \neq i}^{N} q_j \right) U + (1 - q_i)(U - K). \tag{1}$$

Here, q_i is the probability that player i chooses D and $1 - q_i$ for choosing C. The number of players is indicated by N. The probability that all players $j \neq i$ choose D is denoted by $\prod q_i$ and $1 - \prod q_i$ indicates the probability that at least one actor $j \neq i$ chooses C. The first term on the right hand side reflects the payoff of player i when he selects D but benefited when there is at least one volunteer. The second term on the right hand side indicates the payoff of player i if he selects C.

We assume that player i adopts the mixed-strategy which is the best strategy for him. In a mixed-strategy equilibrium, every action played with positive probability must be a best response to other players' mixed strategies. This implies that player i is indifferent between choosing C and D, so a small change in the payoff E_i with respect to q_i (i.e., the probability of choosing D) becomes zero:

$$\frac{dE_i}{dq_i} = -U \prod_{j \neq i}^{N} q_j + K = 0. \tag{2}$$

After assuming $q_i = q_j$, we can obtain probability that player i chooses D

$$q_i = \left[\frac{K}{U}\right]^{\frac{1}{N-1}} = \beta^{\frac{1}{N-1}}, \tag{3}$$

where $\beta = K/U$ is cost ratio, which can be interpreted as the risk of volunteering. Accordingly, the probability that player i chooses C is given as

$$p_i = 1 - q_i = 1 - \beta^{\frac{1}{N-1}}. \tag{4}$$

The probability that at least one player selects C is denoted by p^*, i.e.,

$$p^* = 1 - q_i^N = 1 - \beta^{\frac{N}{N-1}}. \tag{5}$$

Equations 4 and 5 show good agreement with the bystander effect, see Fig. 1. Figure 1(a) shows a decreasing trend of p_i as the number of players N increases, inferring that players are less likely to volunteer seemingly because they believe other players will volunteer. Note that the social pressure from other players is not considered here, so the existence of volunteers does not affect on players' behavior. Figure 1(b) presents the trend of p^* which reflects the chance that an injured person is rescued. As the number of players N increases, the value of p^* approaches to a certain value, $1 - \beta$.

2.2   Social Force Model

According to the social force model [15], the position and velocity of each pedestrian i at time t, denoted by $x_i(t)$ and $v_i(t)$, evolve according to the following equations:

$$\frac{d x_i(t)}{dt} = v_i(t) \tag{6}$$

and

$$\frac{d v_i(t)}{dt} = f_{i,d} + \sum_{j \neq i} f_{ij} + \sum_B f_{iB}. \tag{7}$$

In Eq. (7), the driving force term $f_{i,d} = (v_d e_i - v_i)/\tau$ describes the tendency of pedestrian i moving toward his destination. Here, v_d is the desired speed and e_i is a unit vector indicating the desired walking direction of pedestrian i. The relaxation time τ controls how quickly the pedestrian adapts one's velocity to the desired velocity. The repulsive force terms f_{ij} and f_{iB} reflect his tendency to keep certain distance from other pedestrian j and the boundary B, e.g., wall and obstacles. A more detailed description of the social force model can be found in previous studies [15–18].

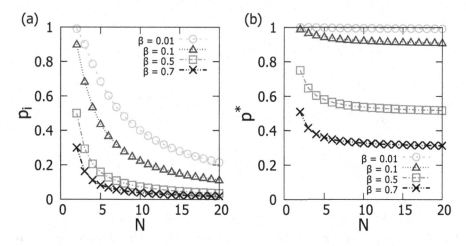

Fig. 1. Bystander effect on helping behavior: (a) p_i, the probability that player i volunteers to rescue an injured person and (b) p^*, the probability that an injured person is rescued.

2.3   Numerical Simulation Setup

Our agent-based model consists of helping behavior model and movement model. The helping behavior model computes the probability that a passerby would help an injured person based on the volunteer's dilemma game. The movement model calculates the sequence of pedestrian positions for each simulation time step. Our agent-based model was implemented from scratch in C++.

Each pedestrian is modeled by a circle with radius $r_i = 0.2$ m. $N_0 = 100$ pedestrians are placed in a 10 m×10 m room indicated by a yellow shade area in Fig. 2. Pedestrians are leaving the room through an exit corridor which is 5 m long and 2 m wide. The place of safety is set on the right, outside of the exit corridor. There are N_i injured persons who need a help in escaping the room and $N = N_0 - N_i$ passersby who are ambulant. Some passersby might turn into volunteers who are going to approach to and then rescue the injured persons. The number of volunteers is determined based on the volunteer's dilemma game presented in Sect. 2.1.

The volunteer's dilemma game is updated for each second. We assumed that the volunteer's dilemma game is a macroscopic behavior like goal selection and path navigation patterns [19]. In line with Heliövaara et al. [6], each passerby can play the volunteer's dilemma game a few times during the whole simulation period. With the update frequency of one time per second, most of passersby play the volunteer's dilemma game up to ten times before they leave the room. A passerby can decide whether he will volunteer to rescue an injured person within a range of 3 m. Once the volunteer decides to rescue the injured person, then he shifts his desired direction walking vector e_i toward the position of injured person. Once the volunteer reaches the injured person, he will flee to the place of safety with the injured person after a preparation time of 5 s.

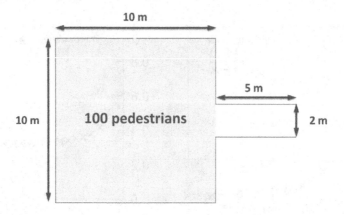

Fig. 2. Schematic depiction of the numerical simulation setup. 100 pedestrians are placed in a 10 m×10 m room indicated by a yellow shade area. Pedestrians are leaving the room through an exit corridor which is 5 m long and 2 m wide. The place of safety is set on the right, outside of the exit corridor. (Color figure online)

The pedestrian movement is updated with the social force model in Eq. (7). The passersby move with the initial desired speed $v_d = v_{d,0} = 1.2$ m/s and with relaxation time $\tau = 0.5$ s, and their speed cannot exceed $v_{max} = 2.0$ m/s. Until now, the speed of volunteers rescuing the injured persons is often assumed by the modelers, like the work of Von Sivers et $al.$ [10,11]. We applied speed reduction factor $\alpha = 0.5$ to the volunteers rescuing the injured persons, so they move with a reduced desired speed $v_d = \alpha v_{d,0} = 0.6$ m/s. Following previous studies [20–22], we discretized the numerical integration of Eq. (7) using the first-order Euler method:

$$v_i(t + \Delta t) = v_i(t) + a_i(t)\Delta t, \tag{8}$$
$$x_i(t + \Delta t) = x_i(t) + v_i(t + \Delta t)\Delta t. \tag{9}$$

Here, $a_i(t)$ is the acceleration of pedestrian i at time t which can be obtained from Eq. (7). The velocity and position of pedestrian i is denoted by $v_i(t)$ and $x_i(t)$, respectively. The time step Δt is set as 0.05 s.

3   Results and Discussion

Fig. 3 shows snapshots of our agent-based model simulations. Open black circles indicate injured persons and full dark circles show volunteers helping the injured persons. Gray circles represent the passersby. If the helping cost is low, all the injured persons are likely to be rescued, as shown in Fig. 3(a). However, if the helping cost is high (i.e., high β), then some injured persons might not be rescued, see the red dotted circle in Fig. 3(b). By systematically changing the value of N_i and β, we observed different patterns of collective helping behaviors summarized

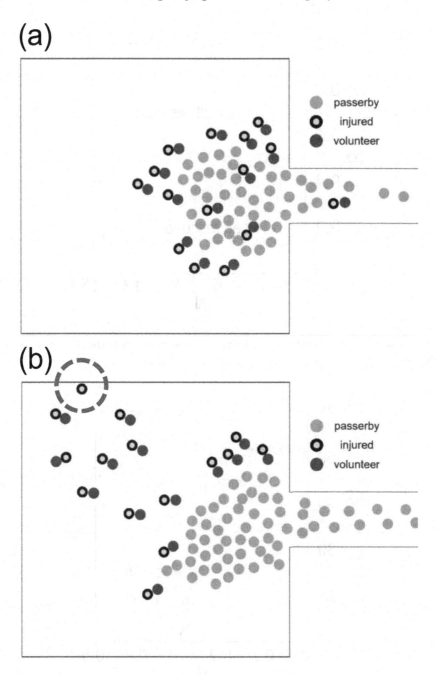

Fig. 3. Snapshots of helping behavior in a room evacuation scenario: (a) all the injured persons are rescued in case of $N_i = 15$ and $\beta = 0.1$, and (b) some injured persons are not rescued (in the red dotted circle) in case of $N_i = 15$ and $\beta = 0.2$. Open black circles indicate injured persons and full dark circles show volunteers helping the injured persons. Gray circles represent the passersby. (Color figure online)

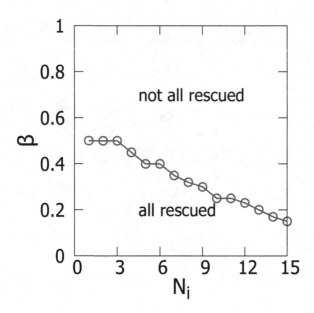

Fig. 4. Schematic phase diagram of collective helping behavior in the room evacuation scenario.

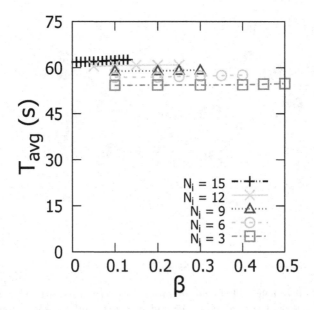

Fig. 5. Average total evacuation time T_{avg} as a function of the number of injured persons N_i and β.

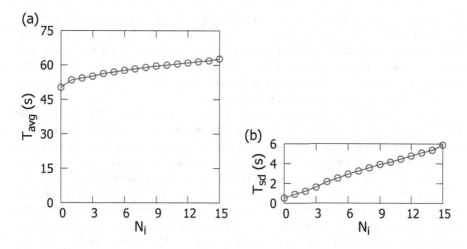

Fig. 6. Total evacuation time in case of $\beta = 0.1$ against the number of injured persons N_i: (a) average T_{avg} and (b) standard deviation T_{sd}.

in the schematic phase diagram (see Fig. 4). For each parameter combination (N_i, β), we performed 30 independent simulation runs.

We also looked into the impact of cost ratio β and the number of injured persons N_i on the total evacuation time T. The total evacuation time T is defined as the length of period from the start of evacuation to the moment when the last evacuee leaves the exit corridor. We measured the average and standard deviation of the total evacuation time, i.e., T_{avg} and T_{sd}, based on the values of total evacuation time T obtained over 30 independent simulation runs for each parameter combination (N_i, β). Figure 5 indicates that change in the value of β does not make a noticeable different to the average total evacuation time T_{avg}. This is seemingly because β only affects the probability that a passerby turns into a volunteer. As indicated in Fig. 6(a), the average total evacuation time T_{avg} increases as the number of injured persons N_i grows. Having more injured persons indicates that there are more volunteers who move in the reduced desired speed, so the total evacuation time increases due to the volunteers rescuing the injured persons. In addition, the standard deviation of total evacuation time T_{sd} increases as the number of injured persons N_i grows, in that the difference in evacuation time among evacuees gets larger.

4   Conclusion

We have numerically investigated helping behavior among evacuees in a room evacuation scenario. Our simulation model is based on the volunteer's dilemma game reflecting volunteering cost and social force model simulating pedestrian movement. We characterized collective helping behavior patterns by systematically controlling the values of cost ratio β and the number of injured pedestrians

N_i. For low cost ratio values, one can expect that all the injured pedestrians are rescued by volunteers. For high cost ratio values, on the other hand, it was observed that not all the injured persons can be rescued. When the number of injured persons is large, a low value of cost ratio yields a result that all the injured pedestrians are rescued. A schematic phase diagram summarizing the collective helping behavior patterns is presented.

A very simple room evacuation scenario has been used in order to study the fundamental role of helping behavior in the evacuation especially the number of evacuated pedestrians. In this study, the severity of injuries are assumed to be the same for all the injured persons, so each injured person can be rescued by a volunteer. According to patient triage scale in Singapore [23,24], patients can be categorized based on the severity of injuries and the desired number of volunteers are different for various types of injuries. Future work can reflect the impact of patient injury levels in collective helping behavior by assuming different number of required volunteers for each patient. This study can be extended from the perspective of game theory. As stated in Diekmann's study [14], it can be interesting to introduce different values of collective good U and volunteering cost K to each passerby. By doing that, we can reflect personal difference in willingness to volunteer in emergency evacuations. In addition, one can imagine that the value of U can be changed depending on the number of injured persons and volunteers. For instance, the values of U are different when there are one injured person, one volunteer and two injured persons, three volunteers. Evolutionary game [4] can be also introduced in order to reflect behavioral changes of passersby influenced by the existence of volunteers, which might be observable in emergency evacuations.

Acknowledgements. This research is supported by National Research Foundation (NRF) Singapore, GOVTECH under its Virtual Singapore program Grant No. NRF2017VSG-AT3DCM001-031.

References

1. Guo, R.Y., Huang, H.J., Wong, S.C.: Route choice in pedestrian evacuation under conditions of good and zero visibility: experimental and simulation results. Transport. Res. Part B: Methodol. **46**(6), 669–686 (2012)
2. Abdelghany, A., Abdelghany, K., Mahmassani, H., Alhalabi, W.: Modeling framework for optimal evacuation of large-scale crowded pedestrian facilities. Euro. J. Oper. Res. **237**(3), 1105–1118 (2014)
3. Yanagisawa, D., Nishinari, K.: Mean-field theory for pedestrian outflow through an exit. Phys. Rev. E **76**(6), 061117 (2007)
4. Hao, Q.Y., Jiang, R., Hu, M.B., Jia, B., Wu, Q.S.: Pedestrian flow dynamics in a lattice gas model coupled with an evolutionary game. Phys. Rev. E **84**(3), 036107 (2011)
5. Shi, D.M., Wang, B.H.: Evacuation of pedestrians from a single room by using snowdrift game theories. Phys. Rev. E **87**(2), 022802 (2013)
6. Heliövaara, S., Ehtamo, H., Helbing, D., Korhonen, T.: Patient and impatient pedestrians in a spatial game for egress congestion. Phys. Rev. E **87**(1), 012802 (2013)

7. von Schantz, A., Ehtamo, H.: Spatial game in cellular automaton evacuation model. Phys. Rev. E **92**(5), 052805 (2015)

8. Johnson, N.R.: Panic at "The Who concert stampede": an empirical assessment. Soc. Problems **34**(4), 362–373 (1987)

9. Drury, J., Cocking, C., Reicher, S.D.: The nature of collective resilience: survivor reactions to the 2005 London bombings. Int. J. Mass Emergencies Disasters **27**, 66–95 (2009)

10. von Sivers, I., Templeton, A., Köster, G., Drury, J., Philippides, A.: Humans do not always act selfishly: social identity and helping in emergency evacuation simulation. Transport. Res. Procedia **2**, 585–593 (2014)

11. von Sivers, I., et al.: Modelling social identification and helping in evacuation simulation. Saf. Sci. **89**, 288–300 (2016)

12. Lin, G.W., Wong, S.K.: Evacuation simulation with consideration of obstacle removal and using game theory. Phys. Rev. E **97**(6), 062303 (2018)

13. Diekmann, A.: Volunteer's dilemma. J. Conflict Resolution **29**(4), 605–610 (1985)

14. Diekmann, A., Przepiorka, W.: "Take one for the team!" Individual heterogeneity and the emergence of latent norms in a volunteer's dilemma. Soc. Forces **94**(3), 1309–1333 (2016)

15. Helbing, D., Molnár, P.: Social force model for pedestrian dynamics. Phys. Rev. E **51**, 4282–4286 (1995)

16. Johansson, A., Helbing, D., Shukla, P.: Specification of the social force pedestrian model by evolutionary adjustment to video tracking data. Adv. Complex Syst. **10**, 271–288 (2008)

17. Kwak, J., Jo, H.-H., Luttinen, T., Kosonen, I.: Collective dynamics of pedestrians interacting with attractions. Phys. Rev. E **88**, 062810 (2013)

18. Viswanathan, V., Lee, C.E., Lees, M.H., Cheong, S.A., Sloot, P.M.A.: Quantitative comparison between crowd models for evacuation planning and evaluation. Euro. Phys. J. B **87**(2), 1–11 (2014). https://doi.org/10.1140/epjb/e2014-40699-x

19. Zhong, J., Cai, W., Luo, L., Zhao, M.: Learning behavior patterns from video for agent-based crowd modeling and simulation. Autonomous Agents Multi-Agent Syst. **30**(5), 990–1019 (2016). https://doi.org/10.1007/s10458-016-9334-8

20. Zanlungo, F., Ikeda, T., Kanda, T.: Social force model with explicit collision prediction. EPL (Europhys. Lett.) **93**(6), 012811 (2011)

21. Zanlungo, F., Ikeda, T., Kanda, T.: Potential for the dynamics of pedestrians in a socially interacting group. Phys. Rev. E **89**, 012811 (2014)

22. Kwak, J., Jo, H.-H., Luttinen, T., Kosonen, I.: Jamming transitions induced by an attraction in pedestrian flow. Phys. Rev. E **96**, 022319 (2017)

23. Parker, C.A., Liu, N., Wu, S.X., Shen, Y., Lam, S.S.W., Ong, M.E.H.: Predicting hospital admission at the emergency department triage: a novel prediction model. Am. J. Emergency Med. **37**(8), 1498–1504 (2019)

24. Singapore General Hospital (SGH) Emergency Care. https://www.sgh.com.sg/patient-care/visiting-specialist/pages/emergency-care.aspx. Accessed 3 Feb 2020

Learning Mixed Traffic Signatures
in Shared Networks

Hamidreza Anvari$^{(\boxtimes)}$ and Paul Lu

Department of Computing Science, University of Alberta, Edmonton, AB, Canada
{hanvari,paullu}@ualberta.ca

Abstract. On shared, wide-area networks (WAN), it can be difficult to characterise the current traffic. There can be different protocols in use, by multiple data streams, producing a mix of different traffic signatures. Furthermore, bottlenecks and protocols can change dynamically. Yet, if it were possible to determine the protocols (e.g., congestion control algorithms (CCAs)) or the applications in use by the background traffic, appropriate optimisations for the foreground traffic might be taken by operating systems, users, or administrators.

We extend previous work in predicting network protocols via signatures based on a time-series of round-trip times (RTT). Gathering RTTs is minimally intrusive and does not require administrative privilege. Although there have been successes in using machine learning (ML) to classify protocols, the use cases have been relatively simple or have focused on the foreground traffic. We show that both k-nearest-neighbour (K-NN) with dynamic time warp (DTW), and multi-layer perceptrons (MLP), can classify (with useful accuracy) background traffic signatures with a range of bottleneck bandwidths.

Keywords: Classification · TCP · Protocol selection · Wide-area networks · High-performance network · Fairness · Shared network

1 Introduction

Knowledge about the state of a data network can be used to achieve high performance. For example, knowledge about the protocols in use by the background traffic might influence which protocol to choose for a new foreground data transfer. Unfortunately, global knowledge can be difficult to obtain in a dynamic, distributed system like a wide-area network (WAN).

Previously, we introduced a machine-learning (ML) approach to network performance, called *optimization through protocol selection (OPS)* [2]. Using local round-trip time (RTT) time-series data, a classifier predicts the mix of protocols in current use by the background traffic. Then, a decision process selects the best protocol to use for the new foreground transfer, so as to maximize throughput while maintaining fairness. We showed that a protocol oracle would choose TCP-BBR [7] for the new foreground traffic if TCP-BBR is already in use in the

© Springer Nature Switzerland AG 2020
V. V. Krzhizhanovskaya et al. (Eds.): ICCS 2020, LNCS 12137, pp. 524–537, 2020.
https://doi.org/10.1007/978-3-030-50371-0_39

background, for proper throughput. Similarly, the protocol oracle would choose TCP-CUBIC [12] for the new foreground traffic if only TCP-CUBIC is in use in the background, for fairness.

However, it was unclear if that result would generalize further. For example, only one bottleneck bandwidth was considered in our first empirical evaluation. On a real network, the bottleneck bandwidth can change dynamically due to different network routing and paths.

After gathering more empirical signatures with different bottleneck bandwidths (Fig. 1b), we show that the applicability of k-nearest-neighbour (K-NN) with dynamic time warping (DTW) remains intact, even with a mix of training data from different bottlenecks. That result is consistent with the ability of DTW to abstract out scaling differences in time-series data.

In our previous study [2], K-NN and MLP were comparable in accuracy. We advocated for the use of K-NN in practice due to the relative simplicity of K-NNs, and due to the explainability of K-NN's predictions and mis-predictions. In other words, if a K-NN makes an incorrect prediction, one can reason about why K-NN was wrong by examining the training data vectors that were the nearest neighbours of the query vector (a.k.a. time-series). In contrast, if an MLP makes an incorrect prediction, it is harder to explain the mistake. MLPs have a multitude of parameters and weights within the layers and perceptrons that make explanations more difficult.

This empirical study shows that even with multiple and/or changing bottleneck bandwidths (i.e., mixed traffic signatures), OPS continues to be possible with either K-NN or MLP. In the future, we plan to explore the use of ML, whether with K-NN, MLP, LSTM, or some other algorithm, to predict other properties of network traffic. So far, OPS has been about selecting an appropriate CCA at the operating system level. But, if the user had an ML oracle that correctly predicted other network characteristics, then the user might choose different application-level strategies (e.g., different video compression algorithm to make bandwidth-vs-quality trade-offs). Or, in the long term, network administrators might be able to identify situations in which the best optimisation is to selectively move some traffic to a different network route (e.g., an overlay network, or a private network).

2   Experimental Setup

We implemented a controlled testbed to allow us to vary different network configurations (Fig. 1). This controlled network extends the testbed from our previous work at 1 Gb/s speeds [3] by adding 10 Gb/s network interface cards (NICs) and switches, enabling us to vary between 1 Gb/s and 10 Gb/s for the native link rate, while keeping the possibility of emulating varying bottleneck bandwidth, $BtlBW$, and end-to-end latency, RTT.

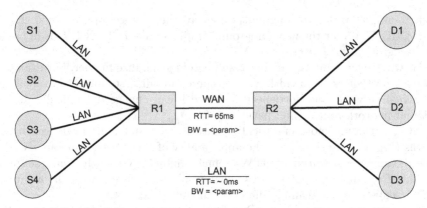

(a) Dumbbell Network Topology (Based on our previous study [3])

Configuration	LAN		WAN	
	BW	RTT	BW	RTT
C1	1 Gb/s	~ 0 ms	500 Mb/s	65 ms
C2	10 Gb/s	~ 0 ms	250 Mb/s	65 ms
C3	10 Gb/s	~ 0 ms	500 Mb/s	65 ms
C4	10 Gb/s	~ 0 ms	1 Gb/s	65 ms
C5	10 Gb/s	~ 0 ms	2 Gb/s	65 ms

Node(s)	CPU (Model/Cores/Freq.)	RAM
$S1, D1$	AMD Opt. 6134 / 8 / 2.30	32 GB
$S2, D2$	Intel Ci3-6100U / 4 / 2.30	32 GB
$S3, D3$	Intel Ci3-6100U / 4 / 2.30	32 GB
$S4$	AMD E2-1800 / 2 / 1.70	8 GB
$R1, R2$	AMD A8-5545M / 4 / 1.7	8 GB

(b) Testbed Configuration Scenarios   (c) Nodes Configuration

Fig. 1. Testbed architecture

2.1   Logical View

Our controlled testbed consists of a dumbbell topology with 7 end-nodes grouped into 2 virtual LANs, and 2 virtual edge routers (i.e., R1, R2) connected over an emulated WAN of bandwidth BtlBW and RTT (Fig. 1a). In the literature, the dumbbell topology is often used as a simplified representation of a shared network.

For the WAN link, the Dummynet [6] network emulator is used on nodes R1 and R2 to control the desired bandwidth, delay, and router queuing properties. The emulated configurations we used in this study are summarised in Fig. 1b. The end-to-end propagation delay (base RTT) for all the scenarios is set to 65 ms for the WAN connection, as found on many medium-to-large WANs; the RTT for LAN connections is negligible. The LAN link bandwidth is either 1 Gb/s or 10 Gb/s, both native link rates available to the nodes. The bottleneck shared bandwidth for WAN (BtlBW) varies from 250 Mb/s up to 2 Gb/s.

The router buffer size at R1 and R2 are fixed at 6 MB. This buffer size accounts for about 0.5 BDP for the highest BtlBW, and up to 2 BDP for the lowest BtlBW (BDP=Bottleneck-Bandwidth × Delay (RTT)). Setting the buffer size to less than 1 BDP, so-called shallow buffering, could result in the under-utilisation of network bandwidth. In contrast, setting the buffer to large sizes could result in an effect called buffer-bloat where the users experience extremely long delays due to long queuing delays at the router. However, the existing version of the Dummynet software for Linux platform has an internal limitation that prevents us from setting the buffer to sizes larger than 6 MB. But this fixed size is sufficiently large to enable senders to saturate the BtlBW for WAN in all configurations.

2.2 Physical View

The testbed is implemented as an overlay on top of a physical network. All the nodes in our testbed are physically located in a dedicated cluster, all running Linux distribution CentOS 6.4 using kernel version `4.12.9-1.el6.elrepo.x86-64`. The hardware configuration of the nodes are provided in Fig. 1c. All the nodes are equipped with two network interface cards of 1 Gb/s and 10 Gb/s native rates. There are two network switches connecting all the nodes in parallel, at 1 Gb/s and 10 Gb/s rates accordingly. There is no interference from other traffic because the cluster is isolated from other networks.

2.3 Data-Transfer Scenarios

For all experimental scenarios (Fig. 1b), the sender nodes on the left-hand-side LAN (S1, S2, S3, and S4) act as traffic generators, sending bulk data over the WAN link. The receiver nodes are on the right-hand-side LAN (nodes D1, D2, and D3). We run two data-transfer tasks simultaneously, between S1-D1 and between S2-D2. During the transfers, regular network pings (to measure RTT) are conducted between S3-D3 (Sect. 5.1). S4 is not normally used for these experiments. Both streams and the pings must travel across the bottleneck link R1-R2. The RTT time series are presented in Fig. 5 and form the basis for our machine-learning experiments in Sect. 5.

Of course, networks often carry multiple streams of traffic, but we start our evaluation with the simpler case of two streams, to make it easier to control interactions.

3 Shared Network: Does Background Traffic Matter?

Networks are not always private. Bandwidth-sharing networks are still the common case for a large number of research and industry users. The workload on shared networks tends to be highly dynamic. As a result, estimating the network condition and available resources (bandwidth, etc.) is one of the challenging tasks in bandwidth-sharing networks. There are a number of studies in this area, investigating the possibility of estimating the available bandwidth in high-speed networks [20,25].

In addition to bandwidth estimation, estimating the network workload and the type of background traffic on the network could also affect the performance of data transfer tasks. In one study, the effect of background traffic on distributed systems has been investigated [23]. Also, in our previous work we have investigated and shown the counter-intuitive performance of some well-known tools and protocols depending on the type of background traffic in the network [1–3]. Hence, obtaining knowledge about the background traffic would allow better and more-efficient adjustments of the network configurations, and choosing appropriate data-transfer tools and protocols.

In this section, we further investigate the impact of background traffic on the performance of TCP CCAs, as well as on a few high-performance data transfer tools.

3.1 Case 1: TCP CCAs Interaction as Background Traffic

Here we briefly review the Transmission Control Protocol (TCP) in terms of its congestion control algorithm (CCA). We review CUBIC and BBR, two popular CCA schemes. CUBIC [12] is the the default TCP CCA deployed on most Linux-based hosts. BBR [7] is a newer CCA. Some studies show that both CUBIC and BBR manifest unfair bandwidth utilisation under various circumstances [18,19].

To further investigate the interoperability of CUBIC and BBR, as well as their impact on the other traffic sharing the same bandwidth, we have run an experiments, using CUBIC and BBR over a shared network. The results are provided in Fig. 2. Due to space limit we only present resluts for C4 configuration.

For each of two possible background traffic, (1) Constant TCP (Fig. 2a), (2) Square-wave pattern TCP Stream (Fig. 2b), we run 4 separate data-transfer tasks as foreground traffic (different bars in Fig. 2):

1. **iperf.** A single TCP stream using the standard iperf tool.
2. **iperf-8.** A combination of 8 parallel TCP streams for data transfer.
3. **iperf-16.** A combination of 16 parallel TCP streams for data transfer.
4. **UDT.** A well-known UDP-based high-performance data transfer tool.

For both background traffic and foreground data-transfer tasks we have examined both CUBIC and BBR to study their interaction and impact on each other. For example, the notation C-C in Fig. 2 specifies that CUBIC is the CCA for the background traffic, as well CUBIC is the CCA for the foreground traffic (whether it is iperf, iperf-8, iperf-16, but NOT UDT since UDT is based on UDP). Similarly, the notation C-B specifies that CUBIC is the CCA for the background traffic, but BBR is the CCA for the foreground traffic (iperf, iperf-8, iperf-16, but NOT UDT).

As shown in Fig. 2, for all the possible mixtures of parameter configurations, there is a significant variation in the observed throughput performance while each tool runs along with another TCP CCA. Both CUBIC and BBR are able to utilize available bandwidth while running in isolation. However, BBR has a significant negative impact on CUBIC in all combinations, regardless of running as a background or foreground stream. CUBIC stream(s) suffer from extreme starvation when running along with BBR streams on a shared networks. In contrast, all-CUBIC and all-BBR scenarios, while not perfect, are considerably fairer in sharing bandwidth compared to heterogeneous combination of the two CCAs.

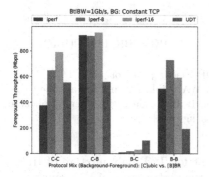 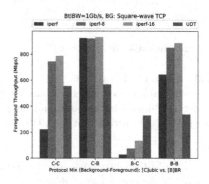

(a) Background Traffic: Constant TCP Stream   (b) Background Traffic: Square-wave pattern TCP Stream

Fig. 2. Impact of different TCP CCAs as background traffic on each other

3.2 Case 2: TCP- vs. UDP-Based Background Traffic

In last section we investigated the impact of single TCP stream of varying CCA algorithm on the foreground traffic. In this section, we further expand our observations for more complicated patterns of background traffic, where background traffic is either a TCP stream or a UPD-based data transfer task as follows: No background (no_bg), constant TCP (bg_const), TCP- and UDP-based square-wave pattern cycling on and off for 10 s (bg_tcp1 and bg_udp1), and an UDP-based square-wave traffic generated by RBUDP [14] data transfer tool. The results are depicted in Fig. 3.

Similar to previous section, here we have studied the possible combinations of CUBIC and BBR where the data transfer tasks are TCP streams. The results here show that further than TCP CCA, the type of background traffic being TCP or UDP stream will differently impact the foreground data transfer performance. While for simple TCP-based background traffic the GridFTP tool seems to perform better than UDT, when UDP-based streams are added as the background traffic the UDT performs relatively better than GridFTP.

3.3 Case 3: Burstiness of the Background Traffic

In the last section we observed that TCP- and UDP-based background traffic could imply considerably different performance for the foreground data transfer

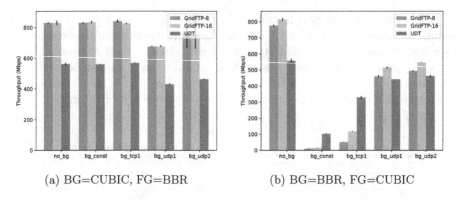

(a) BG=CUBIC, FG=BBR   (b) BG=BBR, FG=CUBIC

Fig. 3. Impact of various TCP- and UDP-based protocols on TCP CCAs. BtlBW=1 Gb/s is the bottleneck bandwidth of WAN link. BG and FG are the CCAs used for TCP-based background and foreground traffic respectively where applicable. (Listed BG CCA is irrelevant for UDP-based background cases.)

tasks. In addition, we have already seen that for TCP-only mixture of traffic, different CCAs could have different impacts the performance of foreground traffic. In this section we will see that when the background traffic is only formed of UDP-based streams, there are still more attributes that could impact the performance of data transfer tasks. For this experiment, we conduct a bursty UDP stream as background traffic, represented by an square-wave pattern. The burstiness of the UDP stream is varying from 1-s ON and 1-s OFF (denoted by bg_udp(1,1)), to 1-s ON and 10-s OFF (denoted by bg_udp(1,10)). The results are provided in Fig. 4. Again, the burstiness of the background traffic could significantly impact the performance of data transfer tasks, where CUBIC is more vulnerable to this burstiness than BBR (Fig. 4a and Fig. 4b respectively)

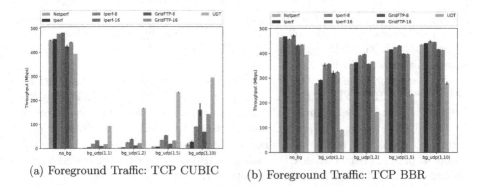

(a) Foreground Traffic: TCP CUBIC

(b) Foreground Traffic: TCP BBR

Fig. 4. Impact of bursty UDP-based background traffic on CUBIC and BBR foreground. Background traffic is generated using *iperf* tool in UDP mode.

4 End-to-End Traffic Probing: Intrusive vs. Non-intrusive

In Sect. 3 we investigated and shown the impact of background traffic, from several aspects, on the performance of foreground traffic and data transfer tasks. With that knowledge, here we will review and discuss the possible network signals that might be used at end-nodes to probe the shared bandwidth and identify the type of background traffic. Such a knowledge would later be utilised to decide on an appropriate set of tools and protocols for transferring data over the network.

Estimating the type and mixture of network traffic is performed from a local, non-global perspective, viewing the network as a black box. In such scenarios one well-known technique is to send probing packets to the network and based on the received response form a model as a proxy for the global network view. Ping and TraceRoute are two widely used tools that utilise this technique.

While end-to-end probing is a flexible and user-level approach, depending on the type of probing to conduct it might face challenges due to its impact on network performance. Sending probing packets to the network would mean that part of network resources will become busy processing non-real data packets for helping an end-node to obtain information about the network. This technique is historically discouraged or blocked by network administrator and infrastructure providers. The challenges for TraceRoute tool and it's numerous variations to deal with imposed restrictions is an evident example of such discouragement.

The two end of network probing spectrum are Intrusive and Non-Intrusive network signalling.

Intrusive Probing. Intrusive probing consists of sending bursts of data traffic to the network in order to estimate the available bandwidth and resources. While this probing policy could result in more accurate predictions, it implies wasting a significant portion of network bandwidth and resources for this probing process. As such, computer networks usually search for such probing workload and apply restrictions, using techniques such as traffic policing [11].

Non-intrusive Probing. At the other end of the spectrum, the most ideal type of network probing does not involve sending any artificial packets to the network. Instead, the end-node would solely rely on the organic signals it receive from network as a result of transferring real user traffic on the network. Despite it benefits and desired behaviour, it is not a practical method for probing network resources in most scenarios. Firstly, solely relying on the organic acknowledgement and signals would limit our exposure to the network and only gives us inconsistent messages which could be very challenging to impossible to draw any conclusion on them. Secondly, such network signals are usually being consumed by the lower layers in the networking stack and are being discarded before handing data over to the user-space. Hence, there is little opportunity to utilise non-intrusive signalling techniques for investigating network characteristics, including the type of background traffic.

A Practical Trade-off: Minimally Intrusive Probing. As a practical trade-off between intrusive and non-intrusive methods, we aim to use a *minimally* *intrusive* probing method. Such a method would enable us to develop sufficient insight about the network, and at the same time would impose minimum impact on the network resources and so will be more probable to be allowed to be carried over and across the networks. In this study, we use RTT as our main network probing facility, studying it variation over time as a potential signature to identify the mixture of background traffic on the network. In particular, in the following section we periodically probe the end-to-end RTT using the `ping` tool to form time-series of RTT values over time. We then use these RTT time-series as a proxy for predicting the mixture of traffic on the network.

5   Learnability of TCP-based Traffic Signatures

In this section, we investigate the feasibility of building classification models to predict the type of background traffic using end-to-end probed RTT time-series. As a proof-of-concept, for this study we investigate the learnability of a mixture of TCP-only traffic on the network (Sect. 3.1). In particular, we consider the following 6 distinct classes of TCP-based traffic mix to be represented by the prediction model:

1. **No Traffic (B0-C0).** No active data communication on the network.
2. **Single CUBIC (B0-C1).** A single TCP CUBIC stream running on the network.
3. **Single BBR (B1-C0).** A single TCP BBR stream running on the network.
4. **Double CUBIC (C1-C1).** Two TCP CUBIC streams running on the network.
5. **Double BBR (B1-B1).** Two TCP BBR streams running on the network.
6. **CUBIC and BBR (B1-C1).** Two TCP streams running on the network: one CUBIC and one BBR.

Intuitively, for the RTT time-series to be used as the input for prediction, they should hold two qualities:

1. Distinct Patterns between traffic signatures: the RTT time-series for different traffic mixtures should be reasonably distinct in order to be trainable with a reasonable accuracy.
2. Repetitive Pattern within each time-series: in order to be able to train a classifier to generalise well to unseen cases of RTT time-series, repetitive patterns should exist in each traffic mixture signature over time.

We have conducted a series of experiments running the above six classes of traffic mixtures, probing end-to-end RTT in the periods of one second. The sample signature of RTT time-series are provided in Fig. 5. The provided samples, intuitively, hold both desired qualities of interclass distinctive patterns and intraclass repetitive patterns.

In what follows we review the process of building training dataset, training prediction models, and studying the accuracy and usability of the trained models.

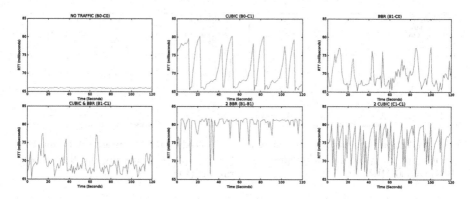

Fig. 5. A Sample of RTT Time-Series Data

5.1 Building a Training Dataset of RTT Time-Series

Using our controlled experimental setup presented in Sect. 2, we gathered RTT time-series, with a 1-s sampling rate, over a one hour period of time. To smooth out the possible noise in the measured RTT value, each probing step consists of sending 10 ping requests to the other end-host, with 1 ms delay in between, and the average value is recorded as the RTT value for that second. We repeated this experiment for the six traffic classes listed above. In addition, to further investigate how well the time-series prediction generalises across varying bottleneck bandwidth configurations, we repeated the same experiment for all the configurations listed in Fig. 1b.

For each probing scenario, we conducted the corresponding CUBIC or BBR traffic stream between pairs of (S1,D1) and (S2,D2) pair of nodes; at the same time, we used the (S3,D3) node pair to conduct the periodic RTT probing and form the RTT time-series for that particular class.

For all data-transfer tasks, we used the *iperf* tool (http://software.es.net/iperf/) for generating TCP traffic of the desired CCA.

To prepare the gathered data for training process, we partition the one-hour RTT time-series into smaller chunks. For this purpose we re-used the partitioning software we developed as part of our previous study [2], partitioning the data based on a given parameter w, representing the RTT time-series length.

5.2 Training Classification Models

For this study, for training classifiers we use k-nearest-neighbours (K-NN), and a multi-layer perceptron (MLP) neural network [5,13]. As for K-NN, we use it along with the well-known dynamic time warping (DTW) distance metric, for the time-series data. DTW was originally used for speech recognition. It was then proposed to be used for pattern finding in time-series data, where Euclidean distance results in a poor accuracy due to possible time-shifts in the time-series [4,15]. For MLP, we use a network with 2 dense hidden layers of

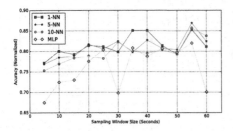 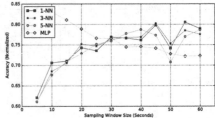

(a) Prediction Accuracy for models trained over a single Configuration (BtlBw=500 Mb/s, RTT=65ms)

(b) Prediction Accuracy for models trained over the mixture of all Configurations (Fig. 1b)

Fig. 6. Learnability of TCP-based Traffic Signatures. Accuracy of K-NN w/ DTW and MLP models

$w * 1.5$ and w nodes, both with ReLU activation function. The output layer applies SoftMax activation function. To avoid overfitting *Dropout* technique is uniformly applied through the layers.

To make an appropriate decision about the parameter value w, length of RTT time-series entries, we did a parameter-sweep experiment where we calculated the classification accuracy for all the classifiers, varying parameter w from 5 s to 60 s, with a 5 s step.

To better estimate accuracy while avoiding overfitting, we use five-fold cross-validation. The reported results are the average accuracy over the five folds on the cross-validation scheme. Figure 6 represents the average accuracy per window size w (in seconds), calculated for three variations of K-NN and MLP models.

Figure 6a presents prediction accuracy for the models trained for a single network configuration. Figure 6b shows the prediction accuracy when the models are trained using the mixed traffic signatures of all configurations.

For both single-configuration and mixed-configuration scenarios, K-NN with DTW yields a better accuracy in most cases. Since K-NN compares against real data points, it is very efficient in making more accurate prediction on average. 1-NN in particular offers the highest accuracy between tested K-NN variations. In contrast, MLP shows variable accuracy, highly sensitive to the parameter w. On the one hand, in Fig. 6a, MLP has lower accuracy than K-NN across different values of w. On the other hand, in Fig. 6b, MLP has the highest accuracy for w below 25 s, but MLP becomes less accurate than K-NN for larger values of w. Recall that, given our methodology, the number of training data instances decreases as w increases.

Although the preference of K-NN vs. MLP depends on the choice of w, the experiment does show that a ML classifier is capable of being accurate in the range of 0.75 to 0.85 for mixed traffic signatures. This accuracy level, being achieved using our relatively simple classification models, confirms our hypothesis on learnability of end-to-end TCP-based traffic signatures using RTT time-series. According to our other study, this level of accuracy is sufficient to improve on the throughput and fairness of the TCP-based data transfers on a shared network, running a mixture of CUBIC and BBR CCAs [2]. Further studying the extensibility of this approach to other mixtures of TCP- and non-TCP-based traffic, as well as adopting more sophisticated classification models such as RNN and LSTM deep neural networks [16], form the next steps of this study to be pursued in the future work.

6   Other Related Work

ML techniques have been used for designing or optimising network protocols. RemyCC uses simulation and ML to create a new TCP CCA, via a decentralised partially-observable Markov decision process (dec-POMDP) [24]. Performance-oriented Congestion Control (PCC) is another recent study where an online learning approach is incorporated into the structure of the TCP CCA. [9]. Online convex optimisation is applied to design a rate-control algorithm for TCP, called Vivace [10]. Another recent approach has been to apply deep reinforcement learning techniques for constructing CCA algorithms [17].

Another line of work includes applying ML techniques to discover network properties. Estimating the available bandwidth in high-speed networks [20,25], identifying TCP CCA in traffic traces [8,21], and predicting TCP unique behaviours and behaviour anomalies [22] are among the topics in this category.

7   Concluding Remarks

We investigated the learnability of traffic signatures of background traffic in shared networks. Currently, the signatures are RTT time-series data based on minimally intrusive end-to-end probing. We gathered a labelled training dataset of 6 different classes of TCP-based background traffic, on a testbed that emulates a shared WAN. Different classifier models were trained, using the signatures. We performed a simple parameter sweep of the available bandwidth of the bottleneck in the testbed. Such prediction models could prove useful for a variety of use-case scenarios, including protocol selection for a data transfer, and network tuning and optimisation.

For future work, we will consider adding more complex traffic mixes, including UDP-based traffic, high-performance data transfer tools (e.g., GridFTP), bursty traffic, and more. We hypothesise that by having sufficiently large training datasets, more sophisticated classification algorithms such as deep neural networks and LSTM might be used.

Acknowledgement. Thank you to Jesse Huard for the original implemenation of MLP.

References

1. Anvari, H., Lu, P.: Large transfers for data analytics on shared wide-area networks. In: Proceedings of the ACM International Conference on Computing Frontiers, pp. 418–423. CF 2016, ACM, New York (2016). https://doi.org/10.1145/2903150. 2911718, http://doi.acm.org/10.1145/2903150.2911718

2. Anvari, H., Huard, J., Lu, P.: Machine-learned classifiers for protocol selection on a shared network. In: Renault, É., Mühlethaler, P., Boumerdassi, S. (eds.) MLN 2018. LNCS, vol. 11407, pp. 98–116. Springer, Cham (2019). https://doi.org/10. 1007/978-3-030-19945-6_7

3. Anvari, H., Lu, P.: The impact of large-data transfers in shared wide-area networks: an empirical study. Procedia Comput. Sci. **108**, 1702–1711 (2017). https://doi.org/10.1016/j.procs.2017.05.211, http://www.sciencedirect. com/science/article/pii/S1877050917308049, international Conference on Computational Science, ICCS 2017, 12-14 June 2017, Zurich, Switzerland

4. Berndt, D.J., Clifford, J.: Using dynamic time warping to find patterns in time series. In: Proceedings of the 3rd International Conference on Knowledge Discovery and Data Mining, pp. 359–370. AAAIWS 1994, AAAI Press (1994). http://dl.acm. org/citation.cfm?id=3000850.3000887

5. Bishop, C.M.: Pattern Recognition and Machine Learning (Information Science and Statistics). Springer, Heidelberg (2006)

6. Carbone, M., Rizzo, L.: Dummynet revisited. SIGCOMM Comput. Commun. Rev. **40**(2), 12–20 (2010). https://doi.org/10.1145/1764873.1764876

7. Cardwell, N., Cheng, Y., Gunn, C.S., Yeganeh, S.H., Jacobson, V.: BBR: congestion-based congestion control. Queue **14**(5), 5020–5053 (2016). https://doi. org/10.1145/3012426.3022184

8. Chen, X., Xu, S., Chen, X., Cao, S., Zhang, S., Sun, Y.: Passive TCP identification for wired and wireless networks: a long-short term memory approach. CoRR abs/1904.04430 (2019). http://arxiv.org/abs/1904.04430

9. Dong, M., Li, Q., Zarchy, D., Godfrey, P.B., Schapira, M.: PCC: re-architecting congestion control for consistent high performance. In: 12th USENIX Symposium on Networked Systems Design and Implementation (NSDI 2015), pp. 395–408. USENIX Association, Oakland, CA (2015). https://www.usenix.org/conference/ nsdi15/technical-sessions/presentation/dong

10. Dong, M., et al.: PCC vivace: online-learning congestion control. In: 15th USENIX Symposium on Networked Systems Design and Implementation (NSDI 2018), pp. 343–356. USENIX Association, Renton, WA (2018). https://www.usenix.org/ conference/nsdi18/presentation/dong

11. Flach, T., et al.: An internet-wide analysis of traffic policing. In: Proceedings of the 2016 ACM SIGCOMM Conference, pp. 468–482. SIGCOMM 2016, Association for Computing Machinery, New York (2016). https://doi.org/10.1145/2934872. 2934873

12. Ha, S., Rhee, I., Xu, L.: Cubic: a new TCP-friendly high-speed TCP variant. SIGOPS Oper. Syst. Rev. **42**(5), 64–74 (2008). https://doi.org/10.1145/1400097. 1400105

13. Hastie, T., Tibshirani, R., Friedman, J.: The Elements of Statistical Learning. SSS. Springer, New York (2009). https://doi.org/10.1007/978-0-387-84858-7

14. He, E., Leigh, J., Yu, O., DeFanti, T.A.: Reliable blast UDP: predictable high performance bulk data transfer. In: Proceedings of the IEEE International Conference on Cluster Computing, pp. 317. CLUSTER 2002, IEEE Computer Society, Washington, DC (2002). http://dl.acm.org/citation.cfm?id=792762.793299

15. Hsu, C.J., Huang, K.S., Yang, C.B., Guo, Y.P.: Flexible dynamic time warping for time series classification. Procedia Comput. Sci. **51**, 2838–2842 (2015). https://doi.org/10.1016/j.procs.2015.05.444, http://www.sciencedirect.com/science/article/pii/S1877050915012521, International Conference On Computational Science, ICCS 2015
16. Ismail Fawaz, H., Forestier, G., Weber, J., Idoumghar, L., Muller, P.A.: Deep learning for time series classification: a review. Data Mining Knowl. Disc. **33**(4), 917–963 (2019). https://doi.org/10.1007/s10618-019-00619-1
17. Jay, N., Rotman, N., Godfrey, B., Schapira, M., Tamar, A.: A deep reinforcement learning perspective on internet congestion control. In: Chaudhuri, K., Salakhutdinov, R. (eds.) Proceedings of the 36th International Conference on Machine Learning. Proceedings of Machine Learning Research, vol. 97, pp. 3050–3059. PMLR, Long Beach, California, USA, 09–15 June 2019. http://proceedings.mlr.press/v97/jay19a.html
18. Kozu, T., Akiyama, Y., Yamaguchi, S.: Improving RTT fairness on cubic TCP. In: 2013 First International Symposium on Computing and Networking, pp. 162–167, December 2013. https://doi.org/10.1109/CANDAR.2013.30
19. Ma, S., Jiang, J., Wang, W., Li, B.: Towards RTT fairness of congestion-based congestion control. CoRR abs/1706.09115 (2017), http://arxiv.org/abs/1706.09115
20. Mirza, M., Sommers, J., Barford, P., Zhu, X.: A machine learning approach to TCP throughput prediction. IEEE/ACM Trans. Netw. **18**(4), 1026–1039 (2010). https://doi.org/10.1109/TNET.2009.2037812
21. Mishra, A., Sun, X., Jain, A., Pande, S., Joshi, R., Leong, B.: The great internet TCP congestion control census. Proc. ACM Meas. Anal. Comput. Syst. **3**(3), 45 (2019). https://doi.org/10.1145/3366693
22. Papadimitriou, G., Kiran, M., Wang, C., Mandal, A., Deelman, E.: Training classifiers to identify TCP signatures in scientific workflows. In: 2019 IEEE/ACM Innovating the Network for Data-Intensive Science (INDIS), pp. 61–68, November 2019. https://doi.org/10.1109/INDIS49552.2019.00012
23. Vishwanath, K.V., Vahdat, A.: Evaluating distributed systems: does background traffic matter? In: USENIX 2008 Annual Technical Conference, pp. 227–240. ATC 2008, USENIX Association, Berkeley, CA, USA (2008). http://dl.acm.org/citation.cfm?id=1404014.1404031
24. Winstein, K., Balakrishnan, H.: TCP ex machina: computer-generated congestion control. In: Proceedings of the ACM SIGCOMM 2013 Conference on SIGCOMM, pp. 123–134. SIGCOMM 2013, ACM, New York, NY, USA (2013). https://doi.org/10.1145/2486001.2486020
25. Yin, Q., Kaur, J.: Can machine learning benefit bandwidth estimation at ultra-high speeds? In: Karagiannis, T., Dimitropoulos, X. (eds.) PAM 2016. LNCS, vol. 9631, pp. 397–411. Springer, Cham (2016). https://doi.org/10.1007/978-3-319-30505-9_30

A Novel Metric to Evaluate In Situ Workflows

Tu Mai Anh Do[1]([✉]), Loïc Pottier[1], Stephen Thomas[2],
Rafael Ferreira da Silva[1], Michel A. Cuendet[3], Harel Weinstein[3],
Trilce Estrada[4], Michela Taufer[2], and Ewa Deelman[1]

[1] USC Information Sciences Institute, Marina Del Rey, CA, USA
{tudo,lpottier,rafsilva,deelman}@isi.edu
[2] University of Tennessee at Knoxville, Knoxville, TN, USA
{sthoma99,mtaufer}@utk.edu
[3] Weill Cornell Medicine, Cornell University, New York, NY, USA
{mac2109,haw2002}@med.cornell.edu
[4] University of New Mexico, Albuquerque, NM, USA
estrada@cs.unm.edu

Abstract. Performance evaluation is crucial to understanding the behavior of scientific workflows and efficiently utilizing resources on high-performance computing architectures. In this study, we target an emerging type of workflow, called *in situ* workflows. Through an analysis of the state-of-the-art research on *in situ* workflows, we model a theoretical framework that helps characterize such workflows. We further propose a lightweight metric for assessing resource usage efficiency of an *in situ* workflow execution. By applying this metric to a simple, yet representative, synthetic workflow, we explore two possible scenarios (Idle Simulation and Idle Analyzer) for the execution of real *in situ* workflows. Experimental results show that there is no substantial difference in the performance of both the *in transit* placement (analytics on dedicated nodes) and the *helper-core* configuration (analytics co-allocated with simulation) on our target system.

Keywords: Scientific workflow · In situ model · Molecular dynamics · High-performance computing

1 Introduction

High performance computing (HPC) is mainstream for enabling the execution of scientific workflows which are composed of complex executions of computational tasks and the constantly growing data movements between those tasks [14]. Traditionally, a workflow describes multiple computational tasks and represents data and control flow dependencies. Moreover, the data produced by the scientific simulation are stored in persistent storage and visualizations or analytics are performed post hoc. This approach is not scalable, mainly due to the fact that scientific workflows are becoming increasingly compute- and data-intensive at

© Springer Nature Switzerland AG 2020
V. V. Krzhizhanovskaya et al. (Eds.): ICCS 2020, LNCS 12137, pp. 538–553, 2020.
https://doi.org/10.1007/978-3-030-50371-0_40

extreme-scale [13]. Storage bandwidth has also failed to keep pace with the rapid computational growth of modern processors due to the stagnancy of I/O advancements [7,16]. This asymmetry in I/O and computing technologies, which is being observed in contemporary and emerging computing platforms, prevents post hoc processing from handling large volumes of data generated by large-scale simulations [2]. Therefore, storing the entire output of scientific simulations on disk causes major bottlenecks in workflow performance. To reduce this disparity, scientists have moved towards a new paradigm for scientific simulations called *in situ*, in which data is visualized and/or analyzed as it is generated [2]. This accelerates simulation I/O by bypassing the file system, pipelining the analysis, and improving the overall workflow performance [4].

An *in situ* workflow describes a scientific workflow with multiple components (simulations with different parameters, visualization, analytics, etc.) running concurrently [4,16], potentially coordinating their executions using the same allocated resources to minimize the cost of data movement [2]. Data periodically produced by the main simulation are processed, analyzed, and visualized at runtime rather than post-processed on dedicated nodes. This approach offers many advantages for processing large volumes of simulated data and efficiently utilizing computing resources. By co-locating the simulation and the analysis kernels, *in situ* solutions reduce the global I/O pressure and the data footprint in the system [7]. To fully benefit from these solutions, the simulation and the analysis components have to be effectively managed so that they do not slow each other down. Therefore, in this paper we study and characterize two specific categories of *in situ* workflows, namely helper-core and *in transit* workflows [2]. Specifically, in the *helper-core* workflow, the analysis component is placed on the same node as the simulation; while in the *in transit* workflow, simulation data is staged to a dedicated node where the analysis is allocated. Workflows are required to capture the individual behavior of multiple coupled workflow components (i.e., concurrent executions of overlapped steps with inter-component data dependency). Throughout the use of a theoretical framework, this paper aims to provide guidelines for the evaluation and characterization of *in situ* workflows. We target a widely-used class of workflows, namely large-scale molecular dynamics (MD) simulations. We argue that the proposed solutions and the lessons learned from the proposed synthetic *in situ* workflows can be directly translated into production *in situ* workflows. This work makes the following contributions:

1. We discuss practical challenges in evaluating next-generation workflows;
2. We define a non-exhaustive list of imperative metrics that need to be monitored for aiding the characterization of *in situ* workflows. We model a framework for *in situ* execution to formalize the iterative patterns in *in situ* workflows—we develop a lightweight approach that is beneficial when comparing the performance of configuration variations in an *in situ* system; and
3. We provide insights into the behaviors of *in situ* workflows by applying the proposed metric in characterizing an MD workflow.

2   Background and Related Work

In Situ Workflows Monitoring. Many monitoring and performance profiling tools for HPC applications have been developed over the past decade [5,12]. With the advent of *in situ* workflows [3], new monitoring and profiling approaches targeting tightly-coupled workflows have been studied. LDMS [1] is a loosely-integrated scalable monitoring infrastructure that targets general large-scale applications that delivers a low-overhead distributed solution, in contrast to TAU [12], which provides a deeper understanding of the application at a higher computational cost. SOS [17] provides a distributed monitoring platform, conceptually similar to LDMS but specifically designed for online *in situ* characterization of HPC applications. ADIOS [9], the next-generation IO-stack, is built on top of many *in situ* data transport layers, e.g. DataSpaces [6]. Savannah [7], a workflow orchestrator, has been leveraged to bundle a coupled simulation with two main simulations, multiple analysis kernels, and a visualization service [4]. These works mainly focus on providing monitoring schemes for *in situ* workflows. This paper instead proposes a novel method to extract useful knowledge from the captured performance data.

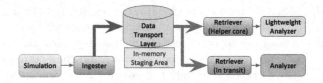

Fig. 1. A general *in situ* workflow software architecture.

In Situ Data Management. FlexAnalytics [19] optimizes the performance of coupling simulations with *in situ* analytics by evaluating data compression and query over different I/O paths: memory-to-memory and memory-to-storage. A large-scale data staging implementation [18] over MPI-IO operations describes a way to couple with *in situ* analysis using a non-intrusive approach. The analytics accesses data staged to the local persistent storage of compute nodes to enhance data locality. Our work mainly focuses on in-memory staging and comprehensively characterizes memory-to-memory transfer using RDMA for both within a compute node (*helper-core*) and across nodes (*in transit*).

3   General *In Situ* Workflow Architecture

In Situ Architecture. In this work, we propose an *in situ* architecture that enables a variety of *in situ* placements to characterize the behavior of *in situ* couplings. Although we focus on a particular type of *in situ* workflows (composed of simulation and data analytics), our approach is broader and applicable to a variety of *in situ* components, for example, several simulations coupled together. The *in situ* workflow architecture (Fig. 1) features three main components:

Table 1. Selected metrics for *in situ* workflows characterization

Name	Definition	Unit
MAKESPAN	Total workflow execution time	s
TIMESIMULATION	Total time spent in the simulation	s
TIMEANALYTICS	Total time spent in the analysis	s
TIMEDTL	Total time spent in data transfers	s
TIMESIMULATIONIDLE	Idle time during simulation	s
TIMEANALYTICSIDLE	Idle time during analysis	s

- A *simulation* component that performs MD computations and periodically generates data in the form of atomic coordinates.
- A data transport layer (DTL) that is responsible for efficient data transfer.
- An *analyzer* component that applies several analysis kernels to the data received periodically from the simulation component via the DTL.

On the data path "simulation-to-analyzer" (1) the *ingester* ingests data from a certain data source and stores them in the Data Transport Layer (DTL) and the (2) *retriever*, in a reverse way, gets data from the DTL to perform further operations. These two entry points allow us to abstract and detach complex I/O management from the application code. This approach enables more control in terms of *in situ* coupling and is less intrusive than many current approaches. The ingester synchronously inputs data from the simulation by sequentially taking turns with the simulation using the same resource. The ingester is useful to attach simple tasks to preprocess data (e.g., data reduction, data compression). The architecture allows *in situ* execution with various placements of the retriever. A helper-core retriever co-locates with the ingester on a subset of cores where the simulation is running—it asynchronously gets data from the DTL to perform an analysis. As the retriever is using the *helper-core* placement, the analysis should be lightweight to prevent simulation slowdown. An in transit retriever runs on dedicated resources (e.g., staging I/O nodes [19]), receives data from the DTL and performs compute-intensive analysis tasks. We compare *helper-core* and *in transit* retrievers in detail in Sect. 6.

In Situ Workflow Metrics. To characterize *in situ* workflows, we have defined a foundational set of metrics (Table 1). As a first metric, it is natural to consider the makespan, which is defined as three metrics corresponding to time spent in each component: the simulation, the analyzer, and the DTL. The periodic pattern enacted by *in situ* workflows may impose data dependencies between steps of coupled components, e.g. the analyzer may have to wait for data sent by the simulation to become available in the DTL for reading. Thus, we monitor the idle time of each individual component. In this work, we use TAU to capture this information and we focus on how to use these data to characterize the *in situ* workflows.

4   *In Situ* Execution Model

4.1   Framework

In traditional workflows, the simulation and the post-processing analyzer are typical components, in which the post-processing follows the simulation in a sequential manner. Let a *stage* be a part of a given component. The simulation component (S) is the computational step that produces the data and (W) is the I/O stage that writes the produced data; The analytics component (R) is the DTL stage that reads the data previously written and (A) is the analysis stage.

However, *in situ* workflows exhibit a periodic behavior: S, W, R, and A follow the same sequential order but, instead of operating on all the data, they operate only on a subset of it iteratively. Here, this subset of data is called a *frame* and can be seen as a snapshot of the simulation at a given time t. Let S_i, W_i, R_i, and A_i be respectively, the simulation, the write, the read and the analysis stage at step i, respectively. In other words, S_i produces the frame i, W_i writes the produced frame i into a buffer, R_i reads the frame i, and A_i analyzes the frame i. Note that, an actual simulation computes for a given number of steps, but only a subset of these steps are outputted as frames and analyzed [15]. The frequency of simulation steps to be analyzed is defined by *stride*. Let n be the total number of simulation steps, S the stride, and m the number of steps actually analyzed and outputted as frames. We have $m = \frac{n}{S}$. However, we model the *in situ* workflow itself and not only the simulation time (i.e., execution times of S_i, W_i, R_i, and A_i are known beforehand), thus the value of the stride does not impact our model. We set $S = 1$ (or $m = n$), so every step always produces a frame.

Execution Constraints. To ensure work conservation we define the following constraint: $\sum_{i=0}^{m} S_i = S$ (m is the number of produced frames). Obviously, we have identical constraints for R_i, A_i, and W_i. Similarly to the classic approach, we have the following precedence constraints for all i with $0 \leq i \leq m$:

$$S_i \rightarrow W_i \rightarrow R_i \rightarrow A_i. \tag{1}$$

Buffer Constraints. The pipeline design of *in situ* workflows introduces new constraints. We consider a frame is analyzed right after it has been computed. This implies that for any given step i, the stage W_{i+1} can start if, and only if, the stage R_i has been completed. Formally, for all i with $0 \leq i \leq m$:

$$R_i \rightarrow W_{i+1}. \tag{2}$$

Equations 1 and 2 guarantee that we buffer at most one frame at a time (Fig. 2). Note that, this constraint can be relaxed such that up to k frames can be buffered at the same time as follows, $R_i \rightarrow W_{i+k}$, where $0 \leq i \leq m$ and $1 \leq k \leq m$. In this work, we only consider the case $k = 1$ (red arrows in Fig. 3).

Idle Stages. Due to the above constraints, the different stages are tightly-coupled (i.e, R_i and A_i stages must wait S_i and W_i before starting their executions). Therefore, idle periods could arise during the execution (i.e., either the simulation or the analytics must wait for the other component). We can characterize two different scenarios, Idle Simulation and Idle Analysis in which idle time occurs. The former (Fig. 2(a)) occurs when analyzing a frame takes longer to complete compared to a simulation cycle (i.e., $S_i + W_i > R_i + A_i$). The later (Fig. 2(b)) occurs when the simulation component takes longer to execute (i.e., $S_i + W_i < R_i + A_i$). Figure 3 provides a detailed overview of the dependencies among the different stages. Note that, the concept of *in situ* step is defined and explained later in the paper.

Intuitively, we want to minimize the idle time on both sides. If the idle time is absent, then it means that we reach the idle-free scenario: $S_i + W_i = R_i + A_i$. To ease the characterization of these idle periods, we introduce two idle stages, one per component. Let I_i^S and I_i^A be, respectively, the idle time occurring in the simulation and in the analysis component for the step i. These two stages

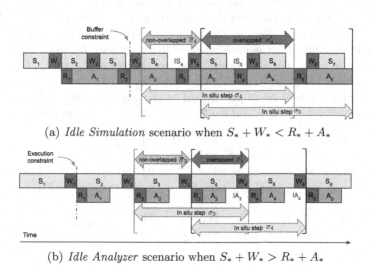

(a) *Idle Simulation* scenario when $S_* + W_* < R_* + A_*$

(b) *Idle Analyzer* scenario when $S_* + W_* > R_* + A_*$

Fig. 2. Two different execution scenarios for *in situ* workflow execution.

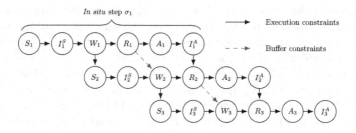

Fig. 3. Dependency constraints within and across *in situ* steps. (Color figure online)

represent the idle time in both components, therefore the precedence constraint defined in Eq. 1 results in:

$$S_i \to I_i^S \to W_i \to R_i \to A_i \to I_i^A. \tag{3}$$

4.2   Consistency Across Steps

This work is supported by the hypothesis that every execution of *in situ* workflows under the above constraints will reach a consistent state after a finite number of warming-up steps. Thus, the time spent on each stage within an iteration can be considered constant over iterations. Formally, there exists j where $0 \leq j < m$ such that for all i where $j \leq i \leq m$, we have $S_i = S_j$. The same holds for each stage, W_i, R_i, A_i, I_i^S, and, I_i^A. This hypothesis is confirmed in Sect. 6, and in practice, we observe that the cost of these non-consistent steps is negligible. Our experiments showed that, on average, $j \leq 3$ for one hundred steps ($m = 100$). Therefore, we ignore the warming steps and we consider $j = 0$. For the sake of simplicity, we generalize *in situ* consistency behavior by denoting $S_* = S_i$ for all $i \geq j$. We also have similar notations for R_*, A_*, I_*^S, I_*^A and W_*. This hypothesis allows us to predict the performance of a given *in situ* workflow by monitoring a subset of steps, instead of the whole workflow. From the two constraints defined by Eq. 3 and Eq. 2, and our hypothesis, we define:

$$S_* + I_*^S + W_* = R_* + A_* + I_*^A. \tag{4}$$

The Idle Simulation scenario is when $I_*^A = 0$, and $I_*^S = 0$ for Idle Analyzer scenario. Let I_* be the total idle time for an *in situ* step, using Eq. 4 we derive:

$$I_* = I_*^S + I_*^A = \begin{cases} R_* + A_* - S_* - W_*, & \text{if } I_*^A = 0 \\ S_* + W_* - R_* - A_*, & \text{if } I_*^S = 0 \end{cases} = |S_* + W_* - (R_* + A_*)|. \tag{5}$$

4.3   *In Situ* Step

The challenge behind *in situ* workflows evaluation lies in collecting global information from multiple components (in our case, the simulation and the analytics) and use this information to derivate meaningful characteristics about the execution. The complexity of such a task is correlated to the number of steps the workflow is running and the number of components involved. By leveraging the consistency hypothesis in Eq. 4, we propose to alleviate this cost by proposing a metric that does not require data from all steps. The keystone of our approach is the concept of *in situ* step. Based on Eq. 3, the *in situ* step σ_i is determined by the timespan between the beginning of S_i and the end of I_i^A. The *in situ* step concept helps us to manipulate all the stages of a given step as one consistent task executing across components that can potentially run on different machines.

Different *in situ* steps overlap each other, so we need to distinguish the part that is overlapped (σ_i') from the other part ($\overline{\sigma}_i$). Thus, $\sigma_i = \overline{\sigma}_i + \sigma_i'$. For example, in Fig. 2(a), to compute the time elapsed between the start of σ_4 and the end of σ_5, we need to sum the two steps and remove the overlapped execution time σ_4'. Thus, we obtain $\sigma_4 + \sigma_5 - \sigma_4'$. This simple example will give us the intuition behind the makespan computation in Sect. 4.4.

The consistency hypothesis insures consistency across *in situ* steps. We denote σ_* as the consistent *in situ* step (i.e, $\forall i, \sigma_i = \sigma_*$), while σ_*' and $\overline{\sigma}_*$ indicate, respectively, the overlapped and the non-overlapped part of two consecutive *in situ* steps. Thus, $\sigma_* = \overline{\sigma}_* + \sigma_*'$. To calculate the makespan, we want to compute the non-overlapped step $\overline{\sigma}_*$. As shown in Fig. 2, the non-overlapped period $\overline{\sigma}_*$ is the aggregation of all stages belonging to one single component in an *in situ* step: $S_* + I_*^S + W_*$ and $R_* + A_* + I_*^A$. Thus, we have two scenarios, if $I_*^S = 0$ then $\overline{\sigma}_* = S_* + W_*$, otherwise $\overline{\sigma}_* = R_* + A_*$. Hence, $\overline{\sigma}_* = \max(S_* + W_*, R_* + A_*)$.

4.4 Makespan

A rough estimation of the makespan of such workflow would be the sum of the execution time for all the stages (i.e, sum up the m *in situ* steps σ_i). But, recall that *in situ* steps interleave with each other, so we need to subtract the overlapped parts:

$$\text{MAKESPAN} = m\,\sigma_* - \sigma_*'\,(m-1) = m\,\overline{\sigma}_* + \sigma_*'. \tag{6}$$

From Eq. 6, for m large enough, the term σ_*' becomes negligible. Since *in situ* workflows are executed with a large number of iterations, then $\text{MAKESPAN} = m\,\overline{\sigma}_*$. This observation indicates that the non-overlapped part of an *in situ* step is enough to characterize a periodic *in situ* workflow. Using our framework and these observations, we define a metric to estimate the efficiency of a workflow.

4.5 *In Situ* Efficiency

Based on our *in situ* execution model, we propose a novel metric to evaluate resource usage efficiency E of an *in situ* workflow. We define efficiency as the time wasted during execution—i.e., idle times I_*^S and I_*^A. This metric considers all the components (simulation and the analysis) for evaluating in situ workflows:

$$E = 1 - \frac{I_*}{\overline{\sigma}_*} = 1 - \frac{|S_* + W_* - (R_* + A_*)|}{\max(S_* + W_*, R_* + A_*)}. \tag{7}$$

This efficiency metric allows for performance comparison between different *in situ* runs with different configurations. By examining only one non-overlapped *in situ* step, we provide a lightweight approach to observe behavior from multiple components running concurrently in an *in situ* workflow. Note that, this model and the efficiency metric can be easily generalized to any number of components.

5 Molecular Dynamics Synthetic Workflow

MD is one of the most popular scientific applications executing on modern HPC systems. MD simulations reproduce the time evolution of molecular systems at a given temperature and pressure by iteratively computing inter-atomic forces and moving atoms over a short time step. The resulting trajectories allow scientists to understand molecular mechanisms and conformations. In particular, a trajectory is a series of *frames*, i.e. sets of atomic positions saved at fixed intervals of time. The *stride* is the number of time steps between frames considered for storage or further *in situ* analysis. For example in our framework, for a simulation with 100 steps and a stride of 20, only 5 frames will be sent by the simulation to the analysis component. Since trajectories are high-dimensional objects, scientists use well-chosen collective variables (CVs) to capture important molecular motions. Technically, a CV is defined as a function of the atomic coordinates in one frame. An example of a complex CV that we will use in this work is the Largest Eigenvalue of the Bipartite Matrix (LEBM). Given two amino acid segments A and B, if d_{ij} is the Euclidean distance between C_α atoms i and j, then the symmetric bipartite matrix $B_{AB} = [b_{ij}]$ is defined by $b_{ij} = d_{ij}$ if $i \in A, j \in B$ or $i \in B, j \in A$ and 0 otherwise. Johnston et al. [8] showed that the largest eigenvalue of B_{AB} is an efficient proxy to monitor changes in the conformation of A relative to B. To study complex behavior of coupling between the MD simulation and the analysis component in exhaustively discussed parameter space, we have designed a synthetic *in situ* MD workflow (Fig. 4). The Synthetic Simulation component extracts frames from previously computed MD trajectories instead of performing an actual, compute-intensive MD simulation. This synthetic workflow is an implementation of the general and abstract software architecture proposed in Sect. 3.

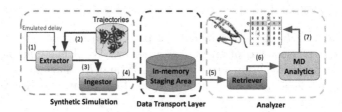

Fig. 4. Synthetic Workflow: the Extractor (1) sleeps during the emulated delay, then (2) extracts a snapshot of atomic states from existing trajectories and (3) stores it into a synthetic frame. The Ingestor (4) serializes the frame as a chunk and stages it in memory, then the Retriever (5) gets the chunk from the DTL and deserializes it into a frame.

Synthetic Simulation. The *Synthetic Simulation* emulates the process of a real MD simulation by extracting frames from trajectories generated previously by an MD simulation engine. The Synthetic Simulation enables us to tune and manage many simulation parameters (discussed in detail in Sect. 6) including the number of atoms and strides, which helps the Synthetic Simulation mimic the behavior of real molecular dynamics simulation. Note that, since the Synthetic Simulation does not emulate the computation part of the real MD simulation, it mimics the behavior of the I/O processes of the simulation. Thus, we define the *emulated simulation delay*, which is the period of time corresponding to the computation time in the real MD simulation. In order to estimate such delay for emulating the simulation time for a given stride and number of atoms, we use recent benchmarking results from the literature obtained by running the well-known NAMD [10] and Gromacs [11] MD engines. We considered the benchmarking performance for five practical system sizes of 81K, 2M, 12M atoms from Gromacs [11] and 21M, 224M atoms from NAMD [10] to interpolate to the simulation performance with the desired number of atoms (Fig. 5(a)). The interpolated value is then multiplied by the stride to obtain the delay (i.e., a function of both the number of atoms and the stride).

In Sect. 6, we run the synthetic workflow with 200K, 400K, 800K, 1.6M, 3.2M, and 6.4M protein atoms. Figure 5(b) shows the emulated simulation delay when varying the stride for different numbers of atoms. The stride varying between 4–16K delivers a wide range of emulated simulation delay, up to 40 s.

Data Transport Layer (DTL). The *DTL Server* leverages DataSpaces [6] to deploy an in memory staging area for coupling data between the Synthetic Simulation and the Analyzer. DataSpaces follows the publish/subscribe model in terms of data flow, and the client/server paradigm in terms of control flow. The workflow system has to manage a DataSpaces server to manage data requests, keep metadata, and create in memory data objects.

Analyzer. The *Analyzer* plays the role of the analytics component in the synthetic *in situ* workflow. More specifically, the *Retriever* subscribes to a chunk from the in memory staging area and deserializes it into a frame. The MD

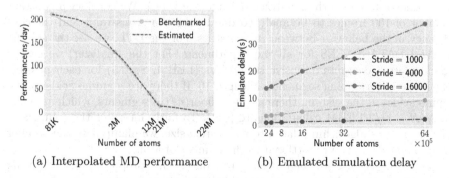

(a) Interpolated MD performance   (b) Emulated simulation delay

Fig. 5. MD benchmarking results from the literature obtained by using 512 NVIDIA K20X GPUs. The results are interpolated to obtain the (a) estimated performance and then combined with the stride to synthesize the (b) emulated simulation delay.

Analytics then performs a given type of analysis on this frame. Recall that, in our model, only one frame at a time can be store by the DTL (see Fig. 3). We leverage DataSpaces built-in locks to ensure that a writing operation to the in memory staging area can only happen when the reading operation of the previous step is complete (constraint model by Eq. 2). Thus, the Analyzer is instructed by DataSpaces to wait for the next chunk available in the in memory staging area. Once a chunk has been received and is being processed, the Synthetic Simulation can send another chunk to the Analyzer.

6   Experiments and Discussions

For our experiments we use Tellico (UTK), an IBM POWER9 system that includes four 32-core nodes (2 compute nodes) with 256GB RAM each. Each compute node is connected through an InfiniBand interconnect network. We target three component placements: (1) helper-core—where the Synthetic Simulation, the DataSpaces server, and the Analyzer are co-located on the same compute node; (2) *in transit S-S* where the Synthetic Simulation and the DataSpaces server are co-located on one node, and the Analyzer runs on another node; and (3) *in transit A-S* where the Analyzer and the DataSpaces server are co-located on one node, and the Synthetic Simulation runs on a dedicated node. Note that the Synthetic Simulation only runs on one physical core as it mimics the behavior of a real simulation. On the other hand, the Analyzer assigns bipartite matrices (see Sect. 5) to multiple processes, so the CV calculation is improved by parallel processing. Since we experimented to designate different numbers of Analyzer processes, we fix that number at 16 processes (number of cores of an IBM AC922) to attain good speed up and to fit the entire Analyzer in a compute node.

Table 2 describes the parameters used in the experiments. For the Synthetic Simulation, we study the impact of the number of atoms (the size of the system) and the stride (the frequency at which the Synthetic Simulation component sends a frame to the Analyzer through the DTL). We consider a constant number of 100 frames to be analyzed due to the time constraint and the consistency in the behavior between *in situ* steps. For the DTL, we use the staging method DATASPACES for all the experiments. For the Analyzer, we choose to calculate a compute-intensive set of CVs (LEBM, Sect. 5) for each possible pair of non-overlapping segments of length 16. If there are n amino acids (alpha amino acids) in the system, there are $N = floor(n/16)$ segments, which amounts to $N(N-1)/2$ LEBM calculations ($\mathcal{O}(n^2)$). To fairly interpret the complexity of this analysis algorithm related to the system size, we manipulate the number of amino acids to be proportional to the number of atoms.

We leverage user-defined events to collect the proposed metrics using TAU [12] (Sect. 3). We focus on two different levels of information, the workflow level and the *in situ* step level (the time taken by the workflow to compute and analyze one frame). At the *in situ* step level, each value is averaged over three runs for each step. At the workflow level, each value is averaged over all

Table 2. Parameters used in the experiments

	Parameter	Description	Values used in the experiments
Synthetic simulation	#atoms	Number of atoms	$[2 \times 10^5, 4 \times 10^5, 8 \times 10^5,$ $16 \times 10^5, 32 \times 10^5, 64 \times 10^5]$
	#strides	Stride	$[1000, 4000, 16000]$
	#frames	Number of frames	100
Data transport layer	SM	Staging method	DATASPACES
Analyzer	CV	Collective variable	LEBM
	lsegment	Length of segment pairs	16

in situ steps at steady-sate and then averaged over the three runs. We also depict the standard deviation to assess the statistical significance of the results. There are two levels of statistical error: for averages across the 3 trials at the *in situ* step level, and for averages over 94 *in situ* steps (excluding the three first steps and the three last steps) in each run at the workflow level.

6.1   Experimental Results

In Situ *Step Characterization*. We study the correlation of individual stages in each *in situ* step. Due to lack of space, the discussion is limited to a subset of the parameter space as the representative of two characterized idle-based scenarios. Figure 6 shows the execution time per step for each component while varying the stride. Confirming the consistency hypothesis across steps discussed in Sect. 4.2, we observe that the execution time per step is nearly constant, except for a few warm-up and wrap-up steps. Figure 6(a) falls under the *Idle Simulation* (IS) scenario, as the I_i^S stage only appears in the Synthetic Simulation step. Similarly in Fig. 6(b), we observe the *Idle Analyzer* (IA) scenario because of the presence of I_i^A. These findings verify the existence of two idle-based scenarios discussed in Sect. 4.1. Since both the Synthetic Simulation and Analyzer are nearly synchronized, we also underline that the execution time of a single step for each component is equal to each other. This information confirms the property of *in situ* workflow in Eq. 4. Overall, we can observe that the I/O stages (W_i and R_i) take an insignificant portion of time compared to the full step. This negligible overhead verifies the advantage of leveraging in-memory staging for exchanging frames between coupled components.

***Execution Scenarios*.** We study the impact of the number of atoms, stride, and components placement on the performance of the entire workflow and each component for different scenarios. Figure 7 shows that the workflow execution follows our model (Fig. 2). While increasing the number of atoms, which increases the simulation time and the chunk size, the total idle time I_* decreases in the IS

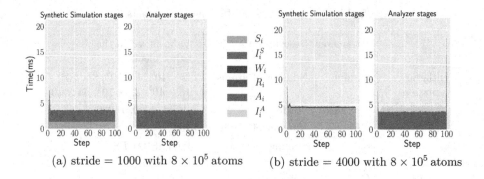

(a) stride = 1000 with 8×10^5 atoms   (b) stride = 4000 with 8×10^5 atoms

Fig. 6. Execution time per step for each component with a *helper-core* placement (all component on the same node). The Synthetic Simulation stages are on the left and the Analyzer stages are on the right (lower is better).

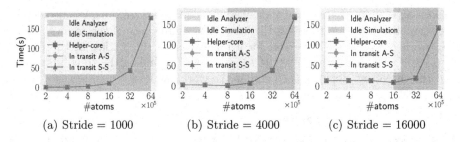

(a) Stride = 1000   (b) Stride = 4000   (c) Stride = 16000

Fig. 7. Detailed idle time I_* for three component placements at different strides when varying the number of atoms (lower is better).

scenario, and increases in the IA scenario. Every *in situ* step exhibits a similar pattern in which at a certain system size the workflow execution switches from one scenario to another. We denote this point as the equilibrium point. We notice that with larger stride, the equilibrium point occurs at larger system sizes. The equilibrium point of the stride $4,000$ occurs at #atoms = 8×10^5, but at #atoms = 16×10^5 with the stride of $16,000$. In terms of component placement comparison, at first glance, there is no big difference in total idle time of the three placements (see Sect. 6.1).

Estimated Makespan. The goal is to verify the assertion made by Eq. 6 stating the makespan of an *in situ* workflow can simply be expressed as the product of the number of steps and the time of one step $m\overline{\sigma}_*$. A typical MD simulation can easily feature $>10^7$ number of *in situ* steps, thus a metric requiring only a few steps to be accurate is interesting. Figure 8 demonstrates the strength of our approach to estimate the MAKESPAN (maximum error ~5%) using *in situ* steps and the accuracy of our model. *In situ* workflows run with a larger number of steps, monitoring the entire system increases the pressure and slows down the execution. Thus, without failures and external loads, only looking at a single non-overlapped step results in a scalable, accurate, and lightweight approach.

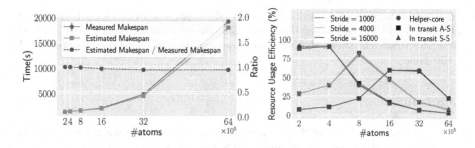

Fig. 8. *Left:* MAKESPAN is estimated from $100\,\overline{\sigma}_*$ using the *helper-core* component placement with stride 16000, the yellow region represents the error. Ratio of *Estimated* MAKESPAN to *Measured* MAKESPAN uses the second y-axes on the right (close to 1 is better). *Right:* Resource usage efficiency (higher is better). (Color figure online)

Resource Usage Efficiency. We utilize the efficiency metric E given by Eq. 7, to evaluate an *in situ* configuration within the objective to propose a metric that allows users to characterize *in situ* workflows. Figure 8-*right* shows that component placements have a small variation in resource efficiency. We infer that the placement of components is not a decisive factor in coupling performance. This result is consistent with the previous finding in Sect. 6.1. The efficiency E increases and reach a maximum in the IS scenario, and decreases after this maximum in the IA scenario. Thus, an *in situ* run is most efficient at the equilibrium point, where $E \approx 1$. If a run is less efficient and classified as the IA scenario, it has more freedom to perform other analyses or increase the analysis algorithm's complexity. In the IS scenario, the simulation is affordable to larger system size.

7 Conclusions

In this study, we explored the challenges of evaluating next-generation *in situ* workflows. We have provided an analysis of *in situ* workflows by identifying a set of metrics that should be monitored to assess the performance of these workflows on HPC architectures. We have designed a lightweight metric based on behavior consistency across *in situ* steps under constrained *in situ* execution model. We have validated the usefulness of this proposed metric with a set of experiments using an *in situ* MD synthetic workflow. We have compared three different placements for the workflow components, a *helper-core* placement and two *in transit* placements in which the DTL server co-locates with different components. Future work will study different models where the constraints are relaxed, for example where the workflow allows to buffer multiple frames in memory instead of one currently. Another promising research line is to extend our theoretical framework to take into account multiple analysis methods. In this case, the time taken by the analysis could vary regarding the method used.

Acknowledgments. This work is funded by NSF contracts #1741040, #1740990 and #1741057; and DOE contract #DE-SC0012636. We are grateful to IBM for the Shared University Research Award that supported the purchase of IBM Power9 system used in this paper.

References

1. Agelastos, A., et al.: LDMS: a scalable infrastructure for continuous monitoring of large scale computing systems and applications. In: SC 2014 (2014)
2. ASCR Workshop on In Situ Data Management (2019)
3. Bauer, A.C., et al.: In situ methods, infrastructures, and applications on high performance computing platforms. In: Computer Graphics Forum, vol. 35 (2016)
4. Choi, J.Y., et al.: Coupling exascale multiphysics applications: methods and lessons learned. In: 2018 IEEE 14th International Conference on e-Science (e-Science) (2018)
5. DeRose, L., Homer, B., Johnson, D., Kaufmann, S., Poxon, H.: Cray performance analysis tools. In: Resch, M., Keller, R., Himmler, V., Krammer, B., Schulz, A. (eds.) Tools for High Performance Computing, pp. 191–199. Springer, Heidelberg (2008). https://doi.org/10.1007/978-3-540-68564-7_12
6. Docan, C., et al.: DataSpaces: an interaction and coordination framework for coupled simulation workflows. Cluster Comput. **15**, 163–181 (2012). https://doi.org/10.1007/s10586-011-0162-y
7. Foster, I., et al.: Computing just what you need: online data analysis and reduction at extreme scales. In: Rivera, F.F., Pena, T.F., Cabaleiro, J.C. (eds.) Euro-Par 2017. LNCS, vol. 10417, pp. 3–19. Springer, Cham (2017). https://doi.org/10.1007/978-3-319-64203-1_1
8. Johnston, T., et al.: In situ data analytics and indexing of protein trajectories. J. Comput. Chem. **38**, 1419–1430 (2017)
9. Lofstead, J.F., et al.: Flexible IO and integration for scientific codes through the adaptable IO system (adios). In: 6th International Workshop on Challenges of Large Applications in Distributed Environments (2008)
10. NAMD performance. https://www.ks.uiuc.edu/Research/namd/benchmarks/
11. Páll, S., Abraham, M.J., Kutzner, C., Hess, B., Lindahl, E.: Tackling exascale software challenges in molecular dynamics simulations with GROMACS. In: Markidis, S., Laure, E. (eds.) EASC 2014. LNCS, vol. 8759, pp. 3–27. Springer, Cham (2015). https://doi.org/10.1007/978-3-319-15976-8_1
12. Shende, S.S., Malony, A.D.: The Tau parallel performance system. Int. J. High Perform. Comput. Appl. **20**(2), 287–311 (2006)
13. da Silva, R.F., et al.: A characterization of workflow management systems for extreme-scale applications. Future Gener. Comput. Syst. **75**, 228–238 (2017)
14. Taylor, I.J., Deelman, E., Gannon, G.B., Shields, M. (eds.): Workflows for e-Science: Scientific Workflows for Grids, vol. 1. Springer, London (2007). https://doi.org/10.1007/978-1-84628-757-2
15. Thomas, S., et al.: Characterizing in situ and in transit analytics of molecular dynamics simulations for next-generation supercomputers. In: 15th eScience (2019)
16. Vetter, J.S., et al.: Extreme heterogeneity 2018 - productive computational science in the era of extreme heterogeneity: report for DOE ASCR workshop on extreme heterogeneity. Technical report, LBNL, Berkeley, CA, USA, December 2018

17. Wood, C., et al.: A scalable observation system for introspection and in situ analytics. In: 2016 5th Workshop on Extreme-Scale Programming Tools (ESPT) (2016)
18. Wozniak, J.M., et al.: Big data staging with MPI-IO for interactive x-ray science. In: 2014 IEEE/ACM International Symposium on Big Data Computing (2014)
19. Zou, H., et al.: FlexAnalytics: a flexible data analytics framework for big data applications with I/O performance improvement. Big Data Res. **1**, 4–13 (2014)

Social Recommendation in Heterogeneous Evolving Relation Network

Bo Jiang[1], Zhigang Lu[1,2(✉)], Yuling Liu[1,2], Ning Li[1], and Zelin Cui[1]

[1] Institute of Information Engineering, Chinese Academy of Sciences, Beijing, China
{jiangbo,luzhigang,liuyuling,liujunrong}@iie.ac.cn
[2] School of Cyber Security, University of Chinese Academy of Sciences,
Beijing, China

Abstract. The appearance and growth of social networking brings an exponential growth of information. One of the main solutions proposed for this information overload problem are recommender systems, which provide personalized results. Most existing social recommendation approaches consider relation information to improve recommendation performance in the static context. However, relations are likely to evolve over time in the dynamic network. Therefore, temporal information is an essential ingredient to making social recommendation. In this paper, we propose a novel social recommendation model based on evolving relation network, named SoERec. The learned evolving relation network is a heterogeneous information network, where the strength of relation between users is a sum of the influence of all historical events. We incorporate temporally evolving relations into the recommendation algorithm. We empirically evaluate the proposed method on two widely-used datasets. Experimental results show that the proposed model outperforms the state-of-the-art social recommendation methods.

Keywords: Social recommendation · Dynamic evolving · Relation network · Network embedding

1 Introduction

The last decades have witnessed the booming of social networking such as Twitter and Facebook. User-generated content such as text, images, and videos has been posted by users on these platforms. Social users is suffering from information overload. Fortunately, recommender systems provide a useful tool, which not only help users to select the relevant part of online information, but also discovery user preference and promote popular item, etc. Among existing techniques, collaborative filtering (CF) is a representative model, which attempt to utilize the available user-item rating data to make predictions about the users preferences. These approaches can be divided into two groups [1]: memory-based and model-based. Memory-based approaches [2,8,16] make predictions based on the similarities between users or items, while model-based approaches [3,9]

V. V. Krzhizhanovskaya et al. (Eds.): ICCS 2020, LNCS 12137, pp. 554–567, 2020.
https://doi.org/10.1007/978-3-030-50371-0_41

design a prediction model from rating data by using machine learning. Both memory-based and model-based CF approaches have two challenges: data sparsity and cold start, which greatly reduce their performance. In particular, matrix factorization based models [13,19] have gained popularity in recent years due to their relatively high accuracy and personalized advice.

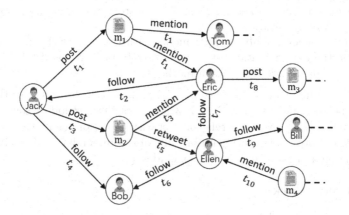

Fig. 1. An illustration of social connections that can change and evolve over time.

Existing research works have contributed improvements in social recommendation tasks. However, these approaches only consider static social contextual information. In the real world, knowledge is often time-labeled and will change significantly over time. Figure 1 shows the entire social contextual information over time which can be derived from links on social networks. User *Jack* post message m_1, which mention users *Tom* and *Eric*, at time point t_1. Subsequently, user *Jack* post message m_2, which mention user *Eric* again, at time point t_3. Meanwhile, message m_2 is retweeted by user *Ellen* at time point t_5. We observe that new social action is often influenced by historical related behaviors. In addition, historical behaviors have an impact on current action over time, and the impact strength decreases with time. On the other hand, we notice that the evolving relation network is very sparse, which greatly reduce the recommendation performance. In order to deal with data sparsity, we leverage network embedding technology, which has contributed improvements in many applications, such as link prediction, clustering, and visual.

In this work, we propose a novel social recommendation model based on evolving relation network, named SoERec, which leverages evolving relation network and network embedding technique. The proposed method explicitly models the strength of relations between pair of users learned from an evolving relation network. To efficiently learn heterogeneous relations, network embedding is employed to represent relation into a unified vector space. We conduct experiments on two widely-used datasets and the experimental results show that our proposed model outperforms the state-of-the-art recommendation methods.

The main contributions of this paper are as follows:

- We construct a dynamic, directed and weighted heterogeneous evolving network that contains multiple objects and links types from social network. Compared with static relation graph, the evolving graph can precisely measure the strength of relations.
- We propose a novel social recommendation model by jointly embedding representations of fine-grained relations from historical events based on heterogeneous evolving network.
- We conduct several analysis experiments with two real-world social network datasets, the experimental results demonstrate our proposed model outperforms state-of-the art comparison methods.

The rest of this paper is organized as follows. Section 2 formulates the problem of social recommendation. Section 3 proposes the method of social recommendation based on evolving relation network to recommend the candidate users. Section 4 presents experimental results of recommendation. Finally, Sect. 5 reviews the related work and Sect. 6 concludes.

2   Related Work

We briefly review the related works from two lines in this section: one on network embedding and the other on social recommendation.

Network Embedding. Network embedding has been extensively studied to learn a low-dimensional vector representation for each node, and implicitly capture the meaningful topological proximity, and reveal semantic relations among nodes in recent years. The early-stage studies only focus on the embedding representation learning of network structure [14,21,24,29]. Subsequently, network node incorporating the external information like the text content and label information can boost the quality of network embedding representation and improve the learning performance [4,7,20,22,23,28]. Network embedding indeed can alleviate the data sparsity and improve the performance of node learning successfully. Therefore, this technique has been effectively applied, such as link prediction, personalized recommendation and community discovery.

Social Recommendation. Recommender systems are used as an efficient tool for dealing with the information overload problem. Various methods of social recommendation have been proposed from different perspectives in recent years including user-item rating matrix [15], network structure [11], trust relationship [5,10,18,27], individual and friends' preferences [6,12], social information [25] and combinations of different features [19,26]. The above social recommendation methods are proposed based on collaborative filtering. These methods all focus on fitting the user-item rating matrix using low-rank approximations, and also use all kinds of social contextual information to make further predictions. Most of the studies that use both ratings and structure deal with static snapshots of networks, and they don't consider the dynamic changes occurring over users' relations. Incorporating temporally evolving relations into the analysis can offer useful insights about the changes in the recommendation task.

3   Problem Statement

The intuition behind is that there are two basic accepted observations in a real world: (1) The current behavior of user is influenced by all his/her historical patterns. (2) A behavior with an earlier generation time has a smaller influence on the user's current behavior, while the one with a later generation time has a greater influence. Therefore, we first formally define the concept of *Evolving Relation Network*, as follows:

Definition 1 (*Heterogeneous Evolving Network*). *A heterogeneous evolving network can be defined as $G = (N_u \cup N_i, E)$, where N_u is the set of vertices representing users, and N_i is the set of vertices representing items, and E is the set of edges between the vertices. The types of edges can be divided into user-user and user-item relationships with temporal information. Hence, G is a dynamic, directed and weighted heterogeneous evolving network.*

From the definition, we can see that each edge not only is an ordered pair from a node to another node, but also has a weight with time-dependent. In order to measure the strength of relations between two nodes objects in the heterogeneous evolving network G, we introduce the concept of *evolving strength*, which is formally defined as follows:

Definition 2 (*Evolving Strength*). *Given an event $e(\psi, t)$ where ψ is an event type (e.g., post, mention, follow, etc.) and t is the timestamp of e. An event sequence Γ between two nodes is a list of events $\{e_1, e_2, \cdots, e_n\}$, ordered by their timestamps $\{t_1, t_2, \cdots, t_n\}$, where $t_i \leq t_j (1 \leq i < j \leq n)$. An event corresponding to an edge. Thus, the strength of evolving relations denoted by F is the sum of individual event influence.*

We formulate the problem of *social recommendation* as a ranking based task in this work, as follows:

Definition 3 (*Social Recommendation Problem*). *Given a heterogeneous evolving network G at time t, and a target user u_i, and a candidate set of items Ψ, we aim to generate a top K ranked list of items $\Omega \in \Psi$ for u_i at time $t + 1$ according to the target user's preference inferred from historical feedbacks.*

4   The Proposed Social Recommendation Model

4.1   Probabilistic Matrix Factorization

Let $R \in \mathbb{R}^{M \times N}$ be the rating matrix with M users and N items. The (i, j)-th entry of the matrix is denoted by R_{ij} that represent the rating of user i for item j. $U \in \mathbb{R}^{K \times M}$ and $V \in \mathbb{R}^{K \times N}$ be user and item latent feature matrices respectively, where K is the dimension of latent factors. The preference of i-th user is represented by vector $U_i \in \mathbb{R}^{K \times 1}$ and the characteristic of j-th item is

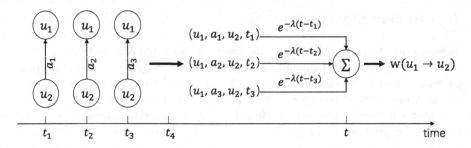

Fig. 2. The evolving strength of relation between pair of users over time.

represented by vector $V_j \in \mathbb{R}^{K \times 1}$. The dot product of U and V can approximate the rating: $\hat{R} \approx U^T V_j$. Recommendation based on Probabilistic Matrix Factorization (PMF) [15] solve the following problem

$$\min_{U,V} \sum_{i=1}^{M} \sum_{j=1}^{N} W_{ij}(R_{ij} - U_i^T V_j)^2 + \gamma(\|U\|_F^2 + \|V\|_F^2) \tag{1}$$

where $W \in \mathbb{R}^{M \times N}$ is a weight matrix. In this work, we set $W_{ij} = 1$ if $R_{ij} \neq 0$ and 0 otherwise. $\gamma > 0$ is the regularization parameter. $(\|U\|_F^2 + \|V\|_F^2)$ can avoid overfitting, $\|\cdot\|_F$ denotes the Frobenius norm of the matrix.

4.2   Modeling Relation Strength

Incorporating the knowledge from present and historical behavior data can accurately measure the strength of influence, as shown Fig. 2. In this work, we model the strength of relation between users as a sum of the influence of each event by multiplying a weight. The weight is calculated by a function, called *decay function*. Since the influence between users can't be less than zero in social networks, the weight ranges from 0 to 1 and decreases with the event's existing time. Thus, we formalize the *decay function* $d_{ij}(t)$ with timestamped information as follows:

$$d_{ij}(t) = e^{-\lambda(t - t_i)} \tag{2}$$

where t is the current time, t_i is the generation time of historical event, and λ is a parameter which controls the decay rate. Through the analyses in the following experiments in the paper, we set the parameter λ as 0.6.

Based on the influence of historical events, we can measure the current strength of social relation between users as follows:

$$w_{ij} = \sum_{t_i=0}^{T} d_{ij}(t_i) \cdot I_{e(\psi, t_i)} \tag{3}$$

where $I_{e(\psi, t_i)}$ is a parameter which controls the weight of different events. To simplify the model, we assume that the importance of any events is equal.

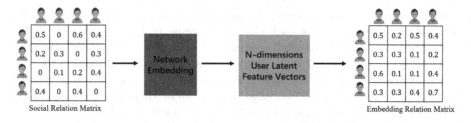

Fig. 3. The network embedding representation learning of user relation network.

4.3   Embedding Relation Network

The learned evolving relation network has three characteristics: (1) a weighted and directed graph; (2) a sparsity graph; (3) heterogeneous information network. In order to learn the evolving relation network, we employ large-scale information network embedding (LINE) [17] model to simultaneously retain the local and global structures of the network. In particular, we leverage the LINE model to learn users' embedded representations of the evolving relation network the first-order proximity and the second-order proximity. As shown Fig. 3, the detailed process is demonstrated as follows.

User Relation with First-Order Proximity. The first-order similarity can represent the relation by the directly connected edge between vertices. We model the joint probability distribution of users u_i and u_j as the first-order similarity $p_1(u_i, u_j)$. The similarity can be defined as follows:

$$p_1(u_i, u_j) = \frac{1}{1 + exp(-\overrightarrow{u}_i^T \overrightarrow{u}_j)} \tag{4}$$

where $\overrightarrow{u}_i \in \mathbb{R}^d$ is the low-dimensional vector representations of vertices u_i. The empirical distribution between vertices u_i and u_j is defined as follows:

$$\hat{p}_1(u_i, u_j) = \frac{w_{ij}}{W} \tag{5}$$

where $W = \sum_{(u_i,u_j) \in E} w_{ij}$, and w_{ij} is the relation strength of the edge (u_i, u_j) measured by Eq. (3). To preserve the first-order proximity in evolving relation network, we use the KL-divergence to minimize the joint probability distribution and the empirical probability distribution as follows:

$$O_1 = -\sum_{(u_i,u_j) \in E} w_{ij} log p_1(u_i, u_j) \tag{6}$$

User Relation with Second-Order Proximity. The second-order proximity assumes that vertices sharing many connections to other vertices are similar to each other. In this work, we assume that two users with similar neighbors have high similarity scores between them. Specifically, we consider each user vertex as a specific "ontext", and users with similar distributions over the "contexts" are assumed to be similar. Thus, each user vertex respectively plays two roles: the user vertex itself and the specific "context" of other user vertices. We introduce two vectors $\overrightarrow{u}_i$ and $\overrightarrow{u}'_i$, where $\overrightarrow{u}_i$ is the representation of u_i when it is treated as a vertex, and $\overrightarrow{u}'_i$ is the representation of u_i when it is treated as a specific "context". For each directed user edge (u_i, u_j), we firstly define the probability distribution of "context" u_j generated by user vertex u_i as follows:

$$p_2(u_j|u_i) = \frac{exp(\overrightarrow{u}'^T_j \overrightarrow{u}_i)}{\sum_{k=1}^{\mathcal{K}} exp(\overrightarrow{u}'^T_k \overrightarrow{u}_i)} \tag{7}$$

where $\mathcal{K}$ is the number of user vertices or "contexts". The empirical distribution of "contexts" u_j generated by user vertex u_i is defined as:

$$\hat{p}_2(u_j|u_i) = \frac{w_{ij}}{d_i} \tag{8}$$

where w_{ij} is the weight of the edge (u_i, u_j) as the same, and d_i is the out-degree of vertex u_i, i.e. $d_i = \sum_{k \in N(i)} w_{ik}$, with $N(i)$ as the set of out-neighbors of u_i.

To preserve the second-order user relation, the following objective function is obtained by utilizing the KL-divergence:

$$O_2 = - \sum_{(u_i, u_j) \in E} w_{ij} log p_2(u_j|u_i) \tag{9}$$

Combining First-Order and Second-Order Proximities. To embed the evolving network by preserving both the first-order and second-order proximities, LINE model can minimize the objective functions O_1 and O_2 respectively, and learns two low-dimensional representations for each user vertex. Then, the two low-dimensional representations are concatenated as one low-dimensional feature vector to simultaneously preserve the local and global structures of evolving relation network. Finally, each user vertex u_i is represented as $\overrightarrow{U} \in \mathbb{R}^{d_1+d_2}$.

4.4 Evolving Relation Embedding Recommendation Model

Incorporating simultaneously user's explicit relation and implicit relation can boost the ability of social recommendation. As mentioned above, LINE model can learn users' embedded representations, where first-order proximity correspond to the strength of explicit relation and second-order proximity correspond to the strength of implicit relation. Hence, the fine-grained relation measure can better predict user ratings by also encoding both the first-order and second-order relationships among users.

After performing the LINE model, we can obtain users' embedded presentations. We then measure the fine-grained relations among users on the basis of the inner product of the presentations as follows:

$$s_{ij} = \frac{\overrightarrow{u}^T_i \overrightarrow{u}_j}{\|\overrightarrow{u}_i\| \|\overrightarrow{u}_j\|} \tag{10}$$

where $\vec{u}_i$ and $\vec{u}_j$ denote the low-dimensional feature representations of users u_i and u_j, respectively. In this work, relation strength w_{ij} can be viewed as a coarse-grained relation value between users u_i and u_j. Compared to coarse-grained measure, the fine-grained measure s_{ij} is more informative, and can effectively distinguish the importance of recent and old events among users. In other words, the fine-grained measure can deduce the strength of latent relation based on neighborhood structures while two users have no explicit connections.

The fact of matter is that user decision making is influenced by his/her own preferences and close friends in real-world situations. Specifically, on the one hand, users often have different preferences for different items. On the other hand, user are likely to accept their friends' recommendations. Thus, we assume that the final rating of user u_i for item v_j is a linear combination between the user's own preference and his/her friends' preferences, where the rating can be defined as follows:

$$\hat{R}_{ij} = \eta U_i^T V_j + (1 - \eta) \sum_{k \in S(u_i)} s_{ik} U_k^T V_j \tag{11}$$

where $S(u_i)$ is the set of most intimate friends of user u_i. In the above equation, the first item corresponds to the prediction rating based on their own preferences, while the second item corresponds to the prediction rating based on the preferences of his/her friends, and η is a parameter that controls the relative weight between user's own preferences and friends' preferences.

The ratings of users to items are generally represented by an ordered set, such as discrete values or continuous numbers within a certain range. In this work, without loss of generality, the differences in the users' individual rating scales can be considered by normalizing ratings with a function $f(x)$:

$$f(x) = \frac{x - R_{min}}{R_{max} - R_{min}} \tag{12}$$

where R_{max} and R_{min} represent the maximum and minimum ratings, respectively. $f(x)$ values can be fell in the $[0, 1]$ interval. Meanwhile, we use the logistic function $g(x) = 1/(1 + e^{-x})$ to limit the predicted ratings $\hat{R}_{ij}$ within the range of $[0, 1]$.

Based on this, the task of social recommendation is likewise to minimize the predictive error. Hence, the objective function of the evolving relation embedding recommendation algorithm is formalized as:

$$\mathcal{L} = \frac{1}{2} \min_{U,V} \sum_{i=1}^{M} \sum_{j=1}^{N} W_{ij} (R_{ij} - g(\alpha U_i^T V_j + (1 - \alpha) \sum_{k \in S(u_i)} s_{ik} U_k^T V_j))^2$$
$$+ \frac{\gamma}{2} (\|U\|_F^2 + \|V\|_F^2) \tag{13}$$

where $S(u_i) = \{k | s_{ik} \geq \epsilon\}$ is the set of most intimate friends of user u_i, and the parameter ϵ is the threshold of the close relation value.

We adopt stochastic gradient descent (SGD) to solve the local minimum solution of $\mathcal{L}$, and learn the latent feature vectors U_i and V_j. The partial derivatives of the objective function $\mathcal{L}$ with respect to U_i and V_j are computed as:

$$
\begin{aligned}
\frac{\partial \mathcal{L}}{\partial U_i} &= \alpha \sum_{j=1}^{N} W_{ij} g'(\alpha U_i^T V_j + (1-\alpha) \sum_{k \in S(u_i)} s_{ik} U_i^T V_j) V_j \\
&\times (g(\alpha U_i^T V_j + (1-\alpha) \sum_{k \in S(u_i)} s_{ik} U_i^T V_j) - R_{ij}) \\
&+ (1-\alpha) \sum_{j=1}^{N} \sum_{p \in S(u_i)} W_{ij} g'(\alpha U_p^T V_j + (1-\alpha) \sum_{q \in S(p)} s_{pq} U_q^T V_j) \\
&\times (g(\alpha U_p^T V_j + (1-\alpha) \sum_{q \in S(p)} s_{pq} U_q^T V_j) - R_{pi}) s_{pi} V_j + \gamma U_i
\end{aligned}
\tag{14}
$$

$$
\begin{aligned}
\frac{\partial \mathcal{L}}{\partial V_j} &= \alpha \sum_{i=1}^{M} W_{ij} g'(\alpha U_i^T V_j + (1-\alpha) \sum_{k \in S(u_i)} s_{ik} U_i^T V_j) U_i \\
&\times (g(\alpha U_i^T V_j + (1-\alpha) \sum_{k \in S(u_i)} s_{ik} U_i^T V_j) - R_i) \\
&\times (\alpha U_i + (1-\alpha) \sum_{k \in S(u_i)} s_{ik} U_k) + \gamma V_j
\end{aligned}
\tag{15}
$$

where $g'(x) = e^{-x}/(1 + e^{-x})^2$ is the derivative of the logistic function $g(x)$.

5 Experiments

In this section, we first describe experimental datasets and metrics. We then present the baselines and the experiments settings. Finally, we give the experimental results and analyze them.

5.1 Datasets

To evaluate the proposed model, we use two real-world datasets for this task: Weibo and Last.fm.

Weibo Dataset[1]. The data is collected from Sina Weibo, which is the most popular microblogging platform in China. It includes basic information about messages (time, user ID, message ID etc.), mentions (user IDs appearing in messages), forwarding paths, and whether containing embedded URLs or event keywords. In addition, it also contains a snapshot of the following network of users (based on user IDs).

---

[1] https://www.aminer.cn/influencelocality.

Table 1. Statistics of the datasets.

Dataset	Weibo	Last.fm
#User	840,432	1,892
#Item	30,000	17,632
#User-user relations	154,352,856	12,717
#User-item relations	355,754	92,834
Density	0.014%	0.71%

Last.fm Dataset[2]. This dataset has been obtained from Last.fm online music system. Its users are interconnected in a social network generated from Last.fm "friend" relations. Each user has a list of most listened music artists, tag assignments, i.e. tuples [user, tag, artist], and friend relations within the dataset social network. Each artist has a Last.fm URL and a picture URL.

For two datasets, the user-user relations are constructed from following or bi-directional friendships between social network users, user-item relations are constructed from the user posting or listening behavior. The statistics of two datasets are summerized in Table 1.

5.2 Evaluation Metrics

We use the mean absolute error (MAE), root mean square error (RMSE) and the average precision of top-K recommendation (Average P@K) to evaluate the performance of recommendation algorithms. According to their definition, a smaller MAE/RMSE or bigger Average P@K value means better performance. For each dataset, {40%, 80%} are selected randomly as training set and the rest as the test set. We will repeat the experiments 5 times and report the average performance.

5.3 Comparison Algorithms

In order to evaluate the effectiveness of our proposed recommendation algorithm, we select following recommendation algorithms as comparison methods:

- PMF [15]: The method adopts a probabilistic linear model with Gaussian distribution, and the recommendations are obtained only by relying on the rating matrix of users to items.
- SoRec [11]: The method integrates social network structure and the user-item rating matrix based on probabilistic matrix factorization. However, the algorithm ignore the temporal changes of relations between users.
- RSTE [10]: The model fuses the users' tastes and their trusted friends' favors together for the final predicted ratings. Similarly, the method doesn't consider the changes of trust relations over time.

---

[2] http://ir.ii.uam.es/hetrec2011/datasets.html.

- SocialMF [5]: The model integrates a trust propagation mechanism into PMF to improve the recommendation accuracy. However, the algorithm represents the feature vector of each user only by the feature vectors of his direct neighbors in the social network.
- TrustMF [27]: The model proposes social collaborative filtering recommendations by integrating sparse rating data and social trust network. The algorithm can map users into low-dimensional truster feature space and trustee feature space, respectively.
- SoDimRec [19]: The model adopts simultaneously the heterogeneity of social relations and weak dependency connections in the social network, and employs social dimensions to model social recommendation.

Table 2. Performance comparisons of different recommender models.

Dataset	Method	MAE(40%)	MAE(80%)	RMSE(40%)	RMSE(80%)
Weibo	PMF	0.9963	0.9110	1.0346	0.9474
	SoRec	0.9602	0.8957	1.0158	0.9329
	RSTE	0.9319	0.8515	1.0023	0.9301
	SocialMF	0.9044	0.8232	0.9778	0.9168
	TrustMF	0.8879	0.8031	0.9465	0.8885
	SoDimRec	0.8528	0.7884	0.9304	0.8757
	SoERec	**0.8249**	**0.7495**	**0.9128**	**0.8655**
Last.fm	PMF	1.0582	1.0292	1.2691	1.1306
	SoRec	1.0442	1.0996	1.2009	1.0971
	RSTE	1.0386	0.9936	1.1716	1.0876
	SocialMF	1.0299	0.9869	1.1546	1.0801
	TrustMF	1.0076	0.9804	1.1408	1.0718
	SoDimRec	0.9967	0.9768	1.1211	1.0639
	SoERec	**0.9851**	**0.9617**	**1.1092**	**1.0590**

The optimal experimental settings for each method were either determined by our experiments or were taken from the suggestions by previous works. The setting that were taken from previous works include: the learning rate $\eta = 0.001$; and the dimension of the latent vectors $d = 100$. All the regularization parameters for the latent vectors were set to be the same at 0.001.

5.4   Experimental Results

Comparisons of Recommendation Model. We use different amounts of training data (40%, 80%) to test the algorithms. Comparison results are demonstrated in Table 2, and we make the following observations: (1) Our proposed approach SoERec always outperforms baseline methods on both MAE and

RMSE. The major reason is that the proposed framework exploits heterogeneity of social relations via time dimension and network embedding technique. (2) Recommendation systems by exploiting social relations all perform better than the PMF method only by using user-item rating matrix in terms of both MAE and RMSE. (3) Among these relation-aware recommendation methods, leveraging more indirect relations method generally achieves better performance than only using direct connections methods. In a word, social relations play an important role in context-aware recommendations.

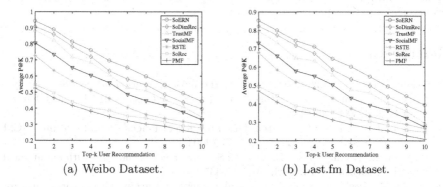

(a) Weibo Dataset.   (b) Last.fm Dataset.

Fig. 4. The overall average P@K score of each method with different K.

Top-K User Recommendation. Figure 4 summarizes the user recommendation performance for the state-of-the-art methods and the proposed model. Generally speaking, it can be shown from the figure that the average P@K value decreases gradually along with the increasing number of K. Besides, we can also observe on both datasets that: Firstly, the proposed method consistently perform better than baseline methods, indicating that the considering cross-time evolving graph embedding by SoERec model can be recommended the more appropriate users than recommendation models without considering time dimension. Secondly, trust-based algorithms (TrustMF, SocialMF and RSTE) consistently perform better than non-trust based benchmarks (SocRec, PMF). It is because trust-based algorithms can fully exploit the network structure, which tackles the incomplete, sparse and noisy problem. Finally, among the different recommendation methods, considering heterogeneous network (SocDimRec and SoERec) significantly performs better than the other methods.

6   Conclusion

In this paper, we propose a novel social recommendation model by incorporating cross-time heterogeneity network of relations. We construct an evolving heterogeneous relation network with timestamp information based on multiple objects and links types. The evolving graph can learn more accurate user relations.

We then use network embedding technique to encode the latent feature spaces of relations into the objective function. To demonstrate the effective of the proposed model, we construct extensive experiments. The experimental results reveal that our proposed method outperforms the state-of-the-art baseline methods.

Acknowledgement. This work is supported by Natural Science Foundation of China (No. 61702508, No. 61802404), and National Social Science Foundation of China (No. 19BSH022), and National Key Research and Development Program of China (No. 2019QY1303). This work is also supported by the ProgramofKeyLaboratoryofNetworkAssessmentTechnology,theChineseAcademyofSciences;Programof BeijingKeyLaboratoryofNetworkSecurityandProtectionTechnology.

References

1. Adomavicius, G., Tuzhilin, A.: Toward the next generation of recommender systems: a survey of the state-of-the-art and possible extensions. IEEE Trans. Knowl. Data Eng. **6**, 734–749 (2005)
2. Deshpande, M., Karypis, G.: Item-based top-n recommendation algorithms. ACM Trans. Inf. Syst. (TOIS) **22**(1), 143–177 (2004)
3. Gopalan, P., Hofman, J.M., Blei, D.M.: Scalable recommendation with hierarchical poisson factorization. In: UAI, pp. 326–335 (2015)
4. Grover, A., Leskovec, J.: node2vec: scalable feature learning for networks. In: Proceedings of the 22nd ACM SIGKDD International Conference on Knowledge Discovery and Data Mining, pp. 855–864. ACM (2016)
5. Jamali, M., Ester, M.: A matrix factorization technique with trust propagation for recommendation in social networks. In: Proceedings of the Fourth ACM Conference on Recommender Systems, pp. 135–142. ACM (2010)
6. Jiang, M., et al.: Social contextual recommendation. In: CIKM, pp. 45–54 (2012)
7. Kipf, T.N., Welling, M.: Semi-supervised classification with graph convolutional networks. arXiv preprint arXiv:1609.02907 (2016)
8. Linden, G., Smith, B., York, J.: Amazon. com recommendations: item-to-item collaborative filtering. IEEE Internet Comput. **7**(1), 76–80 (2003)
9. Liu, N.N., Yang, Q.: Eigenrank: a ranking-oriented approach to collaborative filtering. In: Proceedings of the 31st Annual International ACM SIGIR Conference on Research and Development in Information Retrieval, pp. 83–90. ACM (2008)
10. Ma, H., King, I., Lyu, M.R.: Learning to recommend with social trust ensemble. In: Proceedings of the 32nd International ACM SIGIR Conference on Research and Development in Information Retrieval, pp. 203–210. ACM (2009)
11. Ma, H., Yang, H., Lyu, M.R., King, I.: SoRec: social recommendation using probabilistic matrix factorization. In: Proceedings of the 17th ACM Conference on Information and Knowledge Management, pp, 931–940. ACM (2008)
12. Ma, H., Zhou, D., Liu, C., Lyu, M.R., King, I.: Recommender systems with social regularization. In: WSDM, pp. 287–296. ACM (2011)
13. Mnih, A., Salakhutdinov, R.R.: Probabilistic matrix factorization. In: Advances in Neural Information Processing Systems, pp. 1257–1264 (2008)
14. Ou, M., Cui, P., Pei, J., Zhang, Z., Zhu, W.: Asymmetric transitivity preserving graph embedding. In: Proceedings of the 22nd ACM SIGKDD International Conference on Knowledge Discovery and Data Mining, pp. 1105–1114. ACM (2016)

15. Salakhutdinov, R., Mnih, A.: Probabilistic matrix factorization. In: Advances in Neural Information Processing Systems (NIPS), pp. 1257–1264 (2007)
16. Sarwar, B.M., Karypis, G., Konstan, J.A., Riedl, J., et al.: Item-based collaborative filtering recommendation algorithms. In: WWW, vol. 1, pp. 285–295 (2001)
17. Tang, J., Qu, M., Wang, M., Zhang, M., Yan, J., Mei, Line: large-scale information network embedding. In: Proceedings of the 24th International Conference on World Wide Web, pp. 1067–1077. International World Wide Web Conferences Steering Committee (2015)
18. Tang, J., Gao, H., Liu, H.: mTrust: discerning multi-faceted trust in a connected world, pp. 93–102 (2012)
19. Tang, J., et al.: Recommendation with social dimensions. In: Thirtieth AAAI Conference on Artificial Intelligence (2016)
20. Tu, C., Liu, H., Liu, Z., Sun, M.: Cane: context-aware network embedding for relation modeling. In: Proceedings of the 55th Annual Meeting of the Association for Computational Linguistics (Volume 1: Long Papers), vol. 1, pp. 1722–1731 (2017)
21. Tu, C., Wang, H., Zeng, X., Liu, Z., Sun, M.: Community-enhanced network representation learning for network analysis. arXiv preprint arXiv:1611.06645 (2016)
22. Tu, C., Zhang, W., Liu, Z., Sun, M., et al.: Max-margin deepwalk: discriminative learning of network representation. In: IJCAI, pp. 3889–3895 (2016)
23. Tu, C., Zhang, Z., Liu, Z., Sun, M.: Transnet: translation-based network representation learning for social relation extraction. In: Proceedings of the 26th International Joint Conference on Artificial Intelligence (IJCAI) (2017)
24. Wang, D., Cui, P., Zhu, W.: Structural deep network embedding. In: Proceedings of the 22nd ACM SIGKDD International Conference on Knowledge Discovery and Data Mining, pp. 1225–1234. ACM (2016)
25. Wang, M., Zheng, X., Yang, Y., Zhang, K.: Collaborative filtering with social exposure: a modular approach to social recommendation. In: Thirty-Second AAAI Conference on Artificial Intelligence (2018)
26. Wang, X., Lu, W., Ester, M., Wang, C., Chen, C.: Social recommendation with strong and weak ties. In: CIKM, pp. 5–14. ACM (2016)
27. Bo Yang, Y., Lei, J.L., Li, W.: Social collaborative filtering by trust. IEEE Trans. Pattern Anal. Mach. Intell. 39(8), 1633–1647 (2016)
28. Yang, C., Liu, Z., Zhao, D., Sun, M., Chang, E.Y.: Network representation learning with rich text information. In: Proceedings of the 24th International Joint Conference on Artificial Intelligence (IJCAI), pp. 2111–2117 (2015)
29. Yang, J., Leskovec, J.: Overlapping community detection at scale: a nonnegative matrix factorization approach. In: Proceedings of the 6th ACM International Conference on Web Search and Data Mining, pp. 587–596. ACM (2013)

DDNE: Discriminative Distance Metric Learning for Network Embedding

Xiaoxue Li[1,2], Yangxi Li[3], Yanmin Shang[1($\boxtimes$)], Lingling Tong[3], Fang Fang[1],
Pengfei Yin[1], Jie Cheng[4], and Jing Li[4]

[1] Institute of Information Engineering, Chinese Academy of Sciences, Beijing, China
{lixiaoxue,shangyanmin,fangfang,yinpengfei}@iie.ac.cn
[2] University of Chinese Academy of Sciences, Beijing, China
[3] National Computer Network Emergency Response Technical Team, Beijing, China
liyangxi@outlook.com, tonglong300@sina.com
[4] State Grid Information & Telecommunication Co., Ltd. (SGIT), Beijing, China
{chengjie,jingli}@sgcc.com.cn

Abstract. Network embedding is a method to learn low-dimensional representations of nodes in networks, which aims to capture and preserve network structure. Most of the existing methods learn network embedding based on *distributional similarity* hypothesis while ignoring *adjacency similarity* property, which may cause distance bias problem in the network embedding space. To solve this problem, this paper proposes a unified framework to encode distributional similarity and measure adjacency similarity simultaneously, named *DDNE*. The proposed DDNE trains a siamese neural network which learns a set of non-linear transforms to project the node pairs into the same low-dimensional space based on their first-order proximity. Meanwhile, a distance constraint is used to make the distance between a pair of adjacent nodes smaller than a threshold and that of each non-adjacent nodes larger than the same threshold, which highlight the adjacency similarity. We conduct extensive experiments on four real-world datasets in three social network analysis tasks, including network reconstruction, attribute prediction and recommendation. The experimental results demonstrate the competitive and superior performance of our approach in generating effective network embedding vectors over baselines.

Keywords: Network embedding · Social network · Metric learning

1 Introduction

Network embedding, as one of network representation learning methods, has been successfully applied in a wide variety of network-based analysis tasks, such as link prediction, network reconstruction, node classification, etc. Different from traditional adjacency matrix representation, which suffers from high dimensionality and data sparsity, network embedding aims to represent each node in a given network as a vector in a low-dimensional latent space.

© Springer Nature Switzerland AG 2020
V. V. Krzhizhanovskaya et al. (Eds.): ICCS 2020, LNCS 12137, pp. 568–581, 2020.
https://doi.org/10.1007/978-3-030-50371-0_42

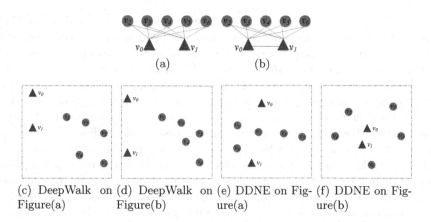

(a)   (b)

(c) DeepWalk on (d) DeepWalk on (e) DDNE on Fig- (f) DDNE on Fig-
Figure(a)   Figure(b)   ure(a)   ure(b)

Fig. 1. Global adjacency similarity and distance bias: Figure 1(a) and Fig. 1(b) are the small part of the given network. Figure 1(c) to Fig. 1(f) are the embedding visualization of Figure 1(a) and Fig. 1(b). Each point indicates one neighbor of ego-node (triangle).

In order to well preserve the structure of a given network, existing researches encode local proximity, and inherent properties to learn network embedding [1,6,8]. Typically, *Node2vec*, *DeepWalk* and *Line* [2,7,9] approximate nodes' local proximity, including the first- and second-order proximity, via random walks or neural network models with specific objective functions. The essence is to learn the vector representation of a node by predicting its neighborhood, which is inspired by the word embedding principle. Based on this principle, the vector representation satisfies the *distributional similarity* property of network, i.e. nodes with similar neighborhoods are closer in the network embedding space.

In practical applications, there is another fundamental property of network besides *distributional similarity*, called *adjacency similarity*. *adjacency similarity* means that a pair of nodes are similar in some aspects. For example, in the link prediction task, node pairs with higher similarity are more likely to be considered as adjacent nodes. In the label propagation task, the adjacent nodes are considered sharing the common labels. So, adjacent nodes should be closer than non-adjacent ones in the network embedding space. However, in most of previous embedding learning methods, this adjacency similarity is ignored, which may generate **distance bias** in the network embedding space [4].

Figure 1 shows an example of what is the distance bias. In Fig. 1(a), node v_0 and v_1 share the same neighbors, but there is no link between them. In contrast, we add a link between node v_0 and v_1 in Fig. 1(b). As a result of previous method (taking DeepWalk as an example), the distance between v_0 and v_1 in Fig. 1(a) is smaller than that in Fig. 1(b) in the network embedding space (shown in Fig. 1(c)). However, if adjacency similarity is taken into consideration, the distance between v_0 and v_1 in Fig. 1(a) would be larger than that in Fig. 1(b) (shown in Fig. 1(e)). We call **this inaccurate estimation of distance between two nodes as a distance bias problem.**

To address the distance bias problem, we propose a novel node embedding method to simultaneously preserve the distributional similarity and adjacency similarity property of the network. This model consists of two modules: the *Node Encoder* and the *Distance Metric-learner*. For a given network, Node-encoder encodes the first-order proximity of the nodes using a neural network model. In the input layer, each node is represented as a sequence of its neighbors, and then it goes through multiple non-linear transformation in hidden layers. Because different neighbors contribute to similarity measurement differently, we adopt the attention mechanism to adaptively assign weight to different neighbors. The output is node embedding representation, and nodes with common neighbors will gain similar encoding. The Distance Metric-learner measures the distance of pair-wise node embedding vectors, which aims to assign the adjacent nodes-pair a smaller distance to highlight the adjacency similarity. For this purpose, we use a well-designed objective function to pull the node toward its neighbors, and push non-adjacent nodes further away. Based on this, the structure of the network would be preserved better in the embedding space.

To verify the effectiveness of our approach, we conduct experiments through network reconstruction, attribute prediction and recommendation tasks on four real-world datasets. We take five state-of-the-art embedding algorithms as comparative methods. The experimental results show that our approach is able to not only solve the distance bias problem, but also outperform comparative methods in all above tasks, especially in network reconstruction.

In summary, the main contributions of this paper are as follows:

- We analyze the distance bias problem in traditional network embedding methods, which is induced by disregarding the adjacency similarity property.
- We propose a discriminative distance metric learning method to preserve the adjacency similarity property of networks and improve the effectiveness of node representations.
- We evaluate our method on three tasks over four datasets and experimental results show that our approach achieves a significant improvement.

2   Proposed Approach

In this section, we present the details of the proposed network embedding based on neural network and attention mechanism. Firstly, we briefly introduce the definition of the problem. Then we discuss the details of the proposed discriminative distance metric learn model DDNE. Finally, we present some discussion and implementation of our objective function.

2.1   Preliminaries

Notations. Given a network $G = (V, E)$, $V = \{v_1, ..., v_V\}$ represents the set of nodes and $E = \{e_{ab}\}_{a,b=1}^{n}$ represents the set of edges. We define the adjacency matrix of G as $X = [X_{ab}]$, where $X_{ab} = 1$ if v_a and v_b linked by an edge, otherwise, $X_{ab} = 0$. Accordingly, given a node pair v_a, v_b in the network, X_{ab} is also the adjacency relation label of this node pair. D is a diagonal matrix, where D_{aa} represents the degree of v_a.

Distributional Similarity. In this paper, the distributional similarity describes the relationship between node and its first-order proximity. For a node v_i, $N(v_i)$ denotes a set of nodes directly connected to v_i. The distributional similarity of v_i is generated by its neighbors $N(v_i)$, which means that nodes with similar neighborhoods are closer in the network embedding space.

Adjacency Similarity. In the network embedding space, the learned embedding vectors of two nodes are expected closer if they are adjacent. Accordingly, for each node pair, if $X_{ij} = 1$, there exists a larger adjacency similarity than those without adjacency relation.

Network Embedding. Given a network denoted as $G = (V, E)$, network embedding aims to learn a mapping function $f : v_i \rightarrow u_i \in R^d$, where $d \ll |V|$. The objective of our method is to make the distance between adjacent node pair closer than those node pairs without adjacency relation in the embedding space, while the distance between node pairs with similar neighbors(distributional similarity) is also closer in this space.

2.2 DDNE

Framework. In this paper, we propose a Discriminative Distance metric learning framework to perform Network Embedding (DDNE), as shown in Fig. 2. In details, this framework consists of two modules: Node Encoder and Distance Metric-learner. Node-encoder encodes each node into embedding vector based on its first-order proximity, while the Distance Metric-learner measure the distance of pair-wise node embedding vectors with some constraints to eliminate the distance bias problem.

Node Encoder. Formally, for ego-node v_i with T neighbors $N(v_i)$, his neighbors can be modelled as a sequence which is the input vector of neural network:

$$R(v_i) = (r_i^1, ..., r_i^T),$$

where r_i^t denotes network structure information about t−th neighbor, which is lookup from the adjacency matrix of network X. For a given node v_i, the output of hidden layer $h_i^t(t = 1, ..., T)$ encode the representation of v_i's t-th neighbor.

$$h_i^{(k)t} = \sigma(W^{(k)}r_i^t + b^{(k)}), \tag{1}$$

where k denotes the k-th layer of our neural network, $k \in (1, K)$. Then the first-order proximity of node v_i is:

$$[h_i^{(K)1}, ..., h_i^{(K)T}]$$

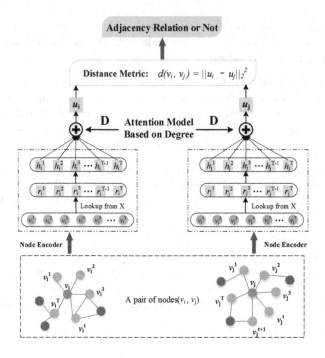

Fig. 2. Framework of the DDNE model.

The embedding vector of node is a sum of its first-order proximity elements, which makes sure that nodes with common neighbors will gain similar encoding (local distributional similarity). As different neighbors contribute to similarity measurement differently, as assumed in Sect. 1, we adopt the attention mechanism to adaptively assign weight to different neighbors.

$$z_i^t = \sigma(W_p h_i^t) \tag{2}$$

$$\alpha_i^t = \frac{exp(z_i^t \times \frac{D_{ii}}{|D_{ii} - D_{tt} + \beta|})}{\sum_{t'=1}^{T} exp(z_i^{t'} \times \frac{D_{ii}}{|D_{ii} - D_{t't'} + \beta|})} \tag{3}$$

$$u_i = \sum_{t=1}^{T} \alpha_i^m \cdot h_i^m \tag{4}$$

where W_p is a trained projection matrix; D_{ii} and D_{tt} are the degree of node v_i and v_i^t, and u_i is the embedding vector of node v_i

In the attention phase, we calculate the weight α_i^t of neighbor v_i^t by Eq. (3), which makes sure that the weight is larger when the degree of v_i^t is comparable with v_i. Node embedding vector u_i is computed by Equation (4). The advantage of our attention model is that it can dynamically learn the weight of each neighbor according to its degree with the ego-node (same, large or low).

Distance Metric-Learner. Embedding vectors generated by distributional similarity based methods may generate distance bias problem. That is to say, the distance between non-adjacent nodes is closer than adjacent nodes, which does not conform with reality. In order to eliminate this problem, we measure the adjacency similarity using distance metric learning method, which aims to pull the distance between adjacent nodes-pair closer to highlight the adjacency similarity. For this purpose, we propose a distance constraint to restrict the distance margin between node pair with adjacency relation (***positive node pair***) and node pair without adjacency relation (***negative node pair***). Based on this, the adjacency similarity would be measured and the distance bias problem in the embedding space would be eliminated, as shown in Figure 3.

Distance Constraint: The distance between positive pair should be smaller than a pre-specified margin m ($m > 0$) and the negative pair should be larger than m. Thus, the constraint between margin m and $d(v_i, v_j)$ is that:

$$D_{ij} = \begin{cases} d(v_i, v_j) < m & \text{if positive node pair: } X_{ij} = 1 \\ d(v_i, v_j) > m & \text{if negative node pair: } X_{ij} = 0 \end{cases} \tag{5}$$

where $d(v_i, v_j)$ is

$$d(v_i, v_j) = \|u_i - u_j\|_2^2$$

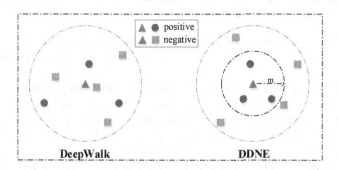

Fig. 3. Distance Constraint: there are two embedding spaces which are generated by DeepWalk and DDNE respectively. We sample seven nodes in each space, where one of them is the ego-node (purple triangle) and others are the neighbors (red circle) or non-adjacent nodes (green rectangle) of the ego-node. In DDNE, a distance constraint is used to make the distance between a pair of adjacent nodes smaller than a threshold and that of each non-adjacent nodes larger than the same threshold. Under this constraint, our method eliminates the bias problem existed in DeepWalk. (Color figure online)

Objective Function. Formally, we transform the above distance constraint into the following optimization problem:

$$l_{ij} = \begin{cases} max(0, d(v_i, v_j) - m)^2 & X_{ij} = 1 \\ max(0, m - d(v_i, v_j))^2 & X_{ij} = 0 \end{cases} \tag{6}$$

where m and $d(v_i, v_j)$ are represent the distance margin and the distance between v_i and v_j, respectively. The loss function of Eq. (5) pulls the nodes toward v_i's neighbor, and pushes non-adjacent nodes further away.

Accordingly, for all node pairs, the objective function of DDNE is that

$$\mathcal{L} = \frac{1}{2N} \sum_{ij} l_{ij} \tag{7}$$

3 Experiment

In this section, we firstly introduce datasets and baseline methods in this work. We then evaluate our proposed methods in three network analysis tasks: network reconstruction, attribute prediction and recommendation. Finally, we analyze the quantitative experimental results and investigate the sensitivity across parameters.

3.1 Datasets and Baseline Methods

Datasets. We conduct our experiments on four networks, including two social networks, one citation network and one recommendation network. Table 1 shows the detailed information about those four networks. The description of those four networks are shown as following:

- **Google+**[1] is one of social networks. In which, nodes represent users and each has gender, university title, job title, last-name and workspace as its attribute.
- **Sina**[2] is the social network. In this network, users have attributes such as following number, self-introduction, constellation, age and location.
- **DBLP**[3] is a citation network in which nodes refer to papers and edges represent the citation relationship among papers. Each paper has attributes like title, authors, publication venue and abstract.
- **Movieslens**[4] is a recommendation network in which nodes refer to users and movies respectively and edges represent viewing record between users and movies. Each user has age, gender and occupation as its attribute information.

---

[1] https://snap.stanford.edu/data/index.html.
[2] https://www.weibo.com/.
[3] https://snap.stanford.edu/data/index.html.
[4] https://grouplens.org/datasets/movielens/100k/.

Table 1. The statistic of datasets

Data	Nodes	Edges	Categories
Google+	3,126	22,829	7
Sina	29,418	800,174	8
DBLP	244,021	4,354,534	9
Movieslens	943	100,000	4

Baseline Methods. We compare DDNE with the following baseline methods:

- **SDNE** [13] is the best topology-only network embedding method, which introduces an auto-encoder algorithm to learn the node embedding vector and considers the first-order and second-order proximities information.
- **LINE** [9] is a popular topology-only network embedding method, which also considers the first-order and second-order proximities information.
- **DeepWalk** [7] is a topology-only network embedding method, which introduces the Skip-gram algorithm to learn the node embedding vector.
- **GraphGAN** [10] is a topology-only network embedding method, which introduces the GAN network to learn the node embedding vector.
- **DDNE** is our proposed method using neural network (NN or LSTM) to model the distributional similarity and distance metric learning to model the adjacency similarity, which include **DDNE$_{NN}$** and **DDNE$_{LSTM}$**.
- **Sigmoid**: In this method nodes are represented by the local proximity through neural network (NN or LSTM) and the network structure is preserved through the sigmoid loss function, which includes **S$_{NN}$** and **S$_{LSTM}$**.

Parameter Setup. For all datasets, the dimension of the learned node embedding vector d is set to 128. In SDNE method, parameters are set to the same as given in the original paper. In DeepWalk method, the parameters are set as following: window size $w = 15$, walks per node $r = 70$ and walk length $t = 35$. In LINE, we set $negative = 8$ and $samples = 10$ million. And in DDNE method the parameters are set as $margin = 1$ and learning rate $\eta = 0.01$.

3.2 Distance Bias Analysis

We utilize the embedding space generated by various network embedding methods to analyze the distance bias problem. Then we sample 100 positive node pairs in each network dataset and 100 negative node pairs. The average Euclidean distance is used to evaluate the performance of each embedding methods, as shown in Fig. 4.

From Fig. 4, we can see some phenomenon:

- Compared with baselines, DDNE can guarantee consistency of the phenomenon that the distance between positive node pairs is closer than the distance between negative node pairs on different datasets. For example, with

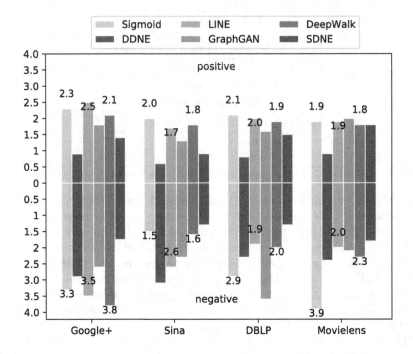

Fig. 4. Average Euclidean distance on networks.

the sigmoid method, the distance between positive node pairs on Sina dataset is 2.0 but the distance between negative node pairs is 1.5, this distance bias is obviously contrary to cognition. Similarly, LINE and SDNE on DBLP, Deep-Walk on Sina will also result in distance bias.

- In contrast, the distances between positive node pairs with DDNE are smallest on all datasets, which means that the embedding vectors obtained by DDNE can better reflect the network structure.

3.3   Network Reconstruction

As the embedding of a network, the learned embedding feature space is expected to well reconstruct the network. Generally, a good network embedding method should ensure that the learned node's embedding vectors can preserve the original network structure. That is also the reason why we conduct this experiment. We use a social network Sina and a paper citation network DBLP as embedding networks. Given a network, we use different network embedding methods to learn the node embedding vectors in feature space R^d. The network reconstruction task is reconstructing the network edges based on distances between nodes in the feature space R^d. We denote the probability of existing edges between v_i and v_j is that:

$$p_{i,j} = \frac{1}{1 + e^{d(u_i,u_j)}} \tag{8}$$

Besides, a threshold β is pre-defined and an edge e_{ij} will be created if $p_{i,j} > \beta$. As the existing edges in the original network are known and can serve as the positive label-data, while the equal amount node pairs which do not exist edges are generated and can serve as the negative label-data. We can evaluate the reconstruction performance of different embedding methods. The accuracy is used as the evaluation metrics and the result is presented in Fig. 5.

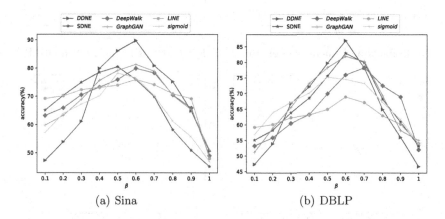

(a) Sina   (b) DBLP

Fig. 5. Precision of network reconstruction on Sina and DBLP.

From Fig. 5, we can see that DDNE achieves the best performance when $\beta = 0.6$, which improves the accuracy by 6% at most comparing to the best baseline SDNE. In addition, our method is more sensitive to the pre-defined threshold β, which indicates that DDNE preserves the network structure better than other methods because there is a clearly distance margin between positive node pairs and negative node pairs in the embedding space generated by DDNE.

Parameter Sensitive Analysis. DDNE has two major parameters: the dimension of embedding vector d and the margin m. We only present the result on Sina and DBLP and omit others due to space limitation. In this experiment, d varies from 10 to 300 and m varies from 0.1 to 2. Figure 6(a) and 6(b) show the accuracy resulted by our method with different embedding dimension. When the embedding dimensions grow, the performance firstly increases significantly, and then does not change drastically for both $DDNE_{NN}$ and $DDNE_{LSTM}$ as the dimension rose to $d = 128$. Besides, on DBLP, the accuracy even increases significantly when the embedding dimensions increases. The figure also shows that $DDNE_{NN}$ beat the best performance and is able to obtain a fairly better accuracy when $d = 128$.

The margin m also influence the attribute prediction performance, as shown in Fig. 6(c) and 6(d). The $DDNE_{NN}$ is more sensitive to the margin value m. This is largely due to the fact that in $DDNE_{NN}$, the local proximity is encoded by the neural network but not the sequence model. The former could encode local

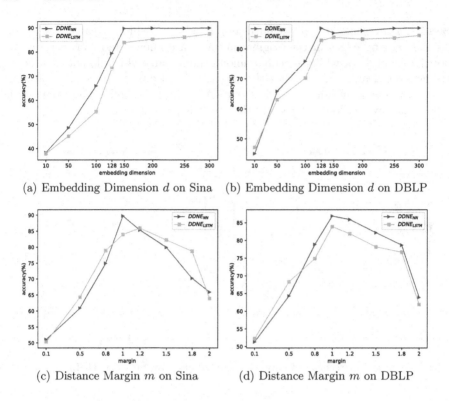

(a) Embedding Dimension d on Sina   (b) Embedding Dimension d on DBLP

(c) Distance Margin m on Sina   (d) Distance Margin m on DBLP

Fig. 6. DDNE parameter analysis

proximity better because, for the neighborhood, the sequence characteristics do not obvious. Thus, the influence of m is manifested in the $DDNE_{NN}$.

3.4   Attribute Prediction

We utilize the vectors generated by various network embedding or social network embedding methods to preform profile prediction task. User always cancel their attribute information or no attributes were filled in because personal attributes often involve users' privacy issues, which results in a problem that user's essential information can not be obtained directly. Thus, attribute prediction task can solve this problem and we treat this task as a classification problem. In our experiment, the embedding vector of each node (user) is treated as its feature vector, and then we use a linear support vector machine model to return the most likely category(value) of the missing attribute. For each dataset, we predict occupation. The training dataset consists of α- portion nodes which are randomly picked from the network, and the rest of users are the test data.

Occupation Prediction. We make the experiment about occupation prediction. The result of this experiment is shown in Table 2.

Table 2. Precision of occupation prediction (%)

Methods	α	SDNE	DeepWalk	LINE	GraphGAN	$DDNE_N$	$DDNE_L$	S_N	S_L
Google+	30%	52.3	46.8	40.9	52.2	**54.6**	54.3	51.4	51.3
	40%	51.9	46.2	50.1	55.4	**56.3**	55.9	53.1	53.2
Sina	30%	62.9	61.9	64.8	**65.4**	**65.4**	63.9	61.8	61.5
	40%	65.3	63.1	66.2	67.1	**68.3**	67.5	65.2	66.0
Movieslens	30%	56.4	53.9	55.6	57.3	**59.2**	58.7	56.1	56.2
	40%	58.5	55.2	57.8	59.9	**61.3**	60.5	58.2	58.9

One can see that DDNE also outperforms other embedding methods. Comparing to the best baseline GraphGAN, our method improve the accuracy by 2.4% at most. Besides, $DDNE$ performs better than Sigmoid which demonstrates that the effective of our distance metric leaning objective function can help preserve the network structure better.

3.5   Recommendation

In this section, we concentrate on the recommendation task and conduct the experiment on Movieslens and DBLP datasets. Given a snapshot of the current network, recommendation tasks refers to recommend new item (movies or papers) that will be picked by users or added in the future time. In order to process this recommendation task, we remove a portion of existing links from the input network. Based on the residual network, nodes embedding vectors are learned by different embedding methods respectively. Node pairs in the remove edges are considered as the positive samples. We also randomly sample the same number of the node pairs which are not connected as the negative samples. Positive and negative samples form the balanced data set. Given a node pair in the sample dataset, we compute the cosine similarity as the score function. The higher node pair score, the greater the possibility of being recommended. Area Under Cur (AUC) is used to evaluate the consistency between the labels and the similarity scores of the samples.

Table 3. AUC of recommendation

Methods	SDNE	DeepWalk	LINE	GraphGAN	$DDNE_N$	$DDNE_L$	S_N	S_L
Movieslens	65.9	62.3	63.6	64.8	**79.8**	78.9	71.2	71.9
DBLP	78.7	75.2	76.1	76.8	**88.9**	88.2	81.0	80.6

From Table 3, we can see that DDNE performs best in both movies and papers recommendation. Compared to the SDNE, DDNE improves the AUC score by 13.9% in Movieslens and 10.2% in DBLP, which demonstrates the effectiveness of DDNE in learning good node embedding vectors for the task of recommendation.

4   Related Work

Network embedding aims to learn a distributed representation vector for each node in a network. Most of existing works can be categorized into three categories: matrix factorization based, random walking based and deep learning based methods. Matrix factorization based methods first express the input network with a affinity matrix into a low-dimensional space using matrix factorization techniques, including singular value decomposition which seeks a low-dimensional projection of the input matrix, and spectral decomposition (eigendecomposition) [3] which uses a spectral decomposition of the graph Laplacian to compute the low-dimensional representation of input data. However, matrix factorization based methods rely on the decomposition of the affinity matrix, which is time-consuming when the data is large real-world networks.

Random Walk is an optimization algorithm in Graph, which can compute the globally optimal solution. As the first attempt, DeepWalk [7] introduces the word2vec algorithms(skip-gram) into learn the embedding of nodes in graph. Another famous work is Node2vec [2], which is a variant of Deepwalk. The most difference between those two is that node2vec changes random walk into biased random walk, and then it can select the next node in an heterogeneous way.

The last category is Deep learning based methods. Tang et al. propose LINE [9], which optimizes a carefully designed objective function through maximizing edge reconstruction probability. SDNE [13] is a deep network embedding method based on auto-encoder, which captures the highly non-linear network structure and exploits the first-order and second-order proximities to preserve the network structure. GraphGAN [10] is a framework that unifies generative and discriminative thinking for network embedding. DKN [12] learns knowledge graph embedding by TransX. The author used a CNN framework for combining word embedding and entity embedding and present an attention-based CTR prediction model meanwhile. SHINE [11] is a network embedding on signed heterogeneous information network, which is also based on auto-encoder.

5   Conclusion

In this paper, we introduce discriminative distance metric learning method to solve the distance bias problem. To adopt the adjacency similarity property, our model is able to preserve the network structure more efficiently. Experiments on three network analysis tasks verified the effectiveness of our approach. In the future work, we will research more deeply on the node encoder. On one hand, we will compare with other deep neural network models, such as CNN or deep RNN. On the other hand, we will try to integrate distributional similarity and adjacency similarity simultaneously in the node encoding phase.

Acknowledgement. This work was sponsored by the National Key R&D Program of China (NO. 2018 YFB1004704), the National Natural Science Foundation of China (U1736106).

References

1. Farnadi, G., Tang, J., Cock, M.D., Moens, M.: User profiling through deep multi modal fusion. In: Proceedings of the Eleventh ACM International Conference on Web Search and Data Mining, WSDM 2018, Marina Del Rey, CA, USA, 5–9 February 2018, pp. 171–179 (2018)
2. Grover, A., Leskovec, J.: node2vec: scalable feature learning for networks. In: Proceedings of the 22nd ACM SIGKDD International Conference on Knowledge Discovery and Data Mining, San Francisco, CA, USA, 13–17 August 2016, pp. 855–864 (2016)
3. Kipf, T.N., Welling, M.: Semi-supervised classification with graph convolution networks. CoRR (2016). abs/1609.02907
4. Feng, R., Yang, Y., Hu, W., Wu, F., Zhang, Y.: Representation learning for scale-free networks. In: Proceedings of the Thirty-Second AAAI Conference on Artificial Intelligence, New Orleans, Louisiana, USA, 2–7 February 2018 (2018)
5. Levy, O., Goldberg, Y.: Neural word embedding as implicit matrix factorization. In: Advances in Neural Information Processing Systems 27: Annual Conference on Neural Information Processing Systems 8–13 December 2014, Montreal, Quebec, Canada, pp. 2177–2185 (2014)
6. Li, C., et al.: PPNE: property preserving network embedding. In: Candan, S., Chen, L., Pedersen, T.B., Chang, L., Hua, W. (eds.) DASFAA 2017. LNCS, vol. 10177, pp. 163–179. Springer, Cham (2017). https://doi.org/10.1007/978-3-319-55753-3_11
7. Perozzi, B., Al-Rfou, R., Skiena, S.: Deepwalk: online learning of social representations. In: The 20th ACM SIGKDD International Conference on Knowledge Discovery and Data Mining, KDD 2014, New York, NY, USA 24–27 August 2014, pp. 701–710 (2014)
8. Ribeiro, L.F.R., Saverese, P.H.P., Figueiredo, D.R.: struc2vec: learning node representations from structural identity. In: Proceedings of the 23rd ACM SIGKDD International Conference on Knowledge Discovery and Data Mining, Halifax, NS, Canada, 13–17 August 2017, pp. 385–394 (2017)
9. Tang, J., Qu, M., Wang, M., Zhang, M., Yan, J., Mei, Q.: LINE: large-scale information network embedding. In: Proceedings of the 24th International Conference on World Wide Web, WWW 2015, Florence, Italy, 18–22 May 2015, pp. 1067–1077 (2015)
10. Wang, H., et al.: Graphgan: graph representation learning with generative adversarial nets. In: Proceedings of the Thirty-Second AAAI Conference on Artificial Intelligence, New Orleans, Louisiana, USA, 2–7 February 2018 (2018)
11. Wang, H., Zhang, F., Hou, M., Xie, X., Guo, M., Liu, Q.: SHINE: signed heterogeneous information network embedding for sentiment link prediction. In: Proceedings of the Eleventh ACM International Conference on Web Search and Data Mining, WSDM 2018, Marina Del Rey, CA, USA, 5–9 February 2018, pp. 592–600 (2018)
12. Wang, H., Zhang, F., Xie, X., Guo, M.: DKN: deep knowledge-aware network for news recommendation. In: Proceedings of the 2018 World Wide Web Conference on World Wide Web, WWW 2018, Lyon, France, 23–27 April 2018, pp. 1835–1844 (2018)
13. Wang, D., Cui, P., Zhu, W.: Structural deep network embedding. In: Proceedings of the 22nd ACMSIGKDD International Conference on Knowledge Discovery and Data Mining, San Francisco, CA, USA, 13–17 August 2016, pp. 1225–1234 (2016)

Extracting Backbone Structure of a Road Network from Raw Data

Hoai Nguyen Huynh[1]($\boxtimes$)(iD) and Roshini Selvakumar[2]

[1] Institute of High Performance Computing, Agency for Science,
Technology and Research, Singapore, Singapore
huynhhn@ihpc.a-star.edu.sg
[2] Raffles Institution, Singapore, Singapore

Abstract. The representation of roads as a networked system of nodes and edges has attracted significant interest in the network literature, generating a large number of studies over the years. Such representation requires a proper identification of what constitute an edge or a node. Intuitively, nodes represent the junctions where roads intersect, and edges are the road segments connecting these junctions. In practice, however, such simplified presentation is not trivial to achieve due to extra details of individual roads. In this paper, we present a set of novel and efficient computational techniques based on elementary geometry and graph theory that can be employed to obtain the essential structure of a road network, while also retaining the crucial geometry of roads, such as shape and length. This is done by dissecting the network into clusters of nodes of degree other than 2 and curves, which contain consecutive nodes of degree 2, connecting these clusters. These clusters of nodes and curves will be collapsed into cleaned nodes and paths, respectively in a simplified mathematical graph. We apply this method to obtain the simplification of road network in Punggol new town in the northeast region of Singapore, and show that application of network analyses such as centrality measures could be performed in a more meaningful and concise manner than on the original road network.

Keywords: Road network · Clustering · Geometrical graph

1 Introduction

Modern computers with various architectures developed in recent years have enabled the storage and processing of a large amount of data, especially the spatial ones. Such development has led to a rapid growth in the availability of (open) urban data for almost every corner on the Earth's surface. The available geospatial data allow us to study different aspects of urban systems, or cities, including their physical elements of road networks. The representation of roads as a networked system of nodes and edges has attracted significant interest in the network literature, generating a large number of studies over the years (see *e.g.* [5, 7, 10, 15, 16, 18]).

ⓒ Springer Nature Switzerland AG 2020
V. V. Krzhizhanovskaya et al. (Eds.): ICCS 2020, LNCS 12137, pp. 582–594, 2020.
https://doi.org/10.1007/978-3-030-50371-0_43

The available geospatial data on road networks, typically exist in the digital format of a shapefile [8] that contains points listed in a spatially sequential order to store the geometrical information of the road lines defined by the points. Although the data format is convenient in storing spatial features of point or lines for roads, the network information concerning the connections among points or intersections between roads is generally not available and requires further processing before studies such as network analysis could be performed [7]. On the other hand, the detailed geometrical information of roads unpremeditatedly makes the application of network analysis on the raw data challenging. The extra elements included in the data, such as multiple (clustered) points when two roads intersect (*e.g.* slip road or filter lane) or multiple lines representing the road segments (*e.g.* lanes) between a pair of junctions, create spatially redundant information that hinders network analyses such as centrality measures, since a single intersection may contain many unnecessary and noisy data points.

As a result, the presentation of roads as networks requires a proper identification of what constitute an edge or a node to ensure that accurate models can be created to study road networks in a meaningful way. In the literature, various methods and algorithms have been proposed to simplify or generalise road networks by removing redundant information and performing merging processes to properly represent nodes and edges in the network. At the simplest level, intermediary nodes along a road line are treated redundant and removed from the network as they are only intermediate to provide connections between a pair of junctions [6]. For more sophisticated simplification, a number of approaches have been proposed including the combination of nodes' degree with road length and road attributes [17,21] or making use of the proximity of network nodes to facilities [14,19]. Alternatively, with the additional availability of dynamic traffic data, main structure of urban road networks have been extracted based on their usage through the determination of importance (or grade) of roads using vehicle trajectories [23] or traffic flow pattern [20,22].

The above-mentioned approaches are either too simplistic (only removing intermediary nodes) or require additional data which may not always be available. In this work, we propose a procedure to simplify a road network from its raw geospatial data using a novel set of techniques based on elementary geometry and graph theory. The simplification would yield the essential structure of the road network, which could be represented as a mathematical graph object. In this approach, we only use spatial information of points along road lines, omitting other road attributes such as name or grade as these may be incorrect or inconsistent at times. The problem we tackle, therefore, boils down to extracting the backbone structure of a network of pure geometric connections. This structure of the road network in an urban system would subsequently allow the studies of other aspects of the system, such as its morphology, beyond the road network itself. In the remainder of this paper, we first describe our proposed procedure for simplification of a road network. After that, we apply the procedure to the network of roads in an area in Singapore and illustrate the value of the simplified network, especially when applying network analyses such as centrality measures.

2   Data and Methods

2.1   Data

OpenStreetMap (OSM) [3] is a collaborative source that is free and features a map, from which geospatial data for different regions of the world can be downloaded. The data come in a number of file formats, with `shapefile` being one of the most popular ones. For the analysis in this work, we utilise shapefiles of road data from Nextzen service [2], which provides up-to-date extracts of OSM data, mostly for major cities in the world. In this format, the data contain entries of road chunks with sequential lists of points along the roads, which could be converted to network format by creating according nodes and edges.

2.2   Simplification of Road Networks

After obtaining the data, we proceed with a network of roads, in which a node represents a point on a road line and an edge connects a pair of successive points along the line. Depending on the level of detail in the data, a road could be represented by a single or multiple road lines, corresponding to the number of lanes of the road. A node could belong to multiple road lines, for example when two roads intersect or merge. However, very often, two road lines cross one another without sharing any common node in the data. In this work, we will treat a road network as a planar graph, *i.e.* no two segments on the same plane can cross without having a common node.

General Framework. The general framework we employ to simplify a road network is depicted in Fig. 1 and involves four steps. Firstly, the graph is planarised by identifying the pairs of segments that cross one another and adding the points of intersection as new nodes in the graph. In the second step, the graph is cleaned by removing the dangling paths of length shorter than 30 m as these are considered dead ends, which don't lead to a place different from the other nodes. Next, the graph is dissected into curves, which are lines of consecutive nodes of degree 2, and clusters of nodes of degree other than 2. Finally, each cluster is collapsed into a single node and the curves connecting the same pair of clusters are merged into a clean path between the newly collapsed nodes in the simplified graph. These four steps are repeated until no further simplication could be made to the final graph.

Graph Planarisation. A road network provides connections among places in an area on the Earth's surface. Therefore, it is reasonable to consider a road network a 2-dimensional object. This property, however, cannot be taken for granted as very often the data doesn't contain information on the points of intersection when roads pass through each other. Without these points, it would not be possible to properly obtain all the useful information from the network

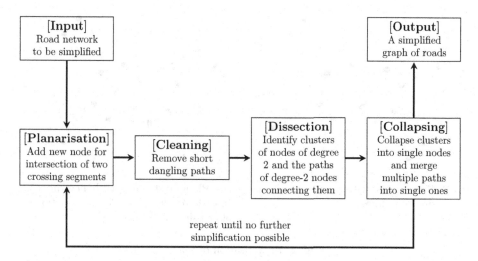

Fig. 1. Procedure for simplification of a road network.

regarding the connections it provides. Therefore, it is imperative that we planarise the graph such that no edges intersect one another, but rather meet only at their end points. In order to carry out the planarisation, intersection points were found and added as nodes (see Fig. 2). In reality, even real roads can lay on top of one another without crossing, it is still suitable and necessary to ensure a road network is planar (see Sect. 3.3 for an argument).

Graph Cleaning. An actual road network can contain short dead ends that branch out from a main road, for example, to provide access to a particular building which might have side entrance. While such access is necessary in reality, the branch does not make a significant difference in terms of connectivity between places. As such, dangling paths of length shorter than 30 m is trimmed off the graph, reducing extra unnecessary junctions in the road network (see Fig. 3).

Graph Dissection. In order to simplify the road network, it is vital that we took out any redundant information that would not contribute to a greater understanding of any one intersection and its connections to other intersections. For example, a junction or road intersection may contain at least one point for a simple intersection that has a 3- or 4-way approach. Most simple intersections tend to have at least four points since the roads have more than one lane. Another example of a junction that may contain redundant information is a junction that contains ramps in addition to the simple intersections. Moreover, roads that have multiple lanes and that intersection would result in one point for every lane intersection.

We notice that a network of roads could be dissected into two groups of elements. The first group contains the chunks of degree-2 nodes that connect from a

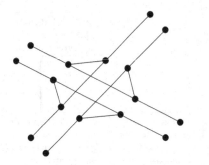

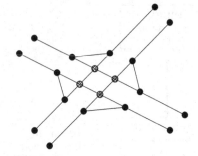

(a) Raw roads with non-intersecting crossing road lines.

(b) Planarised roads with new nodes (crosshatched) added for intersections of crossing road lines.

Fig. 2. Planarisation of roads.

non-degree-2 node to another non-degree-2 node. These chunks can be uniquely and effectively identified by picking a degree-2 node and start tracing in opposite directions (as the node has only 2 neighbours) until a non-degree-2 node is reached at both ends. The idea behind this approach is that if all connecting roads, characterised by paths of degree-2 nodes, were removed, it would leave clusters of (connected) nodes that can be considered as corresponding road junctions since there are no road chunks connecting nodes within them (see Fig. 4). These clusters would form the second group of elements in the road network.

Node and Curve Collapsing. After dissecting the road network into two groups of non-degree-2-node clusters and degree-2-node chunks, we collapse them into simple nodes and curves, respectively, to construct the simplified network.

Nodes. For each cluster, all the points are collapsed into a single node located at their (geometric) centroid and added as a new node to represent the corresponding junction in the simplified network (see Fig. 5). By replacing the points with their centroid, the representative node contains contribution from all member nodes.

Curves. We apply the same concept to curves, *i.e.* to get the representative curve that could be considered "average" of all the pertaining curves. To do that, we notice that after the node collapsing above, all the curves share the same end points, which are also the end points of the averaged curve. It is reasonable to argue that the general direction of the curves is set by these two end points. Therefore, we use a line passing these two points as our reference line. We then find the projection of all points in the curves onto this reference line. The total number of projections on the reference line is the sum of number of points in all curves, unless some points in two different curves share the same projection.

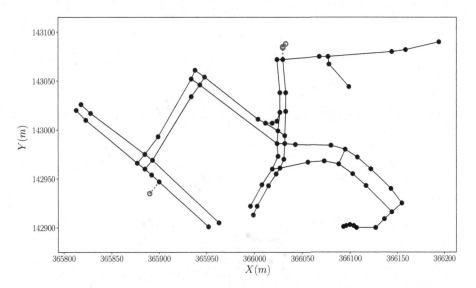

Fig. 3. Removal of short dangling paths. A dangling path starts from a degree-1 node and ends at the first non-degree-2 node encountered, upon tracing along the path. In this illustration, the dotted red paths have length shorter than 30 m and are to be trimmed off. After the operation, the nodes where those short dangling paths terminate (and get removed) will have 1 degree less. (Color figure online)

For every projection identified, a line perpendicular to the reference line is plotted to find its intersections with all the curves. The centroid of these intersections will contribute as a point on the averaged curve, in the same order as the projections on the reference line (see Fig. 6).

3   Results and Discussion

In order to apply the simplification procedure devised above, we report a case study of analysis of roads in the planning area of Punggol in the northeast region of Singapore. Punggol in the last 20 years has been developed into a new town with 11 districts, spanning a total area of $9.57 \, km^2$ [1]. The simplified network after applying the procedure removes the extra details of roads by collapsing complex junctions into single nodes and merging multiple curves between the same pair of nodes into a thin simple path, to obtain the essential structure of the road network (see Fig. 7). This simplified network, however, still retains the overall geometrical properties of the original network such as the shape and (average) length of the paths between places.

3.1   Network Attributes

In Table 1, we compare some basic attributes of the network of roads in Punggol before and after simplification. It can be seen that a large number of spatially

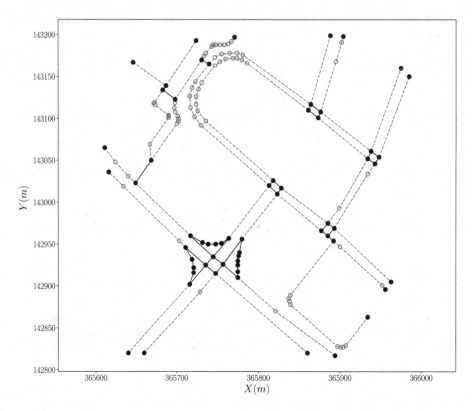

Fig. 4. Dissection of a road network into two types of elements, clusters of non-degree-2 nodes and chunks of degree-2 nodes connecting the clusters. The paths tracing the chunks are dashed in the plot with its nodes being hollow circles. The clusters are depicted with solid lines and filled circles. (Color figure online)

redundant nodes have been removed (and/or replaced) in the simplification process, leaving only 3,265 necessary nodes compared to the original 6,168 or a gross 47% reduction. Similarity, a large number of edges have also been cleaned off in the operation, from 6,917 down to 3,401, which is more than two times slimming down. More importantly, it should be noted that the number of degree-2 nodes reduces significantly, due to merging of multiple paths between junctions.

The simplication also brings attention to high-degree nodes, of 5 or more, which indicate the emergence of proper junctions at which multiple roads from different directions meet. In the original presentation, such nodes are very rare as junctions tend to comprise a bulky set of degree-4 nodes, which arise from a pair of intersecting road segments, even when more than two roads cross each other.

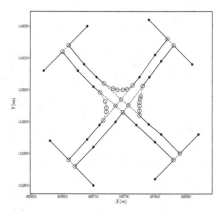

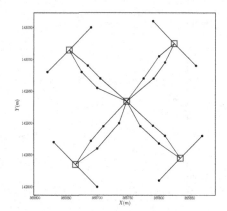

(a) Original road network with clusters of nodes highlighted in red. Their nodes are plotted in hollow circle and the intra-cluster edges are dashed. Inter-cluster edges are solid.

(b) Simplified road network is obtained by replacing the clusters of nodes with their respective centroid. The newly added centroid nodes are marked with hollow square markers, while other nodes are intact and presented as black filled circles.

Fig. 5. Simplification of clusters of nodes at road junctions. (Color figure online)

3.2 Centrality Measures

With the simplified network, we can apply common network analysis techniques such as centrality measures. Four measures that are commonly used for spatial networks are degree, betweenness, closeness and eigenvectors [4,7]. As shown in Table 1, degree centrality of a junction cannot be accurately understood in the original road network as most points at a junction have either degree 3 or 4 when two road segments intersect. In the simplified network, the degree centrality correctly reflect the number of connections that a junction has to others, which refers to the number of directions one can travel from a node.

Table 1. Comparison of basic network properties before and after simplification.

Feature	Original	Simplified
Number of nodes	6,168	3,265
Number of edges	6,917	3,401
Number of nodes of degree 2	4,573 (74.14%)	2,881 (88.24%)
Number of nodes of degree 3	926 (15.01%)	137 (4.20%)
Number of nodes of degree 4	411 (6.66%)	59 (1.81%)
Number of nodes of degree 5	2 (0.03%)	25 (0.76%)
Number of nodes of degree more than 5	0 (0%)	19 (0.58%)

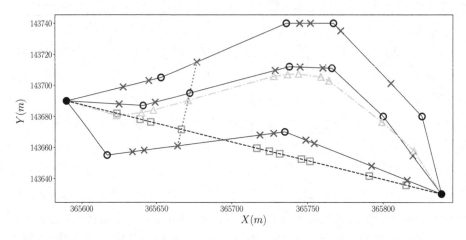

Fig. 6. Collapse of multiple curves into a single "averaged" curve. All the curves share the same end points, which are presented in two filled circle markers (●). Hollow circle markers (○) are other points in the curves, which are connected by solid black lines. The dashed line between two end points is the reference line. Every hollow circle marker has its projection on the reference line, which is shown as a hollow green square marker (□). For each of these projections, a perpendicular line is plotted to find intersections with all the curves, which are marked blue crosses (×). For each of the sets of intersections (including the curve point itself, in hollow circle marker), the centroid is found (red triangle marker (△)) and constitutes a point of the averaged curve. (Color figure online)

Another benefit of having a simplified network structure compared to the original one is that spatial distribution of importance measures like betweenness centrality is more meaningful and properly represents significance of places on a map. As an illustration, the top 50 nodes in the network of roads in Punggol, Singapore, are shown in Fig. 8 for both original and simplified network. It could be seen that in the simplified network, the nodes with highest betweenness centrality measures locate at the major crossroads in the area. However, in the original network, such nodes are mostly located in clusters near the centre part of the network, which is due to redundancy of nodes at the same place and paths between places in the unsimplified network. This highlights the fact that the simplification is important before analysis of spatial networks could be meaningfully applied. This is to remove the spatial redundancy encoded in the original data.

3.3   Discussions

For simplicity of the procedure, we use a simple edge to represent each road segment in the network, with no specification of direction of the edge. In other words, we treat all roads as two-way, with flow possible in either direction, although this does not always hold true in the real world. Nevertheless, this would not affect the main objective in this study of modelling the geometric properties

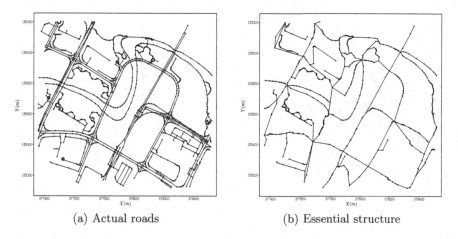

(a) Actual roads   (b) Essential structure

Fig. 7. Representation of network of roads in a zoom-in area in Punggol, Singapore, before and after applying the simplification.

of the network of roads, which essentially concerns the spatial pattern of roads that provide connections between places. In that context, roads are considered in more generic network sense, where junctions are taken to be representative nodes and (abstract) links are established to indicate the flow among the nodes, be it people, materials or information. This is particularly relevant in the context of urban morphology studies, which deal with the spatial organisation of an urban system [11,12], whose structure is primarily laid out by the skeleton of the road network. Yet, for other, more detailed and conventional, purposes such as routing of vehicles, the directionality of roads remains important, which is, however, beyond the scope of this paper.

For the same reason that we're only concerned with the geometrical presentation of road networks that depicts the connections between places, it can also be argued that road network can be reasonably considered planar for the purpose of this study. In reality, a road network may not strictly be planar due to existence of flyover or tunnel. Yet, a flyover or tunnel is usually part of a system at a junction that come with the lanes leading traffic from the flyover of tunnel to the road below or above it. Hence, planarisation of such crossing (creating a point of intersection) does not introduce invalid information to the network. The new point of intersection would still be in the same junction and eventually get merged in the simplification process.

In the process of developing the simplification procedure reported in this study, we came across an approach which purely looks at the spatial location of nodes at road junctions. We find it beneficial to discuss this plausible approach as well as its limitation, to highlight the effectiveness and superiority of the reported method.

This point-based approach utilises the clustering algorithm DBSCAN (Density-Based Spatial Cluster of Applications with Noise) [9] (see also

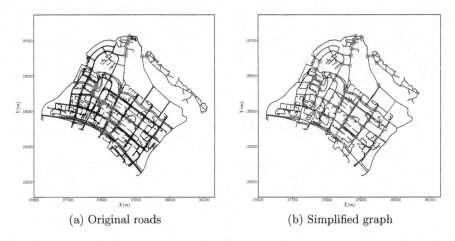

(a) Original roads   (b) Simplified graph

Fig. 8. Spatial distribution of top 50 nodes (shown as hollow red square markers) with highest betweenness centrality values in the network of roads in Punggol, Singapore, before and after applying the simplification. (Color figure online)

continuum percolation [12,13]) in order to identify the points that belong to the same junction. This approach is based on the fact that points of a road junction are in close proximity to one another. The main concept of DBSCAN is to scan for cluster of points in different regions, taking the density of points into account. A distance parameter can be set for DBSCAN, which will determines the maximum distance between a point A and any other point B for B to be considered in the same cluster or neighbourhood as A. Using this algorithm, clusters of junction nodes in a road network can be easily identified, based on their proximity.

Despite its simplicity, this approach suffers multiple limitations that have to be overcome before any useful result can be obtained. Firstly, the main limitation of DBSCAN would be that the input parameter of the maximum distance between any 2 points to be considered in the same cluster is difficult to determine, but significantly influences the result. The problem with using DBSCAN in this context is that each intersection is unique and contains points distributed in different manners, which makes it harder or even not possible to determine one single parameter for an entire road network region. For instance, in one simple regular intersection, points are relatively close together since the roads are parallel and reasonably of close distance to one another. However, road lanes inevitably have different lengths and widths, and there is no one distance that could fit every single intersection in the real world road networks. This problem becomes even more evident when we consider road intersections that contain ramps, slip lanes, staggered junctions or other specialised types of junctions. It would then only mean that a different distance has to be determined for each intersection, which makes the problem counter-intuitive. DBSCAN is used for intersections, but the complexity of the different types of intersections makes it vital to identify the type of intersection and where they are before DBSCAN is even used.

4   Conclusions

In this paper, we present a procedure to extract the backbone structure of spatially embedded networks, such as road networks. The procedure involves simplification of multiple nodes present at the same junction due to multiple lines connecting the same pair of places in the raw data. This is achieved by dissecting the networks into clusters of nodes of degree other than 2 and chunks of nodes of degree 2 serving as paths between the clusters. Applying this simple yet effective technique, we obtain a simplified structure of road network in the Punggol planning area in the northeast region of Singapore. We show that network analysis such as centrality measures on the simplified network can be performed in a more concise and meaningful manner than on the original network.

Acknowledgement. This research is supported by National Research Foundation (NRF) Singapore (Award NRF2018AU-SG01). RS thanks the support of the A*STAR Science Award (JC) under the 2019 A*STAR Student Research Attachment Programme (RAP).

References

1. Data.gov.sg. https://data.gov.sg/
2. Nextzen metro extracts. https://www.nextzen.org/metro-extracts/index.html
3. OpenStreetMap. https://www.openstreetmap.org
4. Barthelemy, M., Bordin, P., Berestycki, H., Gribaudi, M.: Self-organization versus top-down planning in the evolution of a city. Sci. Rep. **3**(1), 2153 (2013)
5. Barthélemy, M., Flammini, A.: Modeling urban street patterns. Phys. Rev. Lett. **100**(13), 138702 (2008)
6. Boeing, G.: OSMnx: new methods for acquiring, constructing, analyzing, and visualizing complex street networks. Comput. Environ. Urban Syst. **65**, 126–139 (2017)
7. Brookes, S., Huynh, H.N.: Transport networks and towns in Roman and early medieval England: an application of pagerank to archaeological questions. J. Archaeol. Sci. Rep. **17**, 477–490 (2018)
8. ESRI: ESRI shapefile technical description (1998). https://www.esri.com/library/whitepapers/pdfs/shapefile.pdf
9. Ester, M., Kriegel, H.P., Sander, J., Xu, X.: A density-based algorithm for discovering clusters in large spatial databases with noise. In: Proceedings of the Second International Conference on Knowledge Discovery and Data Mining, KDD 1996, pp. 226–231 (1996)
10. Gudmundsson, A., Mohajeri, N.: Entropy and order in urban street networks. Sci. Rep. **3**(1), 3324 (2013)
11. Huynh, H.N.: Continuum percolation and spatial point pattern in application to urban morphology. In: D'Acci, L. (ed.) The Mathematics of Urban Morphology, pp. 411–429. Birkhäuser, Springer Nature, Cham (2019). https://doi.org/10.1007/978-3-030-12381-9_18
12. Huynh, H.N.: Spatial point pattern and urban morphology: perspectives from entropy, complexity, and networks. Phys. Rev. E **100**(2), 022320 (2019)
13. Huynh, H.N., Makarov, E., Legara, E.F., Monterola, C., Chew, L.Y.: Characterisation and comparison of spatial patterns in urban systems: a case study of U.S. cities. J. Comput. Sci. **24**, 34–43 (2018)

14. Kim, Y., Fukuyasu, H., Yamamoto, D., Takahashi, N.: A road generalization method using layered stroke networks. In: Proceedings of the 3rd ACM SIGSPA-TIAL International Workshop on Location-based Recommendations, Geosocial Networks and Geoadvertising - LocalRec 2019, Chicago, Illinois, pp. 1–10. ACM Press (2019)

15. Marshall, S.: Streets and Patterns. Routledge, London (2004)

16. Rosvall, M., Trusina, A., Minnhagen, P., Sneppen, K.: Networks and cities: an information perspective. Phys. Rev. Lett. **94**(2), 028701 (2005)

17. Tian, J., Xiong, F., Lei, Y., Zhan, Y.: Revising self-best-fit strategy for stroke generating. In: Harvey, F., Leung, Y. (eds.) AdAdvances in Geographic Information Science, pp. 183–192. Springer, Cham (2015). https://doi.org/10.1007/978-3-319-19950-4_11

18. Xie, F., Levinson, D.: Measuring the structure of road networks. Geogr. Anal. **39**(3), 336–356 (2007)

19. Yamamoto, D., Murase, M., Takahashi, N.: On-demand generalization of road networks based on facility search results. IEICE Trans. Inf. Syst. **E102.D**(1), 93–103 (2019)

20. Yu, W., Zhang, Y., Ai, T., Guan, Q., Chen, Z., Li, H.: Road network generalization considering traffic flow patterns. Int. J. Geogr. Inf. Sci. **34**(1), 119–149 (2020)

21. Zhang, Q.: Road network generalization based on connection analysis. In: Fisher, P.F. (ed.) Developments in Spatial Data Handling, pp. 343–353. Springer, Heidelberg (2005). https://doi.org/10.1007/3-540-26772-7_26

22. Zhang, W., Wang, S., Tian, X., Yu, D., Yang, Z.: The backbone of urban street networks: degree distribution and connectivity characteristics. Adv. Mech. Eng. **9**, 168781401774257 (2017)

23. Zhou, C., Li, W., Jia, H.: Road network generalization based on float car tracking. ISPRS. Int. Arch. Photogramm. Remote Sens. Spatial. Inf. Sci. **XLI–B4**, 71–77 (2016)

Look Deep into the New Deep Network: A Measurement Study on the ZeroNet

Siyuan Wang[1,2](✉) (iD), Yue Gao[1,2], Jinqiao Shi[1,3], Xuebin Wang[1,2], Can Zhao[1,2], and Zelin Yin[1,2]

[1] Institute of Information Engineering, Chinese Academy of Sciences, Beijing, China
{wangsiyuan,gaoyue,wangxuebin,zhaocan,yinzelin}@iie.ac.cn
[2] School of Cyber Security, University of Chinese Academy of Sciences, Beijing, China
[3] Beijing University of Posts and Telecommunications, Beijing, China
shijinqiao@bupt.edu.cn

Abstract. ZeroNet is a new decentralized web-like network of peer-to-peer users created in 2015. ZeroNet, called an open, free and robust network, has attracted a rising number of users. Most of P2P networks are applied into file-sharing systems or data-storage system and the characteristics of these networks are widely investigated. However, there are obvious differences between ZeroNet and conventional P2P networks. Existing researches rarely involve ZeroNet and the characteristics and robustness of ZeroNet are unknown. To tackle the aforementioned problem, the present study measures the ZeroNet peer resources and site resources separately, and at the same time, proposes collection methods for both. No like other simulation experiments, the experiments of this paper are set on real-world environment. This is also the first measurement study about ZeroNet. Experimental results show that the topology of the peer network in ZeroNet has scarce edges, short distances and low clustering coefficients, and its degree distribution exhibits a special distribution. These indicate that the peer network of ZeroNet has poor robustness and the experimental results of the ZeroNet resilience verify this issue. In addition, this paper represents an improved peer exchange method to enhance the robustness of the ZeroNet. We also measure the topology characteristics, languages, sizes and versions of the sites in ZeroNet. We find that the size of the sites and the client version are also the reasons for the low robustness of the ZeroNet.

Keywords: ZeroNet · P2P · Measurement · Network · Peer · Site

1 Introduction

With BitTorrent-based architecture and Bitcoin-cryptography-based account [3], Zeronet has the characteristics of no single point of failure, uncensorable, and free [2]. Recent years, ZeroNet, with its advantages, has attracted a rising number of users. Current studies of ZeroNet is lacking, but the P2P networks have been extensively studied. [6,10,12,16] study the P2P network from the perspective of topology, while [5,7,9] paid more attention to the robustness of the P2P network.

© Springer Nature Switzerland AG 2020
V. V. Krzhizhanovskaya et al. (Eds.): ICCS 2020, LNCS 12137, pp. 595–608, 2020.
https://doi.org/10.1007/978-3-030-50371-0_44

However, there are significant differences between the ZeroNet and traditional P2P networks. Most of P2P networks are applied into file-sharing systems or data-storage systems, but ZeroNet is essentially a site publishing platform. Sites, which shared in ZeroNet, are small and connected. It is different from conventional P2P network and it may cause changes in the topology and robustness. To the best of our knowledge, there is no relevant measurement on ZeroNet yet.

The goal of this paper is to track the aforementioned problems. In this work, the communication protocol of ZeroNet was analyzed and the protocol-based resource collection method was proposed. On this foundation, we model and analyze the peers network and sites network separately. For the peers network of ZeroNet, three important properties are taken into consideration: the degree distribution, small world characteristics and the graph resilience to failures or targeted attacks. For the sites network, in addition to the above characteristics, languages, sizes and versions of the sites in ZeroNet are also measured.

The contribution of this paper can be summarized as follows:

- By analyzing the communication protocol of ZeroNet, the protocol-based resource collection method was proposed. On this basis, we estimate the number of peers and sites in the whole network.
- The present study finds that the peer network in ZeroNet has scarce edges, short distances and low clustering coefficients, and its degree distribution unlike other P2P networks exhibits irregular distribution. This is closely related to the peer exchange mechanism and site size of the ZeroNet and they reduce the robustness of the ZeroNet.
- An improved node exchange method is proposed in this paper which has the ability to increase the site access success rate.
- This study also obtains the topology characteristic, language, version and size distribution of the site network in ZeroNet. Experimental results show the language distribution of ZeroNet is basically the same as that of Dark Web and Surface Web. The small size of sites and low client version are also the reasons for the low robustness of the ZeroNet.

2   Background

ZeroNet, created in 2015, is a decentralized network similar to website and composed of peer-to-peer users. ZeroNet is a BitTorrent-based structure, which also transfers site files in a similar manner. ZeroNet is essentially a site publishing platform. Sites can be accessed through an ordinary web browser where ZeroNet application acts as a local webhost for such pages. In addition, trackers and peer exchange(PEX) message from the BitTorrent network are used by ZeroNet to negotiate connections between peers.

2.1   ZeroNet Components

There are three fundamental components in ZeroNet network: peer, tracker and site.

- peer: Each peer refers to a running zero network client (differentiated by ip and port).
- tracker: A ZeroNet tracker is a special type of server that assists in the communication between peers using the BitTorrent protocol. Tracker keeps track of where site copies reside on peer machines, which ones are available at time of the client request, and helps coordinate efficient transmission and reassembly of the copied file.
- site: Sites in ZeroNet are like files in the BitTorrent network. The client obtains the peers holding the site from the tracker or PEX. Then download the site files from the obtained peers for browsing. Instead of having an IP address, sites are identified by a public key (specifically a bitcoin address). The private key allows the owner of a site to sign and publish change, which propagate through the network.

2.2 ZeroNet Protocols

There are four crucial protocols in ZeroNet network: PEX, Announce, ListModify and StreamFile.

- PEX: Peer exchange(PEX) is a protocol which is a part of ZeroNet site file sharing protocol. It allows a group of peers that are collaborating to share a given file to do so more swiftly and efficiently. PEX greatly reduces the reliance of peers on a tracker by allowing each peer to directly update others in the swarm as to which peers are currently in the swarm.
- Announce: Announce is used to request peers of site from tracker server. When the client sends an Announce message to the tracker server, the tracker will return to the client other nodes that it knows are holding the site and record the client as the site holder.
- ListModify: The ListModify protocol can check the update status of the site without downloading the site. When the node receives the ListModify message, it will return the latest update time of the site.
- StreamFile: Zeronet utilizes the StreamFile protocol to download site files. When the peer receives the StreamFile message, it will transmit the site file data over the TCP link.

3 Related Work

A lot of measurement studies are performed on P2P network [13], such as [11, 14] etc. [6,10,12,16] conducted researches on the topology of P2P network. [6, 16] measure the typology of BitTorrent in real and simulated environments, respectively, [12] detects peers and statistics the proportion of nodes contributing to the Ethereum and [10] measures the extent of decentralization of Bitcoin and Ethereum. However, The purpose and use of Zeronet is completely different from them. ZeroNet is essentially a site publishing platform rather than a file sharing system. The files shared in ZeroNet are sites which have a very small size and it may cause changes in the topology.

[5, 7, 9] analyzed the resilience and robustness of the P2P network. [9] analyzes the robustness of Bitcoin under the AS-level attack. [7] verify the influence of sybil attack to P2P network. [5] measures the impact of neighbor selection to the performance and resilience of P2P. ZeroNet has attracted a large number of users due to its declared high robustness, but its true robustness is unknown.

Based on BitTorrent architecture, sites in ZeroNet are censor-resistant, studies related to the dark web [4] is also very relevant to the ZeroNet. For example, the most relevant work is [15] which crawls the sites in Tor [8] and measures some of the properties and privacy of the sites in Tor.

4   Peer Resource Measurement and Analysis

This section gives a brief description of the work of peer resources. The work is mainly divided into two parts: the peer resource collection and the topology analysis. By the experiment of resilience, we find that the robustness of ZeroNet is poor. In addition, the improved peer exchange method is proposed.

4.1   Peer Resource Collection

By the way of deploying a collection system, we have collected most of peers in ZeroNet. The peer coverage ratio and the average connection number also be calculated to verify that the vast majority of peers in ZeroNet are detected by us. Figure 1 depicts the structure of peer network in ZeroNet. The larger the size of the nodes in the graph, the more neighbors there are.

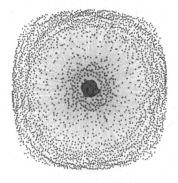

Fig. 1. Structure of peer network in ZeroNet

Resource Collection Method. For the sake of collecting the peers of ZeroNet as much as possible, our collection system comes up with two components: active part and passive part. The active part is faster and more targeted, while passive part has the advantages of avoiding the extra network burden and discovering nodes behind NATs. The active part contains obtaining peer information from trackers by Announce messages and from peers by PEX messages. Each epoch every site will be utilized to construct PEX message and Announce message to obtain peers from other peers and trackers. The passive part is deployed by encapsulating modified ZeroNet client into docker container, so that one computer can run multiple ZeroNet clients. Modified ZeroNet clients will record the peers who build connections with us.

Resource collection runs from November 11th to 13th. 2 active-collecting machines and 15 passive-collecting machines with 90 passive clients are deployed.

Coverage Ratio. The peer coverage ratio $p(n)$ is the collection system detecting the percentage of peers and it represents the discovering ability of the system. Borrowing from the method used in a, $p(n)$ is evaluated as follows:

$$\Delta C = C(m+1) - C(m) \tag{1}$$

$$\Delta C = N(1 - p(n)) \tag{2}$$

$$p(n) = 1 - \frac{C(m+1) - C(m)}{N} \tag{3}$$

Where $C(m)$ is the total number of distinct peers discovered after mth epoches, ΔC is the number of newly discovered peers and N is the number of all distinct peers (identified by the IP address and the listening port). Time cost of one epoch is 5 s. As shown in Fig. 2, after 40 epoches, the number of new peers and distinct peers discovered by the server per epoch is basically the same. The collection system averagely finds 303.74 distinct peers and 1.33 of them are new, so it can be estimated that the peer coverage ratio is nearly 99.56% and snapshot is taken every 100 s.

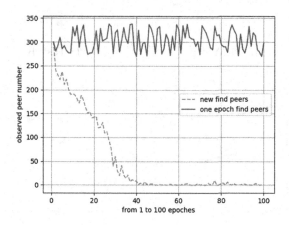

Fig. 2. The number of new peers and distinct peers discover by the server per epoch.

This study also compares the number of the average connections of each peer with the modified ZeroNet client deployed by us, we find the numbers are similar.

4.2   Topology Analysis

In this part, we select peer networks of four sites to perform their respective topological measurements and these sites have different sizes and different types. Some of their basic properties are shown in Table 1. The topology of the ZeroNet is represented by an undirected graph $G = (V, E)$. V denotes the set of ZeroNet peers and E denotes the set of connections of peers. For example, $(u, v) \in E$ means that peer u in the peer list of peer v. Degree distribution, small word and resilience, three topological properties of peer network in ZeroNet are measured.

Degree Distribution. Degree distribution is one of the most important metrics which represents the structure of peer network in ZeroNet. Existing research found degree distribution usually have close relation with the resilience and routing latency of the network.

The degree distribution of the peer cluster is shown in Fig. 3 where x-axis is the node rank in descending order of degrees and y-axis is the degree. Unlike most P2P networks, measured degree distribution of peer cluster is not very

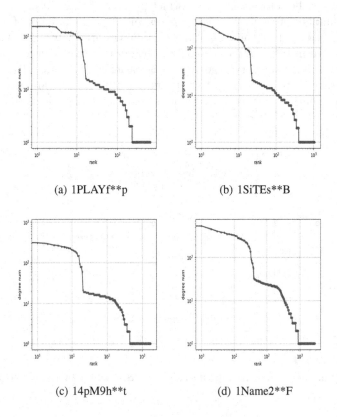

(a) 1PLAYf**p   (b) 1SiTEs**B

(c) 14pM9h**t   (d) 1Name2**F

Fig. 3. Degree distribution of peer network

Table 1. Peer small world characteristics

Site	Type	Nodes	Edges	cof	L
1PLAYf**p	Real network	662	1792	0.1644	3.1302
	Simulative network	662	1861	0.0107	3.9510
1SiTEs**B	Real network	1046	3354	0.2684	2.8348
	Simulative network	1046	3377	0.0051	3.9269
14pM9h**t	Real network	1711	4690	0.1882	2.9061
	Simulative network	1711	4698	0.0027	4.5669
1Name2**F	Real network	2791	9978	0.1425	3.1947
	Simulative network	2791	9867	0.0027	4.2776

consistent with the power-law distribution. Lots of peers in ZeroNet have a low degree which indicates that ZeroNet may not be robust like other P2P networks.

Small World. Small world property is characterized by dense local clustering or cliquishness of connections between neighboring nodes. Most of P2P networks have been verified to be a small-world, because PEX or PEX-similar Mechanism helps peers know each other within a short distance. Our goal is to verify whether peer network in ZeroNet exhibits small-world properties though its PEX mechanism is weak.

Though all connections of peer network in ZeroNet can't be obtained, we also can get an approximate value of the shortest path length L between peers (most of peers and connections are detected).

$$L = \frac{2\sum_{i,j=1}^{i,j=N} dis_{ij}}{N(N-1)} \tag{4}$$

where dis_{ij} is the length of shortest path between node i and node j.

The clustering coefficient cof is calculated by the Eq. (5) and Eq. (6):

$$cof_v = \frac{2 * E_v}{d_v(d_v - 1)} \tag{5}$$

$$cof = \sum_{v=1}^{N} cof_v \tag{6}$$

where d_v is the count of the neighbors of node v, E_v is the count of connections between the the neighbors of node v. In order to clearly measure the level of the average shortest distance and clustering coefficient of peer network in ZeroNet, an analogous network with the same count of nodes and the same count of edges was simulated. The calculating results of these networks are revealed in Table 1.

The peer network in ZeroNet has a short peer distance and high clustering coefficient which are orders of magnitude larger than analogous random networks. So it is a small-world network.

Resilience. For a robust network, when some nodes leave the network, the network should still maintain good connectivity. The resilience of peer network in ZeroNet is examined in two different ways of node removal: random removal, and particularly removing the highest-degree nodes first.

Figure 4 shows the proportion of remaining nodes in the largest connected component of four peer networks when facing random deletion and targeted deletion modes. This figure clearly illustrates peer networks in ZeroNet are extremely robust to random peer removals (after removing 22–55% peers). However, when we particularly remove the highest-degree nodes first, it quickly becomes very fragmented (after removing 8–22% peers). This means that the resilience of the ZeroNet is very poor. The peer network is centered on a small number of nodes. When these nodes leave the network, it will seriously affect the robustness of the network. We can also find that the larger the cluster network, the stronger the resilience of the cluster, on the contrary, the weak resilience.

4.3   Reason Analysis and Suggestion

Through the analysis of the characteristics of the peer networks in ZeroNet, we find that the degree of some nodes in the ZeroNet is very low, the network is highly centralized and has significant small-world characteristics. The possible reasons for the current characteristics of ZeroNet are as follows: (1) As the presence of NATs and firewall, lots of peers can't be connected. There are few connections between them. (2) No DHT in ZeroNet and high dependence on tracker. No DHT has resulted in a single form and inadequate of node exchange in ZeroNet. As tracker in ZeroNet can also be a peer, a large number of sites can be downloaded from these peers without PEX operation to get more peers. So node network is obviously centered on tracker. (3) Different from the file sharing P2P network, the file resources in ZeroNet are very small (see the site resource measurement and analysis section). When a user visits a site, the site download time is very short and the seeders required are very small. So the connections between nodes are sparse.

If more online nodes can be found, the success of ZeroNet visits can be significantly improved, thus improving the robustness of ZeroNet. This paper puts forward an improved peer exchange mechanism for ZeroNet client. By similar methods, we have collected far more sites than clients can access. We start a daemon thread to continuously perform PEX with online nodes, so as to make up for the deficiency of ZeroNet native PEX operations. Specific steps are to start a daemon thread and do the following: (1) Collect all currently online nodes. (2) Use the most popular sites to PEX with all online nodes to get new nodes. (3) Verify that the nodes are found to be linear. (4) Repeat (2), (3).

In this way, the importance of trackers in the ZeroNet is weakened. Experiments have found that even if trackers are blocked, we can still download most of the sites in ZeroNet by crawler with improved peer exchange mechanism. Therefore, our method can improve the connectivity of the network and enhance the robustness of the peer network in ZeroNet.

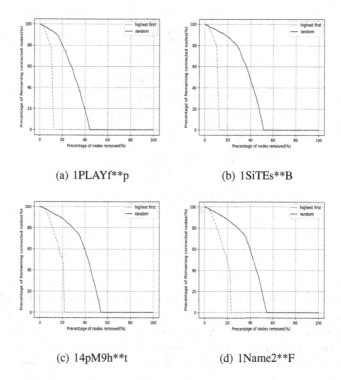

(a) 1PLAYf**p

(b) 1SiTEs**B

(c) 14pM9h**t

(d) 1Name2**F

Fig. 4. The proportion of remaining nodes in the largest connected component changes with the proportion of removed nodes

5   Site Resource Measurement and Analysis

Sites in ZeroNet are equivalent to files in BT network. However, there is link relationship between sites that files in BT don't have. In this section, we will introduce how to collect site resources and the characteristic of sites in ZeroNet. Figure 5 depicts the typology of site network in ZeroNet. The more connections to other ZeroNet sites there are, the larger the size of the site in the graph.

5.1   Site Resource Collection

The crawler is specifically designed according to the ZeroNet protocol, which is fast and light than regular Dark Web crawler [15]. Furthermore, we check for site updates via ListModify messages without downloading the site. The sites download from the nodes found by peer collection module through the StreamFile message. When a new site or changed site is downloaded, we will retrieve the new ZeroNet site from it.

Experiment lasted from November 1, 2019 to December 1, 2019. More than 14000 sites in ZeroNet and downloaded more than 1300 online sites were found by using the above method.

604   S. Wang et al.

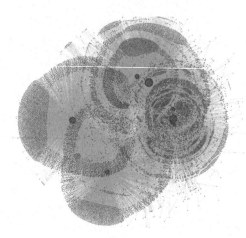

Fig. 5. Typology of site network in ZeroNet

From the crawl results, it can be seen that even if the characteristics of ZeroNet make website always online (as long as the node downloading the website is online), there are still a large number of which cannot be accessed. One of the most important reasons is that a large number of sites are created, and very few sites can be popular. These sites are hardly visited and there will be no peers serving them.

Coverage Ratio. The most important way to publish a new site in ZeroNet is by publishing it to the popular sites. [15] collects site data by using a headless browser to simulate user clicks which can collect all sites that clients can access. We use both the method in [15] and our proposed site collection method for resource collection. 98.7% of the sites collected by [15] is covered by our method, and 45.3% of the sites covered by our method cannot be collected. The sites collected by our proposed method can cover 98.7% of the sites collected by [15] and an additional 45.3% sites which cannot be collected by method [15] are collected. Existing data can not give the exact number of current coverage ratio, but we have verified that our site collection method can collect far more sites than the ZeroNet client can visit. Figure 6 plots the distinct sites and new sites found by our crawler. After 10 days of data collection, the number of stations found daily has stabilized. Our collection method can collect most of online site.

5.2 Topology Analysis

Similar to the peer network topology measurement, we used the same method to measure the topological characteristics of the site network: degree distribution and small-world characteristics. The resilience of site network is not measured because it is meaningless, because most popular sites will not be offline, as long as one peer holding the site is online. Undirected graph $G = (V, E)$ of site

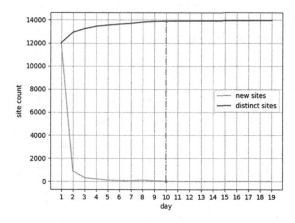

Fig. 6. Distinct sites and new sites found

topology is defined as follows: The nodes in the graph are the sites in the network (distinguish by site address) and the edges donate the links in the pages of sites.

The measurement results show that ZeroNet has characteristics of the small world which is similar to the surface network and the dark network.

The degree distribution of site network is shown in Fig. 7 and the small world property is shown in Table 2.

Table 2. Site small world characteristics

Type	Nodes	Edges	cof	L
Real network	14116	29418	0.33866	3.62390
Simulative network	14116	29418	0.00015	6.88375

Except for the end of the curve, the site degree distribution shows a clear power-law distribution and a large number of sites without external links exist in ZeroNet. The possible reason is that the cost of creating a site in ZeroNet is low, no server costs are required, and a large number of humble sites are created (only a "Hello!" in pages).

By comparison with the random graph, it is found that site network in ZeroNet has a high clustering coefficient and a short average shortest distance. Especially the clustering coefficient is many orders of magnitude higher than the random graph. Hence, site network in ZeroNet is obviously centralized and has small world property. We also verify the type of central site and found that most sites of them are navigation sites. Because ZeroNet is based on the BT architecture, popular sites in Zeronet will not go offline due to a single point of failure. So the centralization of the site network will not reduce the robustness of the ZeroNet.

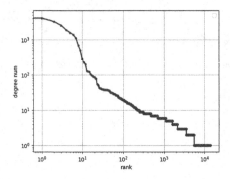

Fig. 7. Degree distribution of site network

5.3 Content Analysis

In this subsection, we measured the distribution of languages, versions and sizes within the active ZeroNet sites.

Languages. This study also measured the distribution of languages of the active ZeroNet sites. We use the same method as [15] by Google Translate API to get the language type of the web pages. Each website may be classified into multiple languages.

We found the sites of ZeroNet in 58 different languages. Among them, English, Russian, Chinese, German, French and Spanish are the most used languages, accounting for 66.2%, 17.3%, 5.5%, 3.4%, 3.1% and 2.3% respectively. Except that Chinese sites account for a relatively high proportion, the rankings of the other five languages are basically consistent with the surface network [1] and the dark web [15].

Version and Size. In this subsection, unique characteristics of ZeroNet Sites are measured, such as the versions of the site and the distribution of site sizes in ZeroNet. Version is the client version which site was published in and site size is the sum of all files in the site directory.

Version and size information of sites in ZeroNet are obtained by the online site file *content.json* download by crawler. As of December 1, 2019, the latest version of ZeroNet is 0.7.1. ZeroNet client has no automatic update mechanism. The top site-publish versions are shown in Fig. 7. 77.6% sites are published by the client with lower than 0.6.* version which suggests most popular and online sites are created long time ago. Only 12.2% sites are published by newest client. Even if the new version of the client of ZeroNet has made a lot of optimizations, there are still a large number of users passively updating the client. This may have a bad influence on the robustness of ZeroNet (Fig. 8).

The distribution of site sizes in ZeroNet is as follows, 95.2% sites are less than 10 MB, 3.8% sites are less than 100 MB and greater than 10 MB, 1.0% sites are

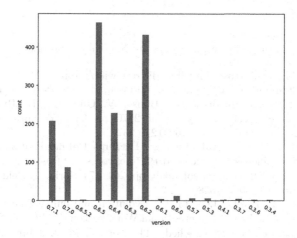

Fig. 8. Version distribution of ZeroNet

greater than 100 MB. Most of the ZeroNet sites are smaller than 10 MB, which is also caused by the default site size limit (10 MB). The small site size makes the site download time very short, and the PEX between peers will be insufficient which also reduces the robustness of ZeroNet.

6   Conclusion

In this paper, we first find the resource collection method of ZeroNet. On this basis, we measure and analyze the peer network and site network of the ZeroNet.

The experimental results reveal that the peer network of the ZeroNet has obvious small world characteristics, while the robustness of the node network of the ZeroNet is poor, especially when a large number of high-level nodes leave the network, the connectivity of the network becomes extremely poor. We analyze that this is mainly caused by the tracker and the peer exchange mechanism of ZeroNet. Therefore, we propose an improved peer exchange mechanism, so that even in the face of most trackers unavailable, the ZeroNet can still be visited.

This article also fist measure the site of ZeroNet. The results show that the degree distribution of the site network in the ZeroNet is a power law distribution, which also has the characteristics of small worlds. The content of the sites in ZeroNet are also measured from three angles: languages, sizes, and versions. The measurement results show that English, Russian, Chinese, German, French and Spanish are the most popular languages among the ZeroNet sites. The site size of ZeroNet is generally small, and most of the online sites are created using lower ZeroNet client versions. This is also one of the reasons why ZeroNet is less robust.

References

1. Usage of JavaScript websites. https://w3techs.com/technologies/
2. ZeroNet home page. https://zeronet.io/
3. ZeroNet Wikipedia. https://en.wikipedia.org/wiki/zeronet
4. Biddle, P., England, P., Peinado, M., Willman, B.: The darknet and the future of content protection. In: Becker, E., Buhse, W., Günnewig, D., Rump, N. (eds.) Digital Rights Management. LNCS, vol. 2770, pp. 344–365. Springer, Heidelberg (2003). https://doi.org/10.1007/10941270_23
5. Chun, B.-G., Zhao, B.Y., Kubiatowicz, J.D.: Impact of neighbor selection on performance and resilience of structured P2P networks. In: Castro, M., van Renesse, R. (eds.) IPTPS 2005. LNCS, vol. 3640, pp. 264–274. Springer, Heidelberg (2005). https://doi.org/10.1007/11558989_24
6. Dale, C., Liu, J., Peters, J.G., Li, B.: Evolution and enhancement of bittorrent network topologies. In: 16th International Workshop on Quality of Service, IWQoS 2008, University of Twente, Enschede, The Netherlands, 2–4 June 2008, pp. 1–10 (2008)
7. Dinger, J., Hartenstein, H.: Defending the Sybil attack in P2P networks: taxonomy, challenges, and a proposal for self-registration. In: Proceedings of the First International Conference on Availability, Reliability and Security, ARES 2006, The International Dependability Conference - Bridging Theory and Practice, Vienna University of Technology, Austria, 20–22 April 2006, pp. 756–763 (2006)
8. Dingledine, R., Mathewson, N., Syverson, P.: Tor: The second-generation onion router. Technical report, Naval Research Lab, Washington DC (2004)
9. Feld, S., Schönfeld, M., Werner, M.: Analyzing the deployment of Bitcoin's P2P network under an as-level perspective. In: Proceedings of the 5th International Conference on Ambient Systems, Networks and Technologies (ANT 2014), the 4th International Conference on Sustainable Energy Information Technology (SEIT-2014), Hasselt, Belgium, 2–5 June 2014, pp. 1121–1126 (2014)
10. Gencer, A.E., Basu, S., Eyal, I., van Renesse, R., Sirer, E.G.: Decentralization in Bitcoin and Ethereum networks. In: Meiklejohn, S., Sako, K. (eds.) FC 2018. LNCS, vol. 10957, pp. 439–457. Springer, Heidelberg (2018). https://doi.org/10.1007/978-3-662-58387-6_24
11. Hei, X., Liang, C., Liang, J., Liu, Y., Ross, K.W.: A measurement study of a large-scale P2P IPTV system. IEEE Trans. Multimedia 9(8), 1672–1687 (2007)
12. Kim, S.K., Ma, Z., Murali, S., Mason, J., Miller, A., Bailey, M.: Measuring Ethereum network peers. In: Proceedings of the Internet Measurement Conference 2018, IMC 2018, Boston, MA, USA, 31 October–02 November 2018, pp. 91–104 (2018)
13. Milojicic, D.S., et al.: Peer-to-peer computing (2002)
14. Peng, Z., Pallelra, R.R., Wang, H.: On the measurement of P2P file synchronization: Resilio Sync as a case study. In: 25th IEEE/ACM International Symposium on Quality of Service, IWQoS 2017, Vilanova i la Geltrú, Spain, 14–16 June 2017, pp. 1–2 (2017)
15. Sánchez-Rola, I., Balzarotti, D., Santos, I.: The onions have eyes: a comprehensive structure and privacy analysis of tor hidden services. In: Proceedings of the 26th International Conference on World Wide Web, WWW 2017, Perth, Australia, 3–7 April 2017, pp. 1251–1260 (2017)
16. Majing, S., Zhang, H., Du, X., Fang, B., Mohsen, G.: A measurement study on the topologies of BitTorrent networks. IEEE J. Sel. Areas Commun. 31(9–Suppl.), 338–347 (2013)

Identifying Influential Spreaders On a Weighted Network Using HookeRank Method

Sanjay Kumar[1,2(✉)], Nipun Aggarwal[2], and B. S. Panda[1]

[1] Computer Science and Application Group, Department of Mathematics,
Indian Institute of Technology Delhi, Hauz Khas 110016, New Delhi, India
[2] Department of Computer Science and Engineering, Delhi Technological University,
Main Bawana Road, New Delhi 110042, India
sanjay.kumar@dtu.ac.in

Abstract. Influence maximization is a significant research problem that requires the selection of influential users who are capable of spreading information in the network such that it can reach to a large number of people. Many real-world networks like Road Network, Email Networks are weighted networks. Influence maximization on weighted networks is more challenging than an unweighted network. Many methods, such as weighted-degree rank, weighted-voteRank, weighted-eigenvalue rank, and weighted-betweenness rank methods, have been used to rank the nodes in weighted networks with certain limitations. In this manuscript, we propose a Hooke's law-based approach named HookeRank method to identify spreaders in a weighted network. We model edge weights as spring constants. The edges present in the network are modeled as springs, which are connected in series and parallel. They elongate by a distance under the effect of a given constant force following Hooke's law of elasticity, and this is the equivalent propagation distance between nodes in the network. The proposed the model finds relevant influential nodes, that can propagate the information to other nodes. A higher HookeRank score implies the greater influential capability of the node in the network. We compared our proposed algorithm with state-of-the-art models and found that it performs reasonably well on real-life data-sets using epidemic spreading Susceptible-Infected-Recovered model.

Keywords: Influence maximization · Hooke's law · Information diffusion · Social networks · Weighted networks

1 Introduction

Many real-world networks, like online social networks, transportation networks, email networks, collaboration networks, and many others are complex weighted networks [1]. These networks can be modeled as graphs, $G = (V, E, W)$ where V represents a set of nodes, E denotes edges between nodes, and W represents the

© Springer Nature Switzerland AG 2020
V. V. Krzhizhanovskaya et al. (Eds.): ICCS 2020, LNCS 12137, pp. 609–622, 2020.
https://doi.org/10.1007/978-3-030-50371-0_45

edge weight. The weight of a connection between two nodes usually depends on the exchange of services, intensity, or duration [2]. If two nodes are frequently interacting or if they have high interactions, then information or diseases are more likely to be transferred between them [3]. Complex networks have a large number of nodes, and interaction between nodes is usually complicated. The evolution of complex networks has led to the establishment of many useful applications like influence maximization, viral-marketing, and information propagation [4]. Influence maximization [5] is a technique to select some constant number of nodes as seed nodes which are capable of spreading information by "Word-of-Mouth" analogy, after knowing the information source. Therefore, influence maximization finds great value in the business. The influence maximization has many applications, for example, controlling the proliferation of messages and rumors, positioning influential researchers, and discovering social leaders. The process of influence maximization consists of two activities, first, identifying the seed spreaders and the second the information diffusion phase. In the study of disease transmission, numerical models are helpful in understanding the spread and control of epidemics. The circulation of the information in the complex network is very similar to the epidemic spreading, and many popular methods for modeling information diffusion are based on the epidemics spreading [6,7]. In epidemiology, mathematical models play a role as a tool in analyzing the spread and control of infectious diseases [8]. Many researchers have successfully applied epidemic spreading models to information propagation in complex networks to estimate the final spread of the information originating from the source nodes.

Most of the influence maximization models on weighted-networks are merely extensions of the algorithms counterparts on unweighted networks by introducing edge weight into the models. Numerous other models have considered standard graph-theoretical features based approach for identifying the important spreaders. In this paper, we propose a Hooke's law of elasticity based approach named HookeRank method to identify spreaders in a weighted network. Our algorithm considers the influence of nodes in a setting where edges are modelled as springs and edge-weights are modelled as elasticity coefficients. We model edge weights as spring constants. The edges present in the network are modeled as springs, which are connected in series and parallel. They elongate by a distance under the effect of an assumed constant force following Hooke's law of elasticity, and this is the equivalent propagation distance between nodes in the network. The contributions of our work are as follows:

1. We propose a novel method based on Hooke's Law of Elasticity in complex weighted networks to find the influential spreaders.
2. We model the equivalent weight between indirectly connected nodes in a weighted network.
3. The proposed algorithm is an improved method of selection of influential nodes on real-world data-sets.

The rest of the paper is organized as follows: Sect. 2 consists of the related work in this field. Section 3 presents the data-sets and the models for information diffusion used in this paper. In Sect. 4, we describe the methodology of our novel method and simulation on a toy network. Section 5 discusses the simulation of our algorithms on various real-world networks. Finally, Sect. 6 concludes our paper.

2 Traditional Centralities

The initial models in the field of influence maximization have majorly been innovations in the field of unweighted networks where all edges are equally important. In real-world networks, these edges are associated with weights that need to be considered while analyzing the strength of these connections, during a cycle of information diffusion. When we consider these aspects of topology, we can gather insights into what is most beneficial for the maximization of information diffusion. The early advances in weighted networks were through centralities like DegreeRank used for unweighted networks, by additional weighing of these edges to achieve a weighted DegreeRank [9]. In a similar pattern, multiple centralities were eventually derived from unweighted networks, evolving into a method for weighted graphs through mathematical adjustments leading to weighted algorithms. Betweenness centrality considers the shortest path of a node in an unweighted graph, and it was extended for weighted version giving the weighted-betweenness centrality [10,11]. Based on the notion of the voting scheme, researchers have proposed influence maximization algorithm in unweighted as well as weighted networks where the nodes getting the highest votes in each round gets selected as spreader nodes [12–14]. The h-index is a measure of the impact of researchers based on the number of citations received, and by augmenting edge weight, Yu et al. proposed a weighted h-index centrality [15]. Weighted-eigenvector centrality applicable in a weighted network is based on the fact that a node is important if its neighbors are also famous and finds the centrality for a node as a function of the centrality of its neighbors [16]. Eades [17] suggested to model the edges of the network as springs to draw graphs by minimizing potential energy. This method was later refined by Fruchterman et al. [18], where they model nodes as electrical charges and edges as connecting springs. The electrical charges make these nodes repel each other. One of the most popular algorithms for drawing graphs is Kamada and Kawai's method, which models the edges of the graph as springs acting following Hooke's Law [19,20]. The method optimizes the length of the spring between any two nodes by minimizing a global cost function. We argue the applicability of the spring-based model to measure the centrality of the nodes and to find influential spreaders.

3 Datasets and Performance Metrics

3.1 Information Diffusion Model

SIR Model: In this paper, we utilize the susceptible-infected-recovered (SIR) model as the data diffusion model [21]. This model divides nodes into three categories Susceptible (S), infected (I), and recovered (R). Susceptible nodes are supposed to receive data from their infected neighbor nodes. The information starts from a subset of the network nodes with the spreading parameter (β), and recovery rate (γ). In the SIR model, initially, all nodes, except seed nodes, are in a susceptible state. After each progression, the infected nodes affect their susceptible neighbors with a likelihood of β. Infected nodes at the next timestamp enter the recovered stage with a likelihood of γ. When arriving at the recovered stage, they are no longer prone and can't be infected again.

3.2 Performance Metrics

The final infected scale $(f(t_c))$:- It is a measure of the final spread of the information originating from the chosen seed nodes at the end of SIR simulations. The final infected scale is the final number of recovered users that passed through the chronological advancements from susceptible, infected, and finally, to recover during the information diffusion process. There are two ways to measure this criterion, first $f(t_c)$ is plotted against time t, which shows us the propagation of the information on the network as time proceeds. Secondly, $f(t_c)$ is plotted against the different fraction of spreaders, which shows us the propagation of the information on the network as the number of spreaders taken by the algorithm initially is changed.

3.3 Datasets Used

Table 1. Real world data-sets for simulation

Dataset Name	Description	Nodes	Edges
PowerGrid [22]	An undirected weighted network containing information about the power grid of the Western States of the United States of America	4941	6954
Facebook-like weighted network [23]	This undirected weighted dataset originates from an online community for students at the University of California, Irvine	1899	20297
US Airports [24]	An undirected weighted network of the 500 busiest commercial airports in the United States	500	28237

4   Methodologies

In this section, we present the mapping of the edges present in the network as springs, which are connected in series and parallel, and describe the proposed HookRank method. The discrete part of individual connections of springs is discussed in detail. In a weighted network, weights generally mean that the higher the weight, the stronger is the connection. The same is true for our network as well since we know from Hooke's law that more is the spring constant, less is the displacement from the spring. Now this new unit of distance between any two Nodes. This distance is the actual distance between these nodes when the information diffusion is to be considered. By normalizing these distances, we can model not only the approximate form but also the equivalent of all the different paths that exist between any two pairs of nodes. When we model this, individual graphs for each node are created, and these nodes can now be evaluated based on the amount of information they can propagate. We will consider that a constant force $F_0 = 1$ continuously acts on the node that is chosen as a seed node. Now the seed node is connected to every other node with a spring constant of $k = w_{ij}$ where k is the spring constant of a spring that connects node i and node j with a weight of w_{ij}. Now using a breadth-first traversal (BFS), calculate the new spring between the edges, if there exist, multiple edges between the springs, on different levels, they must be added since they are definitely parallel. When traversing from the node of one level to another, use the series combination to generate equivalent springs and to calculate most probable distances and finally propagating information through each of these nodes to find the maximum amount of spreading that takes place (Fig. 1).

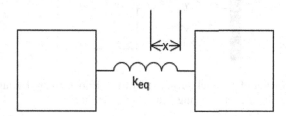

Fig. 1. Calculating the heuristic distance between two nodes based on weights of the edges, being modeled using springs and evaluating to a single spring, following Hooke's Law of Elasticity

Parallel: When Springs are placed in parallel, they end up as a joint spring with the total elasticity of a new spring of a spring constant that can be modeled using the fact that the spring is definitely much stiffer.

Series: It is possible to add the contributions of the springs in series. The Springs in series make a more flexible spring that tends to elongate more than the springs that are previously involved in the complete connection.

4.1   Distance Calculation

When two springs of different spring constants, k_1 and k_2 respectively are placed in series with each other, we get (Fig. 2):

$$1/k_{eq} = 1/k_1 + 1/k_2 \tag{1}$$

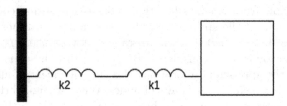

Fig. 2. Demonstration of serial springs following Hooke's law. Series edges occur when more than one node occurs in the path.

When two springs of different spring constants, k_1 and k_2 respectively are placed in parallel with each other, we get (Fig. 3):

$$k_{eq} = k_1 + k_2 \tag{2}$$

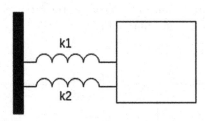

Fig. 3. Demonstration of parallel springs following Hooke's law. Parallel edges occur when more than one path of reaching the same node exists.

This means in an actual network is that springs in series are stiffer if the strength of ties in the individual connections is strong. It also implies that more connections from one node to another, add up a single connection, as seen in case of parallel connections. Now the equivalent distance between any of these nodes, under a constant force is given as:

$$x = f/k \tag{3}$$

Assuming $f = 1$, without loss of generality, we can easily see that the full measure of distance in this network is relative. A breadth-first search similar method is used for the nodes to find the equivalent value of k between all indirect neighbors.

4.2   Proposed Algorithm

Based on the notion of modeling edges as springs, we compute the centrality of the nodes in the network. Edge weights in the network are the spring-constants of these modeled springs. The proposed HookeRank method uses the following steps:

1. For each node, we perform a breadth-first search to all other nodes and calculate the distance of each node in series from the nearest neighbor.
2. In the BFS traversal, when a new node is encountered, we add its distance from its parent using a series combination as in Eq. 2 and add this neighbor to the queue.
3. All the nodes that occur again in the BFS traversal are assumed to be in a parallel connections and add up to the spring constant according to Eq. 1.
4. When for a given node, the queue is completely processed, we get its equivalent tree.
5. Calculate the HookeRank value for the node by finding the weighted average degree of the given node in the equivalent tree.
6. This procedure gives us the HookeRank value for each node, and the result is stored in a dictionary containing the node and its HookeRank value.
7. As, the objective of influence maximization is to select top c nodes, where c is a constant. Here, the top c nodes are the nodes having the maximum HookeRank score values in the ranking, and such nodes can be chosen as the influential spreaders.

4.3   Time Complexity of the Proposed Algorithm

The time complexity of the HookeRank method consists of initialization of spring constants and selection of the node with the highest number of the closest neighbor score and finding the level order traversal with respect to all the nodes (Step 1 to Step 4). Overall this makes the complete algorithm bounded by $O(V(E+V))$. This time complexity, however, reduces because we know that if an equivalent spring from A to B has the constant k, then spring from B to A has the same spring constant (Step 6).

5   Results and Analysis

We perform the experiment of the proposed HookeRank method along with the contemporary centrality measures like weighted- degree, weighted betweenness centrality, weighted eigenvector centrality, and weighted voteRank. The investigation has been performed on a toy network and three real-world networks of different nature, application, and size that are listed in Table 1. We use the SIR model to compute the final infected scale, $f(t_c)$, as a function of spreaders fraction and final infected scale in terms of increasing timestamps. The results were averaged over SIR 100 simulations. For simplicity and to maintain consistency in the analysis for all data-sets, we chose infection rate (β) as 0.01, meaning that when a node is infected, then it can infect 1% of its neighbors randomly.

5.1   Simulation of the Proposed Algorithm On a Toy Network

Here, we simulate the working of the proposed HookeRank method using a toy network, as depicted in Fig. 6. The network is a weighted graph with edge weights representing the stiffness constant of the spring (Fig. 4).

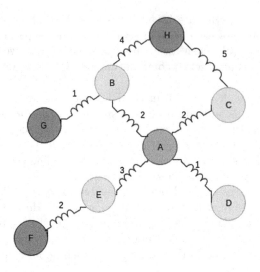

Fig. 4. A sample Weighted Network where the edges are modelled as springs and the spring constant is analogous to the weight of the edge

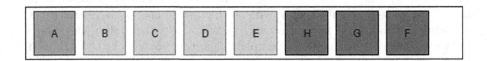

Fig. 5. The run of a breadth-first search with respect to A as the starting node

Let us consider the steps of the algorithm for a sample node A to understand the working of this algorithm. We take a FIFO queue and put A inside the queue. Now, the distance to A is 0. Now let's compute the spring constant for the first neighbors B, C, D, E using the direct connection of the spring. These are then popped out, and their neighbors are pushed into the queue. The equivalent spring constant is found through a series connection through the parent. In the case of multiple parents at the same level, the constant is found using a parallel combination, as in the case of node H. For the node A, the simulation and calculation of the equivalent distances of all the other nodes by using a breadth-first search are performed. Its immediate neighbors are processed first, and so on, the different layers are highlighted in a level order fashion, as shown in Fig. 5. We can now compute the value of all the neighbors of A, as performed in the

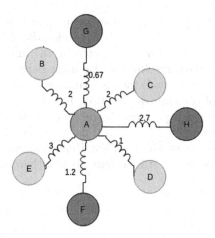

Fig. 6. The final network with respect to node A. Notice that certain indirect neighbors became direct neighbors under the action of both series and parallel connections.

above section. The computation will result in a graph similar to Fig. 6. Notice the connections and the spring constants. Thus, the average spring constant for A becomes, $12.57 / 8 = 1.52$, and thus, an elongation averaging 1.57 are taken, giving us a HookeRank value of 1.57 for the node A.

Table 2. Results of HookeRank Score of various nodes in the toy Network

Node	A	B	C	D	E	F	G	H	HookeRank-value
A	0.0	2.0	2.0	1.0	3.0	1.2	0.67	2.7	1.57
B	2.0	0.0	3.2	0.67	1.2	0.75	1.0	4.0	1.60
C	2.0	3.2	0.0	0.67	1.2	0.75	0.76	5.0	1.69
D	1.0	0.67	0.67	0	0.75	0.69	0.4	0.52	0.58
E	3.0	1.2	1.2	0.75	0.0	2.0	0.55	1.4	1.26
F	1.2	0.75	0.75	0.69	2.0	0.0	0.4	0.82	0.83
G	0.67	1.0	0.76	0.4	0.55	0.4	0.0	0.8	0.57
H	2.7	4	5	0.52	1.4	0.82	0.8	0.0	**1.90**

A similar calculation can be performed for each of the nodes, and their equivalent HookeRank value is calculated, as given in Table 2. Based on the value of the HookRank score, node H is elected as the top spreader in the toy network.

5.2 Simulation of the Proposed Algorithm on Real-Life Networks

Figure 7, Fig. 8, and Fig. 9 depicts the final infection scale $(f(t_c))$ with respect to the percentage of spreaders for three real-life data-sets with infection rate (β)

as 0.01. We consider the percentage of influential spreaders as the seed nodes in the range of 2%, 4%, 6%, 8%, and 10% to plot the final infection scale. In Fig. 7, note that the number of nodes affected by the infection is maximum for HookeRank on the US-Airports Network for most percentages of spreaders. In Fig. 8, HookeRank greatly exceeds the performance of other algorithms towards increasing the count of the spreader fraction. In the weighted PowerGrid data, shown in Fig. 9, HookeRank performs better than most other algorithms from an early stage. In the weighted PowerGrid data, shown in Fig. 10, the increase in the number of spreaders results in WVoteRank becoming marginally close to HookeRank, but our algorithm still performs better than all other algorithms in the simulation.

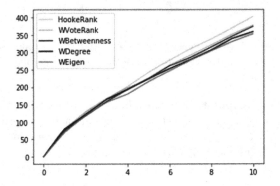

Fig. 7. The infection scale with respect to the percentage of spreaders on US-Airports network with $\beta = 0.01$.

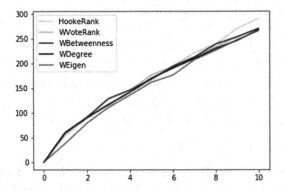

Fig. 8. The infection scale with respect to the percentage of spreaders on Facebook-like weighted network with $\beta = 0.01$.

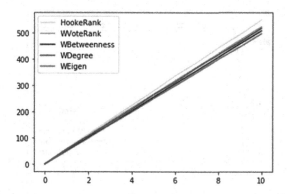

Fig. 9. The infection scale with respect to the percentage of spreaders on US PowerGrid network with $\beta = 0.01$.

Figure 10 presents the final infection scale ($f(t_c)$) with respect to the increasing timestamps with infection rate (β) as 0.01 and top 7% influential as seed nodes on US PowerGrid network. Figure 11 shows the final infection scale ($f(t_c)$) with respect to the increasing timestamps with infection rate (β) as 0.01 and top 5% influential as seed nodes on US PowerGrid network. Figure 12 displays the final infection scale ($f(t_c)$) with respect to the increasing timestamps with infection rate (β) as 0.01 and top 5% influential as seed nodes on Facebook-like weighted network.

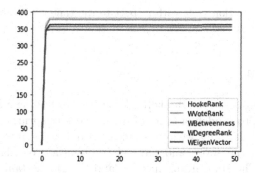

Fig. 10. The final infection scale with respect to the time on US PowerGrid Network with $\beta = 0.01$ and $\rho = 7\%$.

From above results on three real-life networks, it is evident that HookeRank performs better than state-of-the-art methods like weighted-degree centrality, weighted-betweenness centrality, weighted-eigenvector centrality, and weighted-voteRank, and also consistently outperforms recent methods like WVoteRank in terms of final infected scale with respect to time t and spreader fraction p on real-world networks as depicted in Table 1.

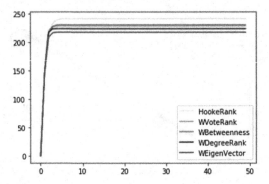

Fig. 11. The final infection scale with respect to the time on US-Airports Network with $\beta = 0.01$ and $\rho = 5\%$.

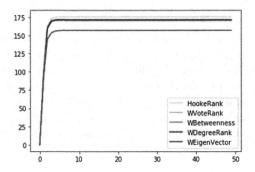

Fig. 12. The final infection scale with respect to the time on Facebook-like weighted Network with $\beta = 0.01$ and $\rho = 5\%$.

6   Conclusion

In this paper, we proposed the HookeRank method for finding influential nodes in weighted networks by modeling edges of the network as springs and edge weights as spring constants. Initially, we found a measure of the distance between indirect neighbors through the series and parallel combination of edges, by modeling them as springs. The HookeRank method and the HookeRank distance can be used to gain a better understanding of the topology in a complex weighted network. By finding the Hookerank value of the nodes, our method locates the top spreaders in the given real-world network to reach a large number of people in the network to maximize the spread of the information. The proposed algorithm incorporates both the local and global properties of a node in the measurement of its spreading capability. We performed the simulation of the proposed method along with contemporary methods on three real-life data-sets taking the basis of evaluation as the final infected scale. The proposed influence maximization algorithm performs considerably well and is effective in real-life scenarios.

References

1. Chen, G., Wang, X., Li, X.: Fundamentals of Complex Networks: Models, Structures, and Dynamics. Wiley, Singapore (2014)
2. Newman, M.E.: Analysis of weighted networks. Phys. Rev. E **70**(5), 056131 (2004)
3. Valente, T.W.: Network models of the diffusion of innovations (No. 303.484 V3) (1995)
4. Kempe, D., Kleinberg, J., Tardos, É.: Maximizing the spread of influence through a social network. In: Proceedings of the Ninth ACM SIGKDD International Conference on Knowledge Discovery and Data Mining, pp. 137–146 (2003)
5. Chen, W., Wang, C., Wang, Y.: Scalable influence maximization for prevalent viral marketing in large-scale social networks. In: Proceedings of the 16th ACM SIGKDD International Conference on Knowledge discovery and Data Mining, pp. 1029–1038. ACM (2010)
6. Wang, W., Tang, M., Stanley, H.E., Braunstein, L.A.: Unification of theoretical approaches for epidemic spreading on complex networks. Rep. Prog. Phys. **80**(3), 036603 (2017)
7. Sun, Y., Liu, C., Zhang, C.X., Zhang, Z.K.: Epidemic spreading on complex weighted networks. Phys. Lett. A **378**(7–8), 635–640 (2014)
8. Pastor-Satorras, R., Castellano, C., Van Mieghem, P., Vespignani, A.: Epidemic processes in complex networks. Rev. Mod. Phys. **87**(3), 925 (2015)
9. Opsahl, T., Agneessens, F., Skvoretz, J.: Node centrality in weighted networks: generalizing degree and shortest paths. Soc. Networks **32**(3), 245–251 (2010)
10. Prountzos, D., Pingali, K.: Betweenness centrality. ACM SIGPLAN Not. **48**(8), 35 (2013)
11. Wang, H., Hernandez, J., Van Mieghem, P.: Betweenness centrality in a weighted network. Phys. Rev. E: Stat. Nonlin. Soft Matter Phys. **77**(4 Pt 2), 046105 (2008)
12. Zhang, J., Chen, D., Dong, Q., Zhao, Z.: Identifying a set of influential spreaders in complex networks. Sci. Rep. **6**(1) (2016). Article number: 27823
13. Sun, H., Chen, D., He, J., Ch'ng, E.: A voting approach to uncover multiple influential spreaders on weighted networks. Phys. A **519**, 303–312 (2019)
14. Kumar, S., Panda, B.S.: Identifying influential nodes in social networks: neighborhood coreness based voting approach. Physica A: Stat. Mech. Appl. 124215 (2020)
15. Yu, S., Gao, L., Wang, Y.F., Gao, G., Zhou, C., Gao, Z.Y.: Weighted H-index for identifying influential spreaders. arXiv preprint arXiv:1710.05272 (2017)
16. Bihari, A., Pandia, M.K.: Eigenvector centrality and its application in research professionals' relationship network. In: 2015 International Conference on Futuristic Trends on Computational Analysis and Knowledge Management (ABLAZE) (2015)
17. Eades, P., Lin, X.: Spring algorithms and symmetry. Theoret. Comput. Sci. **240**(2), 379–405 (2000)
18. Fruchterman, T.M., Reingold, E.M.: Graph drawing by force-directed placement. Software Prac. Experien. **21**(11), 1129–1164 (1991)
19. Kamada, T., Kawai, S.: An algorithm for drawing general undirected graphs. Inf. Process. Lett. **31**(1), 7–15 (1989)
20. Slaughter, W.S.: The Linearized Theory of Elasticity. Birkhäuser (2001)
21. Hethcote, H.W.: The mathematics of infectious diseases. SIAM Rev. **42**(4), 599–653 (2000)

22. Colizza, V., Pastor-Satorras, R., Vespignani, A.: Reaction-diffusion processes and metapopulation models in heterogeneous networks. Nat. Phys. **3**, 276–282 (2007)
23. Opsahl, T., Panzarasa, P.: Clustering in weighted networks. Soc. Networks **31**(2), 155–163 (2009). https://doi.org/10.1016/j.socnet.2009.02.002
24. Watts, D.J., Strogatz, S.H.: Collective dynamics of small-world networks. Nature **393**, 440–442 (1998)

Community Aware Models of Meme Spreading in Micro-blog Social Networks

Mikołaj Kromka[(✉)], Wojciech Czech, and Witold Dzwinel

Department of Computer Science, AGH University of Science and Technology, WIET,
Kraków, Poland
{mkromka,czech,dzwinel}@agh.edu.pl

Abstract. We propose the new models of meme spreading over social network constructed from Twitter *mention* relations. Our models combine two groups of diffusion factors relevant for complex contagions: network structure and social constraints. In particular, we study the effect of perceptive limitations caused by information overexposure. This effect was not yet measured in the classical models of community-aware meme spreading. Limiting our study to hashtags acting as specific, concise memes, we propose different ways of reflecting information overexposure: by limited hashtag usage or global/local increase of hashtag generation probability. Based on simulations of meme spreading, we provide quantitative comparison of our models with three other models known from literature, and additionally, with the ground truth, constructed from hashtag popularity data retrieved from Twitter. The dynamics of hashtag propagation is analyzed using frequency charts of *adoption dominance* and *usage dominance* measures. We conclude that our models are closer to real-world dynamics of hashtags for a hashtag occurrence range up to 10^4.

Keywords: Complex network · Information spreading · Network dynamics

1 Introduction

Spreading memes over social networks became the subject of interdisciplinary research in many fields of science. From linguistics and sociology to statistical physics, we observe numerous works analyzing sociolinguistic phenomenon of virality, trying to explain key mechanisms of meme propagation and predict its future popularity [3,15,18].

In common understanding, meme is a piece of information or a unit of cultural transmission, which replicates over population. In the context of social media the meme could be a link, hashtag, phrase, image or video. The viral is defined as longstanding meme with the high probability of reposting. The process of meme proliferation, typically different than epidemic spreading, is therefore described as complex contagion. The deviation from simple epidemic patterns is caused

© Springer Nature Switzerland AG 2020
V. V. Krzhizhanovskaya et al. (Eds.): ICCS 2020, LNCS 12137, pp. 623–637, 2020.
https://doi.org/10.1007/978-3-030-50371-0_46

by multiple social factors and cognitive constraints such as homophily, confirmation bias, social reinforcement, triadic closure or echo chambers [15,17]. In this work, we consider spreading of hashtags - specific string-type concise memes used extensively to tag microblog Twitter posts and decorate them with additional semantics, personal affect or extended context. The substrate for hashtag propagation is formed by social network of Twitter users and posts, connected by three types of relations: *follow, retweet* and *mention.*

Modeling meme dynamics is an interesting topic, essential not only in broadly described marketing and business analysis but also more recently in public security sector, where the new challenges related to elections manipulation, fake news, terror attacks or riots were raised [4]. What factors should be considered in modeling hashtag propagation and recognizing virality? The recent studies suggest that while the content and related affect seems natural driver of virality, the social influence [8] and network structure [6,18] play more important role in meme diffusion process. This is particularly visible for hashtags, which are intended to be ultra concise and short.

The impact of network structure on meme dynamics is typically modeled as a social reinforcement effect, for which the probability of meme adoption increases with the number of exposures [17]. Motivated by recent blog posts about overused memes, we argue that social reinforcement mechanism should be supplemented with negative feedback loop, which reflects that the users can be annoyed with the meme, when they see it too frequently. We hypothesise, that by adding overexposure effect to simulation, we will achieve meme dynamics closer to one observed for real tweets. To prove our hypothesis we first, in Sect. 2, introduce basic definitions and describe five new models of hashtag propagation over micro-blog network. These models use the baseline structural factors known from other works [14,17,18], but additionally take into account that multiple exposures of the meme can have both positive and negative effect on an adoption. In Sect. 3, we present how one-day Twitter network dump with the real hashtags was used for analyzing our models and comparing them with the classical models described in [17]. The results of experiments, visualizations and quantitative model comparison are presented in Sect. 4. The wider context of meme spreading is described in Sect. 5, where the overview of related works is presented. We conclude in Sect. 6, by presenting summary of our findings and discussing limitations of our approach.

The key contributions of this work are new models of hashtag spreading, which take into account not only community structure and social reinforcement but also meme overexposure. Five different mechanisms of incorporating overexposure are proposed. We also perform validation of the new models against the models described in [17], based on Twitter data.

2   Spreading Models

Herein we introduce definitions required in the rest of this work and describe meme spreading models including five new models combining the influence of communities and cognitive limitations of individuals.

2.1   Networks with Communities

We define an undirected, static graph as a tuple $G = (V, E)$ where V is a set of vertices and $E = \{\{v_1, v_2\} : v_1, v_2 \in V\}$ is set of edges. We also define a community as a subset of vertices $C_i \subseteq V, i \in \{1, ..., n\}$, where $n \in \mathbb{N}$ is a number of all communities. We assume that the community structure is fixed in time. The set of all communities $C = \{C_i\}$ must contain all nodes from the graph: $\bigcup C_i = V, C_i \in C$. Similarly to [17,18], we assume that the communities are disjoint, meaning that one vertex is a member of exactly one community $\forall v \in V \; \exists! C_i : v \in C_i$, or in other words $\forall u, v \in V; C_i, C_j \in C, C_i \neq C_j : u \in C_i \wedge v \in C_j \Rightarrow u \neq v$.

2.2   Spreading Process

The defined graph is a base network on which the meme spreading processes take place. Spreading itself can be defined as an iterative function taking any known or unknown parameters of the network and returning state of the spreading. This requires assumption about similar timescales of all spreading processes, which is not always true - the same number of tweets can be produced really quickly during an intensive discussion and rather slowly in a marketing campaign. For our work it is enough to assume that spreading process uses the knowledge about the topology of the network, community structure and a state of spreading from the previous iteration.

We borrow the language from epidemiology and define generation of a tweet as an infection. To be more formal, we define a state of the spreading S_t at given time $t \in T \subset \mathbb{N}$ as a subset of infected nodes $S_t \subseteq V$. If the spreading process is more complex, i.e. there are more possible states of the node, then new disjoint subsets of nodes must be defined. The spreading process is then defined as a function $f_s : S_t \times G \times C \to S_{t+1}$.

For this work we assume that at any given iteration t only one node can be infected and we can infect the same node multiple times during the process. We define infected node $v_t \in V$ as: $S_t = S_{t-1} \cup \{v_t\}$. Then, the spreading process function can be simplified to a sequence of nodes infected (v_t) in each iteration $f_s : S_{t-1} \times G \times C \to \{v_t\}$. For describing spreading processes, we define neighborhood of a set of nodes as all nodes from its hull that are not in that set $S \subseteq V : N(S) = \{v_1 : \exists v_2 : \{v_1, v_2\} \in E \wedge v_1 \notin S\}$. We also define inclusive neighborhood of a set $S \subseteq V$ as $N_{incl}(S) = N(S) \cup S$.

2.3   Spreading Models

In this section, we provide spreading process functions for baseline models (M1 to M4 from [17]) and for five models proposed by us.

Random Sampling Model (M1). In the random model, at each iteration, we randomly choose the infected node with a uniform probability. It assumes that the network topology does not affect the spreading. This model will be used as a baseline for calculating spreading metrics.

$$v_t = rand_t(V) \tag{1}$$

Network Structure Model (M2). Here we assume that the network structure has an impact on spreading. This effect is reflected by choosing a random, already infected vertex and then selecting one of its topological neighbors:

$$v_t = rand(\{v : v \in N_{incl}(\{u\}); u = rand(S_{t-1})\}) \tag{2}$$

In addition, with constant probability $p_{restart} = 0.15$, the process is restarted from a random node, as in Eq. (1).

Social Reinforcement Model (M3). At each step of the spreading process, the node with most infected neighbors is infected. If there is no unique vertex satisfying this requirement, then a random one (from the all with the same number of infected neighbors) is selected. The restart process is included similarly to the model M2.

$$v_t = rand(\{v : v \in N_{incl}(S_{t-1}) \wedge argmax_v(N(v) \cap S_{t-1})\}) \tag{3}$$

Homophily Model (M4). Homophily model uses community structure to define nodes having similar interests. We choose randomly an already infected vertex and then, select randomly its neighbor from the same community. The restart process is included in M2 and M3.

$$v_t = rand(\{v : v \in N_{incl}(\{u\}) \wedge u, v \in C_i; u = rand(S_{t-1})\}) \tag{4}$$

Unique Information Overexposure Model (M5). In meme spreading in social networks some of hashtags are so unique, that they unlikely appear in multiple places of the graph but they are still affected by social reinforcement. To include that effect, we change the M3 model by removing the restart process and adding a requirement to not infect already infected nodes. To some extent this is similar to Susceptible-Infected-Resistant (SIR) epidemic model, but it is a more directed approach, because we are choosing a vertex with the most infected neighbors. This is an example of simple overexposure, because each node will not be infected multiple times with the same hashtag.

$$v_t = rand(\{v : v \in N(S_{t-1}) \wedge argmax_v(N(v) \cap S_{t-1})\}) \tag{5}$$

Global Increase of Generation Probability Model (M6). One of the cognitive limitations in social interactions is that people get bored, when they see the same meme too many times. To model such mechanism, we infect a node with the most exposures, but if a neighbor vertex was infected multiple times, we count all of these occurrences as exposures. Secondly, the restart probability at each step grows linearly from 0 to 1. This means that initially, the spreading is rapid and not constrained by community structure. After a period of time, when the members of a community are overexposed, new random vertices out of a community are likely to be infected.

This definition requires setting maximum number of iterations, for which the spreading occurs. Similarly to choosing $p_{restart}$, this is strictly connected to finding timescale of a process. In this model, we assume that the spreading lasts as long as the longest spreading time observed in the real data, noted as T_{max}. We define a number of exposures for a given node v at iteration t as:

$$n_{exp}(v, t) = |\{u : u \in N(v) \wedge u = v_\tau, \tau < t\}|. \tag{6}$$

Then, selection of the currently infected node can be represented as:

$$v_t = rand(\{v : argmax_v(n_{exp}(v, t))\}). \tag{7}$$

The restart probability is:

$$p_{restart} = \frac{t}{T_{max}}. \tag{8}$$

Local Increase of Generation Probability Model (M7). Similarly to M6, we increase the restart probability linearly but it is set to 0 after each restart.

$$v_t = rand(\{v : argmax_v(n_{exp}(v, t))\}) \tag{9}$$

The restart probability is:

$$p_{restart} = \frac{i}{T_{max}}, \tag{10}$$

where i is the number of iterations since last restart.

Inverse Exposure Model (M8). This is a more complex overexposure model, where the exposures are counted separately for all nodes. Then, the infected node is selected randomly, with the probability inversely proportional to the number of its exposures:

$$v_t = rand_{ipe}(N_i(S_{t-1})), \tag{11}$$

where $rand_{ipe}$ is function choosing a node with probability inversely proportional to the number of exposures for each node. Restart has a constant probability of

$$p_{restart} = 0.15. \tag{12}$$

Inverse Exposure Model with No Restarts (M9). Similar to M8 model but without a restart mechanism. Choosing the next infected node is described using the same function from Eq. (11).

2.4   Spreading Metrics

To compare different models, we calculate the following metrics. Most of them are taken from [17].

Intra- and Inter-community Activity. The patterns of within- and between-community meme activity are reflecting the influence of these sub-structures on diffusion process and can be used to quantify overall spreading dynamics. In models M2 and M4 the meme transfer happens between a source node (u - infected in one of the previous iterations) and a destination node (v_t). If the nodes belong to the same community $u, v_t \in C_i$, the transition is classified as *intra-community activity* (a_{intra}). If they belong to different communities $u \in C_i \wedge v_t \in C_j \wedge i \neq j$, it is called *inter-community activity* (a_{inter}). To be able to compare these metrics between different networks we normalize it by dividing them by the number of all inter- ($edges_{inter}$) and intra-community edges ($edges_{intra}$) in the network.

$$a_{intra} = \frac{c_{intra}}{edges_{intra}} \tag{13}$$

$$a_{inter} = \frac{c_{inter}}{edges_{inter}} \tag{14}$$

Here c_{intra}, c_{inter} are the numbers of spreading events occurring within the same and between the different communities.

Usage Dominance. We define *usage dominant community* as a community with the biggest number of spreading events: $C_d = \max_{C_i} \sum_{t=0}^{T_{max}} |\{v_t \in C_i\}|$ and a number of spreading events produced in this community as $T_{C_d} = \sum_{t=0}^{T_{max}} |\{v_t \in C_d\}|$. Then *usage dominance* is defined as

$$D_u = \frac{|T_{C_d}|}{T_{max}}. \tag{15}$$

The high values of *usage dominance* indicate that the meme was trapped mostly within one community and its dynamics is more local. The probability of becoming the viral is low.

Adoption Dominance. We define *adoption dominant community* (C_a) as a community with the biggest number of vertices infected by spreading. Note that it is different than C_d for which we count all spreading events, here we count only unique vertices: $C_a = \max_{C_i} |S_{T_{max}} \cap C_i|$. The *adoption dominance* can be defined as:

$$D_a = \frac{|C_a|}{|S_{T_{max}}|}. \tag{16}$$

3   Simulation Details

Each model was run at least twice on the network recreated from real tweets. Ground truth data were obtained from the same set of tweets and analyzed on the same network, to obtain metrics reflecting real-world spreading dynamics.

3.1   Recreated Network

We gathered 2 045 413 tweets using public Twitter API. From that, we selected only the ones containing *mention relation*, which involves at least two users: one source and possibly multiple targets. Each source-target pair was treated as an undirected edge $e \in E$, to form initial graph G. Then, the largest connected component of G was extracted. In our case it consisted of 710 195 vertices and 919 022 edges. This was the base graph, on which the simulations were run. In Table 1, the details of recreated networks are summarised. The visualisation of the graph using IVGA, a fast force-directed method from [5], is presented in Fig. 1.

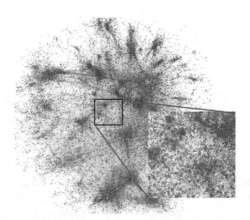

Fig. 1. Visualization of the giant component of sampled Twitter network (*mention relation*) using IVGA algorithm [5]. The color of the node represents the community. (Color figure online)

Table 1. Recreated network details.

$|V|$	$|E|$	$edges_{intra}$	$edges_{inter}$	$|C|$	Max. degree	Triangles
710 195	919 022	782 244	136 778	32 948	391	48 115

Static networks do not completely model temporal relationships occurring between nodes in real networks. In our case the timescales of spreading are much lower than the changes in relationships between users which is why we can assume that they do not change over time. We are using *mentions* relation because in its nature it is bidirectional: the user mentioning another user is aware of the message content as its creator and the mentioned user is notified about that message as well. This is why we can assume that our base network is undirected.

Community Detection. We used *INFOMAP* algorithm to obtain partitioning of the nodes into communities C. Its main advantage is linear time complexity $O(|E|)$, which is practical for large graphs. Secondly, it does not require the number of expected communities as an input. This is not achievable, e.g., for *Label Propagation* algorithm.

The network was split into 32 948 communities, with the smallest communities consisting of just two nodes and the largest containing 5394 nodes. The distribution of communities is presented in Fig. 2. We also ran simulations on the same network with community structure created using label propagation algorithm. The results were similar but we do not present them here because of space constraints.

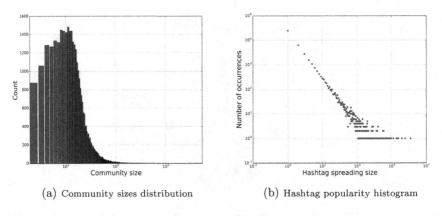

(a) Community sizes distribution   (b) Hashtag popularity histogram

Fig. 2. The frequencies of community sizes detected using *INFOMAP* algorithm on sampled Twitter network (a) and Hashtag popularity (defined as number of occurrences) distribution (b) in logarithmic scale. Broader distribution is visible for highly popular hashtags. It is caused by insufficient time to spread more. With the time passing most of these points would update the trend upwards.

3.2   Ground Truth Data

From all the tweets, we filtered out the ones without a hashtag or with at least one vertex not present in the largest component. For each unique hashtag, the list of tweets using it was extracted and sorted by the tweet's time $t \in T$. To form the ground truth for measuring and comparing hashtag spreading models, we simulated meme propagation on the sample Twitter network by infecting nodes according to their order in that list. The case when two consecutive nodes are not connected is modeled by the restart mechanism described in Sect. 2.

The distribution of hashtag popularity does not completely follow the power law as presented in Fig. 2. Because of the large number of hashtags (41 356) we calculated spreading metrics only for a subset of them. It was chosen so that, the hashtags were distributed evenly in the spreading popularity bins. For our calculations of *usage dominance* and *adoption dominance* we used data for 280 unique hashtags spread across 54578 iterations in total.

4   Results

Based on simulation results, we learned that our models M5–M9 follow the real-world hashtag spreading dynamics more accurately than models M2–M4 (both for unpopular and viral hashtags). For metric calculation, we define hashtags with low a_{intra} as those having its value lower than average. Analogously, we define hashtags with high a_{intra} as those with the value higher than average.

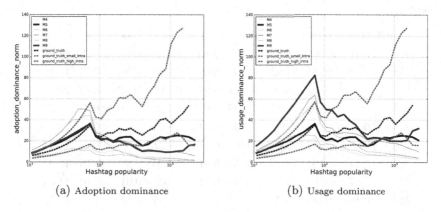

(a) Adoption dominance   (b) Usage dominance

Fig. 3. Adoption (a) and usage dominance values for models M4–M9. Each line represents changes in proportions of hashtags produced inside *adoption* (C_a) and *usage dominant communities* (C_d) to number of all hashtags produced in the network. The results are averaged across 30 logarithmic bins and normalized by the random sampling model (M1). *ground_truth* represents ground truth spreading including all hashtags. The series: *ground_truth_small_intra* and *ground_truth_high_intra* show the results for hashtags with a_{intra} lower and higher than average, accordingly. The standard error slightly exceeds 2% only for one model. For the rest of the models, as well as for the ground truth, its value is below 1%.

The *usage dominance* and *adoption dominance* measures are normalized by dividing them by the values obtained for random sampling model (M1) under the same conditions, similarly to [17]. As our intention was to analyze multi-use memes, rather than rare memes, we removed hashtags with the number of adoptions lower than 10. This also prevents including artifacts like tweets starting to become popular just before end of data collection. Next, we aggregated dominance metrics into 30 logarithmic bins. The results of simulation quantified using *adoption dominance* and *usage dominance* are shown in Fig. 3. Overall, inverse exposure (M9), unique information overexposure (M5) and local increase of generation probability (M7) models are performing the best when comparing to all hashtag ground truth results (*ground_truth*). This is confirmed by both the highest correlation and the lowest error, as shown in Table 2 and Table 3. There is one exception - large error for M9 model in *usage dominance*, which we discuss below. Our models adapting different overexposure mechanisms (especially M7), have generally better results than the baseline M2–M4 models from literature [17]. The critical value used for testing statistical significance of correlation is 0.05. We performed additional tests to verify how well the models reflect dynamics of hashtags divided into two groups: high intra-community activity (specific, trapped memes) and low intra-community activity (generic, viral-candidates).

Table 2. Correlation of non normalized and root mean square error of normalized *adoption dominance* D_a. Highlighted entries represent the best two values in each column.

Model	Ground truth			High a_{intra}			Low a_{intra}		
	corr	p-val	rmse	corr	p-val	rmse	corr	p-val	rmse
M2	0.82	<0.01	24.32	−0.37	0.06	59.63	**0.93**	<0.01	12.14
M3	0.8	<0.01	25.96	−0.37	0.05	61.11	**0.92**	<0.01	13.51
M4	0.91	<0.01	22.27	−0.48	0.01	57.81	0.91	<0.01	**10.72**
M5	**0.92**	<0.01	**9.7**	−0.58	<0.01	**44.57**	0.91	<0.01	**9.33**
M6	0.89	<0.01	23.4	−0.6	0.11	56.72	0.87	<0.01	18.44
M7	0.91	<0.01	18.99	−0.62	0.01	**50.97**	0.86	<0.01	17.9
M8	0.87	<0.01	23.7	−0.45	0.02	59.08	0.91	<0.01	11.64
M9	**0.94**	<0.01	**16.52**	−0.61	<0.01	51.33	0.87	<0.01	11.59

Table 3. Correlation of non normalized and root mean square error of normalized *usage dominance* D_u. Highlighted entries represent the best two values in each column.

Model	Ground truth			High a_{intra}			Low a_{intra}		
	corr	p-val	rmse	corr	p-val	rmse	corr	p-val	rmse
M2	0.84	<0.01	23.76	−0.58	<0.01	60.01	**0.94**	<0.01	11.6
M3	0.86	<0.01	22.69	−0.6	<0.01	58.97	**0.93**	<0.01	10.45
M4	0.88	<0.01	21.08	−0.71	<0.01	57.44	0.84	<0.01	**10.34**
M5	**0.91**	<0.01	**10.08**	−0.7	<0.01	**45.8**	0.92	<0.01	**9.23**
M6	**0.91**	<0.01	20.11	−0.79	<0.01	52.79	0.83	<0.01	19.02
M7	**0.92**	<0.01	**16.85**	−0.78	<0.01	46.7	0.84	<0.01	20.47
M8	0.88	<0.01	19.66	−0,67	<0.01	55.77	0.9	<0.01	10.98
M9	**0.92**	<0.01	24.85	−0.77	<0.01	**45.46**	0.84	<0.01	31.75

Comparing the results of spreading with the ground truth below-average a_{intra} (*ground_truth_small_intra*), we see that existing models slightly outperform our overexposure models when it comes to correlation. This is visible mostly for M2 (correlation 0.93 and 0.94), based on the plots, we observe that this is mostly due to the initial phase of spreading. Errors are still much lower for our models (especially M5). It is worth noting, that correlation and errors are almost on the same level for M5 and M8 models.

The correlations of spreading pattern *ground_truth_high_intra* (higher than average a_{intra}) with all the simulated trends are either statistically insignificant or negative. Nevertheless, based only on the error value, we can see that the overexposure models are mimicking real world spreading more precisely. Much higher absolute values of the errors suggest that this type of spreading is hard to be modeled by all frameworks M1–M9.

A significant difference is observed between *usage* and *adoption dominance* measures for model M9 for initial phase of spreading. As visible in high and similar correlation values (0.94 and 0.92 accordingly), the shape of the function is preserved but the error value is much higher for *usage dominance*, especially for low a_{intra}. The effect is also present in the ground truth data, but with a much smaller amplitude. Most likely this is because the probability of tweeting the same hashtag by the same user is lower in real life compared to the models. The M9 model is more susceptible to that, because it can spread more freely inside a community. This is validated by comparing it with M5, which is a more directed spreading model demonstrating almost no difference between the values of *usage* and *adoption dominance*.

Table 4. Correlation of non normalized and root mean square error of normalized *usage dominance* D_u for highly popular tweets (popularity > 500). Highlighted entries represent the best two values in each column.

Model	Ground truth			High a_{intra}			Low a_{intra}		
	corr	p-val	rmse	corr	p-val	rmse	corr	p-val	rmse
M2	0.79	0.01	33.66	−0.5	0.17	89.51	0.76	0.02	18.31
M3	0.79	0.01	32.03	−0.49	0.18	87.92	0.73	0.02	16.68
M4	**0.8**	0.01	31.39	−0.5	0.17	87.32	**0.78**	0.01	16.04
M5	**0.85**	<0.01	**15.38**	−0.58	0.10	**71.26**	**0.77**	0.02	**1.83**
M6	0.72	0.03	28.7	−0.35	0.36	84.73	0.65	0.06	13.54
M7	0.79	0.01	19.35	−0.49	0.18	75.21	**0.78**	0.01	6.71
M8	0.75	0.02	30.14	−0.44	0.23	86.07	0.69	0.04	14.83
M9	0.78	0.01	**16.38**	−0.44	0.23	**72.16**	0.72	0.03	**5.89**

634   M. Kromka et al.

Because viral or really popular hashtags are the most important in our analysis, in Table 4, we also present *adoption dominance* D_a results for hashtags
with more than 500 spreading events. Good results for M5 model show, that
it is the most robust framework presented in this article. Interestingly, highly
popular hashtags with lower than average a_{intra} are modeled much better with
the inclusion of overexposure. This is visible in low error and high correlation
values for models M5, M7 and M9. From the baseline models from literature,
M4 has the best results.

5   Related Works

In the broad literature of the subject we find the works focusing on different
aspects of meme popularity. These can be grouped into four major categories:
content, social influence, time characteristics and network structure. Content
appeal is typically analyzed for longer posts, for which sentiment, affect or emotional load can be established. In [7], the authors present the study of sample tweets and based on news/non-news classification and sentiment analysis
conclude that negative sentiment for news messages and positive sentiment
for non-news improves retweetability. The influence of psychological arousal
(emotions resulting from a content) on information diffusion was analyzed in
[2]. The authors provide evidence of perceptive claim that the high-arousal
content triggers significant activation of users and boosts meme proliferation.
Exhaustive sociolinguistic analysis of social media communication presented in
[15] reveals important observations regarding meme lifecycle and virality. The
authors emphasize the importance of *phatic communication* in maintaining vital-
ity of social network. This type of communication does not transport quality
content or meaning but still enables emergence of *conviviality*, understood as a
production of social engagement, based on temporal and elastic collectives consuming virals. Simple interactions and networking seem more important drivers
of social network dynamics than sharing valuable information. The comprehensive approach for meme bursts prediction is presented in [1], where features
based on content, network and time are used to determine future meme popularity. The interesting method of meme identification based on content and
time was presented in [13]. The authors represent the content stream dumps in
a form of Graph-of-Words and use *k-core decomposition* algorithm to identify
viral memes.

The notion of social network *influencer* is widely spread in mass culture
and affects modern online marketing strategies causing emergence of semi-
professional content producers on Twitter or Instagram. The importance of seed
agents responsible for initiating information spread over social network was studied in many works including [10, 16]. Cognitive constraints caused by heavy information flows such as limited attention or confirmation bias, forces individuals
to aggressively filter content and optimize evaluation time by adapting options
of others [9]. The high authority of peer providing information and multiple
exposure increases the probability of meme adoption (retweet, mention). From

structural perspective, the influence can be quantified using vertex centrality measures such as *degree, betweenness centrality, clustering coefficient* or *Page Rank*. Nevertheless, the more complete model of meme propagation should also include non-structural external knowledge such as forwarding activity and interests.

Temporal patterns of meme diffusion were also analyzed in the context of process modeling and virality prediction. In [19], the authors used wavelet transform and trajectory clustering techniques to discover six major types of growth-decay characteristics of user attention curves. Different types of time series were identified for professional blogging, micro-blogging or news agencies. Long-term time variations in Twitter hashtag exposure and adoption was studied in [11]. The authors found that timing characteristics highly depend on the topic and frequency of hashtag exposure. The features extracted from the time series of first tweet adopters were used in [18] for classification-based virality prediction. It was also shown, that structural features of social network are more discriminative than time-based features.

The structure of social network was identified as a key component of meme spreading dynamics [3,6,12,14]. In particular, community structures can trap information flows and transform simple epidemic spreading to complex contagion, affected additionally by social mechanisms of homophily, reinforcement or overexposure. Inter-community concentration of early meme adopters is one of the most important features in predicting virality [17,18]. Surprisingly, as presented in [14], the local structural properties of communities are less important than mesoscopic community view. In the context of hashtag propagation over Twitter network, it was shown that spreading of less popular hashstags should be described as complex contagion, subject to homophily and social reinforcement, while virals propagate similar to diseases [17]. Apart from communities, the structure of subgraphs induced by early adopters is also crucial in modeling future meme dynamics. In [3], the authors perform extensive feature engineering for hashtag virality prediction and build classification model, which uses subgraph conductance as one of the most discriminative features. More recently, in [6], the authors analyze structural properties of meme adoption graphs (MAGs) and formulate MAG-based, generic framework for spreading models comparison.

In our work we were inspired by structural models taking into account social reinforcement and homophily [17,18], but we decided to additionally include negative feedback loop related to overused memes.

6 Conclusion

We demonstrated that hashtag spreading is a complex process, which cannot be accurately modeled based on the network structure only. Many factors like *intra-community activity* or popularity play an important role. In general, our models taking into account meme overexposure follow the dynamics of hashtag spreading more precisely, especially, when compared to the models from other works. Unique information overexposure model (M5) is performing the best when compared to the ground truth dynamics obtained from all hashtags.

The good results obtained for M9 model should also be emphasised. This model takes into account overexposure constraint by generating tweet with the probability inversely proportional to the number of exposures. This is the most apparent proof, that the negative feedback loop for the social reinforcement should be considered, when modeling meme spreading.

As visible in large error values and negative or insignificant correlations, hashtags with the higher than average *intra-community activity* are the hardest to reproduce. Interestingly, overexposure models have smaller *rmse* errors compared to social-reinforcement and homophily ones. Spreading with the lower than average *intra-community activity* values is overall best described by the homophily model (M4). This is mostly due to its accuracy for unpopular hashtags. For virals, defined as having popularity greater than 500 spreading events, the M5 and M7 (the local increase of generation probability model) are much more precise. The accuracy of M5 model is the order of magnitude better than one obtained for M2–M4 models.

To have more complete overview of hashtag dynamics, the further work should focus on analysing the timescale of spreading and the stability of solutions when working with missing data.

Acknowledgements. The research presented in this paper was supported by the funds assigned to AGH University of Science and Technology by the Polish Ministry of Science and Higher Education and in part by PL-Grid Infrastructure.

References

1. Bai, J., Li, L., Lu, L., Yang, Y., Zeng, D.: Real-time prediction of meme burst. In: 2017 IEEE International Conference on Intelligence and Security Informatics (ISI), pp. 167–169. IEEE (2017)
2. Berger, J., Milkman, K.L.: What makes online content viral? J. Mark. Res. **49**(2), 192–205 (2012)
3. Bora, S., Singh, H., Sen, A., Bagchi, A., Singla, P.: On the role of conductance, geography and topology in predicting hashtag virality. Soc. Network Anal. Min. **5**(1), 1–15 (2015). https://doi.org/10.1007/s13278-015-0300-2
4. Bright, P.: How the London riots showed us two sides of social networking. Ars Technica (2011)
5. Dzwinel, W., Wcisło, R., Czech, W.: ivga: a fast force-directed method for interactive visualization of complex networks. J. Comput. Sci. **21**, 448–459 (2017)
6. Elsharkawy, S., Hassan, G., Nabhan, T., Roushdy, M.: Modelling meme adoption pattern on online social networks. In: Web Intelligence, vol. 17, pp. 243–258. IOS Press (2019)
7. Hansen, L.K., Arvidsson, A., Nielsen, F.A., Colleoni, E., Etter, M.: Good friends, bad news - affect and virality in Twitter. In: Park, J.J., Yang, L.T., Lee, C. (eds.) FutureTech 2011. CCIS, vol. 185, pp. 34–43. Springer, Heidelberg (2011). https://doi.org/10.1007/978-3-642-22309-9_5
8. Muchnik, L., Aral, S., Taylor, S.J.: Social influence bias: a randomized experiment. Science **341**(6146), 647–651 (2013)

9. Qiu, X., Oliveira, D.F., Shirazi, A.S., Flammini, A., Menczer, F.: Limited individual attention and online virality of low-quality information. Nat. Hum. Behav. **1**(7), 0132 (2017)

10. Romero, D.M., Galuba, W., Asur, S., Huberman, B.A.: Influence and passivity in social media. In: Gunopulos, D., Hofmann, T., Malerba, D., Vazirgiannis, M. (eds.) ECML PKDD 2011. LNCS (LNAI), vol. 6913, pp. 18–33. Springer, Heidelberg (2011). https://doi.org/10.1007/978-3-642-23808-6_2

11. Romero, D.M., Meeder, B., Kleinberg, J.: Differences in the mechanics of information diffusion across topics: idioms, political hashtags, and complex contagion on Twitter. In: Proceedings of the 20th International Conference on World Wide Web, pp. 695–704. ACM (2011)

12. Saxena, A., Iyengar, S., Gupta, Y.: Understanding spreading patterns on social networks based on network topology. In: Proceedings of the 2015 IEEE/ACM International Conference on Advances in Social Networks Analysis and Mining 2015, pp. 1616–1617. ACM (2015)

13. Shabunina, E., Pasi, G.: A graph-based approach to ememes identification and tracking in social media streams. Knowl.-Based Syst. **139**, 108–118 (2018)

14. Stegehuis, C., Van Der Hofstad, R., Van Leeuwaarden, J.S.: Epidemic spreading on complex networks with community structures. Sci. Rep. **6**, 29748 (2016)

15. Varis, P., Blommaert, J.: Conviviality and collectives on social media: virality, memes, and new social structures. Multilingual Margins: J. Multilingualism Periphery **2**(1), 31–31 (2015)

16. Weng, L., Flammini, A., Vespignani, A., Menczer, F.: Competition among memes in a world with limited attention. Sci. Rep. **2**, 335 (2012)

17. Weng, L., Menczer, F., Ahn, Y.Y.: Virality prediction and community structure in social networks. Sci. Rep. **3**, 2522 EP (2013). article

18. Weng, L., Menczer, F., Ahn, Y.Y.: Predicting successful memes using network and community structure. In: Eighth International AAAI Conference on Weblogs and Social Media (2014)

19. Yang, J., Leskovec, J.: Patterns of temporal variation in online media. In: Proceedings of the Fourth ACM International Conference on Web Search and Data Mining, pp. 177–186. ACM (2011)

A Dynamic Vote-Rank Based Approach for Effective Sequential Initialization of Information Spreading Processes Within Complex Networks

Patryk Pazura[1]([⊠]), Kamil Bortko[1], Jarosław Jankowski[1],
and Radosław Michalski[2]

[1] Faculty of Computer Science and Information Technology, West Pomeranian
University of Technology, Szczecin, Poland
{ppazura,kbortko,jjankowski}@wi.zut.edu.pl
[2] Department of Computational Intelligence, Wrocław University of Science
and Technology, Wrocław, Poland
radoslaw.michalski@pwr.edu.pl

Abstract. Seed selection is one of the key factors influencing information spread within networks. Whereas most solutions are based on single-stage seeding at the beginning of the process, performance increases when additional seeds are used. This enables the acquisition of knowledge about ongoing processes and activating new nodes for further influence maximisation. This paper describes an approach based on the Vote-Rank algorithm with dynamic rankings for sequential seed selection. The results prove the increased performance of dynamic rankings compared to the static version and show how the frequency of ranking updates affects both performance and computational costs.

Keywords: Information spreading · Vote-Rank · Influence maximization · Seed selection · Sequential seeding · Viral marketing · Social networks

1 Introduction

Information spreading processes are observed in various aspects of social interaction and commercial activity. They are behind social movements [9], viral marketing [21], political campaigns [4], and spread of misleading information [2]. After gathering additional knowledge about their performance, further actions are often taken to change their dynamics and increase or decrease coverage [25]. Apart from solutions focused on seeding at the beginning of the process without any further actions, other solutions gather knowledge from the process and use additional seeds to improve the process including but not limited to sequential seeding [13], seeding scheduling [34], or adaptive seeding [33]. The approach proposed in this paper is based on the selection of sequential seeds with the

© Springer Nature Switzerland AG 2020
V. V. Krzhizhanovskaya et al. (Eds.): ICCS 2020, LNCS 12137, pp. 638–651, 2020.
https://doi.org/10.1007/978-3-030-50371-0_47

use of the Vote-Rank algorithm based on adaptive rankings recomputed before additional seeds are selected. Recomputation uses knowledge gathered about infections within the network and Vote-Rank only considers inactive nodes, not the nodes already activated. This enables selecting only those nodes with higher potential for spreading as seeds, selected within areas not covered by infections. In this study, we compared the results from static Vote-Rank with the proposed dynamic approach and investigated the influence of recomputation frequency on the final outcome. The remainder of the paper is organized as follows: Sect. 2 provides a literature review, the conceptual framework is presented within Sect. 3, followed by experimental results presented in Sect. 4. The results are summarized and the paper is concluded in Sect. 5.

2   Literature Review

Information spread within social networks has received attention from researchers from various disciplines. Studies related to information spread have focused on factors affecting their dynamics, the roles of social ties, network topology, and the roles of the links within the network [26]. Models derived from epidemiology research, like SIS, SIR, and their variants, were initially used for prediction and analysis [19]. Later, more dedicated approaches, like the independent cascade model [20] and linear threshold model [5] considered network structures. They were verified with the use of agent-based simulations and the Monte Carlo method [6] or analytical solutions like mean-field models [32] or branching processes [15]. Research was initially performed mainly on single layer static networks, but in recent years, more attention is being focused on multilayer networks [32] and spreading processes within temporal networks [11,16]. Apart from single processes, multiple processes were analysed with mechanisms related to competition, cooperation, and other forms of interaction [3].

Information spread processes are usually initialised by selected nodes, called seeds, with the use of a dedicated seed selection methods [10]. The influence maximisation problem leads to several challenges and solutions for initial nodes selection [20]. The main goal is to select a set of seeds with high potential to initiate the spread and activate their neighbours. Early approaches were mainly based on heuristics with high degree or other centrality measures like closeness or eigenvector centrality [36]. Apart from simple heuristics, the greedy approach is much more effective, delivering results closer to optimum [20]. Further attempts were made to improve its computational performance with possible applications within larger networks [7]. The number of seeds was analysed to find the minimal effective seed sets [27]. Other solutions considered costs in a form of budgeted solutions [29]. The negative impact of high intensity seeds on users was identified as an over-exposure problem [1]. Another possible goal is limiting overlapping seeds and maximising distance between seeds with the Vote-Rank algorithm, which ranks seed candidates by its votes acquired from direct and indirect neighbours, with higher ranks grouped together with increasing distance from other seeds [38].

Most of earlier solutions focused on selection of seeds at the beginning of the process, without additional actions during the process. Another possibility is spreading seeding over time, with only a fraction of the seeds used at the beginning, in the form of sequential seeding [13], seeding scheduling [34], or adaptive seeding [33]. The main mechanics are based on avoiding selecting nodes with a high potential to be naturally activated by their neighbours as seeds. Sequential seeding can be used to revive stopped processes or add seeds when the processes are still ongoing [14]. This approach was proven to never deliver worse results than single-stage seeding [17]. Extensions showed how performance of sequential seeding is affected by the topology of networks [26], entropy-based centrality [30], and effective degree [12].

3   The Conceptual Framework

Static rankings used for seed selection only at the beginning of the process create the threat of selected seeds becoming victims of the natural diffusion process. Sequential seeding uses the sequence of seeds instead of using all in a single step to deliver a better result due to the potential of a natural diffusion process. This paper presents an approach that improves the Vote-Rank algorithm with the recomputations and the use of nodes that are effective from the perspective of diffusion processes, that are not yet activated. Earlier study showed that seeds generated with Vote-Rank used sequentially deliver better results than with a single stage [13], but only static Vote-Rank generated at the beginning of the process was analysed and used. The approach presented here avoids gathering votes from network nodes that are no longer effective for information spreading, meaning they were already activated and used for information spreading.

3.1   Illustrative Example

Although Vote-Rank is effective for seed selection, the static ranking generated only at the beginning of the process creates the possibility that the initially good candidates with a high number of votes are no longer good candidates for seeding in the next stages. The proposed approach assumes creation of new ranks based on Vote-Rank before seeding actions. During rank computation, voting only considers votes from nodes available for activation. Already active nodes are not considered. As a result, not yet activated seed candidates with a higher number of direct and indirect connections are preferred. To demonstrate the potential performance of the proposed approach, a toy example is presented in Fig. 1. A small network based on nine nodes was generated with the use of the Watts–Strogatz model [37]. A rewiring probability of 0.1 was used and a value of two was assigned to the neighbourhood within which the vertices of the lattice were connected.

Simulation was performed with the use of the coordinated execution proposed in [17], where agent-based simulations processes are not based on randomly generated values in each run, but on values assigned to network edges

$A \rightarrow B$, representing the probability of passing information from node A to B and from B to A. If propagation probability is assigned to whole the process with the value p, then information is passed through all edges with weights $\leq p$. The number of network versions with weights is equal to the planned number of runs. The main advantage of this process is the ability to compare methods within identical conditions. The same approach was used for all simulations here. Figure 1 shows the network used with weights assigned to the edges and simulated spreading according to the independent cascade model [20]. Figure 1 **(A)** shows the process based on sequential seeding with the use of static Vote-Rank and Fig. 1 **(B)** shows the process based on dynamic Vote-Rank. Descriptions are provided within the figure caption together with the mechanics behind the dynamic Vote-Rank approach and the advantage produced by a higher number of activations within the network.

3.2   General Assumptions for Experimental Study

In general, experimental study was planned within two stages, with different experimental plans, goals and assumptions. In the first stage, illustrated in Fig. 2 **(Stage I)**, the main goal was to analyse differences between coverage of processes with the use of static (computed only one at the beginning) and dynamic (recomputed before additional seeding takes place) Vote-Rank. Ranking of nodes is created in step 0 and Vote-Rank VR(0) is created. In n subsequent steps additional seeds are selected from the ranking according to the sequential approach and new activations are performed. Seeding is conducted in revival mode, after process dies out. While nodes ranking based on the Vote-Rank is effective at the beginning of the process, together with ongoing process more and more nodes are activated. Proposed approach is focused on better utilization of the knowledge about activations. Vote-Rank is recomputed before each seeding action. Recomputation is based on reduced network without activated nodes. It is assumed that dynamic raking will deliver better network coverage, what is illustrated in Fig. 2 **(A2)**. At each seeding step i, Vote-Rank is computed and new ranking VR(i) is used.

In the Stage II (Fig. 2) study is focused on effects of frequency of rank recomputation on final coverage and computational costs related to the time of rank calculation. Figure 2 **(B1)** shows approach with recomputations taken in every seeding step. It is expected to be the most effective in terms of coverage, but with highest computational time needed for new rankings creation. Another possibility is to reduce computational time with medium intervals between computations and medium coverage increase Fig. 2 **(B2)**. Together with growing intervals between recomputations, the performance of selected seeds will be dropping. It is considered as lowest performance at Fig. 2 **(B3)**, but still better than for processes based on static rankings.

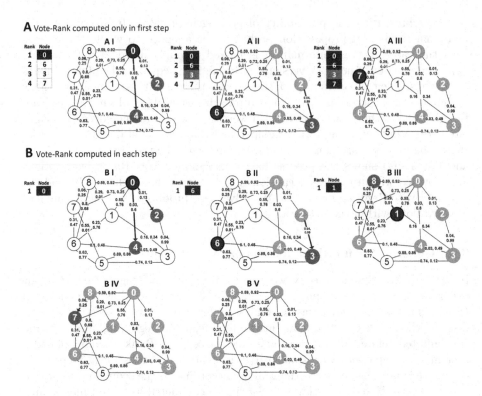

Fig. 1. Results from **(A)** Spreading process based on static Vote-Rank computed only once, at the beginning of the process. Initially computed ranking is used during whole process. **(B)** Information spreading process based on sequential seeding and dynamic Vote-Rank computed in every step when seeding takes place. Both sequential seeding processes use three seeds with single seeds used per simulation step. Contagion process is based on Independent Cascade Model with propagation probability $PP = 0.1$. Every edge has assigned two values of weights, therefore to activate neighbour of node A - suppose that is node B, first value on the edge between A and B must be smaller or equal to propagation probability value. **(A)** Information spreading is initiated by three seeds used in sequence in a form of sequential seeding. Nodes in the network are ranked by Vote-Rank algorithm, with top four nodes presented in the table. First seed is used at the beginning **(AI)**, second seed is used when the process dies out **(AII)**, and the third seed in the same way **(AIII)**. In step **AIII** we select last node as seed. While node 3 from Vote-Rank was activated in a natural way in stage **AIII**, node 7 is selected as a seed. This process ends with the total of 6 activated nodes. **(B)** During this process we compute Vote-Rank at each seeding step. While only one seed is used in each step, only one node is needed with the highest value of Vote-Rank. **(BI)** In first step according to the ranking node 0 is selected as seed. It tries to activate its neighbors with $PP = 0.1$ and nodes 2 and 4 become active, because of appropriate weighs assigned to the edges allowing transmission. **(BII)** In the next step, previous activated nodes infecting further, node number 3 becomes active, and node 6 is activated as a seed. In the last stage, **(BIII)**, node 1 is selected as a seed and it is activating node number 8. As a result 7 nodes within the network are activated.

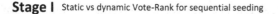

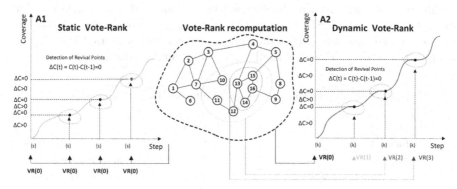

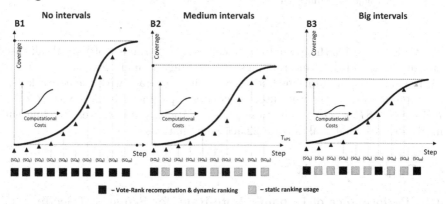

Fig. 2. Two parts of empirical study based on **Stage I:** Comparison of performance of sequential seeding based on static **(A1)** and dynamic ranking of nodes with the use of Vote-Rank algorithm **(A2) Stage II:** Analysis of impact of recomputations frequency on coverage and computational costs with **(B1)** computations in each seeding step **(B2)** medium intervals between computations and **(B3)** big intervals with low frequency and costs of computations (lowest coverage).

3.3 Plan of Experiments

An experimental setup is based on two types of Vote-Rank rankings (static and dynamic) and 10 real networks $N1$–$N10$ from [8, 18, 22, 22–24, 28, 31, 35, 37] respectively, containing from 1133 to 16264 nodes and from 5451 to 146160 edges. All values used for all parameters are presented in Table 1. Simulations were performed ten times per configuration and results were averaged. For Stage I, together with other diffusion parameters we obtained $R \times N \times PP \times SF \times SP$ for simulations when we were seeding each time when diffusion dies out. Propagation probability (PP) represents propagation probability according to

Independent Cascade Model [20]. Each activated node is contacting all not active neighbours and with given probability activates them with only single possible attempt. Seeds fraction (SF) represents the percentage of nodes selected as seeds. Number of seeds per step (SP) represents the number of seeds used in each step of sequential seeding. They are resulting in 4,500 configurations.

Table 1. Networks and parameters of diffusion used in simulations

Symbol	Parameter	No. of variants	Values
R	Ranking type	2	Static Vote-Rank, Dynamic Vote-Rank
N	Network	10	Real networks from various areas
PP	Propagation probability	9	0.1, 0.2, 0.3, 0.4, 0.5, 0.6, 0.7, 0.8, 0.9
SF	Seeds fraction	5	1%, 2%, 3%, 4%, 5%
SP	Seeds per step	5	1, 2, 4, 8, 16
RI	Recomputation interval	5	1, 2, 4, 8, 16

While the goal of the Stage I was to compare performance of Vote-Rank based on the static and dynamic rankings, in the Stage II main goal was to analyse the impact of recomputations frequency on final coverage with the network. Five recomputation intervals were used. It creates experimental space with $R \times N \times PP \times SF \times SP \times RI$ combinations for simulations when we were seeding with fixed intervals what makes total 22,500 combinations.

4   Empirical Study

4.1   Performance of Dynamic Vote-Rank for Sequential Seeding

Overall analysis compared the dynamic Vote-Rank based approach with static Vote-Rank for sequential seeding with results presented in Fig. 3 (**A**). All simulation cases of dynamic approach results with not worse, and in most cases higher coverage than static approach. The finest obtained improvement is at the level of 40%. Comparing the value of the Wilcoxon test there is a statistically valid ($p < 0.05$) difference between results from spreading coverage for static and dynamic rankings. The value of Hodges-Lehmann estimator at the level 27.564 indicates a significant improvement in result continuity.

Figure 4 (**A**) and (**B**) show all simulation cases in terms of (**A**) seeds per step and (**B**) propagation probabilities. For propagation probability the most noticeable difference is for PP = 0.1. Further tendency is visible - the higher propagation probability, the lower increase of performance.

In the next stage the role of propagation probability, number of seeds and network was analysed. Figure 5 (**A**) shows a systematic decline of performance as PP increases. The value of increase of coverage from over 8% for PP = 0.1 to nearly 2% for PP = 0.9 is observed. Wilcoxon test was used to analyse results for propagation probability (PP). For PP = 0.1 Hodges-Lehmann estimator was

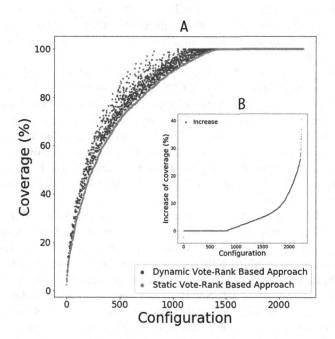

Fig. 3. **(A)** The comparison of coverage for dynamic Vote-Rank Based Approach and Static Vote-Rank based approach. **(B)** The increase of the coverage obtained by Dynamic Vote-Rank Based Approach.

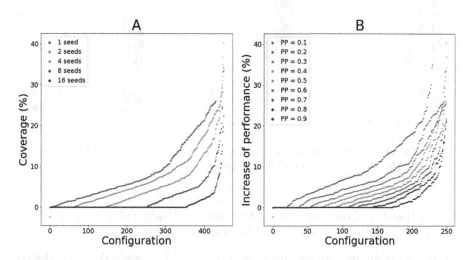

Fig. 4. Results from all simulation cases for **(A)** The number seeds per step with sorted by performance value **(B)** Propagation probabilities sorted by performance.

obtained at the level 11.69 while for PP = 0.9 it was at the lower level 5.37. Statistical significance of results was confirmed with p < 0.05. Here it is also visible a two-fold drop in the difference. The values oscillate mostly in the range from 8.4 to 11.69.

Similar tendency is visible in Fig. 5 (**B**), i.e. for number of seeds per step. At the value of 1 it is 8%, then at 2 about 7% and almost twice to the level of about 4% for the number of seeds 4. A decrease to the level of 1% occurs for 16 number of seeds. Based on the Wilcoxon test, taking into account the number of seeds per step, differences for dynamic and static rankings are visible. A downward trend in the statistical difference is observed, which is statistically significant for each number of seeds per step. We see a two-fold difference of Hodges-Lehmann estimator from 17.11 to 8.8 (for SP = 2: 17.11; for SP = 4: 15.33; for SP = 8: 12.46; for SP = 16: 8.8).

Figure 5 (**C**) shows differences in results for used network. With networks N1 and N2, it maintains the increase of coverage level above 9%, followed by a decrease to about 1% for network N3. Network 4 maintains a level close to 10%. Starting from network N5 to N10, we see a clear decline to 2–4%. Here we can divide the fall into two groups: a large decrease in network from N5 to N10 and a increase in networks N1, N2, N4. Wilcoxon test was also used for comparing results for pairs of dynamic and static rankings for all used networks N1–N10. Statistical significance (p < 0.05) was obtained. The range of results is from 5.37 for network N3 and up to 10.01 for network N2. Other results reach values close to 9.00, so they are closer to the network with the best result.

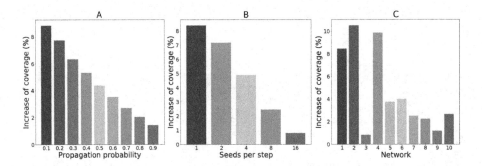

Fig. 5. Difference in averaged increase of coverage for every case of (**A**) Propagation probability, (**B**) Seeds per step and (**C**) Network

4.2 Adjusting Computations Frequency - Trade Off Between Computational Costs and Coverage Increase

In this stage of analysis we considered pros and cons of dynamic Vote-Rank based approach in terms of seeding percentage and recomputation interval to figure out how it affect on coverage performance and computational time. We assumed as seeds per step (SP) number of seeds for every stage of sequential seeding process.

In terms of number of seeds (SP), in Fig. 6 (**A**) is shown how the number of seeds per step (SP) affected coverage performance for each recomputation interval. The lowest coverage performance was observed for 16 seeds per step and interval with value 16, while for 1 seed per step with recomputation in each step the highest coverage performance was obtained. Mean value of coverage performance for 1 seed group is 77.79, for 2 seeds is 76.29, for 4 seeds is 75.75, for 8 seeds is 74.96, for 16 seeds is 70.85. It means that coverage performance decreases along with the number of seeds per step and growing interval between recompuations. Regarding how number of seeds per step and interval affects computational time, we show it in Fig. 6 (**B**), the longest calculation time was observed for 16 seeds with the highest interval, and was equal to 75.25 s. The shortest calculation time was observed for 1 seed, and was 10.99 s. For 2 seeds was 11.72 s, for 4 seeds was 14.76 s, for 8 seeds was 22.09 s. As we can infer, adding more seeds isn't profitable. Both computational time and coverage performance fare worse than with lower number of seeds.

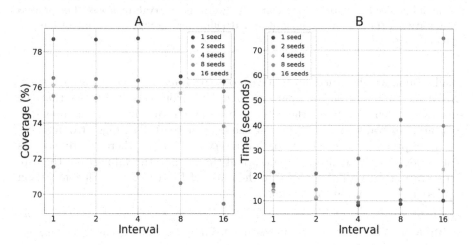

Fig. 6. Results from all simulation cases with showed how the number of seeds affects (**A**) the coverage and (**B**) computational time.

In terms of recomputation interval (RI) we carried out analysis to find out how recomputation interval effects on the efficiency. In Fig. 7 (**A**) is showed coverage performance with each of the colors representing results for different number of seeds per step, while in Fig. 7 (**B**) computational time is showed. In Fig. 7 (**A**) as we can see tendency that the greater we set interval, the smaller coverage is obtained. We can also notice relationships concerning seeds per step similar to those in Fig. 7 (**A**) and (**B**). The most effective combinations values of recomputation interval and seeds per step is small value of these both. When it comes to Fig. 7 (**B**) computational time, for smaller recomputation interval, there is no need to calculate a rank for steps forward. Consequently we calculate

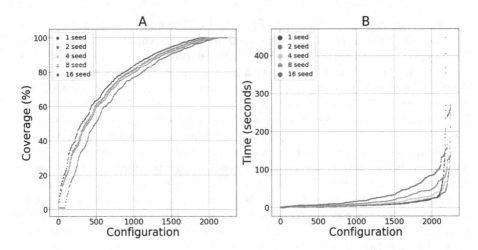

Fig. 7. **(A)** Results from all configurations for different recomputation intervals and the number of seeds per stage **(B)** Relation between the interval steps of seeding process computational time for all configurations. Results were grouped by value of interval step and averaged.

smaller rank, which turned out to have a positive impact on computational time. Analyzing the intergroup comparison using Wilcoxon tests we can see that the smallest differences between the intervals are showed when interval 1 is compared to 4, and interval 2 to 4. They are at the level of 14.61. The biggest differences are more pronounced when comparing intervals 8 to 16, where the differences reach about 35.24, i.e. over two and a half times, than the values from the top. On average, the range of results is in the range of 14.61 to 21.41. All comparison results are presented in the Table 2.

Table 2. Wilcoxon two variants test divided into static and dynamic rankings

Comparison based on group analysis									
Differences	1 vs 2	1 vs 4	1 vs 8	1 vs 16	2 vs 4	2 vs 8	2 vs 16	4 vs 8	4 vs 16	8 vs 16
between	26.68	14.61	17.99	21.28	14.91	17.44	20.76	19.11	21.41	35.24
intervals									

5   Conclusions

The main goal of this study was to analyse the effects of seed selection for sequential seeding with the use of dynamic rankings generated with the Vote-Rank algorithm. In the typical approach, network nodes are ranked once at the beginning of the process and seeds are selected according to their rank. Together with ongoing spreading processes within network changes, nodes with high potential for seeding at the beginning may no longer be effective. This occurs, for example, if a high fraction of their neighbours are already activated. In the proposed approach, nodes are ranked with the use of a network reduced by already activated nodes. Votes are gathered only from nodes able to be activated. The results demonstrated the performance of the proposed approach with a revival mode when additional seeding occurs after the process dies out. The results were dependent on network characteristics and the increase in performance when compared to the static version was above 10%. The best results were observed for low propagation probabilities. High performance was observed for a low number of seeds used in each step, with best result for one seed per step. Recomputation frequency increased the performance, with the best results obtained for recomputation in every step, but this results in higher computational costs. In many cases, larger intervals between recomputations still improved the performance with lower computational costs.

From the perspective of real applications, it is observed that recent marketing solutions focus on adaptive approaches with the use of knowledge gathered from earlier stages of campaigns. The same can be applied to viral marketing with more natural strategies based on spreading budgets and seeds allocation over the time. It creates potential for dynamic Vote Rank usage, with the ability to cope with large networks, same like it was proved for its static version.

The presented findings provide several future directions for adaptive seeding and usage of knowledge from network states observed when information spreading occurs. Future work could extend the proposed approach toward a more adaptive version with the ability to estimate the time when the recomputation should be performed to maximise the outcome. Another direction is modification of the vote counting method with the use of information about activations within the network.

Acknowledgments. This work was supported by the National Science Centre, Poland, grant no. 2016/21/B/HS4/01562.

References

1. Abebe, R., Adamic, L.A., Kleinberg, J.: Mitigating overexposure in viral marketing. In: Thirty-Second AAAI Conference on Artificial Intelligence (2018)
2. Acemoglu, D., Ozdaglar, A., ParandehGheibi, A.: Spread of (mis) information in social networks. Games Econ. Behav. **70**(2), 194–227 (2010)
3. Bharathi, S., Kempe, D., Salek, M.: Competitive influence maximization in social networks. In: Deng, X., Graham, F.C. (eds.) WINE 2007. LNCS, vol. 4858, pp. 306–311. Springer, Heidelberg (2007). https://doi.org/10.1007/978-3-540-77105-0_31
4. Boichak, O., Jackson, S., Hemsley, J., Tanupabrungsun, S.: Automated diffusion? Bots and their influence during the 2016 U.S. presidential election. In: Chowdhury, G., McLeod, J., Gillet, V., Willett, P. (eds.) iConference 2018. LNCS, vol. 10766, pp. 17–26. Springer, Cham (2018). https://doi.org/10.1007/978-3-319-78105-1_3
5. Chen, W., Yuan, Y., Zhang, L.: Scalable influence maximization in social networks under the linear threshold model. In: 2010 IEEE International Conference on Data Mining, pp. 88–97. IEEE (2010)

6. Frias-Martinez, E., Williamson, G., Frias-Martinez, V.: An agent-based model of epidemic spread using human mobility and social network information. In: 2011 IEEE Third International Conference on Privacy, Security, Risk and Trust and 2011 IEEE Third International Conference on Social Computing, pp. 57–64. IEEE (2011)
7. Goyal, A., Lu, W., Lakshmanan, L.V.: Celf++: optimizing the greedy algorithm for influence maximization in social networks. In: Proceedings of the 20th International Conference Companion on World Wide Web, pp. 47–48. ACM (2011)
8. Guimera, R., Danon, L., Diaz-Guilera, A., Giralt, F., Arenas, A.: Self-similar community structure in a network of human interactions. Phys. Rev. E **68**(6), 065103 (2003)
9. Harlow, S.: It was a facebook revolution: exploring the meme-like spread of narratives during the egyptian protest. Revista de comunicación **12**, 59–82 (2013)
10. Hinz, O., Skiera, B., Barrot, C., Becker, J.U.: Seeding strategies for viral marketing: an empirical comparison. J. Mark. **75**(6), 55–71 (2011)
11. Holme, P.: Temporal network structures controlling disease spreading. Phys. Rev. E **94**(2), 022305 (2016)
12. Jankowski, J.: Dynamic rankings for seed selection in complex networks: balancing costs and coverage. Entropy **19**(4), 170 (2017)
13. Jankowski, J., Bródka, P., Kazienko, P., Szymanski, B.K., Michalski, R., Kajdanowicz, T.: Balancing speed and coverage by sequential seeding in complex networks. Sci. Rep. **7**(1), 891 (2017)
14. Jankowski, J., Bródka, P., Michalski, R., Kazienko, P.: Seeds buffering for information spreading processes. In: Ciampaglia, G.L., Mashhadi, A., Yasseri, T. (eds.) SocInfo 2017. LNCS, vol. 10539, pp. 628–641. Springer, Cham (2017). https://doi.org/10.1007/978-3-319-67217-5_37
15. Jankowski, J., Michalski, R., Kazienko, P.: The multidimensional study of viral campaigns as branching processes. In: Aberer, K., Flache, A., Jager, W., Liu, L., Tang, J., Guéret, C. (eds.) SocInfo 2012. LNCS, vol. 7710, pp. 462–474. Springer, Heidelberg (2012). https://doi.org/10.1007/978-3-642-35386-4_34
16. Jankowski, J., Michalski, R., Kazienko, P.: Compensatory seeding in networks with varying avaliability of nodes. In: 2013 IEEE/ACM International Conference on Advances in Social Networks Analysis and Mining (ASONAM 2013), pp. 1242–1249. IEEE (2013)
17. Jankowski, J., Szymanski, B.K., Kazienko, P., Michalski, R., Bródka, P.: Probing limits of information spread with sequential seeding. Sci. Rep. **8**(1), 13996 (2018)
18. Joshi-Tope, G., et al.: Reactome: a knowledgebase of biological pathways. Nucleic Acids Res. **33**(suppl-1), D428–D432 (2005)
19. Kandhway, K., Kuri, J.: How to run a campaign: optimal control of sis and sir information epidemics. Appl. Math. Comput. **231**, 79–92 (2014)
20. Kempe, D., Kleinberg, J., Tardos, É.: Maximizing the spread of influence through a social network. In: Proceedings of the Ninth ACM SIGKDD International Conference on Knowledge Discovery and Data Mining, pp. 137–146. ACM (2003)
21. Leskovec, J., Adamic, L.A., Huberman, B.A.: The dynamics of viral marketing. ACM Trans. Web (TWEB) **1**(1), 5 (2007)
22. Leskovec, J., Kleinberg, J., Faloutsos, C.: Graph evolution: densification and shrinking diameters. ACM Trans. Knowl. Disc. Data (TKDD) **1**(1), 2 (2007)
23. Leskovec, J., Mcauley, J.J.: Learning to discover social circles in ego networks. In: Advances in Neural Information Processing Systems, pp. 539–547 (2012)

24. Ley, M.: The DBLP computer science bibliography: evolution, research issues, perspectives. In: Laender, A.H.F., Oliveira, A.L. (eds.) SPIRE 2002. LNCS, vol. 2476, pp. 1–10. Springer, Heidelberg (2002). https://doi.org/10.1007/3-540-45735-6_1

25. Libai, B., Muller, E., Peres, R.: Decomposing the value of word-of-mouth seeding programs: acceleration versus expansion. J. Mark. Res. **50**(2), 161–176 (2013)

26. Liu, Q., Hong, T.: Sequential seeding for spreading in complex networks: influence of the network topology. Phys. A **508**, 10–17 (2018)

27. Long, C., Wong, R.C.W.: Minimizing seed set for viral marketing. In: 2011 IEEE 11th International Conference on Data Mining, pp. 427–436. IEEE (2011)

28. Newman, M.E.: Scientific collaboration networks. I. Network construction and fundamental results. Phys. Rev. E **64**(1), 016131 (2001)

29. Nguyen, H., Zheng, R.: On budgeted influence maximization in social networks. IEEE J. Sel. Areas Commun. **31**(6), 1084–1094 (2013)

30. Ni, C., Yang, J., Kong, D.: Sequential seeding strategy for social influence diffusion with improved entropy-based centrality. Physica A: Stat. Mech. Appl. 123659 (2019)

31. Opsahl, T.: Why anchorage is not (that) important: Binary ties and sample selection (2011). http://toreopsahl.com/2011/08/12/why-anchorage-is-not-that-important-binary-tiesand-sample-selection. Accessed Sept 2013

32. Sahneh, F.D., Scoglio, C., Van Mieghem, P.: Generalized epidemic mean-field model for spreading processes over multilayer complex networks. IEEE/ACM Trans. Netw. (TON) **21**(5), 1609–1620 (2013)

33. Seeman, L., Singer, Y.: Adaptive seeding in social networks. In: 2013 IEEE 54th Annual Symposium on Foundations of Computer Science, pp. 459–468. IEEE (2013)

34. Sela, A., Ben-Gal, I., Pentland, A.S., Shmueli, E.: Improving information spread through a scheduled seeding approach. In: Proceedings of the 2015 IEEE/ACM International Conference on Advances in Social Networks Analysis and Mining 2015, pp. 629–632. ACM (2015)

35. Šubelj, L., Bajec, M.: Software systems through complex networks science: review, analysis and applications. In: Proceedings of the First International Workshop on Software Mining, pp. 9–16. ACM (2012)

36. Wang, X., Zhang, X., Zhao, C., Yi, D.: Maximizing the spread of influence via generalized degree discount. PLoS ONE **11**(10), e0164393 (2016)

37. Watts, D.J., Strogatz, S.H.: Collective dynamics of 'small-world' networks. Nature **393**(6684), 440 (1998)

38. Zhang, J.X., Chen, D.B., Dong, Q., Zhao, Z.D.: Identifying a set of influential spreaders in complex networks. Sci. Rep. **6**, 27823 (2016)

On the Planarity of Validated Complexes of Model Organisms in Protein-Protein Interaction Networks

Kathryn Cooper[1]([⊠]), Nathan Cornelius[1], William Gasper[1], Sanjukta Bhowmick[2], and Hesham Ali[1]

[1] College of Information Science and Technology, University of Nebraska at Omaha, Omaha, NE, USA
kmcooper@unomaha.edu
[2] College of Engineering, University of North Texas, Denton, TX, USA

Abstract. Leveraging protein-protein interaction networks to identify groups of proteins and their common functionality is an important problem in bioinformatics. Systems-level analysis of protein-protein interactions is made possible through network science and modeling of high-throughput data. From these analyses, small protein complexes are traditionally represented graphically as complete graphs or dense clusters of nodes. However, there are certain graph theoretic properties that have not been extensively studied in PPI networks, especially as they pertain to cluster discovery, such as planarity. Planarity of graphs have been used to reflect the physical constraints of real-world systems outside of bioinformatics, in areas such as mapping and imaging.

Here, we investigate the planarity property in network models of protein complexes. We hypothesize that complexes represented as PPI subgraphs will tend to be planar, reflecting the actual physical interface and limits of components in the complex. When testing the planarity of known complex subgraphs in *S. cerevisiae* and selected mammalian PPIs, we find that a majority of validated complexes possess this planar property. We discuss the biological motivation of planar versus nonplanar subgraphs, observing that planar subgraphs tend to have longer protein components. Functional classification of planar versus nonplanar complex subgraphs reveals differences in annotation of these groups relating to cellular component organization, structural molecule activity, catalytic activity, and nucleic acid binding. These results provide a new quantitative and biologically motivated measure of real protein complexes in the network model, important for the development of future complex-finding algorithms in PPIs. Accounting for this property paves the way to new means for discovering new protein complexes and uncovering the functionality of unknown or novel proteins.

Keywords: Planar graphs · PPI networks · Protein complexes · DDI networks

© Springer Nature Switzerland AG 2020
V. V. Krzhizhanovskaya et al. (Eds.): ICCS 2020, LNCS 12137, pp. 652–666, 2020.
https://doi.org/10.1007/978-3-030-50371-0_48

1 Introduction

1.1 A Brief History and Motivation

In the early stages of bioinformatics research, many studies focused on data generation approaches along with standard analysis of this data, to take advantage of the rapid advancement of biomedical technologies. The lack of data availability in the early days of bioinformatics meant that every attempt was made to take full advantage of all available data. These large, aggregated databases make datasets from multiple research groups and experiments available but has also led to certain practices that impedes the quality of the data if attention is not paid to details of the dataset provenance. Such practices include aggregation of data collected under different experimental conditions or incorporating relationships obtained via prediction rather than observed experiments.

Recently, with the massive explosion of available data in the bioscience and medical domains, the attention has shifted towards a focus on validation, data quality, and in-depth data analysis. To achieve these objectives, there is a need to develop advanced validation mechanisms to assess the quality of the large currently available biological data. We posit that an important step in this direction is to study underlying properties or features associated with current datasets and use these futures to validate the various databases and assess the quality of their data items. In this work, we explore how studying the underlying structural properties of biological networks can lead to a better understanding of the nature of the network data. In particular, we look into the impact of the physical aspects that are associated with protein interaction networks and how the physical restrictions of the interactions enforce certain properties in such networks. Our primary hypothesis is that protein complexes are likely to form planar underlying structures when represented as a subgraph of a protein-protein interaction network, particularly if their domains or subcomponents are large. Proving such hypothesis will open the door to a new direction in utilizing the large amount of data associated with biological networks and objectively assess their quality.

1.2 Overview of Network Modeling of PPIs

Modeling of protein-protein interaction (PPI) networks has grown in popularity since 1999 with the advancement of open source community databases for sharing PPI data, a rapidly growing body research on the link between network models and biological functionality (Barabasi and Albert 1999; Barabasi and Oltvai 2004; Jeong et al. 2001), and the development of algorithms and tools for clustering proteins to identify common functionality (Barabasi and Albert 1999; Barabasi and Oltvai 2004; Jeong et al. 2001; Brohee and van Helden 2006). A number of popular algorithms designed specifically for clustering proteins from PPI networks are now available, including (but certainly not limited to) ClusterONE for finding overlapping protein complexes (Nepusz, Yu and Paccanaro 2012), HC-Pin for functional complex discovery (Wang et al. 2011), Altaf-Ul-Amin's 2006 algorithm for detecting complexes in large PPI networks (Altaf-Ul-Amin et al. 2006), PRODISTIN for prediction of cellular function in PPI complexes (Brun, Herrmann and Guénoche 2004), as well as MCODE (Bader and Hogue 2003), MINE (Rhrissorrakrai and Gunsalus 2011), and SPICi (Jiang and Singh 2010). All of these aforementioned approaches are a part of a large majority of clustering algorithms built for protein-protein interaction networks that use a density measure or function to some extent to identify clusters or complexes within a protein-protein interaction network. While nearly all of the aforementioned literature notes explicitly in their work that density is not the only factor with weight in clustering edges in a protein-protein interaction network, a majority of algorithms can simplify protein complex identification with the justification that complexes are represented as densely connected clusters in a PPI network. This is typically done using a hard-clustering approach (Pu et al. 2007), but performance is mixed.

1.3   3D Structure of Protein Complexes *in Vivo*

Inherently, any clustering algorithm that uses density as a major component of its algorithm makes an assumption that a denser subgraph is the desired outcome, which may not always be the case. As a protein complex grows in size (in length of protein complex components and/or number of interaction partners), it becomes more and more unlikely that all components of a protein complex will have space to physically interface with one another. Inherently, a protein chain in its tertiary or quaternary form can typically only be bound to one partner per interface at a time (Keskin et al. 2008). It is known that the stability of protein-protein interactions can be measured by affinity as transient or permanent if they are part of a non-obligate PPI complex (Acuner Ozbabacan et al. 2011). Further information is known about the stability and permanence of protein-protein interactions; for example, interactions between homodimeric proteins tend to be more stable in their PPI interfaces than heterodimers (Jones and Thornton 1996) and also tend to be easier to predict (Keskin et al. 2008). One reasoning behind this is that the interfaces of heterodimers tend to be flatter than homodimers (Jones and Thornton 1996).

Note that a PPI network is only a model. For example, due to the nature of the techniques used to infer PPIs at the systems level (such as tandem affinity purification, mass spectrometry, or older techniques such as the Y2H experimental system,), a protein complex as it is found within its quaternary form in the cellular machine may not necessarily be accurately represented by the PPI network. Many of these techniques present a protein of interest (bait) and determine through affinity which other proteins (prey) interact with it outside of their normal functioning in the cell, meaning that the PPIs measured represent physical interactions but not their spatial arrangement or temporal stability (Uetz et al. 2000). Therefore, a number of factors, such as protein interactor length, binding affinity, experimental system used to determine the interaction, and stability of the interaction may or may not be represented in a PPI network.

1.4   Planarity in Graph Theory

The term "planar graph" denotes a well-known graph theoretic property indicating that a graph is planar if it can be embedded on a plane without having its edges cross. This notion differs from "planarity" that has been used to describe shape and size of a protein's interface with another within its 3D structure (Janin, Bahadur and Chakrabarti 2008;

Jones and Thornton 1996). Henceforth, when referring to planarity or planar graphs, we refer to the graph theoretic definition, as in Definition 1 below.

Definition 1. *A graph* $G = (V, E)$ *has a planar embedding if it can be drawn on a plane without crossing any of its edges. A graph is planar if it has at least one planar embedding (West 1996).*

In this paper, we assume complexes are represented within PPI networks as an induced subgraph $G = (V, E)$ where G is a simple graph, meaning it contains neither self-loops nor multiple edges, and edges representing interaction relationships are binary ($0 =$ does not exist, $1 =$ exists). Subgraphs are not required to be connected graphs. (See the example given in Fig. 1).

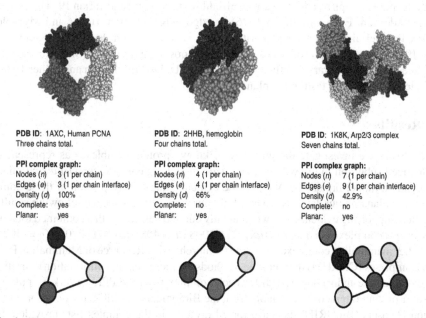

Fig. 1. Three examples of 3D protein structures and a dummy graph model of their PPI representation. The top row shows a given protein complex with its different protein components highlighted with a different color; the bottom row provides an example of how that complex might be represented graphically. Note that interactions/edges in the graphical model are drawn where there is a physical interface within the 3D protein structure. In the middle, we provide dummy examples of measures of number of nodes, edges, and density, as well as planarity and completeness of each complex. (Color figure online)

Interestingly, there appears to be no prior research into the planarity of subgraphs representing protein complexes mined from protein-protein interaction networks. A 2010 model submitted to arXiv notes that while some interactions are too complex to be reliably represented in the "protein – edge – protein" format of the PPI, it is possible to model the relationship between PPI network topology and relative protein abundance on the assumptions that there exists a subset of protein interactions tend to be flat, stable, and ordered (Heo, Maslov and Shakhnovich 2010). However, a search for applications of planar graphs reveals no prior research in biological networks.

1.5 Characterizing a Graph as Planar

There exist several algorithms for testing whether a given graph is planar. The most well-known ones use a direct application of Kuratowski's basic planarity theorem which states that a graph is planar if and only if it does not contain K_5, $K_{3,3}$, or any of their subdivisions as an induced subgraph. Note that K_5 denotes a complete graph (clique) of five vertices and $K_{3,3}$ denotes a complete bipartite graph of six vertices with three vertices in each set. However, Kuratowski's method is expensive to test in practice, particularly for large graphs. Linear time planarity testing algorithms include expanding a smaller planar graph by adding paths (path addition, Hopcroft and Tarjan 1974), vertices (vertex addition, Even and Tarjan 1976) or edge (edge addition, Boyer and Myrvold 2004). Parallel algorithms for planarity testing have also been developed (Klein and Reif 1988). Several graph softwares such as the Boost Graph Library (Siek, Lumsdaine and Lee 2002) and Library of Efficient Data Types (LEDA) (Mehlhorn and Näher 1989) include algorithms for testing the planarity of graphs.

2   Results

In this work we investigate the planarity of known protein complexes as represented by induced subgraph in the well-characterized model organism *S. cerevisiae* and other mammalian model organisms. We provide evidence that a large portion of these complexes are planar in our datasets. To highlight our work, we provide the following results and their supporting evidence for two manually curated datasets with a combined total of 808 known complexes in *S. cerevisiae*, CYC2008 ($n = 408$) and YHTP2008 ($n = 400$), and other mammalian complexes from the Comprehensive Resource of Mammalian Protein Complexes (CORUM) dataset (See Methods for more detailed information). Briefly, we extracted the induced subgraph for each complex from the PPI network by pulling all intra-protein interactions available from the Biological General Repository for Interaction Datasets (BioGRID) database for all proteins in the complex lists provided by

Table 1. The average lengths in amino acid (AA) residues for all proteins in planar and nonplanar subgraphs in *S. cerevisiae* datasets CYC2008 and YHTP2008 is given below. This table also includes the absolute value of difference (Δ) in averages between planar and nonplanar protein lengths. An unpaired Wilcoxon Rank Test was performed on the lengths of the proteins in each dataset (planar vs. nonplanar) and the averages are significantly different (p-value <<< 0.001).

Dataset	Avg. protein length (AA)		Δ (AA)	P-value
	Planar	Non-planar		
CYC2008	546.62	463.39	83.24	1.17 E−07
YHTP2008	598.90	520.72	78.18	2.13 E−06

the datasets. Only interactions that are classified as "physical" were analyzed to reflect the spatial nature of the interaction, so only "physical" experimental system edges were kept.

2.1 Protein Complexes as a Graph Tend to Be Planar

We applied a planarity checking algorithm (see Methods) to the 3,129 validated complex subgraphs from yeast and other model organisms to characterize each one as either "planar" or "nonplanar". We find that 2,619 (83.6%) were planar graphs, and the remaining 510 subgraphs (16.3%) were nonplanar. Further, for each subgraph in our dataset, 100 random graphs with the same number of nodes (n) and edges (m) were also evaluated for planarity. Interestingly, we observe that 99.38% of planar subgraphs maintained their planar quality even when edges were randomly shuffled within their structure. This consistency would imply that the planar nature of the subgraphs is primarily a result of size and density. We hypothesize that this relationship between planarity of an induced subgraph and complex size may be a result of the inherent properties of the interactors in the complex.

The length of a protein involved in a planar subgraph is longer on average than the length of a protein involved in a nonplanar complex. Each subgraph used is made up of a list of ORF ids and interactions. There were 506 planar ORFs and 1,415 nonplanar ORFs in the CYC2008 dataset, and 854 planar ORFs and 1,223 nonplanar ORFs in the YHTP2008 dataset. Lengths in AA residues for proteins involved in both planar and nonplanar subgraphs were retrieved from the Saccharomyces Genome Database using their ORF IDs. Lengths of these proteins were compared, and on average, proteins in planar subgraphs tended to be ~78 to 83 AA longer than proteins involved in nonplanar subgraphs (in the YHTP2008 and CYC2008, respectively) as shown in Table 1. The differences in means were found to be significant (p-value < 0.001) in both datasets using a Wilcoxon Rank unpaired test. However, it can be argued that any subgraph with $n = 4$ proteins or less will automatically be planar as there is a planar embedding for all iterations and subgraphs of a K_4 graph. When we examine the planarity of only subgraphs with 5 or more nodes in all datasets we find that only 31.22% of subgraphs total are planar (combined dataset, $n = 236$), and the remaining subgraphs ($n = 520$, 68.78%) are not planar, as shown in Table 2. Unfortunately, this result is not significant by a paired t-test (p-value > 0.01) and so does not provide sufficient evidence to speculate on the biological motivation, if any, versus circumstantial or coincidental planar quality of subgraphs. We can speculate, however, that as subgraph size (by node count) grows, it is likely that a subgraph will lose its planar quality, further investigated "Density in validated protein subgraphs in S. cerevisiae" section.

2.2 Function of Proteins Involved in Planar and Nonplanar Subgraphs in Yeast

A measured difference in planar versus nonplanar subgraphs leads one to question if the planar quality of subgraphs in these *S. cerevisiae* datasets is biologically motivated, circumstantial, or coincidental. To further probe this question, we annotated the planar and nonplanar datasets for CYC2008 and YHTP2008 using the PANTHER Functional Classification Tool for the GO Biological Process tree, the GO Molecular Function tree,

Table 2. Count of valid planar and nonplanar subgraphs in the datasets where the number of nodes is greater than or equal to 5. Planar/nonplanar column refers to those subgraphs labeled as such by our algorithm. Each column has a count for the number of subgraphs characterized as such, and the percent of the total that it represents for that dataset.

Table 2	Planar		NonPlanar		
Dataset	Count	%	Count	%	Total
CYC2008	11	10.48%	94	89.52%	105
YHTP2008	18	20.69%	69	79.31%	87
Bovine	1	100.00%	0	0.00%	1
Dog	0	0.00%	0	0.00%	0
Human	156	30.83%	350	69.17%	506
Mouse	36	85.71%	6	14.29%	42
Rabbit	0	0.00%	0	0.00%	0
Rat	14	93.33%	1	6.67%	15
Total	**236**	**31.22%**	**520**	**68.78%**	**756**

and the PANTHER Protein Class ontology. Here, we report those annotations with a strong representation (>5% of hit against input) and/or annotations which differed (not necessarily significantly) between planar and nonplanar subgraphs. The goal of this exercise was to determine if there were biologically motivated differences on a broader level between proteins involved in planar versus nonplanar subgraphs. We observed that there were specific annotations within each classification that showed differences between planar and nonplanar graphs (Fig. 2).

Specifically, we observe differences in the GO Biological Process result for *cellular component organization or biogenesis*, where planar-involved proteins have a lower representation than nonplanar involved proteins. We also see a minor difference in the GO Biological Process result for *response to stimulus* (GO:0005198), where planar-involved proteins have a higher annotation rate than non-planar-involved proteins. When examining the GO Molecular Function result, there is a larger difference between planar proteins (3.4% and 5.4% for CYC2008 and YHTP2008, respectively) and nonplanar proteins (14.9% and 12.9% for CYC2008 and YHTP2008, respectively) in the *structural molecule activity* annotation. Per the Gene Ontology website, this annotation is defined as "the action of [the] molecule contributes to the structural integrity of a complex or its assembly within or outside of a cell." Interestingly, this would imply that proteins found in nonplanar subgraphs in yeast are more likely to play a role in *structural molecule activity*. We also observe a higher rate of planar proteins annotated with the GO term *catalytic activity* (GO:0003824), (35.5% and 42.0% in planar CYC2008 and YHTP2008 versus 26.2% and 22.9% in nonplanar CYC2008 and YHTP2008, respectively). This annotation is typically given to molecules that catalyze biochemical reactions. Finally, we observe a difference in annotation rates in the PANTHER Protein Class annotation for *nucleic acid binding* (PC00171), with rates of 17.5% and 14.5% for planar CYC2008 and YHTP2008, respectively, compared to 31.1% and 31.6% for nonplanar CYC2008 and YHTP2008. This annotation designates molecules that bind to nucleic acids (i.e. DNA or RNA), which would imply that proteins involved in planar subgraphs are less likely to engage DNA or RNA binding compared to their nonplanar counterparts.

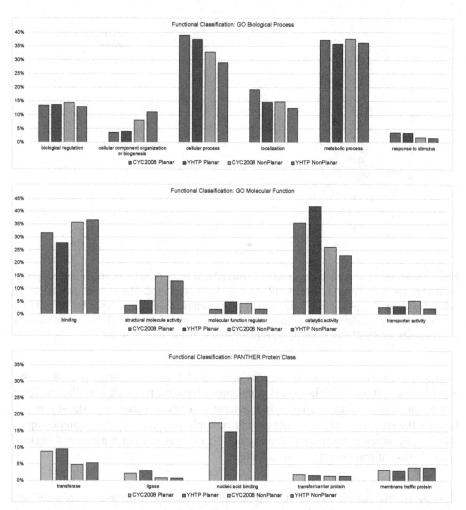

Fig. 2. Selected PANTHER functional classification results for GO Biological Process (top), GO Molecular Function (middle), and PANTHER Protein Class (bottom). The x-axis represents the annotation label or name given, and the y-axis represents the % of input against hit, or effectively the number of proteins in the given dataset labeled with that annotation versus the total number of proteins in the dataset.

2.3 Density in Validated Protein Subgraphs in *S. Cerevisiae*

When comparing the relative size of each complex, we find that there appears to be a natural boundary for planar subgraphs in terms of node size. In Fig. 3, we plot the number of nodes (x-axis) against the edge density (y-axis) for each cluster and include planar and nonplanar labels. Although the number of planar subgraphs far outweighs the number of nonplanar subgraphs, it is apparent in both datasets that the more nodes a subgraph has, the less likely it is to be planar. In both datasets, there are no subgraphs with more than $n = 16$ nodes that are planar. The average edge densities for all subgraphs with enough

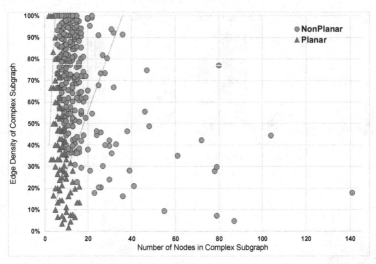

Fig. 3. Scatterplots of node count (x-axis) versus edge density (y-axis) of planar and nonplanar complexes for all *S. cerevisiae* and CORUM complexes combined. The plots do not reflect the volume of planar complexes (there are far more planar complexes than nonplanar, but they all have similar node size and density and as such overlap in the graph). Both plots show that the majority of complexes are small - there are no planar complexes beyond n = 16 nodes.

entries to measure statistical significance and $n > 4$ nodes is reported in Table 3. We also observe that for both planar and nonplanar subgraphs, there are a not-insignificant number of known, validated subgraphs that have lower than average edge density, which furthers the argument while density should certainly play a role determining complex membership when performing clustering on PPI networks, using it alone will exclude some portion of real complexes in the data.

Table 3. The average edge density of complexes in all evaluated datasets where there are enough complexes to measure statistical significance. Average edge density in complex subgraphs with $n = 5$ nodes or more only is reported, with associated p-value for a Student's T-test of unequal variance between means between planar and nonplanar complex subgraphs.

Dataset	Average edge density of complex subgraphs		
	Non-planar	Planar	P-value
CYC2008	89.72%	77.10%	0.0175
Human (CORUM)	79.33%	49.18%	8.3326 E−33
Mouse (CORUM)	72.88%	26.26%	0.0021
YHTP2008	77.86%	63.57%	0.0168

Table 4. Count of valid planar and nonplanar subgraphs in the 3did DDI datasets, where a subgraph consists of individual protein complexes, nodes represent domains with a protein, and edges represent interactions between domains. Each column has a count for the number of subgraphs characterized as such, and the percent of the total that it represents for that dataset

	Planar		Nonplanar		Totals
	Count	%	Count	%	
All complexes	733	84.64%	133	15.36%	866
Complexes with >4 DDIs	231	63.46%	133	36.54%	364

2.4 Planarity of Domain-Domain Interactions

It could be argued that the planarity or lack thereof in protein subgraphs can be attributed to the domain-domain interactions of the proteins themselves, not the entirety of the protein. Domain-domain interactions (DDIs) are the physical contact points for protein-protein interactions, where one protein component of a complex may have many interactions, and the domains of a protein are where proteins physically interface with themselves and other subcomponents. Therefore, we captured all known and validated DDIs for *S. cerevisiae* from the 3did dataset (https://3did.irbbarcelona.org/) and examined the planarity of known DDI's within a validated RSCB PDB protein complex. We find that regardless of inclusion of 'small' complexes ($\leq$4 DDIs or less), the majority of complexes have DDIs that form planar subgraphs, the opposite of what is found with examining PPI complexes (Table 4).

We re-examined complex subgraphs from *S. cerevisiae* at the DDI level, identifying 352 PSCB PDB complexes with known DDIs and their corresponding planarity. In this work, the length of the DDI in amino acid residues is measured, and the results show that planar complexes have a longer physically interacting regions (35.4 AA residues on average, $n = 173$) than nonplanar ones (29.9 AA residues on average, $n = 176$). The difference between the means is statistically significant ($p < 0.0005$ using an unpaired Student's t-test with unequal variance). These results suggest that on average, planar interactions at the complex level correspond with longer DDI interactions.

3 Methods

3.1 Data Download and Pre-processing

We chose to begin our study of planarity in PPI networks in *Saccharomyces cerevisiae* due to the extensive body of research on PPIs in the organism itself (Fields and Song 1989; Ho et al. 2002; Krogan et al. 2006; Schwikowski, Uetz and Fields 2000; Uetz et al. 2000; Von Mering et al. 2002a; Von Mering et al., 2002b), including the sentinel paper by Jeong et al. in 2001 examining centrality and essentiality in yeast PPIs (Jeong et al. 2001). We used two datasets of protein complexes described by Pu et al. 2007, curated through a multi-step procedure of clustering densely connected subunits of the yeast PPI network, and mapping to a high-quality consolidated PPI network (Pu et al. 2007).

The result of this work is two catalogs of protein complexes in yeast, the first focusing on literature-validated, heteromeric protein complexes derived from small-scale experimentation (CYC2008, $n = 408$) and the second focusing on complexes derived from high-throughput assays (YHTP2008, $n = 400$) with interactions supported by literature (Pu et al. 2008). These complexes and their components given as ORF id numbers are available as node lists from http://wodaklab.org/cyc2008 in multiple file formats and were downloaded in September 2018. We also included the Comprehensive Resource of Mammalian Protein Complexes (CORUM) non-redundant complex dataset downloaded on June 27, 2019 from their website https://mips.helmholtz-muenchen.de/corum/#download. This website contains over 4,000 validated protein complexes from *H. sapiens*, *B. Taurus* (bovine), *C. familiaris* (dog), *M. musculus* (mouse), *R. norvegicus* (rat), and *O. cuniculus* (rabbit). The complexes from the Wodak and CORUM datasets were then mapped to their respective protein-protein interaction networks downloaded from BioGRID's August 2018 release (3.4.164, file BIOGRID- ORGANISM-3.4.164.tab.zip) to elucidate their network structure. One subgraph of this PPI was generated for each *S. cerevisiae* protein complex in the YHTP2008 and CYC2008 datasets, and the same was performed for all *H. sapiens* datasets in the *H. sapiens* BioGRID PPI, for the *C. familiaris* (dog) dataset and the *C. familiaris* PPI, and so on. Complex subgraphs were generated in the following manner: First, the set of the proteins involved in each individual complex were extracted from the two complex datasets. Then, for each set of proteins, the interaction network was searched for edges such that both nodes coincident with the edge were in the given complex. Duplicate edges and self-loops were removed from this network before evaluation of planar structure. Edges are undirected. Edges and nodes not explicitly named in the PPI catalogs were removed. The resulting simple subgraph induced by this process was then extracted and stored for further analysis. The result was a total of 3,129 protein subgraphs (as separate connected network components) with nodes and edges as they exist in PPI format.

We also collected domain-domain interactions for proteins for which high-resolution three-dimensional structures are known in *S. cerevisiae* using the 3did database. We downloaded all 4,451 PDB IDs for complexes in yeast on January 16, 2020. The DDI's contained in the 3did dataset were mapped to their PDB ID using Pfam domains. Each PDB ID represented one complex with domains represented as nodes and domain-domain interactions mined from 3did represented edges. These complex subgraphs were then analyzed for planarity by our algorithm implementation.

3.2 Planarity Testing

The planarity of the subgraphs was tested using the Boyer and Myrvold planarity test (Boyer et al. 2004), an $O(n)$ planarity test based on embeddings via edge addition and Kuratowski subdivisions. This algorithm returns a result of "True" if the graph G given as input is planar and "False" if it is not. In addition to the testing of the planarity of the subgraphs themselves, for each individual subgraph a series of 100 random graphs with the same number of nodes (n) and edges (m) were also evaluated for planarity. These random graphs were created by generating m random edges where the endpoints of each edge were randomly chosen from the set of all n nodes. The generated edges were filtered to prevent duplicate edges and self-loops, resulting in m unique, unordered pairs of distinct nodes. Our code for checking the planarity of subgraphs has been made available at https://github.com/ndcornelius/complex-graphs.

3.3 Validation for Nonplanar Subgraphs

The complete interaction datasets for the model organism datasets were downloaded from BioGrid on May 20, 2019 from the version 3.5.172 release archive. ORF ids were used to identify nodes and all other data included was stored as node or edge attributes. We only wanted to investigate physical interactions so we removed any "genetic" Experimental System types. As an example, the *S. cerevisiae* network as downloaded, after removal of genetic interactions, self-loops, multiple edges, and direction, included 6,313 nodes and 110,596 edges (0.56% edge density). There was a total of 17 different types of physical experimental systems included in the BIOGRID filtered network that resulted, and all 17 measure a physical interaction of protein to protein or RNA with varying levels of quality based on experimental system (types available upon request).

3.4 Functional Analysis of *S. Cerevisiae* Subgraphs with >4 Nodes

The four sets of ORF ids from both datasets (CYC2008, planar and nonplanar as well as YHTP2008, planar and nonplanar) was analyzed to characterize functionality with the online PANTHER Classification system (version 14.1) using their functional classification tool using the *S. cerevisiae* reference genome. This tool reports a number of measures, including an annotation label or name according to the ontology being used, the accession number of that annotation, and the "% hit against input", or the number of IDs in the input against the total number of IDs in the input. We performed functional classification of all four subsets across five ontologies: GO Biological Process, GO Molecular Function, GO Cellular Component, PANTHER Protein Class, and PANTHER Pathway. In this work, we report those annotations with a strong representation (>5% of hit against input) and/or annotations which differed (not necessarily significantly) from planar to nonplanar subgraphs. The goal of this exercise was to identify any broad functional differences between planar/non-planar complexes quantitatively.

3.5 Comparison of Protein Length in Planar vs. Nonplanar Subgraphs

Subgraphs were sorted into two types (planar and nonplanar) and gene lists (using ORF as an id) for each complex were generated using in-house Python scripts. Thus, we were able to compile a list of all ORF ids for proteins involved in planar and nonplanar subgraphs for both the CYC2008 and YHTP2008 datasets. We used the *Saccharomyces* Genome Database (www.yeastgenome.org) to pull protein lengths for all ORF ids in planar and nonplanar subgraphs. Average protein lengths (in AA residues) for each group were calculated, and within datasets, length of proteins involved in planar and nonplanar subgraphs were compared using an unpaired Wilcoxon rank-sum test.

4 Discussion

Bioinformatics as a scientific discipline has gone through various stages of maturity in the last few decades. In its next stage, it is anticipated that rigorous validation and verification studies will play significant roles in solidifying major Bioinformatics findings and will increase their impact in advancing biomedical research. The reported work of this paper represents a step in this direction by employing biologically motivated concepts to analyze and measure of the quality of the widely-used biological networks.

Subgraph density has long been a measure of importance when determining the functional potential of a network structure in protein-protein interaction networks. While there is no doubt that density plays a role in finding complexes in protein-protein interaction networks, there are other underlying physical properties of proteins in complex that can be revealed with application of more advanced graph theoretic concepts. In this work, we have applied a planarity checking algorithm to 2 datasets of known PPI complexes in *S. cerevisiae* and found that a majority of protein subgraphs possess this planar property. We have identified a relationship between this planar property that may be linked to physical and spatial constraints of protein interactions at the cellular level and should be investigated with further studies. In the reported results, we find that in the broad majority of planar subgraphs, the planar embedding is not random. We also find that proteins in planar subgraphs tend to be 78–83 amino acid residues longer than proteins in nonplanar subgraphs. We do identify some functional properties of these subgraphs that differ between planar and nonplanar proteins. However, this is a preliminary study and we do realize that further work is needed to determine if this difference is significant. In the future, we plan to expand our research to more high confidence PPI datasets and more model organisms to further confirm our original hypothesis, that *due to the physical nature of protein interactions, protein subgraphs are likely to form planar underlying structures, particularly if their domains or subcomponents are large.*

This research is important for the study of protein-protein interaction networks for several reasons. First, it offers a new structural measure that is readily identifiable from the network structure, without any biological annotation or input. This could allow for the improvement or development of protein complex finding algorithms by uncovering subgraphs that were previously undiscoverable because they were not necessarily dense (for example, having 40% edge density versus 75%), but have this planar component. Secondly, it opens the door to further analysis of structure of domain-domain interactions, a subfield of protein-protein interaction research; we preliminarily find that DDI networks in yeast also maintain this planar component, perhaps even more stringently. Thirdly, it allows for the re-use and re-analysis of existing PPI datasets with the justification that this planar property may reveal previously unknown or partially known protein complexes, opening the door for discovery from our existing community databases. We look forward to expanding our proposed work and investigating further this interesting planar property in PPI networks.

References

Acuner Ozbabacan, S.E., Engin, H.B., Gursoy, A., Keskin, O.: Transient protein–protein interactions. Protein Eng. Des. Sel. **24**(9), 635–648 (2011)

Altaf-Ul-Amin, M., Shinbo, Y., Mihara, K., Kurokawa, K., Kanaya, S.: Development and implementation of an algorithm for detection of protein complexes in large interaction networks. BMC Bioinform. 7(1), 207 (2006)

Bader, G.D., Hogue, C.W.: An automated method for finding molecular complexes in large protein interaction networks. BMC Bioinform. 4, 2 (2003)

Barabasi, A.L., Albert, R.: Emergence of scaling in random networks. Science (New York) 286(5439), 509–512 (1999)

Barabasi, A.L., Oltvai, Z.N.: Network biology: understanding the cell's functional organization. Nat. Rev. Genet. 5(2), 101–113 (2004). https://doi.org/10.1038/nrg1272

Boyer, J.M., Myrvold, W.J.: On the cutting edge: Simplified O (n) planarity by edge addition. J. Graph Algorithms Appl. 8(2), 241–273 (2004)

Brohee, S., van Helden, J.: Evaluation of clustering algorithms for protein-protein interaction networks. BMC Bioinform. 7, 488 (2006). https://doi.org/10.1186/1471-2105-7-488

Brun, C., Herrmann, C., Guénoche, A.: Clustering proteins from interaction networks for the prediction of cellular functions. BMC Bioinform. 5(1), 95 (2004)

Even, S., Tarjan, R.E.: Computing an st-numbering. Theor. Comput. Sci. 2(3), 339–344 (1976)

Fields, S., Song, O.: A novel genetic system to detect protein–protein interactions. Nature 340(6230), 245 (1989)

Heo, M., Maslov, S., Shakhnovich, E.I.: Protein abundances and interactions coevolve to promote functional complexes while suppressing non-specific binding. arXiv Preprint arXiv:1007.2668 (2010)

Ho, Y., et al.: Systematic identification of protein complexes in saccharomyces cerevisiae by mass spectrometry. Nature 415(6868), 180 (2002)

Hopcroft, J., Tarjan, R.: Efficient planarity testing. J. ACM (JACM) 21(4), 549–568 (1974)

Janin, J., Bahadur, R.P., Chakrabarti, P.: Protein–protein interaction and quaternary structure. Q. Rev. Biophys. 41(2), 133–180 (2008)

Jeong, H., Mason, S.P., Barabasi, A.L., Oltvai, Z.N.: Lethality and centrality in protein networks. Nature 411(6833), 41–42 (2001). https://doi.org/10.1038/35075138

Jiang, P., Singh, M.: SPICi: a fast clustering algorithm for large biological networks. Bioinformatics (Oxford, England) 26(8), 1105–1111 (2010). https://doi.org/10.1093/bioinformatics/btq078

Jones, S., Thornton, J.M.: Principles of protein-protein interactions. Proc. Natl. Acad. Sci. 93(1), 13–20 (1996)

Keskin, O., Gursoy, A., Ma, B., Nussinov, R.: Principles of protein–protein interactions: what are the preferred ways for proteins to interact? Chem. Rev. 108(4), 1225–1244 (2008)

Klein, P.N., Reif, J.H.: An efficient parallel algorithm for planarity. J. Comput. Syst. Sci. 37(2), 190–246 (1988)

Krogan, N.J., et al.: Global landscape of protein complexes in the yeast S. cerevisiae. Nature 440(7084), 637 (2006)

Mehlhorn, K., Näher, S.: LEDA a library of efficient data types and algorithms. Paper presented at the Int'l Symposium on Mathematical Foundations of Computer Science, pp. 88–106 (1989)

Nepusz, T., Yu, H., Paccanaro, A.: Detecting overlapping protein complexes in protein-protein interaction networks. Nat. Methods 9(5), 471 (2012)

Pu, S., Vlasblom, J., Emili, A., Greenblatt, J., Wodak, S.J.: Identifying functional modules in the physical interactome of saccharomyces cerevisiae. Proteomics 7(6), 944–960 (2007)

Pu, S., Wong, J., Turner, B., Cho, E., Wodak, S.J.: Up-to-date catalogues of yeast protein complexes. Nucleic Acids Res. 37(3), 825–831 (2008)

Rhrissorrakrai, K., Gunsalus, K.C.: MINE: module identification in networks. BMC Bioinform. 12, 192 (2011). https://doi.org/10.1186/1471-2105-12-192

Schwikowski, B., Uetz, P., Fields, S.: A network of protein–protein interactions in yeast. Nat. Biotechnol. 18(12), 1257 (2000)

Siek, J., Lumsdaine, A., Lee, L.: The Boost Graph Library. Addison-Wesley, Boston (2002)

Uetz, P., et al.: A comprehensive analysis of protein–protein interactions in S cerevisiae. Nature **403**(6770), 623 (2000)

Voevodski, K., Teng, S., Xia, Y.: Finding local communities in protein networks. BMC Bioinform. **10**(1), 297 (2009)

Von Mering, C., et al.: Comparative assessment of large-scale data sets of protein–protein interactions. Nature **417**(6887), 399 (2002a)

Wang, J., Li, M., Chen, J., Pan, Y.: A fast hierarchical clustering algorithm for functional modules discovery in protein interaction networks. IEEE/ACM Trans. Comput. Biol. Bioinform. (TCBB) **8**(3), 607–620 (2011)

West, D.B.: Introduction to Graph Theory. Prentice Hall, Upper Saddle River (1996)

Fig. 1. Image of 1AXC (Gulbis, J.M., Kelman, Z., Hurwitz, J., O'Donnell, M., Kuriyan, J.: Structure of the C-terminal region of p 21WAF1/CIP1 complexed with human PCNA. Cell **87**(2), 297–306 (1996)) created with Protein Workshop (Moreland, J.L., Gramada, A., Buzko, O.V., Zhang, Q., Bourne, P.E.: The molecular biology toolkit (MBT). BMC Bioinform. **6**, 21 (2005))

Fig. 1. Image of 2HHB (Fermi, G., Perutz, M.F., Shaanan, B., Fourme, R.: The crystal structure of human deoxyhaemoglobin at 1.74 Å resolution. J. Mol. Biol. **175**(2), 159–174 (1984)) created with Protein Workshop (Moreland, J.L., Gramada, A., Buzko, O.V., Zhang, Q., Bourne, P.E.: The molecular biology toolkit (MBT). BMC Bioinform. **6**, 21 (2005))

Fig. 1. Image of 2HHB (Robinson, R.C., et al.: Crystal structure of Arp2/3 complex. Science **294**(5547), 1679–1684 (2001)) created with Protein Workshop (Moreland, J.L., Gramada, A., Buzko, O.V., Zhang, Q., Bourne, P.E.: The molecular biology toolkit (MBT). BMC Bioinform. **6**, 21 (2005))

Towards Modeling of Information Processing Within Business-Processes of Service-Providing Organizations

Sergey V. Kovalchuk[1]([✉]), Anastasia A. Funkner[1], Ksenia Y. Balabaeva[1], Ilya V. Derevitskii[1], Vladimir V. Fonin[2], and Nikita V. Bukhanov[1,3]

[1] ITMO University, Saint Petersburg, Russia
{kovalchuk,funkner.anastasia,k_balabaeva,ivderevitckii}@itmo.ru
[2] PMT Online, Moscow, Russia
fonin@pmtonline.ru
[3] LLC "GazpromNeft STC", Saint Petersburg, Russia
bukhanov.nv@gazpromneft-ntc.ru

Abstract. The paper presents an ongoing research project aimed towards the development of an approach to modeling of information spreading in complex organizations during providing service to customers and performing domain-specific business processes (BP). The approach is based on the idea of information spreading modeling and includes the view of this process in three main levels (physical, informational, and technical). The multi-layer view enables the integration and automation of obtained models in order to resolve domain-specific tasks of BP understanding, optimization, and extension. Two domain-specific case studies from healthcare and HR-management were considered within the approach proposing a way to resolve existing problems.

Keywords: Business process · Information processing · Information system · Complex network

1 Introduction

The idea of information processing as a paradigm for a description of complex systems attracts the recent attention of scientific society in various applications ranging from a description of complex physical systems as a source of information to analyzing mutual information in complex social systems [13]. One of the interesting areas with a specific multi-layer view to information processing is the information process in a large organization that involves social, informational, organizational, financial, and other types of resources interconnected within a set of business processes (BP). The most interesting class of such systems is an organization that provides a service to a customer. Taking a broad view of the idea of service, it could cover healthcare, education, social service, legal regulation, and many other systems that involve human agents on both sides (customer and

V. V. Krzhizhanovskaya et al. (Eds.): ICCS 2020, LNCS 12137, pp. 667–675, 2020.
https://doi.org/10.1007/978-3-030-50371-0_49

provider). One can consider many such systems on different levels, each of which has its complexity and own ways of information processing interpretation. This may include a) information storing and processing in (often distributed) information systems like corporate information systems or domain-specific information systems (like medical information systems, MIS) [5]; b) explicit and implicit (performed without direct regulations and rules) BP implementation in a complex domain-specific environment of an organization [11]; c) complex network of agents and resources of different types and roles as a medium for information and knowledge exchange [15]. Within our ongoing study, we are aimed at the development of a holistic approach for modeling and optimization of existing BPs in such multi-level complex systems by considering it through the lens of information processing.

2   Related Works

The proposed approach has a certain level of multidisciplinarity, but two key groups of related should be mentioned due to high maturity. This includes BP modeling and information systems/processes management.

Modeling of business processes is a mature area with a well-grounded theoretical and technological background. One of the important tools here is process mining [1] adopted to multiple areas and provided the foundation for modeling and simulation of BPs [16]. Nevertheless, the area has multiple open issues, including management of complex multi-aspect process modeling [10], complexity [4,8], and interpretability [11] of BP models, various uncertainty [8,12] and performance [9] assessing and management, etc.

The complexity of information processing within an information system attracts certain attention of researchers and practitioners in the area of business and IT management. One of the directions developed in this area is business - IT alignment, which hires various concepts to bring systematization into the management procedures. The approaches include conceptual joining of business-process management and IT management [14], co-evolutionary approaches [2], complex adaptive systems [17], etc. Still, the investigations in this area are mainly focused on planning and management from strategic, operational, social points of view.

Although the mentioned areas are well-developed high complexity of considered processes, existing open issues, and multiple levels of consideration in BPs within the presented problem bring certain limitations to the direct application of existing solutions.

3   Basic Concepts

We are focused on a specific class of systems which mainly include organizations (or department within an organization) providing a service to a customer (or to other departments). One can identify the following characteristics of such a system:

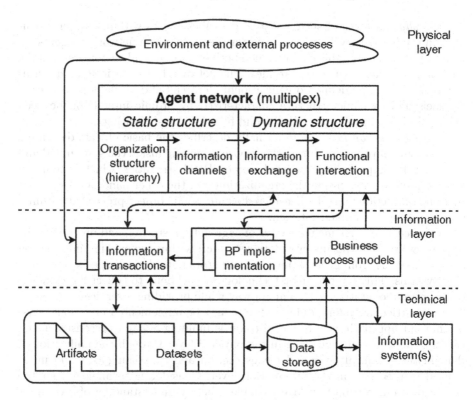

Fig. 1. Information processing in a complex organizational structure.

1. Social interaction within the business processes which include human actors as a subject of BPs and as a target of them.
2. Multi-level, multi-purpose interaction between the agents within the context of BP implementation (including long-distance interaction).
3. High uncertainty of internal or external (within the environment) processes that define or/and influence BPs.
4. Multiple levels of regulation usually include official rules, formal and informal recommendations, common practices etc.
5. Individual strategies, experience, goals, and behavior models of the actors.
6. Diverse, limited, weekly structured data are reflecting the ongoing BP implementation, which commonly includes sparse observations, human-readable artifacts (documents, instructions, messages, etc.) stored in various information systems (including enterprise information systems, EIS).

As a result, such an information processing system obtains characteristics of complex systems [3]. Within our study, we consider such a system as a multi-layer structure with a focus on information processing (see Fig. 1). This structure includes three main layers reflecting different points of view to the information processing.

The physical layer of such a system describes agents within a system. Usually, on the highest level, one can consider the system as a network of agents (in terms of organization, this could include both employees and customers) and the environment where the system resides. The network has a static structure usually defined by a predefined organizational hierarchy and available information channels and dynamic structure characterized by atomic interaction between agents during information exchange and functional co-operating.

The information layer more formally describes the basic objects describing the structure of processes during the service providing inside the system. BP and its particular implementations are based on BPM semantics, which is common to describe processes inside the organization. On the other hand, a set of information transactions may be considered as an atomic object proceeded within a system during BP implementation.

The technical layer includes ways of aggregation, storing, and processing the appearing artifacts (documents, data sets, etc.) within the existing information systems ranging from general-purpose data storage to domain-specific information systems. This layer forms a technological backbone to collect life evidence of the system operating in a common way while providing a service.

In a particular system, different layers may be implemented in various ways. Within our approach, we are aimed towards building a way to reconstruct and estimate information processing in the physical layer through the unification of data in the technical layer and reconstruction and controlling BPs and information transactions in the information layer. Considering technologies for the transition from two bottom layers enable a better understanding of a complex system in the physical layer and optimization of its functional characteristics through the explicit modification of available business-processes and the introduction of new elements in EIS (e.g., in the form of decision support systems). Interconnection of all three layers opens a way to automate BP optimization in a continuous way adapting to the changing of the BPs, their effectiveness (due to the environment changing), or goal criteria defined as a target for organization improvement.

4   Case Studies

Currently, we consider several domains that may be mapped onto the proposed idea, including healthcare, human resource management, education, legal regulation, etc. Each of the areas has its own agent roles, structure, and nature of BPs, architecture, and functional characteristics of information systems, sources of uncertainty and process variation, behavioral models, and strategies of agents. Within the following sub-sections, we provide a preliminary analysis of several cases form existing projects in the domains of healthcare and HR management.

4.1   Distributed Medical Information System

Remote monitoring of chronic patients is an important task aimed to reveal the peculiarities of chronic disease development (which is usually slowly developed and weakly observed process) and treatment effectiveness. We participate in a project performed by PMT Online for developing of a country-scale home monitoring system for chronic disease patients [6]. One of the experimental setup developed within the project was home monitoring of arterial hypertension patients, which perform self-measurements of blood pressure, and the data was transferred to the system via GSM network. The setting was deployed in 83 hospitals in 19 regions of Russia. During January 2019, it collected measurements for about 7k AH patients treated by 433 physicians (with an average of 16 patients per physician) with about 9 core controlling operators. The agents form a network shown in Fig. 2a, which can be considered as information transfer channels. The system (see the architecture illustrated in Fig. 2b) filter the measurement events according to the rule-based procedure into several categories depending on the urgency of physician intervention (initial, with three levels: green, yellow, and red). One of the important questions was how we could tune the filtering rules to reduce the load of the system agents involved in the event processing (operators, experts, physicians) as even in the basic experimental setting, the system generates about 70k events per month? A key requirement here is keeping false-negative triggering as low as possible and optimize false-positive triggering at the same time. A way for tuning the procedure is updating the triggering rules (see Trigger #1 and Trigger #2 in Fig. 2b), which controls information transfer from agent to agent.

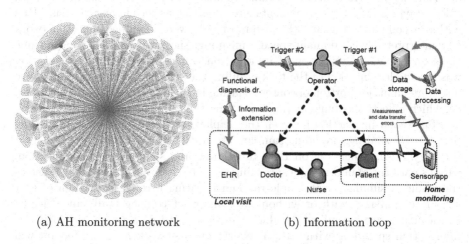

(a) AH monitoring network   (b) Information loop

Fig. 2. Information processing in a complex organizational structure.

From the point of view of the proposed approach, the physical layer includes a network of patients, physicians, experts, and operators interconnected with different information channels, including MIS-based data transfer, phone calls, messaging, etc. The BP for data processing is quite straight-forward (see Fig. 2b) but is subject to control and optimization. The technical layer includes distributed MIS integrated local MISes in hospitals. Having in mind the idea of information transfer, we are developing the simulation version of such a system, including a patient's blood pressure dynamics [7] and controllable state prediction [6] to discover the best triggering rules to reduce redundant information flow.

4.2 Implicit Business Processes in Company Network

A large company usually implements a multitude of BPs during its regular activity. This includes explicit BPs defined in the standards, rule books, recommendations, etc. and implicit BPs defined by best practices, common sense, basic organization principles, etc. Within such a complex structural and functional environment, single employee activity may vary significantly. A research question raised within a project performed with GazpromNeft STC was how we could describe and understand a complex behavior of an employee having diverse artifacts within a "digital trail" of his/her activity within a common business process. The goal was to develop a DSS for HR management, taking into account a multitude of factors (formal, functional, informal, behavioral, etc.) influencing effectiveness, skills, (self-)assessment of an employee. Could we identifying advance under- and over-qualified persons, persons under the stress conditions, persons to be promoted or fired, etc.

One of the original practices of the company was a self-defined description of activities within existing BPs. Analyzing a set of 869 such semis-structured descriptions we've identified an organizational structure of the company (Fig. 3a, colors stays for different departments' areas) as well as the closeness of basic BPs within a company (Fig. 3b, 21 clusters of BPs were identified in the network and interpreted with domain-specific meaning, shown in colors). At the same time, an important reflection of information transfer and BP implementation was the implementation of the task of filling the self-defined description. We've discovered a major portion of content sharing in such documents (see Fig. 3c and d), which reflects the dynamic (having a timestamp for the provided documents) and functional interaction between the employees on company scale (including peer-to-peer sharing, providing samples, etc.). All three networks have different topology. Also, depending on BPs, functional, and structural interconnection, various patterns of information spreading could be discovered. E.g., Fig. 3e shows the result of simulation of simple random iterating through the graph of text sharing considering with or without filtering exact copying ("all" and "deg+" in the figure correspondingly), which can be treated as a rough estimation of information spreading within the network. Further work in this direction will be aimed towards the identification of information pathways according to the objective and subjective (self-defined) BPs structure.

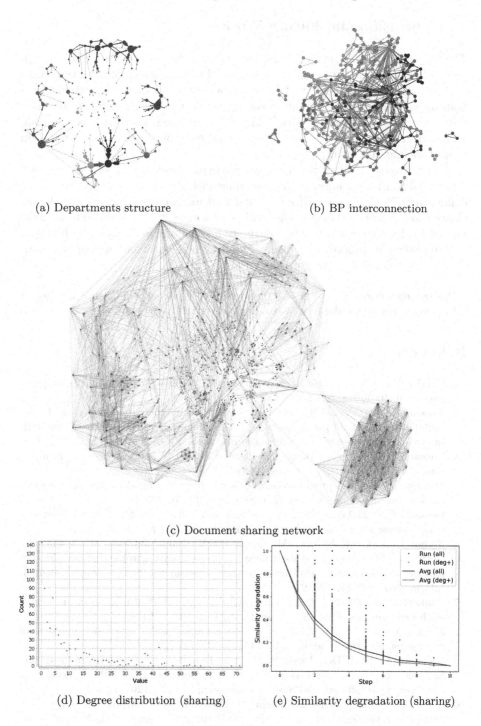

(a) Departments structure

(b) BP interconnection

(c) Document sharing network

(d) Degree distribution (sharing)

(e) Similarity degradation (sharing)

Fig. 3. Information processing in a complex organizational structure.

5   Conclusion and Future Work

The presented work is currently ongoing. Within the proposed conceptual app-roach of multi-layered information spreading in the organization, we've consid-ered several domain-specific tasks defining a way to understand emergent phe-nomena within the information spreading networks. Having defined BPs as a basic structure for dynamic processing of information, there still is a certain variation in BP implementation. The selected examples include case studies in healthcare and HR management areas.

Further research directions include three main directions. First, the proposed concept of multi-layer information spreading with BP as a core objective will be defined more formally and without a detailed explanation. Second, the deeper elaboration of proposed case studies will be performed to assess the mutual infor-mation in the system's elements, information spreading pathways, and finally, a way to resolve the proposed problems. Finally, seeking problems and applications in other domains will be performed.

Acknowledgments. This work was supported by the Ministry of Science and Higher Education of Russian Federation, Goszadanie No. 2019-1339.

References

1. van der Aalst, W.: Process Mining, 2nd edn. Springer, Heidelberg (2016). https://doi.org/10.1007/978-3-662-49851-4
2. Amarilli, F., Van Vliet, M., Van Den Hooff, B.: An explanatory study on the co-evolutionary mechanisms of business IT alignment. In: Proceedings of the 38th International Conference on Information Systems, Seoul 2017, pp. 1–22 (2018)
3. Boccara, N.: Modeling Complex Systems. Graduate Texts in Physics. Springer, New York (2010). https://doi.org/10.1007/978-1-4419-6562-2
4. Cardoso, J.: Business process control-flow complexity: metric, evaluation, and val-idation. Int. J. Web Serv. Res. (IJWSR) 5(2), 49–76 (2008)
5. Delone, W.H., McLean, E.R.: The DeLone and McLean model of information sys-tems success: a ten-year update. J. Manag. Inf. Syst. 19(4), 9–30 (2003)
6. Elkhovskaya, L., Kabyshev, M., Funkner, A., Balakhontceva, M., Fonin, V., Kovalchuk, S.: Personalized assistance for patients with chronic diseases through multi-level distributed healthcare process assessment. Stud. Health Technol. Inform. **261**, 309–312 (2019)
7. Funkner, A.A., Zvartau, N.E., Kovalchuk, S.V.: Motif identification in vital signs of chronic patients. Procedia Comput. Sci. **156**, 105–113 (2019)
8. Jung, J.Y., Chin, C.H., Cardoso, J.: An entropy-based uncertainty measure of process models. Inf. Process. Lett. **111**(3), 135–141 (2011)
9. Leemans, M., Van Der Aalst, W.M., Van Den Brand, M.G.: Hierarchical per-formance analysis for process mining. In: Proceedings of the 2018 International Conference on Software and System Process, pp. 96–105 (2018)
10. Mendling, J., Baesens, B., Bernstein, A., Fellmann, M.: Challenges of smart busi-ness process management: an introduction to the special issue. Decis. Support Syst. **100**, 1–5 (2017). https://doi.org/10.1016/j.dss.2017.06.009

11. Mendling, J., Reijers, H.A., Cardoso, J.: What makes process models understandable? In: Alonso, G., Dadam, P., Rosemann, M. (eds.) BPM 2007. LNCS, vol. 4714, pp. 48–63. Springer, Heidelberg (2007). https://doi.org/10.1007/978-3-540-75183-0_4

12. Pegoraro, M., Uysal, M.S., van der Aalst, W.M.: Efficient construction of behavior graphs for uncertain event data. arXiv preprint arXiv:2002.08225 (2020)

13. Quax, R., Apolloni, A., Sloot, P.M.A.: Towards understanding the behavior of physical systems using information theory. Eur. Phys. J. Spec. Top. **222**(6), 1389–1401 (2013)

14. Rahimi, F., Møller, C., Hvam, L.: Business process management and IT management: the missing integration. Int. J. Inf. Manag. **36**(1), 142–154 (2016). https://doi.org/10.1016/j.ijinfomgt.2015.10.004

15. Reagans, R., McEvily, B.: Network structure and knowledge transfer: the effects of cohesion and range. Adm. Sci. Q. **48**(2), 240 (2003)

16. Rozinat, A., Wynn, M., van der Aalst, W., ter Hofstede, A., Fidge, C.: Workflow simulation for operational decision support. Data Knowl. Eng. **68**(9), 834–850 (2009). https://doi.org/10.1016/j.datak.2009.02.014

17. Schilling, R.D., Kazem Haki, M., Beese, J., Aier, S., Winter, R.: Revisiting the impact of information systems architecture complexity: a complex adaptive systems perspective. In: Proceedings of the 38th International Conference on Information Systems, Seoul 2017, pp. 1–18 (2018)

A Probabilistic Infection Model for Efficient Trace-Prediction of Disease Outbreaks in Contact Networks

William Qian[1], Sanjukta Bhowmick[2(✉)], Marty O'Neill[2],
Susie Ramisetty-Mikler[2], and Armin R. Mikler[2]

[1] University of Pennsylvania, Philadelphia, USA
bqqian@sas.upenn.edu
[2] University of North Texas, Denton, USA
{sanjukta.bhowmick,Marty.ONeill,Susie.Ramisetty-Mikler,
Armin.Mikler}@unt.edu

Abstract. We propose a novel method which we call the Probabilistic Infection Model (PIM). Instead of stochastically assigning exactly one state to each agent at a time, PIM tracks the likelihood of each agent being in a particular state. Thus, a particular agent can exist in multiple disease states concurrently. Our model gives an improved resolution of transitions between states, and allows for a more comprehensive view of outbreak dynamics at the individual level. Moreover, by using a probabilistic approach, our model gives a representative understanding of the overall trajectories of simulated outbreaks without the need for numerous (order of hundreds) of repeated Monte Carlo simulations.

We simulate our model over a contact network constructed using registration data of university students. We model three diseases; measles and two strains of influenza. We compare the results obtained by PIM with those obtained by simulating stochastic SEIR models over the same the contact network. The results demonstrate that the PIM can successfully replicate the averaged results from numerous simulations of a stochastic model in a single deterministic simulation.

Keywords: Computational epidemics · Outbreak simulation · SEIR model

1 Introduction

Two popular approaches for modeling infectious diseases are the simulation of disease spread through stochastic agent-based modelling; and the use of deterministic meta-population models [1,10]. Stochastic agent-based models represent specific individuals or groups of individuals as *agents*. Each agent's actions are governed by a set of rules which may themselves be functions of each agent's characteristics, or of each agent's environment. Interactions between pairs of agents which emerge as each one follows these rules establishes a contact network through which infectious disease can spread.

The original version of this chapter was revised: the first author's first name contained a typo. The correction to this chapter is available at https://doi.org/10.1007/978-3-030-50371-0_52

© Springer Nature Switzerland AG 2020, corrected publication 2020
V. V. Krzhizhanovskaya et al. (Eds.): ICCS 2020, LNCS 12137, pp. 676–689, 2020.
https://doi.org/10.1007/978-3-030-50371-0_50

Many meta-population models use a system of differential equations to approximate the rate of change of the number of individuals in each disease state (e.g., susceptible, infected, etc.). The mathematical description of the SEIR (Susceptible, Exposed, Infected, and Recovered) framework[1] is given in Fig. 1. S, E, I, and R represent the number of individuals in Susceptible, Exposed, Infected, and Recovered states respectively. The total population is then given by $N = S + E + I + R$. Parameter β is the proportion of contacts between members of S and members of I that lead to disease transmission. Parameter σ is the rate at which the exposed become infected. Parameter γ is the recovery rate at which the infected transition to the recovered state.

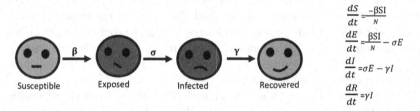

$$\frac{dS}{dt} = \frac{-\beta SI}{N}$$

$$\frac{dE}{dt} = \frac{\beta SI}{N} - \sigma E$$

$$\frac{dI}{dt} = \sigma E - \gamma I$$

$$\frac{dR}{dt} = \gamma I$$

Fig. 1. A pictoral representation of the SEIR model, along with the modeling equations.

Meta-population disease models are computationally efficient due to their deterministic nature. Further, closed form approximations of significant epidemiological parameters such as the basic reproduction number R_0 (i.e. the expected number of secondary cases caused by a single infectious individual in a completely susceptible population) can be derived analytically using meta-population models. However, these models assume a homogeneous mixing rate within a homogeneous population. Thus, they do not take into account the diversity of a population which could lead certain individuals to have more contacts than others.

The stochastic agent-based approach incorporates population heterogeneity which could lead to variations in the numbers of contacts corresponding to each individual. Modeling interactions between pairs of individuals allows for flexibility in dictating specific patterns of behavior for individual agents. These models use stochastic processes to decide which contacts (i.e., edges) represented in the network lead to state transitions of agents (i.e., vertices) from Susceptible to Exposed at each simulated time step. However, due to the reliance on stochastic processes, a single run of an outbreak simulation using these models is not representative of an expected outcome. Thus, these models often require hundreds of repeated trials per unique set of parameters in order to properly estimate trends in the data. The computation required for this repetition of trials limits the scope of the analysis that can feasibly be done using these models. Analysis using stochastic models is complicated further by the fact that it is difficult to

---

[1] Births and deaths among the population are not considered in the model.

derive closed-form expressions for important quantities such as the basic reproduction number R_0 without direct experimentation.

Our work is motivated by the advantages and drawbacks of these two popular epidemiological models. We introduce the Probabilistic Infection Model (PIM), which combines the heterogeneity of the stochastic models with the computational efficiency and deterministic nature of the meta-population models. The **key idea** of PIM is to calculate for each vertex in a contact network, *the probabilities of the four SEIR states associated with that vertex.* To compute the probability function, we leverage the research conducted in escape probabilities by Thomas and Weber [16]. The probabilities for each state and each vertex are compounded over windows of time corresponding to the latent and infectious periods of the given disease. This allows for probabilistic values of different states over time at the individual level and also provides the expected values of the sizes of the SEIR sub-populations corresponding to each state. As an added advantage, our proposed PIM model allows us to compute an expression for $R_0(v_0)$, which yields the value of R_0 for specific single infective individuals in an otherwise susceptible contact network.

We applied our model to a contact network created from class enrollment data from the University of North Texas. We conducted our experiments with three sets of disease parameters and compared the results with those produced by the stochastic models. Our results demonstrate that the PIM simulations are similar to those produced by averaging trials from Monte Carlo models. This similarity is most notable when simulating diseases that are highly infectious.

2   The Probabilistic Infection Model

In this section we describe our proposed Probabilistic Infection Model. In Table 1, we provide a list of the terms that we use in our computations, along with their definitions. In the standard stochastic model, for a given contact event, a vertex selects a single neighbor in the network to simulate a contact. Due to the stochasticity of the model, the simulation must be run multiple times to estimate how population sizes for each SEIR state change over simulated time.

In our probabilistic infection model, contact events occur between adjacent vertices. Thus, all neighbors of a specific vertex have a probability to make a contact. For any given contact event, we set the contact probability per pair of vertices to be proportional to the weight of their corresponding edge. The probability that vertex v will be contacted by vertex u as a result of a single contact expended by u is $\Psi(u,v) = \frac{w(u,v)}{\sum_{x \in N(v)} w(u,x)}$; $w(u,v)$ is the weight of the edge (u,v) and $N(v)$ is the set of neighbors of vertex v. Note that this function is not commutative. The probability of a contact from vertex u to vertex v, will differ from the probability of a contact from vertex v to vertex u, depending on each vertex's number of neighbors and weights of the adjacent edges.

Each time v is contacted by an infectious individual u, there is a transmission probability $T(u,v)$. The probability that vertex v is infected by u on day t as a result of a single contact made by u is then given by

$$\delta_t(u, v) = \Psi(u, v) \cdot I_t(u) \cdot T(u, v) \tag{1}$$

i.e. the product of the probability of contact between u and v, the probability the u is infected on day t, and the transmission probability between u and v.

Table 1. Notation used in equations

Notation	Definition
$S_t(v)$	Probability that a vertex v is susceptible on day t
$E_t(v)$	Probability that a vertex v is exposed on day t
$I_t(v)$	Probability that a vertex v is infectious on day t
$R_t(v)$	Probability that a vertex v is recovered on day t
$N(v)$	Set of neighbors of vertex v
σ_v	The incubation period, time between exposed to infected state, for vertex v
γ_v	The infectious period, time between infected to recovered state, for vertex v
$\Omega_t(v)$	The number of contacts that vertex v makes on day t
$\Psi(u, v)$	Probability that vertex u contacts vertex v as a result of a single contact expended by u
$\delta_t(u, v)$	Probability that vertex v is infected by u on day t as a result of a single contact expended by u
$T(u, v)$	Probability that an infectious vertex u infects vertex v upon contact

Lemma 1. *Given that a vertex v is in the exposed state, i.e. $E_x(v) > 0$ and $I_x(v) = 0$ on day x, v will have $I_t(v) > 0$, i.e. be in an infectious state on day t for some $t > x$, only if it was contacted by an infectious vertex within the critical infection window of $t - (\gamma_v + \sigma_v) + 1$ and $t - \sigma_v$.*

Proof. We note that since each partial infection received by v has a latent period σ_v, the infection probability of v, for a day r prior to day t, will remain unchanged for $t - \sigma_v + 1 \leq r \leq t$. Moreover, because the infectious period is γ_v, any infections that arose from interactions made by v on or before day $t - (\gamma_v + \sigma_v)$ would have expired by day t. Thus, taking these together, the time between $t - (\gamma_v + \sigma_v) + 1$ and $t - \sigma_v$ is the **critical infection window** where an infectious contact will take v to an infectious state on day t.

Figure 2 depicts how this critical window affects the state of the vertex. For ease of explanation, we consider the probabilities in this example to be 0 or 1, which can occur if there is only one successful infectious contact. Consider the vertex v to be in an exposed state ($E_x(v) = 1$) . In Case 1, if an infectious contact occurs within the critical infection window, then v will be in an infected state ($I_t(v) = 1$) on day t. If, Case 2, the infectious contact occurs after the critical infection window then v will remain in exposed state ($E_t(v) = 1$) on day t. If, Case 3, the infectious contact occurs before the critical infection window then v will be in recovered state ($R_t(v) = 1$) on day t.

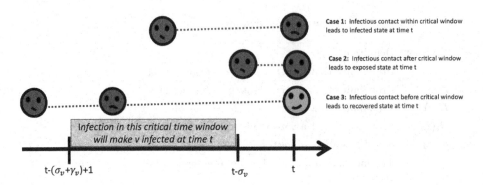

Fig. 2. A pictoral representation of the duration of infections with respect to the critical infection window

2.1 Computing the Probability for Each State

We now derive the expressions for computing the probability of each state for a given vertex v and a day t. We assume at the beginning of the simulation, i.e. at day 0, all vertices are either completely (with 100% probability) in the susceptible state or in the infected state.

Let $\Omega_t(u)$ denote the number of contacts that u makes on day t. The probability of v *not* being infected due to *one* contact made by u on day t is $1 - \delta_t(u, v)$. Taking all neighbors of v, the probability that v is not infected by any of the neighbors is $\prod_{u \in N(v)} (1 - \delta_t(u, v))^{\Omega(u)}$, where we make the approximation that each event where vertex v is not infected by some contact is independent.

Susceptible State: The probability that the vertex is in a susceptible state is the probability that v is not infected by any of the neighbors since day 0 to current day t. Thus;

$$S_t(v) = \prod_{n=0}^{t} \prod_{u \in N(v)} (1 - \delta_n(u, v))^{\Omega_n(u)} \tag{2}$$

Exposed State: Any susceptible vertex that was infected σ_v (the incubation period) days earlier will be exposed. Thus the probability of the exposed state is the probability of being in the susceptible state on day $max(0, t - \sigma_v)$ minus the current probability of the susceptible state on day t.

$$E_t(v) = S_{max(0, t - \sigma_v)}(v) - S_t(v) \tag{3}$$

Infectious State: Any susceptible vertex that was infected $\sigma_v + \gamma_v$ (the incubation period + infectious period) days earlier will be in an infectious state. The probability of the exposed state is the probability of being in the susceptible state on day $max(0, t - \sigma_v)$ minus the current probability of the exposed state on day t.

$$I_t(v) = S_{max(0, t-(\gamma_v + \sigma_v))}(v) - S_{max(0, t-\sigma_v)}(v) \qquad (4)$$

Recovered State: Any susceptible vertex that was infected before the critical infection window $t - (\sigma_v + \gamma_v)$ will have recovered by day t. The probability of the recovered state is 1 minus the probability that the vertex was still susceptible $\gamma_v + \sigma_v$ days prior.

$$R_t(v) = 1 - S_{max(0, t-\gamma_v - \sigma_v)}(v) \qquad (5)$$

The total number of individuals ever infected at the end of an outbreak can be computed by several methods. One method is to take the expected number of recovered individuals by summing over $R_L(v)$ for all v, where L is the last day of the simulation. Another way to approximate this quantity is to integrate the expected number of infected individuals $\sum_{v \in V(G)} I_t(v)$ over time and divide the result by the disease's infectious period to account for over-counting. Since time is counted in discrete steps, this integral can be reduced to a sum.

Thus, given an outbreak of length L in days;

$$\sum_{v \in V(G)} R_L(v) \approx \sum_{n=0}^{L} \sum_{v \in V(G)} \frac{1}{\gamma_v} I_n(v) \qquad (6)$$

This is satisfied in standard Monte Carlo models as well as in our PIM model.

Moreover, using PIM, we can calculate the value of the basic reproduction rate, R_0, for a specific single infective v_0 in a contact network where all other vertices are susceptible, as follows:

$$R_0(v_0) = \sum_{v \in N(v_0)} \left(1 - \prod_{n=0}^{\gamma_{v_0}-1} (1 - T(v_0, v) \Psi(v_0, v)^{\Omega_n(v_0)}) \right) \qquad (7)$$

Here the δ factor is replaced by just the product of transmission and contact probabilities, as $I_n(v_0) = 1$ for $0 \leq n < \gamma_{v_0}$.

2.2 Infection Redundancy Correction

One critical issue in using the PIM model is the effect of infection redundancy. This problem is illustrated in Fig. 3. Consider on day t, vertex v is exposed to the infection $\delta_t(u, v)$ through contact with vertex u. Once v reaches an infected state on day $t + \sigma_v$, it will expose vertex u to the infection $\delta_{t+\sigma_v}(v, u)$. However, note that some of the infections contributing to the value of $I_{t+\sigma_v}(v)$ have originated from u. This will result in u compounding its own probability of being infected, by incurring these redundant infections.

In order to correct this effect, we modify the infection from vertex u to vertex v by correcting each $\delta_t(u, v)$ to only factor in u's probability of being

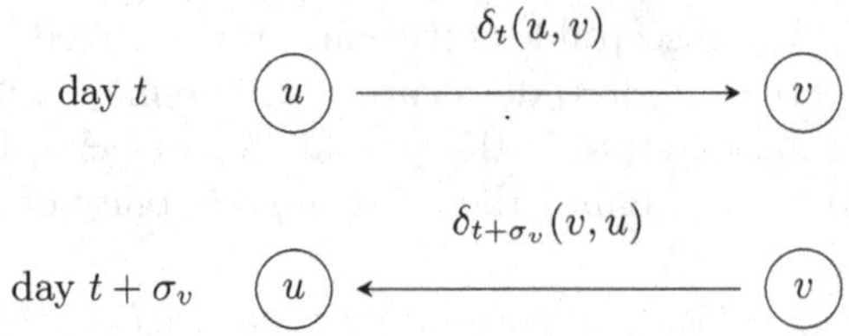

$$\delta_t(u,v)$$

day t u ———————→ v

$$\delta_{t+\sigma_v}(v,u)$$

day $t+\sigma_v$ u ←——————— v

Fig. 3. An illustration of the infection redundancy problem.

infectious as a result of contacts from vertices other than v. This ensures that infections originating from u will not be returned to u by any of u's direct neighbors. Making this correction will improve the accuracy provided by PIM at the expense of computation time.

To calculate this, consider

$$X = \prod_{n=max(0,t-(\gamma_u-\sigma_u-1))}^{max(0,t-\sigma_u)} \prod_{s\in N(u)} (1-\delta_n(s,u))^{\Omega_n(s)}$$

and

$$Y = \prod_{n=max(0,t-(\gamma_u-\sigma_u-1))}^{max(0,t-\sigma_u)} (1-\delta_n(v,u))^{\Omega_n(v)}$$

Then X represents the probability that u was not infected in the critical infectious window by any of its neighbors (using the same logic as calculating for $S_t(v)$ earlier). Y represents the probability that u was not infected in the critical infectious window by vertex v. Since the values are given as products, the ratio of $\frac{X}{Y}$ approximates the probability that u was *not infected* in the critical infectious window by any of its neighbors *and* also discards the effect of infections from v. The probability that u is infected as a result of contacts with vertices other than v is then given by $1-\frac{X}{Y}$. We thus modify the probability that v is infected by u on day t as a result of a single contact made by u to obtain

$$\delta_t(u,v) = \Psi(u,v) \cdot T(u,v) \cdot \left(1 - \frac{\prod\limits_{n=max(0,t-\gamma_u-\sigma_u+1)}^{max(0,t-\sigma_u)} \prod\limits_{s\in N(u)} (1-\delta_n(s,u))^{\Omega_n(s)}}{\prod\limits_{n=max(0,t-\gamma_u-\sigma_u+1)}^{max(0,t-\sigma_u)} (1-\delta_n(v,u))^{\Omega_n(v)}}\right)$$

(8)

where the factor representing the probability that u was infectious on day t has been modified to prevent infection redundancy. We note that this is an approximate correction, as it is still possible for an infection to return to its source after passing through multiple vertices. Since an infection moving down a path of vertices gets exponentially smaller in magnitude as the length of the path increases, it is expected that the effect would be increasingly negligible for higher order corrections.

3   Experimental Results

In this section we present our experimental results of comparing the simulation of PIM with the stochastic Monte-Carlo simulations.

Constructing the Contact Network. Creating a reliable contact network presents a challenge in computational epidemiology [7]. This is because such as traditional methods of determining contacts such as surveys or sensor based tracking cannot scale. Surveys are also affected by recall bias, where part participants may not remember all of their contacts. As a solution to this problem, we observe that many of the daily routines of individuals are based on scheduled activities, such as going to meetings, going to appointments, attending classes etc. Thus if we have information about these scheduled activities we can create a reliable network of most of the frequently occurring contacts. Based on this assumption, we created a contact network of students based on the class-enrollment data for the Fall 2016 semester at the Discovery Park campus of the University of North Texas.

Our data contained information of 3700 students. Each student was assigned a randomly generated id to identify them uniquely, as well as to anonymize the data. The dataset contained the student ids and the classes in which each student was enrolled. Online classes and classes without regular meeting times were excluded. From this data, we constructed a graph where each student was a vertex, and two vertices (students) were connected by an edge if the corresponding students shared a class. The weight of an edge was the average duration of shared class time between the students.

3.1   Experiment Parameters

Experimentation was done with the parameters described in Table 1, and were run with the graph constructed from class-enrollment data. For each vertex v, 3 contacts were given per hour of average time spent in class over all weekdays by the student represented by v. Of the disease-specific parameters, the incubation and infection rates, measles parameters were adapted from [8,15], whereas influenza parameters were adapted from [2,4,6]. Two sets of parameters were chosen for influenza that varied in length of incubation and infectious periods. We used the same values of σ, γ and T for all vertices and edges.

In PIM simulations, a single vertex v_0 was selected to be infected, with $I_n(v_0) = 1$ for $0 \leq n < \gamma$, and $R_n(v_0) = 1$ for $n \geq \gamma$. The remaining vertices were initially completely susceptible. The probability values of the states of each vertex were obtained by computing the functions given in Eqs. 2–5 over the time period. The number of infected individuals at time t in days was determined by summing over $I_t(v)$ for all $v \in V(G)$. We terminated each simulation after day t if outbreak activity was sufficiently small, i.e. the total number of vertices

Table 2. The parameters used in simulations.

Disease parameters

Disease	Incubation period (σ) in days	Infectious period (γ) in days	Transmission probability (T)	Number of contacts ($\Omega_i(v)$) in hours^{-1}
Measles	8	5	.9	3
Influenza 1	2	5	.1	3
Influenza 2	1	3	.1	3

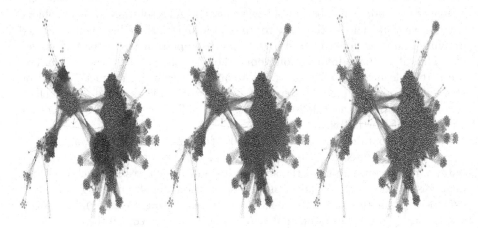

Fig. 4. States of the vertices in the contact network based on the PIM model on day 35. Yellow vertices are fully susceptible, whereas redder vertices have a higher probability of being infected at a given time. Green vertices have a probability of 95% or greater of being recovered. From left to right, the values are for Measles (left), Influenza 1 (middle) and Influenza 2 (right). (Color figure online)

with high probability of exposed and infected states was small. We quantitatively measured this using the following conditions:

$$\sum_{v \in V(G)} E_t(v) + I_t(v)) \leq 0.5$$

$$|\sum_{v \in V(G)} (E_t(v) + I_t(v)) - \sum_{v \in V(G)} (E_{t-1}(v) + I_{t-1}(v))| \leq 0.5$$

The simulations were terminated if both these conditions were satisfied. In addition, simulations were not terminated before day 20. These bounds were selected to ensure that simulations do not end prematurely. Figure 4 shows the state of the vertices in the network as per the PIM model, on day 35. As can be seen, the measles epidemic spreads faster and takes longer time to recover (more red and less green nodes) than the influenza models.

In simulations using the stochastic model, the same graph, seed vertex of infection and parameters were used. 100 trials were run with a seeded random number generator for each of the three disease parameters. Contacts between vertices occurred randomly, with the probability of contact between vertices u and v for any given contact event proportional to $w(u, v)$. Disease transmission occurred with probability T at the time of a successful contact between a susceptible and infectious individual.

3.2 Results

The results demonstrate that PIM produces results most similar to those produced by stochastic Monte Carlo models for diseases that are more highly infectious. As seen in Table 3, the Monte Carlo model and PIM produced similar values for the total number of infected individuals in an outbreak. Additionally, while the peak number of infected individuals and day of peak infection produced by PIM tended to be within one standard deviation of the mean values produced by the Monte Carlo trials, for all disease parameters, PIM outbreaks peaked slightly earlier and higher than the average Monte Carlo trial. This becomes more apparent when the parameters for less infectious diseases are used.

We believe that earlier peaks are observed partially due to an artifact of the stochastic method. In stochastic trials with low parameters, no outbreak of the disease is likely to be observed until multiple days have passed. Outbreak trials with peaks that are lower, occur later and show greater variance in the peak day of infection are observed as a result. This contrasts with PIM, which allows the seed of infection to partially contact multiple neighbors concurrently, possibly causing slightly earlier and higher peaks of infection. In addition, the approximation that events are independent may propel the initial spread of infection at a slightly greater rate, an effect that would be most noticeable for less infectious diseases.

Table 3. A comparison of outbreak attributes between PIM and the averaged values of 100 stochastic simulations. The standard deviation is shown for each averaged value.

	Probabilistic infection model		
Disease	Total infected	Peak infected	Day of peak
Measles	3644.21	1059.10	38
Influenza 1	2930.08	787.61	31
Influenza 2	2077.31	454.38	22–23
	Monte carlo model		
Disease	Total infected	Peak infected	Day of peak
Measles	3647.95 ± 0.22	1021.35 ± 132.12	38.58 ± 2.39
Influenza 1	3011.49 ± 38.04	755.90 ± 72.12	34.03 ± 4.21
Influenza 2	2094.01 ± 109.01	394.72 ± 47.80	27.01 ± 4.60

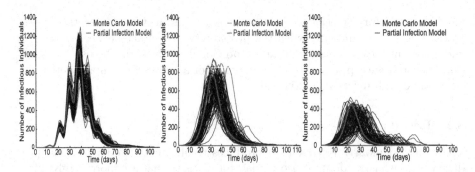

Fig. 5. A comparison between PIM and 100 simulations of the stochastic SEIR model with respect to the number of infectious individuals over the entire simulation. From left to right, the curves are for Measles (left), Influenza 1 (middle) and Influenza 2 (right). (Color figure online)

Figure 5 demonstrates that the attributes of the SEIR curves produced by PIM are similar to those of the average outbreak curves obtained from 100 stochastic trials. This similarity is most notable in simulations of highly infectious diseases, such as when using the parameters for measles; in Fig. 6 left, we show the simulation time series using measles parameters for all four states, showing that the PIM model closely follows the averaged curves of 100 trials of the stochastic model. In addition, we compare the infectious state probability curves of individual vertices produced by the PIM model: Fig. 6 right shows the $I_t(v)$ curves produced by PIM for the seed infected node as well as for 100 vertices that were randomly sampled from the set of initially susceptible vertices for the measles simulation. Most vertices reached their peak probability of being infected around day 38, which is consistent with the peak day of infection given in Fig. 5.

Influence of Correction Parameter. We now test by how much the correction due to redundant infection (as discussed in Sect. 2.2) affects the simulations. Figure 7 shows a comparison between simulations with PIM when correction the probability of vertex v infecting vertex u uses the modified version as in Eq. 7, and one where the original Eq. 1 is used. For each v_0, the percent difference between the peak number of infected individuals produced by PIM with and without correction was less than 0.2%, suggesting that one-level-deep backflow correction is a sufficient approximation.

4   Related Research

Computational epidemics is an active area of research. Several software tools for simulating disease over a population have been developed including EpiSims [9] and DiSimS [5] that use high performance computing, and Broadwick [14] which uses a sequential, but modular framework that can be modified for various disease parameters. Our PIM method can also be implemented to be parallel, and thus can be executed on large networks.

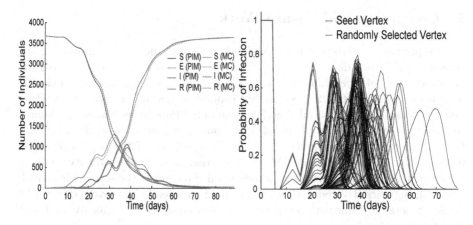

Fig. 6. State of vertices in the measles simulation. Left: Comparison between the number of vertices in each state over time for PIM and the Monte Carlo (MC) method averaged over 100 trials. Right: Probability of infection of 100 randomly selected vertices of the network. The peak occurs around days 35–45.

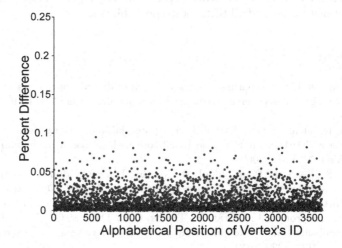

Fig. 7. The percent difference between the peak number of infected individuals is shown for simulations for measles produced by PIM with and without backflow correction for every possible initially infectious v_0.

The challenges of creating reliable contact networks are discussed in [7]. In 2008 [13], a cross-sectional survey on 7,290 participants conducted by different public health institutes or commercial companies was conducted to build a contact network. Another study [12], performed through the 2009 H1N1 flu pandemic on a population of 36 people based on communication using sensors. However, neither of these methods are scalable as compared to our method of utilizing scheduled data. Recent studies have also looked into the dynamic contact networks [3] and the effect of misinformation in developing contact networks [11].

5 Conclusion and Future Work

In this paper, we introduce a probabilistic infection model for simulating the spread of infectious diseases on contact networks. Our model encapsulates the advantages of both deterministic meta-population models as well as stochastic models on contact networks. We further propose a method of obtaining contact networks based on the scheduled activities of individuals in specific environments (e.g., businesses, schools, etc.), and simulate our model on a contact network built from a university's class enrollment data. Comparisons of the results obtained from stochastic modelling and PIM on the contact network of university students demonstrate that our approach produces similar results to the stochastic model, but with significantly reduced computational overhead. Moreover, our model gives a tractable framework for probabilistic analysis of outbreak dynamics at the individual level.

As part of our future work, we will experiment with latent periods, infectious periods and transmission probabilities selected from distributions rather than as static values. In addition, we will pursue further studies of vaccine distribution and other individual-level outbreak intervention strategies by applying PIM's approximations for individual SEIR state-probabilities.

References

1. Ajelli, M., et al.: Comparing large-scale computational approaches to epidemic modeling: agent-based versus structured metapopulation models. BMC Infectious Diseases
2. Balcan, D., et al.: Seasonal transmission potential and activity peaks of the new influenza a (h1n1): a monte carlo likelihood analysis based on human mobility. BMC Med. **7**(1), 45 (2009)
3. Bansal, S., Read, J., Pourbohloul, B., Meyers, L.A.: The dynamic nature ofcontact networks in infectious disease epidemiology. J. Biol. Dyn. **4**(5), 478–489 (2010).https://doi.org/10.1080/17513758.2010.503376, pMID: 22877143
4. Cori, A., Valleron, A.J., Carrat, F., Tomba, G.S., Thomas, G., Boëlle, P.Y.: Estimating influenza latency and infectious period durations using viralexcretion data. Epidemics **4**(3), 132 (2012)
5. Deodhar, S., Bisset, K.R., Chen, J., Ma, Y., Marathe, M.V.: An interactive, web-based high performance modeling environment for computational epidemiology. ACM Trans. Manage. Inf. Syst. **5**(2), 7 (2014). https://doi.org/10.1145/2629692
6. Drewniak, K., Helsing, J., Mikler, A.R.: A method for reducing the severity of epidemics by allocating vaccines according to centrality. In: ACM Conference on Bioinformatics, Computational Biology, and Health Informatics (2014)
7. Eames, K., Bansal, S., Frost, S., Riley, S.: Six challenges in measuring contact networks for use in modelling. Epidemics **10**, 72 – 77 (2015). https://doi.org/ 10.1016/j.epidem.2014.08.006,http://www.sciencedirect.com/science/article/pii/ S1755436514000413, challenges in Modelling Infectious Disease Dynamics
8. Enanoria, W.T., et al.: The effect of contact investigations and public health interventions in the control and prevention of measles transmission: A simulation study

9. Halloran, M.E., et al.: Modeling targeted layered containment of an influenza pandemic in the United States. Proc. Nat. Acad. Sci. **105**(12), 4639–4644 (2008). https://doi.org/10.1073/pnas.0706849105,https://www.pnas.org/content/105/12/4639

10. Henson, S., Brauer, F., Castillo-Chavez, C.: Mathematical models in population biology and epidemiology. Am. Math. Monthly **110**(3), 1 (2003)

11. Holme, P., Rocha, L.E.C.: Impact of misinformation in temporal network epidemiology. Netw. Sci. **7**(1), 52–69 (2019). https://doi.org/10.1017/nws.2018.28

12. Jain, S., Benoit, S.R., Skarbinski, J., Bramley, A.M., Finelli, L.: For the 2009 pandemic influenza A (H1N1) virus hospitalizations investigation team: influenza-associated pneumonia among hospitalized patients with 2009 pandemic influenza A (H1N1) virus-United States, 2009. Clinical Infectious Diseases **54**(9), 1221–1229 (2012). https://doi.org/10.1093/cid/cis197

13. Mossong, J., Hens, N., Jit, M., Beutels, P., Auranen, K., Mikolajczyk, R.T., Massari, M., Salmaso, S., Tomba, G.S., Wallinga, J., Heijne, J.C.M., Sadkowska-Todys, M., Rosińska, M., Edmunds, W.J.: Social contacts and mixing patterns relevant to the spread of infectious diseases. PLoS Med. **5**, 1083–1087 (2008)

14. O'Hare, A., Lycett, S., Doherty, T., Monteiro Salvador, L., Kao, R.: Broadwick: a framework for computational epidemiology. BMC Bioinformatics **17**, 65 (2016). https://doi.org/10.1186/s12859-016-0903-2

15. Ponciano, J.M., Capistrán, M.A.: First principles modeling of nonlinear incidence rates in seasonal epidemics. PLoS Comput. Biol. **7**(2), e1001079 (2011)

16. Thomas, J.C., Weber, D.J.: Epidemiologic Methods for the Study of Infectious Diseases. Oxford University Press, Oxford (2001)

Eigen-AD: Algorithmic Differentiation of the Eigen Library

Patrick Peltzer$^{(\boxtimes)}$, Johannes Lotz, and Uwe Naumann

Informatik 12: Software and Tools for Computational Engineering,
RWTH Aachen University, 52056 Aachen, Germany
info@stce.rwth-aachen.de

Abstract. In this work we present useful techniques and possible enhancements when applying an Algorithmic Differentiation (AD) tool to the linear algebra library Eigen using our in-house AD by overloading (AD-O) tool dco/c++ as a case study. After outlining performance and feasibility issues when calculating derivatives for the official Eigen release, we propose *Eigen-AD*, which enables different optimization options for an AD-O tool by providing add-on modules for Eigen. The range of features includes a better handling of expression templates for general performance improvements as well as implementations of symbolically derived expressions for calculating derivatives of certain core operations. The software design allows an AD-O tool to provide specializations to automatically include symbolic operations and thereby keep the look and feel of plain AD by overloading. As a showcase, dco/c++ is provided with such a module and its significant performance improvements are validated by benchmarks.

Keywords: Algorithmic Differentiation · Linear algebra · Eigen

1 Introduction

In this work, the C++ linear algebra library Eigen[1] is used as a base software implementing linear algebra operations for which derivatives are to be computed using Algorithmic Differentiation (AD) [6,10] by overloading. Derivatives of computer programs can be of interest, e.g. when performing uncertainty quantification [18], sensitivity analysis [1] or shape optimization [4]. AD enables the computation of derivatives of the output of such programs with respect to their inputs. This is done using the tangent model in *tangent mode* or the adjoint model in *adjoint mode*, where the latter is also known as adjoint AD (AAD). In AAD, the program is first executed in the *augmented primal run*, where required data for later use is stored. Derivative information is then propagated through the tape in the *adjoint run*. For both, tangent and adjoint, the underlying original code is called the *primal*, and the used floating point data type and its

---

[1] http://eigen.tuxfamily.org.

© Springer Nature Switzerland AG 2020
V. V. Krzhizhanovskaya et al. (Eds.): ICCS 2020, LNCS 12137, pp. 690–704, 2020.
https://doi.org/10.1007/978-3-030-50371-0_51

variables are called *passive*. Vice versa, code where derivatives are computed and its respective data type and variables are called *active.*

A wide collection of AD tools can be found on the community website[2]. In general, one can divide the available software into *source transformation* and *operator overloading* tools. While source transformation essentially has the potential to create more efficient code, supporting complex language features like they are available in C++ is connected to higher expense for the tool authors. AD by overloading (AD-O) on the other hand can be applied to arbitrary code as long as operator overloading is supported by the programming language. In terms of AD-O, the recorded data in the augmented primal run is referred to as the *tape*, and creating the tape is called *taping*. The propagation in the adjoint run is known as *interpreting* the tape.

Applying an AD-O tool to dedicated libraries poses a significant issue, as by principle they require the usage of an extended floating point data type (from now referred to as the *custom AD data type*). This change in data type is often impractical and breaks hand tuned performance gains [3]. Therefore, software combining AD-O and linear algebra has been realized with, e.g. Adept [7] or the *Stan Math Library* [2], where the latter makes heavy use of Eigen. Eigen allows the direct utilization of AD-O tools due to its extensive use of C++ templates. At a later point in this paper, concrete implementations and benchmarks for AD-O in Eigen will be presented using dco/c++, which is an AD-O tool actively developed by NAG Ltd.[3] in collaboration with RWTH Aachen University.

To our best knowledge, there has not been a work focusing on the application of an AD-O tool to Eigen while preserving the philosophy of plain AD-O. The goal is that the AD-O tool user benefits from optimizations without explicitly being aware of them. Swapping the data type of Eigen and using the AD-O tool as usual should be all that is required to compute derivatives. However, several problems concerning performance and feasibility of the derivative computation will arise from this concept. This work proposes approaches and solutions to overcome them.

The next section provides more background on AD-O and also introduces the concept of symbolic derivatives. Section 3 presents *Eigen-AD* which is a fork of Eigen. It contains several optimizations and improvements for the application of an AD-O tool, which are demonstrated and benchmarked using dco/c++ in Sect. 4. Section 5 summarizes the results and suggests possible future works.

Note than an extended version of this work exists [13]; refer to it for further details.

2   Using AD-O and Symbolic Derivatives

Most of the performance improvements presented at a later point in this paper are based on symbolic differentiation (SD), in which derivatives are evaluated

---

[2] http://www.autodiff.org/.

[3] https://www.nag.co.uk/.

692 P. Peltzer et al.

analytically at a higher level than with AD. This section demonstrates the differences between evaluating derivatives symbolically and with AD-O by using the matrix-matrix product $C = AB$ with $A, B \in \mathbb{R}^{2 \times 2}$ as an example.

Let this specific product kernel be implemented using Eigen as follows:

```
template<typename T>
void matmul(const Matrix<T,2,2>& A, const Matrix<T,2,2>& B,
    Matrix<T,2,2>& C) {
  for(int i=0; i<2; i++)
    for(int j=0; j<2; j++)
      for(int k=0; k<2; k++)
        C(i,j) += A(i,k)*B(k,j);
}
```

Listing 1.1. 2×2 matrix-matrix multiplication kernel.

The primal code is called using the passive data type `double` as template argument T. For an active evaluation, the function must be called using the custom AD data type of an AD-O tool as T. As mentioned in the previous section, the AD-O tool first performs the augmented primal run when in adjoint mode. Both, the `+=` and the `*` operators in line 6, are overloaded by the tool and act as the entry points for taping. The tape is an implementation dependent representation of the *computational graph* of the program, which contains all performed computations and their corresponding partial derivatives. Figure 1 displays the computational graph of the matrix-matrix multiplication kernel in Listing 1.1. Vertices represent variables accessed in the augmented primal run, including temporary instantiations from the `*` operator in line 6 (denoted as z). The edge weights are the partial derivatives of the respective computations. In the adjoint run, the graph is traversed in reverse order, propagating the adjoint value of the output towards the inputs. This is done by multiplying subsequent edge weights and adding parallel edge weights. Effectively, the loops of Listing 1.1 are executed in reverse order; derivatives are computed on *scalar level*.

In contrast to the differentiation of all occurring scalar computations, it may also be possible to rewrite the derivative using matrix expressions so that derivatives are computed on *matrix level*. Staying with the example above, the adjoint propagation on the computational graph in Fig. 1 can be written as follows:

$$\bar{A}_{0,0} = \bar{C}_{0,0}B_{0,0} + \bar{C}_{0,1}B_{0,1} \qquad \bar{B}_{0,0} = \bar{C}_{0,0}A_{0,0} + \bar{C}_{1,0}A_{1,0}$$
$$\bar{A}_{0,1} = \bar{C}_{0,0}B_{1,0} + \bar{C}_{0,1}B_{1,1} \qquad \bar{B}_{0,1} = \bar{C}_{0,1}A_{0,0} + \bar{C}_{1,1}A_{1,0}$$
$$\bar{A}_{1,0} = \bar{C}_{1,0}B_{0,0} + \bar{C}_{1,1}B_{0,1} \qquad \bar{B}_{1,0} = \bar{C}_{0,0}A_{0,1} + \bar{C}_{1,0}A_{1,1}$$
$$\bar{A}_{1,1} = \bar{C}_{1,0}B_{1,0} + \bar{C}_{1,1}B_{1,1} \qquad \bar{B}_{1,1} = \bar{C}_{0,1}A_{0,1} + \bar{C}_{1,1}A_{1,1}$$

$$\Rightarrow \bar{A} = \bar{C}B^T \qquad (1) \qquad\qquad \Rightarrow \bar{B} = A^T\bar{C} \qquad (2)$$

Adjoint values are denoted with a bar. Equations (1) and (2) compute the adjoints of the input data using matrix-matrix multiplications. Using these equations, it is not necessary to tape any computations in Listing 1.1. Instead, the adjoints can directly be computed in the adjoint run on matrix level as long as the values of the input data are available.

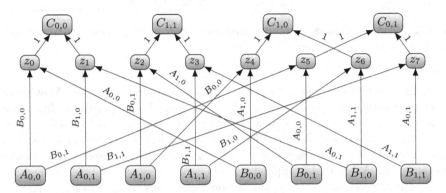

Fig. 1. Computational graph for the 2×2 matrix product kernel displayed in Listing 1.1: Vertices represent variables accessed in the augmented primal run, edges define computation operands and edge weights are the respective partial derivatives. Temporaries z storing the results of the $*$ operator in line 6 may be optimized away by an AD-O tool when taping.

3   Eigen-AD

Applying an AD-O tool to Eigen will lead to severe limitations sooner or later. Eigen comes with optimized kernels, e.g. for 8-byte double precision data. Traditionally, custom AD data types are larger and these optimizations do not work anymore. Regarding AAD application, the complexity of many frequently used linear algebra operations scales cubically with the input dimension. This is the case for, e.g. matrix decompositions or matrix products. Since the memory required by the tape scales roughly linearly with the number of operations required by an algorithm, the tape size can quickly exceed the amount of available RAM and therefore makes an evaluation of the derivatives not feasible at all.

To overcome these issues, the Eigen source code has been adjusted and extended to help optimize the application of AD-O tools. The resulting software has been named *Eigen-AD*. All source code changes are generically written and do not modify the original Eigen API, but provide entry points which can be used by additional modules. Based on that, we have added a generic Eigen-AD base module which provides a clean interface for developers to control and implement optimized operations in their tool specific AD-O tool module. Refer to Fig. 2 for the package architecture.

The existing Eigen test system has been extended so that every Eigen test can also be performed for an AD-O tool's tangent and adjoint data types. Compiling and running the tests successfully ensures compatibility of the AD-O tool

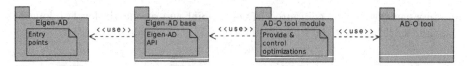

Fig. 2. Eigen-AD package architecture: Authors can implement an AD-O tool module for their AD-O tool.

with all of the tested Eigen functions. The philosophy is that an AD-O tool is able to determine derivatives of all Eigen operations algorithmically, while selected operations are provided with optimized computations for their derivatives. Another aim is to keep the look and feel of AD-O, i.e. optimizations and improvements shall not require a separate interface but be used automatically whenever the AD-O tool is applied.

The next sections present optimization approaches realized in Eigen-AD.

3.1   Nesting Expression Templates

The concept of expression templates was originally proposed to eliminate temporaries when evaluating vector and matrix expressions and to be able to pass algebraic expressions as arguments to functions [16]. The first aspect, also known as *lazy evaluation*, has been complemented by the concept of *Smart Expression Templates* [8] which both are implemented in Eigen.

In the context of AD, using expression templates is especially relevant, since every temporary contributes to the computational graph as it was also demonstrated in Fig. 1 in Sect. 2. In the AAD case, the computational graph needs to be stored in memory and is then traversed in the reverse run, increasing memory and run time requirements with each additional temporary. This can be avoided by constructing expression templates for the right hand side of the assignment and evaluate them altogether. Therefore, some AD-O tools also implement an expression template mechanism, e.g. dco/c++ or Adept [7,12].

When applying such an AD-O tool to Eigen, both expression template engines are nested, where the AD-O tool layer is accessed by the scalar operations of Eigen. This is not an intended use case for Eigen, and therefore Eigen is not aware that it may receive template expressions. The returned template expression is then implicitly casted back to the custom AD data type, resulting in a temporary which must be considered for the derivative evaluation. This destroys the gains originally made by using expression templates in the AD-O tool.

As an example, consider the unary minus operator, implemented in Eigen as a functor named scalar_opposite_op. Its class template parameter Scalar corresponds to the custom AD data type and it is also used in the parenthesis operator as the in- and output types. An assignment of the form $A = -B$, where A and B are Eigen 1×1 matrices containing a single scalar of the adjoint data type, will result in an additional vertex in the computational graph, analogous to the temporaries z of the computational graph in Fig. 1.

When looking at the way the Eigen functors are used, it is not necessary to explicitly prescribe what types they return. Due to Eigen's generic design and as long as the occurring types are compatible – meaning the required casts/specializations/overloads are available – there is no need to force the scalar type at this level. This is a fitting case to use the C++-14 feature of *auto return type deduction* which allows a function to deduce the return type from the operand of its return statement. Therefore, replacing the return type of the functors with the *auto* keyword allows the passing of expression types from the AD-O tool to the Eigen internals. Besides that, it must be ensured that the functors allow arbitrary input types, as they can now be called with expression types as parameters as well.

Evaluating the modified `scalar_opposite_op` functor will avoid the additional vertex in the computational graph. This optimization can be applied to all Eigen scalar functors and also to several Eigen math functions like `sin` or `exp` which are supported by the AD-O tool's expression templates.

3.2   Symbolic Derivatives

As introduced in Sect. 2, mathematical insight can be exploited to evaluate a derivative symbolically. Such an evaluation can be superior to the AD-O solution in terms of performance, run time-wise and also memory-wise in the adjoint case. This observation motivates the inclusion of symbolic derivatives for certain linear algebra routines, yielding a *hybrid* implementation [9].

The Eigen-AD base module provides an interface for AD-O tool developers to implement symbolic derivatives. At the moment, entry points for products as well as for any computation concerning a dense solver are supported. Refer to the Eigen-AD base module technical guide for further information. In the next sections, equations for symbolic adjoints of selected operations are introduced.

SD Dense System Solver. Consider the system of linear equations:

$$A\mathbf{x} = \mathbf{b} \tag{3}$$

where $A \in \mathbb{R}^{n \times n}$ is the system matrix, $\mathbf{b} \in \mathbb{R}^n$ is the right hand side vector and $\mathbf{x} \in \mathbb{R}^n$ is the solution vector. There exists a wide variety of approaches to solve the problem shown in Eq. (3) which make use of decomposing the matrix A into a product of other matrices, e.g. the *LU decomposition*. Eigen offers one dense solver class for each decomposition type.

AAD for the solution of a system of linear equation includes the taping and the interpretation of the decomposition, which yields a run time and memory overhead of $\mathcal{O}(n^3)$. However, when evaluating the adjoints symbolically using Eqs. (4)–(5) as presented in [5], the decomposition is completely excluded from taping and interpreting.

$$A^T \cdot \bar{\mathbf{b}} = \bar{\mathbf{x}} \tag{4}$$

$$\bar{A} = -\bar{\mathbf{b}} \cdot \mathbf{x}^T \tag{5}$$

As it can be seen, the adjoint values of the right hand side vector $\mathbf{b}$ can be determined by solving an additional linear system. By saving the computed decomposition of A in the augmented primal run, it can then be reused in the adjoint run. This reduces the run time and memory overhead for differentiating to $\mathcal{O}(n^2)$ [11].

Symbolic Inverse. Inverting a matrix, i.e. computing

$$C = A^{-1} \tag{6}$$

is implemented in Eigen as a member function of a dense solver. Corresponding adjoints can be computed using Eq. (7) [5].

$$\bar{A} = -C^T \bar{C} C^T \tag{7}$$

Compared to AAD, the memory overhead is reduced to $\mathcal{O}(n^2)$; however, the adjoint run still has a run time overhead of $\mathcal{O}(n^3)$ due to the matrix multiplications.

Symbolic Log-Abs-Determinant. Another member function of the dense solvers is the computation of $x \in \mathbb{R}$ using the log-abs-determinant of a matrix $A \in \mathbb{R}^{n \times n}$ as shown in Eq. (8). Such a computation is relevant for, e.g. computing the log-likelihood of a multivariate normal distribution.

$$x = \log |\det(A)| \tag{8}$$
$$\bar{A} = A^{-T} \bar{x} \tag{9}$$

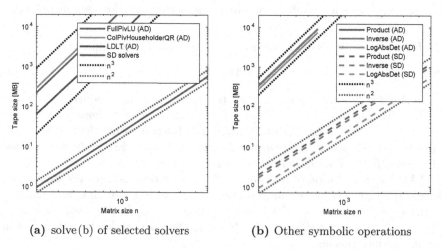

(a) solve(b) of selected solvers   (b) Other symbolic operations

Fig. 3. Tape size comparison between AD and SD solvers: All new symbolic implementations have a memory complexity of $\mathcal{O}(n^2)$ compared to $\mathcal{O}(n^3)$ of the algorithmic versions.

Equation (8) is implemented in Eigen for the QR dense solvers, and adjoints can be computed according to Eq. (9) [14]. The inverse can be computed by reusing the decomposition which was kept in memory for the adjoint run. While the run time overhead is still $\mathcal{O}(n^3)$, the symbolic implementation improves the memory overhead to $\mathcal{O}(n^2)$.

Symbolic Matrix-Matrix Product. For $A \in \mathbb{R}^{n \times m}, B \in \mathbb{R}^{m \times p}, C \in \mathbb{R}^{n \times p}$, the adjoints of the matrix-matrix product in Eq. (10) can be computed using Eqs. (11)–(12) according to [5].

$$C = AB \tag{10}$$
$$\bar{A} = \bar{C}B^T \tag{11}$$
$$\bar{B} = A^T\bar{C} \tag{12}$$

Note that this matches the results derived in Sect. 2 for the 2×2 matrix-matrix product. While differentiating the matrix-matrix product with AAD has a run time and memory complexity of $\mathcal{O}(nmp)$, utilizing the symbolic evaluation reduces the memory overhead to $\mathcal{O}(nm+mp)$. The input matrices A and B must be saved in the augmented primal run and then be multiplied with the adjoint values of the output according to Eqs. (11)–(12) in the adjoint run. Note that the run time complexity can not be improved using the symbolic evaluation.

4   Benchmarks

As mentioned in the beginning of this paper, an Eigen-AD tool module has been implemented for dco/c++. In order to verify the implementation, extensive measurements were made. They were performed using a single thread on an i7-6700K CPU running at 4 GHz with AVX2 enabled and 64GB RAM available for the tape recording, using the g++ 7.4 compiler on Ubuntu 18.04. The respective linear algebra operations were performed for increasing matrix size n using dynamic-sized quadratic matrices $\mathbb{R}^{n \times n}$ and one evaluation of the first-order adjoint model was computed with all output adjoints set to 1. The inverse() results shown here use the underlying PartialPivLU, the logAbsDeterminant() function the FullPivHouseholderQR solver. From now on, the dco/c++ Eigen module is referred to as dco/c++/eigen, and computations which are not using symbolic implementations but only plain overloading are denoted as algorithmic or as AD.

4.1   dco/c++/eigen Benchmarks

In this section, the theoretical considerations from Sect. 3.2 are validated with benchmarks. To emphasize the improvements, reference measurements for the corresponding algorithmic computations without the dco/c++/eigen module are given where appropriate.

Memory Consumption. All symbolic evaluations introduced in Sect. 3.2 lower the memory overhead introduced by an AD-O tool to $\mathcal{O}(n^2)$. In order to visualize this effect, the tape size of dco/c++ has been measured for the algorithmic and for the symbolic implementations and is displayed in Fig. 3. For clarity reasons, only selected dense solvers are shown in Fig. 3a, but similar patterns were measured for the other solvers as well. The symbolic implementations keep a complete primal solver in memory which is accounted with a $n \times n$ matrix on the tape. Since the symbolic logAbsDeterminant() function does not require any additional data, it has the same memory usage as its corresponding solver. The symbolic inverse() function additionally saves the transposed input matrix, the symbolic matrix-matrix product stores the two input matrices.

As it was expected, all new implementations have a memory complexity of $\mathcal{O}(n^2)$, while the algorithmic versions display a cubic behaviour and quickly exceed the amount of available RAM.

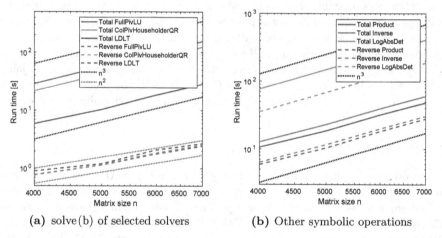

(a) solve(b) of selected solvers   (b) Other symbolic operations

Fig. 4. Total run times and run times of the adjoint run of the new symbolic operations: As described in Sect. 3.2, the symbolic solve(b) function improves the run time complexity of the adjoint run to $\mathcal{O}(n^2)$ and effectively cancels the AD overhead compared to the $\mathcal{O}(n^3)$ primal run time complexity. The other symbolic operations still have the same run time complexity of the adjoint run as the primal code, but are noticeably faster.

Run Time Analysis. Figure 4 visualizes the run time measurements for the symbolic operations, split into total execution time and run time of the adjoint section. As stated in Sect. 3.2, solving a system of linear equations reduces the adjoint run time to $\mathcal{O}(n^2)$, which is confirmed by the measured run times in Fig. 4a. All other symbolic evaluations do not lower the complexity, since a matrix-matrix product or an inverse must be computed in the adjoint run. However, as it can be inferred from the gap between total and adjoint run times in Fig. 4b, the overhead introduced by the adjoint run is rather moderate.

Comparison to Primal Operations. To put the symbolic run times into perspective, the primal run times have been recorded as well. Comparing them both by computing the factor between the respective run times is a good measure to assess the performance of the derivative computation. The results are displayed in Fig. 5.

In contrast to the previous run time analysis, we now compare to the primal code which is highly optimized. Beside the overhead introduced by the AD adjoint section, additional copy instructions are performed in the augmented primal run by the symbolic operations. Due to the convenient fact that the symbolic `solve(b)` evaluation reduces the run time overhead to $\mathcal{O}(n^2)$, all solvers will converge towards a factor of 1 with increasing matrix size, since the ratio is dominated by the $\mathcal{O}(n^3)$ primal code. However, as it can be seen in Fig. 5a, the conversion rate depends on the specific solver. For the other symbolic operations, the run time complexity of the adjoint run can not be improved, which makes a factor of 1 impossible. Instead, the factor depends on the additional computations performed in the adjoint run. For the presented symbolic evaluations, an obtainable factor between 2 and 3 is reasonable. Figure 5b shows a corresponding convergence pattern.

Generically speaking, for very fast primal operations – like the optimized Cholesky `LLT` solver or the matrix-matrix product – it is hard to achieve good factors for smaller input dimensions due to the additional copy overhead introduced by the symbolic implementation. In contrast to that, operations with higher computational costs – like the `JacobiSVD` decomposition or the `logAbsDet()` function – are dominated by the primal code run time-wise and no significant overhead from the AAD code can be measured even for small matrices.

4.2 Comparison to Other AD-O Tools

To put the above given measurements into perspective, it is reasonable to compare them to results from other AD-O tools. The following tools were considered:

- Adept [7]
- ADOL-C [17]
- CoDiPack [15]
- FADBAD++ 2.0[4]
- Stan Math Library [2].

All tools evaluate a matrix-matrix product of two randomly filled $\mathbb{R}^{n \times n}$ matrices. One evaluation of the first order adjoint model is performed using plain AD-O of the Eigen library or using the tool's special API if available. Since Stan only

---

[4] http://www.fadbad.com/fadbad.html.

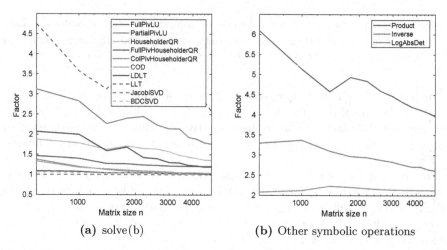

(a) solve(b)   (b) Other symbolic operations

Fig. 5. Run time relation symbolic to primal operations: The run time overhead for the symbolic `solve(b)` implementation vanishes for larger matrices as all factors converge towards 1. For the other symbolic operations, the overhead does not vanish but still appealing factors are achieved.

provides a `grad()` function, its benchmark was modified to compute the scalar value `z = (A*B).sum()` and the corresponding gradient of z. All of the following results were produced using Eigen-AD. However, internal benchmarks have not shown a considerable difference to the standard Eigen version.

It must be said that the shown run times do not imply the feasibility of the tools in general, since they are all designed with different use cases and restrictions in mind. They were utilized to our best knowledge, but no tool specific experts were involved in these measurements. While `dco/c++/eigen` provides its best performance with this setup, we believe that other tools can be optimized by their developers to get similar results. Therefore, the given results only represent the current situation and are likely to change in the future.

The measured run times are displayed in Fig. 6. Note that the notion `dco/c++` refers to plain overloading, and the remark *auto only* describes the usage of the `dco/c++/eigen` module without any symbolic implementations, i.e. only with the optimization from Sect. 3.1 in place. In contrast to that, *full* names the default behaviour when using the module, with the auto return type deduction and symbolic implementations enabled.

All non-specialized tools show the same computational complexity. Differences are non-negligible, though. The feasibility of the auto return type deduction of `dco/c++/eigen` introduced in Sect. 3.1 can be observed, since the smaller amount of temporaries speeds up the computation. In contrast to the other general purpose AD-O tools, Adept also allows the computation of a matrix-matrix product using the `matmul` function from its API. In this case, no Eigen is used but instead the storage types defined by Adept. As it can be expected, this specially designed feature from Adept is faster than the general AD-O tool approach. This

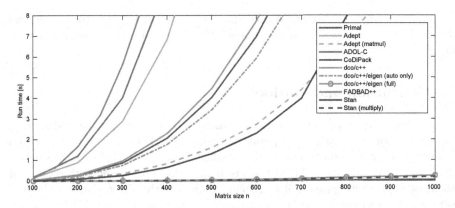

Fig. 6. Run time comparison of different AD-O tools for the matrix-matrix product: Besides plain algorithmic overloading versions, Adept and Stan also offer optimized functions via their API (denoted by parenthesis).

also applies to Stan, although in this case it really is plain AD-O using Eigen storage types. Stan specializes a few Eigen functions such that it internally evaluates optimized matrix-vector products for matrix-matrix products.

Referring to the insight gained in Sect. 3.2, symbolic evaluations have the potential to drastically improve the performance of AD application. In the case of the matrix-matrix product, actual computations are made using the passive data type and profit from all optimizations in Eigen, while only two additional matrix-matrix products need to be evaluated in the adjoint run. This explains why dco/c++/eigen as well as the implementation in the multiply function of Stan drastically outperform all other tools.

Since Stan provides optimized linear algebra functions using Eigen, another benchmark was performed for solving a dense system. Stan offers a mdivide_left(A,b) function to solve a system of linear equations which internally will always use the ColPivHouseholderQR decomposition. Therefore, the algorithmic and the dco/c++/eigen measurements displayed in Fig. 7 also utilize this solver class. While Stan uses the same symbolic evaluation from Eqs. (4)–(5), it performs another decomposition in the adjoint run. dco/c++/eigen on

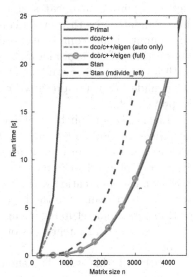

Fig. 7. Algorithmic and symbolic solve(b): Due to saving of the whole decomposition object, dco/c++/eigen achieves a faster run time than Stan.

the other hand keeps the decomposition from the augmented primal run in

P. Peltzer et al.

memory and reuses it later. While Stan keeps the AAD run time overhead
`dco/c++/eigen` at $\mathcal{O}(n^3)$, the implementation in `dco/c++/eigen` improves it to
$\mathcal{O}(n^2)$.

5 Conclusion and Outlook

In this work, we have have outlined challenges which occur when calculating
derivatives for linear algebra operations using an AD-O tool with the Eigen
library. To overcome these issues, the modified library fork Eigen-AD was developed, aimed at authors of AD-O tools to help them improve the performance of
their software when applying it to Eigen. Changes to the Eigen source code were
kept generic and entry points are provided for a general Eigen-AD base module which can be utilized by an individual AD-O tool module via a dedicated
API. Care was taken to realize the improvements via C++ specializations, which
keep the look and feel of plain AD-O. General performance improvements were
made regarding the usage of expression templates by the AD-O tool and specific
operations can now be reimplemented conveniently by an AD-O tool module in
order to provide symbolic implementations.

As a showcase, such a module has been implemented for the AD-O tool
`dco/c++`, where important linear algebra operations like the matrix-matrix product or solving of a linear system are differentiated symbolically. Benchmarks
have validated the theoretical considerations and underlined the improvements
in computational complexity regarding run time and memory usage. It was
shown that AD-O tool modules can cancel the AAD overhead for dense solvers
with a corresponding implementation and comparisons with other AD-O tools
were made to put the produced results into context which further confirmed the
improvements.

Eigen-AD is publicly available[5] and other AD software authors are welcome
to provide a module for their AD-O tool which can be included in the fork as
well as participate in the future development. Investigation into more parts of
Eigen are planned in order to extend the Eigen-AD API. Furthermore, there
has been communication with the Eigen development team and best efforts were
made to keep changes to the Eigen source as general as possible. In combination
with the modular setup regarding the Eigen-AD base module and individual tool
modules, a partial integration of the changes into future Eigen versions should
be discussed.

All in all, this work has shown the potential of adjusting a linear algebra
library to optimize the evaluation of derivatives using an AD-O tool. In the
case of Eigen, relatively small changes to its source code allow to provide a
general API which can be utilized by other AD-O tools and provide a superior
performance compared to ordinary AD-O.

---

[5] https://gitlab.stce.rwth-aachen.de/stce/eigen-ad.

References

1. Bischof, C.H., Bücker, H.M., Rasch, A.: Sensitivity analysis of turbulence models using automatic differentiation. SIAM J. Sci. Comput. **26**(2), 510–522 (2004). https://doi.org/10.1137/S1064827503426723
2. Carpenter, B., Hoffman, M.D., Brubaker, M., Lee, D., Li, P., Betancourt, M.: The Stan Math Library: reverse-mode automatic differentiation in C++. CoRR abs/1509.07164 (2015)
3. Dunham, B.Z.: High-order automatic differentiation of unmodified linear algebra routines via nilpotent matrices. Dissertation, Department of Aerospace Engineering Sciences, University of Colorado at Boulder, Boulder, USA (2017)
4. Gauger, N.R., Walther, A., Moldenhauer, C., Widhalm, M.: Automatic differentiation of an entire design chain for aerodynamic shape optimization. In: Tropea, C., Jakirlic, S., Heinemann, H.J., Henke, R., Hönlinger, H. (eds.) New Results in Numerical and Experimental Fluid Mechanics VI, pp. 454–461. Springer, Heidelberg (2008). https://doi.org/10.1007/978-3-540-74460-3_56
5. Giles, M.B.: Collected matrix derivative results for forward and reverse mode algorithmic differentiation. In: Bischof, C.H., Bücker, H.M., Hovland, P., Naumann, U., Utke, J. (eds.) Advances in Automatic Differentiation, pp. 35–44. Springer, Heidelberg (2008). https://doi.org/10.1007/978-3-540-68942-3_4
6. Griewank, A.: Evaluating Derivatives, Principles and Techniques of Algorithmic Differentiation. No. 19 in Frontiers in Appl. Math., SIAM, Philadelphia, 2000, Second edn. (2008)
7. Hogan, R.J.: Fast reverse-mode automatic differentiation using expression templates in C++. ACM Trans. Math. Softw. **40**(4), 26:1–26:16 (2014). https://doi.org/10.1145/2560359
8. Iglberger, K., Hager, G., Treibig, J., Ruede, U.: Expression templates revisited: a performance analysis of current methodologies. SIAM J. Sci. Comput. **34**(2), 42–69 (2012)
9. Lotz, J.: Hybrid approaches to adjoint code generation with dco/c++. Ph.D. thesis, RWTH Aachen University (March 2016)
10. Naumann, U.: The Art of Differentiating Computer Programs: An Introduction to Algorithmic Differentiation. Society for Industrial and Applied Mathematics, Philadelphia (2012)
11. Naumann, U., Lotz, J.: Algorithmic differentiation of numerical methods: tangent-linear and adjoint direct solvers for systems of linear equations. Tech. rep., RWTH Aachen, Department of Computer Science (June 2012)
12. The Numerical Algorithms Group Ltd. (NAG), Wilkinson House, Jordan Hill Road, Oxford OX2 8DR, United Kingdom: dco/c++ User Guide version 3.2.0 (2017)
13. Peltzer, P., Lotz, J., Naumann, U.: Eigen-AD: algorithmic differentiation of the eigen library (2019). arXiv:1911.12604
14. Petersen, K.B., Pedersen, M.S.: The matrix cookbook (November 2012). version 20121115
15. Sagebaum, M., Albring, T., Gauger, N.R.: High-performance derivative computations using CoDiPack (2017)
16. Veldhuizen, T.: Expression templates. C++ Rep. **7**(5), 26–31 (1995)

17. Walther, A.: Getting started with ADOL-C. In: Combinatorial Scientific Comput-
    ing (2009)
18. Wang, M., Lin, G., Pothen, A.: Using automatic differentiation for compressive
    sensing in uncertainty quantification. Optim. Methods Softw. **33**(4–6), 799–812
    (2018). https://doi.org/10.1080/10556788.2017.1359267

Correction to: A Probabilistic Infection Model for Efficient Trace-Prediction of Disease Outbreaks in Contact Networks

William Qian, Sanjukta Bhowmick, Marty O'Neill,
Susie Ramisetty-Mikler, and Armin R. Mikler

Correction to:
Chapter "A Probabilistic Infection Model for Efficient Trace-Prediction of Disease Outbreaks in Contact Networks" in: V. V. Krzhizhanovskaya et al. (Eds.): *Computational Science – ICCS 2020*, LNCS 12137, https://doi.org/10.1007/978-3-030-50371-0_50

The original version of this chapter was revised. The first author's first name contained a typo. It has been corrected to "William".

The updated version of this chapter can be found at
https://doi.org/10.1007/978-3-030-50371-0_50

© Springer Nature Switzerland AG 2020
V. V. Krzhizhanovskaya et al. (Eds.): ICCS 2020, LNCS 12137, p. C1, 2020.
https://doi.org/10.1007/978-3-030-50371-0_52

Author Index

Aggarwal, Nipun 609
Ahn, Kwangwon 413, 422
Ali, Hesham 652
Alonso, Jorge Blanco 45
Altintas, Ilkay 276
Anvari, Hamidreza 524
Awile, Omar 45
Ayala, Alan 262

Balabaeva, Ksenia Y. 667
Baliś, Bartosz 220
Bazior, Grzegorz 486
Berzins, Martin 175
Bhatia, Ashna 499
Bhowmick, Sanjukta 652, 676
Bortko, Kamil 638
Böser, Sebastian 118
Broeckhove, Jan 385
Bukhanov, Nikita V. 667

Cai, Wentong 513
Cappelli, Carlo 304
Castellano, Maurizio 304
Cebrian, Pau 234
Cheng, Jie 568
Condon, Laura E. 276
Cooper, Kathryn 652
Cornelius, Nathan 652
Cuendet, Michel A. 538
Cui, Zelin 554
Czech, Wojciech 623

D'Ambrosio, Raffaele 59
da Silva, Rafael Ferreira 538
Deelman, Ewa 538
Derevitskii, Ilya V. 667
Di Giovacchino, Stefano 59
Do, Tu Mai Anh 538
Dongarra, Jack 262
Dzwinel, Witold 623

Eberl, Hermann J. 399
Eckhoff, David 499

Enfedaque, Pablo 248
Estrada, Trilce 538
Expósito, Roberto R. 31

Fang, Fang 568
Ferreira, Leonardo 471
Fonin, Vladimir V. 667
Franco, Daniel 191
Funika, Włodzimierz 73
Funkner, Anastasia A. 667

Ganellari, Daniel 290
Gao, Yue 595
Garrido, Daniel 471
Gasper, William 652
Gatta, Roberto 304
Gergel, Victor 17
Gomez-Sanchez, Pilar 191
Guleva, Valentina Y. 432

Haase, Gundolf 290
Haber, Tom 161
Haidar, Azzam 262
Hens, Niel 385
Hieronymus, Maicon 118
Hines, Michael 45
Höb, Maximilian 206
Huber, Dominik 132
Huynh, Hoai Nguyen 582

Ivanchev, Jordan 499

Jacob, João 471
Jankowski, Jarosław 638
Jegatheesan, Thulasi 399
Ji, Guseon 413
Jiang, Bo 554
Jin, Zongze 147

Karp, Artur 445
Kawiak, Andrzej 312, 327
Keegan, Liam 45
Kim, Woo Chang 413

King, James   45
Kitowski, Jacek   73, 220
Knoll, Alois   499
Kolingerová, Ivana   459
Kong, Hyeongwoo   413
Koperek, Paweł   73
Kovalchuk, Sergey V.   667
Kozinov, Evgeny   17
Kranzlmüller, Dieter   206
Kromka, Mikołaj   623
Krzhizhanovskaya, Valeria V.   371
Kumar, Sanjay   609
Kumbhar, Pramod   45
Kuylen, Elise   385
Kwak, Jaeyoung   513
Kwaśniewicz, Łukasz   312, 327

Lamotte, Wim   161
Lee, Nam-Kyoung   422
Lees, Michael H.   513
Lenkowicz, Jacopo   304
León, Betzabeth   191
Leppkes, Klaus   290
Li, Jing   568
Li, Ning   554
Li, Xiaoxue   568
Li, Yangxi   568
Li, Zhenzhen   147
Liesenborgs, Jori   161, 385
Liu, Fan   147
Liu, Yuling   554
Lotz, Johannes   290, 690
Lu, Paul   524
Lu, Zhigang   554
Łukasik, Szymon   445
Luque, Emilio   191

Maijer, Mathijs   342
Maňák, Martin   459
Marchesini, Stefano   248
Maxwell, Reed   276
McSweeney, Thomas   3
Michalski, Radosław   638
Mikler, Armin R.   676
Minev, Peter   102
Moure, Juan Carlos   234
Mu, Weimin   147

Naumann, Uwe   290, 690
Nemeth, Balazs   161

O'Neill, Marty   676
Olschanowsky, Cathie   276
Ong, Marcus E. H.   513
Orzechowski, Michał   220

Pałka, Dariusz   445, 486
Panda, B. S.   609
Parkinson, Dilworth   248
Paszyński, Maciej   102
Pazura, Patryk   638
Peltzer, Patrick   290, 690
Pera, Donato   59
Perciano, Talita   248
Pirola, Ilenia   304
Podsiadło, Krzysztof   102
Pottier, Loïc   538
Presbitero, Alva   371
Purawat, Shweta   276

Qian, William   676
Quax, Rick   371

Ramisetty-Mikler, Susie   676
Rexachs, Dolores   191

Sahasrabudhe, Damodar   175
Schmidt, Bertil   118
Schneider, Piotr   312, 327
Schreiber, Martin   132
Schulz, Martin   132
Schürmann, Felix   45
Selvakumar, Roshini   582
Shang, Yanmin   568
Shi, Jinqiao   595
Silva, Daniel Castro   471
Siwik, Leszek   102
Sloot, Peter M. A.   371
Słota, Renata G.   220
Solak, Esra   342
Song, Fengguang   88
Szkandera, Jakub   459

Tagliaferri, Luca   304
Taufer, Michela   538
Thomas, Stephen   538

Tomov, Stanimire   262
Tong, Lingling   568
Touriño, Juan   31
Treur, Jan   342, 357
Trivedi, Anuradha   248

Vallati, Mauro   304
van Leeuwen, Merijn   357
Veiga, Jorge   31

Walton, Neil   3
Wang, Siyuan   595
Wang, Weiping   147
Wang, Xuebin   595
Wąs, Jarosław   445, 486
Weinstein, Harel   538

Wierzbicki, Adam   312, 327
Wójcik, Grzegorz M.   312, 327
Wolthuis, Kirsten   357

Yang, Dai   132
Yi, Eojin   422
Yin, Pengfei   568
Yin, Zelin   595

Zhao, Can   595
Zhu, Weilin   147
Zhu, Ziyuan   147
Zigon, Bob   88
Zounon, Mawussi   3
Zumbusch, Gerhard   290